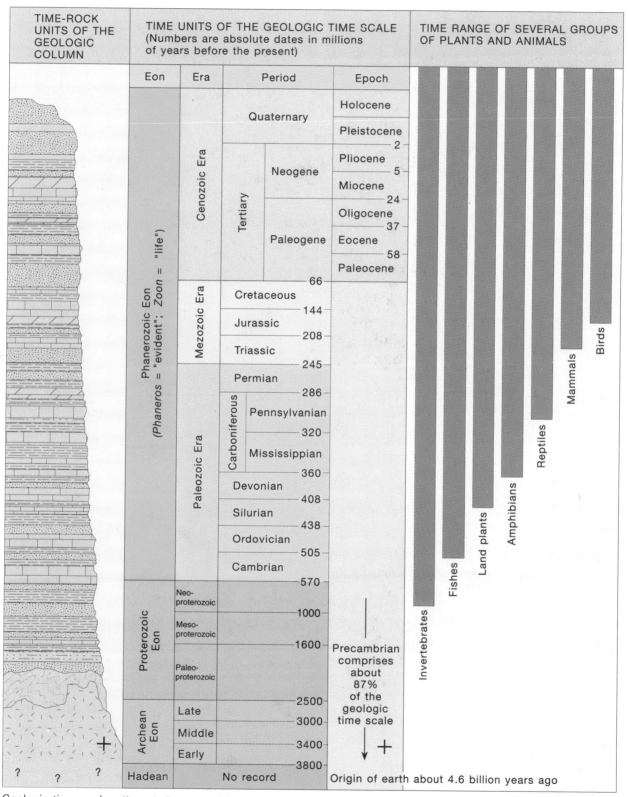

Geologic time scale. *(from A. R. Palmer, The decade of North American geology, geologic time scale, Geology 11:503–504, 1983. Proterozoic divisions are those recommended by the Subcommission on Precambrian Stratigraphy of the International Union of Geological Sciences; K.A. Plumb, New Precambrian time scale, Episodes 14(2):139–140.)*

The Earth Through Time

Painting of Pleurocoelus by Eleanor M. Kish, reproduced with permission of the Canadian Museum of Nature, Ottawa, Canada.

Whenever it is possible to find the cause of what is happening, one should not have recourse to the gods.

POLYBIUS, Greek Historian, Second Century, BCE

The Earth Through Time

FIFTH EDITION

Harold L. Levin

Professor of Geology
Washington University
St. Louis

Saunders Golden Sunburst Series

SAUNDERS COLLEGE PUBLISHING
Harcourt Brace College Publishers

Fort Worth   Philadelphia   San Diego
New York   Orlando   San Antonio
Toronto   Montreal   London   Sydney   Tokyo

Text Typeface: Sabon
Composition: York Graphic Services, Inc.
Publisher: John Vondeling
Developmental Editor: Lee Marcott
Senior Project Editor: Anne Gibby
Copy Editor: Patricia M. Daly
Proofreader: Betty Gittens
Art Director: Caroline McGowan
Text Designer: Rebecca Lemna
Text Layout: Ruth Dudero
Cover Designer: Lawrence Didona
Text Artwork: TASA Graphic Arts, Inc.
Production Manager: Charlene Squibb
Marketing Manager: Angus McDonald

Cover Photocredits: Painting of Dimetrodon entitled "Yellow Sails in the Sunset" by David Peters, Photo
credit © Philip and Karen Smith/Tony Stone Images.
page vii: Blue Valley Benches and the Henry Mountains at sunset. Along the Fremont River, Utah.
(*Scott T. Smith*)
page xi: Badlands at Burns Basin, Badlands National Park, South Dakota. (*Jeff Gnass*)
page xiii: Death Valley National Monument, California. Fall Canyon, Grapevine Mountains.
(*Scott T. Smith*)
page xiv: Cuernos Del Paine, Torres Del Paine National Park, Andes Mountains, Chile. (*Tom Till*)
page xiv: Massive Ordovician sandstone beneath beds of dolomite. (*Hal Levin*)
page xv: Granite Cliffs on Acadia Coast, Mount Desert Island, Acadia National Park, Maine.
(*Jeff Gnass*)
page xv: *Uintatherium,* an Eocene mammal, carved in Salem limestone. Washington University, St. Louis.
(*Hal Levin*)

Printed in the United States of America

The Earth Through Time, Fifth Edition

ISBN: 0-03-005167-3

Library of Congress Catalog Card Number: 95-74830

678901234   032   10 98765432

For Kay,
and for our grandchildren,
Eli, Mollie, Natalie, Emily, Caitlyn, and Candis.

May they have the wisdom to treat the Earth
kindly.

Preface

A friend who knows little of science recently remarked, "at work on yet another textbook revision? Surely things have not changed that much since the last edition!" I explained that we live in a time when the frontiers of Earth science advance with exceptional speed. Rapid progress is catalyzed by technological advances that permit us to gather data that were beyond our capabilities only a few years ago. "The Earth has changed," I said, "but more to the point, our knowledge, and the way we test and interpret that knowledge, has changed. Students merit the most up-to-date information."

The Earth Through Time is for college students taking either their initial course in geology, or their second course in a physical geology–historical geology sequence. Many of these students will not major in a science, but have an understandable interest in the vibrant planet on which they live. The text will help these students understand the many physical, chemical, and biologic events that have shaped their environment, and will provide insight into how scientific questions are resolved. Few would disagree that such matters are important to a student's liberal education. For those planning to complete an academic major in the earth sciences, *The Earth Through Time* provides much of the background information needed for upper-level earth science courses.

We now possess an enormous body of facts about the Earth. Some of these facts must be learned, for one cannot understand processes that have shaped the Earth without them. It is not, however, the objective of *The Earth Through Time* to present Earth history as a compendium of facts. Rather, the text emphasizes how theories have been developed, how they are validated, and how inferences can be made from observations about fossil remains, from the physical and chemical characteristics of rocks, or from the isotopic composition of minerals.

To facilitate an approach that emphasizes understanding over memorization, the first chapter of the fifth edition presents the underlying principles used in deciphering Earth history as developed by such founders of the science as Nicolaus Steno, James Hutton, and Charles Lyell. Time provides the framework for any history, whether cultural or geologic. Therefore, methods used in dating rocks and the development of the geologic time scale are also introduced in this initial chapter.

The second chapter is a survey of common rocks and minerals—the raw materials from which inferences about Earth history are drawn. Some students may have been exposed to this information in an earlier physical geology course. For such students, the chapter will provide a helpful review. Having learned the basic prerequisite information, students can proceed to the core information of historical geology. They can explore the historical significance of sedimentary rocks and fossils, learn about the operation of plate tectonics and sea-floor spreading, and follow the chronologic sequence of topics dealing with the physical and biological events of each geologic era.

The changes in this fifth edition include new photographs and drawings, refashioned explanations, and additions of new information that has recently appeared in the primary literature. Descriptions of the geology of several national parks has been added. In Chapter 3, a broader discussion of Walther's Law, Vail cycles, and facies analysis is presented. The section on the evolution of plants in Chapter 4 is rewritten, and new concepts relating to evolutionary theory introduced. With reorganization of the text, material relating to the origin of the solar system and the beginning of life precede the chapters dealing with Archean and Proterozoic history. A discussion of organisms dependent upon deep sea hydrothermal vents has been added. Chapters 7 through 15 examine the sequential history of the Earth and its inhabitants. Mass extinctions during the Paleozoic and Mesozoic eras, causes of cyclicity in the stratigraphic succession, dinosaur habits, effects of Deccan volcanism, the Messinian event, and hominid evolution are among the many topics that have been revised in this edition.

Not all instructors will have time for the final chapter dealing with "Moons, Meteorites, and Planets." The chapter, however, is largely self-contained. It can be omitted, assigned as reading, or inserted wherever the instructor deems appropriate. The chapter reminds students that the Earth is one of a family of planets that are related in origin.

■ Special Features

Topics related to text discussions have been placed throughout the text as boxes labeled *Commentary*. Their purpose is to further illustrate concepts and spark student interest. Examples include *Sedimentary Way-up Structures; A Tale of Two Deltas; Amber, the Golden Preservative; Riches of Greenstone Belts; The Colossal Ordovician Ash Fall;* and *Is There a Bolide Impact in Our Future?*

Nearly everyone agrees that one of the best ways students can learn about geology is to visit rock and fossil localities in the field. Students usually get a taste of the importance of field observations in field trips taken as part of their geology course. One would hope that non-majors would have an opportunity to continue making such observations on their own. Often the opportunity to do so is provided during vacation trips. For this reason, as well as to provide further illustration of concepts examined in the text, seven special boxes about the geology of national parks are provided. Among these are brief descriptions of the geology of Grand Canyon, Hawaii Volcanoes, Voyageurs, Shenandoah, Acadia, Zion, and Badlands national parks.

■ Help for the Student

Many learning aids have been included to help the student master the content of their geology course. Each chapter begins with an outline that informs the reader about content and sequence. This is followed by introductory paragraphs that provide an overview of what follows and why it is important. A summary is provided at the end of each chapter for review and use in preparation for examinations. Throughout the text, key terms are in **boldface** type. Two kinds of questions are provided with each chapter. A list of **Review Questions** focuses on important themes and text knowledge. They can usually be answered directly from information given in the text. **Discussion Questions** test comprehension and challenge the student to expand on text information. **Supplemental Readings and References** provide more comprehensive information for the student wishing to pursue topics further, perhaps as a basis for term papers.

At the back of the text, students will find an extensive illustrated **Glossary**. All definitions in the glossary are in conformity with the *Glossary of Geologic Terms* published by the American Geological Institute. Should the need arise to check the age or correlation of a rock unit mentioned in the text, the student can refer to the **Formation Correlation Charts** in the Appendices. Also in the Appendices are **A Classification of Living Things, Physiographic Provinces of the United States, Periodic Table of Chemical Elements,** and English/Metric **Convenient Conversion Factors.**

■ Ancillaries

The Earth Through Time is amply supported with materials to facilitate and enhance teaching. These supplementary materials include an **Instructor's Manual** with **Test Bank** written by David T. King, Jr. of Auburn University. It includes detailed chapter outlines, multiple choice questions, and answers to all the end-of-chapter Discussion Questions. To enhance the students' understanding of the text, **A Study Guide** has been prepared by Vicki Harder. It includes Chapter Overviews, Learning Objectives, detailed Chapter Outlines, Summaries, Questions for Review, Terms to Remember, Completion Questions, and True/False Questions. Answers to all True/False Questions and Multiple Choice Questions are provided. **Computerized Test Banks** are available in Macintosh and IBM Windows 3.5 and 5.25 disks.

The **Saunders College Videodisc** includes over 2000 colorful, still images from 10 of Saunders' best-selling Geology, Earth Science, and Geography texts. The videodisc also includes almost an hour of live-action footage. Derived from the Encyclopedia Britannica archives, these moving images feature video clips of landscapes and geological phenomena, along with animated segments that bring geological processes to life.

LectureActive Software accompanies the video-disc. This software allows the instructor to customize lectures by giving quick access to the video clip and still frame data on the videodisc.

A barcode manual is also part of the ancillary package. The manual contains descriptions, bar-codes, and text references for each still image and video clip. This allows the professor to access the images on the videodisc by either using a light pen to scan the barcodes or using the remote control to enter the frame number.

Many aspects of geology are based on observations of exposed strata in the field. Although not as effective as an actual field excursion, an extensive set of 35-mm slides is one way to bring field observations into the classroom. The **Saunders 35-mm Slide Package** includes five hundred 35-mm slides, and is accompanied by **125 overhead transparencies.**

■ *Acknowledgments*

This edition of *The Earth Through Time* owes much to the guidance and wisdom of its reviewers. Their scrutiny of the manuscript and incisive comments have improved the text, and facilitated the endless endeavor to keep it up-to-date. I extend my thanks to all of these earth scientists, including:

Warren Huff
University of Cincinnati

Ernst Kastning
Radford University

David T. King, Jr.
Auburn University

Barun K. Sen Gupta
Louisiana State University

James Stevens
Lamar University

Carl Vondra
Iowa State University

Peter Whaley
Murray State University

Lisa White
San Francisco State University

William Zinsmeister
Purdue University

I would also like to extend my thanks to the following earth scientists who contributed as reviewers to the previous four editions of this book:

Dennis Allen
University of South Carolina, Aiken

William Ausich
Ohio State University

David R. Berry
California State Polytechnic University

William Berry
University of California, Berkeley

Michael Bikerman
University of Pittsburgh

Roger J. Cuffey
University of Pennsylvania

James H. Darrell
Georgia Southern University

Larry E. Davis
Washington State University

William H. Easton
University of Southern California

George F. Engelmann
University of Nebraska at Omaha

Stanley Fagerlin
Southwest Missouri State College

Vicki Harder
Santa Teresa, New Mexico

John A. Howe
Bowling Green State University

John R. Huntsman
University of North Carolina

Allen Johnson
West Chester State College

Gary D. Johnson
University of South Dakota

William M. Jordan
Millersville University of Pennsylvania

Roger Kaesler
University of Kansas

Larry Knox
Tennessee Technical University

Peter Kresan
University of Arizona

Ralph L. Langenheim, Jr.
University of Illinois

W. Britt Leatham
California State University, San Bernardino

Peter B. Leavens
University of Delaware

Joseph Lintz
University of Nevada at Reno

Daniel L. Lumsden
Memphis State University

Donald Marchand
Old Dominion University

William H. Mathews III
University of British Columbia

Dewey McLean
Virginia Polytechnic Institute

Eldridge Moores
University of California, Davis

Peter Nielsen
Keene State College

Cathryn Newton
Syracuse University

Donald E. Owen
Lamar University

John Pope
Miami University

Jennifer Smith Prouty
Corpus Christi State University

Thomas Roberts
University of Kentucky

Thomas W. Small
Frostburg University

Leonard W. Soroka
St. Cloud State University

Calvin H. Stevens
San Jose State University

Kenneth Van Dellen
Macomb Community College

Thomas T. Zwick
Eastern Montana College

The assistance I received from my editors at Saunders College Publishing was indispensable. My friend and publisher John Vondeling has provided enthusiastic support for this and every previous edition. The burden of revision was made lighter because of the efficiency, help, and professionalism of developmental editor Lee Marcott. The formidable task of converting manuscript to printed page fell to project editor Anne Gibby. She cheerfully and efficiently saw to the completion of a multitude of tasks without ever showing impatience at the author's requests to insert last minute changes.

Harold L. Levin
St. Louis
July 1995

Contents Overview

Contents

The Earth Through Time

The Castle, an erosional feature formed by erosion in the Wingate Sandstone, Capitol Reef National Park, Utah. *(Photograph by James Cowlin).*

The face of places and their forms decay;

And that is solid earth that once was sea;

Seas, in their turn, retreating from the shore,

Make solid land, what ocean was before.

OVID, *Metamorphoses*, XV

CHAPTER 1

Introduction to Earth History

We live on the third planet from the sun. Our planet was formed 4.6 billion years ago and since that time has circled the sun like a small spacecraft observing a rather average star. Scarcely 500,000 years ago, primates called *Homo sapiens* evolved on planet Earth. Unlike earlier animals, these creatures with oversized brains and nimble fingers asked questions about themselves and their surroundings. Their questioning has continued to the present day. How was the Earth formed? Why do earthquakes occur? What lies beneath the lands we live on and beneath the floor of the ocean? Even ancient people sought answers for these questions. In frail wooden ships they probed the limits of the known world, fearing that they might tumble from its edge. Their descendants came to know the planet as an imperfect sphere, and they began an examination of every obscure recess of its surface. In harsher regions, exploration proceeded slowly. It has been only within the last 100 years that humans have penetrated the deep interior of Antarctica. Today, except for a few areas of great cold or dense forest, the continents are well charted. New frontiers for exploration now lie beneath the oceans and outward into space.

■ Changing Views of the Third Planet

A major step in the advance of science in the oceanic frontier occurred in the late 1800s with the scientific expeditions of British oceanographic vessels. Promi-

FIGURE 1–1 *H.M.S. Challenger (From the* Report of the Scientific Results of the Exploring Voyage of H.M.S. Challenger during the years 1873–1876. *Narrative, Part II, 1885.)*

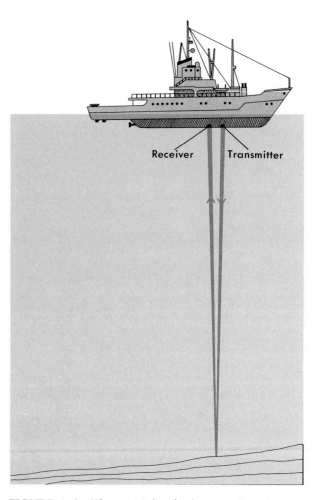

FIGURE 1–2 The principle of echo sounding. A transmitter sends a sound wave, which is reflected back to the surface by the ocean bottom and is picked up by a receiver. By knowing the total time involved and the speed of sound in the ocean (1500 meters per second), water depth can be determined. *(From McCormick, J. M., and Thiruvathukal, J. V., 1981.* Elements of Oceanography, *2nd ed. Philadelphia: Saunders College Publishing.)*

nent among these was the naval ship H.M.S. *Challenger* (Fig. 1–1). Although limited by the relatively primitive technology of their time, the *Challenger* scientists obtained an extraordinary amount of information about water depths, movement of ocean currents, marine life, and the nature of sediments on the ocean floor. The advancement of technology since that time has allowed that base of knowledge to be enormously expanded. Criss-crossing the seas, ships equipped with precision depth recorders that employed the principle of echo sounding (Fig. 1–2) plotted continuous topographic profiles of the sea floors. Related methods using stronger energy sources provided images of layers of rock and sediment beneath the ocean floor. Such cross-sections of the Earth's crust are called seismic profiles (Fig. 1–3). A panoramic view of undersea geology emerged that was more magnificent and complex than anyone had imagined (*see* Fig. 1–4 on pp. 4–5). The magnetic characteristics of the ocean floors were also examined, and the scientific community was startled to learn that the Earth's magnetic polarity had occasionally reversed itself during the past.

To study further the effects of these phenomena on Earth history, a unique deep-sea drilling ship, the *Glomar Challenger,* was constructed and quickly put into operation. With this research vessel, geologists were able to penetrate deeply into the sea floor and bring on deck the sediment and rock needed to decipher the history of the ocean basins. Samples from beneath the sea were combined with interpretations of seismic profiles and magnetic data, stimulating the development of exciting new hypotheses of drifting crustal segments, splitting continents, and continu-

ously changing global geography. The success of the *Glomar Challenger* led to the outfitting of a new drilling ship, the *JOIDES Resolution* (Fig. 1–5), which began operation in 1985.

Exploration of space has also contributed to our new view of the Earth. Intricately engineered spacecraft circled the Earth and set down on the moon. Members of our inquisitive species reached down and held lunar rocks in their gloved hands. Analysis of those rocks has yielded important clues to the early history of the Earth. Satellites and spacecraft continue to circle the Earth, transmitting thousands of images of our planet to scientists below. These im-

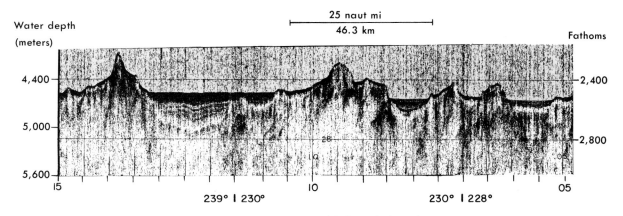

FIGURE 1–3   Continuous seismic profiles showing abyssal hills and abyssal plains along the edge of the Mid-Atlantic Ridge.   *(From Hayes, D. E., and Pimm, A. C., Initial Reports of Deep Sea Drilling Project, 1972. National Science Foundation publication, vol. 14, pp. 341–376.)*

ages have given us an unprecedented synoptic view of a dynamic, vibrant planet, and on a day-to-day basis they have provided information useful in finding mineral deposits, assessing worldwide agricultural productivity, and ascertaining global changes in environmental conditions. Perhaps most important, however, these data have made all of us aware of the stunning beauty and fragility of our planetary home.

FIGURE 1–5   The *JOIDES Resolution,* successor to the *Glomar Challenger.* This oceanographic research vessel is designed for taking drill cores from the bottom of the ocean floor. The vessel is a floating oceanographic research center, with a seven-story laboratory stack that occupies 12,000 square feet. On-board facilities include laboratories for sedimentology, paleontology, geochemistry, and geophysics. The ship can suspend as much as 9000 meters of drill pipe to obtain core samples. *(Courtesy of Ocean Drilling Program, National Science Foundation.)*

■ Geology: Its Physical and Historical Components

Geologists concern themselves with an exceptional variety of scientific tasks and therefore must employ knowledge from diverse fields. Some examine the composition and texture of meteorites and moon rocks. With magnifiers and computers, others scrutinize photographs of planets to understand the origin of the features that characterize their surfaces. Still others are busily unraveling the structure of mountain ranges, attempting to predict the occurrence of earthquakes and volcanic eruptions, or studying the behavior of underground water, glaciers, or streams. Large numbers of geologists search for fossil fuels and the metallic ores vital to our standard of living. They worry, as do you and I, about the fate of humans in a world of diminishing resources. To do their work, geologists draw on the knowledge of astronomy, physics, chemistry, mathematics, and biology. For example, the petroleum geologist must understand the physics of moving fluids, the chemistry of hydrocarbons, and the biology of the fossils used to trace subsurface rock layers. Because geology incorporates the information of so many other scientific disciplines, it can be termed an eclectic science. *Eclectic* is a term useful in describing a body of selected information drawn from a variety of sources. All sciences are eclectic to some degree, but geology is decidedly so.

For convenience of study, the body of knowledge called geology can be divided into **physical geology** and **historical geology.** The origin, classification, and composition of earth materials, as well as the varied processes that occur on the surface and in the deep

FIGURE 1–4   The physical features of the ocean floors and the continents.   *(Copyright © Marie Tharp.)*

interior of the Earth, are the usual subjects of physical geology. Historical geology addresses the Earth's evolution, changes in the distribution of lands and seas, growth and destruction of mountains, succession of animals and plants through time, and developmental history of the solar system. The historical geologist examines planetary materials and structures to discover how they came into existence. He or she

works with the tangible *result* of past events and must work backward in time to discover the *cause* of those events. In this endeavor, geologists employ the usual procedures of science—namely, the collection of observations, the formulation of hypotheses to explain those observations, and the validation of the hypotheses by means of further observations and tests. Thus, observed facts serve both as the basis for building hypotheses and as the ultimate check on their accuracy.

COMMENTARY
Growth Lines and Time

The study of growth lines in ancient corals provides an interesting example of how scientific observations can lead to hypotheses. In the early 1900s, Professor John Wells of Cornell University was musing over the fact that although paleontology (the study of fossils and ancient life forms, from the Greek *palaios*, meaning "ancient") provided a way to determine *relative* geologic age, fossils could not be used in determining absolute geologic time. His thoughts led him to consider the astronomic basis for time and to search for some detectable effect of movement in the Sun–Earth–Moon system on the fossil remains of organisms. On theoretical grounds, geophysicists and astronomers had determined that the present 24-hour period of the Earth's rotation on its axis has not been constant throughout geologic time. Because of tidal friction, the Earth has been slowing down at a rate of about 2 seconds per 100,000 years. Thus, the length of the day has been increasing, and the number of days in a year has been steadily decreasing. On the basis of work completed by earlier researchers, Wells knew that the fine growth lines on the exoskeletons of coral organisms might represent daily growth increments (Fig. A). Thus, a coral might secrete one thin ridge of calcium carbonate each day. In addition to the fine daily growth lines, there are coarser monthly bands (presumably related to monthly breeding cycles during which carbonate secretion was inhibited) and still broader annual bands in corals living in areas of seasonal fluctuations. Wells counted the growth lines on several species of living corals and found that the count "hovers around 360 in the space of a year's

FIGURE A Growth banding displayed by a specimen of the extinct coral *Heliophyllum halli*. The finer lines may represent daily growth increments. There are approximately 200 growth lines per centimeter. Together with the annual bands, the growth increments can be used to estimate the number of days in a year at the time the animal lived.

growth." Proceeding next to fossil corals of successively older geologic age, Wells counted correspondingly larger numbers of growth lines in the yearly

---

■ *The Founders of Historical Geology*

Historical geology is a venerable science. Its beginnings can be traced to the time of classical Greece. Like other sciences, progress in historical geology has been based on the continuous accumulation of knowledge provided by past generations of workers. They have provided the foundations of geology, upon which modern theories and precepts depend. A partial list of early contributors to our understanding of the origin and history of the earth would include Nicolaus Steno, Abraham Gottlob Werner, James Hutton, William Smith, Georges Leopold Cuvier, Alexander Brongniart, Charles Lyell, and Charles Darwin.

Nicolaus Steno

Nicolaus Steno (1638–1687) was a Danish physician who settled in Florence, Italy. There he became physician to the Grand Duke of Tuscany. Since the duke was a generous employer, Steno had ample time to tramp across the countryside, visit quarries, and examine strata. His investigations of sedimentary rocks

FIGURE B The changing length of the day through geologic time. *(From Wells, J. W. 1963. Nature 197:948–950.)*

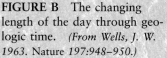

segment of the exoskeleton. For example, on corals that had been determined to be about 370 million years old, Wells found 398 daily growth lines in a yearly increment. This evidence, suggesting that there were nearly 400 days in a year 370 million years ago, correlated astonishingly well with the calculations provided by the astronomers. Historical geologists could now state with considerable confidence that long ago when the first land vertebrates began to appear, the days were shorter but there were more of them in a year. Furthermore, once the relationship between number of days in the year and absolute geologic age had been plotted on a graph (Fig. B), it might be possible to derive the age of a stratum by counting the growth lines on the fossil corals collected from that stratum. Of course, the method would be valid only if the slowdown of the Earth's rotation has been uniform.

As in all forms of geochronology, the Wells method has its limitations. Only exceptionally well-preserved fossils are suitable, and large numbers must be subjected to growth band counts in order to obtain statistically valid results. Many specimens show variations caused by factors in the environment unrelated to ordinary daily changes. Nevertheless, the method has provided an interesting correlation between astronomic and biologic phenomena.

led him to formulate such basic principles of historical geology as superposition, original horizontality, and original lateral continuity

The **principle of superposition** states that in any sequence of undisturbed strata, the oldest layer is at the bottom and successively higher layers are successively younger. It is a rather obvious axiom. Yet Steno, on the basis of his observations of strata in northern Italy, was the first to explain the concept formally. The fact that it is self-evident does not diminish the principle's importance in deciphering Earth history. Furthermore, the superpositional relationship of strata is not always apparent in regions where layers have been steeply tilted or overturned (Fig. 1–6). In such instances, the geologist must examine the strata for clues useful in recognizing their uppermost layer. The way fossils lie in the rock and the evidence of mudcracks and ripple marks are particularly useful clues when one is trying to determine which way was up.

The observation that strata are often tilted led Steno to his **principle of original horizontality.** He reasoned that most sedimentary particles settle from fluids under the influence of gravity. The sediment then must have been deposited in layers that were nearly horizontal and parallel to the surface on which

FIGURE 1–6 Steeply dipping strata grandly exposed in the Himalayan Mountains. It is often difficult to recognize the original tops of beds in strongly deformed sequences such as this. *(Courtesy D. Bhattacharyya.)*

they were accumulating (Fig. 1–7). Hence, steeply inclined strata indicate an episode of crustal disturbance after the time of deposition (Fig. 1–8).

The **principle of original lateral continuity** was the third of Steno's stratigraphic axioms. It pertains to the fact that, as originally deposited, strata extend in all directions until they terminate by thinning at the margin of the basin, end abruptly against some former barrier to deposition, or grade laterally into a different kind of sediment. This observation is significant in that whenever one observes the exposed cross-section of strata in a cliff or valley wall, one should recognize that the strata continue laterally for a distance that can be determined by field work and

drilling. When geologists stand on a sandstone ledge at one side of a canyon, it is the principle of original lateral continuity that leads them to seek out the same ledge of sandstone on the far canyon wall and then to realize that the two exposures were once continuous.

Today, we recognize that Steno's principles are basic to the geologic specialty known as **stratigraphy**, which is the study of layered sedimentary rocks, including their texture, composition, arrangement, and correlation from place to place. Because stratigraphy enables geologic events as recorded in rocks to be placed in their correct sequence, it is the key to the history of the Earth.

FIGURE 1–7 Superpositional sequence of undisturbed, horizontal, mainly Permian strata. Canyonlands National Park, Utah. The Colorado River is in the foreground. *(Photo by Peter L. Kresan.)*

FIGURE 1–8 Steeply dipping shale and sandstone strata. Ouachita Mountains of Oklahoma.

Interpreters of the Geologic Succession

The stratigraphic principles formulated by Steno in the seventeenth century were rediscovered several decades later by other European scientists. Among the most prominent of these early geologists were John Strachey (1671–1743), Giovanni Arduino (1714–1795), Johann G. Lehmann (1719–1767), Georg Füchsel (1722–1776), and Peter Simon Pallas (1741–1811). John Strachey is best remembered for his use of the principles of superposition and original lateral continuity in deciphering the stratigraphic succession of coal-bearing formations in Somerset and Northumberland, England. He clearly illustrated the sequence of formations encountered at the surface and in mines and described the manner in which horizontal strata rested upon the eroded edges of inclined older formations. Years later this type of stratigraphic relationship would be termed *unconformable*.

Whereas John Strachey was particularly interested in the local stratigraphic succession, other naturalists developed a broader, more global view of the geologic succession. In Italy, Giovanni Arduino classified mountains according to the most abundant type of rock that composed them. He defined Primary mountains as constructed of crystalline rocks of the kinds later to be named igneous and metamorphic. Arduino recognized that rocks of the Primary group were likely to be the oldest in a mountain system and were usually exposed along the central axis of ranges. Secondary mountains were constructed of layered, well-consolidated, fossiliferous rocks. Such rocks were later to be named sedimentary. Arduino's Tertiary designation was reserved for unconsolidated gravel, sand, and clay beds as well as lava flows.

Classifications similar to that of Arduino also appeared in the works of the German scientists Lehmann and Füchsel. These men were not rocking chair theorists. Both were excellent field geologists. Füchsel worked chiefly in the mountains of Thuringia, whereas his contemporary Lehmann tramped on the rocks of the Harz and Erz Gebirge. They prepared excellent summaries of the stratigraphic succession in these mountains and further developed a remarkably perceptive understanding of some of the events involved in the making of mountain ranges.

This insight into the history of mountains was improved as a result of the work of a tireless field geologist named Peter Simon Pallas. Under the patronage of Catherine II of Russia, Pallas traveled across the whole of Asia and made careful studies of the Ural and Altai Mountains. He recognized the threefold division of mountains formulated by his predecessors. In addition, Pallas was able to construct a general geologic history of the Urals, and he provided a lucid description of how the rock assemblages change as one travels from the center to the flanks of mountain systems.

Abraham Gottlob Werner

One of the most influential geologists working in Europe near the close of the eighteenth century was Professor Abraham Gottlob Werner (1750–1817). Werner's eloquent and enthusiastic lectures at the Freiberg Mining Academy in Saxony transformed that school into an international center for geologic studies. Werner was a competent mineralogist, and many geologists of his day used his scheme for the identification of minerals and ores. Werner is not, however, remembered as much for his contributions to mineralogy as he is for his interpretation of the geologic history of the Earth. The cornerstone of that interpretation was his insistence that all rocks of the Earth's crust were deposited or precipitated from a great ocean that once enveloped the entire planet. Today we know that some rocks (the group called sedimentary) are indeed often of marine origin. Others, however, are decidedly not formed in water. Because they believed that all rocks had formed in the ocean, Werner and his many followers became known as neptunists (after Neptune, the Roman god of the sea).

Werner envisioned his universal ocean in the earliest stage of Earth history as a hot, steamy body saturated with all the dissolved minerals needed to form the rocks of his oldest division. He called these *Primitive Rocks,* or *Urgebirge.* Most of these rocks would later come to be known as igneous and metamorphic.

The second stage in the Wernerian interpretation of Earth history witnessed a subsidence and cooling of the primitive ocean. It came to resemble the ocean of today. Werner told his students that this change was marked by the deposition of fossil-bearing, well-consolidated, stratified rocks that lay above the Urgebirge. These he designated *Transition Rocks* and suggested that they were deposited when the Earth had passed from an uninhabitable to an inhabitable condition. The fossils proved the planet had become suitable for life. Today, we would refer to these rocks as part of Europe's predominantly sedimentary Paleozoic sequence of strata.

Above the Transition Rocks, Werner noted the occurrence of sandstones, shales, coal beds, very fossiliferous limestones, and occasional layers of a black rock later determined to be basalt. These basalt layers were actually old lava flows. For all of these rocks lying above the Transition Rocks, Werner employed Johann Lehmann's term *Flötzgebirge.* A final term, *Alluvium,* was used for the unconsolidated sand, gravel, and clay that rested on the Flötzgebirge.

Although initially received with great interest and enthusiasm, Werner's ideas were soon to draw criticism. His theory failed to explain what had become of the immense volume of water that once covered the Earth to so great a depth that all continents were submerged. An even greater problem was his insistence that basaltic lava layers such as those in the Flötzgebirge were deposited in precisely the same manner as the enclosing limestones and shales. With visible, indisputable field evidence, geologists such as J. F. D'Aubisson de Voisins (1769-1832) in France clearly demonstrated the volcanic origin of these basaltic layers. Geologists with this opposing view came to be known as plutonists (after Pluto, the Roman god of the Underworld). According to the plutonists, fire rather than water was the key to the origin of igneous rocks. James Hutton of Scotland was a prominent plutonist who clearly stated that rocks like basalt and granite "formed in the bowels of the Earth of melted matter poured into rents and openings of the strata."

James Hutton

James Hutton (1726–1797), an Edinburgh physician and geologist, is remembered not only as a staunch opponent of neptunism but also for his penetrating comprehension of how geologic processes alter the Earth's surface. For Hutton, the Earth was a dynamic, ever-changing place in which new rocks, lands, and mountains arise continuously as a balance against their destruction by erosion and weathering. He took a cyclic view of our planet, as opposed to Werner's more static concept of an Earth that had changed very little from its beginning down to the present time. In addition, Hutton believed that "the past history of our globe must be explained by what can be seen to be happening now." This simple yet powerful idea was later to be named **uniformitarianism** by William Whewell. Charles Lyell (1797–1875) became the principal advocate and interpreter of uniformitarianism. We will speak of this great man again in the pages ahead.

Perhaps because it is so general a concept, uniformitarianism has been reinterpreted and altered in a variety of ways by scientists and theologians from Hutton's generation down to our own. Some of the ideas about what uniformitarianism now implies would seem strange to Hutton himself. If the term *uniformitarianism* is to be used in geology (or any science), one must clearly understand what is uniform. The answer is that the physical and chemical laws that govern nature are uniform. Hence, the history of the Earth may be deciphered in terms of present observations on the assumption that natural laws are invariant with time. These so-called natural laws are merely the accumulation of all our observational and experimental knowledge. They permit us to predict the conditions under which water becomes ice, the behavior of a volcanic gas when it is expelled at the Earth's surface, or the effect of gravity on a grain of sand settling to the ocean floor. Uniform natural laws govern such geologic processes as weathering, erosion, transport of sediment by streams, movement of glaciers, and the movement of water into wells.

Hutton's use of what later was termed uniformitarianism was simple and logical. By observing geologic processes in operation around him, he was able to infer the origin of particular features he discovered in rocks. When he witnessed ripple marks being produced by wave action along a coast, he was able to state that an ancient rock bearing similar markings was once a sandy deposit of some equally ancient shore. And if that rock now lay far inland from a coast, he recognized the existence of a sea where Scottish sheep now grazed.

Hutton's method of interpreting rock exposures by observing present-day processes was given the catchy phrase "the present is the key to the past" by Sir Archibald Geike (1835–1924), a Scot with a brilliant career of discovery and experimentation in geology. The methodology implied in the phrase works very well for solving many geologic problems, but it must be remembered that the geologic past was sometimes quite unlike the present. For example, before the Earth had evolved an atmosphere like that existing today, different chemical reactions would have been prevalent during weathering of rocks. Life originated in the time of that primordial atmosphere under conditions that have no present-day counterpart. As a process in altering the Earth's surface, meteorite bombardment was once far more important than it has been for the last three billion years or so. Many times in the geologic past continents have stood higher above the oceans, and this higher elevation results in higher rates of erosion and harsher climatic conditions compared with intervening periods when the lands were low and partially covered with inland seas. Similarly, at one time or another in the geologic past volcanism was more frequent than at present.

Nevertheless, ancient volcanoes disgorged gases and deposited lava and ash just as present-day volcanoes do. Modern glaciers are more limited in area than those of the recent geologic past, yet they form erosional and depositional features that resemble those of their more ancient counterparts. All of this suggests that present events do indeed give us clues to the past, but we must be constantly aware that in the past, the rates of change and intensity of processes often varied from those to which we are accustomed today and that some events of long ago simply do not have a modern analogue.

In order to emphasize the importance of natural laws over processes in the concept of uniformity, many geologists prefer to use the term actualism as a substitute for uniformitarianism. **Actualism** is the principle that *natural laws* governing both past and present processes on Earth have been the same. Hutton's friend John Playfair never suggested the term actualism, but provided an eloquent statement of it when, in 1802, he wrote the following lines:

> Amid all the revolutions of the globe the economy of Nature has been uniform, and her laws are the only thing that have resisted the general movement. The rivers and the rocks, the seas and the continents have changed in all their parts; but the laws that describe those changes, and the rules to which they are subject, have remained invariably the same.

The eighteenth-century concept of uniformitarianism was not the only contribution James Hutton made to geology. In his *Theory of the Earth,* published in 1785, he brought together many of the formerly separate thoughts of the naturalists who preceded him. He showed that rocks recorded events that had occurred over immense periods of time and that the Earth had experienced many episodes of upheaval separated by quieter times of denudation

and sedimentation. In his own words, there had been a "succession of former worlds." Hutton saw a world of cycles in which water sculpted the surface of the Earth and carried the erosional detritus from the land into the sea. The sediment of the sea was compacted into stratified rocks, and then by the action of enormous forces the layers were cast up to form new lands. In this endless process, Hutton found "no vestige of a beginning, no prospect of an end." In this phrase, we see Hutton's intoxicating view of the immensity of geologic time. No longer could geologists compress Earth history into the short span suggested by the Book of Genesis.

At Scotland's Isle of Arran and at Jedburgh, Scotland, Hutton came across exposures of rock where steeply inclined older strata had been beveled by erosion and covered by flat-lying younger layers (Figs. 1–9A and B). It was clear to Hutton that the older sequence was not only tilted but also partly removed by erosion before the younger rocks were deposited. The erosional surface meant that there was a gap or hiatus in the rock record. In 1805, Robert Jameson named this relationship an **unconformity.** More specifically, Hutton's Jedburgh exposure was an angular unconformity because the lower beds were tilted at an angle to the upper. This and other unconformities provided Hutton with evidence for periods of denudation in his "succession of worlds." Although he did not use the word *unconformity,* he was the first to understand and explain the significance of this feature.

During most of his career, Hutton's published reports attracted only modest attention, and a good part of that attention came from opponents who preferred to follow the views of Abraham Gottlob Werner. To remedy this situation, British scientists who appreciated the value of Hutton's ideas convinced his friend John Playfair, a professor of mathematics and natural philosophy, to publish a summary of and commentary on *Theory of the Earth.* The work by Playfair was published in 1802 under the title *Illustrations of the Huttonian Theory of the Earth.* Whereas Hutton's writing was often complex and difficult to follow, Playfair's text was lucid, easy to read, unburdened by lengthy quotations from foreign sources, and highly persuasive. Indeed, subsequent geologists of the nineteenth century based much of their understanding of Hutton's ideas not on their reading of Hutton's original publications but on the intelligent and convincing phrases of John Playfair.

Hutton died five years before the publication of *Illustrations.* Throughout his life he had been a man absorbed in the investigation of the Earth. He was seen frequently in the field, scrutinizing every rock exposure he happened upon, and soon became so familiar with certain strata that he was able to recog-

FIGURE 1–9A Angular unconformity at Siccar Point, eastern Scotland. It was here that James Hutton first realized the historical significance of an unconformity. The drawings in Figure 1-9B indicate the sequence of events documented in this famous exposure. *(Photograph courtesy of E. A. Hay.)*

nize them at different localities. What he was unable to do well, however, was determine if dissimilar-looking strata were roughly equivalent in age. He had not discovered how to correlate beds that did not have a similar composition and texture (lithology). This problem was soon to be resolved by William Smith (1769–1839).

William Smith

William Smith was an English surveyor and engineer who devoted 24 years to the task of tracing out the strata of England and representing them on a map. Small wonder that he acquired the nickname "Strata Smith." He was employed to locate routes of canals, to design drainage for marshes, and to restore springs. In the course of this work, he independently came to understand the principles of stratigraphy, for they were of immediate use to him. By knowing that different types of stratified rocks occur in a definite

sequence and that they can be identified by their lithology, the soils they form, and the fossils they contain, he was able to predict the kinds and thicknesses of rock that would have to be excavated in future engineering projects. His use of fossils was particularly significant. Prior to Smith's time, collectors rarely noted the precise beds from which fossils were taken. Smith, on the other hand, carefully recorded the occurrence of fossils and quickly became aware that certain rock units could be identified by the particular assemblages of fossils they contained. He used this knowledge first to trace strata over relatively short distances and then to extend over great distances his "correlations" to strata of the same age but of different lithology. Ultimately, this knowledge led to the formulation of the **principle of biologic succession**. This principle stipulates that the life of each age in the Earth's long history was unique for particular periods, that the fossil remains of life permit geologists to recognize contemporaneous deposits around

FIGURE 1–9B Sequence of events indicated by the Siccar Point exposure.

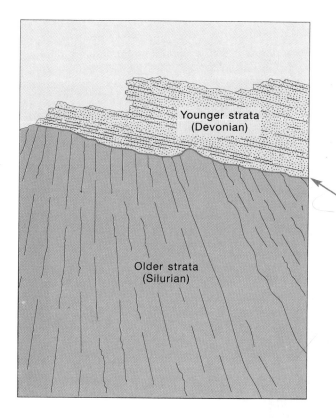

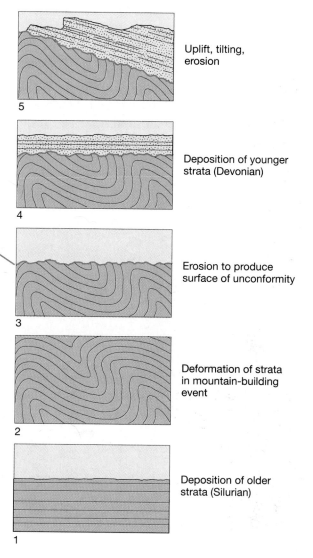

Uplift, tilting, erosion

Deposition of younger strata (Devonian)

Erosion to produce surface of unconformity

Deformation of strata in mountain-building event

Deposition of older strata (Silurian)

the world, and that fossils could be used to assemble the scattered fragments of the record into a chronologic sequence.

Smith did not know why each unit of rock had a particular fauna. This was 60 years before the publication of Darwin's *Origin of Species*. Today, we recognize that different kinds of animals and plants succeed one another in time because life has evolved continuously. Because of this continuous change, or evolution, only rocks formed during the same age contain similar assemblages of fossils.

News of Smith's success as a surveyor spread widely, and he was called to all parts of England for consultation. On his many trips, he kept careful records of the types of rocks he saw and the fossils they contained. Armed with his notes and observations, in 1815 he prepared a geologic map of England and Wales that has remained substantially correct even today. His work clearly demonstrated the validity of the principle of biologic succession.

Cuvier and Brongniart

The use of fossils for the correlation and recognition of formations was not exclusively William Smith's discovery. At the same time that Smith was making his observations in England, two scientists across the English Channel in France were diligently advancing the study of fossils. They were Baron Georges Léopold Cuvier (1769–1832) and his close associate Alexander Brongniart (1770–1847). Cuvier was an expert in comparative anatomy, and with this knowledge he became the most respected vertebrate paleontologist of his day. Brongniart was a naturalist who worked not only on fossil vertebrates but on plants and minerals as well. Together these men established the foundations of vertebrate paleontology. They validated Smith's findings that fossils display a definite succession of types within a sequence of strata and that this succession remains more or less constant wherever found. They also noticed that certain large groupings of strata were often separated by unconformities. As one would pass from one group of strata across the unconformity into the overlying group, a dramatic change in the kinds of animals preserved as fossils was apparent. From this observation, the two French scientists concluded that the history of life was marked by frightful catastrophes involving sudden violent flooding of the continents and abrupt crustal upheavals of stupendous magnitude. The last of these catastrophic episodes was considered to be the Noachian deluge. Cuvier and Brongniart believed that each catastrophe resulted in the total extinction of life and was then followed by the appearance of new animals and plants. Cuvier did not speculate on how each of the many new species originated. Many geologists of the time, including the eminent Charles Lyell, held that geologic history was a uniform and gradual progression and could not accept Cuvier's concept of catastrophism. Thus began a catastrophism versus uniformitarianism controversy that rivaled the earlier neptunist–plutonist debates in scope and passion. Uniformitarianists argued that many seemingly abrupt changes in fossil faunas were caused by missing strata or other imperfections in the geologic record. Other *apparent* breaks in the fossil record were actually not sudden, and the ancestors of each animal group could be found as fossils in underlying beds. However, if we put aside Cuvier's insistence on periodic complete regeneration of life, the central idea of his catastrophism is not invalid. Today, geologists recognize the possibility that rampant volcanism, asteroid impact, or the onslaught of harsh climatic conditions may have caused mass extinctions of life at various times in the geologic past.

Charles Lyell

In the early nineteenth century, the English geologist Charles Lyell (Fig. 1–10) authored the classic work, *Principles of Geology.* This work both amplified the ideas of Hutton and presented the most important geologic concepts of the day. The first volume of this work was printed in 1830. It grew to five volumes and became immensely important in the Great Britain of Queen Victoria. In these volumes one can recognize Lyell's skill in explaining and synthesizing the geologic findings of contemporary geologists. As his friend Andrew C. Ramsey remarked, "We collect the data, and Lyell teaches us the meaning of them." Lyell's *Principles* became the indispensable handbook of every English geologist. In it are amplified many of the principles expressed earlier by Hutton regarding the recognition of the relative ages of rock bodies. For example, Lyell discusses the general principle that a geologic feature that cuts across or penetrates another body of rock must be younger than the

FIGURE 1–10   Sir Charles Lyell.   *(Courtesy of Geological Society of London.)*

rock mass penetrated. In other words, the feature that is cut is older than the feature that crosses it. This generalization, called the **principle of cross-cutting relationships,** applies not only to rocks but also to geologic structures like faults and unconformities. Thus, in Figure 1–11, fault *b* is younger than stratigraphic sequence *d;* the intrusion of igneous rock *c* is younger than the fault, since it cuts across it; and by superposition, rock sequence *e* is youngest of all.

Another generalization to be found in Lyell's *Principles* relates to **inclusions.** Lyell logically discerned that fragments within larger rock masses are older than the rock masses in which they are enclosed. Thus, whenever two rock masses are in contact, the one containing pieces of the other will be the younger of the two.

In Figure 1–12A, the pebbles of granite (a coarse-grained igneous rock) within the sandstone tell us that the granite is older and that the eroded granite fragments were incorporated into the sandstone. In Figure 1–12B, the granite was intruded as a melt into the sandstone. Because there are sandstone inclusions in the granite, the granite must be the younger of the two units.

Charles Darwin

As noted earlier, William Smith and some of his contemporaries were able to recognize that strata were

FIGURE 1–11 An example of how the sequence of geologic events can be determined from cross-cutting relationships and superposition. From first to last, the sequence indicated in the cross-section is deposition of *d*, faulting to produce fault *b*, intrusion of igneous rock mass *c*, and erosion followed by deposition of *e*. Strata labeled *d* are oldest, and strata labeled *e* are youngest.

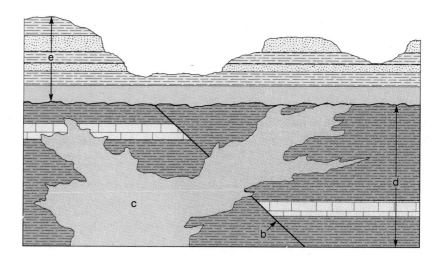

often characterized by particular fossils and that there was a general progression toward more modern-looking assemblages of shells in higher, and thus younger, strata. It was Charles Darwin (1809–1882) who provided a general theory that would account for the changes seen in the fossil record.

As a young man, Darwin had acquired an impressive knowledge of both biology and geology. That knowledge was the basis for his securing an unpaid position as a naturalist aboard the *H.M.S. Beagle*, bound for a five-year mapping expedition around the world. On his return from the voyage in 1836, Darwin (Fig. 1–13) had assembled volumes of notes in support of his theory of evolution of organisms by natural selection. His theory was based on a logical system of observations and conclusions. He observed

that all living things tend to increase their numbers at prodigious rates. Yet in spite of their reproductive potential to do so, no one group of organisms has been able to overwhelm the Earth's surface. In fact, the actual size of any given population remains fairly constant over long periods of time. Because of this, Darwin concluded that not all the individuals produced in any generation can survive. In addition, Darwin recognized that individuals of the same kind differ from one another in various morphologic and physiologic features. From this and the previous observation, he concluded that those individuals with the most favorable variations would have the best chance of surviving and transmitting their favorable traits to the next generation. Darwin had no knowledge of genetics and therefore did not know the cause

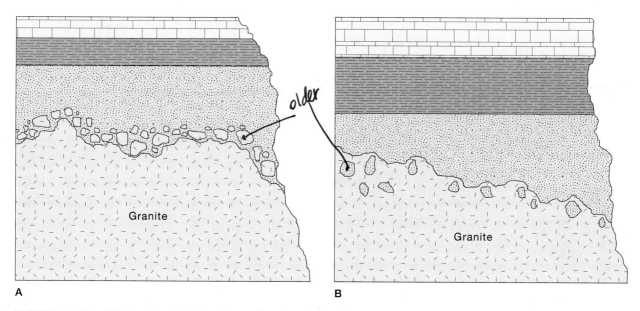

FIGURE 1–12 (A) Granite inclusions in sandstone indicate that granite is the older unit. (B) Inclusions of sandstone in granite indicate that sandstone is the older unit.

FIGURE 1–13 Charles Darwin as a young man. This portrait was made shortly after Darwin returned to England from his voyage around the world on the *H.M.S. Beagle*. Observations made during this voyage helped him formulate the concept of evolution by natural selection.

of the variation that was so important to his theory. Gregor Mendel's 1865 report of experiments in heredity had escaped his attention. In the decades following Darwin's death, geneticists clearly established that the variability essential to Darwin's natural selection is derived from new gene combinations that occur during reproduction and from genetic mutation.

Possibly because he was reluctant to face the controversy that his theory would provoke, Darwin did not publish his findings on his return to England. He did, however, confide in Charles Lyell and the great botanist Joseph Hooker, and these friends urged him to publish quickly before someone else anticipated his discoveries. Yet Darwin continued to procrastinate. Then, in 1858, a comparatively unknown young naturalist named Alfred Russel Wallace sent Darwin a manuscript for review that contained the basic concepts of natural selection. Wallace had conceived of natural selection while on a biological expedition to Indonesia. The idea came to him while suffering with malaria shortly after he read an essay on overpopulation by Thomas Malthus.

Understandably disturbed by Wallace's letter, Darwin sought the help of his friends Lyell and Hooker. Recognizing the importance of giving Darwin the credit he deserved for his discovery and the long years of assembling supporting evidence, the two scientists arranged for a presentation of Darwin's work and Wallace's paper before the Linnaean Society. Thus, the theory of natural selection was credited simultaneously to both scientists. Darwin

now worked at top speed to complete his famous *Origin of Species*. The book was published in 1859. In that volume, Darwin hoped to accomplish two things. The first was to convince the world that evolution had occurred. Organisms had evolved or changed throughout geologic time. The second was to propose a mechanism for evolution. That mechanism was natural selection. Darwin's success in achieving his objectives can be measured by the fact that, within a decade, organic evolution had become the guiding principle in all paleontologic and biologic research. His book changed the way people viewed the world, and for this reason it has been described as one of the greatest books of all time.

Darwin died at his home in Down, England, in 1882. By that time, geologists everywhere were using their knowledge of evolution, biologic succession, superposition, cross-cutting relationships, and inclusions to decipher Earth history.

■ *Time and Geology*

Many people are intrigued by the great age of rocks and fossils. Geology instructors are aware of this interest, for they are often asked the age of rock and mineral specimens brought to them by students and amateur collectors. When told that the samples are tens or even hundreds of millions of years old, the collectors are often pleased but also perplexed. "How can this person know the age of this specimen by just looking at it?" they think. If they insist on knowing the answer to that question, they may next receive a short discourse on the subject of geologic time. It is explained that the rock exposures from which the specimens were obtained have long ago been organized into a standard chronologic sequence based largely on superposition, evolution as indicated by fossils, and actual rock ages in years obtained from the study of radioactive elements in the rock. The geologist's initial estimate of age is based on experience. He or she may have spent a few hours kneeling at those same collecting localities and had a background of information to draw on. Thus, at least sometimes, a geologist can recognize particular rocks as being of a certain age. The science that permits accomplishment of this feat is called <u>geochronology</u>. It is a science that began over four centuries ago when Nicholas Steno described how the position of strata in a superpositional sequence could be used to show the relative geologic age of the layers. As described earlier, this simple but important idea was expanded and refined much later by William Smith and some of his contemporaries. These practical geologists showed how it was also possible to correlate strata. Outcrop by outcrop the rock sequences with their contained fossils were pieced together, one above the other, until a standard geologic time scale based on relative ages had been constructed.

There are two different frames of reference when dealing with geologic time. The work of William Smith and his contemporaries was based on the concept of **relative geologic dating.** It involved placing geologic events and the rocks representing those events in the order in which they occurred without reference to actual time or dates measured in years. Relative geologic time tells us which event preceded or followed another event, or which rock mass was older or younger relative to others. In contrast, **actual geologic dating** expresses the actual age of rocks or geologic events as determined by the decay of radioactive elements.

The Standard Geologic Time Scale

The early geologists had no way of knowing how many time units would be represented in the completed geologic time scale, nor could they know which fossils would be useful in correlation or which new strata might be discovered at a future time in some distant corner of the globe. Consequently, the time scale grew piecemeal, in an unsystematic manner. Units were named as they were discovered and studied. Sometimes the name for a unit was borrowed from local geography, from a mountain range in which rocks of a particular age were well exposed, or from an ancient tribe of Welshmen; sometimes the name was suggested by the kind of rocks that predominated.

Divisions in the Geologic Time Scale

Geologists have proposed the term **eon** for the largest divisions of the geologic time scale. In chronologic succession, the first three eons of geologic time are the **Hadean, Archean,** and **Proterozoic** (Fig. 1–14). The beginning of the Archean corresponds approximately to the ages of the oldest known rocks on Earth. Although not universally used, the term *Hadean* refers to that period of time for which we have no rock record, which began with the origin of the planet 4.6 million years ago. The Proterozoic Eon refers to the time interval from 2500 to 544 million years ago.

The rocks of the Archean and Proterozoic are informally referred to simply as Precambrian. The antiquity of Precambrian rocks was recognized in the mid-1700s by Johann Lehman, a professor of mineralogy in Berlin, who referred to them as the "primary series." One frequently finds this term in the writing of French and Italian geologists who were contemporaries of Lehman. In 1833, the term appeared again when Sir Charles Lyell used it in his formation of a surprisingly modern geologic time scale. Lyell and his predecessors recognized these "primary" rocks by their crystalline character and took their uppermost boundary to be an unconformity that separated them from the overlying—and therefore younger—fossiliferous strata.

All of the remainder of geologic time is included in the **Phanerozoic Eon.** As a result of careful study of superposition accompanied by correlations based on the abundant fossil record of the Phanerozoic, geologists have divided it into three major subdivisions, termed **eras.** The oldest is the **Paleozoic Era,** which we now know lasted about 300 million years. Following the Paleozoic is the **Mesozoic Era,** which continued about 179 million years. The **Cenozoic Era,** in which we are now living, began about 66 million years ago.

As shown in Figure 1–14, the eras are divided into shorter time units called **periods;** and periods in turn can be divided into **epochs.** Eras, periods, epochs, and divisions of epochs, called **ages,** all represent intangible increments of time. They are **geochronologic** units. The actual rocks formed or deposited during a specific time interval are called **chronostratigraphic** units. Table 1–1 indicates the chronostratigraphic units that correspond to geochronologic units. A **system** refers to the actual rock record of a period, whereas a **series** is the chronostratigraphic equivalent of an epoch, and a **stage** represents the tangible rock record of an age. As an example of the way these terms are used, one might correctly speak of climatic changes during the Cambrian *Period* as indicated by fossils found in the rocks of the Cambrian *System.*

Recognition of Geochronologic Units

Geochronologic units bear the same names as the chronostratigraphic units to which they correspond. Thus we may speak of the Jurassic System or the Jurassic Period according to whether we are referring to the rocks themselves or to the time during which they accumulated. Geochronologic terms have come into use as a matter of convenience. Their definition is necessarily dependent on the existence of tangible chronostratigraphic units. The steps leading to the recognition of chronostratigraphic units began with the use of superposition in establishing age relationships. Local sections of strata were used by early geologists to recognize beds of successively different age and, thereby, to record successive evolutionary changes in fauna and flora. (The order and nature of these evolutionary changes could be determined because higher layers are successively younger.) Once the faunal and floral succession was deciphered,

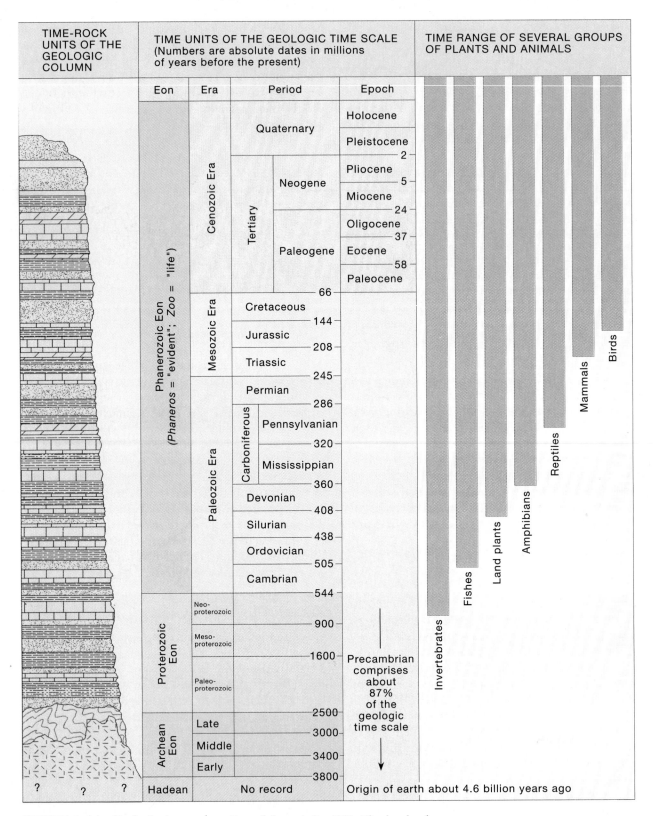

FIGURE 1–14   Geologic time scale.   *(From Palmer, A. R., 1983. The decade of North American geology, geologic time scale, Geology 11:503–504. Proterozoic divisions are those recommended by the Subcommission on Precambrian Stratigraphy of the International Union of Geological Sciences: Plumb, K. A. New Precambrian time scale, Episodes, 14(2):139–140.)*

TABLE 1–1   Hierarchy of Time and
Time-Stratigraphic Terms

Time Divisions	Equivalent Time-Stratigraphic Divisions
Era	Erathem (rarely used)
Period	System
Epoch	Series
Age	State
Chron	Zone (Chronozone)

fossils provided an additional tool for establishing the order of events. They could also be used for correlation, so that strata at one locality could be related to the strata of various other localities. No single place on Earth contains a complete sequence of strata from all geologic ages. Hence, correlation to standard sections of many widely distributed local sections was necessary in constructing the geologic time scale (Fig. 1–15). Clearly, the time scale was not conceived as a coherent whole but rather evolved piece by piece as a result of the individual studies of many geologists. Indeed, for some units at the series and stage level, the process continues even today. The fact that the time scale developed in piecemeal fashion is apparent when one reviews its growth and development.

The Geologic Systems

The Cambrian System   The rocks of the **Cambrian System** take their name from *Cambria*, the Latin name for Wales. Exposures of strata in Wales (Fig. 1–16) provide a standard section with which rocks elsewhere in Europe and on other continents can be correlated. The standard section in Wales is named Cambrian *by definition*. All other sections deposited during the same time as the rocks in Wales are recognized as Cambrian *by comparison* to the standard section.

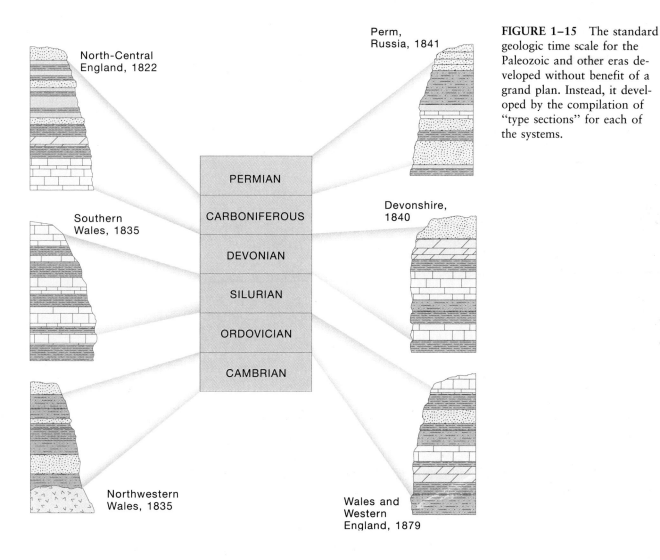

FIGURE 1–15   The standard geologic time scale for the Paleozoic and other eras developed without benefit of a grand plan. Instead, it developed by the compilation of "type sections" for each of the systems.

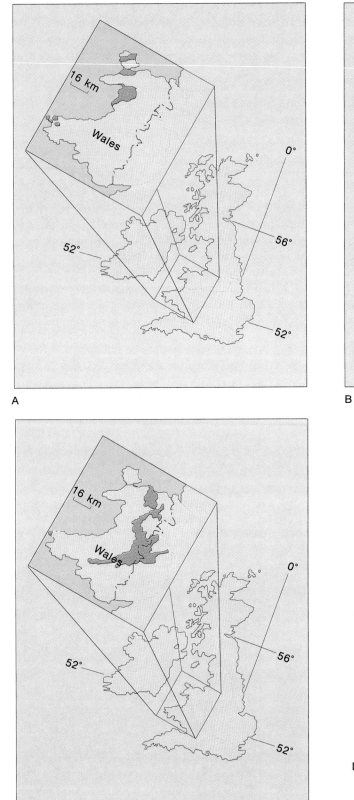

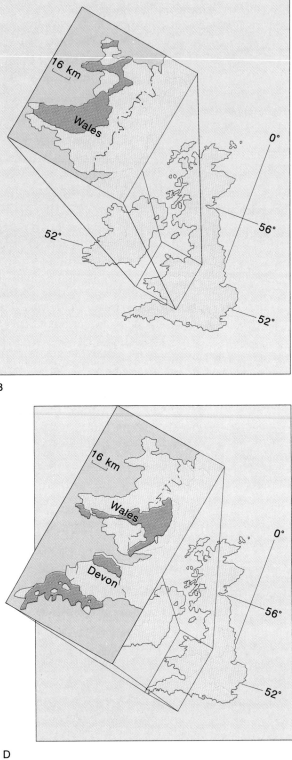

FIGURE 1–16 Outcrop areas for strata of the Cambrian (A), Ordovician (B), Silurian (C), and Devonian (D) systems in Great Britain.

Adam Sedgwick, a highly regarded professor of geology at Cambridge (Fig. 1–17), named the Cambrian in the 1830s for outcrops of poorly fossiliferous dark siltstones and sandstones. The area in north Wales that Sedgwick studied was noted for its complexity, yet he was able to unravel its geologic history on the basis of spatial relationships and lithology.

The Silurian and Ordovician Systems At about the same time that Sedgwick was laboring with outcrops that were to become the Cambrian System, his former student, Sir Roderick Impey Murchison, had begun studies of fossiliferous strata outcropping in the hills of south Wales, Murchison named these rocks the **Silurian System,** taking the name from early inhabitants of western England and Wales known as the *Silures.* In 1835, Murchison and Sedgwick presented a paper, *On the Silurian and Cambrian Systems, Exhibiting the Order in Which the Older Sedimentary Strata Succeed Each Other in England and Wales.* With this publication, the two geologists initiated the development of the early Paleozoic time scale. In the years that followed, a controversy arose between the two men that was to sever their friendship. Because Sedgwick had not described fossils distinctive of the Cambrian, the unit could not be recognized in other countries. Murchison argued, therefore, that the Cambrian was not a valid system. During the 1850s, he maintained that all fossiliferous strata above the "Primary Series" (the old name for Precambrian) and below the Old Red Sandstone (of Devonian age) belonged within the Silurian System. Sedgwick, of course, disagreed, but his opinion that the Cambrian was a valid system did not receive wide support until fossils were described from the upper part of the sequence. The fossils proved to be similar to faunas in Europe and North America. Hence, the Cambrian did meet the test of recognition outside England. Using these fossils as a basis for reinterpretation, the English geologist Charles Lapworth proposed combining the upper part of Sedgwick's Cam-

FIGURE 1–17 Adam Sedgwick, one of the foremost geologists of the nineteenth century. A professor of geology at Cambridge University. Sedgwick is best remembered for deciphering the highly deformed system of rocks in northwestern Wales that he defined as the Cambrian System. He also founded the geologic museum at Cambridge which bears his name. *(Courtesy of the Cambridge Museum, Cambridge, England.)*

brian and the lower part of Murchison's Silurian into a new system. In 1879, he named the system **Ordovician** after the *Ordovices,* an early Celtic tribe. The first three systems of the Paleozoic were thus established (Fig. 1–18).

The Devonian System The **Devonian System** was proposed for outcrops near Devonshire, England (Fig. 1–15) by Sedgwick and Murchison in 1839 (prior to the year of their bitter debate). They based

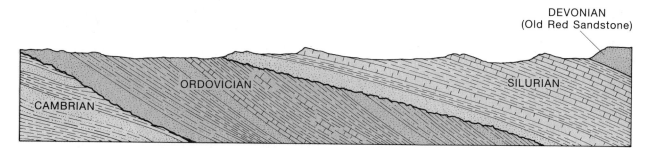

FIGURE 1–18 Generalized geologic cross-section for the Silurian type region. Unconformities separate the Ordovician from the Cambrian and Silurian systems. Silurian strata are inclined toward the east, with more resistant rocks forming escarpments that face toward the west.

their proposal on the fact that the rocks in question lay beneath the previously recognized Carboniferous System and contained a fauna that was different from that of the underlying Silurian and overlying Carboniferous. In their interpretation of the distinctive nature of the fauna, they were aided by the studies of William Lonsdale, a retired army officer who had become a self-taught specialist on fossil corals. Further evidence that the new unit was a valid one came when Murchison and Sedgwick were able to recognize it in the Rhineland region of Europe. The Devonian rocks of Devonshire were also found to be equivalent to the widely known *Old Red Sandstone* of Scotland and Wales.

The Carboniferous System   The term **Carboniferous System** was coined in 1822 by the English geologists William Conybeare and William Phillips to designate strata that included beds of coal in north-central England. Subsequently, it became convenient in Europe and Britain to divide the system into a *Lower Carboniferous* and *Upper Carboniferous*—the latter containing most of the workable coal seams. Two systems in North America, the **Mississippian** and **Pennsylvanian,** are broadly equivalent to these subdivisions. The American geologist Alexander Winchell formally proposed the name Mississippian in 1870 for the Lower Carboniferous strata that are extensively exposed in the upper Mississippi River drainage region. In 1891, Henry S. Williams provided the name Pennsylvanian for the Upper Carboniferous System. The Mississippian and Pennsylvanian are now recognized as subsystems of the Carboniferous.

The Permian System   The **Permian System** takes its name from a Russian province called Perm (from the nearby town of Perm) located on the western side of the Ural Mountains (Fig. 1–19). In 1840 and 1841,

Murchison, in company with the French paleontologist Edouard de Verneuil and several Russian geologists, traveled extensively across western Russia. To his delight, Murchison found he was able to recognize Silurian, Devonian, and Carboniferous rocks by the fossils they contained. As a result, he became even more convinced that groups of fossil organisms succeed one another in a definite and determinable order. Murchison established the new Permian System for rocks that overlay the Carboniferous System and contained fossils similar to those in German strata (the Zechstein beds), which had the same stratigraphic position as the Magnesian Limestone in England. Field studies had previously shown that the Magnesian Limestone rested on Carboniferous strata. Thus Murchison was able to include the Magnesian Limestone within the Permian by correlation. The fossils of the new system appeared distinctly intermediate between those of the Carboniferous below and the Triassic above. In a letter to the Society of Naturalists of Moscow dated October 8, 1841, Murchison stated, "The Carboniferous System is surmounted, to the east of the Volga, by a vast series of marls, schists, limestones, sandstones, and conglomerates to which I propose to give the name 'Permian System.'" Murchison's establishment of the Permian System provides a fine example of the logic employed by early geologists in putting together the pieces of the standard time scale.

The Triassic System   The influence of British geologists in providing names for the system of the Paleozoic is by now obvious. However, their presence is not as evident in the development of Mesozoic nomenclature (Fig. 1–20). The **Triassic,** for example, was applied in 1834 by a German geologist named Frederich von Alberti. The term refers to a threefold division of rocks of this age in Germany. However, because the German strata in the type area are poorly fossiliferous, the standard of reference has been shifted to richly fossiliferous marine strata in the Alps.

The Jurassic System   Another German scientist, Alexander von Humboldt, proposed the term **Jurassic** for strata of the Jura Mountains between France and Switzerland. However, in 1795, when he used the term, the concept of systems had not been developed. As a result, the Jurassic was redefined as a valid geologic system in 1839 by Leopold von Buch.

The Cretaceous System   During the same year that Conybeare and Phillips were defining the Carboniferous, a Belgian geologist named Omalius d'Halloy proposed the term **Cretaceous** (from the Latin *Creta,* meaning "chalk") for rock outcrops in France, Bel-

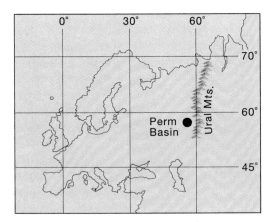

FIGURE 1–19   Location of the Perm Basin in Russia where Sir Roderick Murchison established the Permian System.

FIGURE 1–20 Type areas for the systems of the Mesozoic and Cenozoic.

gium, and Holland. Although chalk beds are prevalent in some Cretaceous exposures, the system is actually recognized on the basis of fossils. Indeed, some thick sections of Cretaceous rocks contain no chalks whatsoever.

The Tertiary System The name **Tertiary** leads us back to the time when geology was just beginning as a science. Giovanni Arduino (1713–1795) suggested a classification with four major divisions: Primary, Secondary, Tertiary, and Quaternary. The Tertiary was derived from his 1759 description of unconsolidated *montes tertiarii* sediments at the foot of the Italian Alps. Later, the Tertiary was more precisely defined, and standard sections for series of the Tertiary were established in France. The **Eocene, Miocene,** and **Pliocene,** for example, were proposed by Charles Lyell in 1832 on the basis of the proportions of species of living marine invertebrates in the fossil fauna. By definition, 3 percent of the fossil fauna of the Eocene still live, whereas the Miocene contained 17 percent, and the Pliocene contained 50 to 67 percent. The term **Oligocene** was proposed by August von Beyrich in 1854, and the term **Paleocene** was proposed 20 years later by Wilhelm Schimper. Other system names are also used in place of the Tertiary. Many geologists now use the terms **Paleogene System** (for the Paleocene, Eocene, and Oligocene) and **Neogene System** (for the Miocene and Pliocene).

The Quaternary System In 1829, the French geologist Jules Desnoyers proposed the term **Quaternary** for certain sediments and volcanics exposed in north-ern France. Although these deposits contained few fossils, Desnoyers was convinced on the basis of field studies that they were younger than Tertiary rocks. In the decade following Desnoyer's establishment of the Quaternary, the unit was further divided into an older **Pleistocene Series,** composed primarily of deposits formed during the glacial ages, and the younger **Holocene Series.**

This brief review describing how geologists drew up a table of geologic time clearly shows a lack of any grand and coherent design. These geologic pioneers were influenced by conspicuous changes in assemblages of fossils from one sequence of strata to another. In many places in Europe they found that such changes frequently occurred above and below an unconformity. The success of their methods is apparent from the fact that, by and large, the systems have persisted and found wide use even to the present day.

Quantitative Geologic Time

Early Attempts at Quantitative Geochronology

After having constructed a geologic time scale on the basis of relative age, it is understandable that geologists would seek some way to assign actual ages in millions of years to the various periods and epochs. From the time of Hutton, leaders in the scientific community were convinced that the Earth was indeed very old, and certainly it was much older than the approximately 6000 years estimated by biblical scholars from calculations involving the ages of post-Adamite generations. But how old was the Earth? And how might one quantify the geologic time scale?

To geologists of the 1800s, it was apparent that to determine the actual age of the Earth or of particular rock bodies, they would have to concentrate on natural processes that continue in a single recognizable way and that also leave some sort of tangible record in the rocks. Evolution is one such process, and Charles Lyell recognized this. By comparing the amount of evolution exhibited by marine mollusks in the various series of the Tertiary System with the amount that had occurred since the beginning of the Pleistocene Ice Age, Lyell estimated that 80 million years had elapsed since the beginning of the Cenozoic. He came astonishingly close to the mark. However, for older sequences, estimates based on rates of evolution were difficult, not only because of missing parts in the fossil record but also because rates of evolution for many taxa were not well understood.

In another attempt, geologists reasoned that if rates of deposition could be determined for sedimentary rocks, they might be able to estimate the time required for deposition of a given thickness of strata.

Similar reasoning suggested that one could estimate total elapsed geologic time by dividing the average thickness of sediment transported annually to the oceans into the total thickness of sedimentary rock that had ever been deposited in the past. Unfortunately, such estimates did not adequately account for past differences in rates of sedimentation or losses to the total stratigraphic section during episodes of erosion. Also, some very ancient sediments were no longer recognizable, having been converted to igneous and metamorphic rocks in the course of mountain building. As a result of these uncertainties, estimates of the earth's total age based on sedimentation rates ranged from as little as a million to over a billion years.

Yet another scheme for approximating the Earth's age was proposed in 1715 by Sir Edmund Halley (1656–1742), whose name we associate with the famous comet. Halley surmised that the ocean formed soon after the origin of the planet and therefore would be only slightly younger than the age of the solid Earth. He reasoned that the original ocean was not salty and that subsequently salt derived from the weathering of rocks was brought to the sea by streams. Thus, if one knew the total amount of salt dissolved in the ocean and the amount added each year, it might be possible to calculate the ocean's age. In 1899, the Irish geologist John Joly attempted the calculation. From information provided by gauges placed at the mouths of streams, Joly was able to estimate the annual increment of salt to the oceans. Then, knowing the salinity of ocean water and the approximate volume of water, he calculated the amount of salt already held in solution in the oceans. An estimate of the age of the oceans was derived from the following formula:

$$\frac{\text{Age of}}{\text{ocean}} = \frac{\substack{\text{Total salt in ocean} \\ \text{(in grams)}}}{\substack{\text{Rate of salt added} \\ \text{(grams per year)}}}$$

Beginning with essentially nonsaline oceans, it would have taken about 90 million years for the oceans to reach their present salinity, according to Joly. The figure, however, was off the mark by a factor of 50, largely because there was no way to account accurately for recycled salt and salt incorporated into clay minerals deposited on the sea floors. Vast quantities of salt once in the sea had become extensive evaporite deposits on land; some of the salt being carried back to the sea had been dissolved, not from primary rocks but from eroding marine strata on the continents. Even though in error, Joly's calculations clearly supported those geologists who insisted on an age for the Earth far in excess of a few million years. The belief

in the Earth's immense antiquity was also supported by Darwin, Huxley, and other evolutionary biologists, who saw the need for time in the hundreds of millions of years to accomplish the organic evolution apparent in the fossil record.

The opinion of the geologists and biologists that the Earth was immensely old was soon to be challenged by the physicists. Spearheading this attack was Lord Kelvin, considered by many to be the outstanding physicist of the nineteenth century. Kelvin calculated the age of the Earth on the assumption that it had cooled from a molten state and that the rate of cooling followed ordinary laws of heat conduction and radiation. Kelvin estimated the number of years it would have taken the Earth to cool from a hot mass to its present condition. His assertions regarding the age of the Earth varied over two decades of debate, but in his later years he confidently believed that 24 to 40 million years was a reasonable age for the Earth. The biologists and geologists found Kelvin's estimates difficult to accept. But how could they do battle against his elegant mathematics when they were themselves armed only with inaccurate dating schemes and geologic intuition? For those geologists unwilling to capitulate, however, new discoveries showed their beliefs to be correct and Kelvin's to be unavoidably wrong.

A more correct answer to the question "How old is the Earth?" was provided only after the discovery of radioactivity, a phenomenon unknown to Kelvin during his active years. With the detection of natural radioactivity by Henri Becquerel in 1896, followed by the isolation of radium by Marie and Pierre Curie two years later, the world became aware that the Earth had its own built-in source of heat. It was not inexorably cooling at a steady and predictable rate, as Kelvin had suggested.

Isotopic Methods for Dating Rocks

The solid Earth is composed of minerals and rocks. **Minerals** are solid, naturally occurring inorganic materials having a definite composition or range of compositions and usually possessing a uniform internal crystal structure. That uniform structure is derived from an orderly internal arrangement of the atoms that combine to make minerals. **Rocks** are solid, cohesive aggregates of the grains of one or more minerals. To understand better the way atoms in rocks and minerals can reveal the numerical age of geologic features, a brief review of the nature of atoms will be useful.

Atoms are the smallest particles of an element that can enter into a chemical reaction. They are also the smallest chemically indivisible particles of an element. An individual atom consists of an extremely minute but heavy nucleus surrounded by rapidly moving **electrons.** The electrons are relatively farther apart than are the planets surrounding our sun; consequently, the atom consists primarily of empty space. However, electrons move so rapidly around the nucleus that they effectively fill the space within their orbits, giving size to the atom and repelling other atoms that may approach.

In the nucleus of the atom are closely compacted particles called **protons,** which carry a unit charge of positive electricity equal to the unit charge of negative electricity carried by the electron. Associated with the protons in the nucleus are electrically neutral particles having the same mass as protons. These are called **neutrons.** Modern atomic physics has made us aware of still other particles in the nucleus. For our understanding of the atom, however, knowledge of protons and neutrons is sufficient. The number of protons in the nucleus of an atom establishes its number of positive charges and is called its **atomic number.** Each chemical element is composed of atoms having a particular atomic number. Thus, every element has a different number of protons in its nucleus.

There are 90 naturally occurring elements that range in atomic number from 1 (for one proton) in hydrogen to 92 (for 92 protons) in uranium (Table 1–2 and Appendix E).

The **mass** of an atom is approximately equal to the sum of the masses of its protons and neutrons. (The mass of electrons is so small that it need not be considered.) Carbon-12 is used as the standard for comparison of mass. By setting the atomic mass of carbon at 12, the atomic mass of hydrogen, which is the lightest of elements, is just a bit greater than 1 (1.008, to be precise). The nearest whole number to the total number of protons and neutrons in an element constitutes its **mass number.** Some atoms of the same substance have different mass numbers. Such variants are called **isotopes.** Isotopes are two or more varieties of the same element that have the same atomic number and chemical properties but differ in mass numbers because they have a varying number of neutrons in the nucleus. By convention, the mass number is noted as a superscript preceding the chemical symbol of an element, and the atomic number is placed beneath it as a subscript. Thus, $^{40}_{20}$Ca is translated as the element calcium having an atomic number of 20 and mass number of 40 (see Table 1–2).

TABLE 1–2 Number of Protons and Neutrons and Atomic Mass of Some Geologically Important Elements

Element and Symbol	Atomic Number (Number of Protons in Nucleus)	Number of Neutrons in Nucleus	Mass Number
Hydrogen (H)	1	0	1
Helium (He)	2	2	4
Carbon-12 (C)*	6	6	12
Carbon-14 (C)	6	8	14
Oxygen (O)	8	8	16
Sodium (Na)	11	12	23
Magnesium (Mg)	12	13	25
Aluminum (Al)	13	14	27
Silicon (Si)	14	14	28
Chlorine-35 (Cl)*	17	18	35
Chlorine-37 (Cl)	17	20	37
Potassium (K)	19	20	39
Calcium (Ca)	20	20	40
Iron (Fe)	26	30	56
Barium (Ba)	56	82	138
Lead-208 (Pb)*	82	126	208
Lead-206 (Pb)	82	124	206
Radium (Ra)	88	138	226
Uranium-238 (U)	92	146	238

*When two isotopes of an element are given, the most abundant is starred.

Radioactivity   The **radioactivity** discovered by Henri Becquerel was derived from elements such as uranium and thorium, which are unstable and break down or decay to form other elements or other isotopes of the same element. Any individual uranium atom, for example, will eventually decay to lead if given a sufficient length of time. To understand what is meant by "decay," let us consider what happens to a radioactive element such as uranium-238 (^{238}U). Uranium-238 has a mass number of 238. The "238" represents the sum of the weights of the atom's protons and neutrons (each proton and neutron having a mass of 1). Uranium has an atomic number (number of protons) of 92. Such atoms with specific atomic number and weight are sometimes termed **nuclides.** Sooner or later (and entirely spontaneously), the uranium-238 atom will fire off a particle from the nucleus called an **alpha particle.** Alpha particles are positively charged ions of helium. They have an atomic weight of 4 and an atomic number of 2. Thus, when the alpha particle is emitted, the new atom will now have an atomic weight of 234 and an atomic number of 90. The new atom, which is formed from another by radioactive decay, is called a **daughter element.**

From the decay of the parent nuclide, uranium-238, the daughter nuclide, thorium-234, is obtained (Fig. 1–21). A shorthand equation for this change is written

$$^{238}_{92}\text{U} \rightarrow ^{234}_{90}\text{Th} + ^{4}_{2}\text{He}$$

This change is not, however, the end of the process, for the nucleus of thorium-234 (^{234}Th) is not stable. It eventually emits a **beta particle** (an electron discharged from the nucleus when a neutron splits into a proton and an electron). There is now an extra proton in the nucleus but no loss of atomic weight because electrons are essentially weightless. Thus, from $^{234}_{90}$Th the daughter element $^{234}_{91}$Pa (protactinium) is formed. In this case, the atomic number has been increased by 1. In other instances, the beta particle may be captured by the nucleus, where it combines with a proton to form a neutron. The loss of the proton would decrease the atomic number by 1.

A third kind of emission in the radioactive decay process is called **gamma radiation.** It consists of a form of invisible electromagnetic waves having even shorter wavelengths than X-rays.

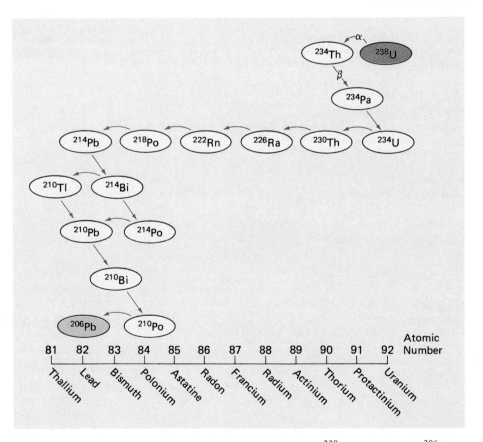

FIGURE 1–21   Radioactive decay series of uranium-238 (^{238}U) to lead-206 (^{206}Pb).

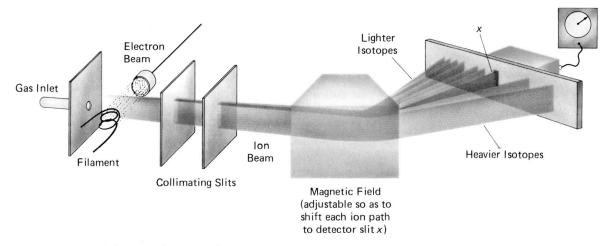

FIGURE 1–22 Schematic drawing of a mass spectrometer. In this type of spectrometer, the intensity of each ion beam is measured electrically (rather than recorded photographically) to permit determination of the isotopic abundances required for radiometric dating.

The rate of decay of radioactive isotopes is uniform and is not affected by changes in pressure, temperature, or the chemical environment. Therefore, once a quantity of radioactive nuclides has been incorporated into a growing mineral crystal, that quantity will begin to decay at a steady rate with a definite percentage of the radiogenic atoms undergoing decay in each increment of time. Each radioactive isotope has a particular mode of decay and a unique decay rate. As time passes, the quantity of the original or parent nuclide diminishes, and the number of the newly formed or daughter atoms increases, thereby indicating how much time has elapsed since the clock began its timekeeping. The beginning, or "time zero," for any mineral containing radioactive nuclides would be the moment when the radioactive parent atoms became part of a mineral from which daughter elements could not escape. The retention of daughter elements is essential, for they must be counted to determine the original quantity of the parent nuclide.

The determination of the ratio of parent to daughter nuclides is usually accomplished with the use of a *mass spectrometer,* an analytic instrument capable of measuring the atomic masses of elements and isotopes of elements. In the mass spectrometer, samples of elements are vaporized in an evacuated chamber, where they are bombarded by a stream of electrons. This bombardment knocks electrons off the atoms, leaving them positively charged. A stream of these positively charged ions is deflected as it passes between plates that bear opposite charges of electricity. The degree of deflection depends on the charge-to-mass ratio. In general, the heavier the ion, the less it will be deflected (Fig. 1–22).

Of the three major families of rocks, the igneous clan is by far the best for isotopic dating. Fresh samples of igneous rocks are less likely to have experienced loss of daughter products, which must be accounted for in the age determination. Igneous rocks can provide a valid date for the time that a silicate melt containing radioactive elements solidified.

In contrast to igneous rocks, the minerals of sediments can be weathered and leached of radioactive components, and age determinations are far more prone to error. In addition, the age of a detrital grain in a sedimentary rock does not give an age of the sedimentary rock but only of the parent rock that was eroded much earlier.

Dates obtained from metamorphic rocks may also require special care in interpretation. The age of a particular mineral may record the time the rock first formed or any one of a number of subsequent metamorphic recrystallizations.

Once an age has been determined for a particular rock unit, it is sometimes possible to use that data to approximate the age of adjacent rock masses. A shale lying below a lava flow that is 110 million years old and above another flow dated at 180 million years old must be between 110 and 180 million years of age (Fig. 1–23). Similarly, the age of a shale deposited on the erosional surface of a 490-million-year-old granite mass and covered by a 450-million-year-old lava flow must be between 450 and 490 million years old. The fossils in that shale might then be used to assign the shale to a particular geologic system or

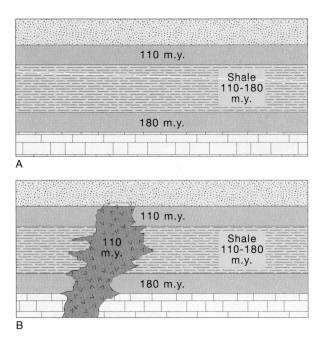

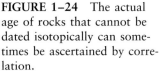

FIGURE 1–23 Igneous rocks that have provided absolute radiogenic ages can often be used to date sedimentary layers. (A) The shale is bracketed by two lava flows. (B) The shale lies above the older flow and is intruded by a younger igneous body. (Note: m.y. = million years.)

series. Then, by correlation, the quantitative age obtained at the initial location (Fig. 1–24, Section A) could be extended to correlative formations at other locations (Fig. 1–24, Section B).

Half-Life One cannot predict with certainty the moment of disintegration for any individual radioactive atom in a mineral. We do know that it would take an infinitely long time for all of the atoms in a quantity of radioactive elements to be entirely transformed to stable daughter products. Experimenters have also shown that there are more disintegrations per increment of time in the early stages than in later stages (Fig. 1–25), and one can statistically forecast what percentage of a large population of atoms will decay in a certain amount of time.

Because of these features of radioactivity, it is convenient to consider the time needed for half of the original quantity of atoms to decay. This span of time is termed the **half-life.** Thus, at the end of the time constituting one half-life, $\frac{1}{2}$ of the original quantity of radioactive element still has not undergone decay. After another half-life, $\frac{1}{2}$ of what was left remains or $\frac{1}{4}$ of the original quantity. After a third half-life, only $\frac{1}{8}$ would remain, and so on.

FIGURE 1–24 The actual age of rocks that cannot be dated isotopically can sometimes be ascertained by correlation.

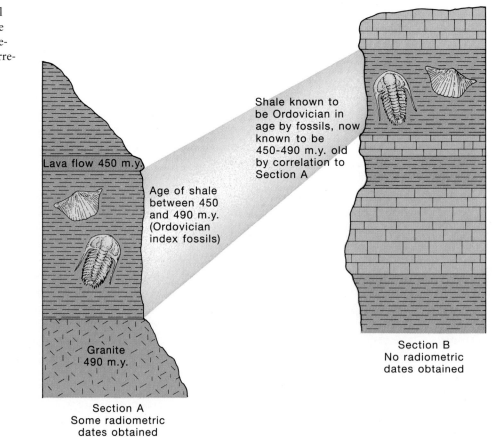

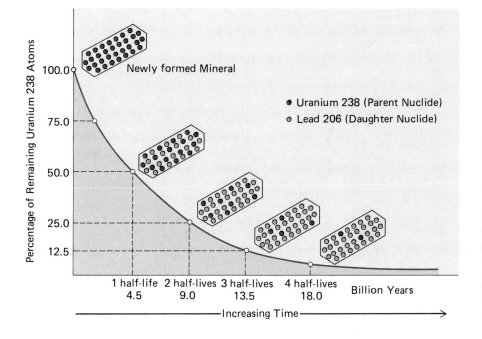

FIGURE 1–25 Rate of radioactive decay of uranium-238 to lead-206. During each half-life, one-half of the remaining amount of the radioactive element decays to its daughter element. In this simplified diagram, only the parent and daughter nuclides are shown, and the assumption is made that there was no contamination by daughter nuclides at the time the mineral formed.

Every radioactive nuclide has its own unique half-life. Uranium-235, for example, has a half-life of 704 million years. Thus, if a sample contains 50 percent of the original amount of uranium-235 and 50 percent of its daughter product, lead-207, then that sample is 704 million years old. If the analyses indicate 25 percent of uranium-235 and 75 percent of lead-207, two half-lives would have elapsed, and the sample would be 1408 million years old (Fig. 1–26).

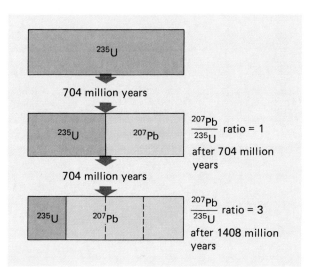

FIGURE 1–26 Radioactive decay of uranium-235 to lead-207.

The Principal Geologic Timekeepers

At one time, there were many more radioactive nuclides present on Earth than there are now. Many of these had short half-lives and have long since decayed to undetectable quantities. Fortunately for those interested in dating the Earth's most ancient rocks, there remain a few long-lived radioactive nuclides. The most useful of these are uranium-238, uranium-235, rubidium-87, and potassium-40 (Table 1–3). There are also a few short-lived radioactive elements that are used for dating more recent events. Carbon-14 is an example of such a short-lived isotope. There are also short-lived nuclides that represent segments of a uranium or thorium decay series.

Uranium–Lead Methods Dating methods involving lead require the presence of radioactive nuclides of uranium or thorium that were incorporated into rocks when they originated. To determine the age of a sample of mineral or rock, one must know the original number of parent nuclides as well as the number remaining at the present time. The original number of parent atoms should be equal to the sum of the present number of parent atoms and daughter atoms. The assumptions are made that the system has remained closed, so neither parent nor daughter atoms have ever been added or removed from the sample except by decay, and that no daughter atoms were present in the system when it formed. The presence, for example, of original lead in the mineral would cause the radiometric age to exceed the true age. Fortunately,

TABLE 1–3   Some of the More Useful Nuclides for Radioactive Dating

Parent Nuclide*	Half-Life (years)†	Daughter Nuclide	Materials Dated
Carbon-14	5730	Nitrogen-14	Organic materials
Uranium-235	704 million (7.04×10^8)	Lead-207 (and helium)	Zircon, uraninite, pitchblende
Potassium-40	1251 million (1.25×10^9)	Argon-40 (and calcium-40)‡	Muscovite, biotite, hornblende, volcanic rock, glauconite, K-feldspar
Uranium-238	4468 million (4.47×10^9)	Lead-206 (and helium)	Zircon, uraninite, pitchblende
Rubidium-87	48,800 million (4.88×10^{10})	Strontium-87	K-micas, K-feldspars, biotite, metamorphic rock, glauconite

*A *nuclide* is a convenient term for any particular atom (recognized by its particular combination of neutrons and protons).

†Half-life data from Steiger, R. H., and Jäger, E., 1977. Subcommission on geochronology: Convention on the use of decay constants in geo- and cosmochronology. *Earth and Planetary Science Letters* 36:359–362.

‡Although potassium-40 decays to argon-40 and calcium-40, only argon is used in the dating method because most minerals contain considerable calcium-40, even before decay has begun.

geochemists are able to recognize original lead and make the needed corrections.

As we have seen, different isotopes decay at different rates. Geochronologists take advantage of this fact by simultaneously analyzing two or three isotope pairs as a means to cross-check ages and detect errors. For example, if the ^{235}U/^{207}Pb radiometric ages and the ^{238}U/^{206}Pb ages from the same sample agree, then one can confidently assume that the age determination is valid.

Isotopic ages that depend on uranium–lead ratios may also be checked against ages derived from lead-207 to lead-206. Because the half-life of uranium-235 is much less than the half-life of uranium-238, the ratio of lead-207 (produced by the decay of uranium-235) to lead-206 will change regularly with age and can be used as a radioactive timekeeper (Fig. 1–27). This is called a lead–lead age, as opposed to a uranium–lead age.

The Potassium–Argon Method   Potassium and argon are another radioactive pair widely used for dating rocks. By means of electron capture (causing a proton to be transformed into a neutron), about 11 percent of the potassium-40 in a mineral decays to argon-40, which may then be retained within the parent mineral. The remaining potassium-40 decays to calcium-40 (by emission of a beta particle). The decay of potassium-40 to calcium-40 is not useful for obtaining numerical ages because radiogenic calcium cannot be distinguished from original calcium in a rock. Thus, geochronologists concentrate their efforts on the 11 percent of potassium-40 atoms that decay to argon. One advantage of using argon is that

it is inert—that is, it does not combine chemically with other elements. Argon-40 found in a mineral is very likely to have originated there following the decay of adjacent potassium-40 atoms in the mineral. Also, potassium-40 is an abundant constituent of many common minerals. However, like all isotopic dating methods, potassium–argon is not without its limitations. A sample will yield a valid age only if none of the argon has leaked out of the mineral being analyzed. Leakage may indeed occur if the rock has experienced temperatures above about 125°C. In specific localities, the ages of rocks dated by this method

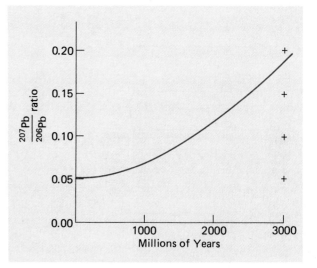

FIGURE 1–27   Graph showing how the ratio of lead-207 to lead-206 can be used as a measure of age.

COMMENTARY
Isotopic Dates for Sedimentary Rocks from Interstratified Ash

The geologic time scale was formulated as a *relative* scale. Since the development of isotopic dating methods, however, geologists have been working to provide numerical ages for the subdivisions of the scale. That effort continues today with periodic improvements and revisions of the absolute ages that calibrate the time scale (see Fig. 1–14). Because the Phanerozoic time scale is based on fossiliferous strata, an often unattainable but ideal procedure would be to date sedimentary rocks directly by using minerals that crystallized in the environment of deposition. Unfortunately, it is far easier to find radioactive nuclides in the minerals of igneous and metamorphic rocks than in sediments. One rather problematic exception is the sedimentary mineral glauconite, which can be dated by the potassium–argon method. Glauconite, however, loses argon on burial and thus is only useful in determining minimum ages. Far greater precision in dating sedimentary rocks can be obtained from radioactive isotopes in the minerals of volcanic ash beds that occur interstratified with sedimentary rocks. Based on uranium–lead ratios obtained from zircon crystals in ash layers, recently improved methods have provided numerical ages for key Paleozoic biostratigraphic boundaries with a precision of better than 1 percent. The analyses also provide ages for fossils in the sedimentary rocks and thereby lend important data for use in studies of rates of evolution, paleoecology, and paleogeography.

Ordovician limestone beds separated by a thin layer of altered volcanic ash called bentonite. The altered ash is in the recessed area beneath the upper massive layer of limestone. It is about 8 inches thick. An age of 453.7 million years was determined for the ash layer by high-resolution U-Pb zircon dating methods.

Reference
Tucker, R. D. et al. 1990. Time-scale calibration by high-precision U-Pb zircon dating of interstratified volcanic ashes in the Ordovician and Lower Silurian stratotypes of Britain. *Earth and Planetary Science Letters*, 100:51–58.

reflect the last episode of heating rather than the time of origin of the rock itself.

The half-life of potassium-40 is 1251 million years (1.251 billion years). As illustrated in Figure 1–28, if the ratio of potassium-40 to daughter products is found to be 1 to 1, then the age of the sample is 1251 million years (1.251 billion years). If the ratio is 1 to 3, then yet another half-life has elapsed, and the rock would have an isotopic age of two half-lives, or 2502 million years (2.502 billion years).

The Rubidium–Strontium Method   The dating method based on the disintegration by beta decay of rubidium-87 to strontium-87 can sometimes be used as a check on potassium–argon dates because rubidium and potassium are often found in the same minerals. The rubidium–strontium scheme has a further advantage in that the strontium daughter nuclide is not diffused by relatively mild heating events, as is the case with argon.

In the rubidium–strontium method, a number of samples are collected from the rock body to be dated. With the aid of the mass spectrometer, the amounts of radioactive rubidium-87, its daughter product strontium-87, and strontium-86 are calculated for each sample. Strontium-86 is an isotope not derived from radioactive decay. A graph is then prepared in which the $^{87}Rb/^{86}Sr$ ratio in each sample is plotted

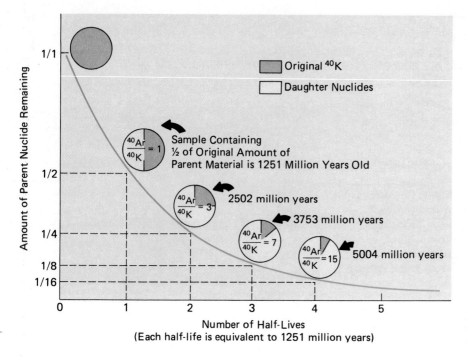

FIGURE 1–28 Decay curve for potassium-40.

against the $^{87}Sr/^{86}Sr$ ratio (Fig. 1–29). From the points on the graph, a straight line that is termed an **isochron** is constructed. The slope of the isochron results from the fact that, with the passage of time, there is continuous decay of rubidium-87, which causes the rubidium-87/strontium-86 ratio to decrease. Conversely, the strontium-87/strontium-86 ratio increases as strontium-87 is produced by the decay of rubidium-87. The older the rocks being investigated, the more the original isotope ratios will have been changed and the greater will be the inclination of the isochron. The slope of the isochron permits a computation of the age of the rock.

The rubidium–strontium and potassium–argon methods need not always depend on the collection of discrete mineral grains containing the required isotopes. Sometimes the rock under investigation is so finely crystalline and the critical minerals so tiny and dispersed that it is difficult or impossible to obtain a suitable collection of minerals. In such instances, large samples of the entire rock may be used for age determination. This method is called **whole-rock analysis.** It is useful not only for fine-grained rocks but also for rocks in which the yield of useful isotopes from mineral separates is too low for analysis. Whole-rock analysis has also been useful in determining the age of rocks that have been so severely metamorphosed that the potassium–argon or rubidium–strontium radiometric clocks of individual minerals have been reset. In such cases, the age obtained from the minerals would be that of the episode of metamorphism, not the total age of the rock itself. The required isotopes and their decay products, however, may have merely moved to nearby locations within the same rock body, and therefore analyses of large chunks of the whole rock may provide valid radiometric age determinations.

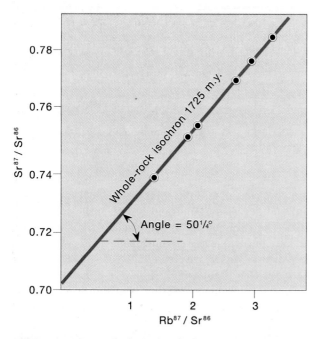

FIGURE 1–29 Whole-rock rubidium–strontium isochron for a set of samples of a Precambrian granite body exposed near Sudbury, Ontario. *(Modified from Krogh, T. E. et al. 1968. Carnegie Institute Washington Year Book 66:530.)*

The Carbon-14 Method Techniques for age determination based on content of radiocarbon were first devised by W. F. Libby and his associates at the University of Chicago in 1947. The method is an indispensable aid to the archeologic research and is useful in deciphering very recent events in geologic history. Because of the short half-life of carbon-14—a mere 5730 years—organic substances older than about 50,000 years contain very little carbon-14. New techniques, however, allow geologists to extend the method's usefulness back to almost 100,000 years.

Unlike uranium-238, carbon-14 is created continuously in the Earth's upper atmosphere. The story of its origin begins with cosmic rays, which are extremely high-energy particles (mostly protons) that bombard the Earth continuously. Such particles strike atoms in the upper atmosphere and split their nuclei into small particles, among which are neutrons. Carbon-14 is formed when a neutron strikes an atom of nitrogen-14. As a result of the collision, the nitrogen atom emits a proton, captures a neutron, and becomes carbon-14 (Fig. 1–30). Radioactive carbon is being created by this process at the rate of about two atoms per second for every square centimeter of the Earth's surface. The newly created carbon-14 combines quickly with oxygen to form CO_2, which is then distributed by wind and water currents around the globe. It soon finds its way into photosynthetic plants because they utilize carbon dioxide from the atmosphere to build tissues. Plants containing carbon-14 are ingested by animals, and the isotope becomes a part of their tissue as well.

Eventually, carbon-14 decays back to nitrogen-14 by the emission of a beta particle. A plant removing CO_2 from the atmosphere should receive a share of carbon-14 proportional to that in the atmosphere. A

FIGURE 1–30 Carbon-14 is formed from nitrogen in the atmosphere. It combines with oxygen to form radioactive carbon dioxide and is then incorporated into all living things.

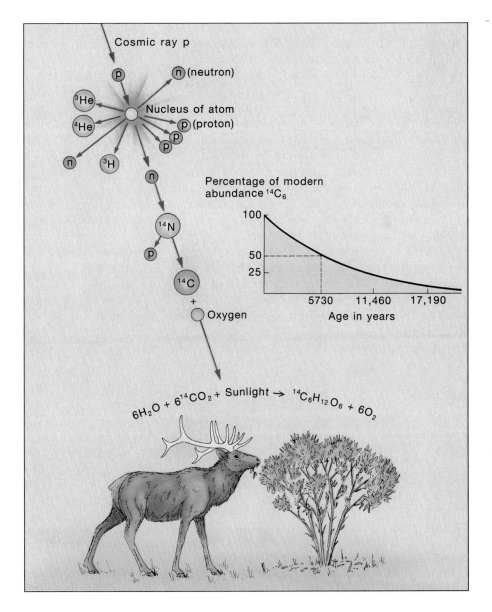

state of equilibrium is reached in which the gain in newly produced carbon-14 is balanced by the decay loss. The rate of production of carbon-14 has varied somewhat over the past several thousand years (Fig. 1–31). As a result, corrections in age calculations must be made. Such corrections are derived from analyses of standards such as wood samples, whose exact age is known from tree ring counts.

The age of some ancient bits of organic material is not determined from the ratio of parent to daughter nuclides, as is done with previously discussed dating schemes. Rather, the age is estimated from the ratio of carbon-14 to all other carbon in the sample. After an animal or plant dies, there can be no further replacement of carbon from atmospheric CO_2, and the amount of carbon-14 already present in the once living organism begins to diminish in accordance with the rate of carbon-14 decay. Thus, if the carbon-14 fraction of the total carbon in a piece of pine tree buried in volcanic ash were found to be about 25 percent of the quantity in living pines, then the age of the wood (and the volcanic activity) would be two half-lives of 5730 years each, or 11,460 years.

The carbon-14 technique had considerable value to geologists studying the most recent events of the Pleistocene Ice Age. Prior to the development of the method, the age of sediments deposited by the last advance of continental glaciers was surmised to be about 25,000 years. Radiocarbon dates of a layer of peat beneath the glacial sediments provided an age of only 11,400 years. The method has also been found useful in studies of groundwater migration and in dating the geologically recent uppermost layer of sediment on the sea floors. Carbon-14 analysis of tissue from the baby mammoth depicted in Figure 4–7 indicates that the animal died 27,000 years ago. The age of giant ground sloth dung recovered from a cave near Las Vegas indicated the presence of these great beasts in Nevada 10,500 years ago. Charcoal from the famous Lascaux cave in France revealed that the artists that drew pictures of mammoths, woolly rhi-noceroses, bison, and reindeer on the walls of the cave lived about 15,000 years ago. In archaeology, dates obtained by the carbon-14 method overturned many cherished concepts by demonstrating that the beginnings of agriculture and urbanization occurred much earlier than had been thought, and the arrival of full civilization was more recent.

Nuclear Fission Track Timekeepers   Nuclear particle fission tracks were discovered in the early 1960s, when scientists using the electron microscope were able to examine the areas around presumed locations of radioactive particles that were embedded in mica. Closer examination showed that the tracks were really small tunnels—like bullet holes—that were produced when high-energy particles of the nucleus of uranium were fired off in the course of spontaneous fission (spontaneous fragmentation of an atom into two or more lighter atoms and nuclear particles). The particles speed through the orderly rows of atoms in the crystal, tearing away electrons from atoms located along the path of trajectory and rendering them positively charged. Their mutual repulsion produces the track (Fig. 1–32). The tracks are only a few atoms in width and are impossible to see without an electron microscope. Therefore, the sample is immersed for a short period of time in a suitable solution (acid or alkali), which rushes up into the tubes, enlarging the track tunnel so that it can be seen with an ordinary microscope.

The natural rate of track production by uranium atoms is very slow and occurs at a constant rate. For this reason, the tracks can be used to determine the number of years that have elapsed since the uranium-bearing mineral solidified. One first determines the number of uranium atoms that have already disintegrated. This number is obtained with the aid of a microscope by counting the etched tracks. Next, one must find the original number of uranium atoms. This quantity can be determined by bombarding the sample with neutrons in a reactor and thereby caus-

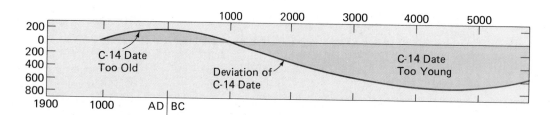

FIGURE 1–31   Deviation of carbon-14 ages to true ages from the present back to about 5000 B.C. Data are obtained from analysis of bristle cone pines from the western United States. Calculations of carbon-14 deviations are based on half-life of 5730 years.   *(Adapted from Ralph, E. K., Michael, H. N., and Han, M. C. 1973. Radiocarbon dates and reality, MASCA Newsletter 9:1.)*

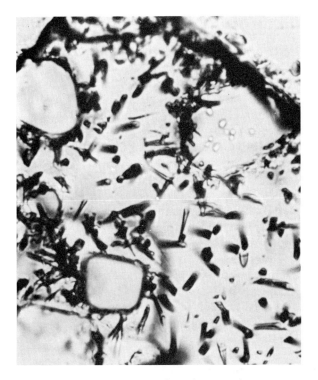

FIGURE 1–32 Fission tracks. These tracks were produced by plutonium-244 in a melilite crystal that was extracted from a meteorite. The small crystals are spinel inclusions. Melilite is a calcium–magnesium aluminosilicate. *(Photograph courtesy of Frank Podosek, Department of Earth and Planetary Sciences, Washington University.)*

ing the remaining uranium to undergo fission. A second count of tracks reveals the original quantity of uranium. Finally, one must know the spontaneous fission decay rate for uranium-238. This information is determined by counting the tracks in a piece of uranium-bearing synthetic glass of known date of manufacture.

Fission track dating is of particular interest to geochronologists because it can be used to date specimens only a few centuries old as well as to date rocks billions of years in age. The method helps to date the period between 50,000 and 1 million years ago—a period for which neither carbon-14 nor potassium–argon methods are suitable. As with all radiometric techniques, however, there can be problems. If rocks have been subjected to high temperatures, tracks may heal and fade away.

■ *The Age of the Earth*

Anyone interested in the total age of the Earth must decide what event constitutes its "birth." Most geologists assume that "year 1" commenced as soon as the Earth had collected most of its present mass and had developed a solid crust. Unfortunately, rocks that date from those earliest years have not been found on Earth. They have long since been altered and converted to other rocks by various geologic processes. The oldest materials known are grains of the mineral zircon taken from a sandstone in western Australia. The zircon grains are 4.1 to 4.2 billion years old. The zircons were probably eroded from nearby granitic rocks and deposited, along with quartz and other detrital grains, by rivers. Other very old rocks on Earth include 3.7-billion-year-old granites of southwestern Greenland, metamorphic rocks of about the same age from Minnesota, and 3.96-billion-year-old rocks from the Northwest territories of Canada (north of Yellowknife, Canada).

Meteorites, which many consider to be remnants of a shattered planet or asteroid that originally formed at about the same time as the Earth, have provided uranium–lead and rubidium–strontium ages of about 4.6 billion years. From such data, and from estimates of how long it would take to produce the quantities of various lead isotopes now found on the Earth, geochronologists feel that the 4.6-billion-year age for the Earth can be accepted with confidence. Evidence substantiating this conclusion comes from returned moon rocks. The ages of these rocks range from 3.3 to about 4.6 billion years. The older age determinations are derived from rocks collected on the lunar highland, which may represent the original lunar crust. Certainly, the moons and planets of our solar system originated as a result of the same cosmic processes and at about the same time.

| *Summary*

Simply stated, geology is the study of all naturally occurring processes, phenomena, and materials of the past and present Earth. Historical geology is that branch of the science concerned particularly with decoding the rock record of the Earth's long history. In the last few decades, advances in technology have added immensely to the store of geologic knowledge.

Interpretation of the new data requires an exceptional understanding not only of geology but also of physics, chemistry, mathematics, and biology. However, the historical inferences that are drawn from the data are frequently derived from the fundamental geologic and paleontologic concepts introduced by Nicolaus Steno (superposition), James Hutton (uniformitarianism), Charles Lyell (cross-cutting relationships, inclusion), William Smith (correlation), Charles Darwin (organic evolution), and others. These scientists formulated the principles by which geologists determine the relative age of rock outcrop, its history of deposition and deformation, and its spatial and chronologic relationship to strata in other regions of the Earth.

In the early stages of its development, geology was totally dependent on *relative* dating of events. James Hutton helped scientists visualize the enormous periods needed to accomplish the events indicated in sequences of strata, and the geologists that followed him pieced together the many local stratigraphic sections by using fossils and superposition. A scale of relative geologic time gradually emerged. Initial attempts to decide what the rock succession meant in terms of years were made by estimating the amount of salt in the ocean, the average rate of deposition of sediment, and the rate of cooling of the Earth. However, these early schemes did little more than suggest that the planet was at least tens of millions of years old and that the traditional concept of a 6000-year-old Earth did not agree with what could be observed geologically.

An adequate means of measuring geologic time was achieved only after the discovery of radioactivity at about the turn of the twentieth century. Scientists found that the rate of decay by radioactivity of certain elements is constant and can be measured and that the proportion of parent and daughter elements can be used to reveal how long they had been present in a rock. Over the years, continuing efforts by investigators as well as improvements in instrumentation (particularly of the mass spectrometer) have provided many thousands of age determinations. Frequently, these numerical dates have shed light on difficult geologic problems, provided a way to determine rates of movement of crustal rocks, and permitted geologists to date mountain building or to determine the time of volcanic eruptions. In a few highly important regions, isotopic dates have been related to particular fossiliferous strata and have thereby helped to quantify the geologic time scale and to permit estimation of rates of organic evolution. Isotopic dating has also changed the way humans view their place in the totality of time.

The isotopic transformations most widely used in determining absolute ages are uranium-238 to lead-206, uranium-235 to lead-207, thorium-232 to lead-208, potassium-40 to argon-40, rubidium-87 to strontium-87, and carbon-14 to nitrogen-14. Methods involving uranium–lead ratios are of importance in dating the Earth's oldest rocks. The short-lived carbon-14 isotope that is created by cosmic ray bombardment of the atmosphere provides a means to date the most recent events in Earth history. For rocks of intermediate age, schemes involving potassium–argon ratios, those utilizing intermediate elements in decay series, or those employing fission fragment tracks are most useful. A figure of 4.6 billion years for the Earth's total age is now supported by ages based on meteorites and on lead ratios from terrestrial samples.

Improvements in numerical geochronology are being made daily and will provide further calibration of the standard geologic time scale in the future. Some of the time boundaries in the scale, such as that between the Cretaceous and Tertiary Systems, are already well validated. Others, such as the boundary between the Paleozoic and Mesozoic, require additional refinement. Additional efforts to incorporate isotopic ages into sections of sedimentary rocks are among the continuing tasks of historical geologists. The usual methods for determining the age of strata involve the dating of intrusions that penetrate these sediments or the dating of interbedded volcanic layers. Less frequently, strata can be dated by means of radioactive isotopes incorporated within sedimentary minerals that formed in place at the time of sedimentation. At present, the best numerical age estimates indicate that Paleozoic sedimentation began about 544 million years ago, the Mesozoic Era began about 245 million years ago, and the Cenozoic commenced about 66 million years ago.

Review Questions

1. What observations were used by Roderick Murchison in the establishment of the Permian System?
2. Explain the difference between a geochronologic term (sometimes called a time term) and a chronostratigraphic term (less formally called a time–rock term)?
3. How did William Smith decide if a bed of rock at a particular location was equivalent to another bed of rock at a distant location?
4. What important component of our modern concept of organic evolution was proposed by Charles Darwin?
5. What criteria might a geologist seek in distin-

guishing the top from the bottom of a stratum that had been forced into vertical alignment or overturned during mountain building?

6. If an intrusion of once molten rock is seen to cut across layers of another rock, which is older?

7. What types of radiation accompany the decay of radioactive isotopes?

8. In attempting an age determination based on the uranium–lead method, why should an investigator select an unweathered sample for analysis?

9. How do the isotopes carbon-12 and carbon-14 differ from one another in regard to number of

protons and neutrons in the nucleus? How is carbon-14 produced in the atmosphere?

10. How do fission tracks originate? What geologic conditions might destroy fission tracks in a mineral crystal?

11. What would be the effect on the isotopic age of a zircon crystal being dated by the potassium–argon method if a small amount of argon-40 escaped from the crystal?

12. State the age of a sample of mummified skin from a prehistoric human that contained 12.5 percent of an original amount of carbon-14.

Discussion Questions

1. Discuss the principles formulated by Steno, Lyell, Smith, and other early geologists in the development of the geologic time scale.

2. Describe the general steps used by geologists and other scientists in their attempt to solve particular problems or understand natural phenomena.

3. What is the meaning of *uniformitarianism*? Cite an example of a process occurring on the earth today that did not occur in the geologic past.

4. Why is the concept of *half-life* necessary? (Why not use "whole life?") What are the half-lives of uranium-238, potassium-40, and carbon-14?

5. Pebbles of a black rock called basalt occur in a sedimentary rock composed of pebbles that is called conglomerate. The pebbles yield an isotopic age of 300 million years. What can be said about the age of the conglomerate? Several miles

away, the conglomerate was cut in a cross-cutting relationship by once molten rock that was found to be 200 million years old. What now can be said about the age of the conglomerate?

6. Has the amount of uranium in the earth increased, decreased, or remained the same over the past 4.6 billion years? What can be said about the amount of lead?

7. How are dating methods involving decay of radioactive elements unlike methods for determining elapsed time by the funneling of sand through an hourglass?

8. What is the advantage of having both uranium and thorium present in a mineral being used for an isotopic age determination?

Supplemental Readings and References

Adams, F. D. 1938. *The Birth and Development of the Geological Sciences.* Baltimore: Williams & Wilkins Co. (Reprinted by Dover Press, 1954.)

Albritton, C. C., Jr. 1980. *The Abyss of Time.* San Francisco, Calif.: Freeman, Cooper & Co.

Berry, W. B. N. 1987. *Growth of a Prehistoric Time Scale.* San Francisco: W. H. Freeman Co.

Dean, D. R. 1992. *James Hutton and the History of Geology.* Ithaca, N.Y.: Cornell Univ. Press.

Eicher, D. L. 1976. *Geologic Time* (2nd ed.). Englewood Cliffs, N.J.: Prentice-Hall, Inc.

Eiseley, L. C. 1959. Charles Lyell, *Sci. Am.* 201(8):98–106. (Offprint No. 846. San Francisco: W. H. Freeman Co.)

Geikie, A. 1905. *The Founders of Geology* (2nd ed.). New York: Macmillan. (Reprinted by Dover Press, 1954.)

Hallam, A. 1983. *Great Geologic Controversies.* Oxford, England: Oxford Univ. Press.

York, D., and Farquhar, R. M. 1972. *The Earth's Age and Geochronology.* Oxford, England: Pergamon Press.

Pearly white, translucent crystals of selenite gypsum ($CaSO_4 \cdot 2H_2O$), Lechuguilla Cave, New Mexico. *(Photograph by Dave Bunnell.)*

What stuff 'tis made of, whereof it is born, I am to learn.

SHAKESPEARE, *The Merchant of Venice*

CHAPTER 2

Earth Materials

Some readers of this book will have taken a prior course in physical geology. These students will have been introduced to rocks and minerals and may use this chapter as a review. For other readers, an examination of earth materials provides essential background for the study of the Earth's origin and the processes that have continuously shaped our planet through its long history. Minerals and rocks (and the fossils within rocks) are the indispensable documents from which the history of the Earth is deciphered.

■ *Minerals*

Minerals are important to all of us. They are the raw materials needed to manufacture the products that sustain our modern technological society. For geologists, the search for new mineral deposits has always been a key objective. Minerals, however, are also of interest to earth scientists because of the evidence they provide about past events. Many minerals form only within a narrow range of physical conditions, and therefore can be used to diagnose the pressures and temperatures involved in the formation of mountain ranges and volcanoes. Some minerals develop exclusively in ocean water and provide evidence of the incursion of seas across former areas of dry land. Others form under conditions of excessive aridity and are used to locate the arid tropical belts of long ago. The magnetic properties of certain minerals provide clues to drifting continents, the widening of oceans, and changes in the planet's magnetic poles. As a further aid to deciphering Earth history, some minerals contain radioactive elements that permit us to determine the actual ages of rocks and structures formed at particular times in the geologic past.

Common Rock-Forming Minerals

Minerals are naturally occurring elements or compounds formed by inorganic processes that have a definite chemical composition or range of compositions as well as distinctive properties that reflect the composition and regular internal atomic structure. Minerals can usually be identified by such physical properties as color, hardness, density, crystal form, and cleavage. **Cleavage** refers to the tendency of some minerals to break smoothly along certain planes of weakness.

More than 3000 minerals have already been discovered and described, but most of these are rarely encountered. For our present purposes, it is important only to consider those minerals that compose the bulk of common rocks. Among these, the most important are the silicate minerals.

Silicate Minerals

About 75 percent by weight of the Earth's crust is composed of the two elements oxygen and silicon (Table 2–1). These elements usually occur in combination with such abundant metals as aluminum, iron, calcium, sodium, potassium, and magnesium to form a group of minerals called the silicates. A single family of silicates, the feldspars, comprises about one-half of the material of the crust, and a single mineral species called quartz represents a sizable portion of the remainder.

Quartz   The principal silicate minerals and some of their properties are listed in Table 2–2. The mineral **quartz** (SiO_2) is one of the most familiar and important of all the silicate minerals. It is common in many different families of rocks. In general appearance, quartz is a glassy, colorless, gray, or white mineral. Quartz is relatively hard (see Mohs scale in the footnote of Table 2–2) and will easily scratch glass. When quartz grows in an open cavity, it may form the hexagonal crystals (Fig. 2–1) that are prized by mineral collectors. More frequently, the crystal faces cannot be discerned because the orderly addition of atoms during crystal growth was interrupted by contact with other growing crystals and results in an aggregate that exhibits crystalline texture. Such common minerals as chert, flint, jasper, and agate are sedimentary varieties of quartz. *Chert* (Fig. 2–2) is a dense, hard, white mineral composed of sub-microcrystalline quartz that forms as a result of the precipitation of silicon dioxide by either biologic or chemical means. *Flint* is the popular name for a dark gray or black variety of chert in which the dark color results from inclusions of organic matter. *Jasper* is an opaque variety of quartz and is red, brown, green, or

FIGURE 2–1   A large crystal of quartz. This specimen is 14 cm (about 5.5 inches) tall.

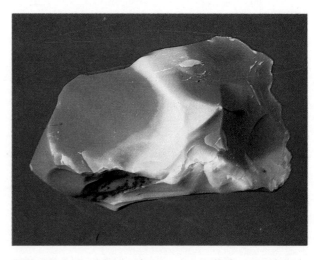

FIGURE 2–2   Chert. This variety (called novaculite) has a grainy texture that makes it useful as a grinding stone.

TABLE 2–1 Abundances of Chemical Elements in the Earth's Crust*

Element and Symbol	Percentage by Weight	Percentage by Number of Atoms	Percentage by Volume
Oxygen (O)	46.6	62.6	93.8
Silicon (Si)	27.7	21.2	0.9
Aluminum (Al)	8.1	6.5	0.5
Iron (Fe)	5.0	1.9	0.4
Calcium (Ca)	3.6	1.9	1.0
Sodium (Na)	2.8	2.6	1.3
Potassium (K)	2.6	1.4	1.8
Magnesium (Mg)	2.1	1.9	0.3
All other elements	1.5	—	—
	100.0	100.0[†]	100.0[†]

*Based on B. Mason, *Principles of Geochemistry*. New York: John Wiley & Sons, 1966. Note the high percentage of oxygen in the Earth's crust.

[†]Includes only the first eight elements.

TABLE 2–2 Common Rock-Forming Silicate Minerals

Silicate Mineral	Composition	Physical Properties
Quartz	Silicon dioxide (silica, SiO_2)	Hardness of 7 (on scale of 1 to 10)*; will not cleave (fractures unevenly); specific gravity: 2.65
Orthoclase feldspar group	Aluminosilicates of potassium	Hardness of 6.0–6.5; cleaves well in two directions; pink or white; specific gravity: 2.5–2.6
Plagioclase feldspar group	Aluminosilicates of sodium and calcium	Hardness of 6.0–6.5; cleaves well in two directions; white or gray; may show striations on cleavage planes; specific gravity: 2.6–2.7
Muscovite mica *(Ferromagnesian minerals)*	Aluminosilicates of potassium with water	Hardness of 2–3; cleaves perfectly in one direction, yielding flexible thin plates; colorless; transparent in thin sheets; specific gravity: 2.8–3.0
Biotite mica	Aluminosilicates of magnesium, iron, potassium, with water	Hardness of 2.5–3.0; cleaves perfectly in one direction, yielding flexible, thin plates; black to dark brown; specific gravity: 2.7–3.2
Pyroxene group	Silicates of aluminum, calcium, magnesium, and iron	Hardness of 5–6; cleaves in two directions at 87° and 93°; black to dark green; specific gravity: 3.1–3.5
Amphibole group	Silicates of aluminum, calcium, magnesium, and iron	Hardness of 5–6; cleaves in two directions at 56° and 124°; black to dark green; specific gravity: 3.0–3.3
Olivine	Silicates of magnesium and iron	Hardness of 6.5–7.0; light green; transparent to translucent; specific gravity: 3.2–3.6
Garnet group	Aluminosilicates of iron, calcium, magnesium, and manganese	Hardness of 6.5–7.5; uneven fracture; red, brown, or yellow; specific gravity: 3.5–4.3

*The scale of hardness used by geologists was formulated in 1822 by Frederich Mohs. Beginning with diamond as the hardest mineral, he arranged the following table:

10 Diamond	8 Topaz	6 Feldspar	4 Fluorite	2 Gypsum
9 Corundum	7 Quartz	5 Apatite	3 Calcite	1 Talc

C O M M E N T A R Y
The Quartz We Wear

Many varieties of the mineral quartz are so attractive when cut and polished that they are highly valued as gems. Single crystals may be cut into faceted geometric shapes that greatly enhance their luster and color. Quartz minerals may also be cut and polished into rounded, convex shapes called cabochons. Among the quartz family gems that are cut from single crystals are amethyst (purple quartz), citrine (yellow quartz), rose quartz, smoky quartz, and rock crystal (colorless quartz). The purple of amethyst is attributed to impurities of colloidal iron. Rose quartz get its color from impurities of molybdenum and often has a cloudy appearance because of myriads of tiny cracks and inclusions. Yellow and brown colors are attributed to defects in the crystal structure of the quartz rather than impurities.

In addition to quartz gems cut from single crystals, many varieties used in jewelry are composed of an aggregate of microcrystalline crystals. The individual crystals are far too small to see except at high magnification. Chalcedony is a common type of microcrystalline quartz that often exhibits a waxy luster. Chalcedony that displays banding is called banded agate. Moss agate is unbanded and contains green or black mosslike inclusions. Carnelian is red microcrystalline quartz, and chrysophase is an attractive green variety. Tiger's-eye quartz is known for its distinctive play of colors that seem to change as the stone is turned in the light. This characteristic, termed chatoyancy, results from light that is reflected off the mineral's fibrous structure. The fibers are relics of asbestos fibers that have been replaced by silica. Jasper is an opaque variety of microcrystalline quartz whose rich red, yellow, or brown colors are the result of a high content of iron oxide.

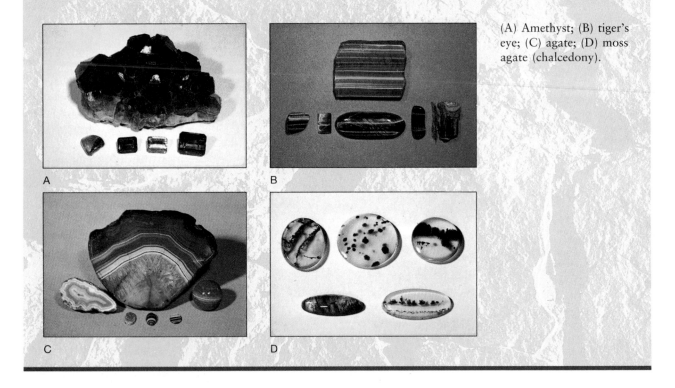

(A) Amethyst; (B) tiger's eye; (C) agate; (D) moss agate (chalcedony).

yellow. *Agate* is a variety of quartz that exhibits distinct banding (Fig. 2–3).

The origin of chert is a complex problem made even more difficult by the fact that different varieties are formed by somewhat different processes. Some cherts are replacements of earlier carbonate rocks. Others appear to have formed as a result of the solution and reprecipitation of silica from the siliceous skeletal remains of organisms. Small amounts of chert may also be precipitated directly from concentrated aqueous solutions.

The Feldspars The **feldspars** are the most abundant constituents of rocks, composing about 60 percent of the total weight of the Earth's crust. There are two major families of feldspars: the orthoclase or potas-

FIGURE 2–3 Agate (banded outer portion) with crystalline quartz in the center.

A

B

FIGURE 2–4 (A) Crystals of orthoclase (potassium feldspar). (B) A cleavage fragment of orthoclase.

sium feldspar group (Fig. 2–4), which comprises the potassium aluminosilicates, and the plagioclase group, which comprises the aluminosilicates of sodium and calcium (Fig. 2–5). Members of the plagioclase group exhibit a wide range in composition—from a calcium-rich end member called *anorthite* ($CaAl_2Si_2O_8$) to a sodium-rich end member called *albite* ($NaAlSi_3O_8$). Between these two extremes, plagioclase minerals containing both sodium and calcium occur. The substitution of sodium for calcium, however, is not random but is governed by the temperature and composition of the parent material. Thus, by examining the feldspar content of a once molten rock, it is possible to infer the physical and chemical conditions under which it originated. Feldspars are nearly as hard as quartz and range in color from white or pink to bluish gray. They have good cleavage in two directions, and the resulting flat, often rectangular surfaces are useful in identification. Those plagioclase feldspars with abundant sodium tend, along with potassium feldspars, to occur in silica-rich rocks such as granite. Calcium-rich plagioclases occur in such rocks as the Hawaiian lavas (called basalts).

Micas Mica is a silicate mineral easily recognized by its perfect and conspicuous cleavage along one directional plane. The two chief varieties are the colorless or pale-colored **muscovite** mica, which is a hydrous potassium aluminum silicate, and the dark-colored

FIGURE 2–5 The variety of plagioclase feldspar known as labradorite. The mineral often displays beautiful blue and gold reflections as well as fine striations on cleavage planes. *(Courtesy of Wards National Science Establishment, Inc., Rochester, New York.)*

FIGURE 2–6 Biotite mica.

FIGURE 2–7 Hornblende. The black scaley flakes on the surface are biotite mica. Specimen is 6 cm long.

biotite mica (Fig. 2–6), which also contains magnesium and iron. Both muscovite and biotite are common rock-forming silicates.

Identification of large specimens of mica is rarely a problem because of its perfect planar cleavage and the way cleavage flakes snap back into place when they are bent and suddenly released. The micas are common constituents of igneous and metamorphic rocks, in which they can be recognized by their shiny surfaces and the ease with which they can be plucked loose with a fingernail or penknife.

Hornblende Hornblende (Fig. 2–7) is a vitreous, black or very dark green mineral. It is the most common member of a larger family of minerals called **amphiboles,** which have generally similar properties. Because of its content of iron and magnesium, hornblende (along with biotite, augite, and olivine) is designated a *ferromagnesian* or *mafic* mineral. Crystals of hornblende tend to be long and narrow. Two good cleavage planes are developed parallel to the long axis and intersect each other at angles of 56° and 124° (Fig. 2–8A).

Augite Just as hornblende is only one member of a family of minerals called amphiboles, **augite** is an important member of the **pyroxene** family, in which many other mineral species also occur. Like hornblende, it is a ferromagnesian mineral and thus dark colored. An augite crystal (Fig. 2–8B) is typically rather stumpy in shape, with good cleavages developed along two planes that are nearly at right angles.

Thus, the cross-section of a crystal appears nearly square rather than rhombic, as in hornblende.

Olivine As you might guess from its name, this glassy-looking iron and magnesium silicate often has an olive green color. It is a common mineral in dark mafic rocks. When not dispersed throughout such rocks, **olivine** may occur as masses of tiny intergrown crystals (Fig. 2–9). Olivine is also an important constituent of stony meteorites. Larger crystals are cut and polished into attractive gemstones called *peridots.*

Clay Minerals Clay minerals are silicates of hydrogen and aluminum with additions of magnesium, iron, and potassium. Their basic structure is similar to that of mica, but because individual flakes are extremely small, their mica-like form can only be seen with the magnification provided by an electron microscope (Fig. 2–10). Unlike many of the silicate minerals already described, clay minerals form as a result of weathering of other aluminum silicate minerals such as feldspars. As a group, clays are known by

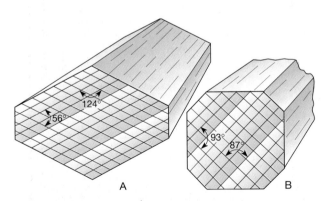

FIGURE 2–8 (A) In crystals of the ferromagnesian mineral hornblende, planes of cleavage intersect at 56° and 124°. (B) In crystals of augite, planes of cleavage intersect at 87° and 93°.

FIGURE 2–9 Olivine.

FIGURE 2–10 Electron micrograph of the clay mineral kaolinite. The flaky, stack-of-cards character of the clay crystals is a manifestation of their silicate sheet structure. Magnified about 2000 times. *(Courtesy of Kevex Corporation.)*

FIGURE 2–11 Calcite, showing the characteristic rhombohedral shape of cleavage pieces. *(Courtesy of Wards Natural Science Establishment, Inc., Rochester, N.Y.)*

their amorphous form, softness, low density, and ability to become plastic when wet. To identify individual species of clay minerals, it is necessary to employ X-ray diffraction techniques.

Nonsilicate Minerals

Approximately 8 percent of the Earth's crust is composed of nonsilicate minerals. These include a host of **carbonates,** sulfides, sulfates, chlorides, and oxides. Among these groups, the carbonates, such as **calcite** and **dolomite,** are the most important. Calcite ($CaCO_3$), which is the main constituent of limestone and marble, forms in many ways. It is secreted as skeletal material by certain invertebrate animals, precipitated directly from sea water, or formed as dripstone in caverns. Calcite is easily recognized by its rhombohedron-shaped cleaved fragments (Fig. 2–11) and by the fact that an application of dilute hydrochloric acid on its surface will cause effervescence.

The term *dolomite* is used both for the carbonate mineral that has the chemical formula $CaMg(CO_3)_2$ and for the rock composed largely of that mineral. (The alternate term *dolostone* is also used occasionally to designate dolomite rock.) In the mineral dolomite, calcium and magnesium occur in approximately equal proportions. The mineral is usually white and is slightly harder than calcite. It will not effervesce when dilute hydrochloric acid is applied to its surface, as does calcite.

Aragonite ($CaCO_3$) is another carbonate mineral that occurs in a different crystal form and more rarely than either calcite or dolomite. Most of us have seen it as the inner "mother-of-pearl" layer of clam shells.

Other common nonsilicate minerals include common rock salt, or **halite** (NaCl), and **gypsum.** Halite is easily recognized by its salty taste and the fact that

FIGURE 2–12   Halite.   *(Courtesy of Wards Natural Science Establishment, Inc., Rochester, N.Y.)*

it crystallizes and cleaves to form cubes (Fig. 2–12). Gypsum is a soft, hydrous calcium sulfate ($CaSO_4 \cdot 2H_2O$). The variety of gypsum called satinspar has a fine, fibrous structure, whereas the variety known as selenite (Fig. 2–13) will split into thin plates. The finely crystalline, massive variety known as alabaster is used widely in carvings and sculpture because of its uniform texture and softness. Halite and the various gypsum minerals are sometimes referred to as **evaporites** because they are often precipitated from bodies of water that have been subjected to intense evaporation.

If you were to make a list of the elements that compose the minerals described thus far, the list would be a surprisingly short one. Only eight elements make up the bulk of these minerals, and only the same eight are abundant in the Earth's crust as well. As shown in Table 2–1, these abundant elements are oxygen, silicon, aluminum, iron, calcium, sodium, potassium, and magnesium.

FIGURE 2–13   A cluster of selenite gypsum crystals. *(Courtesy of Wards Natural Science Establishment, Inc., Rochester, N.Y.)*

■ Rocks

Rock Conversions

We have noted that minerals can provide information about the environment in which they formed. It is therefore possible to know the origin of the rocks containing those minerals. Geologists are agreed on a fundamental division of rocks into three great families according to difference of origin. **Igneous rocks** are those that have cooled from a molten state. **Sedimentary rocks** consist of materials formed from the weathered products of pre-existing rocks that have been transported, deposited, and lithified. **Metamorphic rocks** are any that have been changed from previously existing rocks by the action of heat, pressure, and associated chemical activity. The changes may include a recrystallization of the previous minerals or growth of entirely new minerals.

It is important to remember that the rocks of any of the three major rock families are not immutable. The Earth's crust is dynamic and ever changing. Any sedimentary or metamorphic rock may be partially melted to produce igneous rocks, and any previously existing rock of any category can be compressed and altered during mountain building to produce metamorphic rocks. The weathered and eroded residue of any family of rock can be observed today being transported to the sea for deposition and conversion into sedimentary rocks. These changes can be incorporated into a schematic diagram that is designated the *rock cycle* (Fig. 2–14).

Although rocks are classified into groups that have had similar origin, the identification of rocks is not based on origin but on description. For identification, it is necessary to know general appearance as well as mineral composition and physical properties. **Texture** (size, shape, and arrangement of constituent minerals) and **mineral composition** are essential for identification. Inferences regarding the origin of the rock are based on geologic observations and experimentation.

Igneous Rocks

Igneous rocks constitute over 90 percent by volume of the Earth's crust, although their great abundance may go unnoticed because they are extensively covered by sedimentary rocks. Much of our mountain scenery is sculpted in igneous rocks formed long ago. Recent volcanic activity (Fig. 2–15) provides an often spectacular reminder of the surface processes that produce igneous rocks (the word *igneous* comes from the Latin *ignis*, meaning fire). It is an appropriate name for rocks that develop from cooling masses of molten material derived from exceptionally hot parts of the Earth's interior. **Magma** is the term used to

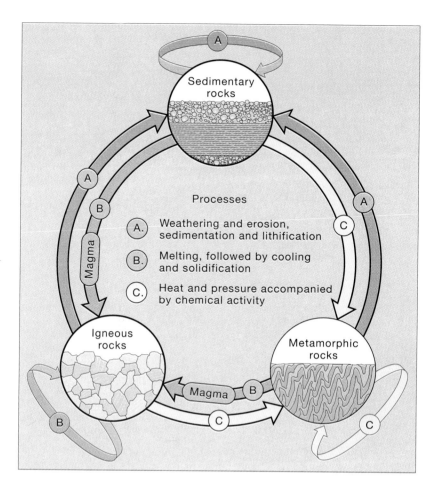

Processes

A. Weathering and erosion, sedimentation and lithification

B. Melting, followed by cooling and solidification

C. Heat and pressure accompanied by chemical activity

FIGURE 2–14 Geologic processes act continuously on Earth to change one type of rock into another.

FIGURE 2–15 Lava flow entering ocean. Kilauea Volcano, 1970. *(Photograph courtesy of D. A. Swanson, U.S. Geological Survey.)*

FIGURE 2–16 Front of an advancing lava flow over an older flow, Kilauea Volcano, Hawaii, February 1990. The advancing flow is breaking into a jagged, rough kind of lava termed *aa*. The underlying older flow exhibits the ropy texture of *pahoehoe* lava. *(Courtesy of J. Plaut.)*

describe this mixture of molten silicates and gases while it is still beneath the surface. If it should penetrate to the surface, it becomes **lava** (Fig. 2–16).

Cooling History of Igneous Rocks

Igneous rocks that formed from magma that had penetrated into other rocks and solidified before reaching the surface are termed **intrusive igneous rocks.** Very large masses of such rocks are sometimes called plutons. Their exposure at the Earth's surface results

A

B

FIGURE 2–17 Differences in the texture of igneous rocks. (A) A coarse-grained intrusive rock called gabbro. (B) A fine-grained igneous rock known as andesite. (C) A porphyry with large phenocrysts of potassium feldspar. It is used here in the ornamental posts outside St. Paul's Cathedral in London.

C

TABLE 2-3 Common Igneous Rocks

Coarsely Crystalline Intrusive Igneous Rocks	Approximate Extrusive Equivalents	Common Silicate Mineral Components	Average Specific Gravity
Granite	Rhyolite	Quartz Potash feldspar Sodium plagioclase (Minor biotite, amphibole, magnetite)	2.7
Diorite	Andesite	Sodium–calcium plagioclase Amphibole, biotite (Minor pyroxene)	2.8
Gabbro	Basalt	Calcium plagioclase Pyroxene Olivine (Minor amphibole, ilmenite)	3.0

from crustal uplift and erosional removal of overlying rocks. In contrast, **extrusive igneous rocks** form from melts that have reached the Earth's surface. This group includes rocks formed from lava erupted from volcanoes or lava that has welled out of fissures.

The grain size of igneous rocks is an index to their history of cooling (Fig. 2–17). Magmas lose heat slowly and retain water. This tends to inhibit formation of crystal nuclei. Thus, there is time and space for the growth of larger crystals around fewer nuclei. In typical intrusive rocks such as granite, diorite, and gabbro (Table 2–3), the intergrown crystals are large enough to be seen readily without magnification. In contrast, extrusive igneous rocks have a finer texture in which crystals are too small to be seen with the unaided eye. The structure of such rocks reflects sudden chilling of molten silicates as they were ejected at the surface of the Earth. When a lava is extruded, there is not enough time for the growth of large crystals. Furthermore, water from the melt is quickly lost to the atmosphere, and without water, myriads of tiny crystals form rather than fewer crystals that can grow to a larger size.

One such extrusive rock is **basalt**, composed of ferromagnesian minerals and tiny rectangular grains of plagioclase feldspars. Obsidian is an extrusive rock that cooled so rapidly that there was insufficient time for crystallization; the melt therefore froze into a glass (Fig. 2–18).

If coarsely crystalline igneous rocks indicate slow cooling and finely crystalline ones indicate rapid cooling, then what would be the cooling history of a rock with large crystals immersed in a very fine-grained matrix? Such rocks are said to have **porphyritic texture.** The large crystals (phenocrysts) were formed slowly at depth and were then swept upward

FIGURE 2-18 The volcanic glass obsidian.

and incorporated in the lava as it hardened at the surface.

Composition of Igneous Rocks

The minerals in an igneous rock reflect the proportions of such elements as silicon, oxygen, aluminum, calcium, iron, magnesium, sodium, and potassium

FIGURE 2–19 The coarse-grained intrusive rock granite. The larger, light-colored grains are potassium feldspar, the dark grains are mainly hornblende and biotite, and the grayish grains are quartz.

present in the magma from which the rock was formed. These eight elements combine in various ways to form feldspars, ferromagnesian minerals, micas, and quartz minerals, which are the constituents of igneous rocks. If the magma is particularly rich in silica, for example, it is likely that quartz will be present in the rock formed by solidification (crystallization) of the melt. In general, intrusive igneous rocks tend to be richer in silica than are extrusive rocks. This observation is related to the fact that greater amounts of silica in a magma increase its viscosity (resistance to flow) so that the thick liquid cannot readily work its way to the surface.

Granite (Fig. 2–19) is a silica-rich, relatively light-colored intrusive rock composed primarily of potassium feldspar, quartz (at least 25 percent), sodium plagioclase, hornblende, and mica. It is derived from magmas so rich in silica that after all chemical linkages with metallic atoms are satisfied, enough silicon and oxygen still remain to form quartz grains. These form an interlocking network with the other minerals to create the crystalline texture of granite.

Another abundant quartz-bearing intrusive rock is called **granodiorite.** Plagioclase is the dominant feldspar in granodiorite. Quartz-bearing igneous rocks such as granite and granodiorite are loosely termed granitic rocks.

Basalt (Fig. 2–20) is a fine-grained extrusive rock derived from a low-silica melt that is rich in iron and magnesium. Basaltic lavas are relatively thin and have been observed to flow with the approximate consistency of motor oil, a fact accounting for the frequency with which they are found at the Earth's surface. Since the silica percentage is low, quartz grains are rarely found in basalt.

There is one final important relationship in the chemistry of magmas and lavas. The silica-rich melts also tend to include ample quantities of potassium and sodium and therefore yield crystals of potassium feldspar, sodium plagioclase, and mica, along with the ubiquitous quartz. Calcium, magnesium, and iron are rather minor constituents of silica-rich magmas, but they increase in abundance in rocks deficient in

A   B

FIGURE 2–20 (A) Hand specimen of basalt. (B) Thin section of basalt viewed with polarized light. Tabular crystals are plagioclase and brightly colored crystals are iron-magnesian silicates. Width of field is 2 mm.

C O M M E N T A R Y
Mount Pelée

Ten kilometers north of the city of St. Pierre on the island of Martinique in the Caribbean Sea is a volcano known as Mount Pelée. The eruption of Mount Pelée in May of 1902 probably qualifies as one of the great natural catastrophes in human history. Eruptions of the volcano began in late March and were characterized by loud explosions and ash showers. The rim of the crater broke on May 5, and water that had been contained within the crater rushed down into the surrounding low coastal areas, incorporating mud and debris and destroying everything in its path. However, the most destructive part of the eruption was yet to come. On May 8, following a series of deafening explosions, a great purplish cloud rose from the top of the mountain and rolled rapidly down the slopes, spreading fanlike as it engulfed the entire town of St. Pierre. Within two minutes all but two inhabitants of the town were killed by the searing heat, gases, and suffocating ash. One of the survivors was a murderer imprisoned in the town's dungeon, and the other a shoemaker whose escape simply defies explanation. In the wake of the hot gases, the entire town was left in flames (see accompanying figure).

A cloud of incandescent dust, ash, and gases such as the one that flowed down the slope of Mount Pelée is known as a **nuée ardente** (fiery cloud). A nuée ardente may take a high toll in lives whenever it passes through a heavily populated area. The extreme mobility of the cloud is caused by expanding hot gases, which contain cinders and ash. The solid particles give the cloud the density required to keep it close to the slope. Cooler air, trapped within the front of the hot moving mass, rapidly expands and increases the turbulence. It is this storm of turbulence that provides the destructive force of nuée ardente.

When the particles of the nuée ardente come to rest, they are so hot that they become welded together, forming a hard igneous rock known as a welded tuff or **ignimbrite.**

The eruption of Mount Pelée did not stop after the destruction of the city. A second nuée ardente swept through the desolated town on May 20, and yet another eruption on August 30 killed 2000 people in the small village of Morne Rouge.

The smoldering ruins of St. Pierre as it appeared on May 14, 1902, six days following the May 8 eruption of Mount Pelée. *(Courtesy of the Institute of Geological Sciences, London.)*

silica. The low-silica rocks utilize these elements in the formation of calcium plagioclases and ferromagnesian minerals. Because of the ferromagnesians and gray plagioclase minerals, low-silica rocks tend to be dark gray, black, or green.

Volcanic Activity

Although deciphering the history of a granitic mass is certainly intellectually stimulating, it is unlikely to evoke the feelings of awe and excitement one experiences when viewing volcanic activity. Volcanoes are basically vents in the Earth's surface through which hot gases and molten rock flow from the Earth's interior. The extrusions may be quiet or explosive. Quiet eruptions are exemplified by the Hawaiian volcanoes and are frequently characterized by truly enormous outpourings of low-viscosity lava (Fig. 2–21). The lava spreads widely to form the gentle slopes of a shield volcano. Explosive eruptions are caused by the sudden release of molten rock driven upward by large pockets of compressed gases. Such explosive eruptions can literally blow the volcanic cone to bits. The catastrophic eruption of Krakatoa in 1883 was heard 5000 kilometers away and was responsible for the death of 36,000 humans. Fortunately, most eruptions are not so violent.

There are, of course, all stages of volcanic activity between quiet and violently explosive. Perhaps in response to changes in the composition of the melt, some volcanoes have even been known to shift from one type of activity to another. Volcanic activity also includes successive outpourings of lava from great fissures to form lava plateaus that extend over thousands of square kilometers. Such fiery floods produced the Columbia and Snake River Plateau as well as the Deccan Plateau of India.

By far, the most abundant kind of volcanic rock is basalt. It underlies the ocean basins, has been built into midoceanic ridges, and has accumulated sufficiently in such places as Hawaii and Iceland to have produced substantial land area.

How and at what depth did this great volume of basalt originate? To answer this question, it is necessary to refer briefly to a model of the Earth's interior that has been formulated by the study of earthquake waves. The model depicts the Earth's basaltic crust as a thin zone averaging about 6 km thick and overlying the mantle of denser olivine- and pyroxene-rich rocks. The boundary between the crust and the mantle is recognized by an abrupt change in the velocity of earthquake waves as they travel downward into the Earth. For many years geologists believed that basaltic lavas originated from the lower part of the basaltic crust. However, several recent lines of evidence suggest that the basaltic lavas may have come from molten pockets of upper mantle material. For example, present-day volcanic activity is closely associated with deep earthquakes that occur within the mantle far beneath the crust. It is likely that fractures produced by these earthquakes could serve as passages for the escape of molten material to the surface. A detailed study of earthquake shocks from particular volcanic eruptions in Hawaii indicates that the erupting lavas were derived from pockets of molten material within the upper mantle at depths of about

FIGURE 2–21 River of lava formed during the 1984 dual eruption of the Mauna Loa and Kilauea volcanoes, Hawaii. *(Courtesy of I. Duncan.)*

100 km. A weak plastic zone (called the "low-velocity zone") in the mantle appears to represent the level at which the lavas originated. The mechanism by which they developed is called partial melting. **Partial melting** is that general process by which a rock subjected to high temperature and pressure is partly melted and the liquid component is moved to another location. At the new location, the separated liquid may solidify into rocks that have a different composition from the parent mass. The word *partial* in the expression *partial melting* refers to the fact that some minerals melt at lower temperatures than others, and so for a time the material being melted resembles a hot slush composed of liquid and still solid crystals. The molten fraction is usually less dense than the solids from which it was derived and thus tends to separate from the parent mass and work its way toward the surface. In this way, melts of basaltic composition separated from denser rocks of the upper mantle and eventually made their way to the surface to form volcanoes.

Many complex and interrelated factors control where in the mantle partial melting may occur or even if it will occur at all. Generally, heat in excess of 1500°C is required, but the precise temperature for melting is also influenced by pressure and the water content of the rock. As pressure increases, the temperature at which particular minerals melt also rises. Thus, a rock that would melt at 1000°C near the surface will not melt in deeper zones of higher pressure until it reaches far greater temperatures. Water has an effect opposite that of pressure, for its presence will allow a rock to start melting at lower temperatures and shallower depths than it would have otherwise. Laboratory experiments indicate that the melting of "dry" mantle rock can occur at depths of about 350 km but that the presence of only a little water can cause partial melting and yield basaltic liquid from depths as shallow as 100 km.

Not all lavas found at the Earth's surface are basaltic. Volcanoes of the more explosive type that are located at the edge of continents around the Pacific and in the Mediterranean extrude a lava called **andesite.** Andesite contains more silica than basalt does, and its lava is more viscous. This greater resistance to flow contributes to the gas containment that precedes explosive volcanic activity. Andesites are considered to be intermediate in silica content between the rocks of the continental crust and those of the oceanic crust of the Earth.

Andesitic rocks may originate in more than one way. Some emplacements result from originally basaltic magmas in which minerals such as olivine and pyroxene form early and settle out, thus leaving the remaining melt relatively richer in silica. This process is called **fractional crystallization.** Other andesites may result from the melting of basaltic ocean crust and siliceous marine sediments as they descend into hot zones of the mantle. The water-rich and silica-enriched melts of andesitic composition might then rise buoyantly and erupt along volcanic island arcs. We will examine this interesting idea more fully in Chapter 5, which deals with plate tectonics.

Sedimentary Rocks

Sedimentary rocks are simply rocks composed of consolidated sediment—particles that are the product of weathering and erosion of any previously existing rock or soil. The components of sedimentary rocks may range from large boulders to the molecules dissolved in water. Sediment is deposited through such agents as wind, water, ice, or mineral-secreting organisms. The loose sediment is converted to coherent solid rock by any of several processes: precipitation of a cementing material around individual grains, compaction, or crystallization. These processes constitute **lithification.**

The most obvious feature of sedimentary rocks is their occurrence in beds or layers called **strata.** Stratification is commonly the result of changes in the conditions of deposition that cause materials of a somewhat different nature to be deposited for a period of time. For example, the velocity of a stream might decrease, causing particles to settle out that might otherwise have stayed in suspension. In another situation, the kind of materials brought into a given depositional site by streams might change, and there would then be a corresponding change in composition of the accumulating layers.

Sandstone, shale, and carbonate rocks (such as limestone) constitute the most abundant sedimentary rocks. **Sandstones** are composed of grains of quartz, feldspar, and other particles that are cemented or otherwise consolidated. **Shale** consists largely of very fine particles of quartz and abundant clay. The carbonates are rocks formed when the carbon dioxide contained in water combines with oxides of calcium and magnesium.

Derivation of Sedimentary Materials

Sedimentary rocks must have originally come from the disintegration and decomposition of older rocks. Commonly, the older rocks are igneous; indeed, these were once the only rocks on Earth. It is therefore instructive to review the manner in which the common components of sedimentary rocks might be derived from an abundant kind of igneous rock, such as granodiorite (Fig. 2–22).

Consider first the quartz in the granodiorite. Quartz will persist almost unchanged during weathering. It is one of the most chemically stable of all the common silicate materials. As the parent rock is gradually decomposed, quartz grains tend to be

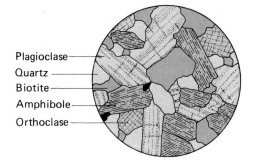

FIGURE 2–22 Sketch of a thin section of granodiorite as viewed through a petrographic microscope. Diameter is about 3 mm.

washed out and carried away to be deposited as sand that will one day become sandstone (Fig. 2–23).

The feldspars decay far more readily than quartz. They are primarily aluminum silicates of potassium, sodium, and calcium. In the weathering process, the last three elements are largely dissolved and carried away by solutions (although some may remain in soils within clay minerals). Ultimately, they reach the sea, where they may stay in solution, or they are deposited as layers of limestone. If large quantities of lake or sea water are evaporated, evaporites such as halite or gypsum may be formed. Of course, not all the feldspars and micas in the granodiorite necessarily decay. Some may persist as detrital grains that become incorporated into sandstones and other sediments.

Biotite is another mineral in the granodiorite source rock. Decomposition of biotite, which is a potassium, magnesium, and iron aluminosilicate, yields soluble potassium and magnesium carbonates, small amounts of soluble silica, and iron oxides. The iron oxides serve to color many sedimentary rocks in shades of brown and red.

Variety Among Sedimentary Rocks

Sedimentary rocks are classified according to their composition and texture. The term *texture* refers to the size and shape of the individual grains and to their arrangement in the rock. A rock that has a **clastic texture** is composed of grains and broken fragments (clasts) of preexisting minerals, rocks, and fossils. Other sedimentary rocks are composed of a network of intergrown crystals and therefore have a crystalline texture.

Clastic Rocks   The fragments of preexisting materials that compose clastic sedimentary rocks range in size from huge boulders to microscopic particles. Particle size is particularly useful in classifying these rocks, which include conglomerates, sandstones, siltstones, and shales. **Conglomerate** (Fig. 2–24) is composed of water-worn, rounded particles larger than 2 mm in diameter. The term **breccia** is reserved for clastic rocks composed of fragments that are angular (Fig. 2–25) but similar in size to those of conglomerates. In **sandstones**, grains range between $\frac{1}{16}$ mm and 2 mm in diameter (see Table 3–1). The varieties of sandstone are then subdivided mainly according to composition. **Siltstones** are finer than sandstones ($\frac{1}{16}$ mm to $\frac{1}{256}$ mm), and **shales** are composed of particles finer than $\frac{1}{256}$ mm. Shales may contain abundant clay minerals, which are flaky minerals that align parallel to bedding planes. As a result, shales characteristically split into thin slabs parallel to bedding planes. This property is termed **fissility.** Rocks lacking fissility but composed of clay-sized particles are called claystones or mudstones.

Carbonate Rocks   The minerals that compose carbonate rocks are primarily calcite, aragonite, and

FIGURE 2–23   Exposure of weathered granodiorite showing accumulation of grains of quartz and partially decomposed feldspar resulting from the disintegration of the parent rock.

FIGURE 2–24   Conglomerate. The pebbles in this particular sample are chert in a carbonate matrix. The largest pebbles are about 2.5 cm in diameter.

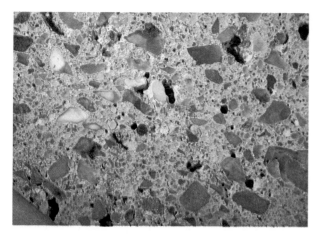

FIGURE 2–25 A polished slab of breccia. Note the angularity of the larger clasts. Area shown is approximately 8 cm across.

dolomite. Both calcite ($CaCO_3$) and aragonite ($CaCO_3$) have the same chemical composition but differ in crystal form. (Aragonite is relatively unstable and is usually converted to calcite.) Calcite is the predominant mineral in limestone. Dolomite— ($CaMg(CO_3)_2$)—predominates in the rock dolomite. (Here both mineral and rock bear the same name.) Carbonate rocks usually also contain variable amounts of impurities, including iron oxide, clay, and particles of sand and silt swept into the depositional environment by currents.

Limestones. The most abundant limestones are of marine origin and have formed as a result of precipitation of calcite or aragonite by organisms and the incorporation of skeletons of those organisms into sedimentary deposits. Inorganic precipitation of carbonate minerals may also form deposits of limestone. The importance of this process is questionable, however, because the precipitation is nearly always closely associated with photosynthetic and respiratory activities of organisms or with the release of tiny particles of aragonite upon the decay of green algae. Strictly speaking, it appears that very few marine limestones are the result of direct chemical precipitation.

After the calcium carbonate has accumulated, it becomes recrystallized or otherwise consolidated into indurated rock that may be variously colored—from white, through tints of brown, to gray. Limestones tend to be well stratified (Fig. 2–26), frequently contain nodules and inclusions of chert, and are often highly fossiliferous (containing fossils). The rock may range in texture from coarsely granular to very fine grained and aphanitic.

In general, limestones consist of one of a combination of textural components such as micrite, carbonate clasts, oolites, or carbonate spar. Micrite is exceptionally fine-grained carbonate mud. Carbonate clasts are sand- or gravel-sized pieces of carbonate. The most common clasts are either bioclasts (skeletal fragments of marine invertebrates) or oöids

FIGURE 2–26 Roadcuts exposing limestone strata are familiar to travelers on interstate highways throughout the Mississippi valley region.

FIGURE 2-27 Oöids. Diameter of field is 2 cm.

(Fig. 2–27), which are spherical grains formed by the precipitation of carbonate around a nucleus. Sparry carbonate is a clear, crystalline carbonate that is normally deposited between the clasts as a cement or has developed by replacement of calcite. These textural categories of limestone as seen through the microscope are shown in Figure 2–28. They permit classification of particular samples as micritic limestone, clastic limestone, oölitic limestone, or sparry (crystalline) limestone.

There are many varieties of limestone. Chalk is a soft, porous variety that is composed largely of calcareous, microscopic skeletal remains of marine planktic (floating) animals and plants. Lithographic limestone is a dense, micritic limestone once widely used as an etching surface in printing illustrations. Some limestones consist almost entirely of skeletal remains of reef corals and other frequently skeletonized marine invertebrates.

Dolomite. As we have already noted, dolomite is a rock composed of the calcium–magnesium mineral of the same name. Like limestone, it occurs in extensive stratified sequences and is not easily distinguishable from limestone. The usual field test for distinguishing dolomite from limestone is to apply cold, dilute hydrochloric acid. Unlike limestone, which bubbles readily, dolomite will effervesce only slightly, if at all.

The origin of dolomites is somewhat problematic. The mineral dolomite is not secreted by organisms during shell building. Direct precipitation from sea water does not normally occur today, except in a few environments where the sediment is steeped in abnormally saline water. Such an origin is not considered adequate to explain the thick sequences of dolomitic rock commonly found in the geologic record. The most widely believed theory for the origin of dolomites is that they result from partial replacement of calcium by magnesium in the original calcareous sediment. However, it is not known for how long or at what time in the history of the rock this dolomitization occurs.

Other Sedimentary Rocks

Chert. We have previously mentioned a form of microcrystalline quartz called chert (SiO_2), noting its occurrence as nodules in limestones (Fig. 2–29). The origin of these nodules is still being debated among petrologists, although the majority believe that they form as replacements of carbonate sediment by silica-rich sea water trapped in the sediment. Some cherts occur in areal-extensive layers and thus qualify as monomineralic rocks. These so-called bedded cherts are thought to have formed from the accumulation of the siliceous remains of diatoms and radiolaria and from subsequent reorganization of the silica into a microcrystalline quartz. Silica from the dissolution of

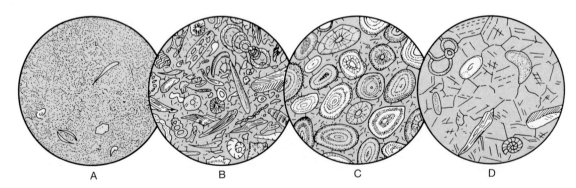

FIGURE 2-28 Textures of limestones as seen in thin section under a microscope. (A) Aphanitic limestone or micrite. (B) Bioclastic limestone with fine-grained sparry calcite as cement. (C) Oölitic limestone. (D) Sparry or crystalline limestone.

FIGURE 2–29 White nodular chert in tan limestone. Fern Glen Formation west of St. Louis, Missouri.

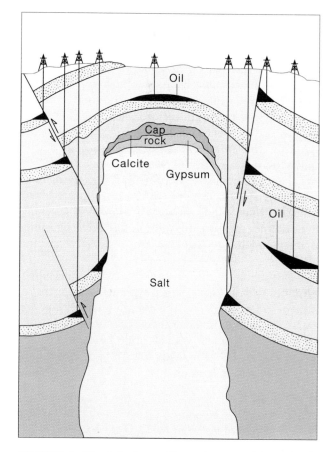

FIGURE 2–30 Salt dome, illustrating possibilities for oil entrapment in domelike structures (top center), by faults, and by pinchout of oil-bearing strata.

volcanic ash is believed to enhance the process; indeed, many bedded cherts are found in association with ash beds and submarine lava flows.

Evaporites. As indicated by their name, evaporites are chemically precipitated rocks that are formed as a result of evaporation of saline water bodies. Only about 3 percent of all sedimentary rocks consist of evaporites. Evaporite sequences of strata are composed chiefly of such minerals as gypsum, anhydrite, halite, and associated calcite and dolomite. Extensive ancient deposits of evaporites are currently being commercially worked in Michigan, Kansas, Texas, New Mexico, Germany, and Israel. The conditions required for precipitation of thick sequences of evaporites include warm, relatively arid climates and a physiographic situation that would provide periodic additions of sea water to the evaporating marine basin. In the Gulf coastal region of the United States, as a result of the pressure of overlying rocks, deeply buried deposits of salt have flowed plastically upward to form underground domes of salt. In the process, the salt has arched overlying strata, thereby producing structures in which petroleum could collect (Fig. 2–30).

Coal. Coal is a carbonaceous rock resulting from the accumulation of plant matter in a swampy envi-

ronment combined with alteration of that plant tissue by both biochemical and physical processes until it is converted to a consolidated carbon-rich material. The biochemical and physical changes may produce a series of products ranging from peat and lignite to bituminous and anthracite coal. For coal to form, plant tissue must be accumulated under water or be quickly buried because vegetable matter, if left exposed to air, is readily oxidized to water and carbon dioxide. With underwater accumulation or quick burial of plant material, a major part of the carbon can be retained.

Metamorphic Rocks

Sir Charles Lyell recognized that igneous or sedimentary rocks, if subjected to high temperature, pressure, and the chemical action of solutions and gases, can be altered to quite different kinds of rocks. Lyell employed the term *metamorphism* (from the Latin *metamorphosis,* meaning change of form) to describe this process. It is still used today to describe alterations in rocks brought about by physical or chemical changes in the environment that are intermediate between those that result in igneous rocks and those that produce sedimentary rocks. Any previously existing rock may be converted to a metamorphic rock, and the changes primarily involve recrystallization of minerals in the rock while it remains in the solid state. In the process of recrystallization, the textural characteristics of the parent rock may be changed while at the same time new minerals develop that are stable under the new conditions of pressure and temperature. New elements need not be introduced; instead, those that are already present are incorporated into different and often denser minerals. Variations in heat and pressure may result in different kinds of metamorphic rocks, even from the same parent material.

Most metamorphic rocks exhibit a layering called **foliation,** which results from the parallel alignment of mineral grains. Whether this foliation is very fine or coarse depends on the size and shapes of the constituent minerals. There are a few nonfoliated metamorphic rocks as well, for many marbles and quartzites do not exhibit foliation.

Metamorphism

Alterations of rock immediately adjacent to igneous intrusions constitute **contact metamorphism** (Fig. 2–31). The changes that occur in the intruded rock are largely the result of high temperatures and the emanation of chemically active vapors that accompany igneous intrusions. Such factors as the size of the magmatic body, its composition and fluidity, and the nature of the intruded rock also influence the kind and degree of contact metamorphism. Important ore deposits are commonly situated in the metamorphosed rock surrounding the intrusives. Examples of such deposits include magnetite and copper ores in metamorphic zones around granite intrusives in the Urals, central Asia, the Appalachian Mountains, Utah, and New Mexico.

Regional or **dynamothermal metamorphism** is a type of rock alteration that is areally extensive and occurs under the conditions of great confining pressures and heat accompanying deep burial and mountain building. In a subsequent chapter, we will discuss how rocks deposited in crustal troughs adjacent to continents may be compressed into mountain systems and thus be regionally metamorphosed. Metamorphic index minerals known to form under specific temperature and pressure conditions are used to decipher the history of growth of these ancient mountainous regions, even when only the roots of the ranges remain.

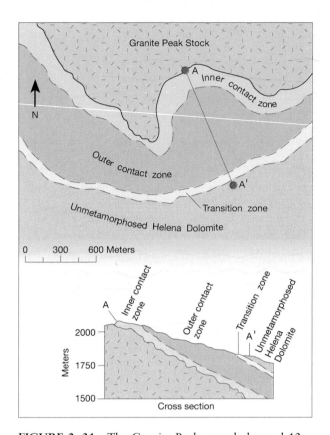

FIGURE 2–31   The Granite Peak aureole located 12 miles east of Lincoln, Montana. An aureole is the zone of contact metamorphism surrounding an igneous intrusion. This metamorphic aureole is developed around a granitic stock that has intruded into dolomite country rock. The intensity of metamorphism diminished outward from the granite margin, and particular assemblages of metamorphic minerals occur within each contact metamorphic zone.   *(Simplified from Melson, W. G. 1966. American Mineralogist 51:404, Fig. 1.)*

Kinds of Metamorphic Rocks

Because any rock can be metamorphosed in a number of different ways, there are hundreds of different kinds of metamorphic rocks. However, for our purposes, we need only consider several that occur extensively at the Earth's surface. It is convenient to divide metamorphic rocks into two groups based on the presence or absence of foliation.

Foliated Metamorphic Rocks

Slate.   In slate, the foliation is microscopic and caused by the parallel alignment of minute flakes of silicates like mica (Fig. 2–32). The planes of foliation are quite smooth, and the rock may be split along these planes of "slaty cleavage." The planes of foliation may lie at any angle to the bedding in the parent rock, and this characteristic helps to differentiate

FIGURE 2–32 These layers of slate from the Kodiak Peninsula of Alaska exhibit well-developed slaty cleavage. Note that the cleavage planes are vertical and are not parallel to bedding, as in the fissility of shale. *(Courtesy of U.S. Geological Survey.)*

slate from dense shale. Slate is derived from the regional metamorphism of shale.

Phyllite. The texture in phyllite is also very fine, although some grains of mica, chlorite, garnet, or quartz may be visible. Phyllite surfaces often develop a wrinkled aspect and are more lustrous than slate. Phyllite represents an intermediate degree of metamorphism between slate and schist. The parent rocks are commonly shale or slate.

Schist. The platy or needle-like minerals in schist are sufficiently large to be visible to the unaided eye; the minerals tend to be segregated into distinct layers. Schists are named according to the most conspicuous mineral present. Thus, there are mica schists (Fig. 2–33), amphibole schists, chlorite schists, and many others. Shales are the usual parent rocks for schists, although some are derived from fine-grained volcanic rocks.

Gneiss. This is a coarse-grained, evenly granular rock. Foliation (Fig. 2–34) results from segregation of minerals into bands rich in quartz, feldspar, biotite, or amphibole. Foliation is coarse and appears less distinct than in schist. High-silica igneous rocks and sandstones are the usual parent rocks for gneisses.

FIGURE 2–33 Steeply inclined mica schists splitting apart along foliation planes. Great Smoky Mountain National Park, North Carolina–Tennessee.

FIGURE 2-34 Coarse (gneissic) foliation developed in a quartz–feldspar–biotite gneiss.

FIGURE 2-36 Hand specimen of an attractive green quartzite.

Nonfoliated Metamorphic Rocks

Marble. A fine to coarsely crystalline rock, marble (Fig. 2–35) is composed of calcite or dolomite and therefore is relatively soft. (It can be scratched with steel.) Marble is derived from limestone or dolomite.

Quartzite. A fine-grained, often sugary-textured rock, quartzite (Fig. 2–36) is composed of intergrown quartz and therefore is very hard. The rock will break through constituent grains; it may be any color. Quartzite is derived from quartz sandstone.

FIGURE 2-35 Marble. On close examination, one can discern the lustrous cleavage surfaces of the calcite crystals. Width of specimen is about 9.0 cm.

Greenstone. A dark green rock, greenstone has a texture so fine that mineral components, except for scattered larger crystals, cannot be seen without magnification. It is derived by the low-grade metamorphism of low-silica volcanic rocks such as basalt.

Hornfels. A very hard, fine-grained rock often studded with small crystals of mica and garnet that have no preferred orientation. Hornfels may form from shale or other fine-grained rocks that are intensely heated at their contact with intrusive igneous bodies.

Historical Significance of Metamorphic Rocks

We have noted that the conditions for metamorphism are developed in regions that have been subjected to intense compressional deformation. Such regions of the Earth's crust either now have, or once had, great mountain ranges. Thus, where large tracts of low-lying metamorphic terrane are exposed at the Earth's surface, geologists conclude that crustal uplift and long periods of erosion leveled the mountains. Metamorphic rock exposures at many localities across the eastern half of Canada represent the truncated foundations of ancient mountain systems.

From studies of the mineralogic composition of metamorphic rocks, it is often possible for geologists to reconstruct the conditions under which the rocks were altered and then to make inferences about the directions of compressional forces, pressures, temperatures, and the nature of parent rocks. Investigators are aided in these studies by the knowledge that specific metamorphic minerals form and are stable within finite limits of temperature and pressure. Maps of **metamorphic facies,** or zones of rocks that formed under specific conditions, can be constructed. Commonly, such maps delineate broad bands of metamorphic rocks, each of which formed under sequentially more intense conditions of pressure and temperature. Imagine a terrane that was once underlain by a thick sequence of calcareous shales, was subjected to compression to produce mountains, and then experienced loss of those mountains by erosion. One might then begin a traverse across the eroded surface on unmetamorphosed shales that were not involved in the mountain building (Fig. 2–37). These shales would contain only unaltered sedimentary minerals. Progressing farther toward the area of most intense metamorphism, one might see that the shales had given way to slates bearing the green metamorphic mineral chlorite. Still farther along the traverse, schists containing such intermediate-grade metamorphic minerals as biotite and garnet would appear. Finally, one might come upon coarsely foliated schists containing kyanite, staurolite, and sillimanite—minerals that develop under high temperature and pressure.

Metamorphic minerals do not always appear in the orderly fashion indicated in the preceding example. Depending on the nature and depth of metamorphism, temperature may increase at a faster or slower rate than does pressure, resulting in the growth of different index minerals. It is also possible for a previously metamorphosed terrane to experience a second, less severe episode of metamorphism. In such cases, a lowering of metamorphic grade might result. Yet another factor is the mineralogic composition of the parent rock and the amount of water present during metamorphism.

Metamorphic rocks may contain a wealth of historical information. A marble containing flakes and veins of chlorite that is dated as 1 billion years old tells the geologist many things. It records the existence of an ancient, somewhat clayey limestone that experienced a relatively low level of metamorphism. Because limestone was the parent rock, the geologist may infer that conditions in the Earth's atmosphere and ocean a billion years ago were similar to those that permit carbonate deposition today.

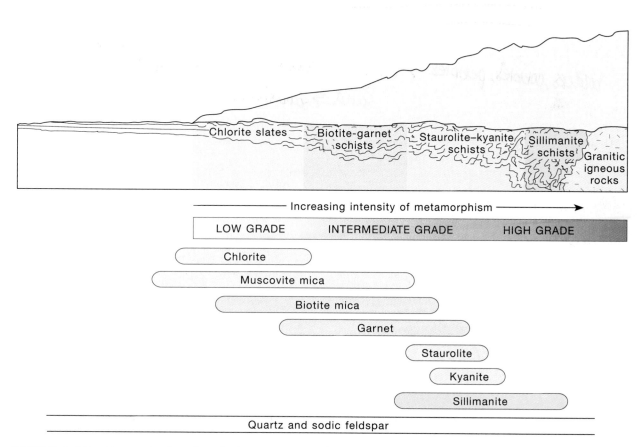

FIGURE 2–37   Changes in the mineralogic composition of a terrane originally underlain by shales following regional metamorphism.

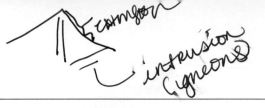

Summary

Rocks are the materials of which the Earth is composed. Rocks are themselves composed of minerals, and minerals, in turn, are constructed of chemical elements. The most common elements in rocks are oxygen, silicon, aluminum, iron, calcium, sodium, potassium, and magnesium. Rocks and minerals are the materials from which geologists make interpretations of ancient environments and past geologic happenings. Silicate minerals are the most important of the rock-forming minerals. Quartz, mica, feldspar, hornblende, augite, and olivine are silicates that were initially crystallized from molten rock. Clay, calcite, dolomite, gypsum, halite, and certain varieties of quartz (such as chert and flint) are formed by processes of weathering and precipitation or deposition at temperatures that prevail at the Earth's surface. Calcite is the main constituent of the carbonate rock limestone, and the mineral dolomite forms a carbonate rock that is also called dolomite. In water bodies that experience intense evaporation, evaporite minerals such as gypsum and halite may be precipitated. Clastic sedimentary rocks are composed of fragments of weathered rock that have been transported some distance from their place of origin. Transported pebbles and cobbles, for example, may be lithified to form the rock called conglomerate, sand grains may form sandstone, and particles of clay may accumulate to form shale.

Rocks are aggregates of minerals. Igneous rocks are those that have cooled and solidified from a molten silicate body. If solidified beneath the Earth's surface, the igneous rocks are intrusive. Granite is an example. Igneous rocks, such as those formed from the lavas flowing from volcanoes, are termed extrusive. Basalt is an example. The texture of igneous rocks provides an index to their cooling history, in that coarse-grained varieties cooled slowly (and are mainly intrusive), whereas fine-grained varieties cooled rapidly (and are mainly extrusive).

Metamorphism refers to all the processes by which previously existing rocks of any kind undergo mineral and textural changes in the solid state in response to changing physical and chemical conditions. The agents of metamorphism include heat, pressure, and chemically active solutions. The basic kinds of metamorphism are contact and regional. Contact metamorphism occurs around the margins of bodies of molten rock. Unlike contact metamorphism, which may be relatively local, regional metamorphism occurs on a large (regional) scale and is usually associated with mountain building. The rocks in the regional belts are well foliated and divisible into distinct metamorphic zones or facies characterized by minerals that formed in response to particular conditions of pressure and temperature.

Review Questions

1. What is a mineral? What characteristics of a true mineral like quartz or feldspar would not be developed in a chunk of solidified glass from a glass manufacturing plant?
2. Why are silicate minerals important in geology? What kinds of silicates might one expect to find in granite? What kinds of silicates might be found in sedimentary rocks?
3. What are the three major categories of rocks, and how are they defined?
4. What type of igneous rock best approximates the composition of the oceanic crust? The continental crust?
5. What inferences can be drawn from the color of igneous rocks? From the grain size of igneous rocks?
6. Of the mineral groups discussed in this chapter, which are particularly characteristic of sedimentary rocks? Which might be particularly effective in causing foliation in metamorphic rocks?
7. List the foliated metamorphic rocks in order of increasingly coarser foliation.
8. List the clastic sedimentary rocks in order of increasingly finer grain size.

Discussion Questions

1. If you were a stone-age human and had to choose between calcite and chert as the material for a spear point, which would you choose? Explain your answer.
2. In a traverse across an ancient metamorphic terrane, how would you recognize the zone where the effects of pressure and heat were once most severe?
3. Discuss the agents of metamorphism, including their interrelationships.
4. With regard to the origin of igneous rocks, discuss what is meant by *partial melting* and *fractional crystallization*.

Supplemental Readings and References

Decker, R., and Decker, B. 1981. *Volcanoes.* San Francisco: W. H. Freeman Co.

Dietrich, R. V., and Skinner, B. J. 1979. *Rocks and Minerals.* New York: John Wiley & Sons.

Ernst, W. G. 1969. *Earth Materials.* Englewood Cliffs, N.J.: Prentice-Hall.

Hochleitner, R. 1994. *Minerals.* Hauppauge, N.Y.: Barron's Press.

Pough, F. H. 1960. *A Field Guide to Rocks and Minerals.* Cambridge, Mass.: Houghton Mifflin Co.

Sinkankas, J. 1964. *Mineralogy for Amateurs.* Princeton, N.J.: Van Nostrand Co.

Dolomite of the Ordovician Joachim Formation forms a resistant ledge over massive quartz sandstone of the St. Peter Formation. The cave was excavated for the extraction of the sandstone used in making glass.

CHAPTER 3

The Sedimentary Archives

Ever since the Earth has had an atmosphere and hydrosphere, sediments have been accumulating on its surface. The sediments, now formed into sedimentary rocks, contain features that tell us about the environment in which they were deposited. By interpreting these bits of evidence in successively higher strata, one can decipher the geologic history of a part of the Earth.

■ *The Tectonic Setting*

Many factors determine the kind of sedimentary rock that will be formed in a particular area. Among these are the method of transport of sedimentary materials; the physical, chemical, and biological processes operating in the place of deposition; the climate under which processes of weathering and erosion take place; and the changes that occur to sediment as it is being converted to solid rock. On a grander scale, the characteristics of an entire assemblage of rocks are also influenced by the tectonics of the region in which deposition takes place. **Tectonics** may be defined as the study of the deformation or structural behavior of a large area of the Earth's crust over a long period of time. For example, a region may be tectonically stable, subsiding, rising gently, or more actively rising to produce mountains and plateaus. Where a source area has recently been compressed and uplifted, an abundance of coarse sediment (forming sandstones and conglomerates) derived from the rugged upland source area will be supplied to the basin. In the geologic past, such a tectonic setting has resulted in the accumulation of great "clastic wedges" of sediment that thinned and became finer away from the former mountainous source area. In other tectonic settings of the past, the source area has been stable and topographically more subdued, so that finer particles and dissolved solids became the most abundant components being carried by streams.

The tectonic setting influences not only the size of clastic particles being carried to sites of deposition but also the thickness of the accumulating deposit. For example, if a former marine basin of deposition had been provided with an ample supply of sediment and was experiencing tectonic subsidence (sinking), enormous thicknesses of sediments might accumulate. Over a century ago, James Hall (1811–1898), an eminent American geologist and paleontologist, recognized that the thick accumulations of shallow water sedimentary rocks in the Appalachian region required that crustal subsidence had accompanied deposition. His reasoning was quite straightforward. It was easy to visualize filling a basin that was 40,000 ft deep with 40,000 ft of sediment. However, where fossils indicated a basin never more than several hundred feet deep, the only way to get tens of thousands of feet of sediment into it would be to have subsidence occurring simultaneously with sedimentation.

In a marine basin of deposition that is stable or subsiding very slowly, the surface on which sedimentation is occurring is likely to remain within the zone of wave activity for a long time. Wave action and currents will wear, sort, and distribute the sediment into broad, blanket-like layers. If the supply of sediment is small, this type of sedimentation will continue indefinitely. Should the supply of sediment become too great for currents and waves to transport, however, the surface of sedimentation would rise above sea level and deltas would form.

It is also important to consider the tectonic framework of entire continents as well as of particular areas. The principal tectonic elements of a continent are **cratons** and **orogenic belts** (Fig. 3–1). Cratons have two components: large areas of exposed ancient crystalline rocks called **shields** as well as surrounding regions called **platforms,** in which these ancient rocks are covered by flat-lying or gently warped layers of sedimentary rocks. Cratons have been undisturbed by mountain-building events since Precambrian time (about 544 million years ago). They comprise the stable interiors of continents.

Orogenic belts are elongate regions that border the craton and have been deformed by compressional forces since Precambrian time. Today, young orogenic belts are recognized by their high frequency of earthquakes and volcanic eruptions, whereas older belts are marked by severely deformed strata, crustal displacements, metamorphic terranes, and huge bodies of intrusive igneous rocks.

The history of the Earth provides many examples of marginal orogenic belts that were once sites for the accumulation of great thicknesses of sediment. Such elongate tracts of sedimentation are found today along the continental shelves and rises. Following long episodes of deposition, the sediments along these tracts may be deformed as a result of an encounter with an oncoming tectonic plate, and a mountain range may form where there was once only a sedimentary basin. These events and their causes will be examined in Chapter 5.

The tectonic setting of deposition largely determines the nature of sedimentary deposits. Conversely, the kind of tectonic setting often can be inferred from a rock's textural and structural features and from its color, composition, and fossils. The tectonic and historical significance of some of these characteristics of sedimentary rocks will be described in the following pages.

■ *Environments of Deposition*

An *environment of deposition* refers to all the physical, chemical, biological, and geographic conditions under which sediments are deposited. Each environment of deposition is characterized by geographic and climatic conditions that modify or determine the properties of sediment that is deposited within it. Thus, the type of sediment becomes the key to the environment. Some sediments, such as chemical precipitates in water bodies, are solely the products of their environment of deposition. Their component minerals formed and were deposited at the same place. Other sediments consist of materials formed elsewhere and transported to the site of deposition. Geologists are keenly interested in the sediment of today's depositional environments because the features they find in the deposits of these modern areas can also be seen in ancient sedimentary rocks. Com-

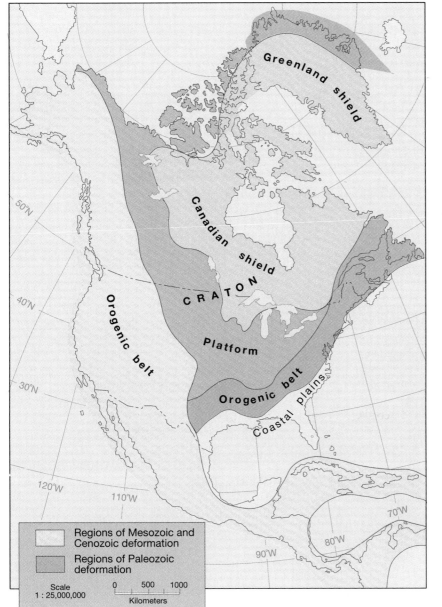

FIGURE 3–1 The craton and orogenic belts of North America.

paring present-day sedimentary deposits to old sedimentary rocks permits one to reconstruct conditions in various parts of the Earth as they were hundreds of millions of years ago.

The Marine Environment

To facilitate discussion, the marine realm can be divided into shallow marine, deep marine, and continental slope environments. The shallow division includes the oceanic topographic regions called continental shelves. **Continental shelves** (Fig. 3–2) are nearly flat, smooth surfaces that fringe the continents in widths that range from only a few kilometers to about 300 km and depths that range from low tide

to about 200 meters. In a geologic sense, these shallow areas are not part of the oceanic crust but resemble the continents in their structure and composition. They are, in fact, the submerged edges of the continents and have a readily apparent continuity with the coastal plains. The outer boundaries of the shelves are defined by a marked increase in slope to greater depths. The smoothness of parts of the continental shelves seems to have been produced in part by the action of waves and currents during the last Ice Age. At that time, sea level was periodically lowered by as much as 140 meters as a result of water being locked in glacial ice. Waves and currents sweeping across the shelves shifted sediment into low places and generally leveled the surface.

C O M M E N T A R Y
A Tale of Two Deltas

Deltas are accumulations of sediment formed by the entrance of a stream into quiet water such as a lake or the sea. The term was proposed about 25 centuries ago by Herodotus, who noted that the Nile Delta had the general shape of the Greek letter Δ. Not all deltas, however, have this shape, for every delta responds differently to depositional and erosional processes that act in opposition to one another. *All deltas are characterized by a restless interplay between the rate of sediment supply brought in by streams and the rate at which that sediment is removed or redistributed by marine processes.*

The Mississippi Delta is termed a bird-foot delta because its many divergent streams (distributaries) extend seaward to form an areal pattern of lowland that resembles the outstretched toes of a bird. In this great delta, the sediment supply vastly exceeds the reworking capabilities of ocean waves and currents. The delta, therefore, builds itself seaward or **progrades** at the mouths of distributaries.

In contrast to the Mississippi Delta, where the amount of sediment brought in by streams greatly exceeds the ability of the marine environment to rework and transport that sediment, the Niger Delta on the west coast of Africa provides an example of a delta in which marine destructional processes are, to a far greater degree, in balance with the supply of sediment. The action of waves and longshore currents along the delta front is capable of reworking newly deposited sand, silt, and clay almost as fast as it is supplied. Over time, however, sedimentation has somewhat exceeded removal so that there has been concentric growth along the entire front of the delta. If the reverse were the case, there would be no delta at all.

The sediments of deltas are rich in organic debris and include many alternate and intersecting bodies of permeable sands and impermeable clays. Because of this, the rocks of ancient deltas have yielded huge volumes of petroleum and natural gas. Subsidence followed by marine incursions and then rebuilding of deltas has resulted in many-leveled sequences of oil reservoirs, such as those in many of the oil fields of the Texas and Louisiana Gulf Coast.

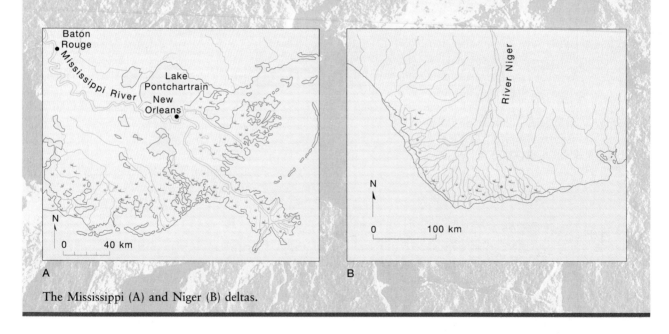

The Mississippi (A) and Niger (B) deltas.

For geologists specializing in the study of sedimentary rocks, the continental shelves hold great interest. All of the sediment eroded from the continents and carried to the sea in streams must ultimately cross the shelves or be deposited on them. Many factors influence the kind of sediment deposited on the shelves, including the nature of the source rock on adjacent land masses, the elevation of source areas, the distance from shore, and the presence of carbonate-secreting organisms. Because of wave turbulence and currents, sediment deposited in the shallow marine environment tends to be coarser than material laid down in the deeper parts of the ocean. Sand, silt, and clay are common. Where there are few continent-

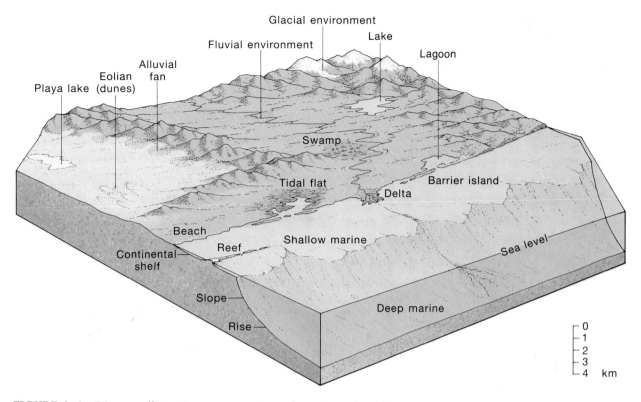

FIGURE 3–2 Diagram illustrating some marine and continental environments of deposition.

derived sediments and the seas are relatively warm, lime muds of biochemical origin may be the predominant sediment. Coral reefs are also characteristic of warm, shallow seas. They remind us of the enormous biological importance of the shallow marine environment. Over most of this realm, sunlight penetrates all the way to the sea floor. Algae and other forms of plant life proliferate. Here one finds the "pastures of the sea" on which, directly or indirectly, a multitude of swimming and bottom-dwelling animals are dependent (Fig. 3–3).

Those areas of the ocean floor that extend from the seaward edge of the continental shelves down to the ocean depths are named the **continental slopes.** Physiographic diagrams of the ocean floor are usually drawn with a large amount of vertical exaggeration, so that the continental slopes appear as steep escarpments. Actually, the inclination of the surface is only

FIGURE 3–3 Marine life flourishes in many areas of the continental shelves, for light is required for the growth of marine plants and plants are the basic components of the food chain that supports marine animals. *(Courtesy of L. E. Davis.)*

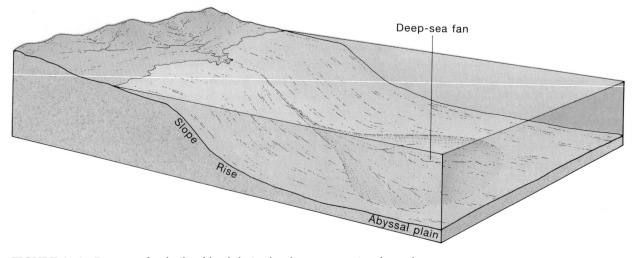

FIGURE 3–4 Deep-sea fan built of land-derived sediment emerging from the lower part of a submarine canyon. Such fans occur in association with such large rivers as the Amazon, Congo, Ganges, and Indus. (Vertical exaggeration 200 : 1.)

3° to 6°. From the sharply defined upper boundary of the continental slope, the surface of the ocean floor drops to depths of 1400 to 3200 meters. At these depths the slope of the ocean floor becomes gentler. The less pronounced slopes comprise the **continental rises** (see Fig. 3–2).

Sediment deposited on slopes and rises is mostly fine sand, silt, and clay. These materials are often transported to sites of deposition by **turbidity currents.** Water in a turbidity current is denser than surrounding water because it is laden with suspended sediment. It therefore flows down the slope of the ocean floor beneath the surrounding clear water. Upon reaching more level areas, the current slows and drops its load of suspended particles. Often the deposits, called turbidites, form submarine fans at the base of the continental slope (Fig. 3–4). In addition to turbidites, slope and rise deposits include fine clay that has settled out of the water column, and large masses of material that have slid or slumped down the slope under the influence of gravity.

In the deep marine environment far from the continents, only very fine clay, volcanic ash, and the calcareous or siliceous remains of microscopic organisms settle to the ocean floor. The exceptions are sporadic occurrences of coarser sediments that are carried down continental slopes into the deeper parts of the ocean by turbidity currents. Coarse sediments may also be dropped into deep water as they are released from melting icebergs.

Transitional Environments

The shoreline of a continent is the transitional zone between marine and nonmarine environments. Here one finds the familiar shoreline accumulations of sand or gravel that we call beaches. Mud-covered **tidal flats** that are alternately inundated and drained of water by tides are also found here, as are deltas. **Deltas** form when streams enter bodies of standing water, undergo an abrupt loss of velocity, and drop their load of sediment. Provided shoreline currents and waves do not remove the deposits as quickly as they are supplied, the delta will grow seaward. In general, progressively finer sediment will be deposited in progressively deeper or quieter water as the current provided by the stream diminishes. At the same time, the stream channel is extended over former deposits, periodically chokes in its own debris, and breaks out to form new branches of the delta.

In addition to deltas, the transitional zone includes such features as **barrier islands,** which are built parallel to shorelines by wave and current action, and **lagoons,** which are often found between the mainland and a barrier island. Swamps are also frequent features of low-lying areas adjacent to the sea.

Continental Environments

Continental environments of deposition include river **floodplains,** alluvial fans, lakes, glaciers, and eolian (wind) environments. The silt, sand, and clay found along the banks, bars, and floodplains of streams are familiar to most of us. In general, stream deposits develop as elongate lenses that are oriented downstream and that grade abruptly from sediment of one particle size to another. In another setting, stream-transported materials may accumulate quickly when a rapidly flowing river emerges from a mountainous area onto a flat plain. The result of the abrupt deposition is an **alluvial fan** (Fig. 3–5). Somewhat quieter deposition occurs in lakes, which are ideal traps for

FIGURE 3–5 Coalescing alluvial fans covering part of the floodplain, Gulf of Suez area, Egypt. The stream is unable to transport the huge amount of debris supplied to it and is dry during part of the year. *(Courtesy of D. Bhattacharyya.)*

sediment. Silt and clay are common lake sediments, although a variety of sediment is possible, depending on water depth, climate, and the character of the surrounding land areas. The **playa lakes** of arid regions are shallow temporary lakes. Periodically they become dry as a result of evaporation. Perhaps the most dramatic environments of deposition are found in areas that have active glaciers. Glaciers have the ability to transport and deposit huge volumes and large fragments of rock detritus. Deposits are characteristically unsorted mixtures of boulders, gravel, sand, and clay (Fig. 3–6). Where such materials have been re-

A

B

FIGURE 3–6 Glacial deposits. (A) In the foreground are glacial gravels being washed and sorted by meltwater. At the center left is a ridge of glacial debris called a *moraine*. The glacier in the background is Athabasca Glacier in Jasper National Park, Canada. The glacier's toe (terminus) can be seen in the center of the photograph. (B) Unsorted glacial debris called *till* in a moraine of a valley glacier, Kenai Peninsula, Alaska.

FIGURE 3–7   Sand dune in the Gran Desierto region of the Sonoran Desert, Mexico. The wind responsible for this dune blew from left to right. Note wind ripples on the windward side.   *(Courtesy of Jeff J. Plaut.)*

worked by glacial meltwater, however, they become less chaotic and resemble stream deposits.

Whereas glaciers can move materials of great size, wind is much more selective in the particle size it can transport. Environments where wind is an important agent of sediment transport and deposition are called **eolian environments** (Fig. 3–7). They are characterized by an abundance of sand and silt, little plant cover, and strong winds. These characteristics typify many desert regions.

■ *Color of Sedimentary Rocks*

We have seen that color in igneous rocks can be used to indicate the approximate amount of ferromagnesian minerals present. Color in sedimentary rock can also provide useful clues to identification. For example, varieties of chert can be identified as flint if they are gray or black, or as jasper if they are red. Color is also useful in providing clues to the environment of deposition of sedimentary rocks. Of the sedimentary coloring agents, carbon and the oxides and hydroxides of iron are the most important.

Black Coloration

Black and dark gray coloration in sedimentary rocks—especially shales—usually results from the presence of compounds containing organic carbon and iron. The occurrence of an amount of carbon sufficient to result in black coloration implies an abundance of organisms in or near the depositional areas as well as environmental circumstances that kept the remains of those organisms from being completely destroyed by oxidation or bacterial action. These circumstances are present in many marine, lake, and estuarine envi-

ronments today. In a typical situation, the remains of organisms that lived in or near the depositional basin settle to the bottom and accumulate. In the quiet bottom environment, dissolved oxygen needed by aerobic bacteria to attack and break down organic matter may be lacking. There also may be insufficient oxygen for scavenging bottom dwellers that might feed on the debris. Thus, organic decay is limited to the slow and incomplete activity of anaerobic bacteria; consequently, incompletely decomposed material rich in black carbon tends to accumulate. In such an environment, iron combines with sulfur to form finely divided iron sulfide (pyrite, FeS_2), which further contributes to the blackish coloration. Such environments of deposition are likely to yield toxic solutions of hydrogen sulfide (H_2S). The lethal solutions rise to poison other organisms and thus contribute to the process of accumulation. Black sediments do not always form in restricted basins. They may develop in relatively open areas, provided the rate of accumulation of organic matter exceeds the ability of the environment to cause its decomposition.

Red Coloration

Hues of brown, red, and green often occur in sedimentary rocks as a result of their iron oxide content. Few, if any, sedimentary rocks are free of iron, and less than 0.1 percent of this metal can color a sediment a deep red. The iron pigments are not only ubiquitous in sediments but are also difficult to remove in most natural solutions.

The iron present in sediment often occurs as either ferrous iron compounds or ferric iron compounds. Ferrous iron oxide (FeO) frequently occurs in oxygen-deficient environments. It is unstable and may slowly oxidize to form ferric iron oxide (Fe_2O_3).

When oxygen is in short supply, ferric iron may be similarly reduced to ferrous iron. Ferric minerals such as **hematite** tend to color the rock red, brown, or purple, whereas the ferrous compounds impart hues of gray and green. Hydrous ferric oxide (**limonite**) is often yellow.

Red Beds

Strata colored in shades of red, brown, or purple by ferric iron are designated **red beds** by geologists. Oxidizing conditions required for the development of ferric compounds are more typical of nonmarine than marine environments; most red beds are floodplain, alluvial fan, or deltaic deposits. Some, however, are originally reddish sediment carried into the open sea. Electron microscope studies of red beds forming today in Baja California indicate that the red coloration developed long after the sediment was deposited. After burial, the decay of clastic ferromagnesian minerals released iron that was oxidized by the oxygen in underground water circulating through the pore spaces. Thus, red coloration may be imparted in the subsurface and may be independent of climate. The paleoenvironmental interpretations one can draw from red beds should be based to a large degree on the associated rocks and sedimentary structures. Red beds interspersed with evaporite layers indicate warm and arid conditions.

Although red beds are more likely to represent nonmarine than marine deposition, they are occasionally interbedded with fossiliferous marine limestones. In such cases, the color may be inherited from red soils of nearby continental areas. Lands located in warm, humid climates often develop such red soils. When the soil particles arrive at the marine depositional site, they will retain their red coloration if there is insufficient organic matter present to reduce the ferric iron to the ferrous state. Otherwise, they will be converted to the gray or green colors of ferrous compounds.

In summary, sedimentary rocks of red coloration may be a product of the source materials, may have developed after burial as a result of a lengthy period of subsurface alteration, or may be the result of subaerial oxidation. Geologists are suspicious of the last possibility because most modern desert sediments are not red unless composed of materials from nearby outcrops of older red beds.

■ *Textural Interpretation of Clastic Sedimentary Rocks*

The size, shape, and arrangement of mineral and rock grains in a rock constitute its **texture.** In addition to the larger grains themselves, the textural appearance of a rock is influenced by the materials that hold the particles together. **Matrix** is one category of bonding material that consists of finer clastic particles (often clay) that were deposited at the same time as the larger grains and that fill the crevices between them. **Cement,** on the other hand, is a chemical precipitate that crystallizes in the voids between grains following deposition. *Silica* (SiO_2) and *calcium carbonate* ($CaCO_3$) are common natural cements. Other cements include *dolomite* ($CaMg(CO_3)_2$), *siderite* ($FeCO_3$), *hematite* (Fe_2O_3), *limonite* ($2Fe_2O_3 \cdot 3H_2O$), and *gypsum* ($CaSO_4 \cdot 2H_2O$).

Texture can provide many clues to the history of a particular rock formation. In carbonate rocks, extremely fine-grained textures such as developed in lithographic limestones probably indicate deposition in quiet water. Fine carbonate muds, which are the source sediment of such rocks, are not likely to settle to the bottom in turbulent water. Whole, unbroken fossil shells confirm the quiet water interpretation. Limestones containing the worn and broken fragments of fossil shells are likely to be the products of reworking by wave action. They are turbulent water deposits.

Size and Sorting of Clastic Grains

Geologists universally use a scale of particle sizes known as the **Wentworth scale** to categorize clastic sediments (Table 3–1). After disaggregation of a rock in the laboratory, the particles can be passed through a series of successively finer sieves, and the weight percentage of each size range in the rock can be determined. It is obvious that a stronger current of water (or wind) is required to move a large particle than a small one. Therefore, the size distribution of grains tells the geologist something about the turbulence and velocity of currents. It can also be an indicator of the mode and extent of transportation. If sand, silt, and clay are supplied by streams to a coastline, the turbulent nearshore waters will winnow out the finer particles, so that gradations from sandy nearshore deposits to offshore silty and clayey deposits frequently result (Fig. 3–8). Sandstones formed from such nearshore sands may retain considerable porosity and provide void space for petroleum accumulations. For this reason, one often finds petroleum geologists assiduously making maps showing the grain size of deeply buried ancient formations to determine areas of coarser and more permeable clastic rock.

One aspect of a clastic rock's texture that involves grain size is sorting. **Sorting** is an expression of the range of particle sizes deviating from the average size. Rocks composed of particles that are all about the

TABLE 3–1   Size Range of Sedimentary Particles

Wentworth scale (in millimeters)	Fractional equivalents (in millimetes)	Particle name
		Boulders
256		
128		Cobbles
64		
32		
16		Pebbles
8		
4		
		Granules
2		
		Very coarse sand
1.0		
		Coarse sand
0.5	1/2	
		Medium sand
0.25	1/4	
		Fine sand
0.125	1/8	Very fine sand
0.0625	1/16	
0.0313	1/32	
0.0156	1/64	
		Silt
0.0078	1/128	
0.0039	1/256	
		Clay

same average size are said to be *well sorted* (Fig. 3–9) and those that include grains with a wide range of sizes are termed *poorly sorted.* Sorting often provides clues to conditions of transportation and deposition. Wind, for example, winnows the dust particles from sand, producing grains that are all of about the same size. Wind also sorts the particles that it carries in suspension. Only rarely is the velocity of winds sufficient to carry grains larger than 0.2 mm. While carrying grains of that size, winds sweep finer particles into the higher regions of the atmosphere. When the wind subsides, well-sorted silt-sized particles drop and accumulate. In general, windblown deposits are better sorted than deposits formed in an area of wave

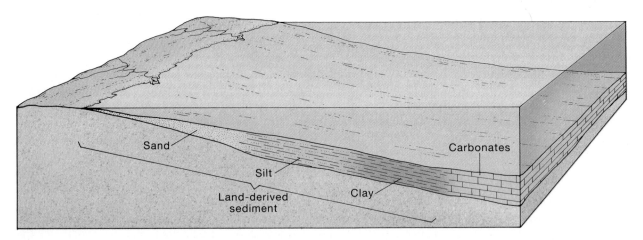

FIGURE 3–8   Idealized gradation of coarser nearshore sediments to finer offshore deposits.

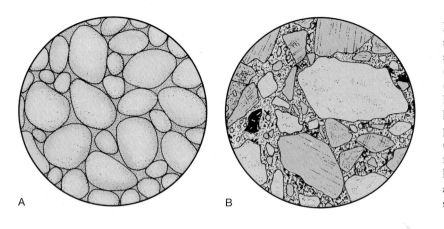

FIGURE 3–9 Sorting of grains in sandstones as seen under the microscope may range from good sorting (A) to poor sorting (B). (A) Quartz (light tan grains) sandstone with carbonate (pink) cement. (B) A sandstone known as graywacke composed of poorly sorted angular grains of quartz (light tan), feldspar (green), and rock fragments (orange). The graywacke lacks cement; spaces between grains are filled with a matrix of clay and silt. Width of fields is 1.5 mm.

FIGURE 3–10 Well-rounded grains of quartz viewed under the microscope. From the St. Peter Formation near Pacific, Missouri. Width of field is 1.85 mm.

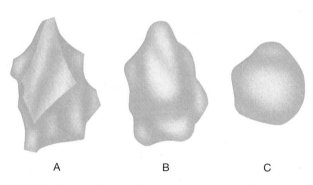

FIGURE 3–11 Shape of sediment particles. (A) An angular particle (all edges sharp). (B) A rounded grain that has little sphericity. (C) A well-rounded, highly spherical grain. *Roundness* refers to the smoothing of edges and corners, whereas *sphericity* measures the degree of approach of a particle to a sphere.

action, and wave-washed sediments are better sorted than stream deposits. It must be kept in mind, however, that if a source sediment is already well sorted, the resulting deposit will be similarly well sorted. Poor sorting occurs when sediment is rapidly deposited without being selectively separated into sizes by currents (see Fig. 3–6B). Poorly sorted conglomerates and sandstones are deposited at the foot of mountains, where stream velocity is suddenly checked. Another example of a poorly sorted conglomerate is **tillite,** a rock deposited by glacial ice containing all particle sizes in a heterogeneous mixture.

Shape of Clastic Grains

The **shape** of particles in a clastic sedimentary rock can also be useful in determining its history. Shape can be described in terms of **rounding** of particle edges and **sphericity** (how closely the grain approaches the shape of a sphere) (Figs. 3–10 and 3–11). A particle becomes rounded by having sharp corners and edges removed by impact with other particles. The relatively heavy impacts between pebbles and granules being transported by water cause rapid rounding. Lighter impacts occur between sand grains in water transport; the water provides a cushioning effect. The result is far slower rounding for sand grains. In conjunction with other evidence, the roundness of a particle can be used to infer the history of abrasion. It is a reflection of the distance the particle has traveled, the transporting medium, and the rigor of transport. It can also be used as evidence for recycling of older sediments.

Arrangement of Clastic Grains

The third element in our definition of texture is the arrangement of the grains in the clastic rock. Geologists examine the rock to ascertain whether the grains are the same size and if they are clustered into zones

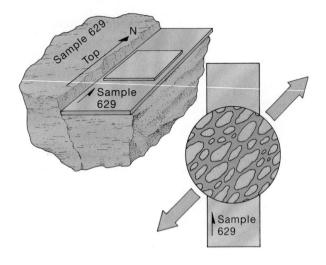

FIGURE 3–12   Grain orientation study. One method of studying grain orientation is to prepare an oriented thin section of a rock whose field orientation has been recorded. Grain orientations are then measured under a microscope equipped with a rotating stage. The angle of the long axis of each elongated grain from the north line is determined. From many individual measurements, a mean orientation—in this example, about N 45°E or S 45°W—is determined and its statistical significance evaluated. A thin section cut perpendicular to bedding might reveal the tilt of the grains and might then be used to determine that the transportation medium flowed northeastward rather than southwestward.

or heterogeneously mixed. These observations may help to determine whether the sediment had been winnowed and sorted by currents or had been dumped rapidly. Such factors as the medium of transport, surface of deposition, and direction and velocity of current control grain orientation. Geologists are particularly interested in studying grain orientation as a means of determining the current direction of the transporting medium. In general, sand grains deposited in moving water tend to acquire a preferred orientation in which the long axes of most of the elongate grains are aligned parallel to the direction of flow. Such particles also tend to be slightly tilted in the upstream direction.

To determine whether preferred orientation exists in a rock unit, geologists must collect many carefully oriented samples (Fig. 3–12) and subject the samples to a statistical analysis of the orientation of constituent grains. Many times, because of weakness or variability of currents, there may be little or no preferred orientation. Also, orientation in windblown sands is almost always less well developed than in water-carried sands. Several grain orientation studies of modern beach sands show that grain orientation in this environment is mainly controlled by backwash; in general, grains become aligned parallel to the shoreline. In river-deposited sands, grain orientation is usually parallel to the elongation of the sand body.

A

B

FIGURE 3–13   Modern and ancient mudcracks. (A) This modern mudcrack formed in soft clay around the margins of an evaporating pond. (B) Mudcracks and wave ripples (caused by wind blowing over a shallow lake) in mudstones of the Oneonta Formation of Devonian age near Unadilla, New York. Divisions on the scale are 1.0 cm.   *(Modern mudcracks courtesy of L. E. Davis; ancient mudcracks courtesy of W. D. Sevon.)*

In glacial sediment, both elongate sand grains and pebbles show a longitudinal orientation parallel to the direction of ice movement. Grain orientation analyses are useful in determining sediment distributional patterns in the geologic past and have occasionally provided important clues to the subsurface location and trend of petroleum-bearing sandstone strata.

■ *Inferences from Sedimentary Structures*

Sedimentary structures are those larger features of sediments that are formed during or shortly after deposition and before lithification. Because particular sedimentary structures result from specific depositional processes, the structures are extremely useful to geologists interested in reconstructing ancient environments. For example, **mudcracks** indicate drying after deposition. These conditions are common on valley flats and in tidal zones. Mudcracks (Fig. 3–13) develop by shrinkage of mud or clay on drying and are most abundant in the subaerial environment. **Cross-bedding** (cross-stratification) is an arrangement of beds or laminations in which one set of layers is inclined relative to the others (Figs. 3–14 and 3–15). The cross-bedding units can be formed by the

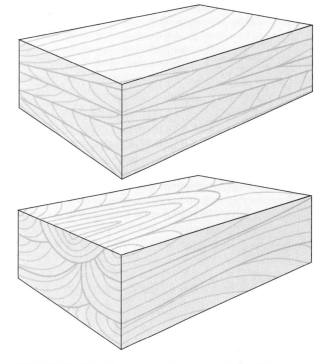

FIGURE 3–14   Two types of cross-bedding. The upper block shows planar laminations, as seen in beach deposits and dunes; the lower block represents trough laminations, as often formed in river channels. The line at the base of each set of laminations represents a surface of erosion that truncates older sets below.

FIGURE 3–15   Cross-bedding (cross-stratification) of the tabular-planar type in the Mountain Lakes Formation of Proterozoic age, Northwest Territories, Canada.   *(Courtesy of G. Ross.)*

COMMENTARY
Sedimentary Way-Up Structures

The principle of superposition tells us that in *undisturbed* strata, the oldest bed is at the bottom and higher layers are successively younger. But what if the strata are deformed and overturned in such a way that the oldest beds are found at the top (Fig. A)? If overturning is not recognized, the geologic sequence of events, the kinds of folds, and features indicating the direction of sediment transport might be misinterpreted. For this reason, geologists working in areas of deformed strata (see Fig. 6–5) carefully scrutinize rocks for indications of the original tops and bottoms of beds. Features providing such information are called **geopetal structures.** Among the more common geopetal structures are some oscillation ripple marks (Fig. 2–22); included fragments; certain types of cross-bedding (Fig. 2–15); graded bedding (Fig. 2–18); mudcracks (Fig. 2–13); bed surface markings such as footprints, trails, and raindrop imprints; sole marks; fossils; and various biologically produced structures (Fig. 3–9).

In the case of oscillation ripple marks (Fig. B), the sharp crests of the ripples normally identify the tops of beds. They point toward the younger beds. If fragments of the rippled rock are found in overlying strata, the overlying bed must be younger and the interpretation is confirmed.

In many kinds of cross-bedding, the cross-beds are concave upward, forming a small angle with beds below and a large angle with beds resting on their truncated upper edges (Fig. C). Way-upness interpretations based on cross-bedding, however, should be confirmed by other geopetal structures, as some cross-beds do not show the concave upward shape.

Graded bedding, in which grains are progressively finer from the bottom to the top of a bed, is another useful geopetal structure. Graded beds are typically formed when fast-moving currents begin to slow, so that large particles are dropped first followed by progressively finer grains.

Mudcracks are geopetal structures formed when mud dries, shrinks, and cracks. The cracks narrow downward, away from the top of the bed. Deposition above mudcracks would fill them, resulting in a corresponding pattern of ridges that identify the bottom of the overlying stratum.

As currents flow across beds of sand, they often erode various kinds of scour marks. An overlying layer of sediment may later fill these depressed markings, forming positive-relief casts in the covering bed (Fig. D). The casts are termed sole markings because they appear on the sole or bottom of the stratum.

Fossils of bottom-dwelling organisms such as corals may also be used to determine way-upness, provided they have been buried in their natural, upright living positions. Some fossils that have been moved by currents may also be useful. For example, the curved shells of clams washed by currents may come to rest in a convex-upward position, as this is hydrodynamically most stable. Finally, many fossil organisms excavated and lived in burrows, such as the U-shaped burrows shown in Figure E. Such structures are often excellent clues to way-upness.

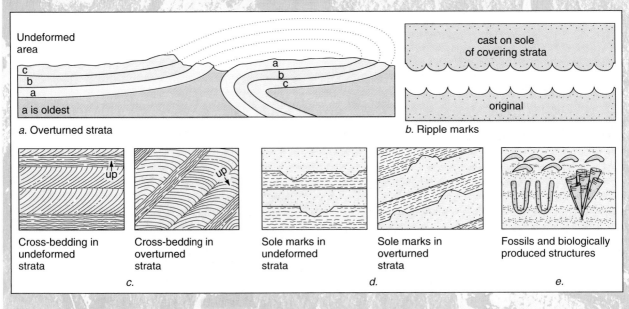

a. Overturned strata

Undeformed area

c
b
a
a is oldest

a
b
c

b. Ripple marks

cast on sole of covering strata

original

Cross-bedding in undeformed strata

Cross-bedding in overturned strata

Sole marks in undeformed strata

Sole marks in overturned strata

Fossils and biologically produced structures

c.

d.

e.

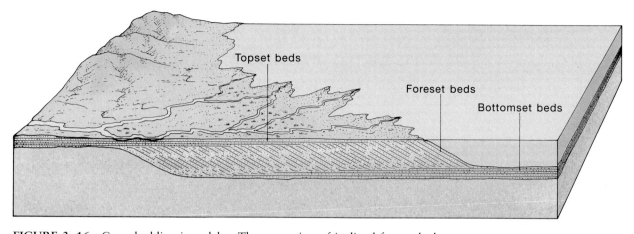

FIGURE 3–16 Cross-bedding in a delta. The succession of inclined foreset beds is deposited over bottomset beds that were laid down earlier. Topset beds are deposited by the stream above the foreset beds.

advance of a delta (Fig. 3–16) or a dune (Fig. 3–17). A depositional environment dominated by currents is inferred from cross-bedding. The currents may be wind or water. In either medium, the direction of the inclination of the sloping beds is a useful indicator of the direction taken by the current. By plotting these directions on maps, geologists have been able to determine the pattern of prevailing winds at various times in the geologic past.

Graded bedding consists of repeated beds, each of which has the coarsest grains at the base and successively finer grains nearer the top (Fig. 3–18). Al-

though graded bedding may form simply as the result of faster settling of coarser, heavier grains in a sedimentary mix, it appears to be particularly characteristic of deposition by the turbidity currents discussed earlier. Turbidity currents are often triggered by submarine earthquakes and landslides that occur along steeply sloping regions of the sea floor. The forward part of the turbidity current contains coarser debris than does the tail. As a result, the sediment deposited at a given place on the sea bottom grades from coarse to fine as the "head" and then the "tail" of the current passes over it. The presence of graded beds may

FIGURE 3–17 Sand being transported by wind action. In this dune at Great Dunes, Colorado, sand is being swept up the windward side (the gentler slope) and deposited on the leeward slope to form the steeper slip face of the dune.

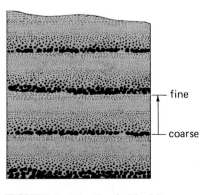

fine

coarse

FIGURE 3–18 Graded bedding.

A   B

FIGURE 3–19   (A) Ripple marks formed in sand along a modern beach.
(B) Ripple marks on a bedding surface of the Tensleep Formation of Pennsylvanian age, Wyoming.   *(Courtesy of L. E. Davis.)*

indicate former turbidity currents. Geologists believe
that turbidity currents frequently characterize unstable, tectonically active environments.

Ripple marks are commonly seen sedimentary
features that developed along the surfaces of bedding
planes (Fig. 3–19). *Symmetric ripple marks* are
formed by the oscillatory motion of water beneath
waves. *Asymmetric ripple marks* are formed by air or
water currents and are useful in indicating the direction of movement of currents (Fig. 3–20). For example, ripple marks form at right angles to current directions; the steeper side of the asymmetric variety
faces the direction in which the medium is flowing.
Although there are instances of ripple marks devel-

oped at great depths on the sea floor, these features
occur more frequently in shallow water areas and in
streams.

■ *Interpretation of Sands and Sandstones*

Among the clastic group of sedimentary rocks, sandstones have been studied in great detail and provide
an extraordinary amount of information about conditions in and near the site of deposition. In particular, the mineral composition of sandstone grains can
be used to identify source areas and to interpret what

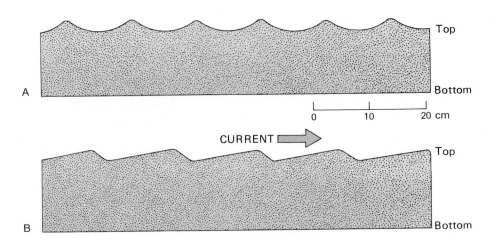

FIGURE 3–20   Profiles of
ripple marks. (A) Symmetric
ripples. (B) Asymmetric ripples.

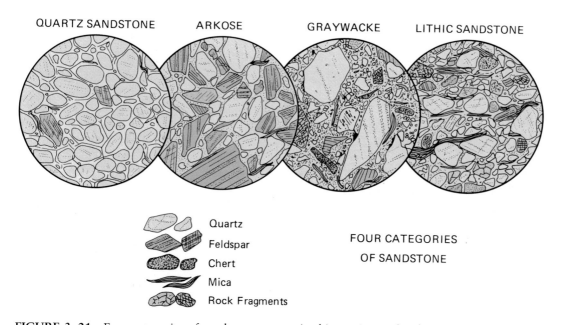

QUARTZ SANDSTONE   ARKOSE   GRAYWACKE   LITHIC SANDSTONE

Quartz
Feldspar
Chert
Mica
Rock Fragments

FOUR CATEGORIES
OF SANDSTONE

FIGURE 3–21 Four categories of sandstone as seen in thin section under the microscope. Diameter of field is about 4 mm.

may have occurred prior to deposition. Often, by closely studying the grains, one can ascertain whether the source material was metamorphic, igneous, or sedimentary. The mineral content also provides a rough estimate of the amount of transport and erosion experienced by the sand grains. Rigorous weathering and long transport tend to reduce the less stable feldspars and ferromagnesian minerals to clay and iron compounds and tend to cause rounding and sorting of the remaining quartz grains. Hence, one can assume that a sandstone rich in these less durable and angular components underwent relatively little transport and other forms of geologic duress. Such sediments are termed *immature* and are most frequently deposited close to their source areas. On the other hand, quartz can be used as an indicator of a sandstone's maturity; the higher the percentage of quartz, the greater the maturity.

In addition to providing an indication of a rock's maturity, composition is an important factor in the classification of sandstones into quartz sandstone, arkose, graywacke, and lithic sandstone (sometimes termed subgraywacke) (Fig. 3–21).

Quartz sandstones (Fig. 3–22) are characterized by a dominance of quartz with little or no feldspar, mica, or fine clastic matrix. The quartz grains are well sorted and well rounded (see Fig. 3–10). They are most commonly held together by such cements as calcite and silica. Chemical cements such as these tend to be more characteristic of "clean" sandstones like quartz sandstones and are not as prevalent in "dirtier" rocks containing clay. The presence of a dense, clayey matrix seems to retard the formation of

chemical cement, perhaps because fine material fills pore openings where crystallization might occur.

Calcite cement may develop between the grains as a uniform, finely crystalline filling, or large crystals may form, and each may incorporate hundreds of quartz grains. Silica cement in quartz sandstone commonly develops as overgrowths on the original grain surfaces.

Quartz sandstones reflect deposition in stable, quiet, shallow water environments such as the shallow seas that inundated large parts of low-lying continental regions in the geologic past or some parts of

FIGURE 3–22 Quartz sandstone. The color in this rock is caused by minor amounts of iron oxide (specimen size approximately 7 × 10 cm).

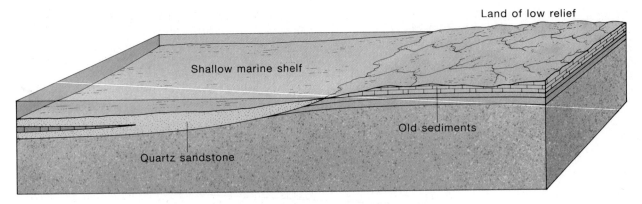

FIGURE 3–23 Idealized geologic conditions under which quartz sandstone may be deposited. There is little tectonic movement in this environment. Water depth is shallow, and the basin subsides very slowly.

our modern continental shelves (Fig. 3–23). These sandstones, as well as clastic limestones, exhibit sedimentary features, such as cross-bedding and ripple marks, which permit one to infer shallow water deposition.

Sandstones containing 25 percent or more of feldspar (derived from erosion of a granitic source area) are called **arkoses** (Fig. 3–24). Quartz is the most abundant mineral, and the angular to subangular grains are bonded together by calcareous cement, clay minerals, or iron oxide. The presence of abundant feldspars and iron imparts a pinkish-gray color to many arkoses. In general, arkoses are coarse, moderately well-sorted sandstones. They may originate as basal sandstones derived from the erosion of a granitic coastal area experiencing an advance of the sea, or they may accumulate in fault troughs or low areas adjacent to granite mountains (Fig. 3–25).

Graywackes (from the German term *wacken*, meaning waste or barren) are immature sandstones consisting of significant quantities of dark, very fine-

FIGURE 3–24 Thin section of an arkose viewed through a petrographic microscope. The clear grains are mostly quartz, whereas the grains that show stripes or a plaid pattern are feldspars. The matrix consists of kaolinite clay and fine particles of mica, quartz, and feldspar. *(Courtesy of the U.S. Geological Survey; photo by J. D. Vine.)*

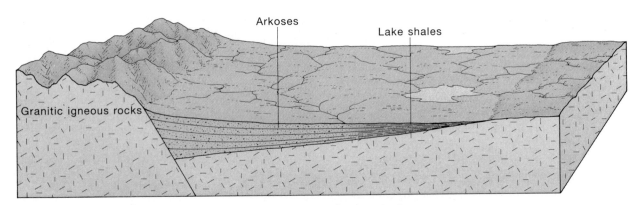

FIGURE 3–25 Geologic environment in which arkose may be deposited.

FIGURE 3–26 Graywacke.

FIGURE 3–27 Thin section of a poorly sorted sandstone (graywacke) observed under petrographic microscope with crossed polarizers. Note wide range of grain sizes. Width of field is 9.0 mm.

grained material (Fig. 3–26). Normally, this fine matrix consists of clay, chlorite, micas, and silt. There is little or no cement, and the sand-sized grains are not in close contact because they are separated by the finer matrix particles. The matrix constitutes approximately 30 percent of the rock, and the remaining coarser grains consist of quartz, feldspar, and rock particles. Graywacke has a dirty, "poured-in" appearance. The poor sorting, angularity of grains (Fig. 3–27), and heterogeneous composition of graywackes indicate an unstable source and depositional

area, in which debris resulting from accelerated erosion of highlands is carried rapidly to subsiding basins. Graded bedding (see Fig. 3–18), interspersed layers of volcanic rocks, and cherts (which may indirectly derive their silica from volcanic ash) further attest to dynamic conditions in the area of deposition. The inferred tectonic setting is unstable, with deposition occurring offshore from an actively rising mountainous region (Fig. 3–28). Graywackes and associated shales and cherts may contain fossils of deep water organisms, indicating deposition at great

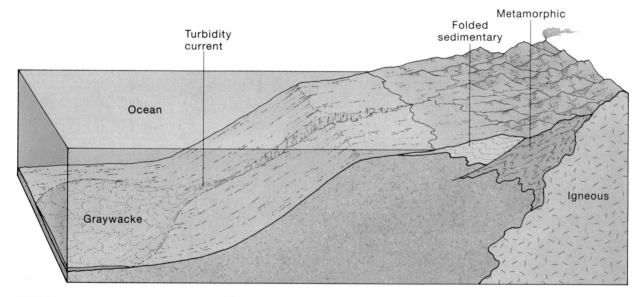

FIGURE 3–28 Tectonic setting in which graywacke is deposited. Frequently graywackes are transported by masses of water highly charged with suspended sediment. Because of the suspended matter, the mass is denser than surrounding water and moves along the sloping sea floor or down submarine canyons as a turbidity current. The graywacke sediment characteristically accumulates in deep-sea fans at the base of the continental slope.

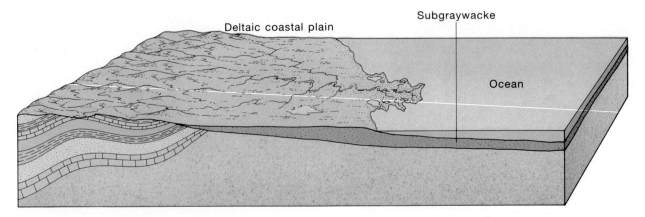

FIGURE 3–29   Deltaic environment in which lithic sandstones may be deposited.

depth. Such shallow water sedimentary structures as cross-bedding and ripple marks are rarely found. To the experienced petrologist, graywackes clearly indicate dynamic, unstable conditions.

Quartz sandstone, arkose, and graywacke are rather distinct kinds of sandstones. A rock that has a more transitional composition and texture is termed a **lithic sandstone** (subgraywacke). In lithic sandstones (see Fig. 3–21), feldspars are relatively scarce, whereas quartz, muscovite, chert, and rock fragments are abundant. There is a fine-grained detrital matrix that does not exceed 15 percent, and the remaining voids are filled with mineral cement or clay. In lithic sandstone, the quartz grains are better rounded and more abundant, the sorting is better, and the quantity of matrix is lower than in graywacke.

The characteristic environments for lithic sandstones are deltaic coastal plains (Fig. 3–29), where lithic sandstones may be deposited in nearshore marine environments or swamps and marshes. Coal beds and micaceous shales are frequently associated with lithic sandstones.

Interpretation of Carbonate Rocks

Limestones are the most abundant of carbonate sedimentary rocks. Although limestone lake deposits do occur, most limestones originated in the seas. Nearly always, the formation of these marine limestones appears to have been either directly or indirectly associated with biologic processes. In some limestones, the importance of biology is obvious, for the bulk of the rock is composed of readily visible shells of molluscs and skeletal remains of corals and other marine organisms. In other limestones, skeletal remains are not present, but nevertheless the calcium carbonate that forms the bulk of the deposit was precipitated from sea water because of the life processes of organ-

isms living in that water. For example, the relatively warm, clear ocean waters of tropical regions are usually slightly supersaturated with calcium carbonate. In this condition, only a slight increase in temperature, loss of dissolved carbon dioxide, or influx of supersaturated water containing $CaCO_3$ "seeds" can bring about the precipitation of tiny crystals of calcium carbonate. Organisms do not appreciably affect temperature, but through photosynthesis, myriads of microscopic marine plants remove carbon dioxide from the water and thus may trigger the precipitation of calcium carbonate. Bacterial decay may also enhance precipitation of calcium carbonate by generating ammonia and thereby changing the alkalinity of sea water. In either case, the precipitate can be considered an inorganic product of organic processes. Carbonate sedimentation today is most rapid in shallow, clear-water tropical marine areas such as the Bahama Banks east of Florida (Fig. 3–30). The carbonate sediments forming today in the Bahama Banks originate in more than one way. Some of them are derived from the death and dismemberment of calcareous algae, such as *Penicillus*, an organism that secretes tiny, needle-like crystals of calcium carbonate. The microscopic shells of other unicellular organisms also contribute to the carbonate buildup. In areas where tidal currents flow across the banks, oöids accumulate. As noted in the previous chapter, oöids are tiny spheres composed of calcium carbonate that are formed when particles roll back and forth on the sea floor and acquire concentric rings of carbonate (Fig. 3–31). Some of the sediment results from the precipitation of tiny crystals of lime from sea water that has been chemically altered by the biologic processes of marine plants. Coarser particles result from the abrasion of the shells of invertebrates or consist of fecal pellets produced by burrowing organisms.

The requirement of warm, clear, shallow seas for the accumulation of modern carbonates seems to apply equally well to ancient deposits. Major sequences of ancient limestones are relatively free of clay and frequently contain an abundance of fossils thought to be representative of shallow, warm seas. Ancient carbonate rocks have developed in a variety of tectonic settings. Thick sections of limestones and dolomites have formed in subsiding basins in West Texas, Alberta, and Michigan. In such areas, optimum conditions for carbonate sedimentation resulted in a rate of accumulation that approximately equaled subsidence. Thick deposits of limestones have also accumulated along the margins of conti-

nents, such as the eastern margin of North America during the Paleozoic Era. Such shelflike areas that lie adjacent to the deep ocean basins are called **carbonate platforms.** During the geologic past, when sea levels were typically higher than today and climates generally warmer, carbonate platforms were more abundant and extensive.

One type of carbonate sedimentary rock that continues to perplex geologists is the magnesium–calcium carbonate rock dolomite. You will recall from Chapter 2 that dolomite is a rock composed primarily of the mineral dolomite, $CaMg(CO_3)_2$. Dolomite is believed to form when magnesium that has been concentrated in sea water replaces a portion of the calcium carbonate in previously deposited calcium carbonate sediment. Supporting this interpretation are fossil shells in dolomite strata that were originally composed of calcite but that subsequently have been changed to dolomite. Today, dolomite formation occurs in only a few areas, usually where evaporation of sea water is sufficiently intense to concentrate magnesium. Yet during parts of the Precambrian, Paleozoic, and Mesozoic eras, dolomites were extensively developed (Fig. 3–32). To account for these ancient deposits, it would appear that extensive areas of evaporative conditions would be required. Recently, geologist David Lumsden discovered a correlation between ancient periods of high dolomite formation and episodes of high sea level. He suggests that during times of eustatic (worldwide) rise in sea level, broad low-lying tracts of the continents were

FIGURE 3–30 Carbonate mud accumulating on the sea floor in the shallow warm waters of the Bahama Banks. *(Courtesy of L. Walters.)*

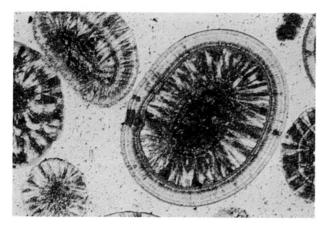

FIGURE 3–31 Thin section view of an oölitic limestone. The oöids are immersed in a cement of sparry (clear) calcite. The large oöid in the center has a maximum diameter of 0.74 cm. *(Courtesy of Graham Thompson and Jon Turk.)*

FIGURE 3–32 Dolomites of Triassic age exposed in the Italian Alps. Dolomites are named after the French geologist Guy de Dolomieu (1750–1801). *(Italian Tourist Board.)*

inundated by shallow seas. Where climatic conditions were favorable, these shallow seas provided the ideal evaporative environments needed to enrich sea water with magnesium. As the magnesium replaced part of the calcium in the calcium carbonate sediment that blanketed the sea floor, calcite was converted to dolomite. When the sea level subsequently fell, the magnesium enrichment process halted and dolomite formation ceased.

Interpretation of Clays and Shales

Shale (Figs. 3–33 and 3–34) is a general term for a very fine-textured, fissile (capable of being split into thin layers) rock composed of clay, mud, and silt. In general, the environmental significance of shales parallels that of the sandstones with which they are associated. Frequently, the silt-size particles in shales are similar in composition and shape to the associated sandstone beds. These silty components can frequently be extracted for study by disaggregating the shale in water and repeatedly pouring off the muddy liquid, retaining the larger particles as a residue.

In shales associated with quartz sandstones, the silt fraction often consists predominantly of rounded quartz grains. Such quartz shales result from the reworking of older residual clays by transgressing shallow seas. Their association with thin, widespread

FIGURE 3–34   Thin section of shale. As indicated here, silt particles are abundant constituents of shale. The brown color results from organic matter that is mixed with the clay, which is the major constituent of shale. *(Courtesy of Graham Thompson and Jon Turk.)*

limestones and quartz sandstones provides evidence for their deposition under stable tectonic conditions.

Feldspathic shales contain at least 10 percent feldspar in the silt size and tend to be rich in the clay mineral kaolinite. Feldspathic shales are frequent associates of arkoses and are presumed to have formed in a similar environment. Such shales are representative of the finer sediment winnowed from coarser detritus and deposited in quieter locations.

Chloritic shales are usually associated with graywackes. As implied by their name, flakes of the green silicate mineral chlorite are common among the silt-sized components. The less flaky particles tend to be angular. Fissility in these rocks is less well developed than in other shales. The clay and silt particles in chloritic shales are generally derived from mountainous, unstable source areas nearby.

A shale type that is the approximate equivalent of a lithic sandstone is designated a micaceous shale. Mica flakes, quartz, and feldspar are all among its silt-sized components. Micaceous shales are deposited under conditions somewhat less stable than the environment for quartz shales. They are particularly characteristic of ancient deltaic deposits.

The clay minerals that occur in shales are complex hydrous aluminosilicates with constituent atoms arranged in silicate sheet structures. Kaolinites, smectites, and illites are the three major groups of clay minerals. Kaolinites are the purest and seem to have a preferred occurrence in terrestrial environments. Smectite may contain magnesium, calcium, or sodium or any combination of these three, whereas potassium is an essential constituent of illites. Illites are the predominant clay mineral in more ancient shales.

FIGURE 3–33   Shale as it appears in the field. *(Courtesy of Graham Thompson and Jon Turk.)*

Unfortunately, positive identification of clay minerals in hand specimens or under the microscope is not possible and can be accomplished only by means of X-ray, thermal, or chemical analyses.

■ *The Sedimentary Rock Record*

Rock Units

William Smith, the British surveyor mentioned in Chapter 1, demonstrated that distinctive bodies of strata could be traced over appreciable distances and therefore could be mapped. He produced an exceptionally fine geologic map of England and Wales in 1815. Smith's map, the first ever made of such high quality and accuracy, was accompanied by a comprehensive table of the rock units encountered in the area. Each of these units was given a particular name, such as the "Clunch Clay," the "Great Oölyte," or the "Cornbrash Limestone." Thus originated the concept of a fundamental unit in geology that was lithologically distinctive, that had recognizable contacts with other units both above and below, and that could be traced across the countryside from exposure to exposure (or in the subsurface from well to well). Smith referred to such a unit as a stratum, but today it is universally known as a **formation** (Fig. 3–35). Furthermore, formations and groupings or subdivisions of formations all constitute rock units. **Rock units,** also called **lithostratigraphic units,** are formally defined as bodies of rock identified by their distinctive lithologic and structural features without regard to time boundaries. Thus rock units are distinctly different from the time–rock or *chronostratigraphic* units defined in the previous chapter. Such features as texture, grain size, clastic or crystalline, color, composition, thickness, type of bedding, nature of organic remains, and appearance of the unit in surface exposures (or in the lithologic record of strata penetrated by wells), are all used to define a rock unit and recognize it in the field. Whereas a chronostratigraphic unit represents a body of rock deposited or

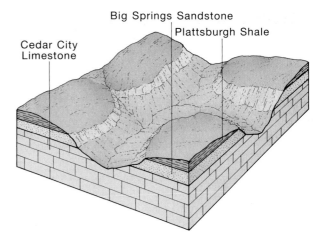

FIGURE 3–35 Formations. The diagram shows three formations. In practice, these formations would be formally named, often after a geographic location near which they are well exposed. For example, the three formations shown here might be designated the Cedar City Limestone, Big Springs Sandstone, and Plattsburgh Shale.

emplaced during a specific interval of time, a rock unit like a formation may or may not be the same age everywhere it is encountered. The nearshore sands deposited by a sea slowly advancing (transgressing) across a low coastal plain may deposit a single blanket of sand (perhaps later to be named the Oriskany Sandstone); however, that sand layer will be older where the sea began its advance and younger where the advance halted (Fig. 3–36).

Naming Rock Units

Formations are given two names: first, a geographic name that refers to a locality where the formation is well exposed or where it was first described, and second, a rock name if the formation is primarily of one lithologic type. For example, the St. Louis Limestone is a carbonate formation named after exposures at St. Louis. When formations have several lithologic types within them, the locality name is simply followed by the term *Formation.*

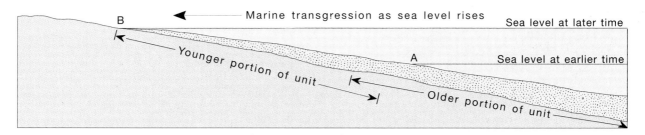

FIGURE 3–36 Diagram showing how the original deposits of a formation may vary in age from place to place.

There are other rock units in addition to formations. Distinctive smaller units within formations may be split out as **members,** and formations may be combined into larger units called **groups** because of related lithologic attributes (or their position between distinct stratigraphic breaks). For example, in Grand Canyon National Park, the Whitmore Wash, Thunder Springs, Mooney Falls, and Horseshoe Mesa rock units are *members* of the massive Redwall Limestone. In another part of the canyon wall, one finds the Tapeats Sandstone, Bright Angel Shale, and Muav Limestone combined to form a larger mappable rock unit known as the Tonto Group (Table 3–2).

Facies

The aforementioned rock terms provide for direct objective mapping of sedimentary beds as well as bodies of metamorphic and igneous rocks. If one is to make inferences about events recorded in rock units, it is useful to employ the term *facies*. A sedimentary **facies** refers to the characteristics or aspects of a rock from which its environment of deposition can be inferred. For example, a body of rock might consist of a bioclastic limestone along one of its lateral margins and micritic limestone elsewhere. Geologists might then delineate a "bioclastic limestone facies" and interpret it as a nearshore part of the rock body, whereas they might interpret the micritic limestone facies as a former offshore deposit. In this case, the distinguishing characteristics are lithologic (rather than biologic); therefore, the facies can be further designated as a **lithofacies.** In other cases, the rock unit may be lithologically uniform, but the fossil assemblages differ and permit recognition of different **biofacies** that reflect differences in the environment. A limestone unit, for example, might contain abundant fossils of shallow water reef corals along its thinning edge and elsewhere be characterized by remains of deep water sea urchins and snails. There would thus be two biofacies developed—one reflecting deeper water than the other. They could, therefore, be designated the "coral" and the "echinoid-gastropod biofacies."

These examples illustrate that facies are clearly the products (sediment, shells of organisms) of particular environments of deposition. Today, as we travel across swamp, floodplain, and sea, we traverse differ-

TABLE 3–2   Rock Units of the Paleozoic Section in Grand Canyon National Park, Arizona

System	Group	Formation	Member
Permian		Kaibab	
		Toroweap	
	Hermit	Coconino	
		Hermit Shale	
Pennsylvanian		Supai	
Mississippian		Redwall Limestone	Horseshoe Mesa
			Mooney Falls
			Thunder Springs
			Whitmore Wash
Devonian		Temple Butte	
Cambrian	Tonto	Muav Limestone	
		Bright Angel Shale	
		Tapeats Sandstone	

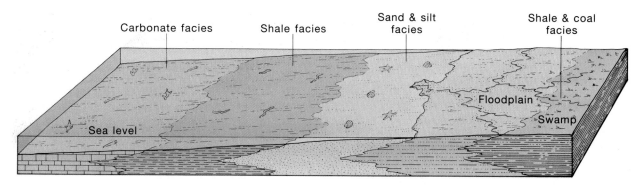

Carbonate facies   Shale facies   Sand & silt facies   Shale & coal facies

Sea level   Floodplain   Swamp

FIGURE 3–37 Sedimentary facies (lithofacies) developed in the sea adjacent to a land area. The upper surface of the diagram shows present-day facies, whereas the front face shows the shifting of facies through time. Notice that bottom-dwelling organisms also differ in environments having different bottom sediment and water depth.

ent environments of deposition (Fig. 3–37). Each of these environments of deposition changes laterally into the adjacent environment of deposition, and each provides its present-day facies that likewise change to adjacent synchronous facies. Geologists record these facies changes on lithofacies and biofacies maps. Because the basis for these maps is time–rock units, such maps provide a view of different facies of essentially the same age. If one were able to make such maps of successively different times, it would become apparent that ancient facies have

shifted their localities as the seas advanced or retreated or as environmental conditions changed.

Consider for a moment an arm of the sea slowly transgressing (advancing over) the land. The sediment deposited on the sea floor may ideally consist of a nearshore sand facies, an offshore mud facies, and a far offshore carbonate facies. As the shoreline advances inland, the boundaries of these facies also shift in the same direction, thereby developing an **onlap** sequence (Fig. 3–38) in which coarser sediments are covered by finer ones. Should the sea subsequently

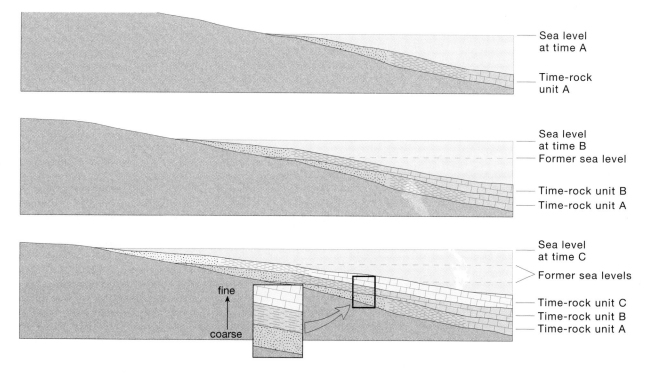

Sea level at time A

Time-rock unit A

Sea level at time B
Former sea level

Time-rock unit B
Time-rock unit A

Sea level at time C
Former sea levels

Time-rock unit C
Time-rock unit B
Time-rock unit A

fine

coarse

FIGURE 3–38 Sedimentation during a transgression produces an onlap relationship in which finer offshore lithofacies overlie coarser nearshore facies (see inset), nearshore facies are progressively displaced away from a marine point of reference, and older beds are protected from erosion by younger beds.

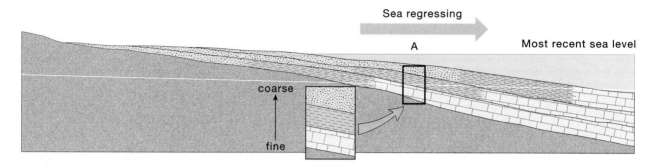

FIGURE 3–39 Sedimentation during a regression produces an offlap relationship in which coarser nearshore lithofacies overlie finer offshore lithofacies, as shown in A. The sandy nearshore facies is progressively displaced toward the marine point of reference. Older beds are subjected to erosion as the regression of the sea proceeds.

begin a withdrawal (regression), the facies boundaries will again move in the same direction as the shoreline, creating as they do so an **offlap** sequence of beds (Fig. 3–39). In offlap situations, coarser nearshore sediment tends to lie above finer sediments. Also, because offlap units are deposited during marine regressions, recently deposited sediment is exposed to erosion, and part of the sedimentary sequence is lost. Study of sequential vertical changes in lithology, such as those represented by offlap and overlap relationships, is one method by which geologists recognize ancient advances and retreats of the seas and chart the positions of shorelines.

Onlap and offlap patterns of sedimentation were recognized as early as 1894 by the German geologist Johannes Walther. Walther observed that the succession of facies occurring laterally is also seen in the vertical succession of facies. Thus, to find what facies are to be encountered laterally from a given locality, one need only examine the vertical sequence of beds at that locality. For example, Section B in Figure 3–40 shows a typical fining upward succession of facies. Point x is in the nearshore silt facies and is overlain by finer shale and then limestone. This same sequence of silt to shale to limestone is seen in moving westward to Section A. Beneath point x is a coarse beach sand, which can be traced laterally (eastward) in Section C. This relationship, in which the vertical succession of facies corresponds to the lateral succession, has been named **Walther's principle.**

If the pattern of sediments spreads seaward from shorelines always graded from nearshore sands to shales and carbonates, as depicted in Figure 3–38, predicting the locations of particular facies in ancient

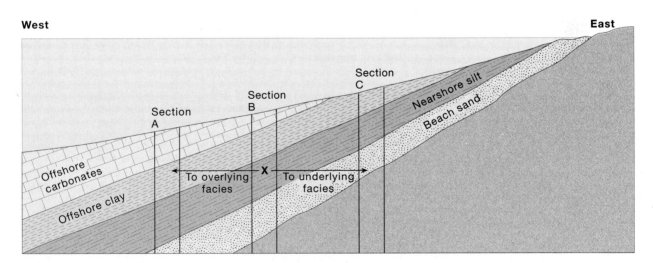

FIGURE 3–40 An illustration of Walther's principle, which states that vertical facies changes correspond to lateral facies changes. *(After Brice, J. C., Levin, H. L., and Smith, M. S. 1993. Laboratory Studies in Earth History, 5th ed., Dubuque, IA: William C. Brown.)*

rocks would be comparatively easy. In reality, however, the task is usually more complex. For example, nearshore sandy facies are not present at all along some coastlines. This may be because little sand is being brought to the coast by streams or because vigorous wave and current action has carried the sand grains away, or possibly because sand grains are trapped in submarine canyons farther up the coast. The nature of sedimentation along a coastline is also controlled by the direction of longshore currents, the location of the mouths of major streams that dump their sedimentary load into the sea, the amount of sediment supplied, the presence of barriers to dispersal of sediment, and whether the agents bringing the sediments to the sea are running water, wind, or glacial ice. Any of these factors complicate the study of facies, but they also provide fascinating problems for the geologist to solve.

The Pervasive Effects of Sea-Level Changes

Whenever a change in sea level occurs that is worldwide and independent of continental tracts, the change is termed **eustatic**. Ice accumulating on the continents during an ice age will cause lowering of sea level because much of the water making up the ice ultimately came from the ocean. Conversely, during warmer episodes, water from melting ice flows back into the ocean, causing a rise in sea level and the landward advance of shorelines along many coasts around the world. In addition to eustatic changes associated with ice ages, upwarping of the floor of an ocean basin or the development of extensive midoceanic ridges can result in a eustatic change in sea level. Midocean ridges are submarine mountain ranges extending for thousands of kilometers across the floors of the major ocean basins. They are composed of basaltic lavas derived from the mantle through a split or rift that widens to admit the basalt. As will be described in Chapter 5, the newly formed basaltic ocean floor moves laterally by a process called sea-floor spreading and at a distance may plunge back into the mantle. The basaltic rocks formed along the midocean ridges are hot and thermally expanded. As a result, they take up space that otherwise would be occupied by water and thus cause a eustatic rise in sea level. When the rate of extrusion of the basalts is rapid, there is likely to be a significant worldwide rise in sea level. Subsequent slower rates of extrusion and spreading would consequently result in lowering of sea level. These changes profoundly influence the geologic history of continental areas, for they determine when the continents are inundated by inland seas, when those seas regress, when there is deposition, and when there is erosion.

The alternate advance and retreat of seas associated with events along midoceanic ridges permit geologists to recognize distinct packages or sequences of strata having erosional boundaries that reflect global cycles of sea-level fluctuations. The cyclic rises and falls of sea level include those called **Vail cycles** after P. R. Vail, who demonstrated their use in lithostratigraphic studies (Fig. 3–41). If the Vail cycles truly reflect eustatic changes in sea level, then they are global and permit worldwide correlation of the package of sediments representing each cycle. Critics of the Vail cycles, however, suggest that at least some of them may have been appreciably affected by vertical movements of coastal regions. Without doubt, changes in the elevation of land areas bordering the ocean can cause effects similar to those resulting from eustatic changes. A coastal tract may experience either tectonic uplift or subsidence. The former is likely to cause a retreat of the sea (offlap) from the rising land area, whereas subsidence might allow the sea to advance.

Precisely how far the sea will advance or retreat during a change in sea level is determined by the amount of change in sea level and the topography of the land. A low-lying, gently sloping terrain would have a much greater area inundated by a small increase in sea level than would a steeply sloping mountainous tract (Fig. 3–42).

Ultimately, the amount of inundation or regression along a coast must be related to the interaction between tectonic movements on the continents and eustatic sea-level changes. If the land area along a continental margin rises at the same time and amount as a eustatic rise in sea level, the tectonic change will cancel out the eustatic one. The effect on shoreline displacement will be minimal. In contrast, if the land adjacent to the ocean is subsiding while sea level rises, a greater marine advance can be expected.

Anyone examining a map depicting one of the world's greatest deltas would quickly recognize that tectonic and eustatic changes are not the only causes of a shift in shorelines. The rapid accumulation of sediment along a coast or at the mouths of rivers will cause land area to be built seaward in the form of a delta. Such seaward migration of the shoreline resulting from nearshore deposition of sediment brought to the sea by rivers is termed **progradation**.

In the geologic past, there have been repeated advances of the sea into low-lying regions of the interiors of continents. Many appear to be directly related to the Vail cycles just discussed. At times these marine transgressions covered as much as two-thirds of North America. The resulting inland seas are termed **epeiric**, meaning "a sea over a continent." In these epeiric seas were deposited the sedimentary rock record of much of the Paleozoic and Mesozoic

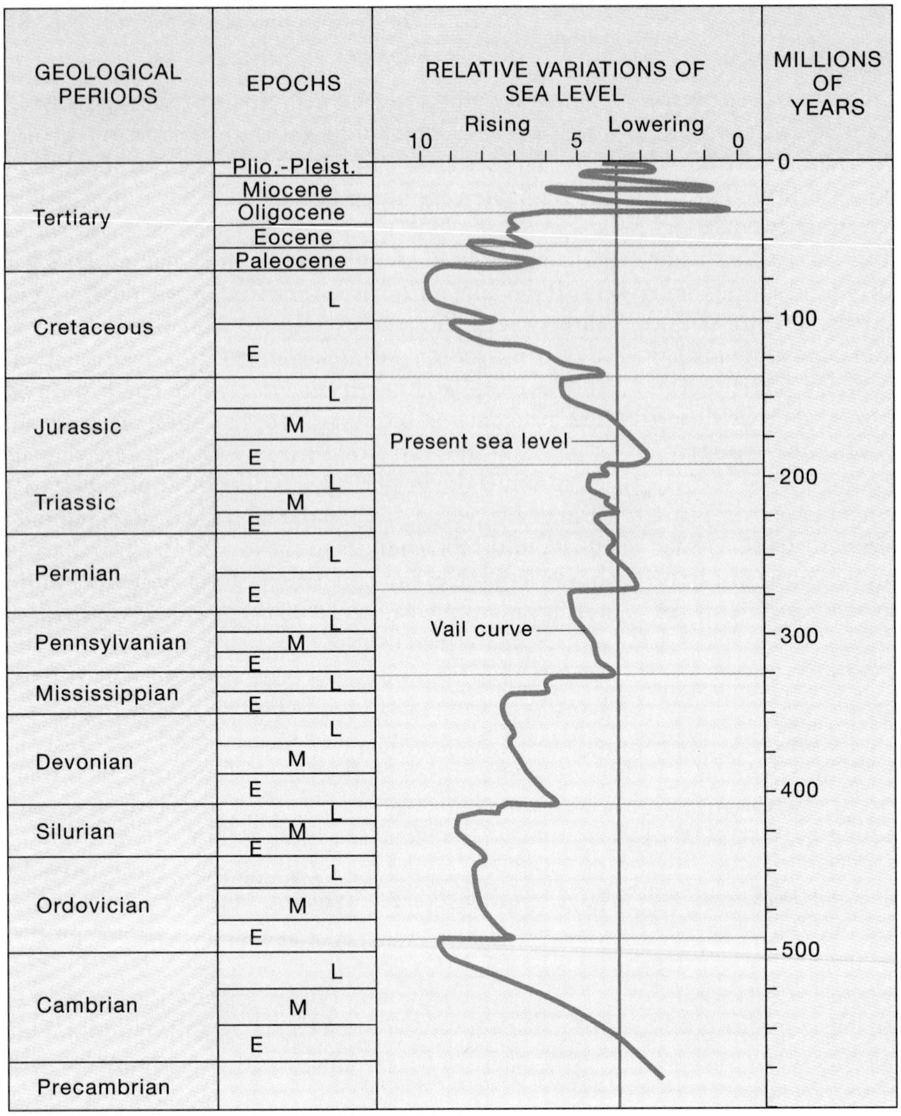

GEOLOGICAL PERIODS	EPOCHS	RELATIVE VARIATIONS OF SEA LEVEL	MILLIONS OF YEARS

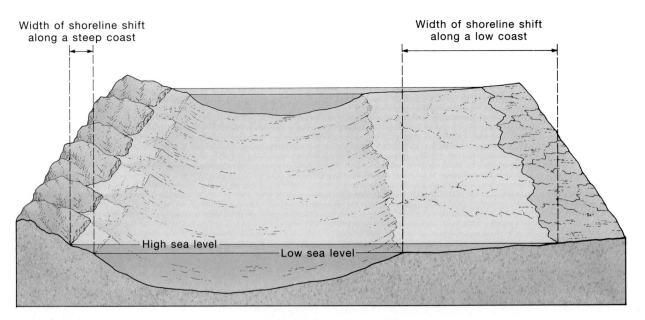

FIGURE 3–41 The Vail sea-level curve of major cycles of sea-level changes. The letters E, M, and L refer to Early, Middle, and Late. *(After Vail, P. R. et al. 1977. American Association of Petroleum Geologists Memoir 26.)*

FIGURE 3–42 A rise or fall in sea level will affect a far greater area along a low coastline than along coastlines composed of highlands that rise steeply adjacent to the sea.

eras. The advance and retreat of the epeiric seas were characteristically rather irregular, often interrupted by partial regressions, and ultimately followed by gradual withdrawal back into the major ocean basins. The average rates of advance and retreat, however, were in no sense catastrophic, for they rarely exceeded a few inches per century—a slow advance indeed, but ample to cause extensive inundation of the continents over the tens of millions of years encompassed by a geologic period.

Correlation

When examining an isolated exposure or rock in a road cut or the bank of a stream, a geologist is aware that the rock may continue laterally beneath the cover of soil and loose sediment and that the same stratum or rock body, or its equivalent, is likely to be found at other localities. The determination of the equivalence of bodies of rock in different localities is called **correlation** (Fig. 3–43). The rock bodies may be equivalent in their lithology (composition, texture, color, and so on), in their age, in the fossils they contain, or in a combination of these attributes. Because there is more than one meaning for the term, geologists are careful to indicate the kind of correlation used in solving a particular geologic problem. In some cases it is only necessary to trace the occurrence of a lithologically distinctive unit, and the age of that unit is not critical. Other problems can be solved only through the correlation of rocks that are of the same age. Such correlations involve time–rock units and

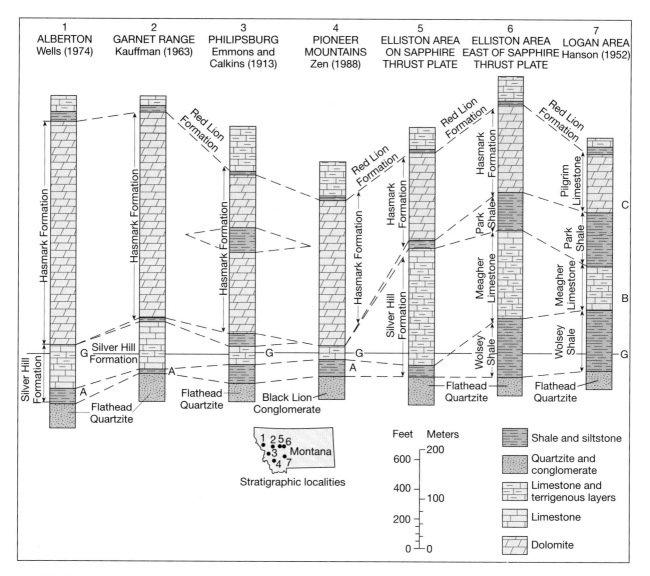

FIGURE 3–43 Correlation of lower Cambrian rock units in western Montana. The letters C, B, G, and A indicate the occurrences of trilobite index fossils *Cedaria, Bathyuriscus, Glossopleura,* and *Albertella. (From Schmidt et al. 1994. U.S. Geological Survey Bulletin 2045.)*

are of the utmost importance in geology. They are the basis for the geologic time scale and are essential in working out the geologic history of any region.

The correlation of strata from one locality to another may be accomplished in several ways. If the strata are well exposed at the Earth's surface, as in arid regions, where soil and plant cover is thin, then it may be possible to trace distinctive rock units for many kilometers across the countryside by actually walking along the exposed strata. In using this straightforward method of correlation, the geologist can sketch the contacts between units directly on topographic maps or aerial photographs. The notations can then be used in the construction of geologic maps. It is also possible to construct a map of the contacts between correlative units in the field by using appropriate surveying instruments.

In areas where bedrock is covered by dense vegetation and a thick layer of soil, geologists must rely on intermittent exposures found here and there along the sides of valleys, in stream beds, and in road cuts. Correlations are more difficult to make in these areas but can be facilitated by recognizing the similarity in position of the bed one is trying to correlate with other units in the total sequence of strata. A formation may have changed somewhat in appearance between two localities, but if it always lies above or below a distinctive stratum of consistent appearance, then the correlation of the problematic formation is confirmed (Fig. 3–44).

A simple illustration of how correlations are used to build a composite picture of the rock record is provided in Figure 3–45. A geologist working along the sea cliffs at location 1 recognizes a dense oölitic limestone (formation F) at the lip of the cliff. The limestone is underlain by formations E and D. Months later the geologist continues the survey in the canyon at location 2. Because of its distinctive character, the geologist recognizes the oölitic limestone in the canyon as the same formation seen earlier along the coast and makes this correlation. The formation below F in the canyon is somewhat more clayey than that at locality 1 but is inferred to be the same because it occurs right under the oölitic limestone. Working upward toward location 3, the geologist maps the sequence of formations from G to K. Questions still remain, however. What lies below the lowest formation thus far found? Perhaps years later an oil well, such as that at location 3, might provide the answer. Drilling reveals that formations C, B, and A lie beneath D. Petroleum geologists monitoring the drilling of the well would add to the correlations by matching all the formations penetrated by the drill to those found earlier in outcrop. In this way, piece by piece, a network of correlations across an entire region is built up.

For correlations of time–rock units, one cannot depend on similarities in lithology to establish equivalence. Rocks of similar appearance have been formed repeatedly over the long span of geologic

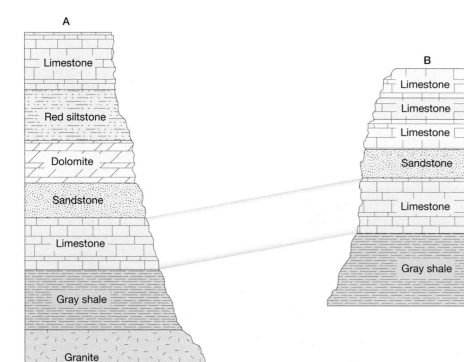

FIGURE 3–44 If the lithology of a rock is not sufficiently distinctive to permit its correlation from one locality to another, its position in relation to distinctive rock units above and below may aid in correlation. In the example shown here, the limestone unit at locality A can be correlated with the lowest of the four limestone units at locality B because of its position between the gray shale and the sandstone units.

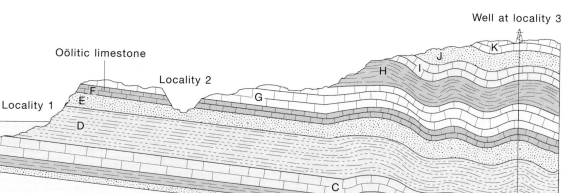

FIGURE 3–45　An understanding of the sequence of formations in an area usually begins with examination of surface rocks and correlation between isolated exposures. Study of samples from deep wells permits the geologist to expand the known sequence of formations and to verify the areal extent and thickness of both surface and subsurface formations.

time. Thus, there is the danger of correlating two apparently similar units that were deposited at quite different times. Fortunately, the use of fossils in correlation may help to prevent mismatching. Methods of correlation based on fossils (biostratigraphic correlation) are fully described in Chapter 4. They are based on the fact that animals and plants have undergone change through geologic time, and therefore the fossil remains of life are recognizably different in rocks of different ages. Conversely, rocks of the same age but from widely separated regions can be expected to contain similar assemblages of fossils.

Unfortunately, there are complications to these generalizations. For two strata to have similar fossils, they would have to have been deposited in rather similar environments. A sandstone formed on a river floodplain would have fossils quite different from one formed at the same time in a nearshore marine environment. How might one go about establishing that the floodplain deposit could be correlated to the marine deposit? In some cases this might be done by physically tracing out the beds along a cliff or valley side. Occasionally, one is able to find fossils that actually do occur in both deposits. Pollen grains, for example, could have been wafted by the wind into both environments. Possibly, both deposits occur directly above a distinctive, firmly correlated stratum such as a layer of volcanic ash. Ash beds are particularly good time markers because they are deposited over a wide area during a relatively brief interval of time. Such key beds are exceptionally useful in establishing the correlation of overlying strata. Finally, the

geologist may be able to obtain the actual age of the strata using radioactive methods, and these values can then be used to establish the correlation.

The correlation of both rock units and chorostratigraphic units from locality to locality within and between continents or between bore holes drilled into the ocean floor is an important component of a branch of geology known as stratigraphy. **Stratigraphy** deals with the study of stratified rocks, their interpretation, and their mutual relationships, description, and identification. Because stratified rocks cover approximately three-fourths of the Earth's total land area and because these strata contain our most readily interpreted clues to past events, stratigraphy forms the essential core of geologic history.

Unconformities

Interpreting the geologic history of an area would be greatly facilitated if the succession of strata was always complete. Unfortunately, such an ideal, uninterrupted sequence of strata is rarely encountered. There are gaps in the geologic record where varying thicknesses of strata have been lost to erosion. We call these breaks in stratigraphic continuity unconformities.

An **unconformity** is a buried surface of erosion that separates two rock masses, the older of which was exposed to erosion prior to the deposition of the younger. The strata, and therefore the geologic record, lost to erosion may encompass tens or even hundreds of millions of years of Earth history.

The three major kinds of unconformities illustrated in Figure 3–46 differ with regard to the orientation of the rocks beneath the erosional surface. Of the types shown, the **angular unconformity** (Fig. 3–47) provides the most readily apparent evidence for crustal deformation. As noted in Chapter 1, James Hutton described a now famous angular unconformity at Siccar Point on the Scottish coast of the North Sea (see Fig. 1–9).

Examples of unconformities are abundant on every continent. Some do not reflect the degree of deformation apparent in the strata at Siccar Point but

FIGURE 3–47 An angular unconformity separates vertical beds of the Precambrian (Proterozoic) Uncompahgre Sandstone from overlying, nearly horizontal Devonian strata of the Elbert Formation at Box Canyon Falls, Ouray, Colorado. *(Photograph by Thomas E. Williams, with permission.)*

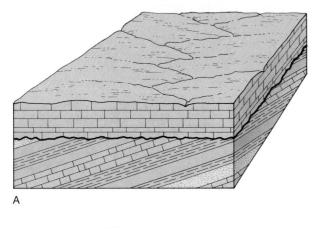

A

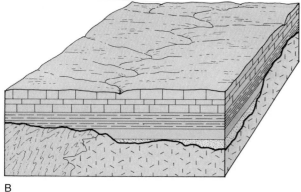

B

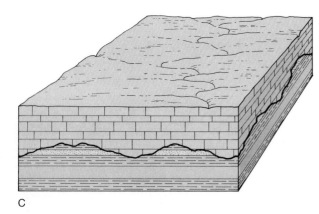

C

FIGURE 3–46 Three types of unconformities. (A) Angular unconformity. (B) Nonconformity. (C) Disconformity.

rather document the simple withdrawal of the sea for a period of time, followed by another marine transgression. The result may be a **disconformity**, in which parallel strata are separated by an erosional surface. The withdrawal and advance of the sea may be caused by fluctuations in the volume of ocean water, but more commonly they are the result of crustal uplift and subsidence. **Nonconformities** are surfaces where stratified rocks rest on older intrusive igneous or metamorphic rocks (Fig. 3–48). In many nonconformities, crystalline rocks were emplaced deep within the roots of ancient mountain ranges that subsequently experienced repeated episodes of erosion and uplift. Eventually, the igneous and metamorphic core of the mountains lay exposed and provided the surface on which the younger strata were deposited.

Although unconformities represent a loss of geologic record, they are nevertheless also useful to geologists. Like lithostratigraphic units, they can be mapped and correlated. They often record episodes of terrestrial conditions that followed the withdrawal of seas. Where regionally extensive unconformities occur, they permit one to recognize distinct sequences or "packages" of strata of approximately equivalent age.

■ *Depicting the Past*

Geologic Columns and Cross-Sections

To aid in the synthesis and interpretation of field observations of sedimentary and other rocks, geologists prepare a variety of maps, graphs, and charts designed to show relationships of rock bodies to one another, their thickness, the manner in which they have been disturbed, and their general composition. The most important graphic devices for communicating such information are columnar sections, cross-sections, and geologic maps.

Columnar sections show the vertical sequence of strata (layers of rock) at any given locality. After geologists have compiled columnar sections at several locations within a region, they may next prepare a *composite geologic column* (Fig. 3–49) that com-

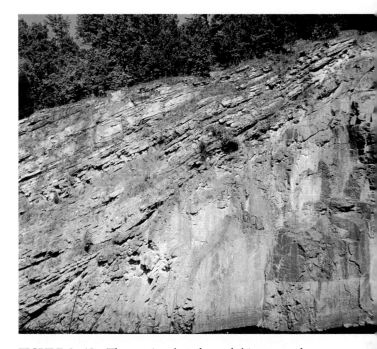

FIGURE 3–48 The erosional surface of this nonconformity is inclined at about 45° and separates Precambrian rhyolitic rock from overlying Upper Cambrian Bonneterre Dolomite. Taum Sauk Mountain, southeastern Missouri. *(Courtesy of D. Bhattacharyya.)*

Composite Geologic Column
Salt River basin
CENOZOIC ERA
QUATERNARY SYSTEM
REC. SERIES
PALEOZOIC
PENNSYLVANIAN
MISSOURIAN
DESMOINESIAN
ATOKAN
MISSISSIPPIAN
OSAGEAN
KINDERHOOKIAN
Alluvium & soil
Glacial deposits & loess
Pleasanton Grp.
Marmaton Grp.
Cherokee Grp.
Cheltenham Fm.
Basal Atokan
Burlington-Keokuk Fm.
Chouteau Fm.
Hannibal Fm.
Limestone
Cherty limestone
Shale
Clay
Sandstone
Coal

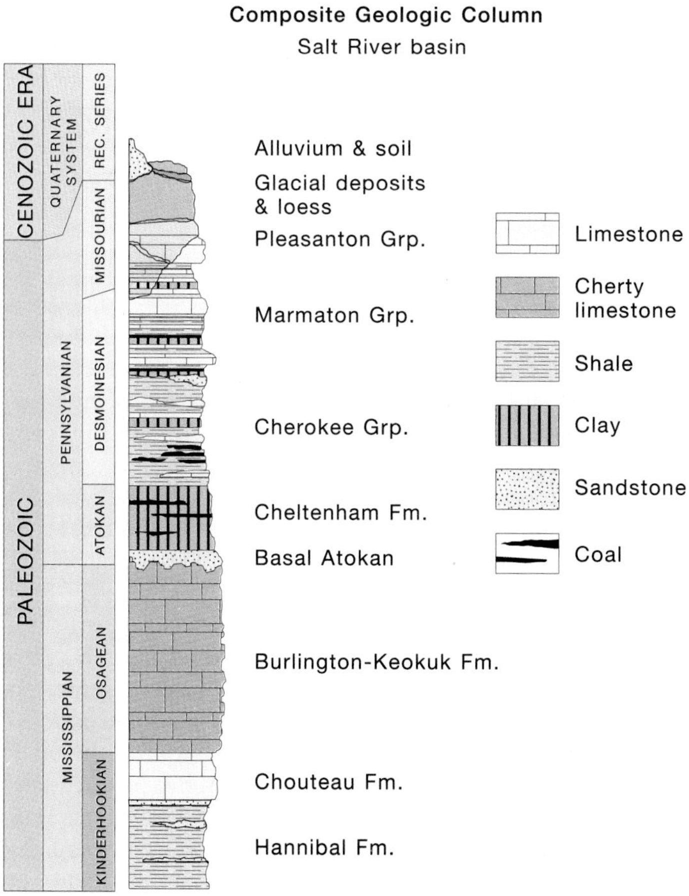

FIGURE 3–49 Example of a composite geologic column.

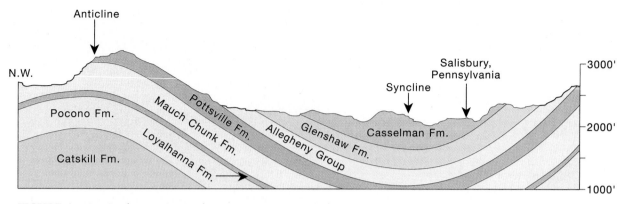

FIGURE 3–50 Geologic structural cross-section across Paleozoic rocks in the Appalachian Mountains, southeastern Pennsylvania, extending northwestward from the town of Salisbury.

bines the data from all of the various individual sections. Individual columnar sections are also used in the construction of cross-sections. As implied by their name, cross-sections show the vertical dimension of a slice through the Earth's crust. Some cross-sections—namely, the stratigraphic type—emphasize the age or lithologic equivalence of strata. The vertical measurements for such stratigraphic cross-sections are made from a horizontal line representing the surface of a definite bed or stratum. That surface is called the *datum.* In nature, the surface used as datum may be inclined or folded, and so stratigraphic sections do not validly show the tilt or position of beds relative to sea level. Stratigraphic cross-sections are most effective in showing the way beds correlate and vary in thickness from exposure to exposure or well to well. To show the way beds are folded, faulted, or tilted, a *structural cross-section* can be prepared (Fig. 3–50). In drawing the structural cross-section, the datum is a level line parallel to sea level, and the tops and bottoms of rock units are plotted according to their true elevations. If the vertical and horizontal scales are similar, the attitude of the beds will be correctly depicted. Many times, however, it is useful to have a larger vertical than horizontal scale to emphasize geologic features.

Geologic Maps

Geologic maps show the distribution of rocks of different kinds and ages that lie directly beneath the loose rock and soil covering most areas of the surface (Fig. 3–51). Assume for a moment that all of this loose material and the vegetation growing on it were miraculously removed from your home state, so that bedrock would be exposed everywhere. Imagine, further, that the surfaces of the formations now exposed were each painted a different color and photo-

graphed vertically from an airplane. Such a photograph would constitute a simple geologic map. In actual practice, a geologic map is prepared by locating contact lines between formations in the field and then plotting these contacts on a base map (Fig. 3–52A, B). Symbols are added to the colored areas to indicate formations and lithologic regions, mineral deposits, and structures such as folds and faults. Once the geologic map is completed, a geologist can tell a good deal about the geologic history of an area. The formations depicted represent sequential "pages" in the geologic record. From the simple geologic map shown in Figure 3–52C, the geologist is able to deduce that there was an ancient period of compressional folding, that the folds were subsequently faulted, and that an advance of the sea resulted in deposition of younger sedimentary layers above the more ancient folded strata.

Paleogeographic Maps

A map showing the geography of a region or area at some specific time in the geologic past is termed a **paleogeographic map.** Such maps are really interpretations based on all available paleontologic and geologic data. The majority of such maps show the distribution of ancient lands and seas (Fig. 3–53). Paleogeographic maps are, at best, of limited accuracy, because since seas advance and retreat endlessly through time, the line drawn at the sea's edge may represent an average of several shoreline positions. They are nevertheless useful for showing general geographic conditions within regions or continents. To prepare a paleogeographic map, one would plot all occurrences of rocks of a given time interval on a map and enclose the area of occurrence in boundary lines (Figs. 3–54A and B). Areas of nonoccurrence may be places of no deposition or places where de-

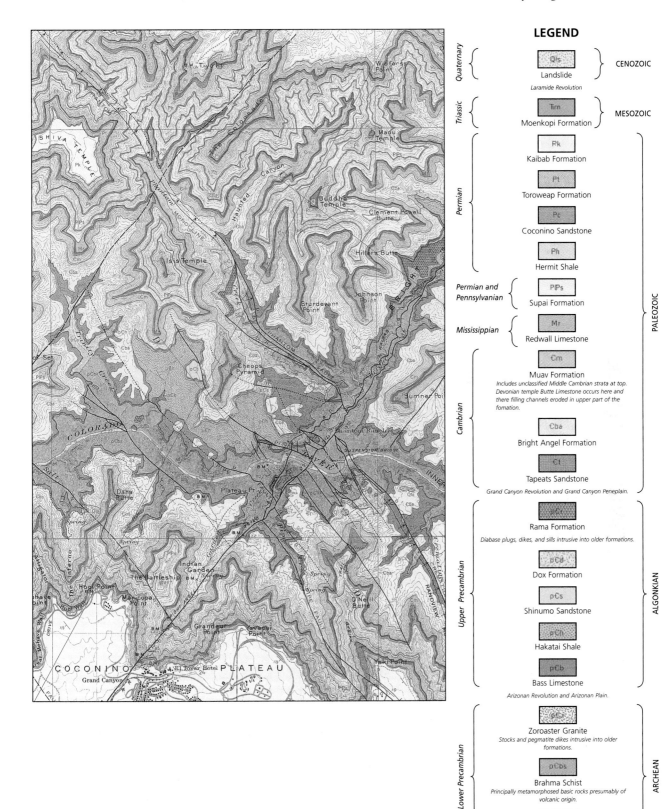

LEGEND

Quaternary {	Q/s — Landslide	} CENOZOIC
	Laramide Revolution	
Triassic {	Ŧrm — Moenkopi Formation	} MESOZOIC
Permian	Pk — Kaibab Formation	
	Pt — Toroweap Formation	
	Pc — Coconino Sandstone	
	Ph — Hermit Shale	
Permian and Pennsylvanian {	P̄Ps — Supai Formation	
Mississippian {	Mr — Redwall Limestone	PALEOZOIC
Cambrian	Ҽm — Muav Formation	

Muav Formation: *Includes unclassified Middle Cambrian strata at top. Devonian temple Butte Limestone occurs here and there filling channels eroded in upper part of the fomation.*

Ҽba — Bright Angel Formation

Ҽt — Tapeats Sandstone

Grand Canyon Revolution and Grand Canyon Peneplain.

ρҼr — Rama Formation
Diabase plugs, dikes, and sills intrusive into older formations.

ρҼd — Dox Formation

ρҼs — Shinumo Sandstone

ρҼh — Hakatai Shale

ρҼb — Bass Limestone

Arizonan Revolution and Arizonan Plain.

Upper Precambrian — ALGONKIAN

ρҼz — Zoroaster Granite
Stocks and pegmatite dikes intrusive into older formations.

ρҼbs — Brahma Schist
Principally metamorphosed basic rocks presumably of volcanic origin.

ρҼv — Vishnu Schist
Meta-sediments with minor meta-volcanics.

Lower Precambrian — ARCHEAN

FIGURE 3–51 Geologic map of part of the Grand Canyon of the Colorado River, Arizona. Note that the outcrop pattern for Paleozoic formations in this area indicates that the strata are horizontal.

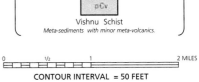

CONTOUR INTERVAL = 50 FEET

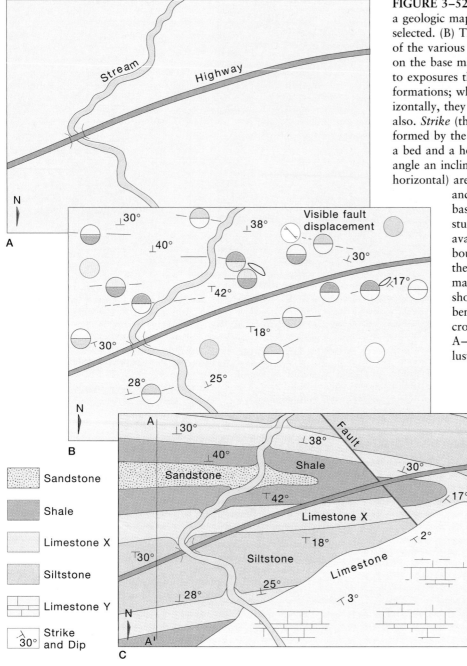

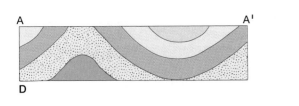

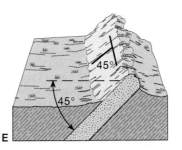

FIGURE 3–52   Steps in the preparation of a geologic map. (A) A suitable base map is selected. (B) The locations of rock exposures of the various formations are then plotted on the base map. Special attention is given to exposures that include contacts between formations; where they can be followed horizontally, they are traced onto the base map also. *Strike* (the compass direction of a line formed by the intersection of the surface of a bed and a horizontal plane) and *dip* (the angle an inclined stratum makes with the horizontal) are measured wherever possible and added to the data on the base map. After careful field study and synthesis of all the available information, formation boundaries are drawn to best fit the data. (C) On the completed map, color patterns are used to show the areal pattern of rocks beneath the cover of soil. (D) A cross-section is shown along line A—A'. (E) A block diagram illustrates strike and dip.

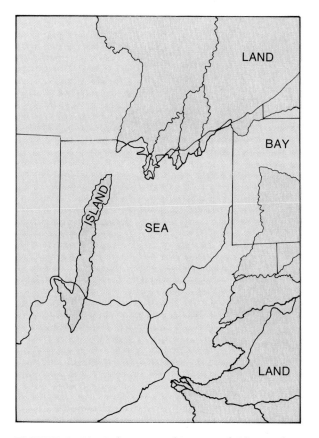

FIGURE 3–53 Paleogeographic map of Ohio and adjoining states during an early part of the Mississippian Period. The data for this study were obtained from outcrops and over 40,000 well records. *(After Pepper, J. F., de Witt, W. J., and Demarest, D. F. 1954.* U.S. Geological Survey Prof Paper 259.)

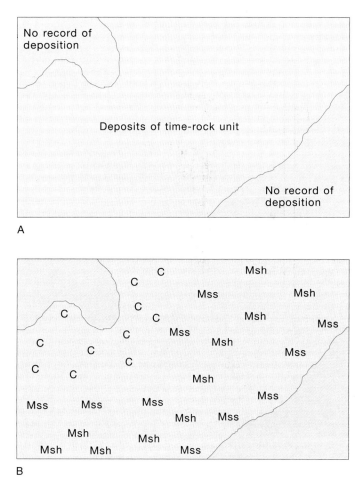

C: Continental sediments containing fossils of fresh water clams and land plants
MSS: Marine sandstone with fossils of marine invertebrates
MSH: Marine shale containing abundant fossils of marine microorganisms

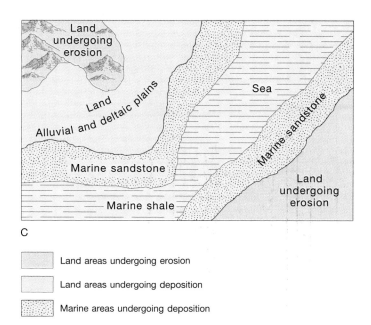

Land areas undergoing erosion

Land areas undergoing deposition

Marine areas undergoing deposition

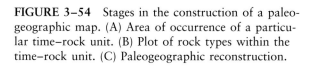

FIGURE 3–54 Stages in the construction of a paleogeographic map. (A) Area of occurrence of a particular time–rock unit. (B) Plot of rock types within the time–rock unit. (C) Paleogeographic reconstruction.

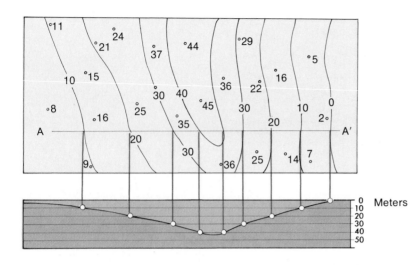

FIGURE 3–55 Diagram illustrating the construction of a simple isopach map in an area of undeformed strata.

posits once existed but were subsequently eroded away. With the help of fossils, the nature of the sediments—that is, whether marine or nonmarine—is determined and plotted on the map. The final step is to complete the paleogeographic reconstruction (Fig. 3–54C).

Isopach Maps

Isopach maps are prepared by geologists to illustrate changes in the thickness of a formation or time–stratigraphic unit. The lines on an isopach map (Fig. 3–55) connect points at which the unit is of the same thickness. On a base map, the geologists plot the thickness of units as they are revealed by drilling or in measured surface sections. Isopach lines are then drawn to conform to the data points. Ordinarily, the upper surface of the unit being mapped is used as the horizontal plane or datum from which thickness measurements are made. An isopach map may be very useful in determining the size and shape of a depositional basin, the position of shorelines, and areas of uplift. Figure 3–56 is an isopach map of Upper Ordovician formations in Pennsylvania and adjoining states. The map indicates a semicircular center of subsidence in southern New York and Pennsylvania in which over 2000 ft of sediment accumulated. The isopach pattern further indicates a highland source area to the southeast.

Lithofacies Maps

Maps constructed to show areal variations in facies can provide additional details and validity to paleogeographic interpretations. Such graphic representations are called **lithofacies maps.** Figure 3–57 is a hypothetical base map of an area subjected to exploratory drilling by oil companies. The logs for the wells are also shown. Geologists first correlate the formations. Then, assuming that the unconformity represents one time plane and the ash bed another, they define the time–rock unit as "X." Paleontologic study of the rocks between the time planes confirms the validity of the time–rock unit. Geologists may now prepare the lithofacies map. Time–rock unit X is missing at well number 11; this may be the result of its not being deposited there or, having been deposited, of its being eroded away. It is logical that the sandy facies was deposited adjacent to a north–south trending shoreline.

Figure 3–58 is a lithofacies map of rocks deposited over 400 million years ago in the eastern United

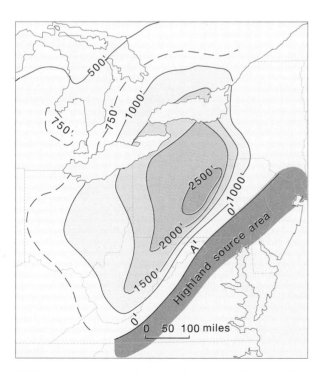

FIGURE 3–56 Isopach map of Upper Ordovician formations in Pennsylvania and adjoining states. *(After Kay, M. 1951. Geological Society of America Memoir No. 48.)*

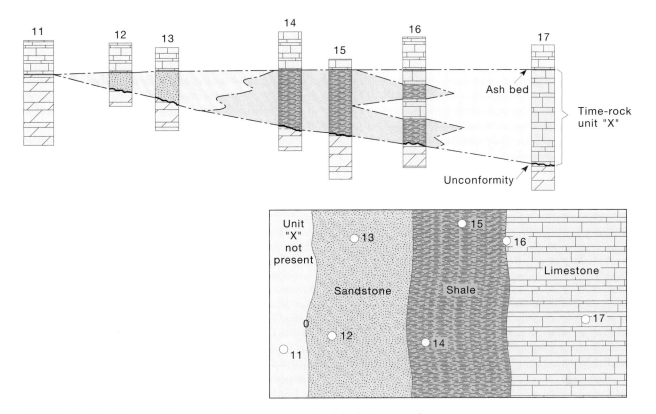

FIGURE 3–57 Diagram illustrating the preparation of a lithofacies map from a subsurface time–rock unit. Well locations are indicated by small circles. Correlation of rock units between wells is indicated by dashed lines. Because of the few control points, the exact position of lithofacies boundaries on this map is somewhat arbitrary.

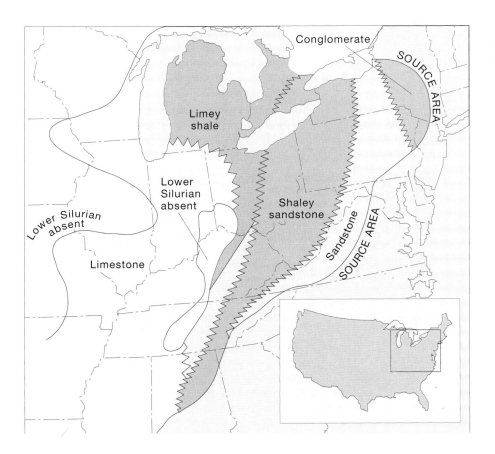

FIGURE 3–58 Lithofacies map of Lower Silurian rocks in the eastern United States. *(After Amsden, T. W. 1955. Bull. Am. Assoc. Petrol. Geol. 39:60–74.)*

States. From this map, one can infer the existence of a highland area that existed at that time along our eastern seaboard and that supplied the coarse clastics. Detrital sediments from the source area become fine, and the section thins as one proceeds westward from the source highlands. Finally, as far west as Indiana, the map indicates that only carbonate precipitates were laid down. The conglomerates were probably the deposits of great alluvial fans built out from the ancient mountain system.

The lithofacies maps just described provide a qualitative interpretation of areal changes in rock bodies. Quantitative lithofacies maps can also be constructed and are frequently used in the study of subsurface formations that are known primarily from well records. By means of contour lines, such maps show the areal distribution of some measurable characteristic of the unit being mapped. For example, contours may be drawn on the percentage of one lithologic component (such as clay) compared to the total unit or on the ratio of one rock type (such as sandstone) to the others within the unit. An isopach map is ordinarily the base map for any of the quantitative maps, since one must know the total stratigraphic thickness of the unit with which individual components are compared.

■ *Importance of Sedimentary Rocks to Society Today*

From neanderthal to astronaut, humans have learned to use sedimentary rocks to improve their lot. Men and women of antiquity fashioned tools from chert and flint, and they later learned the art of manufacturing useful containers from clay. Their successors developed the means of extracting copper and iron from the rocks and from these metals made more sophisticated implements. In addition to the metals they sometimes contain, sedimentary formations include coal seams and contain the oil, gas, and groundwater important to our present way of life.

Mineral Fuels

The mineral fuels—petroleum, gas, and coal—are our most important sedimentary resources. They are essential for heat and power and metal refining; they are also the source of many chemicals useful in the manufacturing of plastics and fertilizers. **Coal** is a brownish-black to black combustible rock that occurs in beds from a few inches to many feet in thickness and is interbedded with shale, sandstone, and other sedimentary rocks. Extensive coal-bearing sequences characterize the Pennsylvanian and Cretaceous systems. Pennsylvanian sequences of strata in

the Appalachians include over 100 individual beds of coal, each composed of the compressed and altered remains of land plants. Coal is abundant in the United States and will become increasingly more important as our reserves of petroleum are consumed.

Commercial accumulations of oil and gas require rather special geologic conditions. These conditions are found almost exclusively in sedimentary rocks. First, there must be a **source rock** for the petroleum. Ordinarily, this is a series of beds rich in the organic remains of unicellular organisms that accumulated along with the particles of sediment. Second, there must be a permeable **reservoir rock** such as sandstone or porous limestone to provide passage and storage for the gas and oil. The reservoir rock is covered by an impermeable **cap rock** to prevent the upward escape of hydrocarbons. Finally, there must be an oil-trapping structure (Fig. 3–59), so that the hydrocarbons can be concentrated in one place.

Petroleum geologists seek out these structures in many ways. Long before drilling programs are begun, geologists conduct a variety of exploratory surveys, the most useful of which are *seismic reflection surveys* (Fig. 3–60). In this technique, seismic waves are sent into the Earth either by the detonation of dynamite in shallow drill holes or by the thumping action of hydraulic vibrators. The latter method is more frequently used today because it is safer and less disturbing to the environment. The depth to the stratum that reflects these waves is determined from the time required for the waves to travel down to a reflecting stratum and back to the surface, where they are detected by portable seismometers. Repeating the process along a predetermined traverse permits one to detect potential oil-bearing geologic structures deep beneath the surface.

Similar seismic exploration can also be accomplished for rock layers beneath the ocean floor by a process called *sub-bottom profiling*. In this process a survey vessel employs devices that emit strong signals produced by air guns or electric sparks. The signals are transmitted to the sea floor, penetrate into it, and are reflected back from the surfaces of sedimentary layers as deep as 10 km below the ocean floor. Hydrophones trailing behind the ship pick up the returned energy as the vessel steams along, and the moving graph paper of the recorder provides an uninterrupted cross-section of what lies deep beneath the bottom of the sea.

Later, as exploratory wells are drilled in an offshore or land area, the lithologic, paleontologic, and geophysical properties of the rocks penetrated are carefully recorded on well logs. Among the instrument-based logs used by geologists, none is more useful than electric logs. They are produced by devices wired to a cylindrical tool called a **sonde**. As the

continued

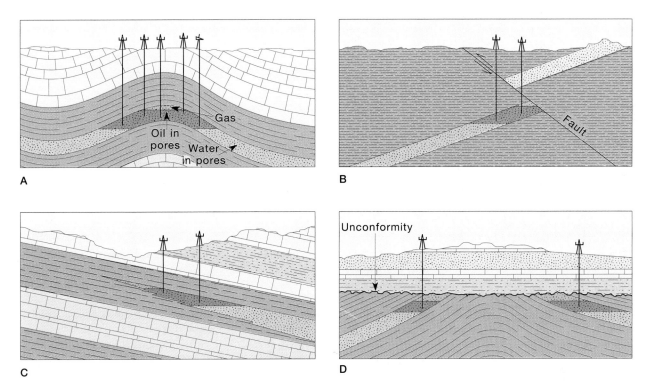

FIGURE 3–59 Idealized cross-sections depicting geologic conditions that may lead to the accumulation of petroleum and natural gas. (A) Anticlinal trap. (B) Fault trap. (C) Stratigraphic trap. (D) Entrapment beneath an unconformity.

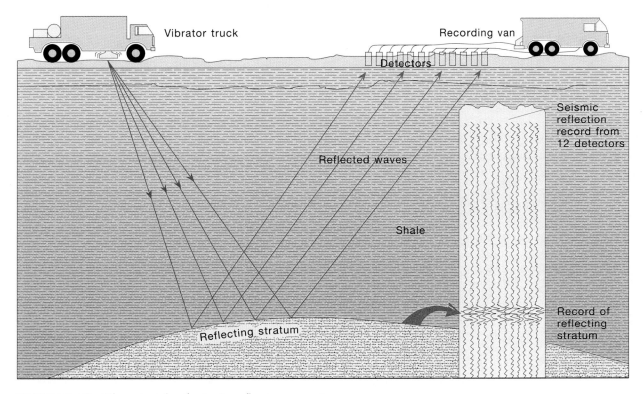

FIGURE 3–60 The principle of seismic reflection surveying.

Grand Canyon National Park

In the year 1540, Hopi indians directed a band of conquistadores to the rim of the Grand Canyon of the Colorado River. One can imagine their sense of wonder as they gazed at this stupendous natural spectacle. Awesome in magnitude and beauty, the great chasm was unlike anything Western humans had ever witnessed. It remains so today, an awesome monument to the erosive force of running water and gravitational downslope transfer of solid rock and weathered debris.

The historical geologist sees the layers of sandstone, shale, limestones, and lava flows exposed in the canyon walls as a great history book that reveals geologic events and changing life over a span of 2 billion years. Rocks of the first chapter are sands and muds that record the presence of a shallow water body in the region. Volcanoes erupted nearby, for the rocks are interbedded with layers of lava and volcanic ash. About 1.7 billion years ago, mountain building deformed and metamorphosed these sediments. In their altered state, they comprise the Vishnu Schist (see Figs. 3–51 and 7–12), seen close up by rafters passing through the canyon's Inner Gorge. The Zoroaster Granite intrudes the Vishnu Schist and can also be seen in the deep clefts of the canyon. An interval of erosion followed the emplacement of the Zoroaster Granite. Then high-silica melts invaded joints and fractures and formed veins and igneous rock, which were later converted to light-colored gneisses. The region was mountainous at this time. Land plants had not evolved, and there was little to retard the forces of erosion. Eventually, the mountains were reduced to lowlands, and their roots can be seen in the canyon walls.

The next chapter in the Grand Canyon story is written in 3700 meters of sedimentary rocks and lava flows spread extensively over the Vishnu Schist. These are the rocks of the Grand Canyon Supergroup (see Fig. 7–13). Following their deposition, the region was subjected to tensional forces that produced north–south trending fault-block mountains. The higher fault blocks became mountains. Debris eroded from these mountains filled intervening downfaulted blocks. The destructional forces of erosion gradually reduced the entire region to a low-lying terrain recorded in geologic history by "the great unconformity" that separates Precambrian from Paleozoic strata. Remnants of the once more extensive Grand Canyon Supergroup are nestled in remaining downfaulted blocks beneath the great unconformity.

Ascending the canyon, we reach rocks of the Paleozoic Era.

continued

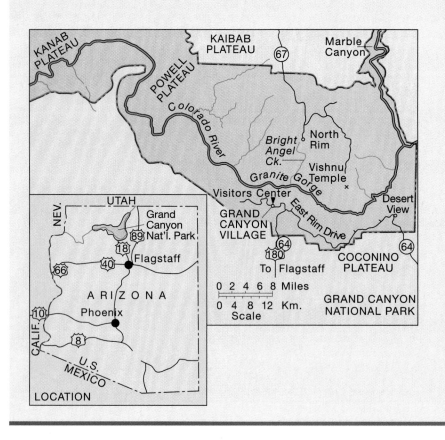

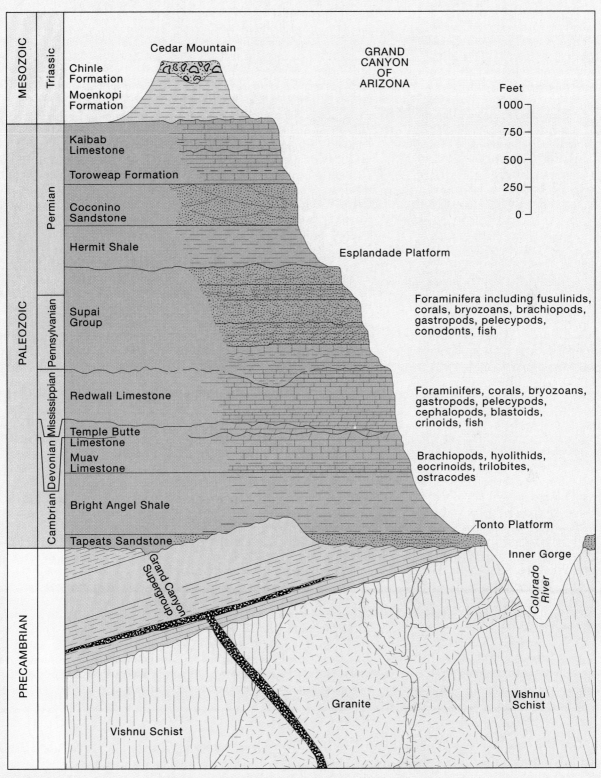

Cedar Mountain

GRAND CANYON OF ARIZONA

MESOZOIC — **Triassic**
Chinle Formation
Moenkopi Formation

Feet
1000
750
500
250
0

PALEOZOIC

Permian
Kaibab Limestone
Toroweap Formation
Coconino Sandstone
Hermit Shale

Esplandade Platform

Pennsylvanian
Supai Group

Foraminifera including fusulinids, corals, bryozoans, brachiopods, gastropods, pelecypods, conodonts, fish

Mississippian
Redwall Limestone

Foraminifers, corals, bryozoans, gastropods, pelecypods, cephalopods, blastoids, crinoids, fish

Devonian
Temple Butte Limestone

Cambrian
Muav Limestone

Brachiopods, hyolithids, eocrinoids, trilobites, ostracodes

Bright Angel Shale

Tonto Platform

Tapeats Sandstone

Grand Canyon Supergroup

Inner Gorge

Colorado River

PRECAMBRIAN

Granite

Vishnu Schist

Vishnu Schist

Generalized Geologic Column for Grand Canyon National Park.   (*From McKee, E. D., The Supai Group of the Grand Canyon, U.S. Geological Survey Professional Paper 1173, 1982.*)

continued

The Paleozoic was a time of repeated inundation and regression of shallow seas. The first of the inundations laid down the nearshore Tapeats Sandstone (see Chapter 8 opening photograph). These sands were followed by the Bright Angel Shale and Muav Limestone as the shoreline shifted eastward. The three formations comprise the Tonto Group. Their sequential change in lithology illustrates the way in which rock units may transgress time boundaries (see Fig. 8–10).

As if pages in our history book had been ripped out, strata of Ordovician and Silurian age are not found in the Grand Canyon. They may have been deposited there, but if so, they have been lost to erosion. Thus, an unconformity (a gap in the stratigraphic record) caps the Cambrian sequence. Above that unconformity, carbonates of the Devonian Temple Butte Limestone were laid down in a shallow sea. Again the sea withdrew, only to return another time during the Mississippian Period. In this Mississippian sea, cherty carbonates of the Redwall Limestone were deposited. Although freshly broken surfaces of the Redwall Limestone are gray, its weathered surface is stained red by iron oxide washed down the face of the precipice from red beds of the overlying Supai Group and Hermit Shale. The Redwall forms bold cliffs that front many of the canyon's promontories. It is richly fossiliferous with the remains of brachiopods, bryozoans, crinoids, and corals.

The withdrawal of the Redwall

Sea is signaled by the presence of estuarian and tidal flat sediments that are part of the Surprise Canyon Formation near the top of the Redwall. Above these sediments of transitional environments, one finds yet another erosional unconformity, and above that surface lie strata of the Pennsylvanian and early Permian Supai Group. The Buddha, Zoroaster, and other spectacular temples in the park are sculpted in the Supai. Nonmarine beds of the Supai exhibit tracks of amphibians, possibly those of reptiles, as well as imprints of ferns. In the overlying Hermit Shale, one can see evidence of nonmarine deposition in the formation's mudcracks and fossils of insects, conifers, and ferns. The floodplains and marshy tracts on which the sediments of the Hermit Shale were deposited were soon covered by migrating dunes of the Permian Coconino Sandstone. The nearly white Coconino sands are cross-bedded. The frosted, well-sorted, well-rounded grains reflect an origin as windblown sediment. Reptiles wandering across the dunes left their footprints in the sand.

Marine limestones and sandstones of the Toroweap Formation rest on the Coconino, recording the advance of a sea over the Coconino dune fields. Above the Toroweap are the bold vertical cliffs of the Kaibab Limestone. This thick and resistant formation forms the surface of the Kaibab Plateau north of the canyon and the Coconino Plateau on the south side. The Kaibab is the final rock unit of the Paleozoic in the Grand Canyon. During the Mesozoic, floodplain sands and silts of

the Triassic Moenkopi Formation, and gravels of the Shinarump Conglomerate, were spread across the region. Near the Grand Canyon, however, all but a few remnants of these formations were swept away by erosion.

Uplift occurred in the Colorado Plateau during the Tertiary Period of the Mesozoic. It was at this time, only about 7 million years ago, that cutting of the canyon began. Thus, although the rocks of the canyon are very old, the canyon itself is geologically young. During the great Ice Age, canyon cutting was probably extraordinarily rapid, as tremendous volumes of glacial meltwater rushed down the Colorado River. The channel deepened, not only as a result of the erosive power of running water itself, but also because of impact and abrasion by cobbles and gravels carried in the rushing currents. Yet if the river alone was the only mechanism for erosion, the canyon would have vertical walls. Weathering and gravitational mass movements of eroded debris downslope broadened the canyon to its present width of over 17 kilometers. The Colorado, like rivers everywhere, acts as a conveyor belt, carrying its own load as well as the debris supplied to its by slides, rockfalls, and other gravity-driven movements that we call mass wasting.

There is much more geology in the Grand Canyon than can be described in these few pages or that can be encompassed in a short visit. If you are fortunate enough to go there, stay the day and watch the sun go down over this magnificent colossus of canyons.

sonde is raised out of the drill hole, it measures the natural electrical properties of the formations with their contained fluids as well as their resistance to an induced flow of electricity. Rocks of different composition and fluid content (oil, fresh water, salt water) provide distinct patterns of lines on the electric log. The correlation of these patterns from well to well often reveals an accurate picture of the configuration

of subsurface rock layers (Figure 3–61), and it is indispensable to modern exploration for oil and the development of oil fields.

Construction Materials and Ores

Other sedimentary material useful to humans include clay for use in ceramics and bricks, limestone for

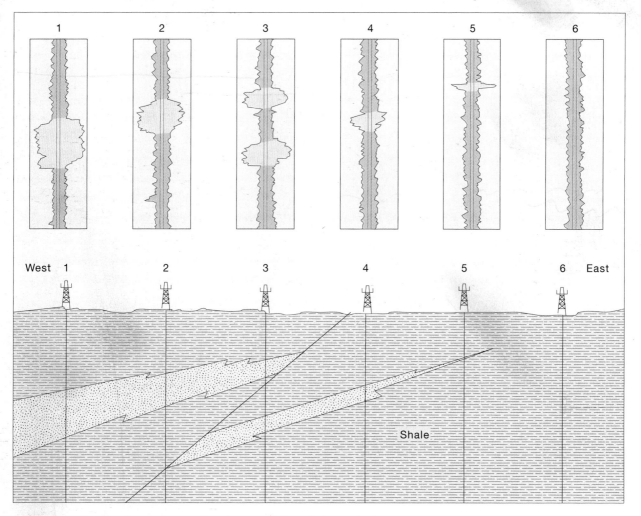

FIGURE 3–61 Six electric logs and an interpretation of the geologic structure penetrated by wells from which the logs were prepared. The prominent "kicks" in the logs record a petroliferous sandstone. From a study of the electric logs, geologists developed an interpretation of a petroliferous sandstone that dips westward, pinches out toward the east, and is disrupted by a reverse fault.

building stones and cement, sand and gravel for concrete and glass, and evaporites for use in the chemical industry. Some sedimentary strata are rich in iron oxide. Those near Birmingham, Alabama, were mined and smelted for several decades. Because minerals such as gold, chromite, diamond, cassiterite (tin oxide), and magnetite are heavier than quartz and other common silicates, they tend to accumulate in quiet areas of streams or to be concentrated by waves as **placer deposits.** Finally, there are sedimentary ores that form in place by the deep chemical decay of a parent rock. The most important metal concentrated in this manner is aluminum. Bauxite, an almost pure hydrous aluminum oxide, is the ore mineral of aluminum. Bauxites developed in once humid, tropical regions as a result of the weathering of source rocks rich in aluminum silicates.

Limited Supply of Mineral Resources

Any commentary on mineral resources would be incomplete without noting that ore bodies and oil fields are limited in extent and are the result of unusual associations of geologic conditions. Furthermore, it is not theory but fact that they are exhaustible and irreplaceable. Their effect upon the standards of living and society is so enormous that every citizen should become involved in political decisions relating to the management of our natural resources.

Summary

Sedimentary rocks represent the material record of environments that once existed on the Earth. For this reason, they are of great importance to the science of historical geology. All sedimentary rocks are formed by the accumulation and consolidation of the products of weathering derived from older rock masses, as well as by chemical precipitation and the accumulation of organic debris. Because of their mode of formation, the composition of sedimentary rocks provides information about source areas. The materials of sedimentary rocks are often transported by wind, water, or ice, or they are carried in solution to be precipitated in a particular environment of deposition. The transporting medium imparts characteristics of texture or composition that can be used to reconstruct the depositional history and tectonic setting. Fossils in sedimentary rocks are splendid environmental indicators. They tell us if strata are marine or nonmarine, if the water was deep or shallow, or if the climate was cold or warm.

Color in a rock provides clues to the chemistry of the depositional medium. In the case of clastic rocks, the size, shape, and arrangement of grains can provide data about the energy of the transporting agent, the distance the grains had traveled, erosional recycling, and the degree to which movements of the Earth's crust had disturbed a basin of deposition. Similar kinds of information are elucidated by such primary sedimentary structures as graded bedding, cross-bedding, and current ripple marks. Sandstones are particularly useful in paleoenvironmental studies. Graywackes, arkoses, quartz sandstones, and lithic sandstones accumulate in particular paleogeographic and tectonic situations. It is the geologist's task to discover the details of those situations by examining the rocks.

Geologists usually divide successions of sedimentary rocks into rock units that are sufficiently distinctive in color, texture, or composition to be recognized easily and mapped. Such rock units are called formations and are not necessarily of the same age throughout their areal extent. A time–stratigraphic unit differs from a rock unit in that it is as assemblage of strata deposited within a particular interval of time. The Cambrian System, for example, is a time–stratigraphic unit including all the rocks deposited in the Cambrian Period.

Within any given time–stratigraphic unit, one may find rocks that vary in composition, texture,

organic content, or other features from adjacent rocks. These rock bodies of distinctive appearance or aspect are called facies. Facies reflect deposition in a particular environmental setting. For example, along a coastline one may find nearshore sand facies that change seaward to shale facies and carbonate facies. As shorelines shift landward or seaward, facies shift accordingly, maintaining their association with a particular local set of environmental conditions.

Shoreline migrations may be the result of such factors as worldwide changes in sea level (eustatic changes), tectonic movements of continental borderlands, or progradation—the seaward advance of the coastline resulting from rapid deposition of sediment brought to the sea by rivers. Progradation is particularly evident in deltas.

The study of facies is of great importance in the development of reconstructions of conditions on earth long ago. Geologists employ various graphic methods to record variations in facies and other attributes of sedimentary rocks. These methods include the preparation of lithofacies, biofacies, geologic, and isopach maps. If examined in chronologic sequence, such maps are useful not only in reconstructing ancient geography but also in providing a picture of the changing patterns of lands and seas to which parts of the Earth have been subjected.

Sedimentary rocks are the source of a great many materials necessary to modern civilization. Some of these materials, such as clay and building stones, are rather commonplace. Others, such as placer deposits of diamonds, chromite, and gold, are more limited in occurrence. Aluminum, a metal derived from a residual mineral named bauxite, is another extensively exploited sedimentary resource. However, the fossil fuels—coal and petroleum—are the most important of all sedimentary materials today. Indeed, oil, coal, and gas power our industrialized world. These resources are present on the Earth in finite quantities and must be carefully managed.

Review Questions

1. What are the three major environments of deposition, and what features in rock sequences might be used to recognize these environments of deposition in ancient rocks?
2. Why are sandstones and siltstones of nonmarine desert environments rarely black or dark gray in color?
3. Under what circumstances might a reddish siltstone containing fossils of marine invertebrates have originated?
4. How does matrix in a rock differ from cement? What kinds of cements can bind a sedimentary rock together? Which cement is most durable?
5. What would be the probable history of a poorly sorted sandstone composed of angular grains immersed in a 30 percent matrix of mud?
6. What is the difference between a lithostratigraphic unit (rock unit) and a chronostratigraphic unit? Give an example of each. How is each defined and recognized?
7. Distinguish between an immature and a mature sandstone.

8. Why, in general, are fine-grained sediments deposited farther from a shoreline than coarse-grained sediments?

9. A geologist examining a sequence of strata notes that limestone is overlain by shale, which in turn is overlain by sandstone. What might this indicate with regard to the advance or retreat of a shoreline?

10. What features or properties of sedimentary rocks are useful in determining the direction of currents of the depositing medium? What features indicate relatively shallow water deposition?

11. What are turbidity currents? How are ancient turbidity current deposits recognized?

12. What sort of paleogeographic information can be inferred from lithofacies maps? From isopach maps?

13. What are the essential natural requirements for the occurrence of an oil pool?

Discussion Questions

1. What is meant by the expression *tectonic setting?* What kind of past tectonic activity can be inferred for a region underlain by a great thickness (over 10,000 meters) of sediment that has been deposited in shallow water (no deeper than 200 meters)?

2. What properties of a sandstone might be used to infer its pre-existing parent rock, its mode of transportation, and its depositional environment?

3. How do quartz sandstones and graywackes differ in composition, texture, matrix and/or cement, and probable environment of deposition?

4. On the Vail curve (Fig. 3–41), how many episodes of major regression are evident during the Paleozoic Era? According to the Vail curve, did episodes of lowering of sea level proceed at the same rate as episodes in which sea level was raised?

5. Examine the lower formations in each of the rock sections correlated in Figure 3–43. Do these sections begin with onlap or offlap? What evidence supports your answer?

6. Refer to the geologic map of part of the Grand Canyon (Fig. 3–48). If you were rafting down this part of the Colorado River, what formation would be visible to you on either side of the river? What is its age? Aside from the Quaternary sediments (river sands, etc.), where are there exposures of the youngest rocks in the area?

7. In Figure 3–58 note the area in which Lower Silurian rocks are absent. How do you account for the absence of these rocks?

8. The petroleum occurrences in the oil traps depicted in Figure 3–59 are all above that part of the reservoir rock that contains water. Why?

9. Figure 3–61 illustrates how a bed may be repeated in vertical section because of faulting. Sketch another way in which the sandstone bed might be repeated. What evidence on the electric log might help you distinguish your configuration from the one depicted in the figure?

Supplemental Readings and References

Davis, R. A. 1978. *Coastal Sedimentary Environments.* New York: Springer-Verlag.

Dunbar, C. O., and Rogers, J. 1957. *Principles of Stratigraphy.* New York: John Wiley & Sons.

Friedman, G. M., Sanders, J. E., and Kipaska–Merkel, D. C. 1992. *Principles of Sedimentary Deposits.* New York: Macmillan.

Hsu, K. J. 1989. *Physical Principles of Sedimentology: A Readable Textbook for Beginners and Experts.* New York: Springer-Verlag.

Kershaw, S. 1991. Way-up structures. *Geol. Today* 4:i–iv.

Lemon, R. R. 1990. *Principles of Stratigraphy.* Columbus, OH: Merrill Publ. Co.

Prothero, D. R. 1990. *Interpreting the Stratigraphic Record,* San Francisco: W. H. Freeman.

Scholle, P. A., and Spearing, D. 1982. *Sandstone Depositional Environments.* Tulsa: American Association of Petroleum Geologists.

Scholle, P. A., Bobout, D. G., and Moore, C. H. 1983. *Carbonate Depositional Environments.* Tulsa: American Association of Petroleum Geologists.

Tucker, M. E. 1981. *Sedimentary Petrology.* New York: John Wiley & Sons.

The trilobite *Isotelus brachycephalus* from strata of Ordovician age near Cincinnati, Ohio. The specimen is 23 cm in length. *(Courtesy of Wards Natural Science Establishment.)*

The search for a fossil may be considered at least as rational as the pursuit of a hare.

WILLIAM SMITH (1769–1839)

CHAPTER 4

The Fossil Record

The Earth is about 4.6 billion years old. Life has been present on it for about 3.5 billion years. The science that seeks to understand all aspects of the succession of plants and animals over that great expanse of geologic time is called **paleontology.** It is a science based on the study of **fossils,** the remains or traces of ancient life. As paleontologists study these relics of former living organisms, they endeavor to learn how these organisms lived and how they grew. What was the organism's true fleshed-out appearance? How did it interact with other organisms and its physical environment? What were its ancestors like, and what changes occurred in its descendants? Why did these changes occur?

113

■ *Preservation*

When one considers the many ways by which organisms are completely destroyed after death, it is remarkable that fossils are as common as they are. Attack by scavengers and bacteria, chemical decay, and destruction by erosion and other geologic agencies make the odds against preservation very high. However, the chances of escaping complete destruction are vastly improved if the organism happens to have a mineralized skeleton and dies in a place where it can be quickly buried by sediment. Both of these conditions are often found on the ocean floors, where shelled invertebrates flourish and are covered by the continuous rain of sedimentary particles. Although most fossils are found in marine sedimentary rocks, they also are found in terrestrial deposits left by streams and lakes. On occasion, animals and plants have been preserved after becoming immersed in tar or quicksand, trapped in ice or lava flows, or engulfed by rapid falls of volcanic ash.

The term **fossil** often implies petrifaction, literally a transformation into stone. After the death of an organism, the soft tissue is ordinarily consumed by scavengers and bacteria. The empty shell of a snail or clam may be left behind, and if it is sufficiently durable and resistant to dissolution, it may remain basically unchanged for a long period of time. Indeed, unaltered shells of marine invertebrates are known from deposits over 100 million years old. In many marine creatures, however, the skeleton is composed of a mineral variety of calcium carbonate called aragonite. Although aragonite has the same composition as the more familiar mineral known as calcite, it has a different crystal form, is relatively unstable, and in time changes to the more stable calcite.

Many other processes may alter the shell of a clam or snail and enhance its chances for preservation. Water containing dissolved silica, calcium carbonate, or iron may circulate through the enclosing sediment and be deposited in cavities such as marrow cavities and haversian canals in bone. In such cases, the original composition of the bone or shell remains, but the fossil is made harder and more durable. This addition of a chemically precipitated substance into pore spaces is termed **permineralization** (Fig. 4–1).

Petrifaction may also involve a simultaneous exchange of the original substance of a dead plant or animal with mineral matter of a different composition. This process is termed **replacement** because solutions have dissolved the original material and replaced it with an equal volume of the new substance (Fig. 4–2). Replacement can be a marvelously precise process, so that details of shell ornamentation, tree

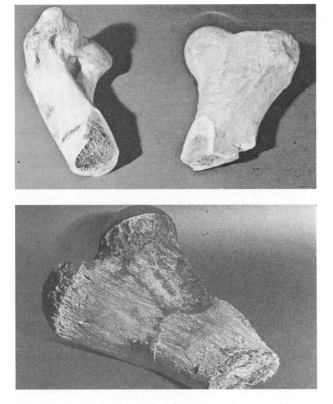

FIGURE 4–1   An example of permineralization. The bone fragments in the upper photograph are part of a cow's femur (upper leg bone). They are unfossilized original bone in which one can readily see the central porous marrow cavity. The bone depicted is a permineralized dinosaur bone in which the porous space has been densely filled with mineral matter.

FIGURE 4–2   The shell (conch) of an extinct marine organism known as an ammonoid. In this Jurassic fossil, the original calcium carbonate skeleton has been replaced with iron sulfide in the form of the mineral pyrite, known to some as fool's gold.

FIGURE 4–3 This fossil seed fern from rocks of Pennsylvanian age has been preserved by carbonization. The frond is approximately 27 cm in length.

rings in wood, and delicate structures in bone are accurately preserved.

Another type of fossilization, known as **carbonization,** occurs when soft tissues are preserved as thin films of carbon. Leaves and tissue of soft-bodied organisms such as jellyfish or worms may accumulate, become buried and compressed, and lose their volatile constituents. The carbon often remains behind as a blackened silhouette (Fig. 4–3).

Fossils may also take the form of molds, imprints, and casts. Any organic structure may leave an impression of itself if it is pressed into a soft material and if that material is capable of retaining the imprint. Commonly, among shell-bearing invertebrates, the shell is dissolved after burial and lithification, leaving a vacant **mold** bearing surface features of the original shell that are opposite of those on the shell itself (ridges are represented by grooves, knobs by depressions, and vice versa). If external features (growth lines, ornamentation) of the fossil are visible, the mold is an *external mold.* Conversely, the *internal mold* shows features of the inside of a shell, such as muscle scars or supports for internal organs. Many invertebrate shells enclose a hollow space that may be either left empty or filled with sediment (Fig. 4–4). The filling is called a *steinkern* (stone core). Finally, molds may be subsequently filled, forming **casts** that faithfully show the original form of the shell (Fig. 4–5).

Although it is certainly true that the possession of hard parts enhances the prospects of preservation, organisms having soft tissues and organs are also occasionally preserved. Insects and even small invertebrates have been found preserved in the hardened res-

FIGURE 4–4 An internal mold (or steinkern) formed by filling of the spiral cavity of an ancient marine snail.

FIGURE 4–5 A cast (above) and mold (below) of a trilobite formed in a nodule of calcareous shale. The trilobite *(Calliops)* is 4 cm long.

FIGURE 4–6 An Eocene insect preserved in amber. The insect is a member of the Order Diptera, which includes flies, mosquitoes, and gnats. Although preservation of insects is uncommon, it is likely that they approached modern levels of abundance and diversity early in the Cenozoic Era. *(Courtesy of W. Bruce Saunders.)*

ins of conifers and certain other trees (Fig. 4–6). X-ray examination of thin slabs of rock sometimes reveals the ghostly outlines of tentacles, digestive tracts, and visual organs of a variety of marine creatures. Soft parts, including skin, hair, and viscera of ice age mammoths, have been preserved in frozen soil (Fig. 4–7) or in the oozing tar of oil seeps. A striking example of preservation of soft parts was discovered in 1984 at Lindow Moss, England, when an excavating machine uncovered the body of a 2000-year-old human. Named Lindow Man (Fig. 4–8), the Briton had been ritually slaughtered (possibly by Druids), stripped, garroted, and then bled. The body was thrown in a bog, where the skin was preserved and the body flattened by the weight of accumulating peat. Only the upper part of the body was recovered; the lower part had been destroyed by the machine before the remains were detected.

Evidence of ancient life does not consist solely of petrifactions, molds, and casts. Sometimes the paleontologist is able to obtain clues to an animal's appearance and how it lived by examining tracks, trails (Fig. 4–9), burrows, and borings. Such markings are called **trace fossils,** and the study of trace fossils is termed **ichnology.** The tracks of an ancient vertebrate animal (Fig. 4–10) may indicate whether the animal that made them was bipedal (walked on two legs) or quadrupedal (walked on four legs), digitigrade (walked on toes) or plantigrade (walked on the flat of the foot), whether it had an elongate or a short body, if it was lightly built or ponderous, and sometimes

FIGURE 4–7 In the summer of 1977, the carcass of this baby mammoth was dug from frozen soil (permafrost) in northeastern Siberia. The mammoth stood about 104 cm tall at the shoulders, was covered with reddish hair, and was judged to be only several months old at the time of death. Dating by the radiocarbon method indicates that death occurred 44,000 years ago. *(Photograph courtesy of Klavdija Novikova, Biologopoczvennyj Institut, Vladivostok, former Soviet Union.)*

whether it was aquatic (with webbed toes) or possibly a flesh-eating predator (with sharp claws).

Trace fossils of invertebrate animals are more frequently found than the tracks of vertebrates, and they are also useful indicators of the habits of ancient creatures. One can sometimes infer if the trace-making invertebrate was crawling, resting, grazing, feeding, or simply living within a relatively permanent dwelling. For example, crawling traces, as might be expected, are linear and show directed movement (Fig. 4–11A). Shallow depressions that reflect the shape of the animal may be resting traces (Fig. 4–11B). Simple or U-shaped structures approximately perpendicular to bedding are often dwelling traces (Fig. 4–11C). Grazing traces (Fig. 4–11D) occur along bedding planes and are characterized by a systematic meandering or concentric and parallel patterns that represent the animal's effort to cover the area containing food in an efficient manner. The three-dimensional counterparts of grazing traces are called feeding traces. Feeding traces are made by animals that consume sediment for the organic nutrients within it. The traces consist of systems of branched or unbranched burrows, as shown in Figure 4–11.

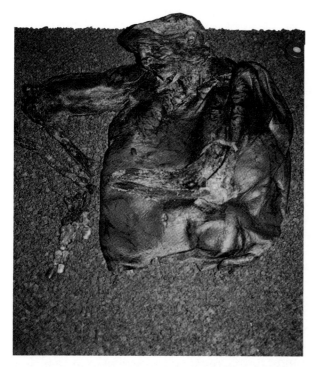

FIGURE 4–8 Preserved torso, arms, and head of the 2000-year-old Lindow Man. This example of preservation of soft tissue was found in a peat bog in 1984 at Lindow Moss, England. The lower half of the body was destroyed by the excavating machine. *(British Museum.)*

FIGURE 4–9 Trace fossil consisting of the trail made by an unknown, possibly wormlike animal about 650 million years ago. The fossil occurs in the Pound Quartzite of South Australia. *(Courtesy of B. N. Runnegar.)*

FIGURE 4–10 Dinosaur trackways arranged to indicate the passage of a biped (identical three-toed imprints), whose tracks were crossed by a quadruped having larger rear than front feet (typical of many quadrupedal dinosaurs). Claw imprints on the biped suggest that it was a predator. What indicates that the quadruped crossed the area after the biped?

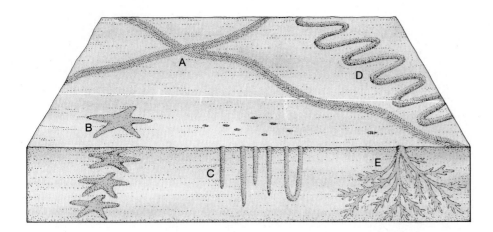

FIGURE 4–11 Traces that reflect animal behavior: (A) crawling traces, (B) resting traces, (C) dwelling traces, (D) grazing traces, and (E) feeding traces.

The Incomplete Record of Life

The fossil record of life is incomplete. If it were a complete and total record, it would include information on all past forms of life for every increment of time and for every place on Earth. Clearly this is unattainable. Only a limited number of animals and plants have been preserved. Others that were preserved have never been exposed to our view by erosion or drilling. Still others have simply not yet been discovered.

The record is more complete for marine life having hard external skeletons and less complete for land life lacking bone or shell. Where burial is rapid and dead organisms are protected from scavengers and agents of decay, the probability of preservation improves. As we have noted, the ocean floor and low-lying land areas where deposition predominates are often favorable places for fossil preservation. Many dinosaur remains, for example, have been discovered in the sands and clays deposited by streams that flowed across the low plains that lay east of the Rocky Mountains during the Cretaceous Period. There were certainly animals and plants living in nearby highland areas as well, but in such places erosion rather than deposition predominated and living things were less likely to escape destruction.

In spite of the many factors that prevent fossilization, the fossil record is remarkably comprehensive. On the basis of this impressive record, paleontologists have been able to piece together a history of past life that is balanced and largely accurate.

■ *The Rank and Order of Life*

The Linnaean System of Classification

Because of the large number of living and fossil animals and plants, random naming would be confusing and inefficient. Realizing this, the Swedish naturalist Carl von Linné, also known as Carolus Linnaeus (1707–1778), established a carefully conceived system for naming animals and plants. The **Linnaean system** uses morphologic structure as the basis of classification and employs binomial nonmenclature at the species level. In this scheme, the first, or **generic,** name was used to indicate a general group of creatures that were visibly related, such as all doglike forms. The second, or trivial, name denoted a definite and restricted group—a species. *Canis lupus,* for example, designates the wolf among all canids. Linné went on to recognize larger divisions such as classes and orders. His groupings were based on traits that seemed most basic or natural; the modern nomenclature system, however, is based on an attempt to be consistent with evolutionary relationships. There must also be consistency in the way organisms are named. For example, it is a rule that no two genera within a kingdom may have the same name. Further, names must be in Latin or Latinized and be printed in italics. The name of the genus must be a single word (nominative singular) and must begin with a capital letter. The species name is an adjective that modifies the generic name (a noun) and therefore begins with a small letter. It must agree grammatically with the generic name. To prevent duplication of names, only the name that is first published is considered valid.

Concepts Involved in Classification

The Species

The species is the fundamental unit in biological classification. A species is a group of organisms that have structural, functional, and developmental similarities and that are able to interbreed and produce fertile offspring. Because members of one species do not breed with members of different species under natural conditions, species exist in reproductive isolation. Species exist during a relatively short period of geologic time, are derived from ancestral populations, and in turn are the population from which new species arise.

In deciding on the validity of giving a species name to a group of seemingly related and similar organisms, biologists have the advantage of being able to determine whether or not the organisms can interbreed. This, of course, is not possible for an extinct population known only from fossils. Thus, the designation of a paleontologic species is unavoidably subjective. The paleontologist must determine the range of variation within the group of organisms being studied and make a logical inference about their inclusion within a designated species. In doing this, one compares the range of variation among individuals of the proposed fossil species with the range of variation within an analogous living species.

Taxonomy

The species is the basic unit in biologic classification, or **taxonomy.** In this system, the various categories of living things are arranged in a hierarchy that expresses levels of kinship. For example, a **genus** (pl. *genera*) is a group of species that have close ancestral relationships; a **family** is a group of related genera; an **order** is a group of related families; a **class** is a group of related orders; a **phylum** (pl. *phyla*) is a group of related classes; and a **kingdom** is a large group of related phyla. To use an example familiar to all, individual humans are members of the Kingdom Animalia, Phylum Chordata, Class Mammalia, Order Primates, Family Hominidae, Genus *Homo,* and species *sapiens.*

There are five kingdoms in the classification system widely used today: the **Monera, Protista** (also called **Protoctista**), **Plantae, Animalia,** and **Fungi** (Fig. 4–12). The Kingdom Monera includes bacteria. Monera means "alone" and refers to the fact that monerans are alone among all the kingdoms in having no membrane-bound nucleus or membrane-bound organelles. Such organisms are termed **prokaryotes.** All organisms that possess membrane-bound nuclei and organelles in their cells are called **eukaryotes.** Monerans are the oldest, most abundant, and simplest of organisms. They populated the Earth well over 3 billion years ago and were responsible for critical changes in the planet's early atmosphere. The Protista are a varied group of mostly unicellular organisms that, unlike the Monera (and similar to the cells of the other kingdoms), have a cell structure that includes a nucleus and organelles. Thus, they are eukaryotes. Protists include animal-like forms that devour food for energy (**heterotrophs**), plantlike photosynthesizers (**autotrophs**), and fungus-like decomposers (**saprophytes**). Members of the Kingdom

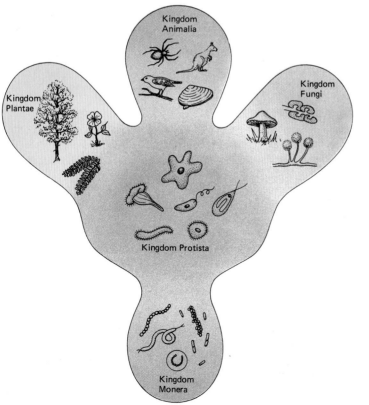

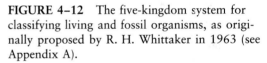

FIGURE 4–12 The five-kingdom system for classifying living and fossil organisms, as originally proposed by R. H. Whittaker in 1963 (see Appendix A).

Plantae are multicellular eukaryotes that typically live on land and make their own food by the process of photosynthesis. They are autotrophs. In contrast, members of the Animalia are heterotrophs. Finally, the Fungi are a kingdom of multicellular eukaryotic organisms that feed on dead or decaying organic material.

The subdivisions and common names for members of the five kingdoms are provided in Appendix A. Even a brief perusal of the listing indicates that biological classification is more than a mere system for cataloging organisms. Because it is based on organic structures and embryologic development, it reflects the broad outlines of evolutionary relationships. In a sense, the classification is a blueprint for constructing the tree of life.

■ *Organic Evolution*

The Roman poet Ovid (43 B.C.E.–A.D. 17) once wrote that "there is nothing constant in the universe, all ebb and flow, and every shape that's born bears in its womb the seeds of change." These words are remarkably relevant when one considers the way life has changed through time as revealed by the fossil record. At times that record indicates change may have been startlingly sudden, whereas at other times evidence shows a more gradual change. In either case, older forms have changed or evolved into newer forms to cope better with changes in their environment. This is not to say that the early, simpler forms of life are gone, for many persist along with their often more complex contemporaries.

The concept that animals and plants have changed through time is not a new one. Anaximander (611–547 B.C.) alluded to the idea 25 centuries ago. This famous Greek taught that life arose from mud warmed in the sun; that plants came first, then animals, and finally humans. Unfortunately, during the Middle Ages, complete faith in the theologic doctrine of special creation of "species" effectively stifled imaginative thinking about evolution. It was not until the eighteenth century in Europe that intellectuals such as Jean Lamarck and Georges Buffon began to challenge the concept of special creation. Buffon was convinced that evolution had occurred, but he offered no explanation of its cause. He believed, however, that the environment was somehow involved in the process that resulted in evolution. Buffon also set down a careful definition of species, noting that species were separate entities that were not able to interbreed successfully with other dissimilar species. The first general theory of evolution to excite considerable public attention was developed by the celebrated naturalist Jean Baptiste de Lamarck in the period between 1801 and 1815. Lamarck's writings were followed a half century later by those of Charles Darwin and Alfred R. Wallace.

Lamarck's Theory of Evolution

The Lamarckian theory of evolution stipulated correctly that all species, including humans, are descended from other species. His theory, however, was based on the mistaken assumption that new structures in an organism appear because of needs or "inner want" of the organisms and that structures once acquired in this way are somehow inherited by later generations. In a similar but reverse fashion, little-used structures would disappear in succeeding generations. For example, Lamarck believed that snakes evolved from lizards that had a strong preference for crawling. Because of this inner need to have long, thin bodies, certain lizards developed such bodies, and because legs became less and less useful in crawling, these structures gradually disappeared. Lamarck's ideas were challenged almost immediately. There was no way to prove by experimentation that such a thing as "inner want" existed. More important, Lamarck's belief that characteristics acquired during the life span of an individual could then be inherited was tested and shown to be invalid. One of these tests consisted of cutting off the tails of mice and then mating them. The offspring of the tailless mice had tails. The tails of the mice in twenty succeeding generations were similarly cut off, and the final generation had tails of the same length as the first. We know that some organisms can "adapt" to environmental conditions in a narrowly limited way within a life span. For example, a fair-skinned woman on the beach can be sunburned by exposure to the sun, but no amount of sunbathing will result in her first child being born with a sunburn. There is no way that somatic (or body) cells can pass characteristics over to reproductive cells and thereby on to the next generation. Thus, the Lamarckian concept of evolution based on use or disuse of organs was discredited.

Darwin's Theory of Natural Selection

Charles Darwin praised Lamarck for his courage in perceiving that the descendants of former creatures have undergone biologic change and are different from their ancestors. As described in Chapter 1, Darwin and his younger contemporary Alfred R. Wallace jointly proposed natural selection as an important mechanism for evolution. Darwin, with his years of careful data collection, and Wallace, with his sudden insight, had discerned that competition for food, shelter, living space, and sexual partners among species with individual variations and surplus reproductive capacity will inevitably result in elimination of the less well fitted and survival of those that are better fitted to their environments.

Darwin's *The Origin of Species by Means of Natural Selection* was published in 1859. It appeared at a time when at least part of the European intellectual atmosphere was more liberal and less satisfied with the theologic doctrine that every species had been independently created. Although the ideas of Darwin and Wallace continued to disturb the religious feelings of some, they nevertheless increasingly acquired adherents among nineteenth- and twentieth-century scientists. Darwin did not consider his views impious. He saw a "grandeur in this view of life with its several powers, having been originally breathed by the Creator into a few forms or into one; and that . . . from so simple a beginning endless forms most beautiful and most wonderful have been and are being evolved."

Mendelian Principles of Inheritance

Every theory has its strong and weak points. In Darwin's theory, the weakness lay in an inability to explain the *cause* of variability in a way that could be experimentally verified. The cause of at least a part of that variability was discovered in the course of elegant experiments on garden peas conducted by a Moravian monk named J. Gregor Mendel (1822–1884). Mendel discovered the basic principles of inheritance. His findings, printed in 1865 in an obscure journal, were unknown to Darwin and unheeded by the scientific community until 1900, when the article was rediscovered. Mendel described the mechanism by which traits are transmitted from adults to offspring. In his experiments with garden peas, he demonstrated that heredity in plants is determined by "character determiners" that divide in the pollen and ovules and are recombined in specific ways during fertilization. Mendel called these hereditary regulators "factors." They have since come to be known as *genes*.

As currently understood, genes are chemical units or segments of a nucleic acid—specifically **deoxyribonucleic acid (DNA)**. As suggested by careful chemical and X-ray studies, the DNA molecule is conceived as two parallel strands twisted somewhat like the handrails of a spiral staircase (Fig. 4–13). The twisted strands are made up of phosphate and sugar compounds and are linked with cross-members composed of specific nitrogenous bases.

The importance of DNA is evident when we realize that it indirectly controls the production of proteins, the essential components of many basic structures and organs. Even the activities of organisms are

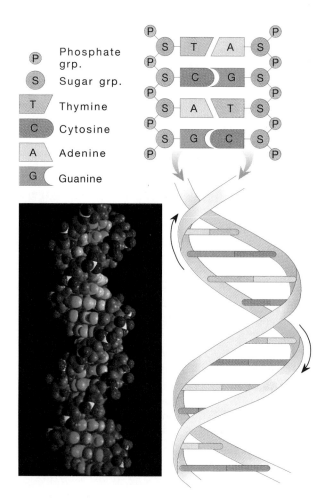

P　Phosphate grp.

S　Sugar grp.

T　Thymine

C　Cytosine

A　Adenine

G　Guanine

FIGURE 4–13 Representations of portions of the deoxyribonucleic acid (DNA) molecule. At the lower left is a computer reconstruction by R. J. Feldmann. The drawing depicts the twisted, double-stranded helix, with side rails composed of alternate sugar (deoxyribose) and phosphate molecules. Each rung of the twisted ladder is composed of one pair of nitrogenous bases. Of these, thymine links to adenine, and cytosine to guanine.

regulated by specific catalytic proteins called enzymes. Without DNA and its products, there would be no life as we know it. Its ability to replicate itself precisely is the basis for heredity, and organic evolution ultimately depends on this remarkable molecule.

The part of the DNA molecule that is active in the transmission of hereditary traits is called a **gene.** In nearly all organisms, genes are linked together to form larger units termed **chromosomes,** the central axis of which consists of a very long DNA molecule comprising hundreds of genes.

Reproduction and Cell Division

Reproduction of an organism may be sexual or asexual or may involve an alternation of sexual and asexual methods. All reproductive methods involve the division of cells. In sexual reproduction there is a union of reproductive or sex cells from separate individuals, whereas in asexual reproduction cells do not unite. The more usual methods of asexual reproduction are binary fission, budding, and spore production. **Binary fission** is found in single-celled organisms such as *Amoeba* that divide to form genetically identical daughter organisms. **Budding** occurs in some unicellular as well as multicellular organisms. In this method, the parent organism simply sprouts a bulge or appendage that may either remain attached to the parent or separate and grow as an isolated individual. The third principal method involving asexual reproduction is the formation of tiny reproductive cells called **spores** by a parent organism such as a seedless plant. The spores are formed by division of special spore parent cells and are shed by the parent organism (called the **sporophyte**) when they mature. When a spore settles onto moist soil, the floor of a water body, or some other suitable surface, it germinates and grows into a tiny plant (called a **gametophyte**), which, in turn, produces male and female sex cells. The union of these cells produces a new plant resembling the original sporophyte.

Organisms that produce asexually have the advantage of increasing their numbers rapidly whenever conditions are favorable. Some disadvantage exists, however, in that they are not able to develop as much variability among their offspring as do sexually reproducing organisms. To understand the reasons for that variability, we need to know some simple facts about chromosomes and their behavior during the development of reproductive cells and during fertilization. We note initially that the kind and number of chromosomes are constant for each species and differ between different species. Except in members of the Kingdom Monera (bacteria, including cyanobacteria), the chromosomes are located in the nucleus of the cells and occur in duplicate pairs. Thus, each chromosome has a homologous mate. In humans, for example, there are 46 chromosomes, or 23 pairs. Cells with paired homologous chromosomes are designated **diploid** cells. In all living things, new cells are being produced constantly to replace worn-out or injured cells and to permit growth. In asexual organisms, and in all the **somatic** or body cells of sexual organisms, the process of cell division that produces new diploid cells with exact replicas of the chromosomal components of the parent cells is called **mitosis** (Fig. 4–14).

In most organisms with sexual reproduction, a second type of division, called **meiosis**, takes place when **gametes** (egg cells or sperm cells) are formed. Meiosis may occur in unicellular organisms, but in multicellular forms it takes place only in the reproductive organs (testes or ovaries). Meiosis consists of two quickly succeeding divisions so as to produce four final daughter cells termed **haploid** because they do not have paired chromosomes. The haploid cells are the gametes or reproductive cells. When two gametes meet during sexual reproduction, the sperm enters the egg to form a single cell, which can now be called the **fertilized egg.** Because two gametes have been combined into one cell, there is now a full complement of chromosomes and genes representing a mix from both parents. The fertilized egg now begins a process of growth by mitotic cell division that eventually leads to a complete organism.

The organism that develops from the union of haploid cells will already have some variability relative to its parents. There is, however, yet another aspect of meiosis that is important in producing variation in offspring. During the initial chromosome division while the chromosomes are still paired, they may break at corresponding places and exchange their severed segments in a process called **crossing over.** The result is an additional mixing of genes.

If one compares the effects of asexual reproduction to sexual reproduction, it is apparent that the latter fosters more rapid evolution. Asexual organisms produce daughter cells that are identical to the parents. Unless there has been a fortuitous alteration of one or more genes (a mutation, as described in the next section), there is little or no change from generation to generation. In contrast, sexual reproduction offers the aforementioned many possibilities for genetic recombination. In addition, sexually reproducing organisms accumulate considerable "invisible" variation in the form of recessive genes. As environmental changes occur, some of these recessive genes will produce traits in the animal or plant that have survival advantages, and new avenues of evolution will be opened.

Mutations

Some of the variation we see among individuals of the same species results from the mixing of genes that occurs during reproduction. However, if genes were never altered, then the number of variations they could produce would be limited. For organisms to evolve a truly new variation, a process is required that will change the genes themselves. That process is known as **mutation.** Mutations can be caused by ultraviolet light, cosmic and gamma rays, and chemicals, including certain drugs. They may also occur spontaneously without a specific causative agent. Mutations may occur in any cell, but their evolutionary impact is greater when they occur in the sex cells, for then succeeding generations will be affected.

To understand how mutations occur, it is useful to re-examine the configuration of DNA (see Fig. 4–

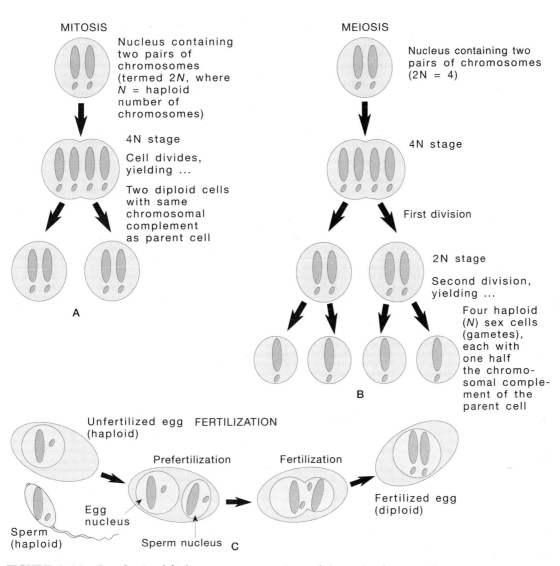

FIGURE 4–14 Greatly simplified summary comparison of the major features of (A) mitosis, (B) meiosis, and (C) fertilization.

13). The steps on the twisted DNA ladder are composed of two kinds of nitrogenous bases: *purines* and *pyrimidines*. Each step is made up of one purine joined to one pyrimidine. The purine called *adenine* is normally coupled with the pyrimidine called *thymine*. Similarly, the purine *guanine* joins with the pyrimidine *cytosine*. Thus, adenine–thymine and guanine–cytosine form **base-pair** steps of the ladder. Genes owe their specific characteristics to a particular order of these base-pair steps. If the order is disrupted, a mutation may result. For example, during cell division, when the twisted strands separate and then proceed to attract bases to rebuild the DNA molecule, there may be a mishap that results in the positioning of a guanine–cytosine base pair in what should have been the location of an adenine–thymine base pair. At that location, the gene is altered and may result in an inheritable change.

Clearly, organic evolution may involve change from at least three different sources: mutations, gene recombinations, and natural selection. Mutations are the ultimate source of new and different genetic material. Recombination (mixing) spreads the new genetic material through the population and mixes the new with the old. Natural selection sorts out the multitude of varying traits, preserving those that by chance are best fitted to a particular environment.

Evolution in Populations

A group of individuals that live close enough together so that each individual has an equal chance to mate with all members of the opposite sex within the group is called a **population**. Populations whose members can interbreed when they come into contact are grouped into that larger biologic entity called species. A species may be composed of a small geographically limited population (such as the sea snail *Retusa obtusa*, which is found only in the waters west of

Great Britain), or the species may include populations with wide geographic distributions (such as the brine shrimp *Artemia salina,* which occurs in saline lakes on all continents except Antarctica and Australia). Because it contains the total genetic composition of an entire group of interbreeding individuals, a population can be considered the basic unit of evolution. Each of the individuals within the population has its own particular combination of genes, and the sum of all these combinations constitutes the **gene pool** of the population. The gene pool is divided up in each generation and partitioned out to the new offspring. Inevitably, in each successive generation, new mutations and genetic combinations occur that manifest themselves in the offspring. As natural selection comes into play, some of these inherited traits will be passed on to the next generation in greater numbers, others in lesser numbers, so that ultimately the gene pool is altered. Thus, a source of evolutionary change becomes the impact of natural selection on the gene pool.

Speciation and Adaptive Radiation

The entire course of evolution depends on the origin of new species. The process by which this occurs is called **speciation.** We have seen that a species is a collection of populations within which there is a free flow of genes. From this definition, it is implicit that there is no gene exchange between two different species. Barriers between different species keep the gene pools separate. These barriers may be reproductive barriers that prevent fertilization, mating, or development of offspring, or they may be geographic barriers, such as islands that separate and isolate land animals. Wherever reproductive or geographic barriers exist, various segments of the population become isolated for many generations. During this period, they are likely to accumulate enough genetic differences so that interbreeding between the segments is no longer possible. Once this has occurred, the altered isolated segments have become different species.

After a new species has appeared, its success will depend on the advantages of its particular attributes relative to the environment. Species may encounter severe competition for food, shelter, living space, and other vital needs, and this may shorten their life span. If, however, a species is able to cope well with its environment, it may increase and persist. The average duration for most fossil invertebrate species has been 5 to 10 million years, a short time from a geologic point of view.

Once a new population has become established, pioneering segments of the population located around the fringes of the habitats may, like their parent species, undergo additional speciation. With in-

numerable successive speciations, diverse organisms characterized by diverse living strategies emerge. This branching of a population to produce descendants adapted to particular environments and living strategies is termed **adaptive radiation.**

The honey creepers of Hawaii provide an illustration of adaptive radiation. These birds, comprising many different species, are believed to have descended from a common ancestor. The most striking differences between the species occur in the sizes and shapes of the beaks, which are adaptively related to the kind of food the birds are dependent on (Fig. 4–15). Some have stout beaks for crushing hard seeds, and others have beaks well adapted for seeking out insects in cracks and crevices or for sucking nectar. All of the these variations in beak morphology are examples of **adaptations,** a word that means the acquisition of heritable characteristics that are advantageous to an individual and a population.

On a higher taxonomic level, one finds adaptive radiations among classes, orders, and families of organisms. All the orders of mammals, for example, had a common origin in an ancestral species that lived during the early Mesozoic. By adapting to different ways of life, the descendant forms came to diverge more and more from the ancestral stock, thereby providing today's rich diversity of mammalian plant eaters, flesh eaters, insect eaters, walkers, climbers, swimmers, and flyers.

Gradual or Punctuated Evolution

A question widely discussed by paleontologists is whether evolution proceeds gradually by means of an infinite number of subtle steps, as Charles Darwin proposed, or if there are sudden sporadic advances during which most of the change in a lineage occurs over a relatively short span of time. Gradual progressive change is referred to as **phyletic gradualism.** *Phyletic* (from *phylogeny*) refers to evolutionary pathways. Thus, proponents of phyletic gradualism believe that change occurs by slow degrees along the evolutionary pathway of a lineage. Where striking evolutionary breaks occur, they are not true breaks at all but places where intermediate or transitional forms are missing because of imperfections in the geologic record. Darwin would insist that the transitional forms once existed but were either not preserved, were destroyed, or have not yet been discovered.

Evidence for phyletic gradualism has been found in several groups of invertebrates, including lineages of well-preserved trilobites from the Ordovician of Wales as well as among the marine planktonic organisms known as foraminifers. In the latter study, the fossil foraminifers were recovered from complete and

FIGURE 4–15 The honey creepers of Hawaii provide a good example of adaptive radiation. When the ancestor of today's honey creepers first reached Hawaii, few birds were present. Succeeding generations diversified to occupy various ecologic niches. Their diversity is most apparent in the way their beaks have become adapted to different diets. Some have curved bills to extract nectar from tubular flowers, whereas others have short, sturdy beaks for cracking open seeds, or pointed bills for seeking out insects in tree bark. *Himatione sanguinea*, at the lower right, has a relatively unspecialized beak and appears to be similar to the original honey creeper ancestor.

uninterrupted well-dated sections of deep-sea cores. The fossils indicated a gradualistic line of descent that produced at least four species of the genus *Globorotalia* over a span of about 8 million years.

Paleontologists who oppose phyletic gradualism argue that the fossil record for the past 3 billion years contains many examples of new groups appearing suddenly in places where there is no evidence of any imperfections in the geologic record. Indeed, they find that the slow and stately advance of evolution can be documented only rarely and that evolutionary progress is more often sporadic than slow and uniform. Although this concept was suggested more than a half century ago, it has recently received new and strong support from paleontologists such as Stephen J. Gould and Niles Eldridge in their research on Pleistocene snails and Devonian trilobites. Gould and Eldridge have suggested the term **punctuated equilibrium** for evolution that consists of fitful sudden advances that punctuate long episodes of little evolutionary progress.

According to this concept, the punctuation or sudden morphologic change that interrupts equilibrium occurs at the periphery of the geographic area occupied by the population. These segments of the population are referred to as **peripheral isolates.**

Gene flow is rapid among the individuals of peripheral isolates, and changes in morphology or physiology leading to speciation may occur within a short span of time. Species do not usually originate in the places where their parental stock exist but rather in boundary zones where variations can be tested against new environmental situations. Should the parent species suffer extinction or severely decline, the new species may move back into the parental domain or they may expand into new territory. With adaptations that are a significant advantage, the new species are likely to enter a period of rapid evolution. This stage is usually followed by a period of moderate evolutionary rates and stability that lasts until the population begins to decline. Subsequently, yet another new group may begin its evolutionary expansion.

The final chapter on the question of punctuated evolution versus phyletic gradualism has not been written. At present, the proponents of punctuated evolution appear to be more numerous than those of phyletic gradualism. Like most controversies in science, however, the answer need not lie totally in one camp, and it is evident that instances of phyletic gradualism can also be recognized in the fossil record of certain groups of plants and animals.

Patterns of Evolution

The retrospective view of life through time provided by the fossil record affords an opportunity to detect overall patterns of evolution. The most familiar way of depicting these patterns is by a diagram called a **phylogenetic tree,** in which various taxa are traced from outermost branches (representing lower taxonomic divisions such as species, genera, or families) through a succession of junctures downward (and backward in time) to the major trunks of the higher taxa. On such a diagram, time is plotted on the vertical axis and biologic change on the horizontal. For the punctuated equilibrium model of evolution (Fig. 4–16A), the phylogenetic tree has short horizontal branches (reflecting sudden morphologic change) that quickly become vertical (to reflect subsequent stability through time). In contrast, the phyletic gradualism model (Fig. 4–16B) shows gently inclined branches that indicate slow and gradual change

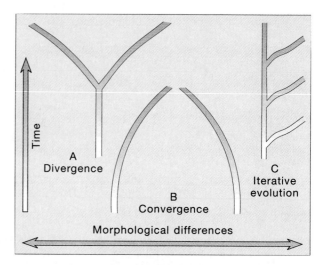

FIGURE 4–17   Three common evolutionary patterns as indicated by morphologic change among different lineages. (A) The *divergence* of two lineages. (B) Two lineages evolve toward morphologic similarity as an example of *convergence.* Presumably, this convergence results from selective pressures to adapt to similar environmental conditions. (C) In *iterative evolution,* morphologically similar forms evolve repeatedly from a root stock.

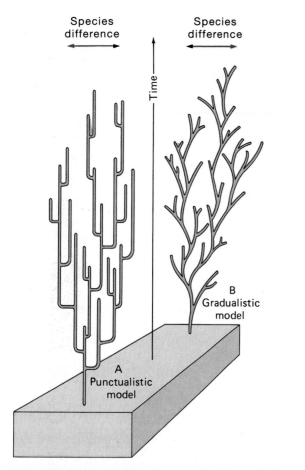

FIGURE 4–16   Phylogenetic trees depicting (A) the punctualistic model and (B) the gradualistic model of evolution. Morphologic change occurs in sideward directions. Time is depicted by the vertical direction. The short horizontal side branches of the punctualistic model depict sudden change, whereas the inclined branches of the gradualistic model suggest slow and uniform change through time.

through time. Branching occurs by gradual deviation of forms.

The phylogenetic tree serves as a model for depicting particular styles of evolution. As is evident from Figure 4–16B, a branch representing a single lineage may simply divide to form two or more lineages. This kind of evolutionary pattern is called **divergence** (Fig. 4–17A). **Convergence** (Fig. 4–17B) occurs when two or more unrelated lineages acquire similar morphologic traits, usually because of adaptations to similar modes of life and environments. A familiar example of convergence is seen in the evolution of superficially similar structures—wings—in birds, bats, and flying reptiles of the Mesozoic known as pterosaurs. **Iterative evolution** refers to the evolution of similar morphology more than once from the parental lineage (Fig. 4–17C). An example is provided by the coiled chambered mollusks known as cephalopods. Three or more times during the history of this class of invertebrates, irregularly coiled forms evolved from the uniformly coiled parental lineage.

Evidence for Evolution

Paleontologic Clues

The proponents of the concept of evolution had a difficult time convincing some of their nineteenth-century contemporaries of the validity of Darwin's theory. One reason was that evolution is an almost imperceptibly slow process. The human life span is

C O M M E N T A R Y
Earbones Through the Ages

Humans and other terrestrial mammals hear because receptor cells in the inner ear generate electrical impulses in response to vibrations caused by sound waves. These waves travel through the external ear canal to the tympanic membrane or ear drum (see accompanying figure). The tympanic membrane vibrates in response, and the vibrations are transmitted through a chain of three small bones located in the middle ear to the sound organ or cochlea of the inner ear. The three bones (auditory ossicles) are hinged to one another in a manner that allows them to act like an amplifier, increasing the force of sound vibrations. The evolution of the ear bones and middle ear cavity are excellent examples of anatomical structures that served a particular function in ancestral animals but are changed in function and form in the descendant animals.

The first of the three auditory ossicles is the **stapes.** One can trace its transformation backward in time to the Silurian, when jaws evolved from cartilaginous or bony supports for gills. The bone that was to become the stapes was originally an upper element of one of these gill supports. In the evolution of fish with true jaws, the gill support just posterior to those that evolved into jaws became a supporting prop between the upper jaw and braincase. It is called the hyomandibular. With the advent of amphibians during the Devonian, the old fish hyomandibular was transformed into the stapes, where it functioned as a transmitter of vibrations from the tympanic membrane to the inner ear. The stapes persisted as the only auditory ossicle throughout the evolutionary history of both amphibians and reptiles. With the advent of mammals, however, two bones in reptiles that had served as the joint between the upper and lower jaws were changed in function to transmit vibrations. They became the **incus** and the **malleus.** Thus, over the past 400 million years or so, the bones that were to evolve into our present hearing apparatus served such diverse functions as supports for the gills, props for the braincase, and articulation for the jaws.

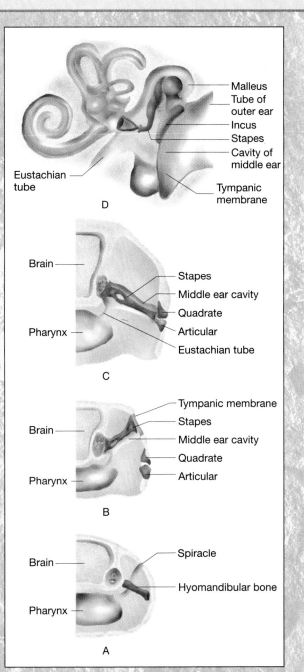

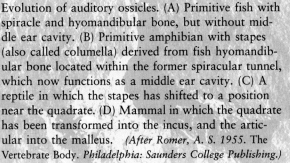

Evolution of auditory ossicles. (A) Primitive fish with spiracle and hyomandibular bone, but without middle ear cavity. (B) Primitive amphibian with stapes (also called columella) derived from fish hyomandibular bone located within the former spiracular tunnel, which now functions as a middle ear cavity. (C) A reptile in which the stapes has shifted to a position near the quadrate. (D) Mammal in which the quadrate has been transformed into the incus, and the articular into the malleus. *(After Romer, A. S. 1955. The Vertebrate Body. Philadelphia: Saunders College Publishing.)*

too short to witness evolutionary changes across generations of plants or animals. Fortunately, we can overcome this difficulty by examining the remains of organisms left in rocks of successively younger age. If life has evolved, the fossils preserved in consecutive formations should exhibit those changes. Indeed, many examples are known of sequential morphologic changes among related creatures during successive intervals of geologic time. The most famous example is provided by fossil horses recovered from formations of the Cenozoic Age (Fig. 4–18). The oldest horses thus far discovered as fossils were about the size of fox terriers. Unlike modern horses, they had four toes on their front feet and three on the rear. As one examines the many branching lineages of the horse "family tree" in successively higher (hence, younger) formations, one is able to see the results of evolutionary change. The animal shows an increase

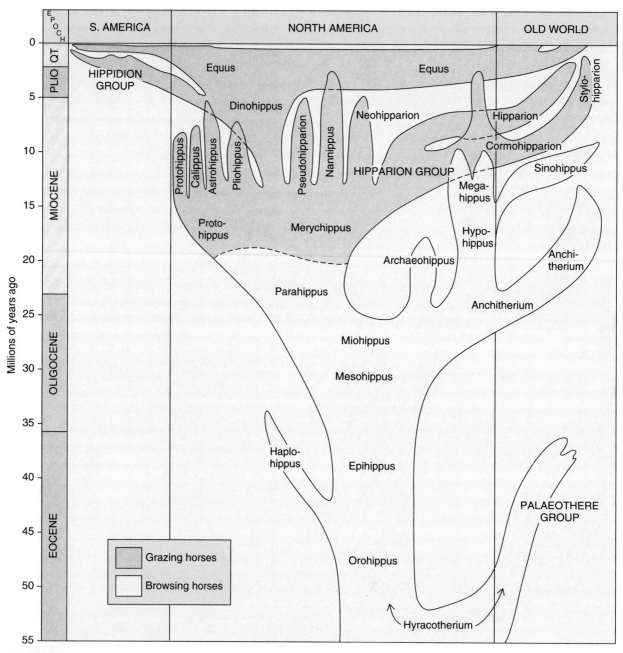

FIGURE 4–18 Phylogenetic tree of horses showing evolutionary relationships among the genera and the transition from browsing horses with low-crowned teeth to grazing horses with high-crowned teeth. *(After MacFadden, B. J., Bryand, J. D., and Mueller, P. A. 1991. Geology, 19:242–245.)*

in size, a reduction of side toes with emphasis on the middle toe, an increase in the height and complexity of teeth, and a deepening and lengthening of the skull.

The change in horse dentition through time provides an interesting example of the important link between environment and organic evolution. During the time that the horse family was evolving, there is paleontologic evidence that grasslands were becoming increasingly widespread in North America and Eurasia. Grass is a rather harsh food for an animal that must feed on plants. The blades contain silica, and because grass grows close to the ground, it is usually coated with abrasive dust. Early members of the horse family lived in forested areas and fed on the less abrasive leaves of trees and shrubs. They were browsers and had the low-crowned teeth of leaf eaters. Later members of the horse family were affected by selection processes that favored variants better able to cope with the problem of tooth wear from eating grasses. Over many generations, they had evolved the high-crowned teeth with complex patterns of enamel that characterize grazing animals (Fig. 4–19). These horses were likely to have been well nourished and hence lived longer lives and were able to produce more progeny. Their evolution must also have affected the evolution of the predators that pursued them and may even have contributed to the evolution of species of grasses that were resistant to damage by grazing. Evolution is an intricate process in which every animal and plant interacts with its neighbors and with the physical environment.

Although fossil remains of horses provide a fine illustration of paleontologic evidence for evolution, hundreds of other examples representing every major group of animals and plants would serve as well. The marine invertebrates known as cephalopods provide exceptionally fine examples of progressive evolutionary changes (Fig. 4–20). Because of this, they are also exceptionally useful in correlation.

Biologic Clues

Supplemental to the paleontologic clues favoring evolution are several persuasive arguments that are more directly biologic in nature. For example, in studies of the comparative morphology of organisms, it is not at all unusual to find body parts that are evidently of similar origin, structure, and development, even though they may be adapted for different functions in various species. For example, in seed plants, leaves are found as petals, tendrils, and thorns. In four-limbed vertebrates, the bones of the limbs may vary in size and shape, but they are fundamentally similar and in similar relative positions in birds, horses, whales, and humans (Fig. 4–21). Such basically similar structures in superficially dissimilar organisms are referred to as **homologous**. The differences in homologous structures are the result of variations and adaptations to particular environmental conditions, but the similarities indicate genetic relationship and common ancestry.

As another line of evidence, biologists point to the existence in animals and plants of useless and usually reduced structures that in other related species are well developed and functional. These **vestigial organs** are the "vestiges" of body parts that were utilized in earlier ancestral forms. Genetic material inherited from those ancestors still produces the vestigial organs, but the importance of these structures to the well-being of the organism has diminished with changes in environment and habits. Humans have over 100 of these vestigial structures, including the appendix, the ear muscles, and the coccyx (tail vertebrae). The vestigial pelvic bones in such different ani-

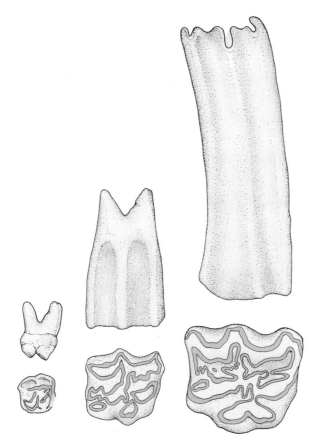

FIGURE 4–19 Development of high-crowned grinding molars in horses. Enamel shown in black. From left to right, *Hyracotherium*, *Merychippus*, and *Equus*. *(From Levin, H. L. 1975.* Life Through Time. *Dubuque, IA: William C. Brown Co.)*

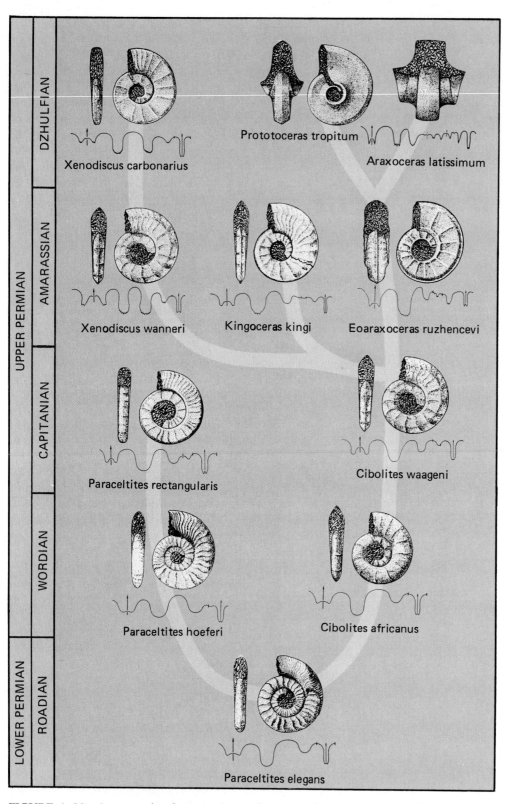

FIGURE 4–20   An example of progressive evolutionary change in a group of Permian ammonoid cephalopods. According to this interpretation, two evolutionary lineages originated from *Paraceltites elegans*, one terminating in *P. rectangularis* and a second producing *Cibolites waageni*. The latter was the ancestral stock for three additional lineages. The curved lines beneath each drawing are tracings of the suture lines; the terms along the left border are the names of Permian stages.   *(From Spinosa, C., Furnish, W. M., and Glenister, B. F. 1975. J. Paleontol. 49(2):239–283.)*

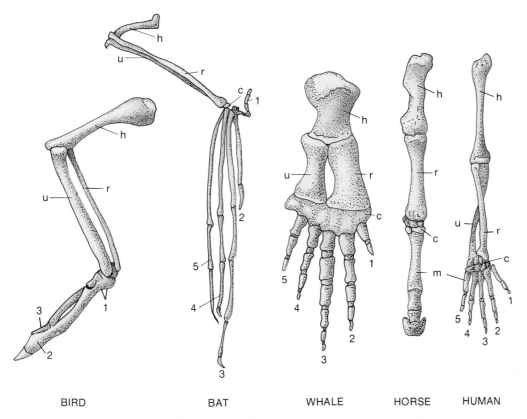

BIRD   BAT   WHALE   HORSE   HUMAN

FIGURE 4–21   Skeleton of right forelimb of several vertebrates to show similarity of structure. Key: *c,* carpals; *h,* humerus; *m,* metacarpals; *r,* radius; *u,* ulna; *1–5,* digits. Limbs are scaled to similar size for comparison.

mals as the boa constrictor and whale (Fig. 4–22) clearly suggest that they were evolved from four-legged animals.

Biologists have provided further evidence of evolution through comparative studies of the embryos of vertebrate animals. In their early stages of develop-ment, embryos of fish, birds, and mammals are strikingly similar. It would seem that all these animals received basic sets of genes from remote common ancestors. These genes control embryologic development for a time. Later in the developmental process, other genes begin to assume control and to cause

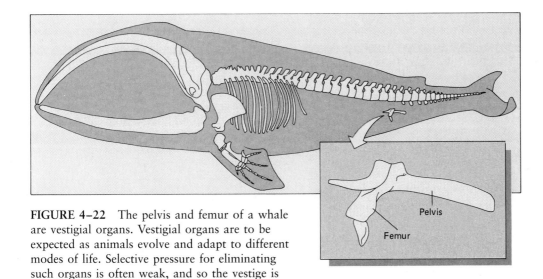

FIGURE 4–22   The pelvis and femur of a whale are vestigial organs. Vestigial organs are to be expected as animals evolve and adapt to different modes of life. Selective pressure for eliminating such organs is often weak, and so the vestige is retained for relatively long periods of time.

Pelvis

Femur

each different species to develop in its own unique way.

Many plants and animals resemble one another not only in structure and embryologic development but in biochemistry as well. Present knowledge indicates that all organisms produce nucleic acids—especially DNA—and that all use a compound called **adenosine diphosphate (ADP)** in life processes that involve energy storage and transfer. Digestive enzymes and hormone secretions are similar in many forms of life. Proteins extracted from corresponding tissues of closely related animals show definite similarities. The antigenic reactions of blood from the various groups of humans have been found to be practically identical or similar to such reactions in the blood of anthropoid apes.

■ *Fossils and Stratigraphy*

Establishing Age Equivalence of Strata with Fossils

One of William Smith's major contributions to geology was his recognition that individual strata contain definite assemblages of fossils. Because of the change in life through time, superposition, and the observation that once species have become extinct they do not reappear in later ages, fossils can be used to recognize the approximate age of a unit and its place in the stratigraphic column. (Such a method for judging the age of a unit would not be possible with inorganic characteristics of strata because they frequently recur in various parts of the geologic column.) Further, because organic evolution is generally global, rocks formed during the same age in identical environments but diverse localities often contain similar fauna and flora if there has been an opportunity for genetic exchange. This permits geologists to match chronologically or correlate strata from place to place. It provides a means for establishing the age equivalence of strata in widely separated parts of the globe.

The Geologic Range

Before fossils could be used as indicators of age equivalence, it was first necessary to determine the relative ages of the major units of rock on the basis of superposition. Geologists began by working out the superpositional sequences locally, and then they added sections from other localities around the world, fitting the segments together into a summary or **composite geologic column** (see Fig. 3–49). The next step was to determine the fossil assemblage from each time–rock unit and to identify the various genera and species. This work began well over a cen-

tury ago and is still very much in progress. Gradually, it became possible to recognize the oldest (or first appearance of) particular species as well as their youngest occurrence (last to appear) stratigraphically. The interval between the first and last appearance of a species constitutes the **geologic range.** Clearly, the geologic range of any ancient organism is not known *a priori* but is determined only by recording its occurrence in numerous stratigraphic sequences from hundreds of locations. Figure 4–23 illustrates one such study that records the geologic ranges of cephalopods (a class of mollusks having

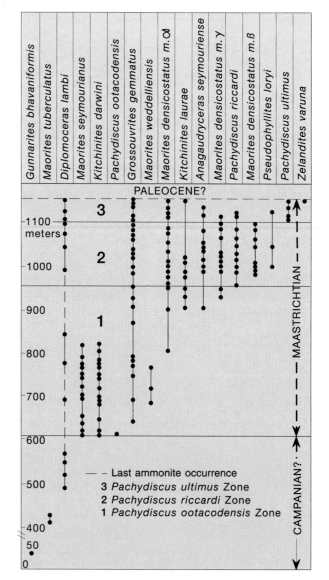

FIGURE 4–23 Example of a range chart showing the ranges of late Cretaceous ammonite cephalopods (chambered mollusks) from the Lopez de Bertodano Formation, Seymour Island, Antarctic Peninsula. The Campanian and Maastrichtian are stages of the Upper Cretaceous. Zonal boundaries are based on the first appearances of the species after which each zone is named. *(From Macellari, E. E. 1986. J. Paleontol. Mem. 18, Part 2.)*

chambered conchs) from a Cretaceous formation in Antarctica. Today, ranges are well known for some species and groups of species but relatively poorly known for others. Fortunately, there are now enough data so that isolated sections of strata in geologically unexplored areas can be located in the composite geologic column by use of their contained fossils.

Identification of Chronostratigraphic Units

The method for identifying time–stratigraphic units is illustrated in Figure 4–24. Geologists working in Region 1 come upon three time–rock systems of strata designated O, D, and M. Perhaps years later in Region 2, they again find units O and D, but in addition, they recognize an older unit—namely, C, below and hence older than O. Finally, while working in Region 3, they find a "new" unit, S, sandwiched between units O and D. The section, complete insofar as can be known from the available evidence, consists of five units that decrease in age as one progresses upward from C to M. The geologists may then plot the ranges of fossil species found in C to M alongside the geologic column. If they should next find themselves in an unexplored region, they might expe-

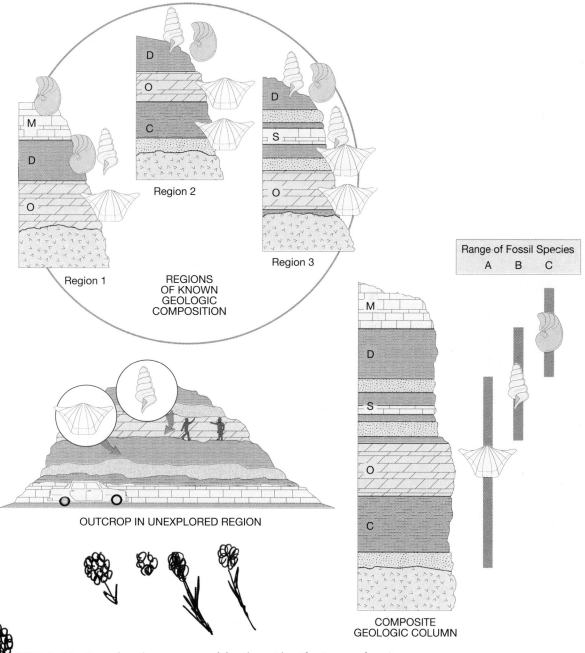

FIGURE 4–24 Use of geologic ranges of fossils to identify time–rock units.

rience difficulty in attempting to locate the position of this rock sequence in the standard column, especially if the lithologic traits of the rocks had changed. However, on discovering a bed containing species **A,** they are at least able to say that the rock sequence might be *C, O,* or *S.* Should they later find fossil species **B** in association with **A,** they might then state that the outcrop in the unexplored region correlates in time with unit *S* in Region 3. In this way, the strata around the world are incorporated into an accounting of global stratigraphy.

Paleontologic Correlation

Of course, the preceding illustration is greatly simplified, and future discoveries may extend known fossil ranges. Geologists must always keep in mind that the chance inability to find key fossils might lead to erroneous interpretations. Fortunately, the chances of making such mistakes are diminished by the practice of using entire assemblages of fossils. Two or three million years from now, geologists might have difficulty in firmly establishing on fossil evidence that the North American opossum, the Australian wallaby, and the African aardvark lived during the same episode of geologic time. However, if they found fossils of *Homo sapiens* with each of these animals, it would indicate their contemporaneity. In this example,

Homo sapiens can be considered the **cosmopolitan species,** for it is not restricted to any single geographic location within the terrestrial environment. The aardvark and wallaby are said to be **endemic species** in that they are confined to a particular area.

In the case of fossilized marine animals, cosmopolitan species have been especially useful in establishing the contemporaneity of strata, whereas endemic species are generally good indicators of the environment in which strata were deposited. Endemic species may slowly migrate from one locality to another. For example, the peculiar screw-shaped marine fossil bryozoan known as *Archimedes* (Fig. 4–25) was endemic to central North America during the Mississippian Period but migrated steadily westward, finally reaching Nevada by Pennsylvanian time and Russia by the Permian Period. Motile larvae of invertebrates that are attached as adults are often widely distributed by ocean currents, thus facilitating the migration of species.

One must assume a tentative attitude in making correlations based on fossils. The validity of a correlation increases each time the sequence of faunal changes is found again at different locations around the world. Geologists must also be aware that all changes are not because of evolution but instead may indicate fauna migrations or shifts in flora that accompanied ancient environmental changes. In this regard, the sudden disappearance of a fossil need not

A   B

FIGURE 4–25   Two specimens of the guide fossil *Archimedes.* Specimen (A) shows only the screwlike axis of this bryozoan animal. In specimen (B), one can see the fragile, lacy skeleton of the colony that is attached to the sharp, helical edge. Specimens are 8 to 10 cm in height.

mean that it became extinct but rather that it moved elsewhere. One might also note that the earliest appearance of a fossil in rocks of a given region might mean it had evolved there; however, it might also signify only that an older species had come into the new locality.

In addition to being on the alert for faunal changes that are related to shifting of ancient environments (as opposed to changes resulting from evolution), paleontologists must also be concerned about the possibility of **reworked fossils.** At various times in the geologic past, weathering and erosion have freed fossils from their host rock. Much as would be the case with any clastic particle, these fossils might then be reworked into younger beds, and the younger strata might then be mistakenly assigned to an older geologic time. Some fossils are particularly resistant to erosion and chemical decay and therefore are more susceptible to reworking. Among such resistant fossils are spore and pollen grains (see Fig. 4–40) and conodonts (see Fig. 4–52). The latter are tiny, toothlike structures composed of calcium phosphate. Conodonts are abundant fossils in some Paleozoic strata.

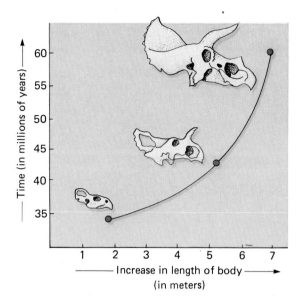

FIGURE 4–26 A study of rate of evolutionary change. In this lineage of horned dinosaurs, the rate of increase in size was initially rapid and subsequently much slower. *(Adapted from Colbert, E. H. 1948. Evolution, 2:145–163.)*

Index Fossils

Many fossils are rare and restricted to a few localities. Others are abundant, widely dispersed, and derived from organisms that lived during a relatively short span of geologic time. Such short-lived but widespread fossils are called **index** or **guide fossils** because they are especially useful in identifying time–rock units and correlating them from one area to another. A guide fossil with a short geologic range is clearly more useful than one with a long range. A fossil species that lived during the total duration of a geologic era would not be of much use in identifying the rocks of one of the subdivisions of lesser duration within that era. It is apparent that the rate of evolution is an important factor in the development of guide fossils. Simply stated, the **rate of evolution** is a measure of how much biologic change has occurred with respect to time. Figure 4–26 provides an example of a change in the rate of evolution among horned dinosaurs. The data indicate that although increase in size was a persistent evolutionary trend, the rate of that change diminished with time.

Obviously, different animals have evolved at different rates. Human beings and their ancestors, for example, are considered to have had a relatively rapid rate of evolution. In general, lineages with rapid rates of evolution provide greater numbers of guide fossils. Usually, rates of evolution can be related to changes in the environment or to the inherent reproductive or genetic characteristics of the evolving populations.

Although index fossils are of great convenience to geologists, correlations and interpretations based on assemblages of fossils are often more useful and less susceptible to error or uncertainties caused by undiscovered, reworked, or missing individual species. Also, by using the **overlapping geologic ranges** of particular members of the assemblage, it is often possible to recognize the deposits of smaller increments of time. The advantage of overlapping ranges can be illustrated with the help of Figure 4–27. Here, rather than using actual ranges of species, larger animal categories are used. It is apparent from this chart that a rock containing both stromatoporoids and goniatites can only be considered middle to late Devonian, whereas the occurrence of members of only one of the fossil groups would not provide as narrow a limit to the age of the rock.

Biostratigraphic Zones

Geologists frequently use the term **biozone** when describing a body of rock that is identified strictly on the basis of its contained fossils. Paleontologic biozones can vary in thickness or lithology and can be either local or global in lateral extent. Without formally naming them, we have already described the two major kinds of biostratigraphic zones: the range zone and the assemblage zone. A **range zone** is simply the rock body representing the total geologic life span of a distinct group of organisms. For example, in Fig-

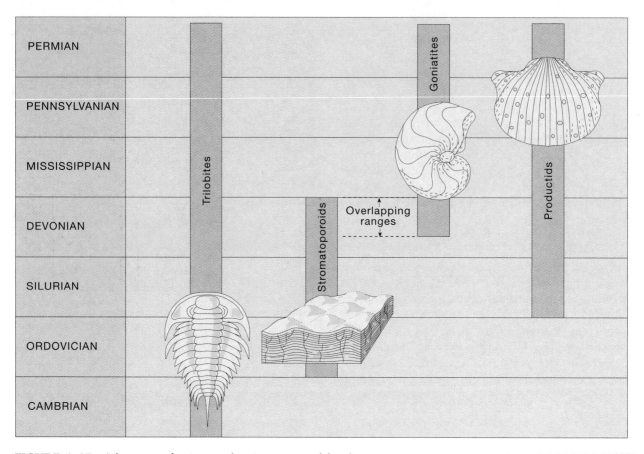

FIGURE 4-27   Advantage of using overlapping ranges of fossil taxa to recognize smaller increments of geologic time. The age of a stratum containing both stromatoporoids and goniatites would be Middle to Upper Devonian.

ure 4–28 the *Assilina* range zone is marked by the first (lowest) occurrence of that genus at point A and its extinction at point B. Geologists may also designate **assemblage zones** selected on the basis of several coexisting taxa and named after an easily recognized and usually common member of the assemblage. In certain areas, however, even though the guide fossil for the assemblage zone may not be present, the other members permit recognition of the zone. Another kind of biostratigraphic zone is the **concurrent range zone.** It is recognized by the overlapping ranges of two or more *taxa* (sing. *taxon*). (The term **taxon** refers to a group of organisms that constitute a particular taxonomic category, such as species, genus, and family.) For example, the interval between X and Y in Figure 4–28 might be designated the *Assilina–Heterostegina* concurrent range zone.

Biozones are of fundamental importance in stratigraphy. They are the basic unit for all biostratigraphic classification and correlation. They are also the basis for time–stratigraphic terms because zones are aggregated into stages, stages into series, and series into systems. The boundaries of these units are usually zonal boundaries.

■ *Fossils as Clues to Ancient Environments*

Paleoecology

Although inert and mute, fossils are vestiges of once lively animals and plants that nourished themselves, grew, reproduced, and interacted in countless ways with other organisms and their physical environment. The study of the interaction of ancient organisms with their environment is called **paleoecology.** Paleoecologists attempt to discover precisely where and how ancient creatures lived and what their habits and morphology reveal about the geography and climate of long ago. This scientific detective work is accomplished in various ways. One can, for example, compare species known only from fossils with living counterparts. The assumption is made that both living and fossil forms had approximately the same needs, habits, and tolerances. One can also examine the anatomy of the fossil and attempt to identify structures that were likely to have developed in response to particular biologic and physical conditions

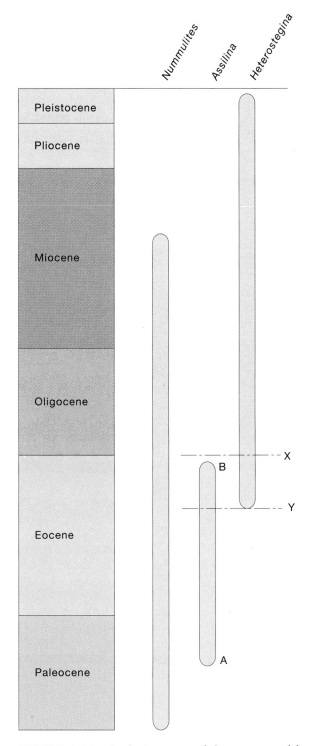

FIGURE 4–28 Geologic ranges of three genera of foraminifers within the family Nummulitidae. The interval between *A* and *B* is the total range of *Assilina*. The interval between *X* and *Y* could be designated the Assilina–Heterostegina concurrent range zone.

in the environment. As noted earlier, such modifications for living in a certain way and performing particular functions are called **adaptations**. For example, the broad, spiny valve of the brachiopod *Marginifera*

FIGURE 4–29 A fossil of the spinose brachiopod *Marginifera ornata*. The shell has been replaced by silica. The enclosing limestone has been dissolved so as to provide excellently preserved fossils with delicate spines intact. The animal rested on its ventral valve (shown here), and its spines provided anchorage. *(Photograph courtesy of Richard E. Grant, U.S. Geological Survey.)*

ornata (Fig. 4–29) was an adaptation that served to support the animal on a sea bed composed of soft mud. In another example, reduction of the skeleton to mere spines in the Silurian trilobite *Deiphon* (Fig. 4–30) was probably an adaptation to provide buoyancy in an animal that fed near the surface of the sea. Coiling in living and fossil cephalopods appears to be an adaptation designed to bring the center of buoyancy above the center of gravity and thereby permit these creatures to keep on an "even keel" while swimming. A list of additional examples of adaptations would be immense, for every creature is a result of hundreds of adaptations. Not all adaptations are morphologic. Biochemical and physiologic adaptations occur as well, although these are more difficult to recognize in fossils.

The language of paleoecology is derived from **ecology**, the study of the present relationships between organisms and their environments. In ecologic studies, one tends to concentrate on the **ecosystem**, which is any selected part of the physical environment together with the animals and plants in it. An ecosystem maybe as large as the Earth or as small as a garden pond. Paleoecologists are particularly interested in the ocean ecosystem because of the richness of the fossil record of marine life. The physical aspects of the ocean ecosystem include the water itself, its dissolved gases (especially carbon dioxide and oxygen), salts (phosphates, nitrates, chlorides, and carbonates of sodium, potassium, and calcium), various organic compounds, turbidity, pressure, light penetration, and temperature.

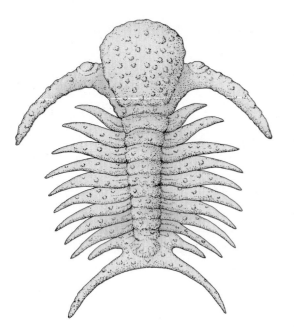

FIGURE 4–30 *Deiphon* was a Silurian trilobite whose extreme spinosity suggests it was a swimmer and floater rather than a bottom dweller, like many of its trilobite cousins. Spinosity increases surface area without adding weight, thus enhancing buoyancy. It is also possible that the swollen anterior structure (glabella) may have been filled with a liquid of low specific gravity, thus providing additional buoyancy. The length of the specimen is 27 mm.

The biologic components of the ocean ecosystem are usually classified according to so-called **trophic** or feeding levels (Fig. 4–31). For example, **producer organisms** such as green plants manufacture compounds from simple inorganic substances by means of photosynthesis. These organisms in the sea are mostly very small (less than 0.1 mm in diameter) and include diatoms and other forms of algae. Producer organisms are eaten by **consumer organisms** such as mollusks, crustaceans, and fish. The primary consumers commonly feed directly on unicellular plants and thus may be further designated as **herbivores.** The secondary consumers that eat the herbivores are called **carnivores.** Tertiary and even quaternary consumers feed on carnivores from lower tropic levels. The ecosystem also contains decomposers and transformers, including bacteria and fungi, which are able to break down the organic compounds in dead organisms and waste matter and produce simpler materials that can be used by the producers. Thus, in an ecosystem, the basic chemical components of life are continuously being recycled. The marine ecosystem also contains parasites, which feed on other organisms without necessarily killing them, and scavengers, which derive their nourishment from dead organisms.

Within any given ecosystem, one finds the specific environments or habitats where certain organisms live. In the ocean, the habitats range from the cold, dark realm of the abyss to the warm, illuminated areas of tropical coral reefs. And within every habitat there are ecologic niches in which particular organisms make their living (Fig. 4–32). Each contains all the biologic, chemical, and physical conditions that permit the organism occupying the niche to survive. Many niches exist in the coral reef habitat. Some are occupied by tiny, tentacled coral animals that build the framework of the reef and feed as carnivores on even smaller crustaceans. Certain marine snails occupy a niche along the surface of algal mats, where

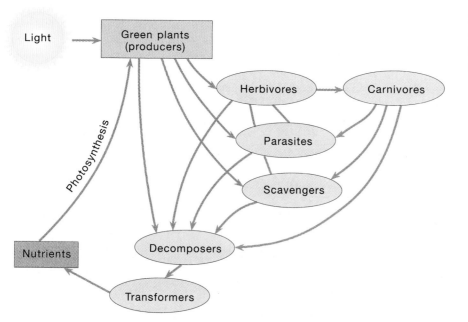

FIGURE 4–31 The movement of materials through an ecosystem. Components within ovals are consumers.

FIGURE 4–32 A Permian patch reef paleocommunity. In this diorama, one can recognize strings of bead sponges (top) and compact sponges (lower left), horn corals (upper left), cephalopods (center and bottom center), and many aggregates of brachiopods. The left-shaped shells to the left of the spiny cephalopod are remains of oysterlike brachiopods *(Leptodus).* *(Courtesy of the National Museum of Natural History.)*

with snail-like sluggishness they graze on films of algae. Behind the reef in areas of soft mud, plump lugworms consume the soupy sediment for its content of organic nutrients. Each creature has its characteristic ecologic niche. Each organism, however, is also a member of a natural assemblage of plants and animals living together in a **paleocommunity.** One of the most interesting and challenging tasks of the paleontologist is to discover the interrelationships between members of the paleocommunity. Which were the producers, herbivores, predators, and decomposers? What elements of the fauna might have been present but are not preserved? And what does the succession of paleocommunities, as seen in underlying and overlying strata, indicate about how the environment changed during the passage of geologic time?

The Marine Ecosystem

To facilitate the study of the ocean ecosystem, ecologists have developed a simple classification of marine environments. It begins with a twofold division of the entire ocean into pelagic and benthic realms. The **pelagic realm** consists of the water mass lying above the ocean floor. It can be divided into a **neritic zone,** which overlies the continental shelves, and an **oceanic zone,** which extends seaward from the shelves (Fig. 4–33).

Within the pelagic realm one finds myriads of small animals and plants that float, drift, or feebly swim. These are **plankton. Phytoplankton** consist of plants and plantlike protists and include algae such as diatoms (see Fig. 12–6) and coccolithophorids (see Fig. 4–51). Planktonic animals constitute **zooplank-**

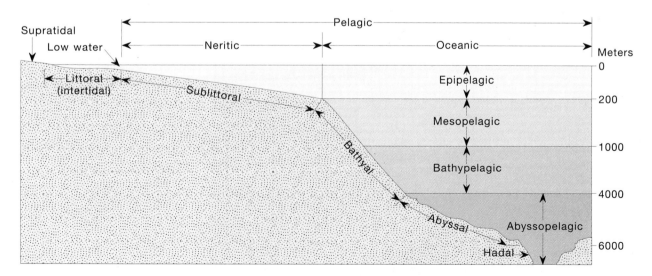

FIGURE 4–33 Classification of marine environments. *(After Hedgpeth, J. W., ed. 1957. Treatise on Marine Ecology and Paleoecology. Geol. Soc. Am. Mem. 67(1):18.)*

FIGURE 4–34 A living planktonic foraminifer similar to those that became abundant during the Cretaceous and Cenozoic. The fine rays are cytoplasm that is extruded through the pores of the tiny calcium carbonate shell. The shell is approximately 85 μm in diameter. *(Manfred Kage/Peter Arnold, Inc.)*

ion and include protists such as radiolaria (see Fig. 14–5), foraminifers (Fig. 4–34), certain tiny mollusks, small crustaceans, and the motile larva of many different families of invertebrates that live the adult stage of their life cycles on the sea floor.

The pelagic realm is also the home of **nekton**, or true swimming animals. Nekton are able to travel where they choose under their own power, and this is clearly advantageous. A swimming creature can search for its food and does not have to depend on food particles carried in chance currents. It can use its mobility to escape predators and can move to more favorable areas when conditions become difficult. The nekton is a diverse group that includes invertebrates such as shrimp, cephalopods, and certain extinct trilobites as well as vertebrates such as fish, whales, and marine turtles.

The second great division of the ocean ecosystem is the bottom or **benthic** realm. It begins with a narrow zone above high tide called the **supralittoral zone.** Several kinds of marine plants and animals have adapted themselves to this harsh environment, where ocean spray provides vital moisture but where drying poses a constant danger. Seaward of the supralittoral zone is the area between high and low tide. This is the **littoral zone.** Like those creatures of the supralittoral area, organisms in the littoral zone must be able to tolerate alternate wet and dry conditions. Some avoid the drying by burrowing into wet sand, whereas others have adapted themselves in various ways to retain body moisture when exposed to air.

Benthic animals and plants are most abundant seaward of the littoral zone in the continuously submerged **sublittoral zone** (Fig. 4–35). This zone extends from low tide levels down to the edge of the continental shelf (about 200 meters deep). Depending on the clarity of the water, light may penetrate to the sea floor in the sublittoral zone, although the base of light penetration is usually slightly less than 200 meters. Various kinds of algae thrive here as well as abundant protozoans, sponges, corals, worms, mol-

FIGURE 4–35 Diorama depicting sublittoral zone organisms living on the floor of the sea covering the central United States during the Silurian Period. Members of this thriving paleocommunity include trilobites, horn and honeycomb corals, algae, branching bryozoan colonies, and small bivalved brachiopods. *(Courtesy of the Milwaukee Public Museum.)*

FIGURE 4–36   Burrows made by infaunal organisms (probably worms) are preserved in the Northview Formation (Mississippian) of Missouri.

lusks, crustaceans, and sea urchins. These animals are never subjected to desiccation. Their adaptations are mainly associated with food gathering and protection from predators. Some of these benthic animals live on top of the sediment that carpets the sea floor and are called **epifaunal.** Others, termed **infaunal,** burrow into the soft sediment for food and protection (Fig. 4–36). Burrowers churn and mix sediment, a process known as **bioturbation.** The result of bioturbation may be destruction of original grain orientations in clastic sediments, rendering them useless for studies of paleocurrent directions. Evidence of cyclic events may be obscured, and magnetic properties of the rock may be altered. On the other hand, bioturbation may provide evidence of the existence of organisms in strata where fossil remains such as shells are rare or absent.

Beyond the continental shelves, the benthic environment is subjected to low temperatures, little or no light, and high pressures. Without light, plants are unable to live at these depths. One encounters this **bathyal** environment from the edge of the shelf to a depth of about 4000 meters. Still deeper levels constitute the **abyssal** environment. The term **hadal** is reserved for the extreme depths found in oceanic trenches. As might be expected, animals are less abundant in the abyssal and hadal environments. Most of these deep-water creatures are scavengers that depend on the slow fall of food from higher levels and predators that feed on the scavengers.

Use of Fossils in Reconstructing Ancient Geography

The geographic distribution of present-day animals and plants is closely controlled by environmental lim-

itations. Any given species has a definite range of conditions under which it can live and breed, and it is generally not found outside that range. Ancient organisms, of course, had similar restrictions on where they could survive. If we note the locations of fossil species of the same age on a map and correctly infer the environment in which they lived, we can produce a paleogeographic map for that particular time interval. One might begin by plotting on a simple base map the locations of fossils of marine organisms that lived at a particular time. This provides an idea of which areas were occupied by seas and might even suggest locations for ancient coastlines. Figure 4–37 shows major land and sea regions during the middle Carboniferous Period. The locations of marine protozoans called *fusulinids* are plotted as open circles. Notice that the lofty ranges of the Rocky Mountains were nonexistent, and their present locations were occupied by a great north–south seaway.

Having obtained a fair idea of the marine regions, one might next give attention to the locations of fossils diagnostic of land areas. The fossilized bones of land animals such as dinosaurs or mastodons would suggest a terrestrial environment, as would their preserved footprints. Fossil remains of land plants, including fossilized seeds and pollen, provide excellent evidence of terrestrial paleoenvironments.

By an analysis of the fossils and the nature of the enclosing sediment, it is often possible to recognize deeper or shallower parts of the marine realm or to discern particular land environments such as ancient floodplains, prairies, deserts, and lakes. River deposits may yield the remains of freshwater clams and fossil leaves. A mingling of the fossils of land organisms and sea organisms might be the result of a stream entering the sea and perhaps building a delta in the process.

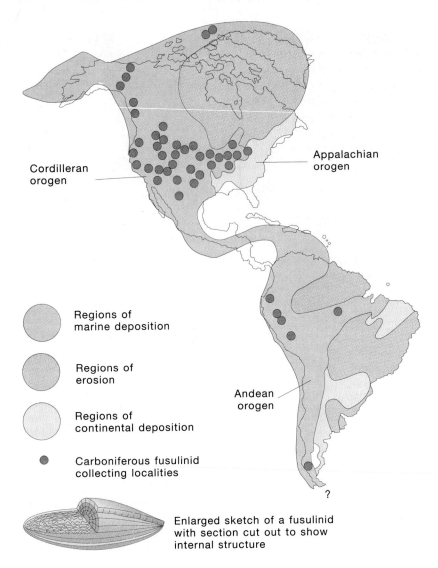

Cordilleran
orogen

Appalachian
orogen

Regions of
marine deposition

Regions of
erosion

Regions of
continental deposition

● Carboniferous fusulinid
collecting localities

Andean
orogen

?

Enlarged sketch of a fusulinid
with section cut out to show
internal structure

FIGURE 4–37 Major land and sea
regions in North and South America
during the Carboniferous Period.
(Adapted from Ross, C. A. 1967. J. Paleontol. 41(6):1341–1354.)

The migration and dispersal patterns of land animals, as indicated by the fossils they left behind, is one important indicator of former land connections as well as mountainous or oceanic barriers that once existed between continents. Today, for example, the Bering Strait between North America and Asia prevents migration of animals between the two continents. The approximately 80 km between the two shorelines (Fig. 4–38), however, is covered by less than 50 meters of water; this might lead one to wonder whether Asia was once connected to North America. The fossil record shows that a land bridge did connect these two continents on several occasions during the Cenozoic Era. The earliest Cenozoic strata on both continents have a fossil fauna uncontaminated by foreign species. Somewhat higher and younger rocks contain fossil remains of animals that heretofore had been found only on the opposite continent. These remains mark a time of land connection. The last such connection may have existed only

14,000 to 15,000 years ago, when Stone Age hunters used the route to enter North America.

Another familiar example of how fossils aid in paleogeographic reconstructions is found in South America. Careful analyses of fossil remains indicate that, in the early Cenozoic Era, South America was isolated from North America; as a result, a uniquely South American fauna of mammals evolved over a period of 30 to 40 million years. The establishment of a land connection between the two Americas is recognized in strata only a few million years old (late Pliocene) by the appearance of a mixture of species formerly restricted to either North or South America. (The migrations were decidedly detrimental to South American species.)

In addition to deciphering the positions of shorelines or locations of land bridges, paleontologists can also provide data that help to locate the equator, parallels of latitude, and pole positions of long ago. It has been observed that in the higher latitudes of the

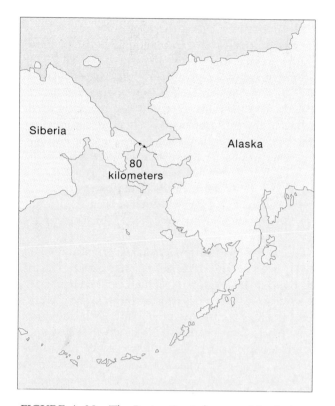

FIGURE 4–39 View of a modern coral reef. Coral reefs harbor an extraordinary diversity of organisms and, in some ways, rival tropical rain forests in their complexity. *(From Villee, C. A. et al. 1989. Biology, 2nd ed. Philadelphia: Saunders College Publishing.)*

FIGURE 4–38 The Bering Strait between North America and Asia. The area between the two continents would become land if sea level were lowered by only 45 meters. Such a lowering of sea level occurred at various times during the Cenozoic Era and permitted migrations across the land bridge.

globe, one is likely to find large numbers of individuals but that these are members of relatively few different species. In contrast, equatorial regions tend to develop a large number of species, but with comparatively fewer individuals within each species. Stated differently, the variety or **species diversity** for most higher categories of plants and animals increases from the poles toward the equator. This is probably related to the fact that relatively few species can adapt to the rigors of polar climates. Conversely, there is a stable input of solar energy at the equator, less duress caused by seasons, and a more stable food supply. These warmer areas place less stress on organisms and provide opportunity for continuous, uninterrupted evolution. Of course, in particular areas, generalizations about species diversity can be upset by special circumstances of altitude, upwelling, or other vagaries of air and water currents.

Another way to locate former equatorial regions (and therefore also the polar regions that lie 90° of latitude to either side) is by plotting the locations of fossil coral reefs of a particular age on a world map. Nearly all living coral reefs (Fig. 4–39) lie within 30°

of the equator. It is not an unreasonable assumption that the ancient reef corals had similar geographic preferences.

Use of Fossils in the Interpretation of Ancient Climatic Conditions

Among the many climatic factors that limit the distribution of organisms, *climate* (especially the temperature component of climate) is of great importance. There are many ways by which paleoecologists gain information on ancient climates. An analysis of fossil spore and pollen grains (Fig. 4–40) often provides outstanding evidence of past climatic conditions. Living organisms with known tolerances can often be directly compared with fossil relatives. Such comparisons are direct examples of the application of **actualism**, a concept introduced in Chapter 1. As noted earlier, corals thrive in regions where water temperatures rarely fall below 18°C, and it is likely that their ancient counterparts were similarly constrained. However, even when a close living analogue is not found, plants and animals may exhibit morphologic traits useful in determining paleoclimatology. Plants with aerial roots, lack of annular rings, and large wood cell structure indicate tropical or subtropical climates lacking in strongly contrasting seasons. Marine mollusks (such as clams, oysters, and snails) with well-developed spinosity and thicker shells tend to occur in warmer regions of the oceans. In the case of particular species of marine protozoans, variation in the average size of individuals or in the direction of coiling can provide clues to cooler or warmer conditions.

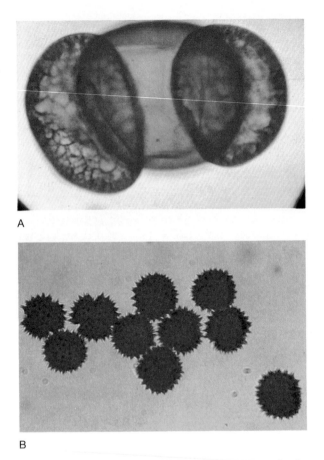

A

B

FIGURE 4–40 Pollen grains. (A) Pollen grain of a fir tree. Note the inflated bladders that make the pollen grain more buoyant. The specimen is about 125 microns in its longest dimension. (B) Pollen grains of ragweed. These grains are about 25 microns in diameter.

An example of environmentally induced changes in shell coiling directions is provided by the planktonic (floating) foraminifer *Globorotalia truncatulinoides* (Fig. 4–41). In the Pacific Ocean, left-coiling tests (shells) of these foraminifers have dominated in periods of relatively cold climate, whereas right-coiling tests predominated during warmer episodes. Another foraminifer, *Neoglobquadrina pachyderma,* exhibits similar changes in coiling directions, not only in the Pacific Ocean but in the Atlantic as well. Such reversals in coiling may occur quickly and over broad geographic areas. For this reason, they are exceptionally useful in correlation studies.

Aside from morphologic changes within species, entire assemblages of foraminifers are widely used in paleoecologic studies. Today, as in the past, benthic (bottom-dwelling) species have inhabited all major marine environments. As a result of the accumulated data on living foraminifers, inferences about salinity and water depths can be made. Fossil foraminifers

have frequently provided the means for recognition of ancient estuaries, coastal lagoons, and nearshore or deep oceanic deposits.

Sometimes, when the overall morphology of a fossil does not provide clues to temperature or climate, the compositions of the skeleton can be used. Magnesium, for example, can substitute for calcium in the calcium carbonate of shelled invertebrates. For particular groups of living marine invertebrates, it has been found that those living in warmer waters have higher magnesium values than do those residing in colder areas. This knowledge has been used to interpret the climate at the times fossil forms were living.

In another kind of biogeochemical study of shell matter, investigators have found that certain organisms build their skeleton of calcium carbonate in both of its two mineral forms: *aragonite* and *calcite.* Moreover, in colder realms, particular species have higher calcite-to-aragonite ratios, whereas their fellow species living in warmer waters have lower calcite-to-aragonite ratios. Secretion of calcium carbonate in the form of aragonite appears to be "easier" at higher temperatures.

There are, however, problems when geochemical means are used to deduce temperature. Aragonite is an unstable mineral that reverts to the more stable calcite. It is nearly absent in Paleozoic rocks. High-magnesium calcite is also unstable. Magnesium may be leached out of the sediment after burial, causing the conversion of high-magnesium calcite to its more stable low-magnesium counterpart.

Oxygen isotope analyses offer yet another way for learning the temperature of ancient seas. When water evaporates from the ocean, there is a fractionation of the oxygen-16 and oxygen-18 in the water. (Oxygen-16, the usual form of oxygen, has an atomic mass of 16, whereas oxygen-18 has an atomic mass of 18.) Oxygen-16, being lighter, is preferentially

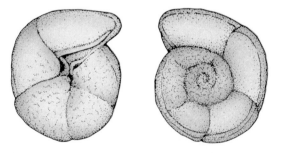

FIGURE 4–41 Shell coiling in the foraminifer *Globorotalia truncatulinoides.* Sketch of both sides of a single left-coiling specimen. (Diameter is about 0.9 mm.) *(From Parker, F. L. 1962. Micropaleontology 8(2):219–254.)*

removed, while the heavier oxygen-18 tends to remain behind. When the evaporated moisture is returned to the Earth's surface as rain or snow, water containing the heavier isotope precipitates first, often near the coastlines, and flows quickly back into the ocean. Inland, the precipitation from the remaining water vapor is depleted in oxygen-18 relative to its initial quantity. If the interior of a continent is cold and contains growing glaciers, the glacial ice will lock up the lighter isotope, preventing its return to the ocean and thereby increasing the proportion of the heavier isotope in sea water. As this occurs, the calcium carbonate shells of marine invertebrates such as foraminifers will also be enriched in oxygen-18 and thereby reflect episodes of continental glaciation.

Even in the absence of ice ages, the oxygen isotope method may be useful as a paleotemperature indicator. As invertebrates extract oxygen from sea water to build their shells, a temperature-dependent fractionation of oxygen-18 occurs between the water and the secreted calcium carbonate in the shell matter. Provided there has been no alteration of the shell, it is possible to find the temperature of the water at the time of shell formation. In a famous early study, the oxygen isotope ratio in the calcium carbonate skeleton of a Jurassic belemnite indicated that the average temperature in which the animal lived was 17.6°C, plus or minus 6°C of seasonal variation. Subsequent studies on both Jurassic and Cretaceous belemnites (Fig. 4–42) provided confirmation of the inferred positions of the poles during the Mesozoic and indicated that tropical and semitropical condi-

tions were far more widespread during late Mesozoic time than they have been in subsequent geologic time.

■ *An Overview of the History of Life*

Setting the Stage

As yet, no fossil evidence has been found of life on Earth during the planet's first billion years. The oldest direct indications of ancient life are remains of bacteria and primitive algae discovered in rocks over 3.5 billion years old. These early fossils are usually considered evidence that the long evolutionary march had begun. However, they also stand at the end of another long and remarkable period during which living things presumably evolved from nonliving chemical compounds. We have no direct geologic evidence to tell us how and when the transition from nonliving to living occurred. What we do have are reasonable hypotheses supported by careful experimentation. Some of these hypotheses will be examined in Chapters 6 and 7, which deal with the Precambrian eons. It will suffice for this brief overview to note that life came into existence early in that great 4-billion-year span of time prior to the beginning of the Cambrian Period.

For most of Precambrian time, living things left only occasional traces. Here and there paleontologists have been rewarded with finds of bacteria, in-

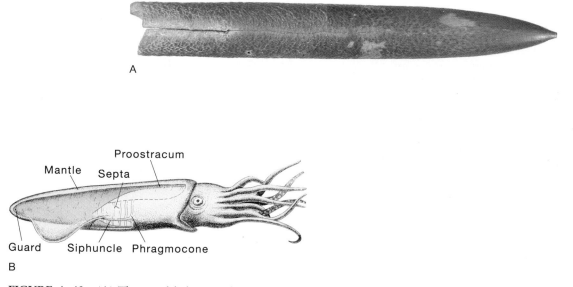

FIGURE 4–42 (A) The usual belemnite fossil consists only of the solid part of the internal skeleton called a guard. The guard here is 9.6 cm long. (B) Belemnites are extinct relatives of squid and cuttlefish. Shown here is an interpretation of the appearance of a belemnite as it would appear if living.

FIGURE 4–43 The circular pods depicted in this mural of the margins of a Precambrian sea are calcareous algal structures called stromatolites. *(Courtesy of the National Museum of Natural History.)*

cluding cyanobacteria (formerly called blue–green algae). Occasionally, filamentous organisms formed extensive algal mats (Fig. 4–43), and these may have produced important quantities of oxygen by photosynthesis. However, all life was at or below the unicellular level until about 1 billion years ago, when the world's first multicellular organisms left their trails and burrows in rocks that would later be called the *Torrowangee Group* of Australia. In rocks deposited about 0.7 billion years ago, fossil metazoans recognizable as worms, coelenterates, and arthropods have been found—albeit rarely—in scattered spots around the world, including the United States, Australia (Fig. 4–44), Canada, England, Russia, China, and South Africa. Thus, near the end of the Proterozoic, the

FIGURE 4–44 *Mawsonites*, a fossil found in the Pound Quartzite of Australia. The formation is late Proterozoic in age. *Mawsonites* is considered to be the mouth end of a jellyfish. *(Courtesy of B. N. Runnegar.)*

stage was set for a wide range of Paleozoic plants and animals.

The Evolutionary History of Animals

Some general observations are possible after a brief scanning of the history of life following the Precambrian. One notes that the principal groups of invertebrates appear early in the paleozoic (Fig. 4–45). Less advanced members of each phylum characterized the earlier geologic periods, whereas more advanced members came along later. Also, most of the principal phyla are still represented by animals today. In our review of fossil creatures, we are not startled by sudden appearances of bizarre or exotic animals and plants. Further, we are able to recognize periods of environmental adversity that caused extinctions. Such episodes were usually followed by much longer intervals of recovery and more or less orderly evolution. One final observation is that there has been a persistent gain in the overall diversity of life down through the ages.

After the Precambrian, life became so varied that it is difficult to describe it both briefly and adequately. We can, however, make some generalizations. Such marine invertebrate animals as trilobites, brachiopods, nautiloids, crinoids, horn corals, honeycomb corals, and twiglike bryozoans were abundant during the Paleozoic Era (Fig. 4–46). The Paleozoic was also the era when fishes, amphibians, and reptiles appeared and left behind a fascinating record of the conquest of the lands. Remains of more modern corals, diverse pelecypods, sea urchins, and ammonoids characterize the marine strata of the Mesozoic Era. However, the Mesozoic is best known as the "age of reptiles," when dinosaurs and their kin domi-

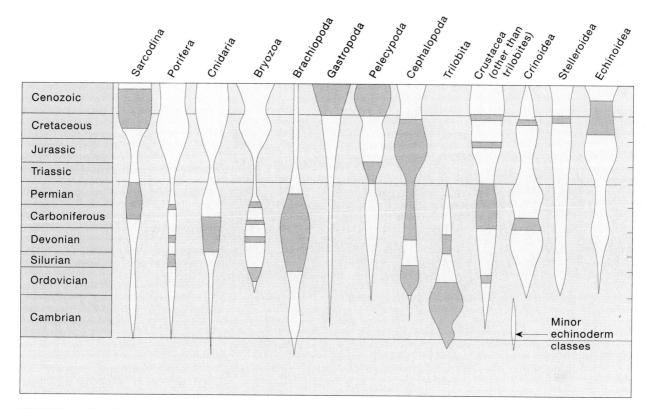

FIGURE 4–45 Geologic ranges and relative abundances of frequently fossilized categories of invertebrate animals. Width of range bands indicates relative abundance. Colored areas indicate where fossils of a particular category are widely used in zoning and correlation.

FIGURE 4–46 A large, straight-shelled cephalopod dominates this scene of an Ordovician sea floor. Also visible are corals, bryozoa, and crinoids. *(Courtesy of the National Museum of Natural History.)*

FIGURE 4–47 Cretaceous carnivorous dinosaurs in combat. *(Courtesy of the American Museum of Natural History.)*

nated the continents (Fig. 4–47). No less important than these big beasts, however, were the rat-sized primitive mammals and earliest birds that skittered about perhaps unnoticed by the "thunder beasts."

The mollusks were particularly well represented in the marine invertebrate faunas of the Mesozoic, and they have continued in importance into the Cenozoic. However, no ammonoid cephalopods have occurred in this most recent era. Rather, rocks of the Cenozoic are recognized by distinctive families of protozoans (foraminifers) and a host of modern-looking snails, clams, sea urchins, barnacles, and encrusting bryozoa. Because the Cenozoic saw the expansion of warm-blooded creatures such as ourselves, it is appropriately termed the "age of mammals."

The Evolutionary History of Plants

Long before the first animal appeared on Earth, there were plants. Plants had their origin among unicellular aquatic protistids of the Precambrian. Among these,

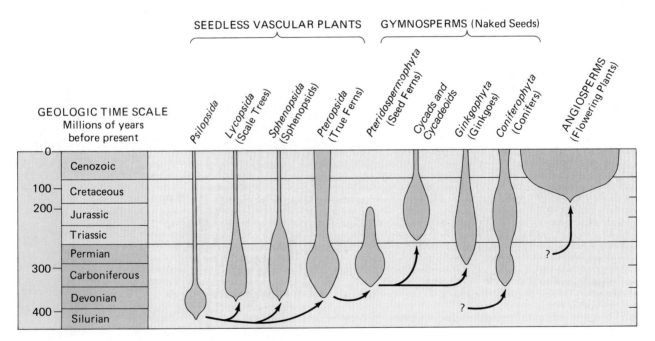

FIGURE 4–48 Geologic ranges, relative abundances, and evolutionary relationships of vascular land plants.

FIGURE 4–49 Foraminifers obtained from the Monterey Formation of Miocene age, California.

the green algae or chlorophytes are the most likely ancestors of vascular land plants. The first invasion of the land was made by plants that reproduced by means of spores (Fig. 4–48). From primitive forms that originated in the Silurian, vascular land plants expanded widely during the Devonian and Carboniferous, forming the bulk of the vegetation in the great coal forests. The most imposing of these spore bearers were the scale trees, whose close-set leaves left a pattern of scalelike scars on trunks and branches. The next group of plants to make their debut were pollen and seed producers. These so-called gymnosperms appeared in the Permian and were widely dispersed during the Mesozoic. Plants that bore seeds *and* flowers had evolved by Cretaceous time. Plants with both seeds and flowers are called angiosperms. They are by far the most abundant plants today and include such familiar trees as oak, maple, sassafras, and birch as well as the grasses that became the primary food source for evolving Cenozoic mammals.

■ *Fossils and the Search for Mineral Resources*

One might be surprised to learn that the majority of paleontologists in the world are not based in museums and universities but rather are industrial scientists using their knowledge of fossils in the search for oil and gas. In their investigations, they generally utilize only very small fossils because these are not likely to be broken by drilling tools. During the drilling process, samples of the rock being penetrated are returned to the surface in the drilling mud or in rock cores. The rock is then broken down, and the microfossils are extracted for study. By far, the most useful

microfossils are the protozoans known as *foraminifers* (Fig. 4–49). Other microfossils that have been used frequently in petroleum exploration include tiny, toothlike structures called conodont elements (see Fig. 4–52); small crustaceans known as **ostracodes** (Fig. 4–50); **pollen grains** and **spores;** and the calcareous remains of the golden brown marine calcareous algae named **coccolithophorids** (Fig. 4–51). Micropaleontologists, specialists in the study of microscopic fossils, prepare subsurface logs that depict the depth from which each species of fossil was obtained. From the accumulated information of many such paleontologic logs, one is able to identify the geologic period of the formations through which the drill penetrated. Particular time–rock units can be recognized by their unique index fossils. Knowledge

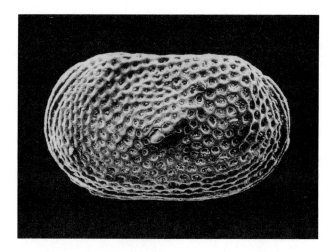

FIGURE 4–50 Electron micrograph of an ostracode recovered from Pennsylvanian shales during the drilling of an oil well. The specimen is 0.32 mm long.

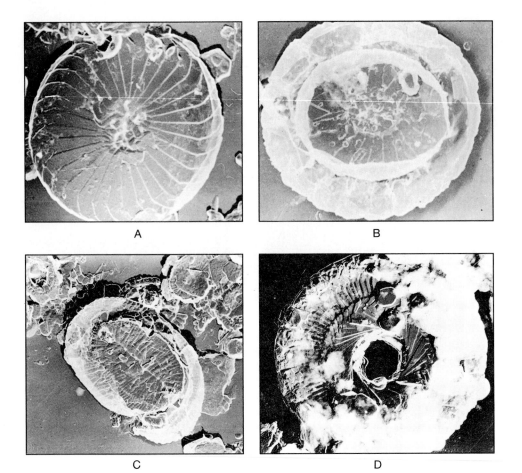

A

B

C

D

FIGURE 4–51 Coccoliths. (A) *Coccolithus leptoporous* (distal side). (B) *Coccolithus leptoporous* (proximal side). (C) *Helicosphaera carteri*. (D) *Cyclococcolithus* sp. For scale, D is 22 micrometers in diameter.

of the location of such key fossils relative to known oil-bearing strata is useful when wells are drilled in unproved territory. Indeed, the microfossils permit correlation from well to well and from area to area across entire petroliferous (oil-containing) provinces. The vertical successions of fossil assemblages in well samples often reflect changes in the depth of water in which the sediment was deposited and can therefore be used to infer tectonic conditions within the petroleum province. Along a single stratigraphic zone, knowledge of the depth of water preferences for microfossils may help in locating the porous reef or strand line sediments that are occasionally saturated with petroleum.

Fossils may also play a role in the exploration of resources other than oil and gas. Mining geologists use fossils to date strata above and below mineralized beds and to help determine directions of movements along faults that offset the ore-bearing strata. With the aid of fossils and careful field work, they are often able to pinpoint the geologic time in which valuable metals were emplaced and then to use that knowl-

edge to seek out ores in unexplored areas. Mapping of ancient algal reefs has shown how these structures may sometimes control the emplacement of metal sulfides.

Sometimes the study of fossils provides unexpected information that is useful in mineral exploration. For example, geologists working in the Appalachian Mountains became intrigued by the fact that a group of fossils called conodont elements exhibited a range of colors from pale yellow to black. Conodont elements (Fig. 4–52) are microscopic hard parts of organisms that lived about 560 to 200 million years ago. They are composed of apatite—the same mineral of which bone is made. The geologists suspected that the color changes were related to deep burial and attendant high temperatures in the sedimentary rocks containing the fossils. By plotting the conodont element color differences on maps, it became apparent that the colors changed systematically in the direction of rocks that had been most deeply buried. In the laboratory, heating of conodont elements provides the same range of colors, from pale yellow (low tem-

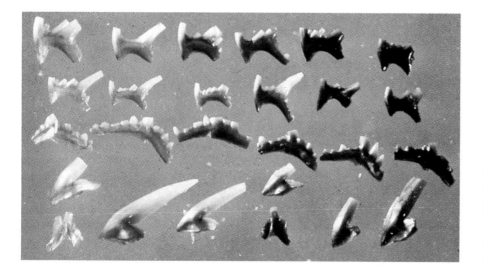

FIGURE 4–52 Conodont elements exhibiting varying degrees of color alteration resulting from deep burial and attendant increases in temperature. *(Courtesy of U.S. Geological Survey.)*

perature) to black (high temperature). Thus, conodont elements provide indices of the depth of burial. These colors are also an aid in exploration, for it has been shown that sedimentary rocks bearing black conodont elements are less likely to yield commercial quantities of oil and gas. This information is now used routinely by oil companies in their search for petroleum.

■ *Speculations About Life on Other Planets*

While reflecting upon the long and complex history of life on Earth, one might drift on to questions of whether or not other planets had—or now have—life as we know it. Is organic evolution a process that is unique to this planet? What properties of the Earth have made it a suitable place for the origin and evolution of life? Could these conditions exist elsewhere in the universe?

To begin with, the Earth is large enough that its gravitational attraction is sufficient to retain an atmosphere. The temperature of most of the Earth's surface is low enough to provide an abundance of water in its liquid form. In addition, that temperature is suitable for chemical reactions required for life processes and is itself a consequence of the size of our sun and its distance from the Earth. Our sun-star also has a life span sufficiently long to permit time for the emergence and evolution of life. Finally, the Earth has always had all of the chemical elements required for life processes.

Life in Our Solar System

We may first consider the possibilities for extraterrestrial life here in our own solar system. All evidence to date indicates that because of either size or distance from the sun, conditions are currently too harsh on neighboring planets to permit the evolution of *higher* forms of life. The Viking mission to Mars sampled the Martian surface for indications of life but found nothing to suggest the existence of present or past organisms on the red planet.

With regard to our moon, scientists had predicted that no evidence of life would be found. The moon lacks such important attributes as an atmosphere and water. The predictions were correct. The moon is sterile. Satellites of other planets in the solar system, however, appear to have a greater potential for harboring life. One of these is Europa, a satellite of Jupiter. Data from the Voyager flybys of Europa suggest that the satellite has a liquid water ocean covered by a thin, fractured crust of ice. It is only conjecture, but perhaps these waters harbor simple forms of life. By analogy, here on Earth certain species of algae and diatoms are known to be living in Antarctica in water that is overlain by impressive thicknesses of ice. They are able to photosynthesize even in the feeble light that has penetrated the overlying ice. Perhaps here and there beneath the icy surface of Europa, somewhat analogous conditions and organisms exist.

Life in the Universe

Thus, we see that the Earth is indeed biologically special when compared with other planetary bodies in our solar system. In contrast, however, it may not be so peculiar at all in the vast realms of the universe. Astronomers estimate that there are about 150 billion stars in our galaxy, and the number of galaxies now appears to be nearly limitless. Of those 150 billion stars in our galaxy, a not unreasonable estimate stipulates that at least 1 billion have planets with size and temperature conditions similar to the Earth's.

These are potentially habitable planets. For the entire known universe (of which our galaxy is but a small part), reputable scientists have estimated that there are as many as 10^{20*} planetary systems similar to our solar system. Such calculations indicate that it is probable that life does exist out there somewhere. Indeed, the universe may be rich in suitable habitats for life—but what kind of life? If we assume such life was formed from the same universal store of atoms and under physical conditions not too dissimilar to those that existed on Earth, then we might very well recognize it as a living thing. But it is highly unlikely that duplicates of humans, cows, or butterflies exist on other planets. There are many variables in the evolutionary interactions of genetics, environment, and time involved in the making of a particular species. To say that the very same mutations, genetic recombinations, and environmental conditions producing a sparrow could occur in precisely the same sequential steps in time on a distant planet seems most improbable.

COMMENTARY
Amber, the Golden Preservative

To many, the term *fossil* brings to mind remains of ancient organisms that have literally been turned to stone. In this chapter, however, a few rare instances of preservation of actual remains have been described. For such preservations, there is nothing superior to the organic substance known as amber (see Fig. 4–6). Insects are the usual organisms preserved in amber, and the bodies of insects, except for being desiccated, are preserved entirely. In addition to insects, spiders, crustaceans, snails, mammal hair, feathers, small lizards, and even a frog have been discovered in amber. By far the most abundant insects preserved in amber are flies, mosquitoes, and gnats of the Order Diptera. Paleontologists deplore the fact that during the late nineteenth century thousands of tons of raw amber were melted down for varnish, causing the loss of untold numbers of exquisitely preserved organisms.

Amber is a fossil resin produced by conifers. It is initially exuded from cracks and wounds in trees and, while still soft and sticky, traps and engulfs insects. The insect's struggle to escape can often be recognized in the swirl patterns around appendages. Later, through evaporation of the more volatile components, the soft resin hardens.

Although amber deposits are known in rocks ranging in age from Lower Cretaceous to the Holocene, the most famous are those found along the coast of the Baltic Sea southwest of the Gulf of Riga near the seaport of Kaliningrad (formerly Konigsberg). The conifer forests in this area were inundated during a marine transgression in early Oligocene time. Marine sediment buried the trees and their contained amber. Subsequently, chunks of amber eroded from the soft clays were distributed across adjacent areas by streams, wave action, and glaciers. Intrigued by their beauty, neolithic families are known to have gathered the rounded yellow and brown pieces of amber. Such prominent philosophers of antiquity as Aristotle, and later Pliny and Tacitus, described amber's physical and chemical properties. As a semiprecious gem, amber was transported along ancient trade routes from the Baltic region to the Mediterranean. Traders called amber "the gold from the north." Then as now, it was used in the manufacture of beads for jewelry. Amber beads, amulets, necklaces, and bracelets have been recovered from Etruscan tombs and from excavations in Mycenae, Egypt, and Rome. However, for modern women adorned in amber jewelry, a word of whimsical caution may be of interest. That person gazing intently at your necklace may be a paleontologist in search of an embalmed mosquito.

Reference
Poinar, G. O., Jr. 1992. *Life in Amber*. Stanford, CA: Stanford University Press.

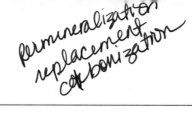

permineralization
replacement
carbonization

Darwin: simple ⟹ complex

Summary

Fossils are the remains or traces of life of the geologic past. The processes that are important in fossilization are varied and include the precipitation of chemical substances into pore spaces (permineralization), molecular exchange for substances that were once part of the organisms with inorganic substances (replacement), or compression of the animal and plant into a thin film of carbonized remains (carbonization). Tracks, trails, molds, and casts are additional varieties of fossils. Paleontologists, the specialists that study fossils, are aware that the fossil record is not complete for all forms of life that have existed on Earth. It is more complete for organisms that had hard parts such as teeth, bone, or shell and for organisms that lived on the floors of shallow seas, where the deposition of sediment is relatively continuous and rapid.

It has been repeatedly demonstrated that particular lineages of fossil organisms exhibit change with the passage of time. This observation is part of the evidence for Darwin's conclusion that life has developed from simple to more complex forms (evolution). Because of this change through time, rock layers from different periods can be recognized and correlated on the basis of the kinds of fossils they contain.

Our modern concept of evolution combines natural selection with hereditary mechanisms for change by means of recombinations of hereditary materials and mutations. The hereditary materials are chromosomes and genes. Chromosomes are located in the nuclei of cells, carry the genes that are composed of DNA, and are responsible for the total characteristics of an organism. A process called meiosis segregates the members of homologous pairs of chromosomes and halves the total number for each gamete. During fertilization, there is a random union of two gametes from parents of unlike sex that brings together assortments of chromosomes (and genes) from two parents, resulting in the production of individuals with different gene combinations. Changes in genes called mutations may occur. Changes may occur also by rearrangement of chromosomes. These changes and redistributions alter the assortment of genes and, hence, the characteristics that are passed on to succeeding generations.

Adaptation is the process by which organisms become fitted to their environments. Commonly, adaptations involve a combination of characteristics, including not only structure but physiology and behavior as well. Different vertebrates, for example, display adaptive modifications of limbs for either swimming or running, or of teeth for various kinds of food. The evolution and spread of a single lineage of organisms into several ecologic niches often results in different forms, each of which is distinctly adapted to a particular habitat. This process of diversification from a central stock is called adaptive radiation.

Within any large adaptive radiation, one finds many examples of an evolutionary pattern called divergence, in which a parent stock splits into two or more distinct lineages. Other evolutionary patterns include convergence, in which two morphologically different groups evolve similarities in morphology as they become adapted to similar environments, and iterative evolution, wherein organisms of similar morphology evolve more than once from the parental lineage.

Charles Darwin viewed evolution as a gradual stately change in lineages through time. This interpretation, called phyletic gradualism, persists today and is probably valid for many groups of organisms. In addition, however, the present generation of paleontologists are finding considerable evidence that evolution entails sudden advances that punctuate periods of relative stability. The process is termed punctuated equilibrium and is thought to occur when new species arise abruptly from isolated populations. The subsequent expansion of new species from these isolated groups results in the sudden faunal or floral changes often seen in the fossil record.

An important goal for paleontologists is to understand the complex relationships between ancient plants and animals and their habitats. This area of paleontology is termed paleoecology. Paleoecology can, in turn, provide information about the distribution of ancient lands and seas, ancient climates, depth of water, natural barriers to migration, and former locations of continents.

Because fossils are so useful in inferring environments of deposition as well as in correlating and determining relative geologic age of strata, they are widely employed by commercial paleontologists engaged in the search for fossil fuels. It is likely that this aspect of paleontology will continue in importance as the search for economically essential minerals intensifies. Paleontologists will continue their work in the future not only in such areas of exploration but possibly also in the detection of once-living things in planets other than the Earth. Statistics suggest the strong probability of present or past life somewhere in the universe.

Review Questions

1. What is a fossil? What are guide fossils? What conditions must a fossil have in order to be considered a guide fossil?

2. What relationship exists between guide fossils and rate of evolution? Between guide fossils and rate of dispersion?

3. Fossil *A* occurs in rocks of Cambrian and Ordovician age. Fossil *B* occurs in rocks that range in age from early Ordovician through Permian. Fossil *C* is found in Mississippian through Permian strata.
   a. What is the maximum possible range of age for a stratum containing only fossil *B*?
   b. What is the maximum possible range for a rock containing both *A* and *B*?
   c. Which is the better guide fossil, *A* or *C*?

4. A chronostratigraphic (time–rock) unit contains a different fossil assemblage at one locality than it does at another located 100 miles away. Suggest a possible cause for the dissimilarity.

5. In drilling for oil, geologists recover Triassic pollen grains in a stratum known to be of Cretaceous age. Explain this occurrence.

6. What are the major stages in the evolution of vascular land plants?

7. What are the usual or most common replacing substances seen in fossilization by replacement?

8. What are the differences between haploid and diploid cells? Between meiosis and mitosis? Between a gymnosperm and an angiosperm?

9. What is meant by the term *adaptation*? Cite an example of an adaptive radiation.

10. What are peripheral isolates? Why do new species commonly arise from peripheral isolates?

11. Why is it that fossils of marine invertebrates are far more abundant than fossils of terrestrial invertebrates?

12. What is convergent evolution? Cite an example.

Discussion Questions

1. How has the science of paleontology contributed to the following?
   a. Validation of the theory of biological evolution
   b. Recognition of geographic changes in the geologic past
   c. Recognition of climatic changes in the geologic past
   d. The search for fossil fuels

2. Defend the hypothesis that, although there may be life outside the Earth somewhere in the universe, there will probably not be humans precisely like ourselves.

3. What were the contributions of Darwin and Mendel to our modern concept of organic evolution?

4. Why is the determination of a fossil species more subjective than the determination of a living species?

5. Using fossils for age correlation is dependent on *a priori* knowledge of their age ranges. How is this knowledge obtained?

6. Distinguish between the concepts of phyletic gradualism and punctuated equilibrium. How did Charles Darwin (and others) account for rapid or abrupt appearances of new species? Which, then, would be a more appropriate geologic section to study for proof of punctuated equilibrium—a continuous set of cores from the floor of the ocean or a section on the continent where there has been repeated uplift and erosion through geologic time?

7. Jean Baptiste de Lamarck's theory of evolution stipulated that structures acquired by individuals during their lives in response to need are then inherited by succeeding generations. Why is this not possible?

8. Evolution is in many ways an opportunistic process in which chance plays an important role. Why is this true?

Supplemental Readings and References

Boardman, R. S., Cheetham, A. H., and Rowell, A. J. (eds.). 1987. *Fossil Invertebrates*. Oxford, England: Blackwell Scientific Publications.

Briggs, D. E. G., and Crowther, P. R. (eds.). 1990. *Paleobiology*. Oxford, England: Blackwell Scientific Publications.

Carroll, R. L. 1988. *Vertebrate Paleonotology and Evolution*. Oxford, England: Blackwell Scientific Publications.

Clarkson, E. N. K. 1993, 3rd ed. *Invertebrate Paleontology and Evolution*. London: Geo. Allen & Unwin, Ltd.

Cowen, R. 1993, 2nd ed. *History of Life*. Palo Alto, CA: Blackwell Scientific Publications.

Darwin, C. 1859. *The Origin of Species* (1963 ed.). Introduction by H. L. Carson. New York: Washington Square Press.

Dodd, R. J., and Stanton, R. J., Jr. 1981. *Paleoecology, Concepts and Applications*. New York: John Wiley.

Donovan, S. K. (ed.) 1992. *The Process of Fossilization*. New York: Columbia University Press.

Eldredge, N., and Gould, S. J. 1972. Punctuated equilibria: An alternative to phyletic gradualism. In *Models in Paleobiology*. Schopf, T. J. M. (ed.) San Francisco: W. H. Freeman.

Gillette, D. D., and Lockley, M. G. (eds.) 1989. *Dinosaur Tracks and Traces*. Cambridge, England: Cambridge University Press.

Gould, S. J. 1982. *The Panda's Thumb*. New York: W. W. Norton & Co.

MacFadden, B. J., Bryant, J. D., and Mueller, P. A. 1991. Sr-isotopic, Paleomagnetic, and biostratigraphic calibration of horse evolution: Evidence from the Miocene of Florida. *Geology* 19:242–245.

Skinner, B. J. (ed.) 1981. *Paleontology and Paleoenvironments*. Readings from *American Scientist*. Los Altos, CA: Wm. Kauffman, Inc.

Stearn, C. W., and Carroll, R. L. 1989. *Paleontology: The Record of Life*. New York: John Wiley.

Stewart, W. N. 1983. *Paleobotany and the Evolution of Plants*. Cambridge, England: Cambridge University Press.

The Andes Mountains as seen from Pehoe Lake, Torres Del Paine National Park, Chile. The Andes formed at a convergent tectonic plate boundary when the westward-moving South American plate collided with the Pacific plate.

(Tom Till.)

The deep interior of the Earth is inaccessible, and no rays of light penetrate to let us see what is below the surface. But rays of another kind penetrate and carry with them their messages from the interior.

Inge Lehmann, 1959[*]

CHAPTER 5

Earth Structure and Plate Tectonics

Before turning to the study of the geologic history of the Earth, it is necessary to understand the basic internal structure of our planet. In this age of artificial satellites and spacecraft, our attention is often directed toward the remarkable discoveries resulting from the exploration of outer space. We tend to forget that there is still much to learn about the "inner space" that lies beneath us. Our deepest wells penetrate only about 10 km of the 6300 km that separate us from the center of the planet. Only seismic waves permit us to "see" into the Earth's interior and to discern its boundaries and physical properties. Those properties bear directly on the forces that have deformed the Earth's crust, that trigger earthquakes, cause volcanic eruptions, and drive gargantuan plates of the planet's outermost

157

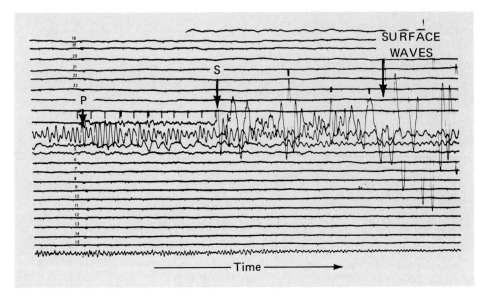

FIGURE 5-1   Record of a magnitude 6 earthquake that occurred in Turkey on March 28, 1964 and was recorded by a vertical component seismograph located in northwestern Canada. Time increases from left to right. Small tick marks represent 1-minute intervals. P means primary waves, S indicates secondary waves. Notice that the first S-waves arrive about 10 minutes after the arrival of the first P-waves.   *(Courtesy of the Earth Physics Branch, Department of Energy, Mines, and Resources, Canada.)*

layer from place to place around the globe. The formation of these plates, how they are formed, how they are destroyed, and how they interact are components of the process called plate tectonics.

■ *Seismic Waves*

The "messages from the interior" mentioned in the epigram at the beginning of this chapter are, of course, seismic waves. Seismic waves permit one to determine the location and thickness and some of the properties of the Earth's internal zones. They are generated when rock masses are suddenly disturbed, as when they break or rupture. Vibrations spread out in all directions from the source of the disturbance. They move outward in waves that travel at different speeds through materials that differ in chemical composition and physical properties. The principal categories of these waves are **primary, secondary,** and **surface.**

Primary waves, or P-waves, are the speediest of the three kinds of waves and therefore the first to arrive at a seismograph station after there has been an earthquake (Fig. 5–1). They travel through the upper crust of the Earth at speeds of 4 to 5 km per second. Near the base of the crust, they speed along at 6 or 7 km per second. In these primary waves, pulses of energy are transmitted in such a way that

the movement of rock particles is parallel to the direction of propagation of the rock itself. Thus, a given particle of rock set in motion during an earthquake is driven into its neighbor and bounces back. The neighbor strikes the next particle and rebounds, and subsequent particles continue the motion. The result is a series of alternate compressions and expansions that speed away from the source of shock. Thus, P-waves are similar to sound waves in that they are *longitudinal* and travel by compression and rarefaction (Fig. 5–2). Vibration is an accordion-like "push–pull" movement that can be transmitted

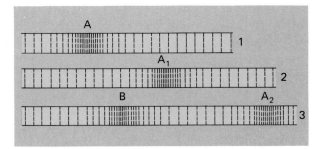

FIGURE 5–2   Movement of primary waves. In 1, the compression has moved the particles closer together at A. In 2, the zone of compression has moved to A_1. In 3, the zone of compression has moved to A_2 and a second compressional zone (B) has moved in from the left.

C O M M E N T A R Y
Recording Earthquakes

It is easy to record the passage of an automobile or train as we stand in one place and observe their movement. We have objects on the ground to use as reference in relation to which the motion can be readily seen. However, during an earthquake, we move in unison with the Earth and there is no stationary frame of reference. To solve this problem, the **seismograph** was developed. Seismographs are the pre-eminent tools for examining the interior of the Earth. They make use of the property of *inertia*, or the tendency of a heavy object to "stay put" as the Earth moves around it. To accomplish this feat, the heavy body is suspended from springs and flexible wires, which serve to absorb Earth movements. The heavy body is able to lag behind, maintaining its position as the Earth moves around it.

To keep the heavy body or weight as independent of Earth motions as possible, there are several different designs of seismographs. For recording vertical movements, the seismograph contains a weight suspended by springs from an overhead support (see accompanying figure). Horizontal movements are recorded by seismographs containing a horizontal pendulum having a weight on a rigid arm supported by a wire. The pendulum is freely attached to a socket joint from the supporting column. In both devices, the weight remains motionless while the support vibrates with the Earth. The difference in motion can be recorded in various ways. One method employs a narrow beam of light emanating from the weight. The beam of light traces a continuous line on photographic paper wrapped around a revolving drum that is fixed to the supporting framework of the seismograph. The drum turns as a clock so that the arrival times and durations of vibrations can be determined.

The accompanying figure illustrates the general principle of the seismograph. Actual seismographs located in earthquake monitoring stations are far more complex. The ray of light used to record vibrations is reflected off a series of mirrors that serve to amplify even very small motions of the ground. Other devices in the seismograph dampen the natural swinging motion of the pendulum and filter out unrelated background motions. To analyze all movements of earthquakes fully, at least two horizontal and one vertical seismograph must be in simultaneous operation.

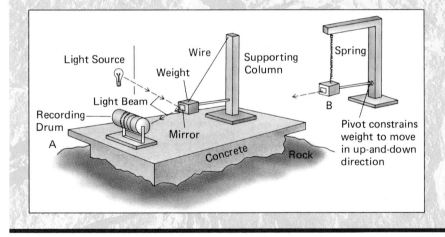

Principle of seismographs recording horizontal (A) and vertical (B) Earth motion. A spot of light on the boom moves across photographic paper on the recording drum as the boom oscillates. (Springs are omitted for clarity.)

through solids, liquids, and gases. Of course, the speed of P-wave transmission will differ in materials of different density and elastic properties. P-waves tend to die out with increasing distance from the earthquake source and will echo or reflect off rock masses of differing physical properties.

Secondary waves, which are also termed S-waves or transverse waves, travel 1 to 2 km per second slower than do P-waves. The movement of rock particles in secondary waves is at right angles to the direction of propagation of the energy (Fig. 5–3). A demonstration of this type of wave is easily managed by tying a length of rope to a hook and then shaking the free end. A series of undulations will develop in the rope and move toward the hook—that is, in the direction of propagation. Any given particle or point along the rope, however, will move up and down in a direction perpendicular to the direction of propagation. It is because of their more complex motion that S-waves travel more slowly than P-waves. They are the second group of oscillations to appear on the seismogram (see Fig. 5–1). Unlike P-waves, secondary waves will not pass through liquids or gases.

Both P- and S-waves are sometimes also termed **body waves** because they are able to penetrate deep into the interior or body of our planet. Body waves travel faster in rocks of greater elasticity, and their speeds therefore increase steadily as they move downward into more elastic zones of the Earth's interior and then decrease as they begin to make their ascent toward the Earth's surface. The change in velocity that occurs as body waves invade rocks of different elasticity results in a bending or refraction of the wave. The many small refractions cause the body waves to assume a curved travel path through the Earth (Fig. 5–4).

Not only are body waves subjected to refraction, but they may also be partially reflected off the surface of a dense rock layer in much the same way as light is reflected off a polished surface. Many factors influence the behavior of body waves. An increase in the temperature of rocks through which body waves are traveling will cause a decrease in velocity, whereas an increase in confining pressure will cause a corresponding increase in wave velocity. In a fluid where no rigidity exists, S-waves cannot propagate and P-waves are markedly slowed.

Surface waves are large-motion waves that travel through the outer crust of the Earth. Their pattern of movement resembles that of waves caused when a pebble is tossed into the center of a pond. They develop whenever P- or S-waves disturb the surface of the Earth as they emerge from the interior. There are actually several different types of motion in surface waves. *Rayleigh surface waves*, for example, have an

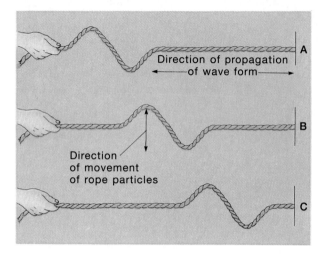

FIGURE 5–3   Analogy of propagation of S-waves by displacement of a rope. In rocks, as in this rope, particle movement is at right angles to the direction of propagation of the wave. A, B, and C show the displacement of the crest from left to right at successive increments of time.

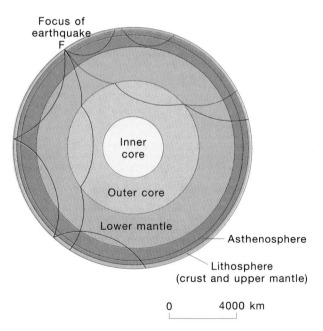

FIGURE 5–4   Cross-section of the Earth showing paths of some earthquake waves. P-waves (black) that penetrate to the core are sharply refracted, as shown in the path from F to 1. S-waves (red) end at the core, although they may be converted to P-waves, traverse the core, and emerge in the mantle again as P- and S-waves. Both P- and S-waves may also be reflected back into the Earth again at the surface.   *(From U.S. Geological Survey publication,* The Interior of the Earth.*)*

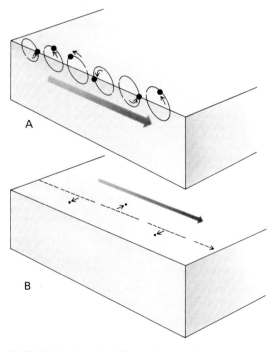

FIGURE 5–5   (A) Elliptical particle motion in Rayleigh surface waves (ground surface shown without displacement). (B) Particle motion in Love wave is in horizontal plane and perpendicular to the direction of wave propagation.

elliptical motion that is opposite in direction to that of propagation, whereas *Love surface waves* vibrate horizontally and perpendicular to wave propagation (Fig. 5–5). Surface waves are the last to arrive at a seismograph station. They are usually the primary cause of the destruction that can result from earthquakes affecting densely populated areas (Fig. 5–6). This destruction results because surface waves are channeled through the thin outer region of the Earth, and their energy is less rapidly dissipated into the large volumes of rock traversed by body waves. Indeed, surface waves may circle the Earth several times before friction causes them to fade. As is the case with ordinary water waves, the motion of surface waves diminishes with depth. It has been demonstrated that the total depth to which surface wave motion can exist is about equal to the distance between two surface wave crests. Thus, surface waves that are 2000 meters from crest to crest will shake a section of crust about 2000 meters thick.

The Divisions of Inner Space

Most of what we know about the Earth's deep interior is derived from the interpretation of countless seismograms documenting recent and past earthquakes around the globe. From such studies, geophysicists have found evidence of the gradual change in rock properties with depth as well as the relatively abrupt boundaries between major internal zones. Boundaries where seismic waves experience an abrupt change in velocity or direction are called **discontinuities.** Two widely known breaks of this kind

FIGURE 5–6 Damage in the Marina district of San Francisco caused by the October 1989 La Prieta earthquake. The first story of this three-story building failed during the earthquake, causing the second story to collapse, leaving only the third story. *(Courtesy U.S. Geological Survey, photograph by D. Perkins.)*

are named, after their discoverers, the Mohorovičić and Gutenberg discontinuities.

The discontinuity discovered by **Mohorovičić** (pronounced Mō-hō-rō-vĭtch′-ĭtz) was based on his observation that seismograph stations located about 150 km from an earthquake received earthquake waves sooner than those nearer to the focus. Mohorovičić reasoned that below a depth of about 30 km, there must be a zone having physical properties that permit earthquake waves to travel faster. That layer is the upper mantle.

In Figure 5–7, the shallow direct waves (A), although traveling at only about 6 km per second, are

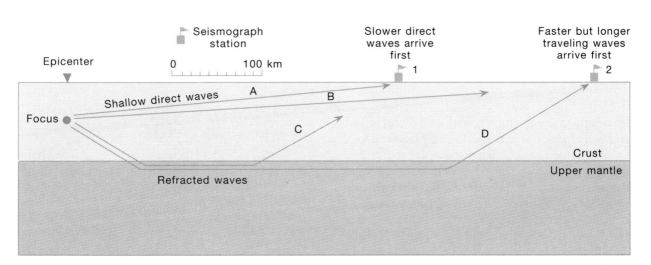

FIGURE 5–7 Mohorovičić's conclusion about the location of the base of the Earth's crust was based on this interpretation of the travel paths of early- and late-arriving seismic body waves. (*Focus* is the true center of an earthquake and the point at which the disturbance originates. *Epicenter* is a point on the Earth's surface above the focus.)

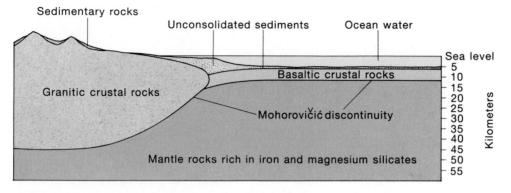

FIGURE 5–8   Generalized cross-section showing location of the Mohorovičić discontinuity.

the first to arrive at the closer seismograph station. At the farther station, however, the deeper but faster traveling wave (D) has caught up and moved ahead of the shallow direct wave and is therefore the first to arrive at seismograph station 2. The situation is analogous to what many of us do when we take a longer highway route to our destination than a more direct route through city streets because the greater speed on the highway allows us to arrive at the destination sooner.

The Mohorovičić discontinuity (or Moho, for short) lies at about 30 or 40 km below the surface of the continents and at lesser depths beneath the ocean floors (Fig. 5–8). It marks the boundary between the crust and the mantle.

The **Gutenberg discontinuity** is located nearly halfway to the center of the Earth at a depth of 2900 km. Its location is marked by an abrupt decrease in P-wave velocities and the disappearance of S-waves. The Gutenberg discontinuity marks the outer boundary of the Earth's core (Fig. 5–9).

The Core

Inferences from Body Waves

As just noted, the boundary of the core was determined by the study of earthquake waves. Seismology has also provided a means for discerning subdivisions of the core and deciphering some of its physical properties. For example, at a depth of 2900 km (the core boundary), the S-waves meet an impenetrable barrier, while at the same time P-wave velocity is drastically reduced from about 13.6 km/second to 8.1 km/second. Earlier, we noted that S-waves are unable to travel through fluids. (Fluids cannot sustain shear.) Thus, if they were to enter a fluid-like region of the Earth's interior, they would be absorbed there and would not be able to continue. Geophysicists believe

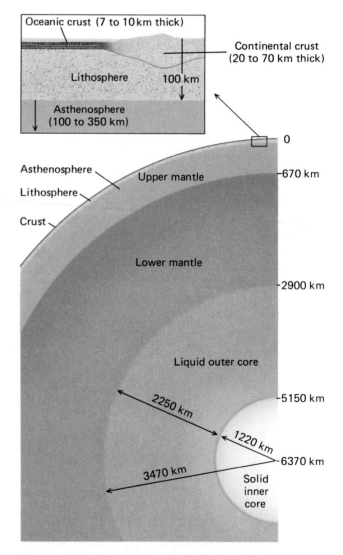

FIGURE 5–9   The structure of planet Earth.   *(From Thompson, G. R., and Turk, J. 1993. Modern Physical Geology. Philadelphia: Saunders College Publishing.)*

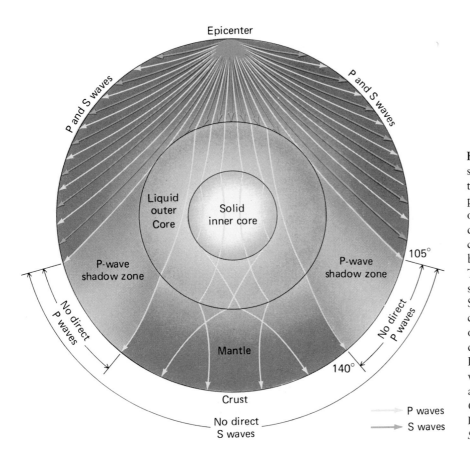

FIGURE 5–10 Refraction of seismic waves as they travel through the Earth. The travel paths bend gradually because of increasing velocity with depth. Abrupt bending or discontinuities occur at the boundaries of major zones. The S-wave shadow zone results from absorption of S-waves in the liquid outer core. P-waves entering the outer core are slowed and bent downward, giving rise to the P-wave shadow zone within which neither P- nor S-waves are received. *(From Thompson, G. R., and Turk, J. 1993.* Modern Physical Geology. *Philadelphia: Saunders College Publishing.)*

this is what happens to S-waves as they enter the outer part of the core. As a result, the secondary waves generated on one side of the Earth fail to appear at seismograph stations on the opposite side, and this observation is the principal evidence for an outer core that behaves as a fluid. The outer core barrier to S-waves results in an *S-wave shadow zone* on the side of the Earth opposite an earthquake. Within the **shadow zone,** which begins 105° from the earthquake focus, S-waves do not appear.

Unlike S-waves, primary waves are able to pass through liquids. They are, however, abruptly slowed and sharply refracted as they enter a fluid medium. Therefore, as primary seismic waves encounter the molten outer core of the Earth, their velocity is checked and they are refracted downward. The result is a *P-wave shadow zone* that extends from about 105° to 140° from the focus (Fig. 5–10). Beyond 140°, P-waves are so tardy in their arrival that they further validate the inference that they have passed through a liquid medium. At the upper boundary of the core, P-waves are also reflected back toward the Earth's surface. Such P-wave echos are clearly observed on seismograms.

The radius of the core is about 3500 km. The inner core is solid and has a radius of about 1220 km, which makes this inner core slightly larger than the

moon. A transition zone approximately 500 km thick surrounds the inner core. Most geologists believe that the inner core has the same composition as the outer core and that it can exist only as a solid because of the enormous pressure at the center of the Earth.

Evidence for the existence of a solid inner core is derived from the study of hundreds of seismograms produced over several years. These studies showed that weak, late-arriving primary waves were somehow penetrating to stations that were within the P-wave shadow zone. Geophysicists recognized that this penetration could be explained by assuming that the inner core behaved seismically as if it were solid.

Composition of the Core

The Earth has an overall density of 5.5 g/cm^3, yet the average density of rocks at the surface is less than 3.0 g/cm^3. This difference indicates that materials of high density must exist in the deep interior of the planet to achieve the 5.5 g/cm^3 overall density. Calculations indicate that the rocks of the mantle have a density of about 4.5 g/cm^3 and that the average density of the core is about 10.7 g/cm^3. Under the extreme pressure conditions that exist in the region of the core, iron mixed with nickel would very likely have the required high density. In fact, laboratory experiments suggest that a highly pressurized iron–

A

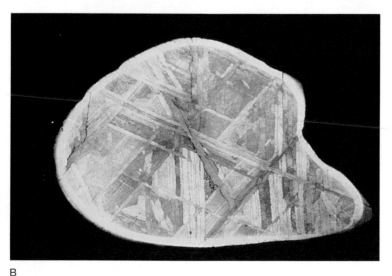

B

FIGURE 5–11 (A) Collecting an iron meteorite on the Antarctic ice. (B) Cut and polished iron meteorite found in Mohave County, Arizona, showing the interesting pattern of interlocking crystals, called the Widmanstatten pattern. *(A, courtesy of NASA; B, from the Washington University collection.)*

nickel alloy might be too dense and that minor amounts of such elements such as silicon, sulfur, or oxygen may also be present to "lighten" the core material.

Support for the theory that the core is composed of iron (85 percent) with lesser amounts of nickel has come from the study of meteorites. Many of these samples of solar system materials are iron meteorites (Fig. 5–11) that consist of metallic iron alloyed with a small percentage of nickel. Some geologists believe that iron meteorites may be fragments from the core of a shattered planet. Their presence in our solar system suggests that the existence of an iron–nickel core for the Earth is plausible.

There is another kind of evidence for the Earth's having a metallic core. Anyone who understands the functioning of an ordinary compass is aware that the Earth has a magnetic field. The planet itself behaves as if there were a great bar magnet embedded within it at a small angle to its rotational axis. Geophysicists believe that the Earth's magnetic field may, in some way, be associated with electric currents. This interpretation is favored by the discovery, over 160 years ago, that a magnetic field is produced by an electric current flowing through a wire. The silicate rocks of the lithosphere and mantle are not good conductors, however, and therefore unlikely materials for the development of electromagnetism. In contrast, iron is an excellent conductor. If scientists are correct in the inference made from density considerations and study of meteorites that the core is mostly iron, then the magnetic field lends credence to that inference.

The Mantle

Materials of the Mantle

As was the case with the Earth's core, our understanding of the composition and structure of the **mantle** is based on indirect evidence. As inferred from seismic data, the mantle's average density is about 4.5 g/cm^3, and it is believed to have a stony, rather than metallic, composition. Oxygen and silicon probably predominate and are accompanied by iron and magnesium as the most abundant metallic ions. The iron- and magnesium-rich rock **peridotite** (Fig. 5–12) approximates fairly well the kind of material inferred for the mantle. A peridotitic rock not only would be appropriate for the mantle's density but also is similar in composition to stony meteorites as well as rocks that are thought to have reached the Earth's surface from the upper part of the mantle itself. Such

suspected mantle rocks are indeed rare. They are rich in olivine and pyroxenes and contain small amounts of certain minerals, including diamonds, that can form only under pressures greater than those characteristic of the crust.

Layers of the Mantle

The mantle is not merely a thick, homogeneous layer surrounding the core but is itself composed of several concentric layers that can be detected by studying earthquake data. In Figure 5–13, note that within the mantle there are three zones of increase in wave velocity. These sudden increases cannot be explained simply as the result of pressure increases with depth. Such increases would be too gradual to cause seismic wave velocity to increase so abruptly. There must, therefore, be some change in the physical nature of the material. One of the zones is encountered at about 400 km and is taken to mark the base of the **upper mantle**. Beneath the upper mantle is the **transition zone**, which extends downward to about

FIGURE 5–12 Peridotite. The green minerals are olivine and spinel, and the black grains are pyroxene. The specimen was collected at Kilbourne Hole maar, Dona Ana County, New Mexico. A *maar* is a crater formed during a violent volcanic explosion. The explosion brought chunks of mantle rock to the Earth's surface.

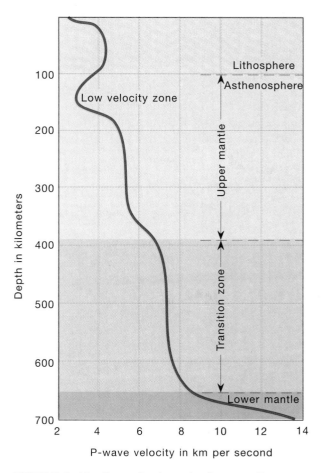

FIGURE 5–13 Generalized graph of average P-wave velocities versus depth in the outer 70 km of the Earth, showing the location of the low-velocity zone.

650 km. The **lower mantle** lies beneath the transition zone and above the core.

The composition of the lower mantle is difficult to infer. A plausible guess that is supported by seismic evidence is that it consists mostly of silicates and oxides of magnesium and iron. In the high-temperature and high-pressure environment of the lower mantle, iron, magnesium, silicon, and oxygen atoms are rearranged into denser and more compact crystals. For example, near the Earth's surface, olivine and pyroxene are relatively stable minerals. However, in the high-temperature and high-pressure environment below a depth of about 400 km, olivine is likely to be converted to a form whose atoms are more closely packed (Fig. 5–14). Similar dense crystals have been formed from olivine in high-pressure laboratory experiments. Rocks composed of such dense materials would be capable of causing the increase in seismic velocities that characterize the mantle.

The upper mantle is of particular importance because its evolution and internal movements affect the geology of the crust. The most remarkable feature of the upper mantle is the **low-velocity zone.** As suggested by its name, this is a region in which there is a decrease in the speed of S- and P-waves (see Fig. 5–13). The low-velocity zone occupies an upper region of the larger **asthenosphere.** The rock here is near or at its melting point, and magmas are generated. Above the asthenosphere is the outer rigid shell of the Earth known as the **lithosphere.** The lithosphere contains the oceanic and continental crust.

Geophysicists believe that seismic waves are slowed in the low-velocity zone not because of a decrease in density, but rather because they enter a region that is in the state of a crystalline–liquid mixture. In such a "hot slush," perhaps 1 to 10 percent of the material would consist of pockets and droplets of molten silicates. Such material would be capable of considerable motion and flow and would serve as a slippery, mobile layer on which overlying lithosphere plates could move.

■ *The Crust*

The **crust** of the Earth is seismically defined as all of the solid Earth above the Mohorovičić discontinuity. It is the thin, rocky veneer that constitutes the continents and the floors of the oceans. The crust is not a homogeneous shell in which low places were filled with water to make oceans and higher places make continents. Rather, there are two distinct kinds of crust that, because of their distinctive compositions and physical properties, determine the very existence of separate continents and ocean basins.

The Oceanic Crust

Beneath the varied topography of the ocean floors lies an **oceanic crust** that is approximately 5 to 12 km thick and has an average density of about 3.0 g/cm^3. Three layers of this oceanic crust can be recognized. On the upper surface is a thin layer of unconsolidated sediment that rests on the irregular surface of the igneous basement layer. The second layer consists of basalts that had been extruded under water (Fig. 5–15). The nature of the deepest layer of oceanic crust is not clear. Many suspect that it is metamorphosed basaltic mantle material that has become somewhat less dense by chemically combining with sea water. In rocks of the oceanic crust, we find a concentration of

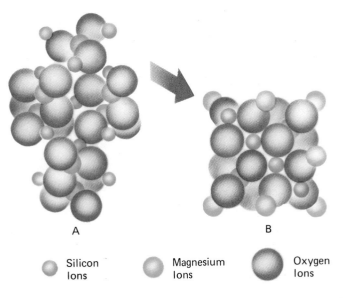

A　　　　　　　B

Silicon Ions　　　Magnesium Ions　　　Oxygen Ions

FIGURE 5–14 High- and low-pressure forms of the mineral olivine. (A) At shallow depths in the mantle, olivine is stable in this form, which is the form of olivine found at the Earth's surface. (B) When the pressure reaches a critical value corresponding to a depth of about 400 km, the molecule collapses into a more dense form in which oxygen ions are more closely packed.

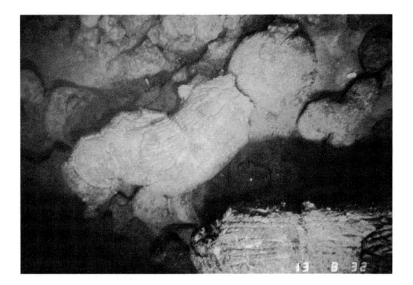

FIGURE 5–15 Pillow structures formed by lavas being extruded on the floor of the Pacific Ocean along the Juan de Fuca ridge, a plate tectonic spreading center west of Oregon. Photo covers 5 meters. *(Courtesy of W. R. Normark and the U.S. Geological Survey.)*

such common elements as iron, magnesium, and calcium. These elements are included in the plagioclase feldspars, amphiboles, and pyroxenes of basalts. The upper mantle is the ultimate source for the lavas that formed the oceanic crust.

The Continental Crust

Properties of the Continental Crust

At the boundaries of the ocean basins, the Mohorovičić discontinuity plunges sharply beneath the thicker **continental crust** (see Fig. 5–8). The depth of the Moho beneath the continents averages about 35 km, although it may be considerably deeper or shallower in particular regions. The continental crust is not only thicker than its oceanic counterpart but also less dense, averaging about 2.7 g/cm^3. As a result, continents "float" higher on the denser mantle than the adjacent oceanic crustal segments. Somewhat like great stony icebergs, the roots of continents extend downward into the mantle.

Origin of the Continental Crust

The condition of balance that exists among segments of the Earth's crust as they come into flotational equilibrium with denser mantle material is called **isostasy.** Because of isostasy, the continental crust stands higher than the denser oceanic crust. Isostasy also explains why mountain ranges experience repeated uplifts following erosional removal of great volumes of rock and soil. As the heavy burden of eroded material is carried away by streams, the ranges rise in much the same way that a boat rises in the water as its cargo is unloaded.

A little "bathtub geology" will help illustrate the concept of isostasy. Assume that you have a large cube of low-density (light) wood, such as pine, and that the pine represents a continent. Another cube of heavy oak is used to represent the denser ocean crust. When placed in a tub of water, the pine will float higher than the oak. Next, place an identical second block of pine precisely on top of the first, and note that the pine blocks will sink somewhat, but the vertical difference between the upper surface of the pine and oak cubes will be greater than before. Thus, both the density of the blocks of material and their size (thickness) are important in determining the level of flotational equilibrium. Imagine further that the upper pine block represents a mountainous region of the continent. Lift the upper pine block off the lower, and notice that the lower block will rise to a new level much as a mountain range might rise when erosional processes have removed its "upper block" and upset the gravitational balance.

The lower density of the continental crust results from its composition. Although it is referred to as being granitic, it is really composed of a variety of rocks that approximate granite in bulk composition. Igneous continental rocks are richer in silicon and potassium and poorer in iron, magnesium, and calcium than igneous oceanic rocks. In addition, extensive regions of the continents are blanketed by sedimentary rocks.

■ Crustal Structures

Continents, mountains, and deep oceans appear to us to be everlasting. The human life span is far too short to permit a view of lands in upheaval, continents splitting and colliding, and ocean basins that expand

FIGURE 5–16   Close-up view of the edge of a fault plane that cuts through beds of sandstone and siltstone near Dry Gulch, Idaho. Beds on the right have slipped downward relative to beds on the left. This is a normal fault. There is a small splinter fault visible just to the left of the larger fault.   *(Courtesy of U.S. Geological Survey.)*

and contract. People are allowed time only to witness an occasional manifestation of these changes, such as an earthquake or volcanic eruption. The effects of the changes, however, are recorded in rocks as they are deformed and broken. Among the effects of Earth movements are faults and folds, a brief discussion of which will provide background for the subsequent topic of plate tectonics.

Faults

A **fault** is a break in the Earth's crustal rocks along which there has been movement (Figs. 5–16 and 5–17). According to the relative movement of the rocks on either side of the plane of breakage, faults can be classified as **normal, reverse,** or **lateral** (Fig. 5–18). Lateral faults can be further designated as *right lateral* or *left lateral* if one looks across the fault zone to see if the opposite block has moved to the right or left. The San Andreas fault in California is a strike–slip fault of the right-lateral type (Fig. 5–19). In normal faults, the mass of rock that lies above the shear plane, called the **hanging wall,** appears to move downward relative to the opposite side or foot wall. Such faults occur where rocks are subjected to tensional forces—forces that tend to stretch the crust. On the other hand, reverse faults exhibit a hanging wall that has moved up relative to the **foot wall.** If we imagine ourselves holding two blocks cut from wood like those of Figure 5–18, we find that they must be shoved together, or compressed, to cause reverse as well as thrust faulting. Thus, regions of the Earth's crust containing numerous reverse faults (and folded strata) are likely to have been compressed at some time in the geologic past. Reverse faults in which the shear zone is inclined only a few degrees from horizontal are termed **thrust faults.** Although examples of all kinds of faults may be found in any mountain belt, compressional structures are decidedly the most prevalent in the world's great mountain ranges.

FIGURE 5–17   A mountain range formed by faulting—the Grand Tetons of Wyoming viewed from Signal Mountain. The Grand Tetons were formed by normal faulting. The western block was uplifted to create the mountains, whereas the east block (represented by the foreground) dropped to form Jackson Hole with its picturesque Jenny Lake. Grand Teton at the extreme left rises over 2000 meters above the valley floor of the downfaulted block.

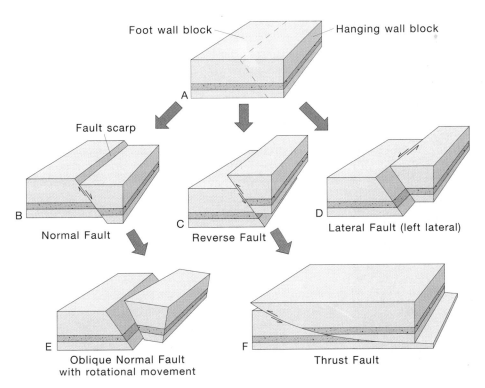

Foot wall block   Hanging wall block

A

Fault scarp

B   Normal Fault

C   Reverse Fault

D   Lateral Fault (left lateral)

E   Oblique Normal Fault with rotational movement

F   Thrust Fault

FIGURE 5–18 Types of faults. (A) The unfaulted block with the position of the potential fault shown by dashed line. In nature, movements along faults may vary in direction, as shown in (E). A thrust fault (F) is a type of reverse fault that is inclined at a low angle from the horizontal.

FIGURE 5–19 Air photo mosaic in which the trace of the right-lateral strike-slip San Andreas fault is clearly visible. Notice the offset streams (arrows). San Luis Obispo County, California. *(Courtesy of U.S. Geological Survey.)*

FIGURE 5–20 Severely folded strata exposed on the north side of the Rhone River valley west of Sierra, Switzerland.

Folds

No less important than faults as evidence for the Earth's instability are the bends in rock strata that are termed **folds** (Fig. 5–20). Sediments are, of course, originally laid down in approximately level layers. As noted in Chapter 1, the perceptive naturalist Nicolaus Steno referred to this observation as the "principle of original horizontality." Steno correctly inferred that if strata were originally horizontal, then folds, flexures, or inclinations of those beds were direct evidence of crustal movement.

The principal categories of folds are anticlines, synclines, domes, basins, and monoclines (Fig. 5–21). **Anticlines** are uparched rocks in which the oldest rocks are in the center and the youngest rocks are on the flanks (Fig. 5–22). **Synclines** are downwardly folded rocks that have youngest beds in the center and oldest rocks on the flanks (Fig. 5–23). Folds tend to occur together, rather like a series of petrified wave crests and troughs. The erosion of anticlines and synclines characteristically produces a topographic pattern of ridges and valleys (Fig. 5–24). The ridges develop where resistant rocks project at the surface, whereas valleys develop along areas underlain by more easily eroded rocks.

Domes and **basins** are similar to anticlines and synclines, except that they have elliptical to roughly circular outcrop patterns. A dome consists of uparched strata in which the beds dip (are inclined) in

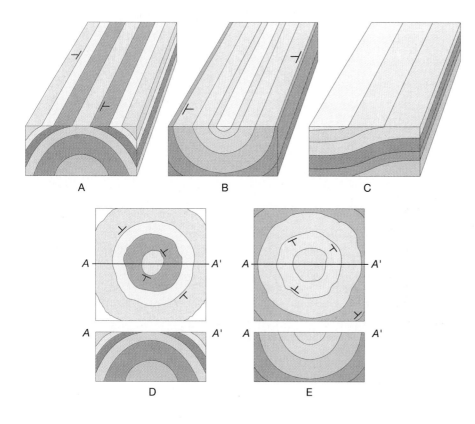

FIGURE 5–21 Types of folds. (A) Anticline. (B) Syncline. (C) Monocline. (D) Map view and cross-section of a dome. Notice that older strata are in the center of the outcrop pattern. (E) Map view and cross-section of a basin. Younger rocks occur in the central area of the outcrop pattern.

FIGURE 5–22 Anticline exposed in the north bank of the Sun River, Teton County, Montana. *(Courtesy of R. R. Mudge and U.S. Geological Survey.)*

all directions away from the center. In contrast, a basin is a downwarp in which beds dip from all sides toward the center of the structure. The erosionally truncated beds of simple domes and basins may form a circular pattern of ridges and valleys. As is the case with anticlines and synclines, older beds are found in the center of truncated domes and younger strata at the center of basins (see Fig. 5–21). Domes and basins vary in size from a few meters to hundreds of kilometers in diameter (see Fig. 8–3).

FIGURE 5–23 Geologic structures often have no relationship to surface topography. Here a downfold in beds (syncline) lies beneath a high ridge. Sideling Hill road cut, 6 miles west of Hancock, Maryland, along U.S. 48, Ridge and Valley Province. *(Courtesy of K. A. Schwartz.)*

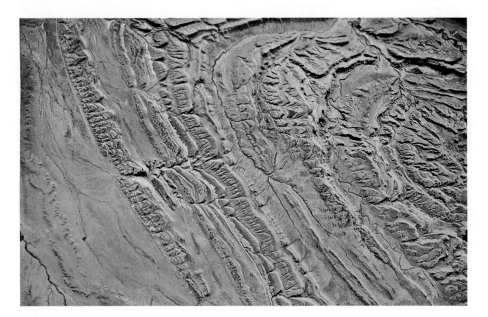

FIGURE 5–24 An anticlinal fold in Wyoming. Because of the sparseness of vegetation and soil cover, rock layers are clearly exposed and differences in resistance to erosion cause layers to be etched into sharp relief. *(Courtesy of U.S. Geological Survey.)*

Although most folds (and particularly those found in mountain belts) are formed by compression, some may develop under a variety of other circumstances. For example, some relatively small folds may result from differential compaction as loose sediment is converted to sedimentary rock, from slumping of mounds of sediment at the time of its accumulation, from an upward protuberance of rock masses from below, from draping over uplifted fault blocks, and from crumpling as great blankets of strata slide over an older rock surface.

■ *Plate Tectonics*

Perhaps more than other scientists, geologists are accustomed to viewing the Earth in its entirety. It is their task to assemble the multitude of observations about the origin of mountains, the growth of continents, and the history of ocean basins into a coherent view of the whole Earth. Exciting and revolutionary discoveries made over the past 4 decades have provided that kind of integrated understanding of our planet. The new view of the Earth's dynamic geology, called **plate tectonics,** is based on the movements of plates of lithosphere driven by convection in the underlying mantle.

As with many scientific breakthroughs, the plate tectonics theory was not fully conceptualized until a foundation of relevant data had been assembled. In the years following World War II, research related to naval operations produced submarine detection devices that also proved useful in measuring magnetic properties of rocks. In the mid-1940s to late 1950s,

the need to monitor atomic explosions resulted in the establishment of a worldwide network of seismometers. This network provided precise information about the global pattern of earthquakes. The magnetic field over large portions of the sea floor was soon to be charted by the use of newly developed and delicate *magnetometers*. Other technologic advances ultimately permitted scientists to examine rock that had been carefully dated by isotopic methods and then to determine the nature of the Earth's magnetic field at the time those rocks had formed. Geologically recent reversals of the magnetic field were soon detected, correlated, and accurately dated. A massive federally funded program to map the bottom of the oceans was launched, and depth information from improved echo depth-sounding devices poured into data-collecting rooms to be translated into maps and charts.

A new picture of the ocean floor began to emerge (see Fig. 1–4). It was at once awesome, alien, and majestic. Great chasms, flat-topped submerged mountains, boundless abyssal plains, and interminable volcanic ranges appeared on the new maps and begged an explanation. How did the volcanic mid-oceanic ridges and deep-sea trenches originate? Why were both so prone to earthquake activity? Why was the mid-Atlantic ridge so nicely centered and parallel to the coastlines of the continents on either side? As the topographic, magnetic, and geochronologic data accumulated, the relationship of these questions became apparent. An old theory called **continental drift** was reexamined, and the new, more encompassing theory of plate tectonics was formulated. It was an idea whose time had come.

Fragmented Continents

It requires only a brief examination of the world map to notice the remarkable parallelism of the continental shorelines on either side of the Atlantic Ocean. If the continents were pieces of a jigsaw puzzle, it would seem easy to fit the great "nose" of Brazil into the re-entrant of the African coastline. Similarly, Greenland might be inserted between North America and northwestern Europe. It is not surprising, therefore, that earlier generations of mapgazers also noticed the fit and formulated theories involving the breakup of an ancient supercontinent.

In 1858 there appeared a work titled *La Création et ses Mystères Devoilés*. Its author, A. Snider, postulated that before the time of Noah and the biblical flood, there existed a great region of dry land. This antique land developed great cracks encrusted with volcanoes, and during the Great Deluge a portion separated at a north–south trending crack and drifted westward. Thus, North America came into existence.

Near the close of the nineteenth century, the Austrian scientist Eduard Suess became particularly intrigued by the many geologic similarities shared by India, Africa, and South America. He formulated a more complete theory of a supercontinent that drifted apart following fragmentation. Suess called that great land **Gondwanaland** after Gondwana, a geologic province in east-central India.

The next serious effort to convince the scientific community of the validity of these ideas was made in the early decades of the twentieth century by the energetic German geophysicist Alfred Wegener. His 1915 book, *Die Entstehung der Kontinente und Ozeane* ("The Origin of the Continents and Oceans"), is considered a milestone in the historical development of the concept of continental drift.

Wegener's hypothesis was straightforward. Building on the earlier notions of Eduard Suess, he argued again for the existence in the past of a supercontinent that he dubbed **Pangea**. That portion of Pangea that was to separate and form North America and Eurasia came to be known as **Laurasia**, whereas the southern portion retained the earlier designation of Gondwanaland. According to Wegener's perception, Pangea was surrounded by a universal ocean named **Panthalassa**, which opened to receive the shifting continents when they began to split apart about 200 million years ago (Fig. 5–25). The fragments of Pangea drifted along like great stony rafts on the denser material below. In Wegener's view, the bulldozing forward edge of the slab might be expected to crumple and produce mountain ranges like the Andes.

The assiduous investigations of Wegener were not to go unchallenged. Criticism was leveled chiefly against his notion that the continents were able to slide along through an unyielding oceanic crust. The eminent and scientifically formidable Sir Harold Jeffries calculated that the ocean floor was far too rigid to allow for the passage of continents, no matter what the imagined driving mechanism. It is now known that Jeffries was correct in asserting that continents cannot—and do not—plow through oceanic crust. Shortly, it will be shown that continents do move, but they move only as passive passengers on large rafts of lithosphere that glide over a comparatively soft and plastic upper layer of the Earth's mantle. Nevertheless, much of the evidence Wegener and

FIGURE 5–25 (A) and (B) Alfred Wegener's view of the Earth as it existed 200 million years ago. At this time there was one ocean (Panthalassa) and one continent (Pangea). (C) By about 20 million years later, the supercontinent had begun to split into northern Laurasia and southern Gondwanaland. (D) Dismemberment continued, and (E) looked this way about 65 million years ago. (F) Further widening of the Atlantic and northward migration of India brings the Earth to its present state.

others had assembled can be used to substantiate both the old and the newer concepts. We will consider this evidence.

Evidence for Continental Drift

The most convincing evidence for continental drift remains the geologic fit of the continents. Indeed, the correspondence is far too good to be fortuitous, even when one considers the expected modifications of shorelines resulting from erosion, deformation, or intrusions following the breakup of Pangea about 200 million years ago. A still closer match results when one fits the continents together to include the continental shelves, which are really only submerged portions of the continents. Such a computerized and error-tested fitting of continents was carried out by Sir Edward Bullard, J. E. Everett, and A. G. Smith of the University of Cambridge (Fig. 5–26). Remarkably, this work showed that over most of the bound-

ary the average mismatch was no more than a degree—a snug fit indeed.

Another line of evidence favoring the drift theory involves sedimentologic criteria indicating similarity of climatic conditions for widely separated parts of the world that were once closely adjacent to one another. For example, in such widely separated places as South America, southern Africa, India, Antarctica, Australia, and Tasmania, one finds glacially grooved rock surfaces and deposits of glacial rubble developed in the course of the late Paleozoic continental glaciation. The deposits of poorly sorted clay, sand, cobbles, and boulders are called **tillites.** They, along with the grooves and scratches on rock surfaces that were apparently beneath the moving ice, attest to a great ice age that affected Gondwanaland at a time when it was as yet unfragmented and lying at or near the south polar region (Fig. 5–27). Furthermore, if the directions of the grooves on the bedrock are plotted on a map, they indicate the center of ice accumulation and the directions in which it moved. Unless the southern continents are reassembled into Gondwanaland, this center would be located in the ocean, and great ice sheets do not develop centers of accumulation in the ocean. Hence, the existence of Gondwanaland seems plausible. In a few instances, oddly foreign boulders in the tillites of one continent are found in the deposits of another continent now located thousands of miles across the oceans. Petrologists are able to trace their parent outcrops to the distant land masses.

There are additional clues to paleoclimatology that can be used to test the concept of moving land masses. Trees that have grown in tropical regions of

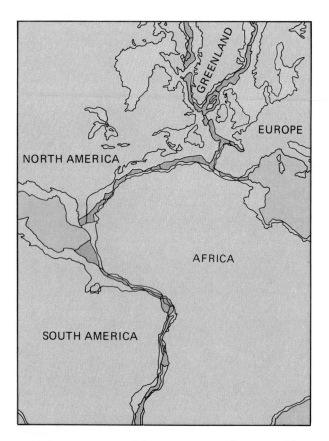

FIGURE 5–26 Fit of the continents as determined by Sir Edward Bullard, J. E. Everett, and A. G. Smith. The fit was made along the continental slope (green color) at the 500-fathom contour line. Overlaps and gaps (shown in orange) are probably the result of deformations and sedimentation after rifting. *(Adapted from Bullard, E. C., et al. 1965. Philos. Trans. R. Soc. Lond. 258:41.)*

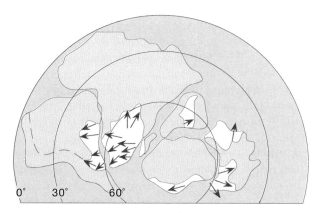

FIGURE 5–27 Reconstruction of Gondwanaland near the beginning of Permian time, showing the distribution of glacial deposits (shown in white). Arrows show direction of ice movement as determined from glacial scratches on bedrock. *(Modified from Hamilton, W., and Kinsley, D. 1967. Geol. Soc. Am. Bull. 78:783–799.)*

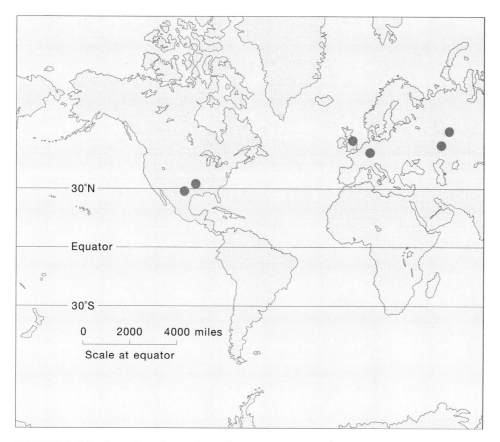

FIGURE 5–28 Location of prominent Permian evaporite deposits.

the globe characteristically lack the annual rings resulting from seasonal variations in growth. Exceptionally thick coal seams containing fossil logs from such trees imply a tropical paleoclimate. The locations of such coal deposits should approximate an equatorial zone relative to the ancient pole position for that age. It is evident that such coal seams now being exploited in northern latitudes must have been moved to those locations from the equatorial zones along which the source vegetation accumulated.

Because of the decrease in solubility of calcium carbonate with rising temperatures, thick deposits of marine limestone also imply relatively warm climatic conditions. Arid conditions, such as those existing today on either side of the equatorial rain belt, can be recognized in ancient rocks by desert sandstones and evaporites. (Evaporites are chemical precipitates such as salt and gypsum that characteristically form when a body of water containing dissolved solids is evaporated.) Today such deposits are ordinarily formed in warm, arid regions located about 30° north or south of the equator. If one believes continents have always been where they are now, then it is difficult to explain the great Permian salt deposits now found in northern Europe, the Urals, and the southwestern United States (Fig. 5–28). The evaporites had probably been

precipitated in warmer latitudes before Laurasia migrated northward.

A similar kind of evidence can be obtained by examining the locations of Permian reef deposits. Modern reef corals are restricted to a band around the Earth that is within 30° of the equator. Ancient reef deposits are now found far to the north of the latitudes at which they had originated.

At least some of the paleontologic support for continental drift was well known to Suess and Wegener. In the Gondwana strata overlying the tillites or glaciated surfaces, there can often be found nonmarine sedimentary rocks and coal beds containing a distinctive assemblage of fossil plants. Named after a prominent member of the assemblage, the plants are referred to as the *Glossopteris* flora (Fig. 5–29). Paleobotanists who have supported the idea of shifting continents have argued that it would be virtually impossible for this complex temperate flora to have developed in identical ways on the southern continents as they are separated today. The seeds of *Glossopteris* were far too heavy to have been blown over such great distances of ocean by the wind.

Another element of paleontologic corroboration for the concept of moving land masses is provided by the distribution in the Southern Hemisphere of fossils

FIGURE 5–29   Fossil *Glossopteris* leaf associated with coal deposits and derived from glossopterid forests of Permian age. This fossil was found on Polarstar Peak, Ellsworth Land, Antarctica.   *(Courtesy of U.S. Geological Survey, J. M. Schopf, and C. J. Craddock.)*

of a small aquatic reptile named **Mesosaurus** (Fig. 5–30). An interpretation, based on its skeletal remains, of this animal's habits as well as the nature of the sediment in which its fossils are found strongly suggests that it once inhabited lakes and estuaries and was not an inhabitant of the open ocean. The discovery of fossil remains of *Mesosaurus* both in Africa and South America lends credence to the notion that

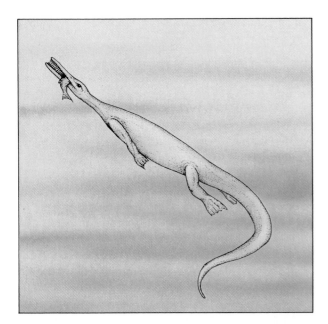

FIGURE 5–30   *Mesosaurus* (about 1 meter long).

these continents were once attached. Perhaps for a time, as they were just beginning to separate, the location of the present coastlines became dotted with lakes and protected bodies of water that harbored *Mesosaurus*. This seems a more plausible explanation than one requiring the creature to navigate the South Atlantic or to wend its way across formidable latitudinal climatic barriers by following the shorelines northward and around the perimeter of the Atlantic. Also, there are no fossil remains to suggest such a perimeter journey.

Before fragmentation of a supercontinent such as Laurasia, one would expect to find numerous similar plants and animals living at corresponding latitudes on either side of the line of future separation. Such similarities do occur in the fossil record of continents now separated by extensive oceanic tracts. Faunas of Silurian and Devonian fishes, for example, are comparable in such now distant locations as Great Britain, Germany, Spitzbergen, eastern North America, and Quebec. These similarities, not only in fishes but also in amphibians, persist into Carboniferous rocks. In Permian and Mesozoic age rocks, one again finds striking similarities in the reptilian faunas of Europe and North America. The fossil evidence implies not only the former existence of continuous land connections but also a uniformity of environmental conditions between the elements of Laurasia that were once at approximately similar latitudinal zones.

When one turns to Cenozoic mammalian faunas, however, one finds the situation to be quite different.

Distinctive faunal elements are evident in Australia, South America, and Africa. Apparently, as the continents became separated from each other by ocean barriers, genetic isolation resulted in morphologic divergence. The modern world's enormous biologic diversity is at least partially a result of evolutionary processes operating on more or less isolated continents. The faunas and floras during the periods prior to the breakup of Pangea were less diverse.

The character, sequence, age, and distribution of rock units have also been examined for insights into concepts of drift. One might presume that locations close to one another on the hypothetic supercontinent would have environmental similarities that would result in resemblances in the kinds of rocks deposited. As indicated by the correlation chart of southern continents (Table 5–1), there is such a similarity in the geologic sections of now widely separated land masses. The sections begin with glacial deposits such as the Dwyka Tillite of Africa and the Itararé Tillites of South America. These cobbly layers are overlain by nonmarine sandstones and shales containing the *Glossopteris* flora and layers of coal. This overlaying may indicate a warming trend from cold glacial to more temperate climates. However, one must be careful with such interpretations, for our most recent continental glaciations occurred under a cool, temperate climate regime. The next higher group of strata shows evidence of deposition under terrestrial conditions with an abundance of alluvial, eolian, and stream deposits that contain fossils of an interesting group of mammal-like reptiles. The up-permost beds of the sequences often include basalts and other volcanic rocks that are of similar age and composition in India and Africa but are slightly younger in South America. The absence of the reptile-bearing Triassic strata and the Jurassic volcanics in Australia is believed to be the result of Australia separating from Africa at an early date—probably early in the Permian—and long before the separation of Africa and South America. Australia must have retained just enough of a connection with Africa to permit the entry of dinosaurs and marsupials.

Yet another way to test the notion of a supercontinent is to see whether geologic structures, such as the trends of folds and faults, match up when now distant continents are hypothetically juxtaposed. Such correlative trends do exist. Folds and faults are often difficult to date, although, if successful, one can establish the contemporaneity of a fault lineage or fold axis now on separated continents. One can also examine the outcrop patterns of Precambrian basement rocks and discern correlative boundaries between now widely separated continents. A folded geosynclinal sequence of Precambrian strata in central Gabon in Africa, for example, can be traced into the Bahia Province of Brazil. Also, isotopically dated Precambrian rocks of west Africa can be correlated with rocks of similar age in northeastern Brazil.

The Testimony of Paleomagnetism

Although Alfred Wegener was unable to provide a satisfactory explanation for what breaks continents

TABLE 5–1   Gondwanaland Correlations

System	Southern Brazil	South Africa	Peninsular India
Cretaceous	Basalt	Marine sediments	Volcanics Marine sediments
Jurassic	(Jurassic rocks not present)	Volcanics	Sandstone and shale Volcanics Sandstone and shale
Triassic	Sandstone and shale with reptiles	Sandstone and shale with reptiles	Sandstone and shale with reptiles
Permian	Shale and sandstones with coal and *Glossopteris* Shale with *Mesosaurus*	Shale and sandstones with coal and *Glossopteris* Shale with *Mesosaurus*	Shale and sandstones with coal and *Glossopteris* Shale
Carboniferous	Sandstone, shale, and coal with *Glossopteris* Tillite	Sandstone, shale, and coal with *Glossopteris* Tillite	Tillite

apart and moves them across the face of the Earth, he had assembled a convincing body of evidence that such events occurred. Discoveries in the late 1950s and 1960s would provide further substantiating evidence. The new information came from the study of magnetism imparted to ancient rocks at the time of their formation and preserved down to the present time. To understand this **paleomagnetism,** as it is called, it is necessary to digress for a moment and consider the general nature of the Earth's present magnetic field.

The Earth's Present Magnetism

It is common knowledge that the Earth has a magnetic field. It is this field that causes the alignment of a compass needle. The origin of the magnetic field is still a question that has not been fully resolved, but many geophysicists believe it is generated as the rotation of the Earth causes slow movements in the liquid outer core. The magnetic lines of force resemble those that would be formed if there were an imaginary bar magnet extending through the Earth's interior. The long axis of the magnet would be the conceptual equivalent of the Earth's magnetic axis, and the ends would correspond to the north and south geomagnetic poles (Fig. 5–31). Although today the geomagnetic poles are located about 11° of latitude from the rotational axis, they slowly shift position. When averaged over several thousand years, the geomagnetic poles and the geographic poles do coincide. If we assume that this relationship has always held true, then by calculating ancient magnetic pole positions from paleomagnetism in rocks we have coincidentally located the Earth's geographic poles. It should be kept in mind, however, that such interpretations are based on the supposition that the rotational and magnetic

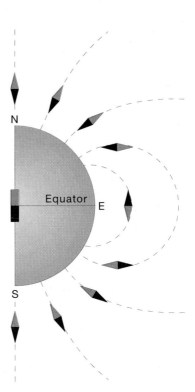

FIGURE 5–32   A freely suspended compass needle aligns itself in the direction of the Earth's magnetic field. The inclination of the needle will vary from horizontal at the equator to vertical at the poles.

poles have always been relatively close together. This seems a reasonable assumption based on the modern condition as well as on paleontologic studies that have shown inferred ancient climatic zones in plausible locations relative to ancient pole positions. Another assumption is that the Earth has always been dipolar. Paleomagnetic studies from around the world thus far support this supposition.

Remanent Magnetism

The magnetic information frozen into rocks may originate in several ways. Imagine for a moment the outpouring of lava from a volcano. As the lava begins to cool, magnetic iron oxide minerals form and align their polarity with the Earth's magnetic field. That alignment is then retained in the rock as its crystallization is completed. In a simple analogy, the magnetic orientations of the minerals responded as if they were tiny compass needles immersed in a viscous liquid. Because they are aligned parallel to the magnetic lines of force surrounding the Earth, they not only point toward the poles (magnetic *declination*) but also become increasingly more inclined from the horizontal as the poles are approached (Fig. 5–32). This

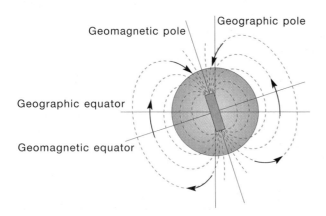

FIGURE 5–31   Magnetic lines of force for a simple dipole model of the Earth's magnetic field.

inclination, when detected in paleomagnetic analysis, can be used to determine the latitude at which an igneous body containing magnetic minerals cooled and solidified.

Magnetism frozen into ancient rocks is called **remanent magnetism.** The type described previously, in which igneous rocks cool past the Curie temperature (also called the *Curie point*) of their magnetic minerals, is further classified as *thermoremanent magnetism.* The **Curie temperature** is simply that temperature above which a substance is no longer magnetic. A few minerals, the most important of which is magnetite (Fe_3O_4), have the property of acquiring remanent magnetism. Magnetite (Fig. 5–33) is widespread in varying amounts in virtually all rocks. The manner in which remanent magnetism is acquired is complicated but can be explained in a general way. Remanent magnetism in a mineral is ultimately due to the fact that some atoms and ions (charged atoms) have so-called magnetic moments; this means that they behave like tiny magnets. The magnetic moment of a single atom or ion is produced by the spin of its electrons. When magnetite takes on its remanent magnetism, the iron ions align themselves within the crystal lattice so that their magnetic moments are parallel. Most igneous rocks crystallize

FIGURE 5–33 Magnetite, an iron oxide mineral that acts as a natural magnet and is capable of picking up pieces of steel. Magnetite may possess polarity like a compass needle and, when suspended by a string, will twist around until its south pole points to the Earth's north pole.

at temperatures in excess of 900°C. At these extreme temperatures, the atoms in the minerals have a large amount of energy and are being violently shaken. The vibrating, shaking atoms are unable to line up until more cooling occurs. Finally, when the Curie temperature is reached (578°C for magnetite), enough energy has been lost to allow the atoms to come into alignment. They will stay aligned at still lower temperatures because the Earth's field is not strong enough to alter the alignment already frozen into the minerals.

Igneous rocks are not the only kinds of Earth materials that can acquire remanent magnetism. In lakes and seas that receive sediments from the erosion of nearby land areas, detrital grains of magnetite settle slowly through the water and rotate so that their directions of magnetization parallel the Earth's magnetic field. They may continue to move into alignment while the sediment is still wet and uncompacted, but once the sediment is cemented or compacted, the depositional remanent magnetism is locked in.

Over the past 2 decades, geophysicists have been measuring and accumulating paleomagnetic data for all the major divisions of geologic time. Their results are partly responsible for the recent revival of interest in continental drift hypotheses. For example, when ancient pole positions were located on maps, it appeared that they were in different positions relative to a particular continent at different periods of time in the geologic past. Either the poles had moved relative to stationary continents or the poles had remained in fixed positions while the continents shifted about. If the poles were wandering and the continents "stayed put," then a geophysicist working on the paleomagnetism of Ordovician rocks in France, for example, should arrive at the same location for the Ordovician poles as a geophysicist doing similar work on Ordovician rocks from the United States. In short, the paleomagnetically determined pole positions for a particular age should be the same for all continents. On the other hand, if the continents had moved and the poles were fixed, then we should find that pole positions for a particular geologic time would be different for different continents. The data suggest that this latter situation is the more valid.

Another way to view the data of paleomagnetism is to examine what are called **apparent polar wandering curves.** The word *apparent* in the expression is necessary, for *polar wandering curves* alone might imply that the poles wander. As we have just noted, this is highly unlikely. The apparent polar wandering curves are merely lines on a map connecting ancient pole positions relative to a specific continent for various times during the geologic past. As shown in Fig-

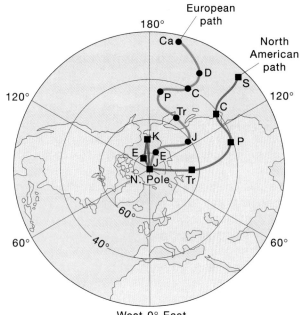

FIGURE 5–34 Apparent paths of polar wandering for Europe and North America. A scatter of points have been averaged to a single point for each geologic period: Ca, Cambrian; S, Silurian; D, Devonian; C, Carboniferous; P, Permian; Tr, Triassic; J, Jurassic; K, Cretaceous; E, Eocene. *(After Bott, M. H. P. 1971.* The Interior of the Earth: Structure and Processes. *New York: St. Martin's Press.)*

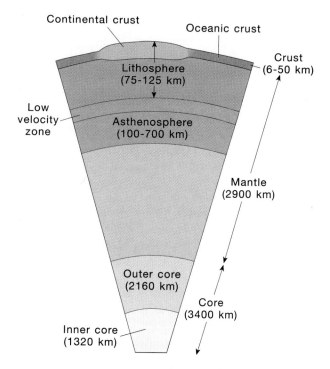

FIGURE 5–35 Divisions of the Earth's interior. (For clarity, the divisions have not been drawn to scale.)

ure 5–34, the curves for North America and Europe met in recent time at the present North Pole. This means that the paleomagnetic data from recently formed rocks from both continents indicate the same pole position. A plot of the more ancient poles results in two similarly shaped but increasingly divergent curves. If this divergence resulted from a drifting apart of Europe and North America, then one should be able to reverse the movements mentally and see if the curves do not come together. Indeed, the Paleozoic portions of the polar wandering curves could be brought into close accord if North America and its curve were to be slid eastward about 30° toward Europe. This sort of information gave new life to the old notion of continental drift. More important, it has become significant in the recent modification of that earlier concept—a modification known as **plate tectonics.**

The Basic Concept of Plate Tectonics

When considered at the general level, plate tectonics is a remarkably simple concept. The **lithosphere,** or outer shell of the Earth (Fig. 5–35), is constructed of

7 huge slabs and about 20 smaller plates that are squeezed in between them. The larger plates (Fig. 5–36) are approximately 75 to 125 km thick. Movement of the plates causes them to converge, diverge, or slide past one another, and this results in frequent earthquakes along plate margins. When the locations of earthquakes are plotted on the world map, they clearly define the boundaries of tectonic plates (Fig. 5–37).

A part of a plate containing a continent would have the configuration shown in Figure 5–38. Plates "float" on the weak, partially molten region of the upper mantle defined earlier as the asthenosphere (from the Greek *asthenos,* meaning "weak"). The asthenosphere is a region of rock plasticity and flowage detected on the basis of changes in seismic wave velocities by the geophysicist Beno Gutenberg.

Plate Boundaries and Sea-Floor Spreading

Central to the idea of plate tectonics is the differential movement of lithospheric plates. For example, plates move apart at **divergent plate boundaries,** which may manifest themselves as midoceanic ridges complete with tensional ("pull-apart") geologic structures. Indeed, the mid-Atlantic ridge approximates the line of separation between the Eurasian and African plate on one side and the American plate on the other (Fig.

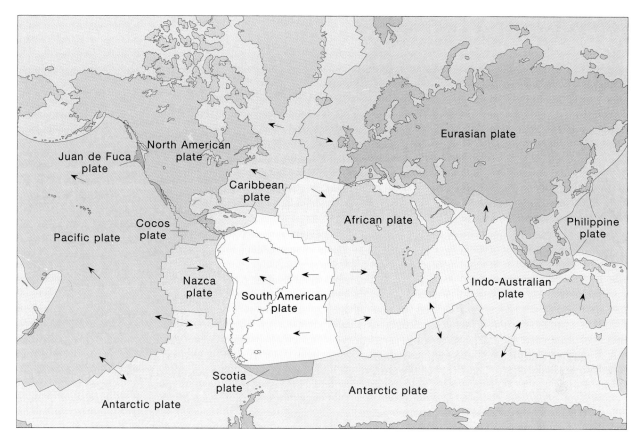

FIGURE 5–36 The Earth's major tectonic plates. Arrows indicate general direction of movement of plates.

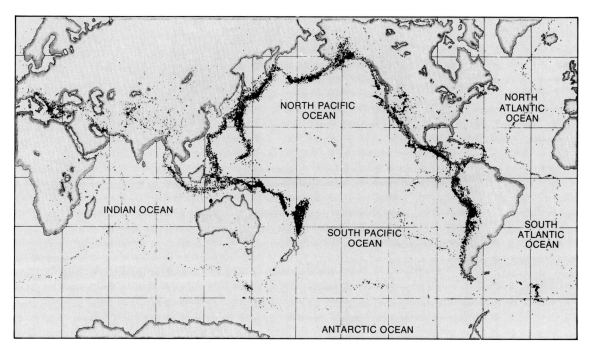

FIGURE 5–37 World distribution of earthquakes. Notice the major earthquake belt that encircles the Pacific and another that extends eastward from the Mediterranean toward the Himalayas and East Indies. The midoceanic ridges are also the site of many earthquakes.

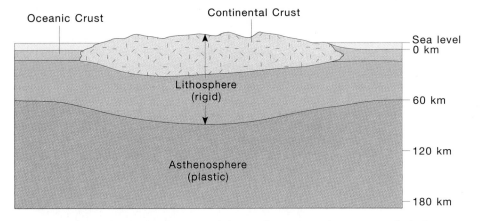

FIGURE 5–38   The outermost part of the Earth consists of a strong, relatively rigid lithosphere, which overlies a weak, plastic asthenosphere. The lithosphere is capped by a thin crust beneath the oceans and a thicker continental crust elsewhere.

5–39). As is to be expected, such a rending of the crust is accompanied by earthquakes and enormous outpourings of volcanic materials that are piled high to produce the ridge itself. The void between the separating plates is also filled with this molten rock, which rises from below the lithosphere and solidifies in the fissure. Thus, new crust (that is, new sea floor) is added to the **trailing edge** of each separating plate

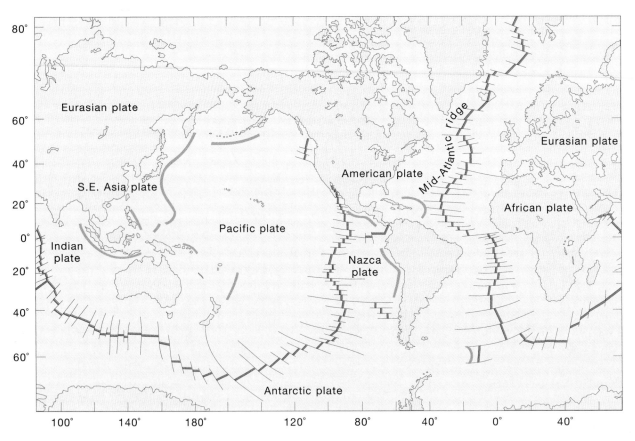

FIGURE 5–39   Locations of midoceanic ridges (red) and trenches (blue). Note fracture zones offsetting the ridges. Dashed lines in Africa represent the East African rift zone.   *(After Isacks, B. et al. 1968. J. Geophys. Res. 73:5855–5899. Copyright, The American Geophysical Union.)*

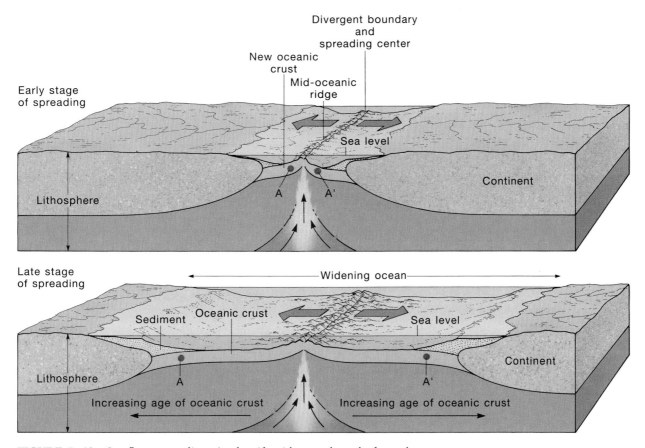

FIGURE 5–40 Sea-floor spreading. As the rift widens and newly formed crust moves away from the ridge axis, the crust stretches and fractures. Some blocks of fractured crust may drop while others remain elevated. Basaltic lavas rise to fill the lower elevations and the space between separating plates. As the lavas solidify, they become part of the trailing edge of a tectonic plate. A and A' are reference points.

as it moves slowly away from the midoceanic ridge (Figs. 5–40 and 5–41). The process has been appropriately named **sea-floor spreading.** Zones of divergence may originate beneath continents, rupturing the overlying land mass and producing such rift features as the Red Sea and Gulf of Aden.

The axis of spreading is not a smoothly curving line. Rather, it is abruptly offset by numerous faults. These features are called **transform faults** and are an expected consequence of horizontal spreading of the sea floor along the Earth's curved surface. Transform faults take their name from the fact that the fault is "transformed" into something different at its two ends. The relative motions of transform faults are shown in Figure 5–42. The ridge acts as a spreading center that exists both to the north and to the south of the fault. The rate of relative movement on the opposite sides along fault segment X to X' depends on the rate of extrusion of new crust at the ridge. Because of the spreading of the sea floor outward

from the ridge, the relative movement is opposite to that expected by ordinary fault movement. Thus, at first glance, the ridge-to-ridge transform fault (Fig. 5–42A) appears to be a left-lateral fault, but the actual movement along segment X–X' is really right lateral. Notice that only the X–X' segment shows movement of one side relative to the other. To the west of X and the east of X', there is little or no relative movement.

If plates are receding from one another at one boundary, they may be expected to collide or slide past other plates at other boundaries. Thus, in addition to the divergent plate boundaries that occur along midoceanic ridges, there are **convergent** and **shear** or **transform boundaries.** Convergent plate boundaries develop when two plates move toward one another and collide. As one might guess, these convergent junctions are characterized by a high frequency of earthquakes. In addition, they are thought to be the zones along which folded mountain ranges

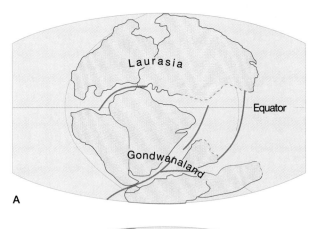

A

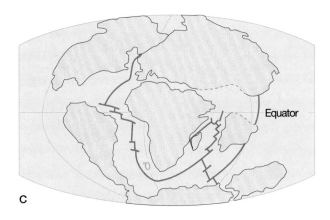

C

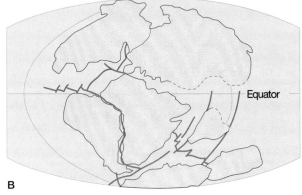

B

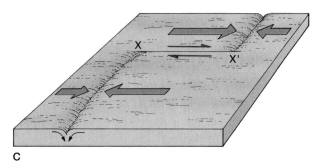

FIGURE 5–41 Disruption of the supercontinent Pangea and dispersal of its continental fragments. Divergent zones indicated by red lines. (A) General conditions at the end of the Triassic, about 180 million years ago. (B) Position of the continents about 135 million years ago. (C) The continental positions at the end of the Cretaceous, about 66 million years ago. *(Based on maps by Dietz, R. S., and Holden, J. C. 1970. J. Geophys. Res. 75.)*

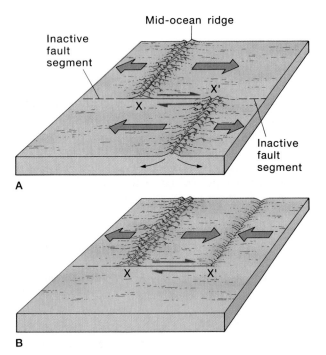

X – X' =
Active fault segment

B

FIGURE 5–42 Three types of transform faults. (A) Ridge–ridge transform. (B) Ridge–trench. (C) Trench–trench.

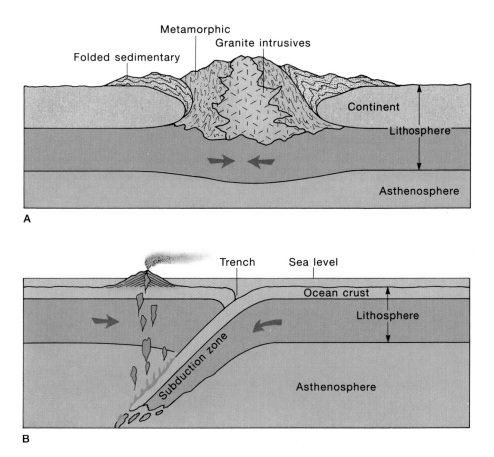

FIGURE 5–43 Two types of convergent plate boundaries. (A) Convergence of two plates, both bearing continents. (B) Convergence of two plates, both bearing oceanic crust.

or deep-sea trenches may develop (Fig. 5–43). The structural configuration of the convergent boundary is likely to vary according to the rate of spreading and whether the leading edges of the plates are composed of oceanic or continental crust. Geophysicists speculate that when the plates collide, one slab may slip and plunge below the other, producing what is called a **subduction zone.** The sediments and other rocks of this plunging plate are pulled downward (subducted), melted at depth, and, much later, rise to become incorporated into the materials of the upper mantle and crust. In some instances, the silicate melts provide the lavas for chains of volcanoes.

An example of a shear plate boundary is the well-known San Andreas fault in California. Along this great fault, the Pacific plate moves laterally against the American plate (Fig. 5–44). Shear plate boundaries are decidedly earthquake prone but are less likely to develop intense igneous activity. They are the active segments of transform faults along which no new surface is formed or old surface consumed.

Crustal Behavior at Plate Boundaries

We have noted that there are three basic kinds of plate boundaries: convergent, divergent, and shear. With regard to the convergent boundaries, there are also three kinds of tectonic behavior, depending on the kinds of crust constituting the margins of the plates that converge. If the leading edge of a plate is composed of continental crust and it collides with an opposing plate of similar composition, the result would be a folded mountain range in which are developed igneous rocks of granitic composition (see Fig. 5–43A). Because continental plate margins are too light and buoyant to be carried down into the asthenosphere, subduction does not occur in this type of collision. Instead, the crust at the plate margins is deformed and may detach itself from deeper zones, and slabs of crust from one plate may be thrust under the other. The zone of convergence between the two plates, recognized by the severity of folding, faulting, and intrusive activity, is termed the **suture zone.**

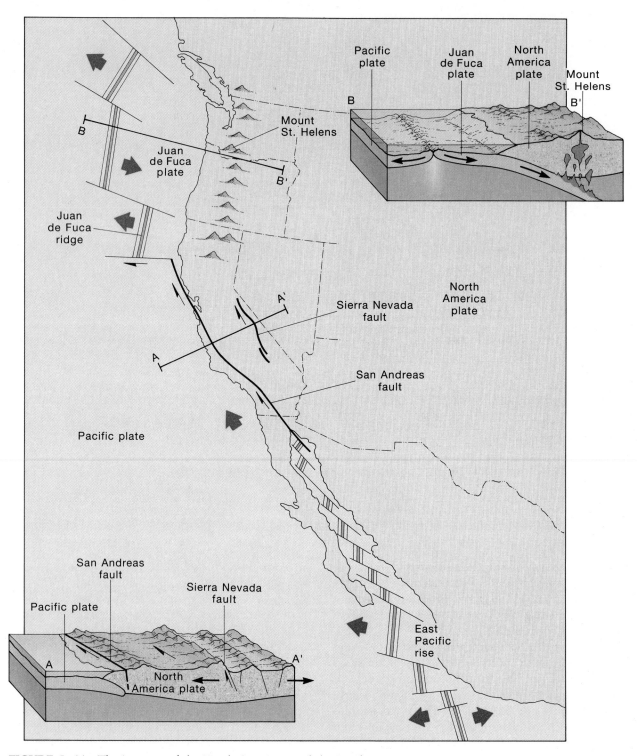

FIGURE 5–44 The juncture of the North American and the Pacific tectonic plates. The double lines are spreading centers. Note the trace of the San Andreas fault. To the north in Oregon and Washington, the small Juan de Fuca plate plunges beneath the North American continent to form the Cascades. *(Courtesy of U.S. Geological Survey.)*

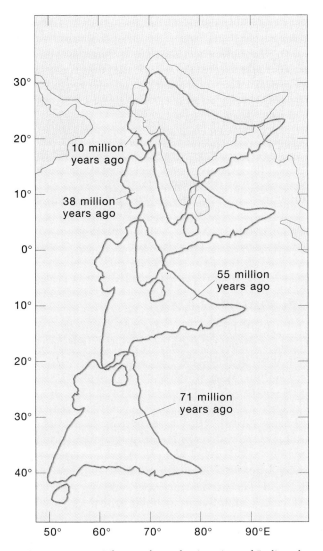

FIGURE 5–45 The northward migration of India relative to Eurasia, showing the position of the continent 71 million years ago (Cretaceous), 55 million years ago (early Eocene), 33 million years ago (Oligocene), and 10 million years ago (late Miocene). The configuration of the northern boundary of India is conjectural. *(From Molnar, P., and Tapponnier, P. 1975. Cenozoic tectonics of Asia: Effects of a continental collision. Science 189(4201):419–426, Copyright 1975, AAAS.)*

A dramatic example of an encounter between two continents occurred during the convergence of India and Eurasia (Fig. 5–45). The mighty Himalayas are the spectacular result of this collision. The study of structures produced during this convergence has provided some interesting insights into the history of continent–continent encounters. It was at one time assumed that once a convergence had occurred, motion between opposing plates would abruptly cease. However, analysis of movement along faults associated with convergence and evidence provided by photographs taken by satellites indicate that this is not true. The plates may continue to converge, albeit at a slower rate, long after making contact with one another. Some of the continued movement is the result of crustal shortening that occurs whenever a flat span of crust is crumpled into a smaller area. The amount of crustal shortening resulting from the India–Eurasia convergence has been estimated at about 1500 km. Geologists can account for 500 to 1000 km of that amount in the folds and thrust faults of the suture zone. The remaining 500 to 1000 km was apparently dissipated along major east–west trending lateral faults in China and Mongolia (Fig. 5–46). Thus, plate convergence may continue long after initial closure by lateral release of plate movement along strike-slip faults (lateral faults) in the region peripheral to the suture zone. Today geologists are finding evidence of similar strike-slip faulting in

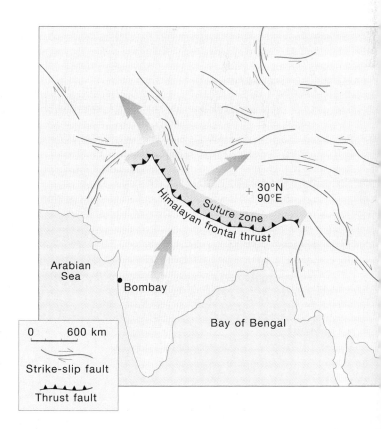

FIGURE 5–46 Major lateral (strike-slip) faults above and adjacent to the Indian suture zone. These faults may represent lateral transfer of some of the movement involved in the India–Eurasia convergence. Geologically recent movement along the faults implies that the plate movement involved in the convergence is still in progress. *(Fault locations from Molnar, P., and Tapponnier, P. 1975. Cenozoic tectonics of Asia: Effects of a continental collision. Science 189(4201):419–426, Copyright 1975, AAAS.)*

older continent–continent convergences such as those that produced the Appalachians and Urals.

The second kind of convergence at plate boundaries involves the meeting of two plates that both have oceanic crust at their converging margins (see Fig. 5–43B). Although the rate of plate movement in an ocean–ocean convergence will affect the kinds of structures produced, it is likely that such locations will develop deep-sea trenches with bordering volcanic arcs such as those of the southwestern Pacific.

Finally, there is the third possibility, which involves the collision of a continental (granitic) plate boundary with an oceanic (basaltic) one. The result of such a collision might be a deep-sea trench located offshore from an associated range of mountains. Numerous volcanoes would be expected to develop, and the lavas pouring from their eruptions would probably be a compositional blend of granite and basalt. The Andes are such a mountain range, and the igneous rock *andesite*, named for its prevalence in the Andes, may represent such a silicate blend.

Regions of ocean–continent convergence (Fig. 5–47) are characterized by rather distinctive rock assemblages and geologic structure. As we have seen, the convergence of two lithospheric slabs results in subduction of the ocean plate, whereas the more buoyant continental plate maintains its position at the surface but experiences intense deformation, metamorphism, and melting. A great mountain range begins to take form as a result of all of this dynamic activity. At the same time, sediments and submarine volcanic rocks along the subduction zone are squeezed, sheared, and shoved into a gigantic, chaotic medley of complexly disturbed rocks called a **mélange.** Within the mélange one finds a distinctive assemblage of deep-sea sediments containing microfossils, submarine lavas, and serpentinized peridotite possibly derived from the upper mantle. Together, these rocks constitute an ophiolite suite (Fig. 5–48). The **ophiolite suites** are splinters of the oceanic plate that were scraped off the upper part of the descending plate and inserted into the crushed forward edge of the continent. Thus, ophiolites mark the zone of contact between colliding continental and oceanic plates. Another clue to the presence of such zones is a distinctive kind of metamorphic rock containing blue amphiboles. These rocks are called **blue schists.** They form at high pressures but relatively low temperatures. This rather unusual combination of conditions is characteristic of subduction zones where the relatively cool oceanic plate plunges rapidly into deep zones of high pressure.

What Drives It All?

If we accept the fact that plates of lithosphere move across the surface of the globe, the next question concerns how they are moved. The propelling mechanism is thought by some scientists to consist of large thermal convection cells produced as mantle material is heated from below, expands, becomes less dense, and slowly rises. On encountering the lithosphere, the flow would diverge laterally, dragging along the overlying slab of lithosphere. Mantle material upwelling along the line of separating currents would join the trailing edges on either side. As the current moved horizontally, it would steadily lose its heat and become denser. Ultimately, it would encounter a similar current moving in an opposing direction, and both viscous streams would descend to be reheated and shunted toward a region of upwelling. Above the descending flow one would find subduction zones and deep-sea trenches, whereas divergent zones and

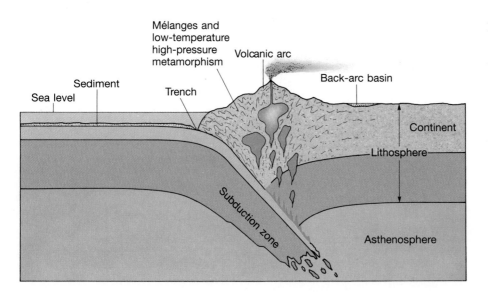

FIGURE 5–47 Convergence of the continental part of a plate with the oceanic part of another plate. The leading edge of the continental plate is crumpled, whereas the oceanic plate buckles downward, creating a trench offshore. The situation generalized here is similar to that off the west coast of South America, where the Nazca plate plunges beneath the South American plate.

Oceanic sediments, including chert, turbidites, and deep-sea oozes.

Basaltic lava with pillow structure

Complex of multiple dike intrusions

Gabbros

Peridotite, serpentine, and other ultramafic rocks

km
0
0.5
1.0

FIGURE 5–48 Idealized section of an ophiolite suite. Ophiolites are thought to be splinters of the ocean floor squeezed into the continental margin during plate convergence.

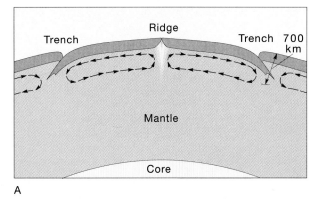

A

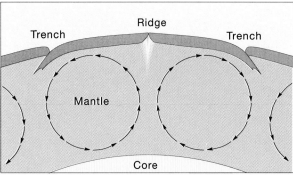

B

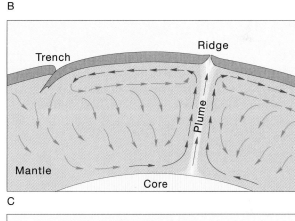

C

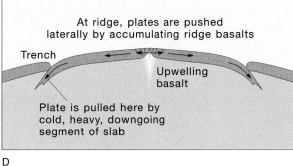

D

FIGURE 5–49 Models proposed for the plate tectonics driving force. (A) Convection model, in which the convection cells are confined to the upper part of the mantle. (B) The convection cells are large and circulate through the entire mantle. (C) The thermal plume model. (D) Sometimes called the push-pull model, plates are pushed laterally at spreading centers and pulled by the plunging, cool, dense leading segment of the plate. A fifth model might be imagined, in which concentric layers of stacked convection cells exist, with the uppermost cells providing the force to move tectonic plates.

spreading centers would mark the locations of the ascending flow.

Several models of convection cell dimensions and distributions have been proposed (Fig. 5–49). One of these confines all convection movement to the relatively thin asthenosphere (Fig. 5–49A), with all of the remaining mantle serving as the heat source. Another suggestion is that convection cells span the entire mantle, from the top of the core to the base of the lithosphere (Fig. 5–49B). Yet a third model favors a stacked convection system with concentric layers of convection cells. Each of these concepts has components that are consistent with existing data, but it is presently too early to state which will ultimately provide the best interpretation. It is also possible that the configuration of convection currents has changed throughout the course of Earth history. Initially, there may have been only one large convecting system that swept together scumlike slabs of continental crust into a single land mass. Later systems may have

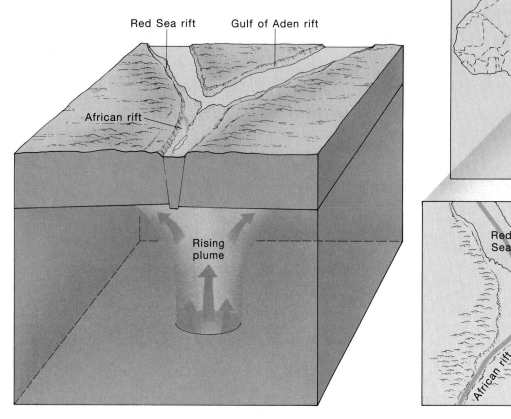

FIGURE 5–50   Rising plumes of hot mantle may cause severe rifts, often forming 120° angles with one another. An example is the Afar triangle, shown at the south end of the Red Sea on the small map of Africa.

been characterized by larger numbers of convection cells located closer to the base of the lithosphere. A concept such as this may account for an apparent increase in continental fragmentations in the Mesozoic compared with earlier eras. This idea was originally advanced by the geophysicist S. K. Runkorn, who postulated that growth of the core of the Earth and consequent shrinkage of the mantle may have brought on the changes in convection cell characteristics. Later information, however, suggests that after the core formed early in the Earth's history, it barely increased in size. This does not, of course, preclude changes in the convection cell patterns for other reasons.

As a mechanism for moving plates of lithosphere, thermal convection may be the best hypothesis thus far advanced. This does not mean that this mechanism should be regarded as an established fact. As yet there has not been an entirely satisfactory way to test the concept, although such currents may be physically plausible, and there is some tenuous evidence that they exist. For example, heat flows from the Earth's interior at a greater rate along midoceanic ridges than from adjacent abyssal plains. Yet analyses of hundreds of measurements of geophysicists at Columbia University suggest that although heat flow

along the mid-Atlantic ridge is 20 percent greater than on the adjacent floors of the ocean basins, it is still not great enough to move the sea floor at the rates suggested by paleomagnetic studies. Another problem relates to the layering in the upper mantle that has been detected in seismologic studies. The mixing that accompanies convectional overturn would seem to preclude such layering.

The search for a mechanism to drive lithospheric plates has produced another convection-related model. In this *thermal plume model* (Fig. 5–49C), mantle material does not circulate in great rolls but rather rises from near the core–mantle boundary in a manner suggestive of the shape and motion of a thundercloud. Proponents of this model suggest that there may be as many as 20 of these thermal plumes, each a few hundred kilometers in diameter. According to the model, when a plume nears the lithosphere, it spreads laterally, doming surface zones of the Earth and moving them along in the directions of radial flow. The center of the Afar triangle in Ethiopia (Fig. 5–50) has been suggested as the site of a plume that flowed upward and outward, carrying the Arabian, African, and Somali plates away from the center of the triangle. Geophysicists have suggested that other such triple rift systems in the world have had a similar origin.

Faced with the uncertainties in all existing theories for plate-moving forces, one should keep an open mind about other, perhaps less popular, hypotheses. Among other mechanisms proposed are "slab-pull" and "ridge-push" (Fig. 5–49D). Slab-pull is thought to operate at the subduction zone, where the subducting oceanic plate, being colder and denser than the surrounding mantle, sinks actively through the less dense mantle, "pulling" the rest of the slab along as it does so. The ridge-push mechanism, which may operate independently of or along with slab-pull, results from the fact that the spreading centers stand high on the ocean floor and have low-density roots. This causes them to spread out on either side of the mid-oceanic ridge and transmit this "push" to the tectonic plate. Other hypotheses suggest rafting of plates in response to stresses induced by differential rotation between the core and mantle, which might then be transmitted to the crust. In the years ahead, studies of rock behavior under high temperature and pressure, geophysical probings of the mantle, and careful field studies will reveal which of the proposed mechanisms can be most confidently accepted.

Tests of Plate Tectonics

We have examined the various lines of evidence supporting Wegener's notion of continental drift. Nearly all of these clues also support the newer concepts embodied in the theory of plate tectonics. The earlier clues were based mostly on evidence found on land. The new theory, with its keystone concept of sea-floor spreading, was developed from evidence gleaned from the sea floor.

In the 1960's, when geophysicists such as Harry Hess of Princeton University began formulating ideas of sea-floor spreading, other scientists were puzzling over findings related to paleomagnetism. The new data were obtained from sensitive magnetometers that were being carried back and forth across the oceans by research vessels. These instruments were able to detect not only the Earth's main geomagnetic field but also local magnetic disturbances or magnetic anomalies frozen into the rocks along the sea floor. Maps were produced that exhibited linear anomaly bands of high- and low-field magnetic intensities parallel to the west coast of North America (Fig. 5–51).

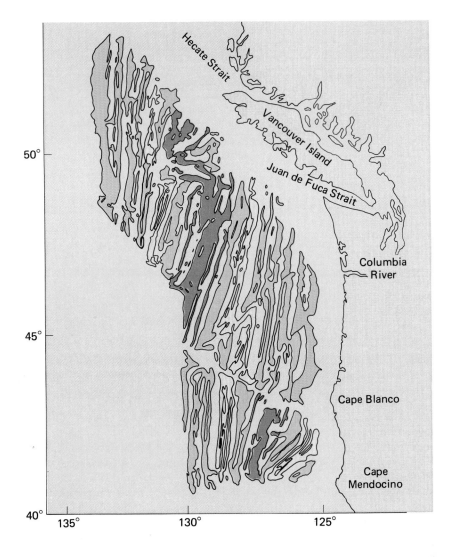

FIGURE 5–51 Magnetic field produced by the rocks on the floor of the northeast Pacific Ocean. Note the symmetry with respect to the ridges (the red areas). The ages of the rocks increase away from the ridges. The color bands are magnetically positive. *(After Raff, A. D., and Mason, R. G. 1961. Magnetic survey of the West Coast of North America, 40° N–52° N. latitude. Bull. Geol. Soc. Am., 72:1267–1270.)*

Surveys of magnetic field strength on traverses across the mid-Atlantic ridge revealed a similar pattern of symmetrically distributed belts of anomalies (Fig. 5–52). Directly over the ridge, the Earth's magnetic field was 1 percent stronger than expected. Adjacent to this zone, the field was somewhat weaker than would have been predicted, and the changes from weaker to stronger and back again often occurred in distances of only a few score miles.

In 1963, F. J. Vine, a research student at Cambridge University, and a senior colleague, Drummond Matthews, suggested that these variations in intensity were caused by reversals in the polarity of the Earth's magnetic field. The magnetometers towed behind the research vessels provided measurements that were the sum of the Earth's present magnetic field strength and the paleomagnetism frozen in the crustal rocks of the ocean floor. If the paleomagnetic polarity was opposite in sign to that of the Earth's present magnetic field, the sum would be less than the present magnetic field strength. This would indicate that the crust over which the ship was passing had reversed paleomagnetic polarity. Conversely, where the paleomagnetic polarity of the sea-floor basalts was the same as that of the present magnetic field, the sum would be greater, and a normal polarity would be indicated.

Since 1963, geophysicists have learned that these irregularly occurring reversals have not been at all infrequent. During the past 70 million years, the Earth's magnetic field has seen many episodes when

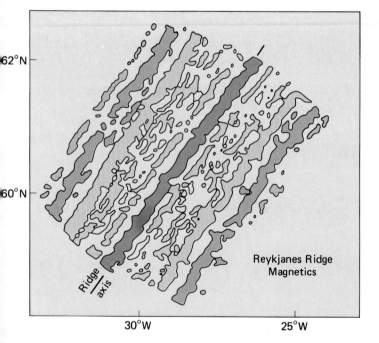

FIGURE 5–52 The magnetic field produced by the ocean-floor rocks over the Reykjanes ridge near Iceland. The outlined areas represent predominantly normal magnetization. The ages of the rocks increase away from the ridge. *(After Heirtzler, J. R., et al. 1966. Magnetic anomalies over the Reykjanes ridge, Deep Sea Res., 13:427–443; and Vine, F. J. 1968. Magnetic anomalies associated with mid-ocean ridges, in Phinney, R. A. (ed). The History of the Earth's Crust. A Symposium, Princeton, NJ: Princeton Univ. Press.)*

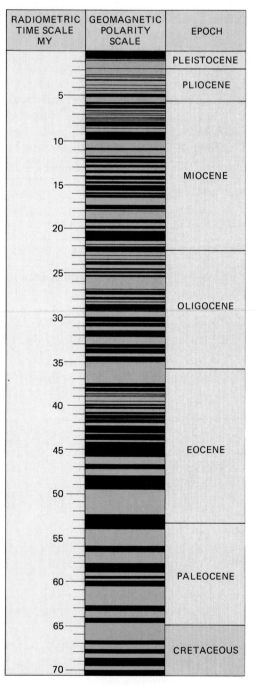

FIGURE 5–53 Reversals of the Earth's magnetic field during the past 70 million years. The field was "normal," as today, during intervals shown in black. *(Modified from Heirtzler, J. R., et al. 1968. J. Geophys. Res. 73:2119–2136. Copyright, The American Geophysical Union.)*

the polarity was opposite to that of today (Fig. 5–53). These changes in polarity were incorporated into the remanent magnetism of the lavas extruded at the midoceanic ridges. The lava would acquire the magnetic polarity present at the time of extrusion and then would move out laterally, as has been previously described. If during that time the Earth's polarity was as it is today, the stripe is said to represent "normal" polarity. If the Earth's normal polarity reversed, the band of extruded lavas that followed behind the previous band would acquire "reverse" polarity as it cooled past the Curie temperature. As the process repeated itself through time, the result would be symmetric, mirror-image patterns of normal and reverse bands on either side of spreading centers such as the midoceanic ridges (Fig. 5–54).

The stripes discovered by Vine and Matthews showed that Harry Hess was correct when he suggested in the early 1960s that the sea floor moves. At that time, however, there seemed to be no feasible way to determine the age of large areas of the ocean floor and thereby ascertain the rate of sea-floor movement. Cores of the basaltic sea floor were not suitable for radiometric dating because basalt is altered, when hot, by contact with sea water. By lucky coincidence, a method for determining the age of the stripes was provided through the work of A. Cox, R. R. Doell, and G. B. Dalrymple in 1963. These geophysicists were investigating the magnetic properties of lava flows on the continents. They were able to identify accurately periodic reversals of the Earth's magnetic field imprinted in superimposed layers of basalt (Fig. 5–55). Many of the layers in these superpositional sequences could be dated by radiometric methods. With such data, the time sequence of magnetic reversals for the past 5 million years was determined. The final step was to correlate these dates obtained from land basalts to the rocks of the sea floor. Confirmation for the ages of some of the ocean basalts was obtained from the study of fossils in overlying sediments.

With knowledge of the age of particular normal or reverse magnetic stripes, it was now possible to calculate rates of sea-floor spreading and sometimes to reconstruct the former positions of continents. Figure 5–56 illustrates how this can be done. For example, if one wishes to know the distance between the eastern coast of the United States and the northeastern coast of Africa about 81 million years ago, one brings together the traces of the two 81-million-year-old magnetic stripes, being careful to move the sea floor parallel to the transform faults, which indicate the direction of movement.

The velocity of plate movement is not uniform around the world. Plates that include large continents tend to move slowly. Their velocity relative to the

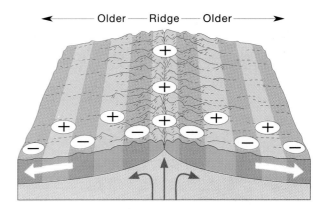

FIGURE 5–54 The normal (+) and reversed (−) magnetizations of the sea floor. Note the symmetry of the magnetizations with respect to the ridge. *(From McCormick, J. M., and Thiruvathukal, J. V. 1981.* Elements of Oceanography, *2nd ed. Philadelphia: Saunders College Publishing.)*

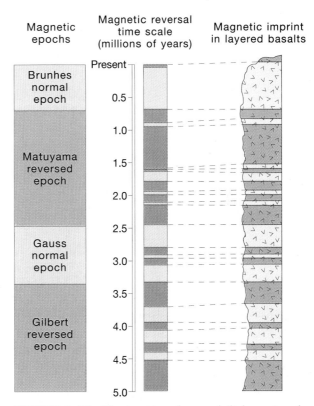

FIGURE 5–55 The imprint of normal (light tan) and reversed (dark tan) magnetic polarity in a composite section of layered basalts from many continental localities. The magnetic reversal time scale is shown in the center column, and at the left are the *magnetic epochs* named after famous investigators of magnetic effects. The magnetic epochs reflect a predominance of one kind of polarity. Within the larger magnetic epochs are *magnetic events* of shorter duration. *(Modified from Cox, A. 1963. Geomagnetic reversals,* Science, *163:237–245.)*

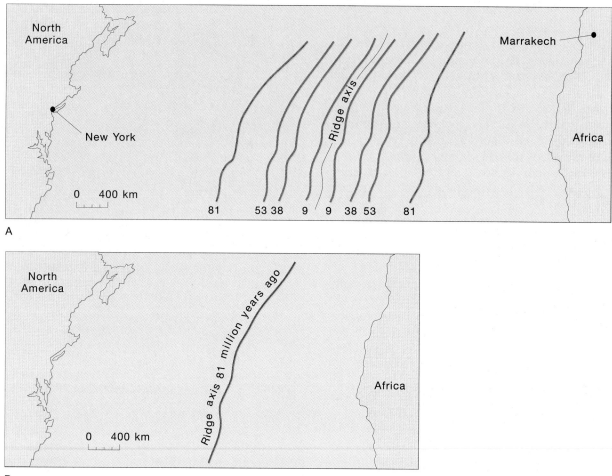

FIGURE 5–56   A method for determining the paleogeographic relations of continents. (A) The present North Atlantic with ages plotted for some of the magnetic stripes. To see the location of North America relative to Africa, say, 81 million years ago, the 81-million-year-old bands are brought together. (B) The result is the view of the much narrower North Atlantic of 81 million years ago.   *(Modified from Pitman III, W. C., and Talwani, M. 1972. Geol. Soc. Am. Bull. 83:619–644.)*

underlying mantle rarely exceeds 2 cm per year. Plates that are largely devoid of large continents have average velocities of between 6 and 9 cm per year. Figure 5–57 provides a summary of the rates and directions of sea-floor spreading as determined by analyses of displacements along transform faults, magnetic anomalies, and other geophysical data.

Evidence of Sea-Floor Spreading from Oceanic Sediment

If newly formed crust joined the ocean floor at spreading centers and then moved outward to either side, then the sediments, dated by the fossil planktonic organisms they contain, could be no older than the surface on which they came to rest. Near a mid-

oceanic ridge, the sediments directly over basalt should be relatively young. Samples of the first sedimentary layer above the basalt but progressively farther away from the spreading center should be progressively older (Fig. 5–58).

In a succession of cruises that began in 1969, the American drilling ship *Glomar Challenger* and subsequent cruises of the JOIDES *Resolution* (see Fig. 1–5) collected ample evidence to prove that the previously stated conclusions on the age distribution of sediments are correct. The studies carried out by scientists associated with the drilling project confirmed earlier assumptions that there were no sediments older than about 200 million years on the sea floor, that the sediments on the sea floor were relatively thin, and that they became thinnest closer to the mid-

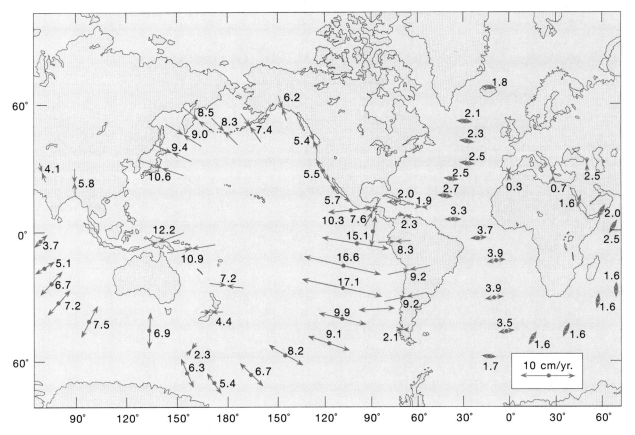

FIGURE 5–57 Rates and directions of sea-floor spreading. Note the more rapid spreading rates in the Pacific. *(Results from Minster, J. B., and Jordon, T. H. 1978. J. Geophys. Res. 83:5331–5354.)*

oceanic ridges. Spreading provided a logical reason for these observations. A given area of oceanic plate surface was simply not in existence long enough to accumulate a thick section of sediments over a long span of geologic time. The sea-floor clays and oozes were conveyed to subduction zones, dragged back down into the mantle, and (as suggested by rock compositional studies) resorbed. Thus, the entire world ocean is virtually swept free of deep-sea sediments every 200 to 300 million years.

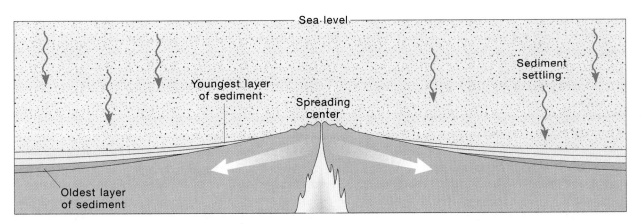

FIGURE 5–58 As a result of sea-floor spreading, sediments near the mid-oceanic ridges are youngest, and sediments directly above the basaltic crust but progressively farther away from the ridge are sequentially older.

C O M M E N T A R Y
Rates of Plate Movement

On examination of the rates at which tectonic plates move away from midoceanic ridges at various places around the Earth, it is immediately apparent that the rate of movement is greater at some places than at others. One reason for this relates to movement on a spherical surface. If the Earth were flat, as some viewed it before the voyages of Columbus, then all locations along a similarly flat plate would be moving away from a midoceanic ridge at about the same velocity. On a sphere, however, the plates are curved. They begin to separate at a location termed a pole of rotation and form an ever-widening split. The effect is somewhat like pulling apart the outer dry layer of an onion. The split in the skin would be widest where it was being pulled apart, and the gap would narrow to a point corresponding to the pole of rotation. On the Earth, plate movement away from the midoceanic ridge near the pole of rotation would be slower because of the lesser distance traversed over a given time interval. Conversely, far from the pole of rotation, plates diverge across a greater distance for the same time interval and hence have greater velocity. One can observe this effect in the North Atlantic, where spreading rates increase from about 0.75 cm per year near Iceland to over six times that amount near the equator. A corollary of this effect is that the width of new oceanic crust also increases with distance from the pole of rotation.

In addition to changes in plate velocity that are a consequence of the Earth's spherical shape, it appears that large expanses of continental crust incorporated into a plate can decrease its velocity. The Pacific and Nazca plates do not carry a heavy load of crust. These plates have higher velocities than the American and Eurasian plates, which are burdened with large continents.

As an example, sea-floor drilling and sampling has thus far been unable to locate sediments in the South Atlantic any older than 150 million years. This knowledge is related to the paleomagnetic calculations that suggest that South America and Africa have been moving apart from one another at a rate of about 4 cm per year. If this rate has been fairly constant, then extrapolating back in time, one may estimate that the two continents should have been in contact about 150 million years ago during the Jurassic. After that time, the two fragments (Africa and South America) moved apart as oceanic basalts filled the gap left behind. In this way, only sediments younger than the data of fragmentation would be present, since there was not an oceanic basin of deposition in existence prior to that time. Similar reconstructions have been made from sea-floor samples and cores elsewhere around the world.

Measurements from Space

The space program has provided yet another method for measuring the movement of tectonic plates. This new method uses laser devices, which emit an intense beam of light of definite wavelength. The beam is capable of traveling across an enormous distance without becoming dispersed. During the *Apollo* missions, astronauts placed three clusters of special reflectors on the moon. Laser beams from Earth are "fired" toward the moon, and by recording the time required for the signal to travel to the moon and return to Earth, one can determine the distance between the laser device and the lunar reflector. The method is accurate to within only 3 cm. Distances from the moon to stations on two moving tectonic plates will, of course, change over a period of time, and such measurements can be used to calculate how fast continents are moving toward or away from one another.

Beaming lasers to artificial satellites from two widely spaced Earth stations, a technique called *satellite laser ranging (SLR),* can also be used to find the distance between two locations on Earth. The Lageos satellite, which circles the Earth at an altitude of 5800 km, is currently being used to reflect laser pulses back to their source on Earth. By accurately measuring the time required for the laser pulse to travel to the satellite and return, the position of the earthbound station is precisely ascertained. A similar measurement made from a second station (perhaps on a different continent) fixes the distance between the two stations. Measurements are made again after a given lapse of time to determine the amount and rate of increase or decrease in distance between the two ground stations.

Another method that provides similar accuracy is called *very long baseline interferometry (VLBI).* In this method, several stations at widely spaced locations on Earth monitor the radio signals from distant quasars (very large redshift, intergalactic, luminous objects). Quasars are so far away from our planet (billions of light-years) that they serve as stationary reference points. The differences in the arrival times of radio signals received from the quasars determine the distance between the ground-based stations, and changes in those distances are then noted for given increments of time, as in the SLR method. Measurements based on monitoring stations in the United States and Europe indicate that the North American and European tectonic plates are moving apart at a rate of 1.9 cm per year and that Hawaii is drifting toward Japan at the relatively rapid rate of 8.3 cm per year. These determinations compare very well with measurements based on the spacing of magnetic bands on the floor of the Atlantic and Pacific oceans.

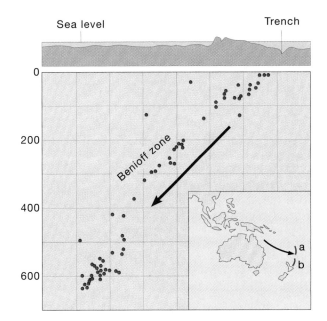

FIGURE 5–59 Vertical cross-section showing spatial distribution of earthquakes along a line perpendicular to the Tonga trench and volcanic island arc. The Tonga trench (*a* on inset map) lies just to the north of the Kermodec trench *(b)*. *(Simplified from Isacks, B., Oliver, J., and Sykes, R. 1968. J. Geophys. Res. 73:5855.)*

Seismologic Evidence for Sea-Floor Spreading

The evidence for the actual movement of lithospheric plates appears to be fairly direct and susceptible to several methods of testing. It is more difficult to prove that plates of the lithosphere are dragged back down into the mantle and resorbed. The best indications that subduction actually does occur come from the study of earthquakes generated along the seismically active colliding edges of plates.

Well known among these studies are the impressive investigations of Hugo Benioff, a seismologist and designer of improved seismographs. Benioff and subsequent investigators compiled records of earthquake foci that occurred along presumed plate boundaries. The charting of the data showed that the deeper earthquake foci occur along a narrow zone that was tilted at an angle of 45° under the adjacent island arc or continent (Fig. 5–59). This inclined seismic shear zone was traced to depths as great as 700 km (435 miles). It was named the **Benioff seismic zone** after its discoverer. These inclined earthquake zones are believed to define the positions of the subducted plates where they plunge into the mantle beneath the overriding plate. The earthquakes are the result of fracturing and subsequent rupture as the cool, brittle lithospheric plate descends into the hot mantle.

In addition to the deep and intermediate focus earthquakes along the Benioff zone, there are intense and often destructive earthquakes at shallow depths where the edges of the two rigid plates press against one another. The 1964 Alaskan earthquake originated at this shallow level. Such quakes are thought to occur along the shear plane between the subducting ocean lithosphere and the continental lithosphere.

Some may result from normal faulting as portions of the lithosphere are arched by the collision.

Gravity Plays a Role

Gravity measured over deep-sea trenches is characteristically lower than that measured in areas adjacent to the trenches. Geophysicists refer to such phenomena as negative gravity anomalies. A **gravity anomaly** is the difference between the observed value of gravity at any point on the Earth and the computed theoretical value. **Negative gravity anomalies** occur where there is an excess of low-density rock beneath the surface. The very strong negative gravity anomalies extending along the margins of the deep-sea trenches can mean only that such belts are underlain by rocks much lighter than those at depth on either side. Because the zone of lower gravity values is narrow in most places, it is believed to mark trends where lighter rocks dip steeply into the denser mantle (Fig. 5–60). Geophysicists assume that the less dense rocks of the negative zone must be held down by some force to prevent them from floating upward to a level appropriate to their density. That force might be provided by a descending convection current, or the plate may be pushed by rising magma along mid-oceanic ridges until it encounters an opposing plate and plunges into the asthenosphere.

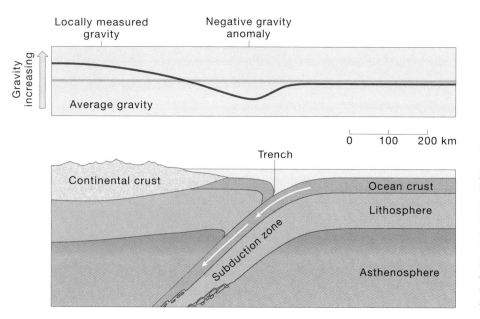

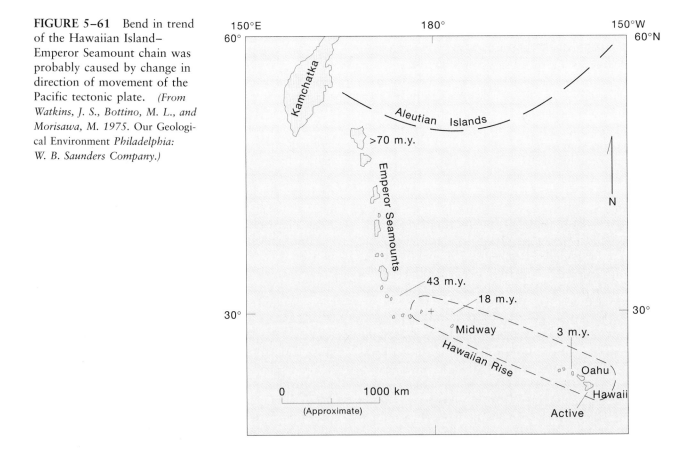

FIGURE 5–60 Diagram showing the variation in gravity over a deep-sea trench. The trench, subducting sediments, and relatively low-density rocks occupy space that would otherwise be filled with denser rocks. Hence, the force of gravity over the trench and subduction zone is weaker than over the Earth generally.

Hot Spots

Anyone examining the superb topographic maps (see Fig. 1–4) of the sea floor that have been produced over the past decade cannot help but notice the chains of volcanic islands and seamounts (submerged volcanoes) that rise in alignment on the sea floor.

Because these volcanoes occur at great distances from plate margins, they must have originated differently from the volcanoes associated with midoceanic ridges or deep-sea trenches. Their striking alignment has been explained as being a consequence of sea-floor spreading. According to this notion, intraoceanic volcanoes develop over a "hot spot" in the astheno-

FIGURE 5–61 Bend in trend of the Hawaiian Island–Emperor Seamount chain was probably caused by change in direction of movement of the Pacific tectonic plate. *(From Watkins, J. S., Bottino, M. L., and Morisawa, M. 1975. Our Geological Environment Philadelphia: W. B. Saunders Company.)*

C O M M E N T A R Y
The Missing Hawaiian Volcano

In this chapter we have described how the Hawaiian chain of volcanic islands originated by passage of the pacific plate over a relatively stable hot spot. In addition to the general alignment of the Hawaiian Islands and Emperor seamounts, however, one can also observe that the volcanoes located along the southern segment of the chain occur sequentially along two lines, as shown in the accompanying figure. Along each line, the volcanoes are fairly evenly spaced, except for a large gap in the western line between the volcanoes Kahoolawe and Hualalai. Geologists have been concerned with this extra space, which seemed to disrupt an otherwise remarkably uniform pattern. Those concerns, however, were put to rest recently when new bathymetric data and geochemical analyses of lava dredged from the sea floor demonstrated the existence of a submerged volcano within what had been assumed was a gap in the chain. The missing volcano was about 70 km long by about 30 km wide, and it had a height of 4 km above the abyssal sea floor. Geologists named the volcano Mahukona after the closest settlement to the volcano on the island of Hawaii.

The question of the gap in the southern Hawaiian chain had been solved, but two questions remained. Why were the volcanoes in a paired sequence, and why did Mahukona become extinct before building itself upward to form an island? In laboratory experiments that model ascending hot spot plumes, the plumes bifurcate as they rise, forming two distinct branches. This may have occurred in the ascending plume that generated the volcanoes of the southern Hawaiian chain, to produce two parallel lines of volcanoes about 35 km apart. One of these volcanoes,

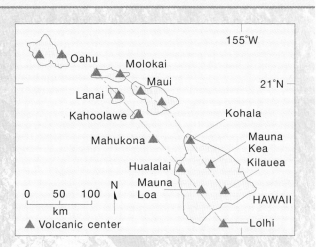

Southern Hawaiian Islands with paired sequence of volcanoes.

Mahukona, did not reach sea level possibly because the branch feeding the western line was smaller or became depleted, causing Mahukona to become dormant before it reached sea level. Other volcanoes on the western line, such as Lanai and Kahoolawe, although they reached sea level, tend to be smaller than volcanoes in the eastern line. These are simple models to explain the characteristics of the Hawaiian chains, and they may be modified with time.

Reference
Garcia, M. O., Kurz, M. D., and Muenow, D. W. 1990. *Geology* 18:1111–1114.

sphere. The hot spot is a manifestation of one of the deep plumes of upwelling mantle rock described earlier. Lava from the plume may work its way to the surface to erupt as a volcano on the sea floor. As the sea floor moves (at rates as high as 10 cm per year in the Pacific), volcanoes form over the hot spot and expire as they are conveyed away. New volcanoes form at the original location. The process may be repeated indefinitely, resulting in a great linear succession of volcanoes that may extend for thousands of kilometers in the direction that the sea floor has moved. The Hawaiian volcanic chain is believed to have formed in this manner from a single source of lava over which the Pacific plate has passed on a

northwesterly course. In support of this concept are radioactive dates obtained from rocks of volcanoes that clearly indicate that those farthest from the source are the oldest.

At the western end of the Hawaiian Islands is a string of submerged peaks called the Emperor seamounts (Fig. 5–61). These submerged volcanoes trend in a more northerly direction than the Hawaiian Islands. However, both the seamounts and the islands are part of a single chain that has been bent as a result of a change in the direction of plate movement. Such an interpretation is supported by age determinations. The oldest of the Hawaiian Islands (near the bend) was formed about 40 million years

(text continues on page 202)

Hawaii Volcanoes National Park

If one wants to see geology in action, there are few places that surpass Hawaii Volcanoes National Park (Fig. A). Within the park, the active volcanoes Mauna Loa and Kilauea periodically provide spectacular fiery fountains and rivers of incandescent lava as evidence that the primordial forces involved in the Earth's development are still at work. Mauna Loa attests to the prodigious ability of volcanoes as land builders. It is the most massive mountain on Earth, rising more than 6 miles (9.6 km) above the floor of the Pacific Ocean.

Broad, gently sloping volcanoes like Mauna Loa and the other volcanoes of the Hawaiian chain are termed *shield volcanoes* because their shape resembles that of the shields of ancient warriors. Their broadly convex shape results from repeated eruptions of highly fluid basaltic lava emerging either from vents or along rift zones.

The Kilauea area of Hawaii Volcanoes National Park is an ideal place to view many of the characteristic features of shield volcanoes. One can easily explore this area along the 11-mile Crater Rim Drive that encircles Kilauea's summit caldera and craters (see accompanying photograph). Along this road, one passes through a lush rainforest environment, a rain shadow desert, lava flows, and well-marked stops and walks to view Sulphur Banks, Steam Vents, the Saggar Museum, Halemaʻumaʻu Crater, Devastation Trail, and Kilauea Ili Crater. On the east side of the drive is the Thurston Lava Tube. Lava tubes are formed when lava breaks through and drains out beneath a solidified crust of basalt. Along the trails leading from Crater Rim Drive, one can find good examples of pahoehoe and aa lava. *Pahoehoe* has a ropy appearance produced in the still-plastic surface scum of lava by the drag of more rapidly flowing lava below. In contrast, *aa* has a blocky, fragmented texture. It forms as the thicker upper layer of lava hardens and is carried in conveyor belt fashion to the front of the flow. There the brittle top layer is broken into jagged chunks that accumulate at the leading edge of the flow. If you were to walk on this jagged material, you might easily guess why Hawaiians call it *aa*.

Basalt is by far the most abundant rock in Hawaii Volcanoes National Park. With successive eruptions, Hawaiian basalts become poorer in silica and more enriched in sodium and potassium. The sodium- and potassium-rich *alkalic* lavas are more viscous. They are found largely near the summits of the shield volcanoes. As is common in many basalts, those of Hawaii contain the mineral olivine. Often one finds large crystals of olivine (phenocrysts) in the basalt, and these are cut and polished for jewelry. The gem name for olivine is *peridot*. Peridot rings and bracelets can be found in almost any Hawaiian shop catering to tourists.

Volcanoes of the Hawaiian Islands are *intraplate* volcanoes. They are located within a tectonic plate rather than along its more dynamic margins. Intraplate volcanoes are in the minority, for more than 75 percent of the Earth's

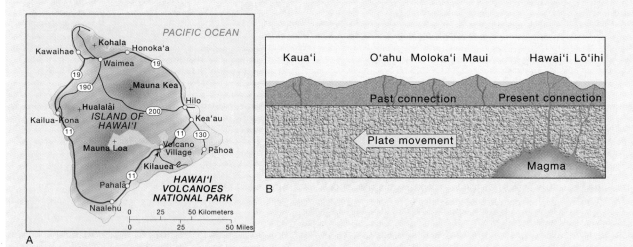

(A) Location map for Hawaii Volcanoes National Park. (B) Cross-section indicating origin of volcanoes of the Hawaiian Island chain as the Pacific plate passed over a hot spot. *(A and B courtesy of the U.S. National Park Service.)*

On the rim of the Kilauea caldera. A pit crater that last erupted in 1967–1968 is visible within the larger caldera. Ama'Uma'U ferns grow in the foreground. *(Photograph by Jeff Gnass.)*

volcanoes are distributed around the edges of the Pacific plate. They form the so-called ring of fire that includes Washington's Mount St. Helens, Alaska's Katmai, Popacatepetl in Mexico, and Japan's Mount Fuji. As discussed in this chapter, the intraplate volcanoes of the Hawaiian chain originated as the Pacific plate moved slowly across a *hot spot* or plume of hot magma rising from the upper mantle. This process accounts for the general alignment of Hawaiian volcanoes. Early eruptions form a seamount, and after thousands of additional eruptions over several hundred thousand years, the volcanic edifice is built above sea level. The island formed in this way continues to grow with subsequent eruptions until movement of the Pacific plate carries it away from the hot spot, severing its connection to the parent magma (Fig. B). Over the past 70 million years, the Pacific plate has carried the Hawaiian Islands northwest of the hot spot at an average rate of 10 cm per year. In recent years, the large island of Hawaii has continued to grow, with periodic contributions of lava from Mauna Loa and Kilauea. Hawaii, however, will not be the final island of the Hawaiian Island chain. To the southeast, the seamount Lo'ihi is building its way to the surface of the sea.

The figure on page 199 shows the location of volcanoes along the southern part of the Hawaiian Island chain. These locations suggest that the volcanoes occur in pairs that are about 35 km apart. Thus, in Hawaii Volcanoes National Park, Mauna Loa has its mate in Kilauea. To account for the paired arrangement of volcanoes, geologists speculate that the plume supplying the volcanoes has two branches, with one branch providing lava to the western volcano and one to the eastern. Kohala at the northern end of Hawaii appears to lack a mate. Actually, however, the mate is present as the seamount Mahukona. Mahukona may have failed to become an island because the branch of the plume supplying the western line of volcanoes was smaller or became depleted.

For scientists, Mauna Loa and Kilauea provide a natural laboratory for on-site study of volcanism. At the Hawaii Volcano Observatory and the University of Hawaii, volcanologists record the periodicity of eruptions, changes in magnetic properties, and variations in the composition and temperature of gases emitted at fumaroles. They record the swelling or tilting of the land surface and carefully consider every tremor associated with subterranean movements of magma. These observations help in finding ways to forecast eruptions and in understanding the behavior of volcanoes everywhere around the world. What we learn from today's Hawaiian volcanoes can be used to interpret the volcanism of long ago and to predict what can be expected in the future.

ago. The seamounts continue the age sequence toward the end of the chain, where the peaks are about 80 million years old. Thus, 40 million years ago the Pacific plate changed course and put a kink in the Hawaiian chain.

Hot spots may not sit still. In the past, they have been assumed to remain fixed in one location as plates slide over them. Recent investigations, however, suggest that hot spots may drift somewhat under the influence of moving currents of mantle material. The analogy would be smoke drifting from a smokestack under the influence of a breeze. Research by geologists Bernard Steinberger and Rick O'Connell indicates that the Hawaiian hot spot is drifting in a direction opposite to that of the Pacific plate, but at only one-tenth of its speed.

The ocean around Hawaii is not the only part of the globe that has hot spots. As indicated in Figure 5–62, hot spots are widely dispersed and occur beneath both continental and oceanic crust. Yellowstone National Park (Fig. 5–63) is over a hot spot that, like the Hawaiian Islands, is in the interior of a continent.

Lost Continents and Alien Terranes

We are accustomed to thinking of continental crust in terms of large land masses such as North America or Eurasia. There are, however, many relatively small patches of continental crust scattered about on the lithosphere. As long ago as 1915, Alfred Wegener described the Seychelles Bank (Fig. 5–64) in the Indian Ocean as a small continental fragment that had broken away from Africa. The higher parts of the Seychelles Bank project above sea level as islands, but many other such small patches of continental crust are totally submerged. Geologists use the term **microcontinents** for these bits of continental crust that are surrounded by oceanic crust. They are recognized by their granitic composition, by the velocity with which compressional seismic waves traverse them (6.0 to 6.4 km/second), by their general elevation above the oceanic crust, and by their comparatively quiet seismic nature.

It is apparent that microcontinents are small pieces of larger continents that have experienced fragmentation. As these smaller pieces of continental

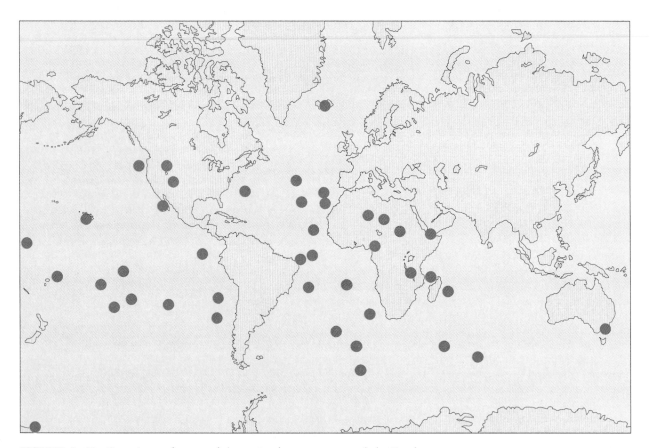

FIGURE 5–62   Locations of some of the major hot spots around the Earth.
(Courtesy of Tom Crough, Department of Geologic and Geophysical Sciences, Princeton University.)

FIGURE 5–63 Mammoth Geyser, Yellowstone Park, Wyoming. Yellowstone is over a hot spot. Surface waters percolating through a system of deep fractures reach the hot rocks below and erupt in columns of hot water and steam.

crust are moved along by sea-floor spreading, they may ultimately converge on the subduction zone at the margin of a large continent. Because they are composed of relatively low-density rock and hence are buoyant, they are a difficult bite for the subduc-

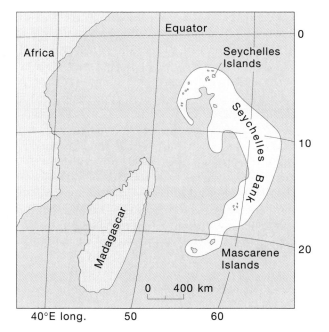

FIGURE 5–64 Location map of the Seychelles Bank.

tion zone to swallow. Their buoyancy prevents their being carried down into the mantle and assimilated. Indeed, the small patch of crust may become incorporated into the crumpled margin of the larger continent as an exotic block, or so-called **suspect terrane.**

It is interesting that geologists found evidence of microcontinents long before the present theory for their origin was formulated. While mapping Precambrian rocks, they came across areas that were incongruous in structure, age, fossil content, lithology, and paleomagnetic orientation when compared to the surrounding geology. It was as if these areas were small, self-contained, isolated geologic provinces. Often their boundaries were marked by major faults. Geologists designated these areas as **alien** or **allochthonous terranes** to indicate that they had not originated in the places where they now rested. Allochthonous terranes have been recognized on every major land mass, with well-studied examples in the northeastern former Soviet Union, in the Appalachians, and in many parts of western North America (Fig. 5–65). In particular, Alaska appears to be largely constructed of allochthonous terranes.

If splinters of continents can be transported on the spreading sea floor, so can pieces of oceanic crust. Particularly in the Cordilleran mountain belt of North America, one finds allochthonous terranes that were apparently microplates of ocean crust containing volcanoes, seamounts, segments of island arcs, and other features of the ocean floor. All of

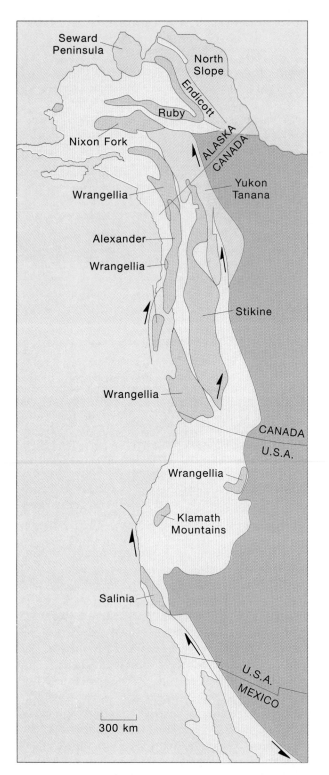

FIGURE 5–65 The larger allochthonous continental terranes in western North America that contain Paleozoic or older rocks. At some time prior to reaching their present locations, these terranes were continental fragments embedded in oceanic crust or microcontinents. Those colored green probably originated as parts of continents other than North America, whereas the pink blocks are possibly displaced parts of the North American continent. *(From Ben-Avraham, Z. 1981. Am. Sci. 69:298.)*

these features were carried to the western margin of North America like passengers on a huge conveyor belt. As the plate that bore them plunged downward at the subduction zone located along the continental margin, volcanoes, seamounts, and the other features were scraped off the descending plate and plastered onto the continental margin. The result was a vast collage of accreted oceanic microplates interspersed with similarly transported microcontinents. It appears that about 50 such allochthonous or exotic terranes exist in the Cordillera, all lying west of the edge of the continent as it existed about 200 million years ago. Clearly this process of successive additions of oceanic and continental microplates can significantly increase the rate at which a continent is able to grow. It is also apparent that active continental margins such as those bordering the Pacific Ocean are likely to grow faster than passive margins because of microplate accretion.

Unraveling the orogenic history of mountain ranges containing multiple allochthonous terranes is an enormously complex task requiring the cooperation of geologists, geophysicists, and paleontologists. Such a collaboration can often yield dramatic results. Recently, paleontologists and geologists working in the Wallowa Mountains of Oregon discovered a massive coral reef of Triassic age. The reef, however, clearly did not belong where it was found. It rested on volcanic rocks whose paleomagnetic orientation indicated that they had solidified from lavas at a latitude considerably closer to the equator. Further investigation indicated that the fossil organisms in the reef were identical to those found in Triassic strata of the Austrian and German Alps. These strata were deposited prior to Alpine mountain building in a seaway called the Tethys that extended from the present east coast of Japan into the Mediterranean region. The similarity between the fossils of Oregon and those of the Alps strongly suggests that the reefs now in Oregon actually grew around the margins of volcanic islands in the Tethys Sea and, in the subsequent 220 million years or so, were transported thousands of kilometers eastward until they collided with the North American continent. By that time, of course, the reef organisms had died, but their fossilized skeletons remained as the reef was accreted onto the western margin of North America.

Another somewhat implausible explanation for the occurrence in Oregon of a Triassic fauna from the Tethys Sea would be the migration of the free-swimming larval stages of reef organisms from island to island across the Pacific until they reached North America. This would require a remarkably continuous string of islands, for larval stages are generally relatively short lived.

◼ *Life on a Traveling Crust*

Of all the tectonic features discussed in this chapter, continents are the easiest to locate in the past. Their shifting positions and changing configurations had far-reaching effects on the Earth's inhabitants. The series of maps depicted in Figure 5–25 shows the continents in various stages of fragmentation since the breakup of Pangea, the supercontinent of Permian–Triassic time that Wegener championed. (Geologists now know that this land mass was preceded by an earlier supercontinent termed Proto-Pangea that fragmented and was reassembled.)

The American geologists J. W. Valentine and E. M. Moores have devised an interesting model that provides a means by which one can more easily imagine the effects on life of such continental fragmentations and reassemblies. Imagine the Earth as having a single supercontinent reminiscent of Pangea centered on the South Pole (rather like Fig. 5–66). In this configuration, the entire coastline of the land mass would be at about the same latitude and thus would have rather similar climatic characteristics. The bottom-dwelling (benthic) shallow-water marine invertebrates that are so frequent in the fossil record would have had ease of migration around the entire supercontinent, because there would have been few climatic barriers. The result would have been a single so-called **shelf province.** Within that province there might have been differences in the topography or composition of the sea floor that might have influenced the kinds of biologic communities that would have lived in particular locations. However, because of the general uniformity of conditions and ease of migration and interaction, global biologic diversity would probably have been relatively low.

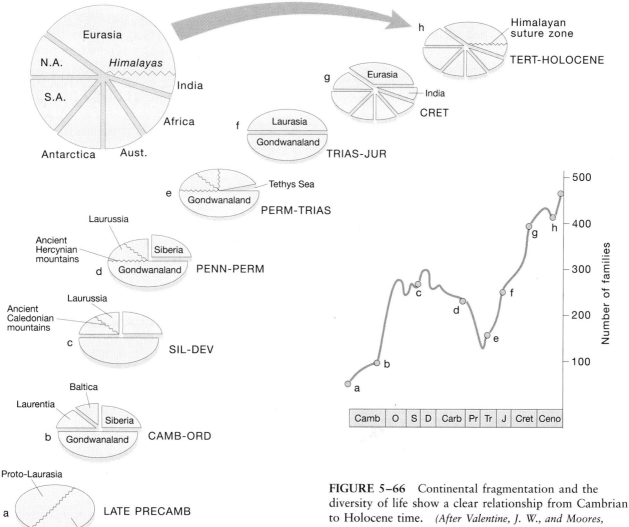

FIGURE 5–66 Continental fragmentation and the diversity of life show a clear relationship from Cambrian to Holocene time. *(After Valentine, J. W., and Moores, E. M. 1972. J. Geol. 80(2):167–184. Copyright © 1972 by the University of Chicago.)*

A single supercontinent would also provide terrestrial vertebrates with relative ease of widespread migration and interaction. This is likely to have caused the development of cosmopolitan land faunas with fewer families and orders of animals than would have been the case if there had been many smaller isolated land areas. Paleomagnetic and paleontologic evidence indicates that although Pangea had begun to divide in the Triassic Era, the rifts did not become effective barriers to dispersal of terrestrial animals until very late in the Cretaceous and Early Cenozoic (Figs. 5–66G and H). Thus, during most of the 200 million years of the Mesozoic "Age of Reptiles," dispersal was a relatively simple matter. In contrast, Cenozoic separations became increasingly more extreme, and many terrestrial segments were soon in positions of relative isolation. During this "Age of Mammals," the earlier supercontinent divided into eight smaller units: North America, Greenland, Eurasia, Africa, India, Australia, South America, and Antarctica. These smaller continents were still further divided by inland seas. Mammals evolving on each continent underwent many separate and different evolutionary radiations and developed distinctive populations. The result was a greater diversity of land vertebrates than the Earth had ever hosted. The Age of Mammals gave rise to over 30 taxonomic orders of mammals in only 65 million years. The previous Age of Reptiles, which lasted over three times as long, saw the evolution of only 20 reptilian orders.

The Cenozoic world is somewhat analogous to the continental configuration depicted in Figure 5–66H. The continents in the model are exceptionally fragmented and widely dispersed. The benthic marine invertebrates that inhabit continental shelves of continents near the equator would probably develop a very high diversity of species because of the general environmental stability there. Continents in higher latitudes would have larger numbers of individuals but relatively less diversity because of the greater rigor of the poleward environments. In terms of the entire Earth, there would be a greater variety of habitats, a larger number of individual centers of biologic radiations, and a corresponding greater increase in global diversity. Today we live in a fragmented world, and life is indeed diverse.

Professors Valentine and Moores are among several Earth scientists who have suggested that unusual episodes of extinctions as well as changes in diversity can be related to plate tectonics. For example, when continents come together to form a supercontinent (Fig. 5–66E), there would be a trend toward extremes of climate, elimination of habitats, regression of inland seas, and loss of barriers to migration. Scores of species would probably perish because of their inability to adjust to these changes, and the fossil record would suggest that an episode of mass extinction had occurred. The survivors and their progeny would probably develop into a rather cosmopolitan world fauna.

We have noted that fragmentation provides multiple centers of biologic diversification. This may well be the reason for the increased variety of life in the geologic periods following the breakup of Proto-Pangea and Pangea. The fragmentation episodes may have been accompanied by worldwide changes in sea level (eustasy) resulting from the growth of midoceanic ridges and the displacement of water that ensued. This is not a wild guess, for it has been calculated that if such ridges today were to vanish, sea level would be lowered approximately 650 meters. During these times of rising sea level, marine waters would spill over onto the low-lying regions of continents, creating vast inland seas ideal for the proliferation of marine invertebrates. These transgressing inland seas would also have a moderating effect on temperature, humidity, and seasonality. The result would be the increase in the number of families of animals depicted in Figure 5–66 for those segments of geologic time in which continents were many and scattered.

The study of the relationship between organisms and their environments is enormously complex. Small changes in faunal elements at lower levels in food pyramids profoundly affect higher levels, often on a global scale. To understand the many variables involved is difficult. To comprehend them from clues in rocks hundreds of millions of years old is far more complex. However, as paleontologists continue to re-examine the data in light of the new theories, they find that the effects of plate tectonics on all the Earth's former inhabitants have been profound.

Summary

The Earth is a complex, dynamic, and delicately balanced planet. The geologic events we witness or see recorded in rocks are ultimately the result of what lies deep beneath the crust. To understand how the Earth works, it is necessary to know the nature of the planetary core and the differentiated zones that surround the core.

Knowledge of the Earth's deep interior is derived from the study of earthquake waves. Among the various kinds of earthquake or seismic waves are primary, secondary, and surface waves. Primary and secondary waves (also called body waves) pass deep within the Earth and therefore are the most instructive. Study of abrupt changes in the velocities of seismic waves at different depths provides the basis for a threefold division of the Earth into a central core; a thick, overlying mantle; and a thin, enveloping crust. Sudden changes in earthquake wave velocities are termed discontinuities.

The core of the Earth, as indicated by the Gutenberg discontinuity, begins at a depth of 2900 km. The core is divided into an inner and outer core. It is likely that the outer core is molten and that the entire core is composed of iron with small amounts of nickel and possibly either silicon or sulfur. The primary evidence for the existence of a liquid outer core is the disappearance of S-waves at a depth of about 2900 km. S-waves are not transmitted through liquids.

Constituting about 80 percent of the Earth's volume, the mantle is composed of iron and magnesium silicates such as olivine and pyroxene. A hot, plastic zone in the mantle is recognized by lower seismic wave velocities. The presence of this low-velocity layer, which is part of the asthenosphere, is an important element in the modern theory of plate tectonics. The asthenosphere may function as a weak plastic layer in which horizontal motions propel the more rigid surface layers as they break apart, collide, or slip past one another.

The seismic boundary that separates the mantle from the overlying crust is the Mohorovičić discontinuity. It lies far deeper under the continents than under the ocean basins. Thus, the continental crust is thicker than the oceanic crust. There are compositional and density differences as well. The continental crust has an overall granitic composition and is less dense than the oceanic crust, which is composed of basaltic rocks. It is likely that earliest patches of continental crust were products of weathering and erosion of pre-existing basaltic crust as well as of conversion of sediment to rocks of granitic character during episodes of metamorphism and melting.

The crust of the Earth is not as firm and stable as we sometimes think. We are reminded of this fact when earthquakes occur and when we observe strata that have been severely deformed and broken by compressional and tensional forces. These forces bend rocks if they are bendable and break them if they are brittle. The result is geologic structures such as faults and folds.

Faults are breaks in crustal rocks along which there has been a displacement of one side relative to the other. According to the directions of that movement, faults are classified as normal, reverse (thrust), lateral or intermediate types. Normal faults result from tensional forces and are recognized by the apparent downward movement of the hanging wall relative to the footwall. In reverse faults, the hanging wall appears to move up relative to the footwall. A thrust fault is merely a variety of reverse fault characterized by a fault plane that has a low angle of inclination relative to a horizontal plane.

No less important than faults as evidence of the Earth's instability are folds, the principal categories of which are anticlines, domes, synclines, basins, and monoclines. Both anticlines and domes have uparched strata but differ in that domes are roughly symmetric, with beds dipping more or less equally away from some central point. Synclines and basins are composed of down-folded strata, with the basin being more or less circular and characterized by beds that dip inward toward the center. In monoclines (Fig. 5–21), strata dip for an indefinite length in one direction and then return to their former, usually horizontal, attitude.

The majority of geologic structures were either directly or indirectly produced by the movement of tectonic plates. They are the products of plate tectonics. According to the concepts embodied in plate tectonics, the crust of the Earth and part of the upper mantle compose a brittle shell called the lithosphere. The lithosphere is broken into a number of plates that override a soft plastic layer in the mantle called the asthenosphere. Heat from within the Earth may create convection currents in the mantle, and such currents may provide a propelling force to move the plates. However, sliding of plates in response to gravity may also prove to be a mechanism for their movement.

Materials for the lithospheric plates originate along fracture zones typified by midoceanic ridges. Along these tensional features, basaltic lavas rise and become incorporated into the trailing edges of the plates, simultaneously taking on the magnetism of the Earth's field as they crystallize. In recent geologic time, this process has provided a record of geomagnetic reversals that in turn permit geophysicists to determine the rates of sea-floor spreading.

As long as they do not have continents on opposing edges, when plates collide, one of them may slide beneath the other and enter the mantle at a steep angle, to be remelted at depth. The zone of collision between plates may be marked by systems of volcanoes, deep-sea trenches, and great mountain ranges. The continental crustal segments of the lithosphere ride passively on the plates. They have a lower density than does oceanic crust and do not sink into subduction zones. This explains why continents are older than the ocean floors and why, when continents collide, they produce mountain ranges rather than trenches. Where continents straddle zones of divergent movements in the asthenosphere, the land mass may break and produce rift features such as the Red Sea and Gulf of Aden in the Afar triangle. These fracture zones may then widen further, resulting in the formation of new tracts of ocean floor.

Aside from the suturing of one large land mass onto another, a continent may grow in either of two ways, both involving mountain building. One process involves orogenic compression and metamorphism of sediments that had accumulated in marginal depositional basins. The second is by accretion of microcontinents and oceanic microplates bearing volcanoes, sea-floor plateaus, and seamounts. These bits and pieces of crust are carried by sea-floor spreading to marginal subduction zones, where they are accreted onto the edge of the continent as allochthonous terranes. The Cordillera of North America appears to be largely composed of a complex collage of such terranes, each having its own distinctive geology that can often be traced to distant source regions.

Many once poorly understood and isolated problems appear to have found an explanation in the theory of plate tectonics. This theory has provided an interrelated network of answers for such questions as how continents grow, why basins of sedimentation experience upheaval to form mountains, what causes the majority of the world's earthquakes and volcanoes, what determines the locations of major mineral deposits, and why the history of life on Earth is interrupted by episodes of crisis or of variation in biologic diversity.

Review Questions

1. If one were to able to drill a well from the North Pole to the center of the Earth, what internal zones would be penetrated?
2. What are the three major categories of seismic waves? Describe their characteristics.
3. What does the presence of an S-wave shadow zone indicate about the interior of the Earth?
4. What is a seismic discontinuity? Where are the Gutenberg and Mohorovičić discontinuities located?
5. How do anticlines (and domes) differ from synclines (and basins) with regard to the age relations of rocks exposed across the erosionally truncated surfaces of these structures?
6. What kind of rock approximates the average composition of the continental crust? The oceanic crust? The mantle?
7. What kind of body waves would be received on a seismograph located 180° from the epicenter of an earthquake?
8. What are the principal categories of faults? What kinds of faults might one find in regions subjected to great compressional forces? What kinds of faults result primarily from tension in the Earth's crust?
9. What is a gravity anomaly? What sort of gravity anomaly might one expect over a subduction zone? A midoceanic ridge?
10. What are folds? What are the principal kinds of folds?
11. Compile a list of items that Alfred Wegener might have used to convince a skeptic of the validity of his theory of continental drift.
12. Do midoceanic volcanoes have lavas of granitic or basaltic composition? What is the composition of lavas in mountain ranges adjacent to subduction zones, such as the Andes? Account for the differences in composition.
13. According to plate tectonics, how did the Himalaya Mountains form? The San Andreas fault? The Dead Sea and Red Sea?
14. According to plate tectonics, where is new material added to the sea floor, and where is older material consumed?
15. A moving tectonic plate must logically have a leading edge, a trailing edge, and sides. Where are the leading and trailing edges of the North American tectonic plate?

Discussion Questions

1. How do the continental crust and oceanic crust differ in composition, thickness, and density? How do these differences relate to isostasy?
2. What is the inferred composition of the mantle? Of the core? On what evidence are these compositions based?
3. If the Earth's internal zonation is the result of heating at an early stage in its history, what were the sources of heat? Was more heat generated during the early Earth's history than today, and if so, why?
4. Explain the concept of isostasy, and relate your explanation to the isostatic adjustments that occur in mountain ranges as they are eroded.
5. What is remanent magnetism? What is its origin? How is it used in finding ancient pole positions? How has remanent magnetism helped validate the concept of plate tectonics?
6. How has the detection of reversals in the Earth's magnetic field been used as support for the concept of sea-floor spreading?
7. How do the alignment and age distribution of volcanic islands in the Pacific Ocean provide evidence of sea-floor spreading?
8. What relationship might thermal convection cells in the mantle have to regions of crustal tension and compression?
9. Why are the most ancient rocks on Earth found only on the continents, whereas only relatively younger rocks are found on the ocean floors?
10. What evidence would one seek to support an interpretation that a particular area in the Appalachians was an allochthonous terrane? How would an allochthonous terrane derived from a microcontinent differ from one derived from a volcanic island arc?

Supplemental Readings and References

Ben–Avraham, Z. 1981. The movement of the continents. *Am. Sci.* 69:291–299.

Bolt, B. A. 1982. *Inside the Earth.* San Francisco: W. H. Freeman & Co.

Dietz, R. S., and Holden, J. 1970. Reconstruction of Pangaea: Break-up and dispersal of continents, Permian to Recent. *J. Geophys. Res.* 75:4939–4956.

Garland, G. D. 1979. *Introduction to Geophysics: Core, Mantle, and Crust,* 2nd ed. Toronto: Holt, Rinehart and Winston.

Jones, P. C., Cox, A., and Beck, M. 1982. The growth of western North America. *Sci. Amer.* 247:78–128.

Moores, E., ed. 1990. *Plate Tectonics: Readings from Scientific American:* New York: W. H. Freeman.

Saleby, J. B. 1983. Accretionary tectonics of the North American Cordillera. *Ann. Rev. Earth & Planet. Sci.* 15:45–73.

Taylor, S. R., and McLennan, S. M. 1985. *The Continental Crust: Its Composition and Evolution.* Oxford, England: Blackwell Scientific Publications.

Valentine, J. W., and Moores, E. M. 1972. Plate tectonics and the history of life in the oceans. *Sci. Amer.* 230:80–89.

van Andel, J. H. 1985. *New Views on an Old Planet.* Cambridge, England: Cambridge University Press.

Wegener, A. 1929 (1966 translations). *The Origin of the Continents and Oceans.* New York: Dover Publications.

Collisions between large bodies that had aggregated in the solar nebula during
the formative stages of the solar system. *(Painting by Don Dixon.)*

What was the Earth like in the very beginning? How does one get from a conglomeratic ball of stardust to a planet with concentric structure—inner and outer core, silicate mantle, crust, oceans, and layered atmosphere?

PRESTON CLOUD, *Oasis in Space,*

W. W. Norton & Co., 1988

CHAPTER 6

The First Two Billion Years

■ *The Earth in Space*

The Earth is one of nine planets that revolve around a rather average star, our sun. It is an approximately spherical planet with a diameter of nearly 13,000 km (8000 mi), with a vast ocean that covers over 71 percent of its surface, and with an atmosphere composed of nitrogen (78 percent) and oxygen (21 percent). The average density of the solid Earth is 5.5 g/cm^3. (For comparison, quartz has a density of 2.6 g/cm^3.) Surface rocks, however, have densities of 2.5 to 3.0 g/cm^3, indicating that high-density material lies in the interior of the Earth. As described in the previous chapter, most of this high-density material is concentrated in the core, which has a diameter of 7000 km (larger than the diameter of the planet Mercury). The core is surrounded by the mantle, which extends from the base of the crust to a depth of 2900 km.

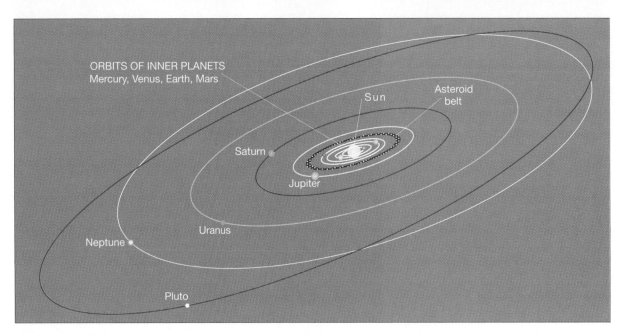

FIGURE 6–1 Orbits of the nine planets in the solar system.

In order of increasing distance from the sun, the planets of our solar system are Mercury, Venus, Earth, Mars, Jupiter, Saturn, Uranus, Neptune, and Pluto. Pluto is considered by many astronomers to be a large satellite that escaped from Neptune. A belt of asteroids orbits the sun in the region between the paths of Mars and Jupiter. This grouping of planets and asteroids around the sun constitutes our solar system (Fig. 6–1). Certainly, ours is not the only such system in the universe. It is true that the suspected planets of other systems are too distant to be detected directly, but we think that they are out there because

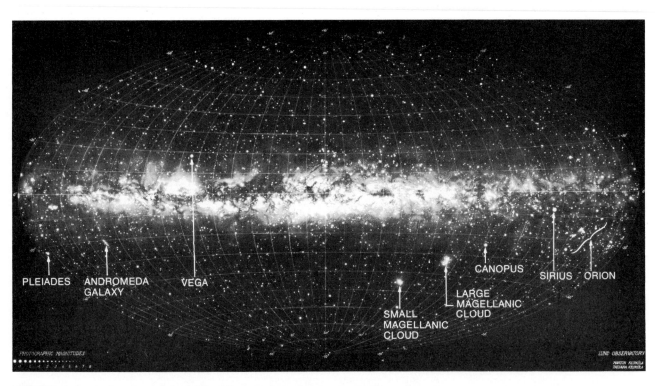

FIGURE 6–2 A drawing of the Milky Way, made under the supervision of Knut Lundmark at the Lund Observatory in Sweden. When we look edge-on toward the center of the galaxy, so many stars are in our line of sight that we see a glow of light in a circular band that appears "milky" in its luminescence. *(Courtesy of Lund Observatory, Sweden.)*

of wobbling motions in distant stars. The wobbling is believed to be caused by the gravitational pull of orbiting, but not visible, planets.

Our solar system is a small part of a much larger aggregate of stars, planets, dust, and gases called a **galaxy.** Galaxies are also numerous in the universe, and their constituents are arranged into several different general forms: tightly packed elliptical galaxies, irregular galaxies, and the more familiar discoidal spiral galaxies with their glowing central bulge and great curving arms.

The galaxy in which our solar system is located is called the **Milky Way galaxy** because as we look toward its dense central bulge, we see a great milky haze of light in the heavens. The Milky Way is a spiral-type galaxy (Fig. 6–2). It rotates slowly in space, completing one rotation about every 240 million years. Our sun is located about two-thirds of the distance (or 26,000 light years) outward from the center of the galaxy to its edge.

■ *The Sun*

For many reasons, the sun is the most important member of our solar system. It is a modest star (in comparison to others in our galaxy) and is composed of hydrogen (about 70 percent) and helium (about 27 percent). The remaining 3 percent consists of heavier elements that exist as gases in the hot interior, where temperatures may exceed 20 million degrees Celsius.

Although an enormous amount of solar energy is intercepted by the Earth, our planet is not roasted but is able to maintain a range of temperatures roughly between −50 and +60°C. The maintenance of this vital temperature range is made possible by three factors. First, because of rotation, the Earth receives energy from the sun on one hemisphere, whereas it returns heat to space over its entire surface. Second, some of the incoming radiation is reflected off the atmosphere and clouds and is directed back into space without ever reaching ground level. Finally, a part of the intercepted radiation is absorbed by the atmosphere and radiated back into space without warming the Earth's surface.

The energy that maintains the sun as a great glowing sphere of gases is derived from a continuous thermonuclear reaction called **fusion.** In the fusion process, hydrogen is changed to helium and excess mass is converted to energy (Fig. 6–3). Each second, the sun transmits an amount of energy equivalent to that which would result from the burning of 25 billion pounds of coal. This energy from the sun is the ultimate force behind the many geologic processes that sculpt the Earth's surface. For example, the sun's rays aid in the evaporation of surface waters, which in turn results in clouds that provide the precipitation required for erosion. Along with the Earth's rotation, the sun's radiation results in winds and ocean currents. Some scientists believe that protracted variations in the heat received from the sun may trigger episodes of continental glaciation or may reduce lush forests to barren wastelands. In company with the moon, the sun helps move the tides. Even primitive people knew the sun's importance and recognized it as the fountainhead and sustainer of life.

■ *Origin of the Universe*

Theories for the origin of the universe must conform to an important astronomical observation called the **red shift.** To understand the red shift, one may recall that light, on passing through a glass prism, is separated into a band of differing colors—a **spectrum.** The spectrum of a star reveals not only the star's composition but whether it is moving toward or away from the Earth and at what speed. For example, if a star is moving away from the Earth, wavelengths of the light from the star are shifted toward the red end of the spectrum. (If moving toward the Earth, the shift would be toward the blue end.) In addition, the greater the velocity of separation, the wider the observed red shift. By 1914, the astronomer W. M.

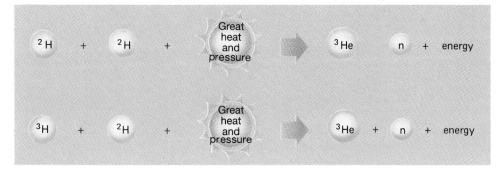

FIGURE 6–3 Examples of two fusion reactions (n = neutron).

Slipher had found 12 galaxies that clearly exhibited red shift. He reasoned that most of the galaxies within his range of observation were moving away from each other at great speed. In 1929, Edwin Hubble made the further discovery that the red shift increased with increasing distances of the galaxies, indicating that the more distant the galaxy, the higher its receding velocity. Thus, red shift indicates that the universe is expanding.

If we were to make a movie of the galaxies moving apart as indicated by the red shift and then play back the film in reverse, it is clear that all galaxies would come together at a single location. This infinitely dense point would then explode. Astronomers have dubbed the great explosion the **big bang.** Calculations based on the amount of expansion that has already occurred in the universe indicate that the big bang occurred between 15 and 18 billion years ago. It marked the instantaneous creation of the present universe. Initial heat in the universe reached about 100 billion degrees Celsius. Atoms could not exist at such high temperatures, so only radiant energy, very light particles called neutrinos, and electrons existed. The formation of protons and neutrons quickly followed. When the universe had cooled to about 1 billion degrees Celsius (only 1.5 to 4.0 minutes after the explosion), fusion reactions produced simple atomic nuclei. It would have been about a million years after the initial explosion before the universe had cooled sufficiently for atoms to form (Fig. 6–4). By about a billion years after the big bang, matter had been collected into galaxies and stars had probably begun to form. In the hot cores of stars, matter was reheated, nuclear reactions ignited, and the synthesis of elements heavier than hydrogen began.

Some astronomers are uneasy about a cosmic evolution that starts with a big bang and ends with galaxies disappearing somewhere out in the farthest

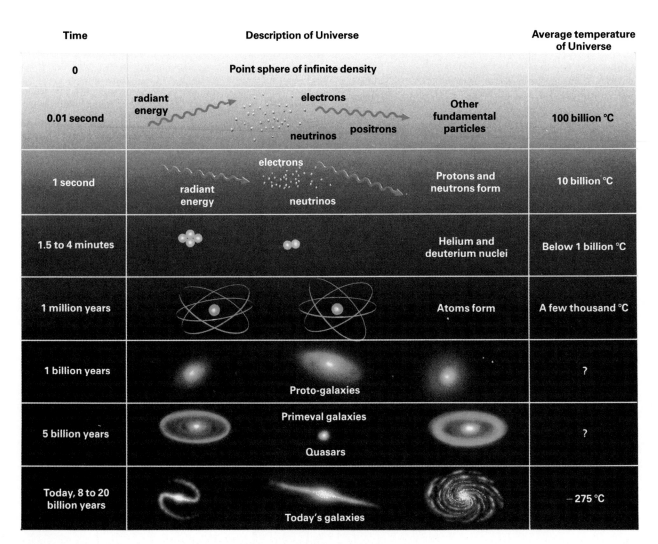

Time	Description of Universe	Average temperature of Universe
0	Point sphere of infinite density	
0.01 second	radiant energy / electrons / neutrinos / positrons / Other fundamental particles	100 billion °C
1 second	radiant energy / electrons / neutrinos / Protons and neutrons form	10 billion °C
1.5 to 4 minutes	Helium and deuterium nuclei	Below 1 billion °C
1 million years	Atoms form	A few thousand °C
1 billion years	Proto-galaxies	?
5 billion years	Primeval galaxies / Quasars	?
Today, 8 to 20 billion years	Today's galaxies	−275 °C

FIGURE 6–4   A brief pictorial summary of the evolution of the Universe. *(Courtesy of G. R. Thompson and J. Turk.)*

reaches of space. They support **steady-state cosmology** of a universe that will continue to expand forever. As expansion progresses, new matter, initially in the form of hydrogen, is formed from combinations of atomic particles in the space between galaxies at about the same rate that older material is receding. In this way, a uniform density of matter in the universe is maintained.

There is also the possibility of combining the big bang and steady-state theories into one theory that stipulates that for every big bang there is a big crunch. According to this so-called **oscillating universe cosmology**, after the great explosion that begins expansion, gravitational forces begin to prevail and cause matter to be drawn back to its place of origin. The expanding universe would thus become a contracting one. In the last stages of contraction, matter would be returning at enormous speed, resulting in compression of all the returned particles into the infinitely dense mass having such extremes of temperature and pressure that stars and their component atomic particles would be disassembled. From this ultradense body would come the next big bang and the initiation of another episode of expansion and contraction. One estimate would put the time required for a complete cycle at 100 billion years.

■ *Birth of the Solar System*

Any account of the origin of the solar system is obliged to conform to certain basic constraints within the system. These include dynamic constraints relating to the movements of planets, compositional constraints, and age constraints.

Dynamic Constraints

The planets of our solar system all revolve around the sun in the same direction and lie approximately in the plane of the sun's rotation. Thus, the solar system is like a disk in shape. The direction taken by the planets as they revolve around the sun is counterclockwise (called **prograde**) when viewed from a hypothetical point in space above the sun's north pole. The direction of rotation of the planets on their axes is also counterclockwise, with the exception of Venus and Uranus. With few exceptions, satellites mimic the movements of the planets they orbit.

Compositional Constraints

As indicated by their differing densities (see Table 6–1) as well as by Earth, moon, and meteorite samples, the planets differ in composition. Mercury, Venus, Earth, and Mars have mean densities of 5.4, 5.2, 5.5,

TABLE 6–1 Size and Density of the Planets

Planet	Diameter (km)	Mean Density (g/cm³)
Mercury	4,878	5.43
Venus	12,104	5.24
Earth	12,756	5.52
Mars	6,794	3.9
Jupiter	142,796	1.34
Saturn	120,000	0.69
Uranus	52,400	1.27
Neptune	50,450	1.58
Pluto	2,400	1.7

and 3.9, respectively. These four planets are small, dense, rocky, and rich in metals. They are termed the **terrestrial planets.** Jupiter, Saturn, Uranus, and Neptune have lesser densities of 1.3, 0.7, 1.3, and 1.6, respectively. They are termed the **Jovian planets** (Fig. 6–5). Like the sun, the composition of Jupiter and Saturn is dominated by hydrogen and helium.

Age Constraints

There are mineral grains in sedimentary rocks known to be at least 4.2 billion years old. The maximum age obtained from lunar samples and from meteorites is 4.6 billion years. It is a determination that agrees with theoretical calculations of the age of the sun. Thus, we can place the birth of the solar system at about 4.6 billion years ago.

The Solar Nebula Hypothesis

When the dynamic, compositional, and age constraints are taken into account, a favored hypothesis is that the solar system was derived from a rotating cloud of dust particles and gases called the **solar nebula.** This basic concept was first suggested in 1755 by Immanuel Kant. About 40 years later, Pierre Laplace further developed Kant's original theory, and in recent years it has been improved with the rich flow of information emanating from the space program.

The modern theory begins with a cold, rarefied cloud of gases and dust particles. This initial material consisted of elements and the chemical compounds formed by combinations of elements. The elements forming the cloud were originally produced by nuclear reactions within stars. The sun, for example, produces about 600 million tons of helium each second from the fusion of hydrogen. Elements heavier than helium are created in stars older than the sun,

FIGURE 6–5   The four giant planets—Jupiter, Saturn, Uranus, and Neptune—all shown to the same scale. (NASA/JPL.)

and in particular in the great explosions that mark the death of stars.

After the elements and their compounds had collected in a region of space, the dust cloud began to rotate in a counterclockwise direction, to contract, and to assume a discoidal shape. About 90 percent of the cloud's mass remained concentrated in the thicker central part. Then, while rotation and contraction were occurring, turbulence began to affect the cloud system. Any departure from the smooth flow of a gas or liquid constitutes turbulence. In the dust cloud it consisted of chaotic swirling and churning movements that were superimposed on the grander primary motion of the entire cloud. Each turbulent eddy was affected by the movement of adjacent eddies, and some served as collecting sites for matter brought to their location by neighboring swirls. When the disk had shrunk to a size somewhat larger than the present solar system, its denser condition permitted condensation of the concentrated knots of matter. Smaller particles merged to build chunks up to several meters in diameter, and the dense swarms swept up finer particles within their orbital paths, thereby

growing in size. This process of accumulation of bits of matter around an initial mass is called **accretion.** Because the particles available for accretion were cold and essentially similar throughout the period of accretion, the process is often further described as **cold homogeneous accretion.** The large globules formed from swarms of accreting bodies, dust, and gases are called **protoplanets.** They were enormously larger than present-day planets. Each rotated somewhat like a miniature dust cloud, and each eventually swept away most of the debris in its orbital path and was able to revolve around the central mass without collision with other protoplanets (Fig. 6–6). The protoplanets' formation from the original dark, cold cloud required an estimated 10 million years.

While the protoplanets were in the process of condensation and accretion, material that had been pulled into the central region of the nebula condensed, shrank, and was heated to several million degrees by gravitational compression. The sun was born. Later, when pressures and temperatures within the core of the sun became sufficiently high, thermonuclear reactions began, pouring out additional en-

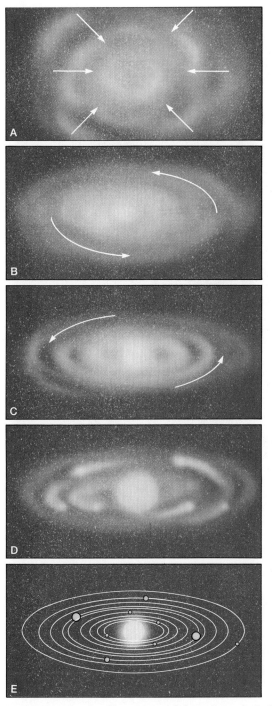

FIGURE 6–6 An illustration of the solar nebula hypothesis for the origin of the solar system. (A) The solar nebula takes form. (B) It begins to rotate, flatten, and contract. (C) The nebula is now distinctly discoidal and has a central concentration of matter that will become the sun. (D) Thermonuclear reactions begin in the primordial sun as material in the thinner area of the disk condenses and accretes to form protoplanets. (E) The late stage in the evolution of the solar system in which planets are fully formed and materials not incorporated in planets are largely swept away by solar wind and radiation. *(From Abell, G. O., Morrison, D., and Wolff, S. C. 1988. Realm of the Universe. Philadelphia: Saunders College Publishing.)*

ergy derived from matter by the nuclear fusion of hydrogen atoms into atoms of helium. (Today, the sun converts about 596 million metric tons of hydrogen into 532 million metric tons of helium each second. The difference represents matter that has been converted to energy.)

As the sun began its thermal history, its radiation ionized surrounding gases in the nebula. These ionized gases in turn interacted with the sun's magnetic lines of force, causing the magnetic field to adhere to the surrounding nebula and drag it around during rotation. At the same time, the nebula caused a drag on the magnetic field and slowed down the sun's rotation. In this way, the old problem of why the sun rotates slowly was explained.

The stream of radiation from the sun, sometimes called **solar wind,** drove enormous quantities of the lighter elements and frozen gases outward into space. This solar force is what causes a comet's tail to show wavy streaming or to be bent away from the sun. As would be expected, the planets nearest the sun lost enormous quantities of lighter matter and thus have smaller masses but greater densities than the outer planets. They are composed primarily of rock and metal that could not be moved away by the solar wind. The Jovian planets were the least affected and retained their considerable volumes of hydrogen and helium.

■ *Differentiation of the Earth*

Cold Accretion Model

If the cold, homogeneous model for accretion is correct, then the Earth would have consisted initially of an unsorted conglomeration of metals and silicates from its center to its surface. The Earth, however, is not now homogeneous throughout. Neither are the other terrestrial planets of the solar system. They, along with the Earth, have experienced **differentiation,** a process whereby a planet becomes internally zoned or layered.

To accomplish differentiation of an initially homogeneous planet, the planet must have been heated and become at least partially melted. Under such conditions, much of the iron and nickel in the planet would percolate downward to form the dense core. The remaining iron and other metals would combine with silicon and oxygen and separate out to form the less dense mantle that surrounds the core. Still lighter components would separate out of the mantle to form an uppermost crustal layer. Geologists do not insist on complete melting for this differentiation to have occurred. Partial melting on a vast scale would be a likely occurrence, as well as solid diffusion of ions mobilized by high temperatures and pressures. Indeed, it would be difficult to account for the present abundance of volatile elements on Earth if the planet had melted completely, for under such circumstances these materials would have been driven off. Even for partial melting and solid diffusion to have occurred, however, an enormous amount of heat would have been required.

The heat needed for differentiation may have been derived from several different processes. Some of it may have been accretionary heat derived from bombardment by particles and meteoric materials when the planet was still actively growing. Much of the heat generated by the great infall of impacting bodies would have been preserved because subsequent showers of material provided blankets of insulating debris. Also, as more and more material accumulated at the surface of the planet, the weight of overburden on underlying zones would have provided heat by gravitational compression. Compression also served to reduce the Earth's size.

Another extremely important source of heat for differentiation would result from the decay of radioactive isotopes incorporated within the materials of the planet during accretion. Much of this radioactively generated heat would have accumulated deep within the Earth, where the poor thermal conductivity of enclosing rocks provided insulation. There was more heat derived from radioactivity in the early days of the Earth's development than today because many of the isotopes having relatively short half-lives had not yet undergone radioactive decay.

Geophysicists estimate that about 1 billion years after the solid Earth formed, there would have been sufficient heat from all these sources to melt iron at depths of 400 to 800 km. One can envision myriads of heavy, molten droplets of iron and nickel gravitating toward the planetary center, displacing lighter materials that had been there (Fig. 6–7). As the mantle began to form, it, too, experienced differentiation, which eventually led to the development of the crust and the exhalation of gases that formed an atmosphere and hydrosphere.

Hot Accretion Model

The alternate hypothesis to the cold accretion model stipulates that the material accumulating to form the planets was extremely hot. The accreting planets, therefore, were not initially homogeneous solid bodies that later differentiated as a result of subsequent heating. On the contrary, it is thought that the internal zonation of planets developed *during,* not after, accretionary accumulation by a process termed **hot heterogeneous accretion** (Fig. 6–8).

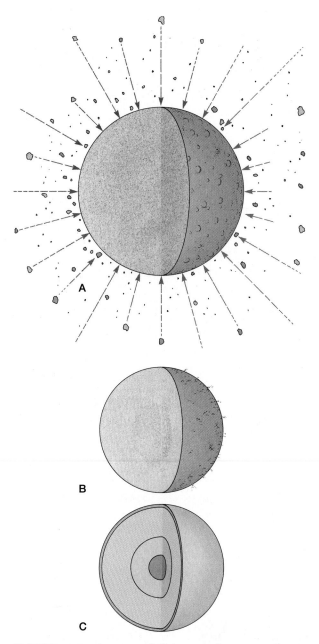

FIGURE 6–7 Conceptual diagrams of stages in the Earth's early history. (A) Representation of the growth of the planet by the aggregation of particles and meteorites that bombarded its surface. At this time, the Earth was composed of a homogeneous mixture of materials. (B) The Earth has lost volume because of gravitational compression. Temperatures in the interior have reached a level at which differentiation has begun. Iron (red drops) sinks toward the interior to form the core, whereas lighter silicates move upward. (C) The result of the differentiation of the planet is evident by the formation of core, mantle, and crust.

The hot heterogeneous model for accretion begins with the solar nebula at a time when its gases were still very hot, possibly reaching temperatures of over 1000°C. It is possible that as it began to cool, heavy, iron-rich compounds condensed, precipitated, and

tinct, subsequent event. It is likely, however, that later heating by meteoric bombardment and radioactive decay could have further propelled the differentiation already established.

Currently, it is not possible to judge which model of differentiation is more valid. It is possible that components of both theories were in operation. Most geologists appear to favor cold homogeneous accretion followed by melting as an explanation for the Earth's internal zonation.

■ *Evolution of the Atmosphere and Hydrosphere*

The Primitive Atmosphere

We have described how the Earth formed by aggregation of small and large bodies that were themselves once part of a nebula of gas and dust. The particles and meteorites incorporated into the growing planet contained significant amounts of gases. Even greater amounts of gas may have been derived from the impact of comets, for these bodies are composed almost entirely of frozen gases, ice, and dust. The **volatiles** (substances easily driven off by heating) either were scattered uniformly throughout the accreting planet by the process of homogeneous accretion or were contained in the impacting bodies that arrived during the late stages of accretion. Among meteorites, volatiles are especially abundant in **carbonaceous chondrites** (Fig. 6–9). These meteorites take their name

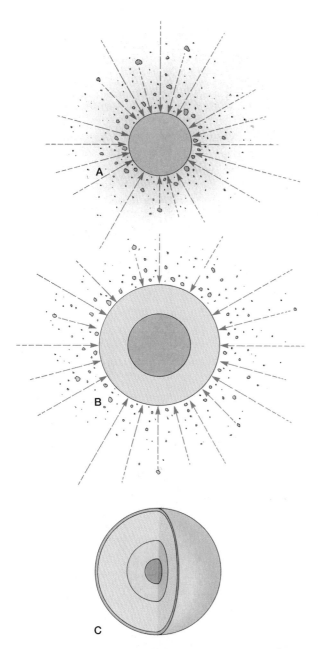

FIGURE 6–8 Origin of the Earth's core according to the hot heterogeneous model of accretion. (A) Primarily iron and nickel condense, collect, and form a core. (B) Silicates envelop the earlier formed core and form a mantle. (C) The mantle differentiates and provides the materials for the crust.

accumulated in masses that would eventually become the iron-rich cores of the terrestrial planets. Next, with continued cooling, lower density silicate minerals began to condense and were swept up by the denser masses formed earlier. The silicates would have accumulated around the core as a thick covering that eventually became the mantle. During the accretionary process, the Earth was hot and probably at least partially molten. In this state, chemical differentiation was able to proceed throughout the period of accumulation. Thus, differentiation was not a dis-

FIGURE 6–9 Small carbonaceous chondrite. This meteorite fell near Murray, Kentucky, in 1950. Scale divisions are in millimeters. *(Washington University collection.)*

from spherical bodies called chondrites that are conspicuously present. In addition to nitrogen, hydrogen, and carbon, some of the dark iron and magnesium silicates in the carbonaceous chondrites contain water. The water and gaseous elements would have been readily released by heat associated with accretion or by the melting and volcanism accompanying the Earth's differentiation.

The process by which water vapor and other gases are released from the rocks that held them and then vented to the surface is termed **outgassing.** Calculations indicate that most of the outgassing of water occurred within the first 1 billion years of Earth history. As evidence of this, the presence of an early ocean is clearly indicated by marine sedimentary rocks dating from as long ago as 3.8 billion years.

The precise chemical nature of the gases released during the early phases of outgassing would have been dependent on the temperatures of the rocks from which the volatiles were derived as well as the state of oxidation of the rocks through which the gases passed. Geochemists believe the outgassed volatiles produced a primitive atmosphere rich in carbon dioxide and water vapor, with lesser amounts of carbon monoxide, hydrogen, and hydrogen chloride. Actually, the volatiles released by modern volcanoes provide an approximation of the composition of the early atmosphere, with the exception that the primitive exhalations probably contained a higher percentage of hydrogen and possibly traces of ammonia and methane as well. The gases from today's volcanoes differ because they are released from relatively young rocks that have been recirculated near the Earth's surface by plate tectonic processes. Uncombined oxygen was not discharged into the early atmosphere, and if small amounts were released from the interior, these amounts would not have escaped immediate recombination with easily oxidized metals such as iron.

One might argue that the gases of our atmosphere were not degassed from the interior but were merely late accumulations of light volatiles left over from the original nebula. The Earth's relative scarcity of inert gases such as argon, neon, and krypton provides strong evidence for opposing this view. These chemically stable gases are too heavy to escape the Earth's gravity. If they were present in gases collected near the end of the accretionary period, they would be unable to escape and would still be here. The inert gases, however, are about a million times less abundant in our present atmosphere than in the sun and other stars. This indicates that our atmosphere is not a residue of gases left over from the solar nebula but was derived from the planet's interior by degassing.

Once at the Earth's surface, the volatiles that had been vented were subjected to a variety of changes (Fig. 6–10). Water vapor was removed by condensation, and seas began to take form. The outgassing of carbon dioxide and other gases resulted in rainfall and ocean water that were considerably more acidic than today. When in contact with rocks, the acidic waters caused rapid chemical weathering and thereby brought calcium, magnesium, and other ions into so-

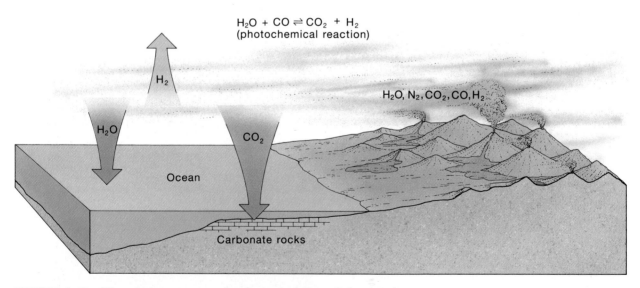

$$H_2O + CO \rightleftharpoons CO_2 + H_2$$
(photochemical reaction)

H_2

H_2O

H_2O, N_2, CO_2, CO, H_2

CO_2

Ocean

Carbonate rocks

FIGURE 6–10 The relative amounts of gases in the primordial atmosphere were different from the abundances vented to the exterior of the Earth during differentiation. Nitrogen tended to be retained in the atmosphere, whereas much of the water vapor was lost to the oceans by condensation. Some CO_2 was combined with calcium and magnesium derived from weathering and extracted to form carbonate rocks; hydrogen, being light, was lost to space.

lution. Much later, when the seas were less acidic and oxygen more prevalent in the atmosphere, these ions would be joined with carbon dioxide to form limestones, the calcareous structures of algae, and the shells of myriads of marine organisms.

An Oxygen-Rich Atmosphere

We have seen that the early atmosphere of the Earth was reducing and nonoxygenic in character. It was followed by an atmosphere rich in oxygen, an atmosphere that was and still is dependent on life for its maintenance. We are reminded of the significance of this atmosphere with every breath we take, yet the earlier nonoxygenic atmosphere was also of vital importance, for in it the earliest forms of life evolved.

The change from an oxygen-poor to an oxygen-rich atmosphere occurred as a result of two processes. The first was dissociation of water molecules into hydrogen and oxygen. Termed **photochemical dissociation,** this process occurs in the upper atmosphere when water molecules are split by high-energy beams of ultraviolet light from the sun. The reaction can be written as follows:

$$2\,H_2O + uv\ radiation \rightarrow 2\,H_2 + O_2$$

Before the Earth had an ozone layer to filter out ultraviolet radiation, some oxygen was probably produced by this process. Today, however, photochemical dissociation does not produce free oxygen at a rate sufficient to balance the loss of the gas by dissipation into space. Another far more important oxygen-generating mechanism came with the advent of life on Earth and, more specifically, with life that had evolved the remarkable capability of separating carbon dioxide into carbon and free oxygen. We now know this process as **photosynthesis.** In the section entitled "Life of the Archean" later in this chapter, we will examine the role of photosynthesis in the evolution of early life forms more fully. It is a process that has made the Earth a truly unique planet in our solar system.

Geologic Clues to the Nature of the Early Atmospheres

The most ancient rocks providing evidence of the atmospheric changes described previously are about 3.5 billion years old. Although they are metamorphic rocks, many of them still contain depositional features indicating that they were originally water-laid sediments. Mineral grains within these old rocks were formed or modified by weathering, and weathering cannot occur without an atmosphere. Further, it is highly probable that the atmosphere was not oxygenic, for there is an obvious absence of oxidized

iron such as hematite. Rather, the sulfide of iron called pyrite is the usual iron mineral found in these rocks. An atmosphere rich in carbon dioxide is acidic, in that carbon dioxide and water combine to form carbonic acid. In such an environment, alkaline rocks like dolomite and limestone do not develop, and this may account for the absence of carbonate rocks from this early stage of Earth history. Chert, a rock more tolerant of such conditions, is commonly found in ancient rock sequences.

Gradual additions of oxygen in the Earth's early atmosphere have often been inferred from the presence of Precambrian rocks called banded iron formations, or BIFs (Fig. 6–11). BIFs exhibit an alteration of rust-red and gray bands. The former are colored by ferric iron oxide (Fe_2O_3), whereas the gray bands are not as rich in oxygen. Banded iron formations first appear in the geologic record about 3 billion years ago and give way to unbanded iron deposits by about 1.8 billion years ago. They may have originated in more than one way. As we have noted, the early Earth was bathed in ultraviolet radiation. Under conditions of intense ultraviolet radiation, insoluble iron hydroxides can be precipitated from sea water that contains iron in solution. The iron hydroxides would then become a component of bottom sediment, later to be converted to ferric iron oxides.

FIGURE 6–11 Banded iron formation. The red bands are hematite and are interbedded with chert. Wadi Kareim, Egypt. *(Courtesy of D. Bhattacharyya.)*

Bacteria may also have played a role in the origin of BIFs. German microbiologists recently discovered that purple nonsulfate bacteria employ photosynthesis to break up dissolved iron carbonate in sea water and precipitate insoluble iron hydroxide. As suggested by the alternation of bands in BIFs, the process may have had a seasonal component. For example, seasonal upwelling of nutrient-rich bottom waters would enhance the bacterial action that produced the iron-rich bands. Such upwelling might also have brought bottom waters rich in dissolved iron to the surface of the sea, where the effects of ultraviolet radiation would be most intense.

Banded iron formations are of enormous economic importance. They provide most of the iron that is mined worldwide, including the great iron ore deposits of the Ukraine, Brazil, Minnesota, Labrador, and western Australia.

In rocks younger than 1.8 billion years, evidence for free oxygen in the atmosphere is seen in red shales, siltstones, and sandstone derived from weathering of older rocks on the continents. One also finds carbonate rocks (dolomites and limestones) of about the same age. If carbon dioxide was still abundant at this time, these carbonate rocks could not have formed.

The Ocean

If the water in the atmosphere was outgassed from the interior, then the bodies of water that accumulated on the surface of the Earth were derived from the atmosphere after it had cooled sufficiently to condense and allow rain to fall. Gradually, the great depressions in the primordial crust began to fill, and oceans began to take form. One cannot help but wonder if the enormous volume of oceanic water could have all come from the interior. Calculations provide a very strong affirmative answer to this question. As noted earlier, vast amounts of water were locked within hydrous minerals that were ubiquitous within the primordial Earth. These waters, "sweated" from the Earth's interior and precipitated onto the uplands, immediately began to dissolve soluble minerals and carry the solutes to the sea. The initial high acidity of the oceans enhanced and speeded the process so that the oceans quickly acquired the saltiness that is their most obvious compositional characteristic. They have maintained a relatively consistent composition by precipitating surplus solutes at about the same rate as they are supplied. Of course, because of its high solubility, sodium remains in sea water longer than do other common elements. However, the fossil record of marine organisms suggests that even this element has not varied appreciably in sea water for at least the past 600 million years.

Having outgassed its present quantity of water early in its history, the Earth has been recycling that water ever since. Water is continuously recirculated by evaporation and precipitation—processes powered by the sun and gravity. Some of the water in the oceans is temporarily lost by being incorporated into hydrous clay minerals that settle to the ocean floors. However, even this water has not permanently vanished, for the sediments may be moved to orogenic belts and melted into magmas that return the water to the surface in the course of volcanic eruptions (Fig. 6–12).

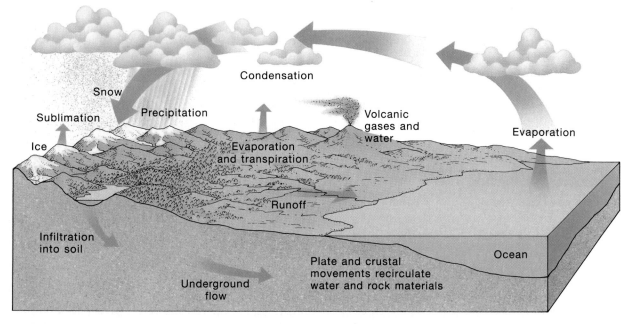

FIGURE 6–12   The constant circulation of water at and near the Earth's surface.

■ *The Archean Eon*

Beneath the Cambrian

When eighteenth-century geologists first began to describe local sequences of strata, they sometimes encountered a primary or basement complex of metamorphic and igneous rocks that lay beneath fossil-bearing sedimentary strata. These older mostly crystalline rocks came to be known as **Precambrian,** a derivative of the term **Cambrian** proposed for overlying younger rocks by Adam Sedgwick in 1835. To most of Sedgwick's contemporaries, correlating and even crudely estimating the age of the apparently unfossiliferous, altered, and deformed Precambrian rocks seemed an impossible task. Nevertheless, geologists recognized that Precambrian terranes were of great importance. They represented an enormous block of geologic time, formed the very cores of continents, and contained rich deposits of iron and other metals. For these reasons, it was inevitable that a few intrepid geologists would devote their lives to deciphering the complex relationships of Precambrian rocks.

One of these pioneers of Precambrian geology was Sir William Logan of the Canadian Geological survey. In the middle 1800s, Logan was able to associate groups of Precambrian rocks according to their superpositional and cross-cutting relationships. To substantiate these relationships, he inferred that those rocks that had experienced the most metamorphism were likely to be the oldest. Modern geologic investigations show that Logan erred in attempting to relate degree of metamorphism to age. Older rocks may escape metamorphism, and younger ones may be radically metamorphosed. Yet in southeastern Canada, where Logan mapped, one could find an older terrane of gneissic rocks as well as younger sequences of less altered metamorphic and sedimentary rocks. Because of this, it seemed reasonable to think of Precambrian time as divisible into an older **Archean Eon** and a younger **Proterozoic Eon.** All of the remainder of geologic time up to the present day was designated the **Phanerozoic Eon.**

Today, all geologists agree that the only reliable basis for correlation of unfossiliferous Precambrian rocks is through isotopic dating techniques. Only after the absolute ages of rocks in a particular region are known can they be placed in correct chronologic order. Radiometric dating of Precambrian rocks has provided a basis for mapping Precambrian terranes and for deciphering their geologic history. The hundreds of dates already obtained permit a tentative calibration of the Precambrian geologic time table (Table 6–2). The beginning of the Proterozoic is placed at about 2.5 billion years ago (Fig. 6–13). The

TABLE 6–2 Time Divisions for the Precambrian

Time in Billions of Years	Time Divisions Followed in This Book*	Events
0.54	Late Proterozoic	Glaciation Grenville orogeny
1.0	1.0	
1.5	Middle Proterozoic	
1.6	1.6	
2.0	Early Proterozoic	Red beds Glaciation
2.5	2.5	Kenoran orogeny
3.	Late Archean	
3.0	3.0	Earliest BIF
3.5	Middle Archean	
3.4	3.4	
Early Archean		
3.8	3.8	Origin of Life
4.0		Oldest sediments
Hadean		Major outgassing Development of internal structure
4.5		Origin of Earth
4.6		

*As recommended by the International Union of Geological Sciences.

Archean is generally considered to have begun about 3.8 billion years ago, a date that is close to the age of the oldest known crustal rocks.

For the earlier interval that began with the birth of the planet, the term **Hadean Eon** is often used. During the Hadean, the Earth was still very hot, convection was vigorous, and steaming vents and volcanoes blistered the planet's cratered surface. It was during the Hadean that the Earth's iron core separated from the silicate mantle and during which the Earth was enveloped in an atmosphere devoid of free oxygen but rich in carbon dioxide, water, and other volcanic gases. It was also a time during which the Earth, its moon, and other planets in the solar system were being vigorously pelted by meteorites and asteroids. Although this episode of bombardment continued for several hundred million years, it has left little direct evidence on the present-day Earth. Meteorites and craters formed by Hadean meteorite showers were destroyed by later episodes of melting and ero-

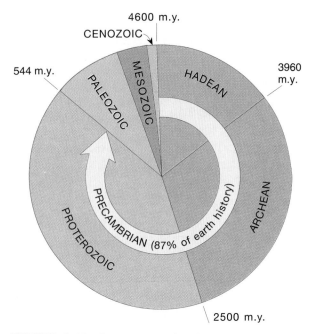

FIGURE 6–13 Proportions of geologic time encompassed by the Precambrian and its divisions, the Hadean, Archean, and Proterozoic Eons.

sion. However, proof of the bombardment is clearly evident on the crater-scarred surface of the moon. Rocks from the lunar highlands are about 4.0 billion years old. At that time, there may have been sufficient heat generated by the vast infall of meteorites to cause melting and consequent resetting of radiometric clocks. Radiometric clocks on the Hadean Earth would have been reset in a similar way.

Shields and Cratons

Although exposures of Precambrian rocks may occur in the deep canyons of some mountains and plateaus, the most extensive exposures are found in broadly upwarped, geologically stable regions of continents called **shields.** Every continent has one or more shields (Fig. 6–14). Extending across 3 million square miles of northern North America is the great Canadian shield. Because continental glaciers of the last ice age have stripped away the cover of soil and debris, Precambrian rocks of the Canadian shield are extensively exposed at the surface.

To the south and west of the Canadian shield, Precambrian rocks are covered by sedimentary sequences of Phanerozoic age. These younger rocks are mostly flat lying or only gently folded. Such stable regions where basement rocks are covered by relatively thin blankets of sedimentary strata are called **platforms.** The platform of a continent, together

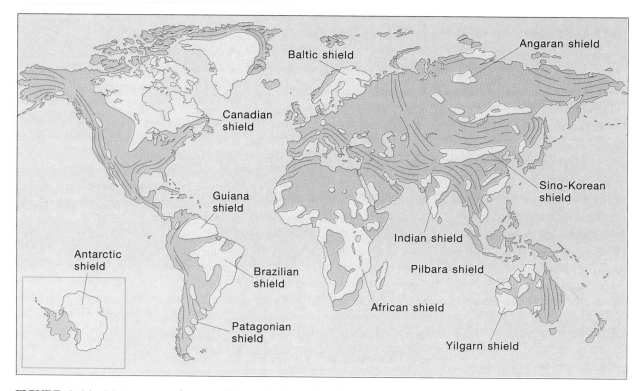

FIGURE 6–14 Major areas of exposed Precambrian rocks (shown in yellow). Trends of mobile belts are indicated by the red lines.

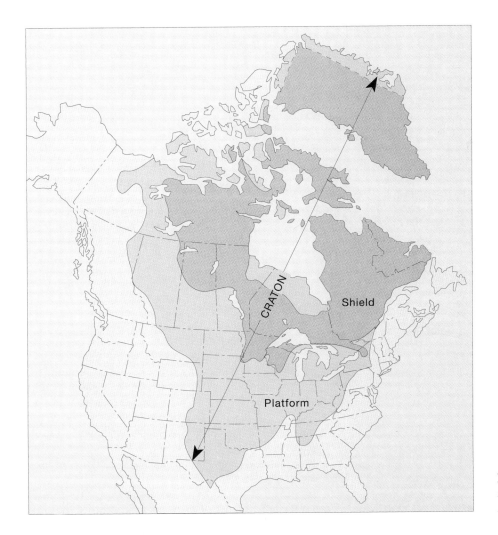

FIGURE 6–15 North American craton, shield, and platform.

with its shield, constitutes that continent's **craton** (Fig. 6–15).

On the basis of differences in the trends of faults and folds, the style of folding, and the ages of component rocks, the Canadian shield has been divided into a number of **Precambrian provinces** (Fig. 6–16). The boundaries of these provinces are often marked by abrupt truncations in structural lineations, or they may be represented by bands of severely deformed rocks of former orogenic belts. In North America, seven Precambrian provinces are recognized: the Superior, Wyoming, Slave, Nain, Hearne, Rae, and Grenville. These provinces were once separate crustal segments that have been consolidated to form the larger North American craton. They are bound together by belts of deformed, metamorphosed, and intruded rocks that mark the location of collision of the various cratonic elements. For this reason, the belts of deformed rocks are called **collisional orogens.** It has been estimated that all of the Archean and Proterozoic elements of the Canadian shield were consolidated by about 1.8 to 1.9 billion years ago.

The Archean Crust

Origin of the Oceanic Crust

The Earth's original crust was a **mafic** crust, meaning that it was dominated by dark iron and magnesium silicates. This mafic crust was derived from an even more mafic (hence **ultramafic**) underlying mantle. To understand the origin of the mafic crust, we must imagine the Earth during its Hadean infancy, when it seethed with heat generated from the decay of short-lived radioactive isotopes, from gravitational forces, and from the infall of meteorites. Even after the final stages of accretion, there was ample heat to melt the upper mantle partially or totally. As a result of the melting, an extensive magma ocean may have covered the Earth's surface. When the surface patches of the magma ocean cooled, they solidified as ultramafic rocks called **komatiites.** Komatiites form at temperatures even greater than the 1100°C required to produce basalt. Thus, they reflect the higher temperature gradients that prevailed during the late Hadean.

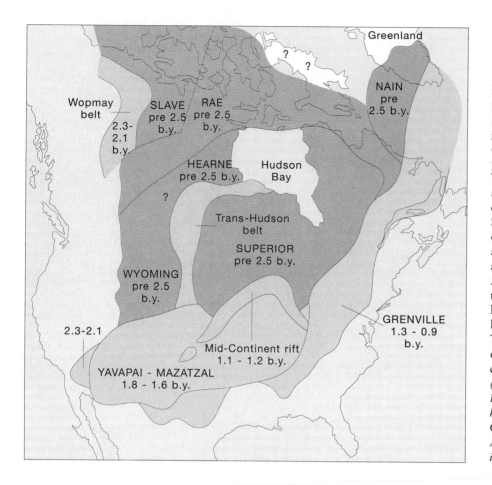

FIGURE 6–16 Precambrian provinces of North America. Archean provinces (those greater than 2.5 billion years old) are in red. Proterozoic provinces are in green. The Trans-Hudson belt is a zone along which rifting and then closure occurred to weld the Superior province to the Wyoming province. The Hearne and Wyoming provinces may actually be a single province. A major episode of rifting that occurred 1.2 to 1.0 billion years ago is represented by the midcontinent rift zone. The Wopmay belt was developed by rifting followed by closure of an ocean basin. *(Adapted from Hoffman, P. F., Precambrian geology and tectonic history of North America,* in Geology of North America, *Vol. A, Ch. 16. Boulder, CO: Geological Society of America, 1989.)*

Even without total melting of the upper mantle to produce a magma ocean, it is probable that extensive partial melting in this zone produced sufficient quantities of magma to form an early crust. As this was occurring, high mantle temperatures and associated vigorous convection probably resulted in some form of plate movement and recycling of surface rocks. Indeed, because komatiites are somewhat denser than hot, partly molten upper mantle, there may have been a tendency for slabs of primordial komatiitic oceanic crust to sink as they were being conveyed laterally by underlying convection currents. With the passage of time, with continued cooling of the mantle, and with partial melting of ultramafic rocks in the upper mantle, komatiitic additions to the surface may have been replaced by mafic (basaltic) magmas. These may have been added to the expanding oceanic crust at spreading centers (Table 6–3).

TABLE 6–3   Summary of the Characteristics of the Earth's Early Oceanic and Continental Crust

	Oceanic Crust	Continental Crust
First appearance	About 4.5 billion years ago	About 4.0 billion years ago
Where formed	Ocean ridges (spreading centers)	Subduction zones
Composition	Komatiite–basalt	Tonalite–granodiorite
Lateral extent	Widespread	Local
How generated	Partial melting of ultramafic rocks in upper mantle	Partial melting of wet mafic rocks in descending slabs

From Condie, K. C., 1989, Origin of the Earth's Crust. *Paleogeography, Paleoclimatology, Paleoecology* 75:57–81, with permission.

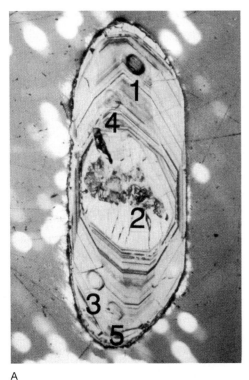

A

B

FIGURE 6–17 (A) Photomicrograph of one of the 3.96-billion-year-old zircon grains extracted from the Acasta Gneiss, Slave province, Northwest Territories of Canada. The grain is 0.5 mm long. Its polished surface has been etched with acid to highlight crystal growth zones. Numbers refer to points selected for analysis. (B) Exposure of the zircon-bearing Acasta Gneiss at the discovery site. *(Courtesy of Sam A. Bowring.)*

Origin of the Continental Crust

In contrast to the mafic nature of the oceanic crust, the continental crust is rich in feldspars, quartz, and muscovite. It is called a **felsic** crust. One of the oldest known patches of felsic crust was recently discovered at a location about 370 km (230 miles) north of the town of Yellowknife in Canada's Northwest Territories. Radiometric dating of zircon grains within these felsic rocks indicates that they are 3.96 billion years old (Fig. 6–17). These old continental rocks are tonalite gneisses. They were formed by metamorphism of **tonalite,** a variety of the igneous rock diorite that contains at least 10 percent quartz. The Isua region of southwestern Greenland also contains very old tonalite gneisses (3.8 billion years old) (Fig. 6–18). The Greenland outcrops are more extensive in area and contain fragments of ultramafic rocks that may be remnants of the Earth's surface that was present before there was any felsic crust. Other patches of extremely old felsic crust have been found in Enderby Land, Antarctica (3.9 billion years old). In addition,

FIGURE 6–18 Archean tonalite gneiss from the Isua region of southwestern Greenland. *(Courtesy of R. F. Dymek.)*

in western Australia, 4.0- to 4.3-billion-year-old detrital zircon grains have been found in Archean sedimentary rocks. Zircon is a relatively common accessory mineral in felsic igneous rocks. The Australian zircons had been eroded from such a rock, transported, and deposited in the fluvial sandstones from which they were extracted.

When we compare the ages obtained from the samples just described to the 4.6-billion-year age of our planet, it is apparent that the Earth had acquired patches of continental (tonalitic and granodioritic) crust a few hundred million years after its formation, or about 4.0 to 4.3 billion years ago. But how did these patches form and grow? Many geologists suggest they originated at subduction zones where slabs of sediment-covered mafic crust plunged into the mantle, there to be partially melted to release felsic components. Because of their lower density, felsic melts ascended to form tonalites and granodiorites. The more mafic materials remained behind in the mantle. By about 3.0 billion years ago, granites formed by a similar mechanism involving partial melting of tonalitic, granodioritic, and sedimentary rock materials.

It is likely that the early Archean segments of continental crust were small, perhaps less than about 500 km (310 miles) in diameter. These initial patches grew larger by several mechanisms that continue to operate today. One such mechanism may have involved the coming together of two tectonic plates

bearing felsic crust at their encountering margins. Compressive deformation and suturing may have occurred to form a larger expanse of continental crust (see Fig. 5–43A). Growth may also have occurred as island arcs and microcontinents were added to the host continent along subduction zones. A third method may have involved the welding of prisms of sediment onto continental margins. This process begins with the weathering and erosion of pre-existing rocks and deposition of the resulting sediment in the ocean adjacent to an upland area. Next, the thick accumulation of sediment is subjected to orogeny caused by an encounter with a converging tectonic plate. The sediment in the prism would have roughly the same bulk composition as granitic igneous rocks. Such material would be converted to felsic crystalline crustal rocks if subjected to intense metamorphism and melting. The repetition of such cycles of deposition and orogeny, along with suturing events and the incorporation of microcontinents, could account for continuing growth of continental crust. As indicated by the known distribution of 3.0- to 2.5-billion-year-old felsic rocks, very large continents had been assembled by late Archean time.

Granulites and Greenstones

In general, Archean rocks of cratons can be grouped into two major rock associations. The first is the **granulite** association. It is composed largely of

---

COMMENTARY
The Riches of Greenstone Belts

At certain locations around the globe, Archean rocks contain important mineral resources. This is especially true of greenstone belts, many of which have contributed significantly to a nation's economic health and supported local economies based on mining and refining ores. It has become apparent that certain kinds of ores occur in particular rock types within the greenstone sequence. As seen in Figure 6–21, rocks near the bottom of the sequence are ultrabasic volcanics and intrusives. Ores of chromite and nickel, as well as asbestos, are mined from these rocks. In the overlying mafic and felsic volcanics, one finds ores of gold, silver, copper, and zinc. At the top of the greenstone "pile" are sedimentary (or metasedimentary) rocks. Here one may find concentrations of manganese, barite, and iron. Sandstones and conglomerates within the sedimentary section may also contain ancient placer deposits

(paleoplacers) that include gold weathered out of older igneous rocks.

The most famous paleoplacer gold deposits in the world are found in the Witwatersrand Supergroup of Late Archean and Proterozoic age in South Africa. As in many of today's placer deposits, the Witwatersrand gold was concentrated in river bars and deltas through the action of water currents. Because particles of gold are heavier than the usual sand and gravel carried by a stream, they tend to be deposited wherever the current has slowed enough to permit their deposition but still has sufficient velocity to carry away lighter rock and mineral grains. Witwatersrand placers are particularly rich where ancient tributaries entered downfaulted, subsiding basins. Since their discovery in 1886, the Witwatersrand gold mines have provided more than half of the world's gold, and they are still producing.

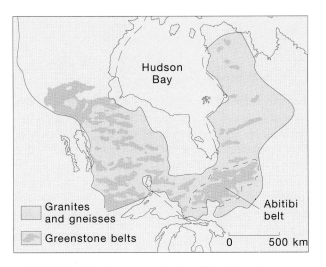

FIGURE 6–19 Greenstone belts of the Superior province. *(After Goodwin, A. M. 1968. Proc. Geol. Assoc. of Canada, 19:1–14.)*

gneisses derived from strongly heated and deformed tonalites, granodiorites, and granites, as well as layered intrusive gabbroic rocks called **anorthosites.** Few, if any, clues about the nature of parent materials can be found in these severely metamorphosed granulite associations.

The second general group of Archean cratonic rocks are **greenstones.** Greenstone associations usually occur in roughly troughlike or synclinal belts (Fig. 6–19). They are prominent features of Archean terranes of all continents. The Abitibi belt of the Superior province is the largest uninterrupted greenstone trend in the Canadian shield (see Fig. 6–19). Like most other greenstone belts, it is composed of basaltic, andesitic, and rhyolitic volcanic rocks along with metamorphosed sediments derived by weathering and erosion of the volcanics. Lavas of greenstone belts exhibit pillow structures that indicate that they were extruded under water (Fig. 6–20).

The most intriguing characteristic of greenstone belts is the roughly sequential transition of rock types from highly mafic near the bottom to felsic near the top (Figs. 6–21 and 6–22). In many, the basal rock masses are komatiites. Overlying basalts are somewhat less mafic and include low-grade metamorphic minerals such as chlorite and hornblende. These green minerals impart the color for which greenstones are named. Proceeding upward in a typical greenstone succession, one finds felsic volcanics above mafic rocks, which are overlain in turn by shales, graywackes, conglomerates, and, less frequently, by banded iron formations. The entire greenstone sequence is often surrounded by granitic intrusive rocks. In the Superior province of Canada, the granitic rocks were emplaced during a deformational event that occurred about 2.6 to 2.7 billion years ago. That event, called the **Kenoran orogeny,** marks the close of the Archean Eon in Canada.

FIGURE 6–20 Exposure of greenstone showing well-developed pillow structures, near Lake of the Woods, Ontario, Canada. The original basalts were extruded under water during Late Archean time. During the Pleistocene Ice Age, glaciers beveled the surface of the exposures and left behind the telltale scratches (glacial striations). *(Courtesy of K. Schultz.)*

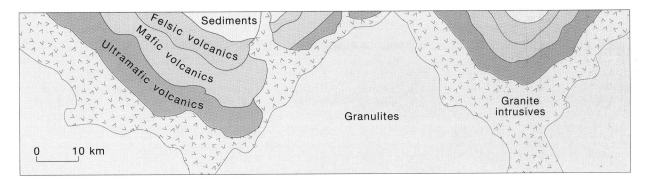

FIGURE 6–21 Generalized cross-section through two greenstone belts. Note their synclinal form and the sequence of rock types from ultrabasic near the bottom to felsic near the top. A late event in the history of the belt is the intrusion of granites. Ultramafic basal layers are particularly characteristic of greenstone belts in Australia and South Africa but do not occur in the Archean of Canada.

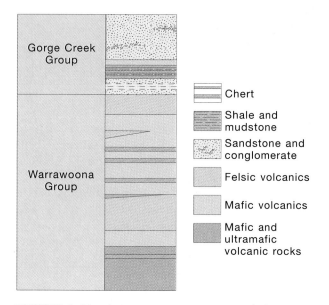

FIGURE 6–22   A representative greenstone belt strati-graphic sequence from the Precambrian shield of western Australia.   *(Adapted from Lowe, D. R. 1980. Ann. Rev. Earth Planet. Sci. 8:145–167.)*

Archean Protocontinents

The upwardly changing sequence from ultramafic or mafic to felsic rocks in greenstone belts records the development of felsic crustal materials from older, more mafic source rocks. The sedimentary rocks that occur most commonly near the top of the greenstone sequence provide clues to the nature of the earliest patches of Archean continental crust. Among the sed-iments are elongate layers of coarse conglomerates containing pebbles of granite. Thus, felsic source rocks were already present at the Earth's surface. Graywackes and dark shales present in the green-stone sedimentary sequence appear to have been deposited in deep-water environments adjacent to protocontinental coasts. In the greenstone sedimen-tary association, we do not find blankets of well-sorted, ripple-marked, cross-bedded sandstones. The existence of broad lowlands that could be flooded by shallow interior seas is therefore unlikely. Nor is there any evidence for wide marginal shelf environ-ments. This absence of blanket sandstones and the presence of graywackes and elongate bodies of con-glomerates suggest that Archean protocontinents were steep-sided, relatively small, narrow land masses with rugged shorelines.

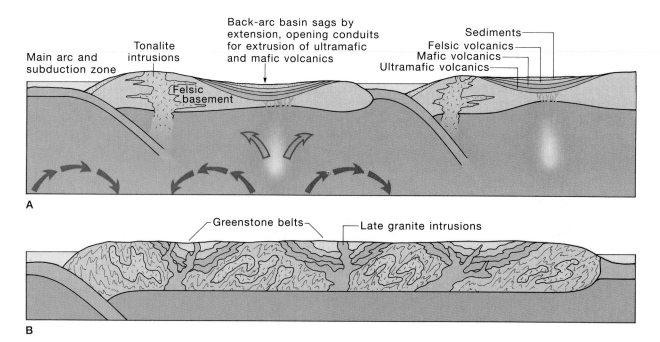

FIGURE 6–23   Plate tectonics model for the development of greenstone belts and growth of continental crust. (A) Plates are in motion, driven by convection cells in the upper mantle. Subduction provides for the emplacement of wedges of oceanic crust and for mixing and melting to provide tonalite intrusions. Behind the main arc, the back arc sags by extension and the greenstone volcanic se-quence is extruded. (B) Compression has occurred to create the greenstone belts with their synclinal form and to aggregate small continental patches into a larger continental mass. Later, granites are intruded in and around greenstone belts. *(Simplified from a model proposed by B. F. Windley, 1984,* The Evolving Continents *2nd ed. New York: John Wiley.)*

Archean Tectonics

The relation between greenstone belts and granulite terranes is currently under active investigation. A plate tectonic model has been proposed to account for the origin of these Archean rocks. Plate tectonics would provide a mechanism capable of supplying emerging continents with mantle-derived materials that could be recycled and partially melted to form additions to continental crust. Certainly, however, Archean plate tectonics would have operated at a far more dynamic level as compared to the Phanerozoic because of the much greater amount of heat still present in the mantle. As compared to Phanerozoic conditions, rates of convection in the mantle would have been far greater, thermal plumes more numerous, midocean ridges longer and more numerous, and volcanism more extensive. Under these conditions, greenstone and granulite associations developed. The first stage would be the growth of volcanic arcs adjacent to subducting plates (Fig. 6–23). Lavas and pyroclastics of the arcs may then have been weathered and eroded to provide sediment to the adjacent trench. In the subduction zones, this wet sediment would be carried downward along with oceanic crust and heated. Partial melting of the mix of materials may have produced increasingly more felsic magmas, and these would have worked their way upward, engulfing earlier volcanics. During or shortly after these events, compressional deformation and metamorphism altered the rocks to form the granulite terranes.

Whereas volcanic arcs were the sites for the birth of granulite associations, the back-arc basins were the probable locations for the development of greenstone belts. In the early stages of that development, ultramafic to mafic lavas ascended through rifts in the back-arcs to form the lower layers of the greenstone sequences. Later extrusions included increasingly more felsic lavas, pyroclastics, and sediments derived from the erosion of adjacent uplands. The final stage of development involved compression of the back-arcs to form the synclinal structures characteristic of greenstone belts. Compression may have occurred simultaneously with the magmatic activity that produced the granitic intrusions that surround the lower levels of greenstone belts.

Archean Rocks

Southern Africa's Barberton Mountain Land and a terrane in western Australia called the Pilbara block contain excellent records of Archean sedimentation. Here one finds large cratons rather than the small protocontinents discussed earlier. These cratons were fully formed and stable as early as 3.0 billion years ago. Because they have been stable since that early

time, sedimentary rocks within them have often escaped severe metamorphism that might otherwise have obscured their environments of deposition. In southern Africa, some of these sedimentary rocks contain gold and have therefore been intensely investigated.

In the greenstone belts of Barberton Mountain Land, the geologic column begins in its lower part with ultramafic and mafic lava flows (Fig. 6–24), followed by pillowed oceanic basalts. These rocks give way upward to felsic lava flows and then graywackes

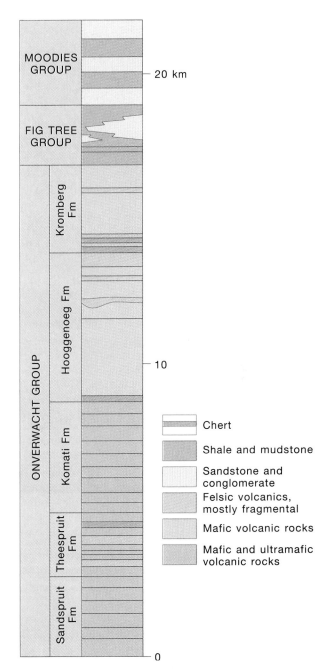

FIGURE 6–24 Stratigraphic sequence in a greenstone belt in Barberton Mountain Land, South Africa. *(After D. R. Lowe 1980. Ann. Rev. Earth Planet. Sci. 8:145–167.)*

Voyageurs National Park

Voyageurs National Park lies along the Minnesota–Ontario border near the southern margin of the Canadian Shield (see location map). The park contains more than 30 lakes surrounded by about 900 rocky islands clothed in forests of fir, spruce, pine, aspen, and birch. Moose, deer, beaver, and wolves make their home on many of the islands. Cars are left at the park's entry stations, and visitors explore the park by boat. Voyageurs National Park takes its name from the colorful French–Canadian trappers who traveled the lakes in long birchbark canoes, carrying pelts to fur company facilities in Grand Rapids, Minnesota.

Rather like a history book in which all but the first and last chapters have been destroyed, the park provides a record of very early and very recent geologic events. Early events can be read from severely deformed and compressed schists and gneisses of Precambrian age. All of the exposed bedrock in the park originated during the Precambrian (see accompanying figure), and most of the formations are Archean. Younger events are revealed in erosional and depositional features resulting from the work of glaciers.

The Archean rocks at Voyageurs National Park include both the granulite and greenstone associations described in this chapter. Mafic volcanics within the green-

Location map of Voyageurs National Park. *(From Voyageurs Official Map and Guide, National Park Service.* Washington, DC: *GPO, 1993, 342398/ 600165.)*

stone association have pillow structures indicating that much of the Archean volcanic activity in this region occurred beneath the sea. Associated metamorphosed graywackes suggest rapid deposition in trenchlike basins.

Near the end of the Archean, the Precambrian Superior Province in which Voyageurs National Park lies was compressed in a massive mountain-building event termed the Kenoran orogeny. One consequence of the high temperatures produced during the orogeny was

the melting of rocks at depth and the emplacement of a granitic batholith (the Vermillion batholith). Pink and gray granites of the batholith can be seen along the southern margin of Kabetogama Lake. Gold veins found in a fault zone at the western edge of the park probably were derived from hydrothermal solutions emanating from the batholith. The gold veins caused a short-lived gold rush in the 1890s.

The last stages of Archean geologic history in this region in-

and claystones probably derived from the erosion of volcanic mountains and cindercones. Pyroclastic beds capped by chert layers also occur. The chert probably formed from the silica of volcanic ash. The general sequence from ultramafic or mafic volcanics to more felsic volcanics to sediments is sometimes repeated, and with each cycle the younger rocks of the sequence are more felsic. Sandstones containing quartz, orthoclase, and granitic detritus are found near the top of the geologic column. These detrital sediments

indicate the presence of nearby granitic source areas exposed during episodes of cratonic uplift. By the time these highly felsic clastic sediments were being deposited, a full range of depositional environments had developed. Deep-water continental slope and rise environments are recorded by the graded turbidites and mudstones of the Fig Tree Group in Barberton Mountain Land. The overlying Moodies Group includes quartz sandstones and shales that were spread across shallow marine shelves, tidal flats, and deltas.

volved erosion of the mountain ranges produced during the Kenoran orogeny. The wearing away of upland areas continued into the Paleoproterozoic. Then, about 2100 million years ago, massive mafic dikes were intruded into the older Archean rocks. The dark rocks of the dikes are clearly seen along the shores of many of the islands.

Rocks younger than Precambrian may have once covered Voyageurs National Park. If so, it is likely that these rocks would have been scoured and scraped away by the great continental ice sheet of the Pleistocene Epoch. As will be described in Chapter 13, this gargantuan blanket of ice covered most of the northern half of North America. The ice was over a mile thick in places. As it advanced slowly southwestward, it was easily able to dislodge huge chunks of bedrock and transport billions of tons of debris. Today many large boulders from distant locations are scattered across the landscape of the Superior Province. They are dubbed glacial erratics. One huge erratic visible on the south shore of Cranberry Lake weighs over 200 tons. Rock fragments within the ice served like the teeth in a rasp, producing glacial scratches and grooves in the underlying Precambrian rocks. When the ice receded, it left behind the glaciated landscape of lakes and islands we know as Voyageurs National Park.

Precambrian rocks exposed along the shore of Lake Kabetogama, Voyageurs National Park, Minnesota. Just offshore is an island in the Kabetogama Narrows. *(Photograph by Jeff Gnass.)*

Intertidal clastic sedimentary rocks of the Pongola Group are above the Moodies rocks, and these are overlain by a remarkable sequence of river-borne conglomerates, sandstones, and shales known as the Witwatersrand Supergroup. The Witwatersrand is notable for several reasons. First, its areal extent (about 103,000 km², or 40,000 square miles), thickness (8 to 10 km, or 5 to 6.3 miles), and nonmarine environment of deposition (fluvial, deltaic, and lacustrine) clearly denote that very large cratons were in existence by about 2.7 billion years ago. Second, the Witwatersrand contains placer gold. Indeed, since the discovery of that gold in 1883, half the world's supply of the precious metal has come from the fluvial and deltaic conglomerates of the Witwatersrand. The particles of gold in these rich placer deposits were derived from the weathering and erosion of mafic volcanics. The volcanics were components of greenstone belts that had been intruded by rising masses of granitic magma.

■ *Life of the Archean*

The splendid diversity of animals and plants on our planet is a consequence of billions of years of chemical and biological evolution that began during the Precambrian. We would like to have fossil documentation of the earliest events in the progression from nonliving to living things, but that kind of direct evidence is lacking. Theories for the origin of life are based on our understanding of present-day biology, coupled with evidence obtained from the study of Archean rocks, meteorites, and planetary compositions.

The Beginning of Life

Systematic progressions from simple to complex structures are a manifest characteristic of both living and nonliving systems. Awareness of such progressions has strongly influenced theories on the origin of life. Subatomic particles such as neutrons, protons, and electrons are built into atoms. Atoms in turn combine to form molecules, and molecules may be joined to form sheets, chains, and a variety of complex molecular structures. Perhaps in a somewhat similar fashion, the atoms of elements essential to organisms (carbon, oxygen, hydrogen, nitrogen, sulfur, and phosphorus) are built into organic molecules. At some stage in the progression, simpler molecules might have added components that would gradually transform them into the complex organic molecules essential to all living things. An analogy might be the manner in which small molecules of nitrogenous bases combine with sugar and phosphorous compounds to form the large deoxyribonucleic (DNA) molecule (see Fig. 4–13). In living organisms, DNA is found chiefly in the nucleus of the cell. It functions in the transfer of genetic characteristics and in protein synthesis. Some theorists suggest that such large and complex molecules might have organized themselves into bodies capable of performing specific functions. Such bodies in unicellular organisms are called **organelles.** Organelles might then have combined to form larger entities that grew, metabolized, reproduced, and mutated. These are the attributes of all living things. Could a similar progression have occurred in the early years of the Archean? We are uncertain about how some of the steps in the progression would have been accomplished, but the basic concept appears to be biochemically feasible.

Preliminary Considerations

The oldest direct evidence of life is found in rocks exposed in Western Australia that are 3.5 billion years old. In these rocks, paleobiologist J. W. Schopf has identified the remains of several different types of microorganisms. Many are strikingly similar to cyanobacteria (formerly called blue–green algae) that are found all around the world today. These fossils strongly suggest that oxygen-producing organisms had already evolved by this early stage in Earth history. The microfossils were found in a chert bed that occurs within the Apex Basalt formation of the Warrawoona Group.

Less direct evidence of the presence of even older life has been found in 3.8-billion-year-old banded iron deposits in southern Greenland. These rocks contain ^{13}C to ^{14}C ratios similar to those in present-day organisms.

As described earlier in this chapter, the Earth accreted from a dust cloud about 4.6 billion years ago, and an initial atmosphere was subsequently generated by volcanic outgassing. Presumably, the gases emitted were similar to those in volcanic eruptions today in that water and carbon dioxide were dominant, with lesser amounts of nitrogen, hydrogen, hydrogen sulfide, sulfur dioxide, carbon monoxide, methane, and ammonia. Note the absence of free oxygen in this list. The planet's earliest organisms must have been anaerobic—that is, they did not require oxygen for respiration. The environment in which they lived lacked a protective atmospheric shield of ozone, which is derived from oxygen and which absorbs ultraviolet radiation from the sun. Perhaps early life escaped these lethal rays by living beneath protective ledges of rock or at greater water depths. Some biologists insist that in the early stages of the development of life, ultraviolet radiation may not have been a hindrance but rather served to impel molecules to interact and form complex associations.

The carbon, hydrogen, oxygen, nitrogen, phosphorus, and sulfur necessary to produce life are among the most abundant elements in the solar system. They were undoubtedly present on the primitive Earth. What had to be accomplished was the enormously intricate organization of those elements into the molecular components of an organism. There are four essential components of life. The first component is protein. **Proteins** are essentially strings of comparatively simple organic molecules called **amino acids.** Proteins act as building materials and as compounds that assist in chemical reactions within the organism. The second of the basic components is **nucleic acids,** such as DNA, mentioned earlier, and ribonucleic acid (RNA). Organic phosphorous compounds provide a third component of life. They serve to transform light or chemical fuel into the energy required for cell activities. An important fourth ingredient for life is some sort of container, such as a cell membrane. The enclosing membrane provides a relatively isolated chemical system within the cell and keeps the various components in close proximity so that they can interact.

From the previous description, it is apparent that amino acids have an important role in the development of the larger and more complex molecules. They are the building blocks of proteins. Two environmental circumstances in the early years of Earth history may have been important in the natural synthesis of amino acids. Prior to the accumulation of the ozone layer in the Earth's upper atmosphere, ultraviolet rays bathed the Earth's surface. As demonstrated in experiments, ultraviolet radiation is capable of separating the atoms in mixtures of water, ammonia, and hydrocarbons and of recombining those atoms into amino acids. A second form of energy capable of accomplishing this feat is electrical discharge in the form of lightning. Either together or separately, lightning and ultraviolet radiation may have stimulated the production of amino acids in the air, in tidal pools, in the upper levels of the oceans, and wherever suitable environmental conditions existed.

Test Tube Amino Acids and a Step Beyond

Scientists in the mid-nineteenth century had succeeded in manufacturing some relatively simple organic compounds in the laboratory. However, it was not until 1953 that the laboratory synthesis of amino acids and other molecules of roughly similar complexity was announced. Stanley Miller, at the suggestion of Harold Urey, performed the now famous experiment. He infused an atmosphere at that time thought to be like that of the Earth's earliest atmosphere into an apparatus similar to the one shown in Figure 6–25. It was a methane, ammonia, hydrogen, and water vapor atmosphere. As the mixture was circulated through the glass tubes, sparks of electricity (simulated lightning) were discharged into the mixture. At the end of only 8 days, the condensed water in the apparatus had become turbid and deep red. Analysis of the crimson liquid showed that it contained a bonanza of amino acids as well as somewhat more complicated organic compounds that enter into the composition of all living things. In additional experiments by other biochemists, it was shown that similar organic compounds could also be produced from gases (carbon dioxide, nitrogen, and water vapor) of the preoxygenic atmosphere. The main requirement for the success of the experiments seemed to be the lack, or near absence, of free oxygen. To the experimenters, it now seemed almost inevitable that amino acids would have developed in the Earth's prelife environment. Because amino acids are relatively stable, they probably increased gradually to

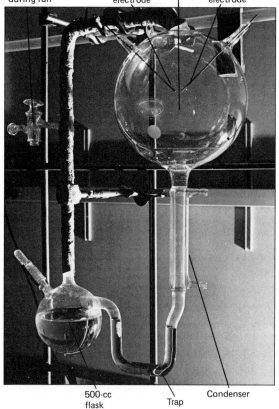

Stopcock for withdrawing samples during run   Tungsten electrode   5-liter flask   Tungsten electrode

500-cc flask   Trap   Condenser

FIGURE 6–25 Stanley Miller and Harold Urey used an apparatus similar to this to replicate what they believed were the conditions of early Earth. An electric spark was produced in the upper right flask to simulate lightning. The gases present in the flask reacted together, forming a number of basic organic compounds. *(Dr. Stanley L. Miller, University of California, San Diego.)*

levels of abundance that would enhance their abilities to join together into more complex molecules.

To come together and form protein-like molecules, amino acids must lose water. This loss can be accomplished by heating concentrations of amino acids to temperatures of at least 140°C. Volcanic activity on the primitive crust would be capable of providing such temperatures. However, the biochemist S. W. Fox discovered that the reaction also occurred at temperatures as low as 70°C if phosphoric acid was present. Fox and his co-workers were able to produce protein-like chains from a mixture of 18 common amino acids. They called these structures **proteinoids** and reasoned that billions of years ago they were the transitional structures leading to true proteins. This is not extravagant conjecture, for Fox was able to find proteinoids similar to those he created in his laboratory among the lavas and cinders

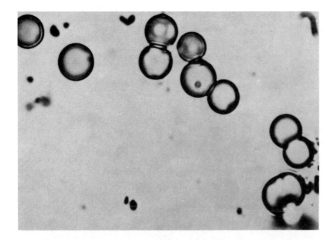

FIGURE 6–26   Proteinoid microspheres formed in hot aqueous solutions of proteinoids that have been permitted to cool slowly. The spheres are about 1 to 2 μm in diameter.   *(Photo by Sidney W. Fox.)*

adjacent to the vents of Hawaiian volcanoes. Apparently, amino acids formed in the volcanic vapors and were combined into proteinoids by the heat of escaping gases.

Hot, aqueous solutions of proteinoids will, on cooling, form into tiny spheres that show many characteristics common to living cells. These **microspheres** (Fig. 6–26), as they are called, have a filmlike outer wall; are capable of osmotic swelling and shrinking; exhibit budding, as do yeast organisms; and can be observed to divide into "daughter" microspheres. They occasionally aggregate linearly to form filaments, as in some bacteria, and they exhibit a streaming movement of internal particles similar to that observed in living cells.

Although complete long-chain nucleic acids have not been experimentally produced under prelife conditions, short stretches of ordered sequences of nucleic acid components have now been produced in the laboratory. Some paleobiologists believe similar sequences could have been formed and accumulated on the surfaces of mineral grains during the Archean. Clay particles, which are much like mica in crystal structure, appear to be particularly favorable sites for the necessary polymerization of organic molecules. Clay minerals contain iron and zinc, which serve as catalysts for polymerization. Further, the tiny flakes of clay have a small electric charge along their planar surfaces that permit them to attract organic molecules or parts of molecules having an opposite charge. The result is not only a concentration of organic molecules on the clay particles but a tendency for those molecules to line up and form molecular chains. Short-chain nucleic acid molecules might have formed in this way during the early stages of the progression toward a living organism.

The Birthplace of Life

There are several good reasons to believe that the earliest organisms originated in the sea. The sea contains the salts needed for health and growth. The waters of the oceans serve as universal solvents capable of dissolving a great variety of organic compounds, and currents in the oceans ceaselessly circulate and mix these compounds. Such constant motion would have favored frequent collisions of the vital molecules and would thereby have increased the probability of their combining into larger bodies. One can imagine the larger bodies adding components, increasing in complexity, and ascending in a thousand infinitesimally small steps from nonliving to transitional things of a biologic netherworld to the first living organisms. Such a sequence could not occur in an environment like that existing on land today. Oxygen and present-day microbial predators would destroy the delicate structures. However, life originated before there were

A

B

FIGURE 6–27   A hydrothermal vent (A) located on a midoceanic ridge in the eastern Pacific. This kind of vent is called a black smoker because the escaping waters are darkened by metallic sulfides precipitated as these waters encounter cold sea water. The nutrient-rich waters of the vents support chemosynthetic bacteria and a variety of invertebrates, including giant tube worms (B).   *(Dudley Foster, Woods Hole Oceanographic Institution.)*

organisms to cause decay and before there was sufficient free oxygen to be troublesome.

The belief that the ocean was the birthplace of life has been with us for many decades. Since the time of Darwin, biologists have spoken of the primordial ocean as a rich organic broth containing all the necessary raw materials for life. Until recently, the more placid, illuminated upper parts of the ocean were considered likely sites for biogenesis. Indeed, the ocean-atmosphere interface is still the best candidate for the site of life's beginning. As another possibility, however, direct observations made from diving submersibles cruising near the surface of midoceanic ridges have provided evidence that life may have originated in total darkness, at abyssal depths, among jets of scalding water rising from submarine hydrothermal vents and volcanic fissures. The evidence for such a stygian beginning of life consists of the presence of actual bacteria called hyperthermophiles that thrive in sea water at temperatures that exceed 100 degrees Celsius. The bacteria are apparently able to live deep within fissures below the actual vents, for they are often ejected in such great numbers that they cloud the surrounding waters. In the absence of light for photosynthesis, the organisms derive energy from chemosynthesis, a process within the cell that causes hydrogen sulfide to react with oxygen and to produce water and sulfur. The chemosynthetic bacteria around vents of the midoceanic ridges form the base of a strange food chain that supports an astonishing variety of invertebrates, including shrimplike arthropods, crabs, clams, and tube worms as much as 10 feet long (Fig. 6–27).

One can envision the steps by which the Earth's first hyperthermophiles originated as beginning with the downward percolation of ocean water through fractures and fissures toward the zone of extremely hot rocks that supply the lavas of midoceanic ridge volcanoes (Fig. 6–28). There the water would be heated to about 1000 degrees Celsius, yet pressure would be so great that it would prevent boiling. The superheated water would react readily with surrounding rocks, extracting elements needed to construct organic molecules. Such molecules would begin to form as the hot waters gradually returned to the surface, cooling as they made their way upward. During that time, amino acids and other organic compounds might be synthesized and combined

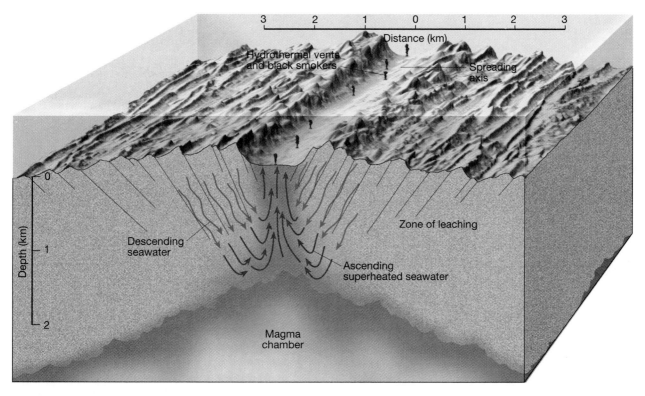

FIGURE 6–28 Conceptual section across the central part of a midoceanic ridge (spreading center) depicting the origin of hydrothermal vents on the ocean floor. Cool water (blue arrows) is gradually heated as it descends toward the hot magma chamber, leaching sulfur, iron, copper, zinc, and other metals from surrounding rocks. The hot water (red arrows) returning to the surface carries these elements upward, discharging them at the hydrothermal springs on the ocean floor, where they support a distinctive community of organisms.

to form the first protocells. Growth, proliferation, and evolutionary processes operating on the protocells might then lead to primitive chemosynthetic bacteria like those seen today in hydrothermal vent environments.

Metabolic Pathways

It is unlikely that the first microorganisms would have evolved a means of manufacturing their own food but rather assimilated small aggregates of organic molecules that were present in the surrounding water. Some would probably have consumed even their developing contemporaries. Today organisms with this type of nutritional mechanism are termed **heterotrophs.** The food gathered by the ancestral heterotrophs was externally digested by excreted enzymes before being converted to the energy required for vital function. In the absence of free oxygen, there was only one way to accomplish this conversion—by **fermentation.** There are many variations of the fermentation process. The most familiar reaction involves the fermentation of sugar by yeast. It is a process by which organisms are able to disassemble organic molecules, rearrange their parts, and derive energy for life functions. A very simple fermentation reaction can be written as follows:

$$\underset{\text{Glucose}}{C_6H_{12}O_6} \rightarrow 2\ \underset{\substack{\text{Carbon} \\ \text{dioxide}}}{CO_2} + 2\ \underset{\text{Alcohol}}{C_2H_5OH} + \text{Energy}$$

Animal cells are also able to ferment sugar in a reaction that yields lactic acid rather than alcohol.

Eventually, the original fermentation organisms would have begun to experience difficult times. By consuming the organic compounds of their environment, they would eventually have created a food shortage; this scarcity, in turn, might have caused selective pressures for evolutionary change. At some point prior to the depletion of the food supply, organisms evolved the ability to synthesize their needs from simple inorganic substances. These were the first autotrophic organisms. Unlike the heterotrophs, **autotrophs** were able to manufacture their own food. The organisms that had developed this remarkable ability saved themselves from starvation. Their evolution proceeded in diverse directions. Some manufactured their food from carbon dioxide and hydrogen sulfide and were the probable ancestors of today's sulfur bacteria. Others employed the life scheme of modern nitrifying bacteria by using ammonia as a source of energy and matter.

However, more significant than either of these kinds of autotrophs were the **photoautotrophs** (Fig. 6–29), which were capable of carrying on photosynthesis. Their gift to life was the unique capability of dissociating carbon dioxide into carbon and free oxy-

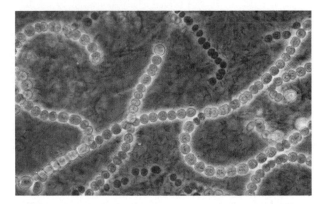

FIGURE 6–29   Photoautotrophs. These organisms, known as *Nostoc,* use light energy to power the synthesis of organic compounds from carbon dioxide and water.   *(Sinclair Stammers/Science Photo Library/Photo Researchers, Inc.)*

gen. The carbon was combined with other elements to permit growth, and the oxygen escaped to prepare the environment for the next important step in the evolution of primitive organisms. In simplified form, the reaction for photosynthesis can be written as follows:

$$6\ CO_2 + 6\ H_2O \xrightarrow{\text{Sunlight}} C_6H_{12}O_6 + 6\ O_2$$

With the multiplication of the photoautotrophs, billions upon billions of tiny living oxygen generators began to change the primeval nonoxygenic atmosphere to an oxygenic one. Fortunately, the change was probably gradual, for if oxygen had accumulated too rapidly, it would have been lethal to developing early microorganisms. Some sort of oxygen acceptors were needed to act as safety valves and to prevent too rapid a buildup of the gas. Iron in rocks of the continental crust provided suitable oxygen acceptors. Eventually, organisms evolved oxygen-mediating enzymes that permitted them to cope with the new atmosphere.

After the surficial iron on the Earth's surface had combined with its capacity of oxygen, the gas began to accumulate in the atmosphere and hydrosphere. Solar radiation acted on atmospheric oxygen to convert part of it to ozone, and ozone in turn formed an effective shield against harmful ultraviolet radiation. Still-primitive and vulnerable life was thereby protected and could expand into environments that formerly had not been able to harbor life. The stage was set for the appearance of aerobic organisms.

Aerobic organisms use oxygen to convert their food into energy. The reaction, which can be considered a form of cold combustion, provides far more energy in relation to food consumed than does the fermentation reaction. This surplus of energy was an important factor in the evolution of more complex forms of life.

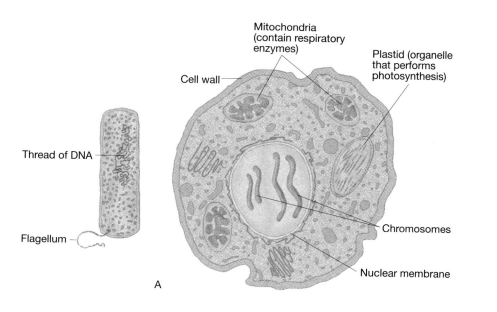

FIGURE 6–30 (A) Comparison of a prokaryotic and eukaryotic cell. Note that the prokaryotic cell lacks a true nucleus, whereas the eukaryotic cell contains a true nucleus and various organelles. Also, prokaryotic cells are smaller, ranging in size from 0.5 to 1.0 μm as compared to 10 to 100 μm for eukaryotic cells. (B) An electron micrograph of the living prokaryote *Bacillus subtilus* that is about to complete cell division (magnification ×15,000). *(B courtesy of A. Ryter.)*

Prokaryotes, Eukaryotes, and Symbiosis

There is no unequivocal evidence to date the transition from chemical or prebiotic evolution to organic evolution. The most accurate statement that can be made in this regard is that the transition occurred at some time prior to 3.5 billion years ago. In rocks of that age, paleontologists have discovered the earliest fossil evidence of life. These oldest fossils belong to a category of microorganisms called **prokaryotes** (Fig. 6–30A). Prokaryotes lack definite membrane-bounded organelles and also do not have a membrane-bounded nucleus in which genetic material is neatly arranged into discrete chromosomes (Table 6–4). Modern prokaryotes do possess cell walls, and most are able to move about. The cyanobacteria (formerly called blue–green algae) are prokaryotes that are capable of photosynthesis (see Fig. 6–29).

TABLE 6–4 Comparison of Some of the Characteristics of Prokaryotic and Eukaryotic Organisms

	Prokaryotes	Eukaryotes
Organisms	Bacteria and cyanobacteria	Protists, fungi, plants, animals
Cell size	1 to 10 μm	10 to 100 μm
Genetic organization	Loop of DNA in cytoplasm	DNA in chromosomes in membrane-bounded nucleus
Organelles	No membrane-bounded organelles	Membrane-bounded organelles (chloroplasts and mitochondria)
Reproduction	Binary fission, dominantly asexual	Mitosis or meiosis, dominantly sexual

COMMENTARY

The Endosymbiotic Theory for the Origin of Eukaryotes

The origin of eukaryotic organisms from prokaryotic ancestors is termed the endosymbiotic theory. The accompanying figure illustrates the steps involved in this important evolutionary theory. At the top of the figure is the original prokaryotic host cell. The host ingests, but does not digest, aerobic bacterial endosymbionts. These survive and reproduce within the host cell so that future generations of the host organism also contain these endosymbionts. At the same time, the host cell provides a safe environment, raw materials, and nutrients. Eventually, the host and endosymbionts develop a mutualistic relationship, and the originally ingested cells lose their ability to function outside the cell. The endosymbionts that were originally photosynthetic bacteria become organelles called chloroplasts and perform the function of photosynthesis. The aerobic bacteria that had been ingested ultimately become mitochondria and function in oxidative metabolism. They are the powerhouses of the cell.

Evidence in support of the endosymbiotic origin of eukaryotes includes observations that mitochondria and chloroplasts possess some of their own genetic apparatus and can, on a limited level, synthesize protein. They can also divide independently of the cell in which they reside. The double membrane enveloping mitochondria and chloroplasts is believed to have formed by invagination of the host cell's outer membrane.

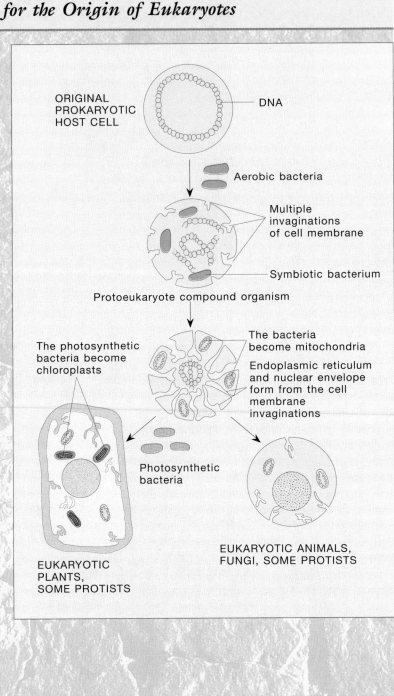

Prokaryotes are asexual and thereby restricted in the level of variability they can attain. In sexual reproduction, as noted in Chapter 4, there is a union of gametes (egg and sperm) to form the nucleus of a single cell, the zygote. The formation of the zygote results in recombination of the parental chromosomes and leads to a multitude of gene combinations among the gametes that give rise to the next generation. In the asexually reproducing prokaryotes, in contrast, a cell divides from the parent and becomes an independent individual containing the identical chromosome complement of the parent cell. Unless mutation intervenes, the number and kinds of chromosomes in individuals produced by asexual reproduction are exactly the same as those in the parent. Thus, the possibilities for variation are more limited than in sexually reproducing organisms. It is probably for this reason that prokaryotes have shown little evolutionary change through more than 2 billion years of Earth history. Nevertheless, such organisms represent an important early step in the history of primordial life.

Evolution proceeded from the prokaryotes to organisms with a definite nuclear wall, well-defined chromosomes, and the capacity for sexual reproduction. These more advanced forms were **eukaryotes** (see Fig. 6–30B). Unlike prokaryotes, eukaryotes contain organelles such as chloroplasts (which convert sunlight into energy) and mitochondria (which metabolize carbohydrates and fatty acids to carbon dioxide and water, releasing energy-rich phosphate compounds in the process).

Biologists believe that the organelles in eukaryotic cells were once independent microorganisms that entered other cells and then established symbiotic relationships with the primary cell. For example, an anaerobic, heterotrophic prokaryote such as a fermentative bacterium might have engulfed a respiratory prokaryote and thereby would have an internal consumer of the oxygen that might otherwise have threatened its existence. A nonmotile organism might acquire mobility by forming a symbiotic association with a whiplike organism like a spirochete. The resulting cell would appear to have a flagellum for locomotion. Natural selection would clearly favor such advantageous symbiotic relationships.

With the development of sexually reproducing eukaryotes, genetic variations could be passed from parent to offspring in a great variety of new combinations. This development was truly momentous, for it led to a dramatic increase in the rate of evolution and was ultimately responsible for the evolution of complex multicellular animals.

The Fossil Record for the Archean

Because of the prevalence of prokaryotic microorganisms in certain Archean sedimentary rocks, the eon might well be called the age of prokaryotes. Among the more widespread indications of these primitive organisms are fossils known as stromatolites. **Stromatolites** are laminar, organic sedimentary structures formed by the trapping of sedimentary particles and precipitation of calcium carbonate in response to the metabolic activities and growth of matlike colonies of cyanobacteria and fewer numbers of other prokaryotes. The study of present-day stromatolites indicates that they develop when fine particles of calcium carbonate settle on the sticky, filamentous surface of cyanobacterial mats to form thin layers (laminae) (Fig. 6–31). The cyanobacterial colonies then grow through the layer and form another surface for the

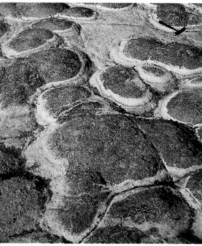

A   B

FIGURE 6–31   (A) Modern columnar stromatolites growing in the intertidal zone of Shark Bay, Australia. Metabolic activities of colonial marine cyanobacteria result in the formation of these structures. Fine particles of calcium carbonate settle between the tiny filaments of the matlike colonies and are bound with a mesh of organic matter. Successive additional layers result in the laminations that are the most distinctive characteristic of stromatolites. (B) Fossil stromatolites from Precambrian rocks exposed in southern Africa. *(Photo A courtesy of J. Ross, and B courtesy of J. W. Schopf.)*

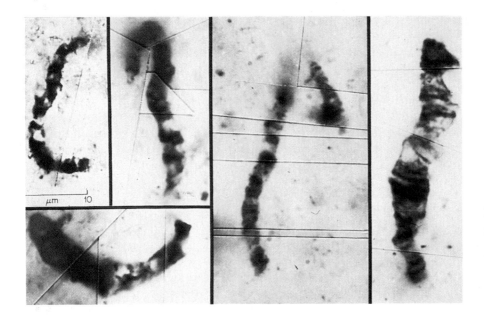

FIGURE 6–32 Filamentous prokaryotic microfossils from 3.5-billion-year-old black cherts of the Archean War-rawoona Group. Pilbara shield of western Australia. *(Courtesy of J. W. Schopf and B. M. Packer.)*

collection of additional fine sediment. Repetition of this process along with precipitation of calcium carbonate produces a succession of laminae. The resulting structures can be cabbage-like, columnar, tabular, or branching in shape.

Stromatolites are more common in Proterozoic rocks than Archean rocks. This may be a reflection of a rarity of the warm shelf environments conducive to the growth of marine stromatolites. The oldest stromatolite-like fossils occur in 3.5-billion-year-old rocks of the Warrawoona Group of Australia's Pilbara shield. The black chert beds in the Warrawoona Group also contain a variety of filamentous fossil prokaryotes (Fig. 6–32). Somewhat younger Archean stromatolites are found in the 3.0-billion-year-old Pongola Group of southern Africa and the 2.8-billion-year-old Bulawayan Group of Australia. The ^{12}C to ^{13}C ratios in the Bulawayan rocks provide evidence of biologic fixation of carbon dioxide by photosynthetic organisms.

Some of the earliest discoveries of Archean microfossils were the result of careful study of thin sections and electron micrographs prepared from rocks of the southern African Fig Tree Group. These rocks are about 3.1 billion years old and consist of variously colored cherts, slates, ironstones, and hard sandstones. Elso S. Barghoorn and J. William Schopf were among the first paleontologists to study the fossils of the Fig Tree Group. They were able to find a number of tiny, double-walled, rod-shaped structures (Fig. 6–33) that had a striking resemblance to modern bacteria. Barghoorn and Schopf named their find *Eobacterium isolatum*—the "isolated dawn bacteria." In their search for larger fossils, the scientists prepared thin sections of the chert and examined these with the optical microscope. The examination disclosed numerous spheroidal bodies very similar to certain cyanobacteria. Because the samples were collected near the town of Barberton, Barghoorn and Schopf named the fossils *Archaeosphaeroides barbertonensis* (Fig. 6–34). The filaments of organic matter and hydrocarbons of organic origin found in Fig Tree rocks are considered evidence of the presence of microscopic life in the Archean. Many beds within the Fig Tree contain so much organic carbon that they are black or dark gray in color. On analysis, the dark layers are found to contain organic compounds regarded as breakdown products of chlorophyll, indicating that photosynthesis was in operation at least 3.0 billion years ago.

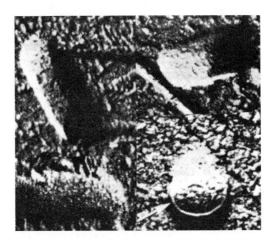

FIGURE 6–33 Electron micrographs of rod-shaped *Eobacterium isolatum*, about 0.6 micron in length. Specimen at lower right has been interpreted as a transverse section through a cell. *(Courtesy of J. W. Schopf.)*

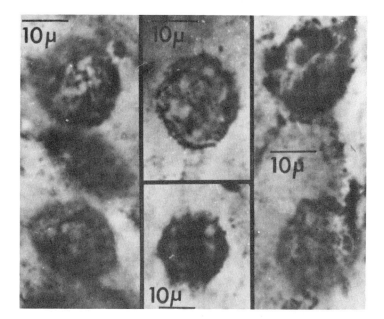

FIGURE 6–34 Spheroidal, alga-like microfossils *(Archaeosphaeroides barbertonensis)* in black chert of the early Precambrian Fig Tree Series. These fossils are 3.1 billion years old. The scale bars are in microns (1 micron is equal to 0.001 mm). *(Courtesy of J. W. Schopf.)*

■ *Mineral Wealth of the Archean*

Archean rocks are rich in metallic ore deposits, including ores of gold, diamonds, iron, copper, chromium, and nickel. Some of these metals were concentrated at the same time that their host rocks were formed, whereas others were introduced at a later date when ore-bearing hot solutions cooled and precipitated their dissolved metals in veins and cavities. Gold is the metal particularly characteristic of the Archean. It occurs in Archean rocks in Africa, North America, Asia, South America, and Australia. Iron, although important locally, is not as abundant in Archean rocks as it is in Proterozoic rocks. Much of the copper, along with zinc and iron, occurs as sulfides that appear to have formed when sea water penetrated hot submarine lavas, reacted with them to extract the metals, and then, on cooling, precipitated the metallic sulfides on the floors of Archean seas. Ores of this type are called *volcanogenic massive sulfide deposits.* They have been observed forming today on the sea floor along the East Pacific rise.

Summary

The long span of time in the history of the Earth that began 3.8 billion years ago after the Hadean and ended 2.5 billion years ago is known as the Archean. Rocks of the Archean are most extensively exposed for study in the broadly upwarped, geologically stable regions of continents known as Precambrian shields. The shields can be divided into a number of Precambrian provinces that tend to be distinctive in age and structural characteristics. The margins of provinces may be marked by belts of deformed rocks called collisional orogens.

The Earth developed its crust during the Archean. Originally, that crust was mafic in composition and was probably formed of extrusions originating from a partially or totally molten underlying mantle. From this original mafic crust, patches of continental crust of a more felsic composition were formed by partial melting of mafic rocks and by passing the weathered products of original mafic terranes through cycles of tectonic and orogenic activity involving melting and metamorphism.

Because of high heat flow from the Earth's interior during the early history of the Archean, it is likely that the crust was frequently disrupted and that continents were small. By late in the Archean, however, large continents had developed by accretion of microcontinents and consolidation of marginal depositional tracts during postdepositional orogenies.

Archean rocks can be grouped into two major rock associations. The first is the granulite association, which consists of gneisses derived from granodioritic and gabbroic rocks. The second is the greenstone association, which typically occurs in synclinal belts within the more extensive granulite terranes. Greenstones consist of mafic or ultramafic volcanics overlain by felsic volcanics and are capped by sedimentary rocks. In theory, both greenstone belts and granulite terranes may have developed in the course of Archean tectonics, in which granulite associations formed along subduction zones adjacent to volcanic arcs and greenstone belts developed in back-arc basins. Archean sedimentary rocks are found chiefly in greenstone belts and consist largely of graywackes, shales, claystones, conglomerates, and cherts.

The Archean was the time when life originated on Earth. There is, of course, no primary evidence of that event in Archean rocks. However, biochemists have shown experimentally that, in an environment largely devoid of free oxygen but containing carbon, hydrogen, nitrogen, and water in which inorganic salts are dissolved, complex organic molecules can be formed when the mixture is activated by ultraviolet light or lightning-induced electrical discharges. These compounds and their derivatives are found in every living organism. They are the molecular basis of all organic structures, heritable information, and energy. A long episode of Precambrian chemical evolution eventually led to the formation of amino acids, sugars, and nucleic acids. These materials somehow acquired the organization and behavior of replicating living systems.

The discovery of bacteria thriving in total darkness at hydrothermal vents on the deep-ocean floor has prompted new theories for the emergence of life. Because of the high temperatures in which these organisms develop, they are called hyperthermophiles. Instead of using energy from the sun as in photosynthesis, the hyperthermophiles use energy released from certain chemical bonds. The process is termed chemosynthesis. Water in the hydrothermal system percolates downward toward magma chambers, where it is superheated and able to dissolve compounds essential for life from surrounding rock. The solutions are then returned to the surface and, on cooling, may form organic molecules and combinations of molecules leading to protocells. In the Archean, the protocells may have evolved into the earliest microorganisms. Hydrothermal systems would have been far more prevalent in the Archean than today, providing ample opportunity for the evolution of chemosynthetic hyperthermophiles.

The first living cells might have subsisted on the chemical energy they derived by consuming other cells and molecules. They were heterotrophs and were followed by cellular organisms (autotrophs) that could use the energy of the sun through photosynthesis. Free oxygen accumulated as a byproduct of photosynthesis, making possible the evolution of organisms that use oxygen to burn their food for energy.

Archean fossils have been found in rocks about 3.5 billion years old and sporadically in younger rocks of Archean age. The organisms preserved are all prokaryotes. They consist of cyanobacteria and other bacteria often associated with the distinctive laminated structures known as stromatolites.

Rocks of the Archean are important sources of such metals as gold, iron, copper, chromium, and nickel. Gold is particularly characteristic of late Archean rocks, whereas the less precious metals occur in volcanogenic massive sulfide deposits that are about 3 billion years old.

Review Questions

1. What are the names of the planets in our solar system? In what galaxy is our solar system located?

2. What is the source of the sun's heat? Given the amount of solar radiation intercepted by the Earth, why is the Earth's surface not hotter than it is?

3. Describe the evidence that indicates that the universe is expanding.

4. What are the features that distinguish the terrestrial planets from the Jovian planets?

5. What is a Precambrian shield? Distinguish between a shield and a craton and between a shield and a platform.

6. Differentiate between the terms *mafic* and *felsic*. What would be an example of a felsic extrusive rock? A felsic intrusive rock? A mafic extrusive rock? Name two minerals most common in felsic rocks and two minerals most common in mafic rocks.

7. Describe the structural configuration and general vertical sequence of rock types in a greenstone belt. In a plate tectonic scenario, where might greenstone belts form and how would they develop?

8. What is the economic significance of the Archean Witwatersrand rocks of Africa? What kind of sedimentary rocks occur in the Witwatersrand Supergroup, and what was their environment of deposition?

9. Why was oxygen not included by Stanley Miller in his famous experiment for artificial production of amino acids?

10. What geologic evidence suggests that free oxygen was beginning to enter the Earth's atmosphere about 3 billion years ago?

11. Differentiate between a prokaryotic organism and a eukaryotic organism; between a heterotroph and an autotroph; between a photosynthetic and a chemosynthetic organism; and between an anaerobic and an aerobic organism.

12. Where, and in what type of rock, do the oldest known fossils occur?

13. What are stromatolites? How do they originate and grow? What role may they have played in the evolution of the Earth's atmosphere?

Discussion Questions

1. Discuss the essential differences between the big bang, steady-state, and oscillating universe hypotheses. If all three of these concepts had equal validity, which would be philosophically most favorable to you and why?

2. Why are Archean rocks more difficult to correlate than rocks of the Phanerozoic? What is the method used to date and correlate most Archean rocks?

3. Discuss how patches of felsic continental crust might have been derived from the mafic rocks that originally covered the Earth.

4. How do komatiites differ from other mafic rock? Where might they be found in greenstone belts? What do they indicate about the temperature gradients present near the surface of the Earth at the time they formed?

5. Discuss the role of symbiosis in the evolution of eukaryotes. What organelles may have originated by symbiosis?

6. In general, what is the inferred sequence of events that theoretically led to the origin of life?

7. What arguments would you use to convince a scientist that his or her hypothesis that life originated in ordinary soil was incorrect?

8. What are hyperthermophiles? What does the presence of hyperthermophiles suggest regarding the possibility of finding microorganisms on other planets or the moons of other planets in the universe?

Supplemental Readings and References

Cone, J. 1991. *Fire Under the Sea.* New York: William Morrow & Co.

Corliss, J. B., Baross, J. A., and Hoffman, S. E. 1981. An hypothesis concerning the relationship between submarine hot springs and the origin of life on Earth. *Oceanaolog. Acta* 57:3219–3230.

Fox, S. W., and Dose, K. 1972. *Molecular Evolution and the Origin of Life.* San Francisco: W. H. Freeman Co.

Holm, N. G. (ed.). 1992. *Origins of Life and Evolution of the Biosphere.* Boston: Kluwer Academic Publishers.

Kershaw, S. 1990. Evolution of the Earth's atmosphere and its geological impact. *Geo. Today* 6(2):55–60.

Lowe, D. R. 1980. Archean sedimentation. *Ann. Rev. Earth Planetary Sci.* 8:145–167.

Margulis, L. 1992. *Symbiosis in Cell Evolution.* 3rd ed. San Francisco: W. H. Freeman Co.

Nisbet, E. G. 1987. *The Young Earth.* Winchester, MA: Allen and Unwin, Inc.

Shock, E. K. 1994. Hydrothermal systems and the emergence of life. *Geotimes* 39(3):13–14.

Schopf, J. W. 1993. Microfossils of the Early Archean Apex Chert: New evidence of the antiquity of life. *Science* 260:640–646.

York, F. 1992. The earliest history of the Earth. *Sci. Am.* 268(1):90–96.

Early Proterozoic graywackes with graded bedding interstratified with shaly limestone, deposited in the aulacogen east of Great Slave Lake, Northwest Territories of Canada. *(Courtesy of P. Hoffman.)*

We crack the rocks and make them ring,

And many a heavy pack we sling;

We run our lines and tie them in,

We measure strata thick and thin,

And Sunday work is never sin,

By thought and dint of hammering.

From the poem *Mente et Malleo,*

written in 1888 by geologist A. F. Lawson

CHAPTER 7

The Proterozoic Eon

The Proterozoic Eon began 2.5 billion years ago and ended only 544 million years ago. This enormous period of time comprises 42 percent of Earth history. The designation of the beginning of the Proterozoic at 2.5 billion years ago is somewhat arbitrary, but it is used because this is about the time when a more modern style of plate tectonics and sedimentation began to be prevalent. Because Proterozoic rocks are less altered, they are easier to study and interpret than rocks of the Archean. Difficulties persist, however, for they are not as abundantly fossiliferous as many Phanerozoic strata.

As described in the preceding chapter, a number of distinct cratonic elements, designated the Superior, Slave, Hearne, Rae, Wyoming, and Nain provinces, had developed in North America during the Archean. These Archean elements were welded together to form a large continent called **Laurentia** during the early Proterozoic. The welding occurred along belts of crustal compression, mountain building, and metamorphism called **orogens** (Fig. 7–1). This process of aggregation and consolidation of the once separate cratonic elements was completed by about 1.7 billion years ago. During the remainder of the Proterozoic, Laurentia continued to grow as a result of accretionary events. Accretion involves additions of crustal materials to continental margins. The additions may consist of sedimentary rocks deposited along the once passive margins of continents that may later be compressed and welded onto the continent. Accretion may instead involve the incorporation of small blocks

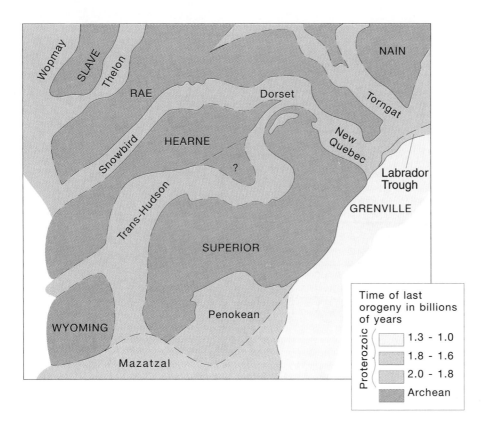

FIGURE 7–1 Expanded and generalized relationships of Proterozoic orogens to Archean provinces. The Wopmay, Penokean, Labrador, and Mazatzal orogens developed by accretion of sedimentary prisms laid down along the margins of Archean cratons. The other orogens are collisional in nature. *(From Hoffman, P. 1989. Precambrian geology and tectonic history of North America, Ch. 16, in Geology of North America. Boulder, CO: Geological Society of America.)*

of crustal material carried to subduction zones adjacent to continents by sea-floor spreading.

The processes of accretion and aggregation of Archean crustal elements were in operation worldwide during the Proterozoic and resulted in the formation of several very large cratons. On and around these cratons, tectonic events occurred and environments of deposition developed that closely resemble those of the succeeding Phanerozoic Eon. By Proterozoic time, wide marginal shelves and epeiric seas existed in which shallow-water clastic sediments and carbonate deposits could accumulate. On the whole, such sediments were rare in the Archean. It is also evident that the Proterozoic was characterized by substantial amounts of lateral plate motion and accompanying subduction, rifting, and sea-floor spreading.

■ *The Paleoproterozoic Record in North America*

Proterozoic Plate Tectonics

In the Canadian shield, accretionary Paleoproterozoic rocks occur primarily in areas that circumvent the older Archean provinces. One such area lies along the western margin of the Slave province in Canada's Northwest Territories. Here, in a belt of deformation, igneous intrusion, and metamorphism called the **Wopmay orogen** (Fig. 7–2), there is evidence of the opening of an ocean basin, sedimentation along resulting new continental margins, and closure of the ocean basin through plate tectonic processes. This sequence of events, well documented in the Phanerozoic, is called a **Wilson cycle**, after J. Tuzo Wilson, one of the pioneers of the plate tectonics theory.

In the region of the Wopmay orogen, evidence for the initial opening of an ocean basin consists of numerous normal (tensional) faults (Fig. 7–3), which created downfaulted or rift valley basins in which alluvial fan and fluvial deposits accumulated. Lavas flowed upward through the fault systems and became interlayered with many of these sediments. Continued opening produced an oceanic tract that slowly widened. As the ocean widened, the western edge of the Slave province became a passive continental margin on which two parallel zones of deposition developed. An eastern zone next to the Slave platform received coastal plain deposits that graded westward (seaward) to shallow marine quartz sandstones. These sandstones were subsequently metamorphosed and are now quartzites. The quartzites pass upward into massive stromatolitic dolomites of the Rocknest

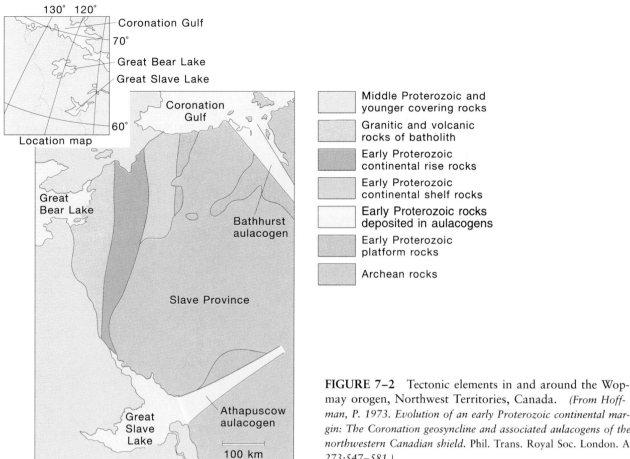

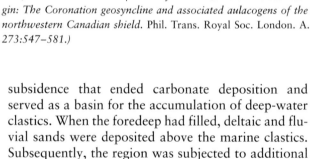

FIGURE 7–2 Tectonic elements in and around the Wopmay orogen, Northwest Territories, Canada. *(From Hoffman, P. 1973. Evolution of an early Proterozoic continental margin: The Coronation geosyncline and associated aulacogens of the northwestern Canadian shield. Phil. Trans. Royal Soc. London. A. 273:547–581.)*

Formation (Fig. 7–4). To the west of these shallow continental shelf deposits, deeper water turbidites of the continental rise were deposited.

After a period of relatively uneventful sedimentation along the Wopmay tract, the more dynamic episode of the Wilson cycle involving ocean closure was initiated. As a result of westward subduction of the leading edge of the Slave plate beneath a microplate approaching from the west, the continental shelf buckled downward, creating a deep elongated area of

subsidence that ended carbonate deposition and served as a basin for the accumulation of deep-water clastics. When the foredeep had filled, deltaic and fluvial sands were deposited above the marine clastics. Subsequently, the region was subjected to additional folding and faulting caused by plate collisions.

These events in the Wopmay orogen closely parallel those that occurred during the late Paleozoic evolution of the Appalachian Mountains. The history of both tracts began with deposition of alluvial deposits

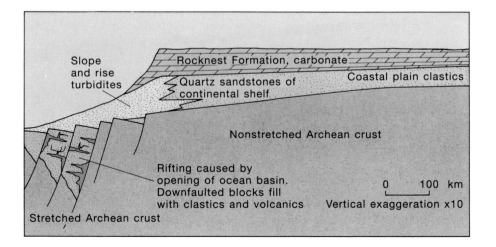

FIGURE 7–3 Relationship of rock units following Paleoproterozoic opening of an ocean basin along the western margin of the Slave craton. *(Simplified from Hoffman, P. 1989. Geology of North America, vol. A:447–512. Boulder, CO: Geological Society of America.)*

A

B

FIGURE 7–4 (A) Dolomite and interbedded shales of the Rocknest Formation, Northwest Territories, Canada. (B) This oblique aerial view impressively illustrates the folding of these dolomitic rocks that resulted from east–west crustal shortening and accretion that occurred during the orogenic phase of the development of the Wopmay orogen.   *(Courtesy of P. Hoffman.)*

in fault-bound basins followed by opening of an oceanic tract; progressed to the deposition of continental shelf and rise sediments along passive continental margins; and ended with ocean closure, deformation, and deposition of nonmarine clastic sediments derived from the erosion of mountains produced during the closure episode.

Proterozoic Aulacogens

Another similarity between Proterozoic tectonics and the tectonics of the Phanerozoic is the development of aulacogens. An **aulacogen** is an inactive rift of a radiating three-rift system that develops over an area of crust that is bulging upward, possibly because of an underlying mantle plume (see Fig. 5–50). As the bulge forms, the stretched crust breaks to form three rift zones that radiate from a common center. Two of

these rifts may widen and become continental margins, whereas the third trends inland away from the continental margin, becomes the "failed arm" of the three-pronged system, and fills with sediment. Two aulacogens have been detected along the western side of the Slave province (see Fig. 7–2). The Athapuscow aulacogen extends northeastward from Great Slave Lake, and the Bathhurst aulacogen trends southeastward from Coronation Gulf.

The Trans-Hudson Orogen

The Wopmay is just one of many orogens that developed along the margins of Archean provinces during the Proterozoic. Another is the **Trans-Hudson orogen**, which separates the Superior province from provinces to the north and west (see Fig. 6–16). Rocks of the Trans-Hudson belt record a historical sequence of initial rifting, opening of an oceanic tract, and subsequent closure along a subduction zone. This closure, with the accompanying severe folding, metamorphism, and intrusive igneous activity, welded the Superior plate to the Hearne and Wyoming plates lying to the west.

A Paleoproterozoic Ice Age

In the region north of Lake Huron, rocks of the Proterozoic are called **Huronian.** They consist mainly of coarse clastics that were eroded from older crystalline rocks. In addition, the Huronian Supergroup includes the **Gowganda Formation,** a rock unit that is notable because of its conglomerates and laminated mudstones of probable glacial origin. Laminations in the mudstones appear to represent regularly repeated summer and winter layers of sediment that are called **varves.** Varved mudstones often form in meltwater lakes that are marginal to ice sheets. The varved sediments in the Gowganda Formation alternate with tillites (consolidated rocks composed of unsorted glacial debris), indicating periodic advances of ice into the marginal lakes. Cobbles and boulders in the tillite have been scratched and faceted, providing evidence of the abrasive action of the ice mass as it moved across the underlying bedrock. The Gowganda Formation is intruded by igneous rocks that have been dated as 2.1 billion years old. They rest on a basement of older crystalline rocks that are 2.6 billion years old. Hence, in North America, glaciation occurred at some time between these two dates. Glacial conditions and accompanying cooler climates were widespread at this time, for rocks similar to the Gowganda are recognized in Finland, southern Africa, and India. This Paleoproterozoic episode of glaciation may have been the first our planet experienced, for the higher thermal regimes of the Archean would have made the accumulation of sufficient ice to form continental glaciers unlikely.

The Animikie Group

Paleoproterozoic rocks surrounding the western shores of Lake Superior include those of the **Animikie Group.** This group of formations is world famous for the bonanza iron ores it contains. The ore minerals are oxides of iron and thus record the presence of free oxygen in the Earth's atmosphere. By implication, the oxygen-generating process of photosynthesis was probably in vigorous operation by this time. Coarse sandstones and conglomerates deposited in shallow water lie near the base of the Animikie Group. These rocks are overlain by cyclic successions of chert, cherty limestone, shales, and banded iron formations. Banded iron formations (BIFs) are also known in

A

B

FIGURE 7–5 Banded iron formations. (A) This banded iron formation is exposed at Jasper Knob in Michigan's Upper Peninsula. (B) Banded iron formations occur worldwide in Proterozoic rocks, as suggested by these at Wadi Kareim, Egypt. *(Photograph B courtesy of D. Bhattacharyya.)*

Archean sequences, but these are not nearly as extensive as those of the Proterozoic (Fig. 7–5). Some of the Canadian deposits are over 1000 meters thick and extend over 100 km. Within the banded iron sequence, there is a formation known as the **Gunflint Chert,** which contains an interesting assemblage of cyanobacteria and other prokaryotic organisms.

The Labrador Trough

East of the Superior province, one finds extensive outcrops of rocks that, like those of the Wopmay orogen, were deposited on the continental shelf,

FIGURE 7–6 Fold belt of the Labrador Trough depicted in a Landsat image made north of Quebec. The belt was once a chain of mountains raised by the collision of two continents about 1.8 to 1.7 billion years ago. Since then, the mountains have eroded, exposing the deeper metamorphic and igneous rocks that were deformed into these folds by the collision. *(Courtesy of NASA.)*

during an orogenic event named the **Hudsonian orogeny.** The Hudsonian orogeny serves as the event that separates Paleoproterozoic from Mesoproterozoic geologic history.

■ *The Mesoproterozoic Record*

Keweenawan Sequence

Among the rocks deposited or emplaced during the Mesoproterozoic were those of the Keweenawan sequence (Fig. 7–7). These rocks rest on either crystalline basement or Animikian strata. They extend from the Lake Superior region southward beneath a cover of Phanerozoic rocks for hundreds of kilometers. Keweenawan rocks consist of clean quartz sandstones and conglomerates as well as basaltic volcanics. The lava flows are well known for their content of native copper. Holes originally formed as gas bubbles in the extruding lava provided some of the voids in which the copper was deposited. In addition, copper fills small joints and pore spaces in associated conglomerates.

The Keweenawan lavas accumulated to a thickness of several thousand meters. Even with so great an outpouring, however, much of the supply of mafic magma was not tapped. It remained beneath the surface, where it crystallized to form the 12,000-meter-thick and 160-km-long mass of the Duluth Gabbro. As noted earlier, basalts are rock types more characteristic of oceanic areas. When such a large quantity of mafic material comes to the surface within the central stable region of a continent, it often signals the presence of a rift zone along which a continent is going to break apart and admit an ocean. The tract where the break begins typically develops tensional faults along which mafic magma rises to the surface to form Keweenawan-like accumulations. Evidence

slope, and a rise of a craton. These rocks occur along a curving, elongate tract known as the Labrador Trough. On the western side of the tract, quartz sandstones, dolomites, and iron formations lie above fluvial and alluvial deposits of fault-bound basins. Pillow lavas, tholeiitic basalts, mafic intrusives, and graywackes occupy an adjacent zone to the east. The eastern zone was subjected to intense folding (Fig. 7–6), metamorphism, and westward thrust faulting

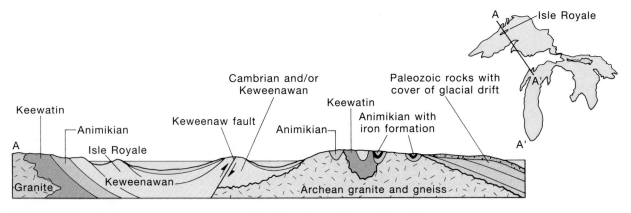

FIGURE 7–7 Geologic cross-section across Lake Superior. *(Adapted from Michigan Geological Survey, Bedrock of Michigan, 1968.)*

from gravity and magnetic surveys, as well as samples from deep drill holes, indicates that the rift zones associated with Keweenawan volcanism developed about 1.2 to 1.0 billion years ago and extended from Lake Superior southward beneath covering sedimentary rocks into Kansas. If this system of rifts had been extended to the edge of the craton, an ocean tract would have formed within the rift system and the eastern United States would have drifted away. Rifting ceased, however, before such a separation could occur.

The Grenville Province

The Grenville province of eastern Canada is the youngest Mesoproterozoic region of the Canadian shield and the last to experience a major orogeny. Exposures of Grenville rocks extend from the Atlantic coast of Labrador to Lake Huron. This, however, is only part of their true extent, for they continue beneath covering Phanerozoic rocks down the eastward side of the United States and westward into Texas (see Fig. 6–16). In the United States, the study of Grenville rocks is difficult, for they have been subjected to the destructive processes that accompanied the building of the Appalachian Mountains during the Paleozoic.

Typical Grenville rocks consist of carbonates and sandstones (Fig. 7–8) that have been metamorphosed and intruded by igneous bodies. In eastern Canada, these rocks have been compressed into overturned

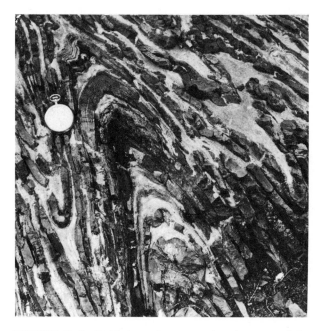

FIGURE 7–8 Folded sandstones and carbonates of the late Proterozoic Grenville Group, Belmont Township, Ontario, Canada. *(Courtesy of the Geological Survey of Canada.)*

folds that have northeasterly axes. Deformation of Grenville sediments occurred 1.2 to 1.0 billion years ago during the **Grenville orogeny.** Two models have been proposed for the cause of this great orogenic event. The first is a plate tectonics model that began with sedimentation along the passive eastern margin of Laurentia. The sedimentation phase was followed by mountain building resulting from collision of another continental block against eastern Laurentia. The alternative model depicts the orogeny as occurring within the continental plate rather than at its margin. In this proposal, a region of subsidence is thought to have formed by ductile stretching of the continental crust over an underlying mantle plume. The depression formed by the stretching became the basin in which Grenville sediments were deposited. Subsequently, the area of stretching shifted laterally, and sediments in the former basin were squeezed against the adjacent cooler and more rigid crust. It is not yet known which of these models is the most valid, but the majority of opinion seems to favor the plate tectonics model.

■ *The Neoproterozoic Record*

The final segment of Proterozoic time is represented in North America by rocks that were deposited in basins and shelf areas surrounding the continent. Most of these sediments were later deformed during Paleozoic orogenic events. In eastern North America, one such event involved rifting and the filling of rift valleys with lava flows and coarse clastics. The sea entered the rift valleys as the beginning of a widening proto-Atlantic ocean, and a new continental margin was established east of the Grenville orogen. The stage was then set for the deposition of the first Paleozoic shelf sediments.

Also in Neoproterozoic rocks, one can find exposures of unsorted boulder beds that are considered by many geologists to be glacial deposits. Such deposits are known in Utah, Nevada, western Canada, Alaska, Greenland, South America, Scandinavia, and Africa. At every locality, the cobbles and boulders exhibit the telltale striations and faceted surfaces that attest to transport by glaciers. This late Proterozoic ice age lasted about 240 million years. It was not the first ice age for the Proterozoic. As described earlier, the Gowganda tillites and correlative formations on other continents reflect widespread glaciation slightly more than 2 billion years ago. Nor would the Neoproterozoic glaciations be the last the Earth would witness, for glacial ice advanced again over major portions of the continents during the Ordovician and Silurian, near the end of the Paleozoic, and again during the Quaternary period of the Cenozoic

Era (Fig. 7–9). Each of these major episodes of glaciation contained shorter intervals during which the ice sheets alternately advanced and regressed. The causes for these oscillations are a matter of great interest to geologists and will be discussed in Chapter 13.

In our standard geologic time scale, the Neoproterozoic is followed by the Cambrian Period of the Paleozoic Era. Nineteenth-century geologists regarded strata that contained the first occurrence of fossils as Cambrian in age. They, of course, had no knowledge of the many animal fossils recently discovered in Neoproterozoic rocks, a few of which are even shell-bearing. These discoveries of the last few decades necessitate a more limited definition of the boundary between the Proterozoic and Phanerozoic. Most geologists now agree that the boundary should be defined by the first occurrence of abundant shelled invertebrates, including archaeocyathids. This definition excludes the relatively rare occurrences of tiny shelled tubular fossils found mostly in continents of the Southern Hemisphere. Also, by insisting that the fossils be shell-bearing, one can exclude strata bearing impressions of soft-bodied animals, such as occur sporadically in latest Neoproterozoic rocks.

■ *Proterozoic Rocks South of the Canadian Shield*

Although outcrops are not as extensive as in Canada, Precambrian rocks also occur south and west of the Canadian shield. Particularly impressive successions are exposed in the Rocky Mountains and Colorado Plateau regions. These rocks have had a complex history that began over 2.5 billion years ago with the evolution of an Archean terrane composed of strongly deformed and metamorphosed granitic rocks. Largely on the basis of the study of remnant patches of volcanic rocks and greenstones in Wyoming, it appears likely that this old Archean mass collided with one or more island arc terranes about 1.7 or 1.8 billion years ago. The line of collision is believed to be marked by shear zones in southern Wyoming and western Colorado characterized by severely crushed and brecciated rocks (Fig. 7–10). Following this orogenic event, there was an extensive episode of magmatism during the Mesoproterozoic. Plutons were emplaced in a broad belt across North America from California to Labrador. The 1.4- to 1.5-billion-year-old plutons are composed primarily of anorthosites and granites. They are **anorogenic,** meaning that they are not associated with an orogenic cycle. This was the first time in the Earth's history that extensive anorogenic magmatism occurred. It is likely that the precondition for this widespread magmatic activity was the existence of a broad region of stable continental crust.

The next event south of the Canadian shield was a widespread episode of rift development about 0.85 to 1.4 billion years ago. Large fault-controlled depressions were formed in which thick sequences of late Precambrian sediments accumulated. They include the Uinta Series of central Utah, the Pahrump Group

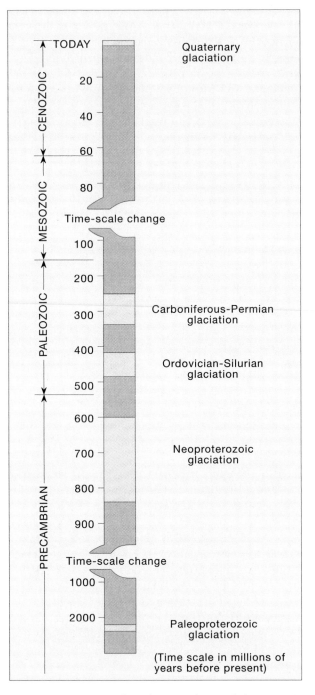

FIGURE 7–9 In its long history, the Earth has experienced several major episodes of widespread continental glaciation. Each of these major episodes contained shorter intervals of advance and retreat of the glaciers.

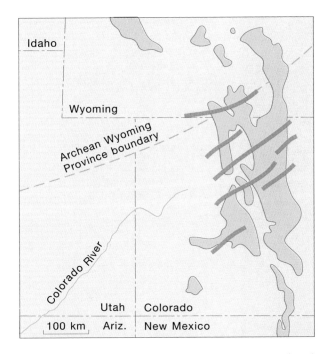

FIGURE 7–10 Shear zones (red) in Wyoming and Colorado developed when the Archean craton collided with island arc terranes during the Paleoproterozoic. Areas of lower Proterozoic outcrops are shown in green.
(Adapted from Karlstrom, K. E., and Bowing, S. A. 1988. Jour. Geology 96:561–571.)

of southeastern California, and the **Belt Supergroup** of Montana, Idaho, and British Columbia. Because of the scenic features with which they are associated, the Belt rocks are of particular interest. In places these rocks are over 12,000 meters thick. Especially impressive are the massive cliff-forming limestones that can be viewed at Waterton Lakes and Glacier National Parks (Fig. 7–11). Although exceptionally thick, the Belt rocks display features such as ripple marks and algal structures, which indicate that they were deposited in relatively shallow water along the passive western margin of North America.

The last event in the late Precambrian history of the western United States was the development of a north–south trending passive margin. At this time the North American plate was moving away from a spreading center located somewhere to the west, and a passive margin marked the continent's trailing edge. The continental shelf was the site of the shallower zone. A westward-thickening wedge of latest Precambrian and Cambrian sediments was deposited on this shelf. The distribution of these sediments was not at all affected by the fault-controlled basins of the earlier episode of deposition.

Precambrian rocks of the Grand Canyon region consist of two distinct units. The lower and older unit is the Vishnu Schist, and the upper unit is the Grand Canyon Series. The Vishnu is a complex body of met-

FIGURE 7–11 Undeformed, horizontal limestones of the Belt Supergroup comprise the upper third of Chief Mountain, located just outside of Glacier National Park, Montana. The lower two-thirds of the mountain consists of deformed Cretaceous beds. A thrust fault separates the two and represents the surface over which the Proterozoic beds were pushed eastward over the weaker underlying Cretaceous strata. The fault is known as the Lewis thrust.

FIGURE 7–12 Vishnu Schist exposed along the walls of the Inner Gorge of the Grand Canyon of the Colorado River, Grand Canyon National Park.   *(Photograph by James Cowlin.)*

amorphosed sediments and gneisses that have been intensely folded and invaded by granites. The granitic intrusions that cut into the Vishnu (and correlative rocks of the southwestern United States) were emplaced 1300 to 1400 million years ago as part of a deformational event named the **Mazatzal orogeny.** A splendid example of an unconformity occurs at the contact between the Vishnu Schist (Fig. 7–12) and the overlying Grand Canyon Supergroup (Fig. 7–13).

The latter unit is Neoproterozoic in age and is itself unconformably overlain by Paleozoic rocks. The Grand Canyon Supergroup correlates with the Belt Supergroup. It consists mostly of clastic rocks (sandstones, siltstones, and shales) that accumulated in a troughlike basin that extended into the craton. The Chuar Group of the Grand Canyon Supergroup contains small, circular carbonaceous structures believed to be the fossil remains of algal spheres.

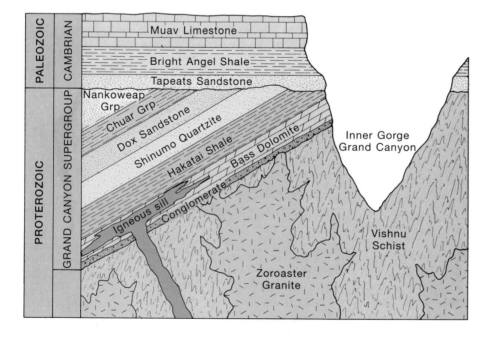

FIGURE 7–13 Vishnu Schist, Grand Canyon Supergroup, and other rocks in the Grand Canyon of the Colorado River.

■ *Precambrian Events Outside North America*

The most significant global event of the Neo-proterozoic was the gathering of continents to form a supercontinent named **Rodinia** (sometimes also called Proto-Pangea to distinguish it from the late Paleozoic supercontinent Pangea). The great ocean that surrounded Rodinia has been named **Mirovia**. It is likely that the Grenville orogeny was caused by continental convergence associated with the assembling of Rodinia.

Like the more familiar later supercontinent of Pangea, Rodinia was not to endure. In the early Cambrian, rifts were already forming between once firmly adjoined land masses. The ocean flooded into the broadening rift zones, one of which separated North America and what is now northern Europe. This Paleozoic ocean is known as **Iapetus.**

The Baltic Shield

Precambrian rocks are found in many of Europe's mountain ranges. However, the most extensive exposures form the surface of the Baltic shield (Fig. 7–14). The northwestern edge of the shield is bounded by an early Paleozoic orogenic belt. In general, there is an east–west trend in the age of rocks composing the Baltic shield. The oldest rocks are of Archean age and are exposed on the eastern side. These are followed by younger Proterozoic rocks in the central region of the shield. The youngest Precambrian rocks are found at the southwestern ends of Norway and Sweden. These rocks are approximately equivalent in age to the Grenville sequence in North America and may have been emplaced at about the time that western Europe began to drift away from Laurentia. A large downfaulted block (graben) near Oslo contains an interesting sequence of fossiliferous Paleozoic strata.

The extremely complex Precambrian history of the Baltic shield has been studied intensively by Swedish and Finnish geologists, who have recognized four great sedimentary and volcanic cycles separated by tectonic movements and batholithic intrusions. The rocks representing the two oldest cycles are mostly schists and gneisses that were once graywackes and volcanics. Conglomerates, ripple-marked quartz sandstones, dolomites, greenstones, slates, and iron ores characterize the third cycle. The most recent rocks escaped severe metamorphism and consist mostly of sandstones and lavas intruded by granites.

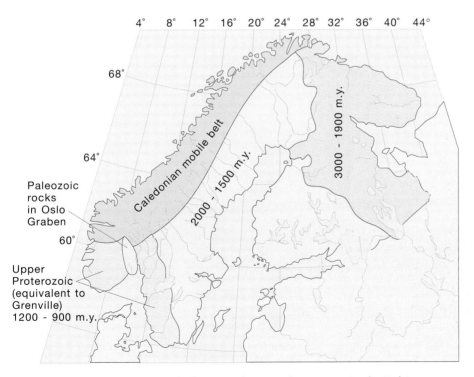

FIGURE 7–14 Region over which Precambrian rocks outcrop in the Baltic shield. Oldest (Archean) rocks are found in the far west, and youngest at the southwestern tips of Norway and Sweden. The Caledonian mobile belt is composed mostly of Paleozoic rocks.

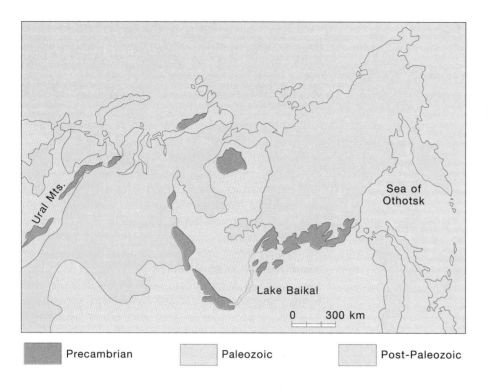

FIGURE 7–15   Areas of Precambrian outcrop in the Angaran shield. Note that vast areas of Precambrian rocks are covered by younger rocks. *(Data from* Geologic Map of U.S.S.R., *1960.)*

The Angaran Shield

The Angaran shield (Fig. 7–15) is more difficult to study than either the Canadian or the Baltic shield. These Precambrian rocks are not exposed over a vast region. Rather, they are revealed as small exposed patches that elsewhere are covered by younger sedimentary rocks of the great Siberian Platform. Covered or not, however, the shield is important as the very nucleus of Asia. To the east and south of this region lay the mobile belts that were later to be fashioned into the Ural and Himalayan mountains. Not unlike the Precambrian of other shields, the Angaran has an older complex of gneisses, schists, and granites that provides evidence of a long and complicated Archean history. There is also a Proterozoic sequence of sedimentary rocks, volcanics, and granitic intrusives that dates from about 1.6 billion years ago. Russian geologists consider these rocks as belonging to a geologic system they have named the **Riphean,** which includes sandstones and limestones of shallow seas. Within the limestones are massive reefs of stromatolites (see Fig. 6–31), the laminar, often moundlike structures produced by cyanobacteria (blue–green algae) that grew in profusion along the carbonate banks of Precambrian seas.

For the very latest Proterozoic rocks, Russian geologists propose the term **Vendian System** (see Fig. 8–6). In the Ukraine where it is best exposed, the

FIGURE 7–16   Regions of exposed Precambrian rocks in South America are shown in light red. The Andean mobile belt is colored green. Yellow areas consist of relatively undisturbed Phanerozoic strata.   *(After Umgrove, J. H. F. 1947.* The Pulse of the Earth. *Nijholt: The Hague.)*

Vendian incorporates strata from the lowest Neoproterozoic glacial deposits upward to the base of the Cambrian. Vendian rocks are notable for the impressions and burrows of soft-bodied metazoans (multicellular animals) they contain. Rarely, tiny tube-shaped fossils whose shells were originally composed of calcium carbonate are found.

South America

The number three seems to characterize the Precambrian geology of South America. There are *three* shields: the Guianan, the Brazilian, and the Patagonian (Fig. 7–16); also, the Precambrian rocks of South America are usually grouped into *three* divisions: Lower, Middle, and Upper Precambrian. As might be expected, gneisses and schists form the oldest (lower) sequence. The Middle Precambrian consists, in large part, of metamorphosed sediments, volcanics, and intrusions that appear to be the product of accretionary evolution. Metamorphic effects are less profound in the Late Precambrian. Slates and phyllites are still in evidence, however, along with quartzites, conglomerates, volcanic flows, and ash beds. Because of a covering of younger sediments over parts of the basement as well as the dense forests that cover many regions, the Precambrian rocks of South America are still not as well known as those of the relatively barren shields of Canada and Scandinavia.

Africa

Precambrian rocks occur over half the surface of Africa and elsewhere lie beneath a veneer of Paleozoic and Mesozoic rocks. Indeed, Africa appears to be essentially a vast platform of abutting shield segments. It is mostly a very stable continent, but along its eastern edge are the remarkable fault structures that produced the African rift valleys. Africa's Precambrian rocks are renowned for their treasure of minerals; exploration for gold, diamonds, copper, uranium, and chromium has resulted in improved understanding of the continent's complex geology. As described in Chapter 6, the Archean of Africa includes thick sections of komatiites and mafic lavas that often appear to be immersed in a sea of granitic intrusions. These igneous rocks constitute the Archean basement of South Africa. In most areas of South Africa, a nonconformity separates the lavas and batholithic granitic masses from the younger sequences of the lower Proterozoic.

By about 1.9 to 1.1 billion years ago, four stable segments of the African craton had been formed (Fig.

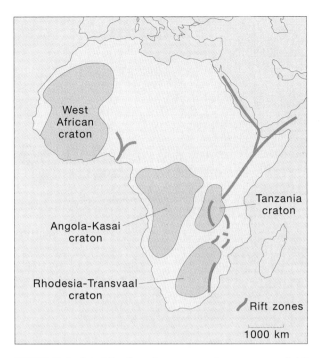

FIGURE 7–17   The four large cratonic segments of Africa.

7–17). One large segment forms much of the western African bulge, two large elements were established in central Africa, and one formed in southern Africa. Belts of metamorphosed granitic crust formed between these crustal segments during the Late Proterozoic and welded them into a unified shield. The orogenic activity accompanying the welding of crustal elements continued until about 400 million years ago and resulted in the continent taking on its present-day outlines.

India

From a geologic point of view, India can be separated into two distinctly different regions. To the north is the Himalayan orogenic belt that developed long after the Precambrian, whereas to the south lies the Indian shield. It is now believed that the shield was brought into juxtaposition with the Himalayan region at a comparatively recent date by northward drifting of what was once an Indian "island continent."

As is the case in North America, geologists have been able to recognize a number of separate provinces within the shield that can be distinguished by their radiometrically determined ages and distinctive

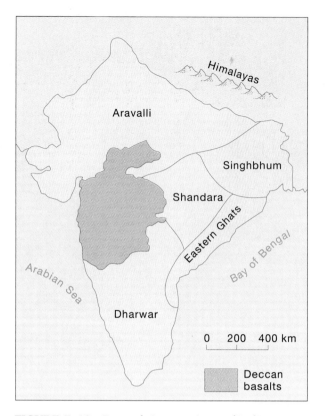

FIGURE 7–18   Precambrian provinces of India.   *(After Nagvi, S. M. and Rogers, J. J. W. 1985. Precambrian Geology of India. Oxford: Oxford University Press.)*

structural characteristics (Fig. 7–18). Five of these provinces are now recognized, and each is believed to be the craton of a separate continental fragment. Orogenic belts and zones of compressional faulting serve to join the cratons together. The joining of component cratons was apparently complete by about 1.5 billion years ago (Middle Proterozoic). Subsequently, the shield was remarkably stable, until in Late Cretaceous time when huge volumes of Deccan basaltic lavas were extruded along numerous fissures. The outpouring of lava may have been in response to the separation of India from the Seychelles Islands. The relative tranquility of India was again disturbed by its collision with Asia only 40 million years ago.

Australia

Australia is composed of Phanerozoic orogenic belts along the eastern side of the continent and consists of shield and platform elements over most of the central and western parts of the continent. The largest area of exposed Archean rocks is in the Yigarn shield in western Australia (Fig. 7–19). North of the Yigarn shield lies the smaller Pilbara shield. Both of these Archean elements contain the greenstone belts and granulite expanses that typify Archean geology on other continents. At least two great episodes of accre-

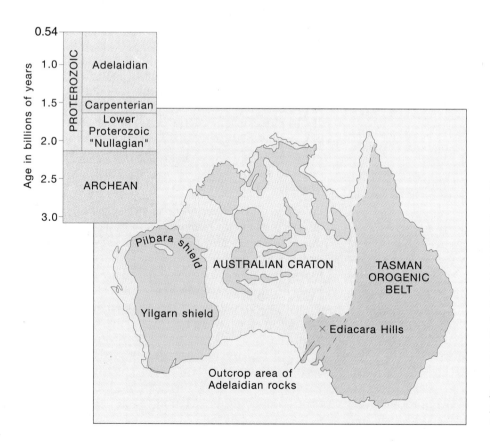

FIGURE 7–19   Central and western Australia consists of a vast cratonic region composed of Precambrian rocks (green) exposed as large patches wherever the cover of mostly horizontal Phanerozoic rocks is absent. The eastern third of the continent contains folded Paleozoic strata of the Tasman orogenic belt. Insert indicates Australian Precambrian time classifications.   *(Generalized from geologic map of Australia, Department of National Development, Canberra, 1958.)*

tion and mountain building are represented in these Archean rocks.

The Proterozoic of Australia includes a thick, widespread sequence of relatively unaltered sedimentary rocks that span a time interval of over 1.5 billion years. Such rocks permit direct comparisons with modern sedimentation processes and provide many clues to paleoclimatic conditions in this part of the Proterozoic world. Included within the sedimentary sequence are extensive shallow marine deposits containing abundant stromatolites. Banded iron deposits of about the same age as those of the Canadian Animikian also occur. Near the top of the Proterozoic sequence are boulder beds deposited by glaciers, varved mudstones of marginal glacial lakes, and glacially striated pavements. Marine sediments slightly younger than the glacial deposits include the Rawnsley Quartzite (a metamorphosed sandstone), well known by geologists because of the impressions of soft-bodied animals it contains.

Antarctica

Beneath the frozen surface of Antarctica lies a Precambrian shield that was poorly understood until the advent of modern techniques for geophysical exploration. Over the past 4 decades, geologists have had an opportunity to examine the bedrock exposed around the margins of the huge Antarctic ice sheet and to visit mountain peaks (nunataks) that pierce the ice cover. Most of the rounded eastern half of Antarctica consists of Precambrian rocks. Archean rocks occupy the eastern margin of the continent, and Proterozoic rocks occur across much of the remainder of the shield. The western part of Antarctica consists of fold belts of late Cretaceous and younger sequences. These mountainous tracts show a marked similarity to those of the Andean orogenic belt of western South America.

■ *The Fossil Record of the Proterozoic*

Life at the beginning of the Proterozoic was not significantly different from that of the preceding late Archean. Blue–green scums of photosynthetic cyanobacteria (Fig. 7–20) constructed filamentous algal mats around the margins of the ocean, and prokaryotes floated in the well-illuminated surface waters of seas and lakes as well. Anaerobic prokaryotes multiplied in environments deficient in oxygen, and these included the thermophiles of deep-sea hydrothermal springs. Stromatolites, which were relatively sparse during the Archean, proliferated around the world during the Proterozoic (Fig. 7–21). Although still

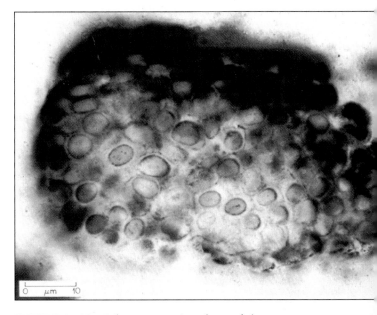

FIGURE 7–20 Paleoproterozoic colony of the cyanobacterium *Eoentophysalis* photographed in a thin section of a Belcher Group stromatolitic chert, Belcher Islands, Canada. *(Courtesy of H. J. Hofmann.)*

FIGURE 7–21 Two-billion-year-old columnar stromatolites from the Kona Dolomite near Marquette, Michigan. *(Courtesy of J. W. Schopf.)*

COMMENTARY
Heliotropic Stromatolites

Stromatolites are among the most abundant kinds of fossils found in Proterozoic rocks. Today, as during the Proterozoic, the photosynthetic cyanophyta that construct stromatolites depend on sunlight for survival and growth. In a study of modern stromatolites, it has been observed that certain species form columnar laminated growths that are inclined toward the sun to gather the maximum amount of light on their upper surfaces. This tendency among organisms to grow toward the sun is called **heliotropism**. Stromatolites, however, may also exhibit growth responses to currents and rising sea level and thus must be examined carefully for evidence that a preferred growth orientation is related to sunlight. Stanley Awramic and James Vanyo (paleobiologist and astronomer, respectively) have studied 850-million-year-old stromatolites from Australia's Bitter Springs Formation for evidence of heliotropism. They believe they found such evidence in species of *Anabaria juvensis*. At the locality studied in central Australia, these stromatolites curve upward in a distinct sine-wave form that appears to record growth that followed the seasonal change in the position of the sun above the shallow sea in which the stromatolitic cyanobacteria lived. In their elongate sinuous form, each sine wave represented a year of growth. The stromatolites grew by the addition of a layer or lamina each day. Thus by counting the laminae in the length of a single sine wave, one can obtain the number of days in the year. The results indicate that there were 435 days in the year 850 million years ago when the stromatolites of the Bitter Springs Formation were actively responding to Proterozoic sunlight.

present today, cyanobacterial mats that construct stromatolitic reefs declined markedly during the early Cambrian. The stromatolitic microbial organisms evidently provided food for newly evolved small shelly fossils and other primitive invertebrates. Extensive overgrazing thus decimated the stromatolites. Even today, stromatolites survive only in environments unsuitable for most grazing invertebrates.

By Mesoproterozoic time, evolution had progressed to the point at which eukaryotes begin to appear in the fossil record. You may recall that single-celled eukaryotes contain a true nucleus enclosed within a nuclear membrane and have well-defined chromosomes and cell organelles. Early single-celled eukaryotes gave rise to multicellular forms and were the probable ancestors of the first unicellular animals. Very likely, the expansion of eukaryotes began about 1.4 billion years ago. It was followed in Neoproterozoic time by the early evolution of multicellular animal life.

Organisms of the Gunflint Chert

As mentioned in the previous chapter, the oldest currently known prokaryotes are those found by paleobiologist J. William Schopf in 3.5-billion-year-old Archean rocks of Western Australia. Schopf reported his find in 1994. These were not the first Precambrian prokaryotes discovered. The initial discovery consisted of Proterozoic fossils detected in 1965 in a now-famous rock unit called the **Gunflint Chert**. This rock unit is exposed along the northwestern margin of Lake Superior. Radiometric dating methods provide an age of 1.9 billion years for the Gunflint Chert. The formation contains abundant and varied flora of so-called thread bacteria and nostocalean cyanobacteria. All the fossils are preserved in chert. Unbranched filamentous forms, some of which are septate, have been given the name *Gunflintia* (Fig. 7–22). More finely septate forms, such as *Animikiea* (Fig. 7–23), are remarkably similar to such living algae as *Oscillatoria* and *Lyngbya*. Other Gunflint fossils, such as *Eoastrion* (literally, the "dawn star"), resemble living iron- and magnesium-reducing bacteria. *Kakabekia* and *Eosphaera* do not resemble any living microorganism, and their classification is uncertain. What does seem certain is that there was an abundance of photosynthetic organisms in the Paleoproterozoic, all actively producing oxygen and thereby altering the composition of the atmosphere. Not only do the Gunflint fossils resemble living photosynthetic organisms, but their host rock contains organic compounds regarded as the breakdown products of chlorophyll.

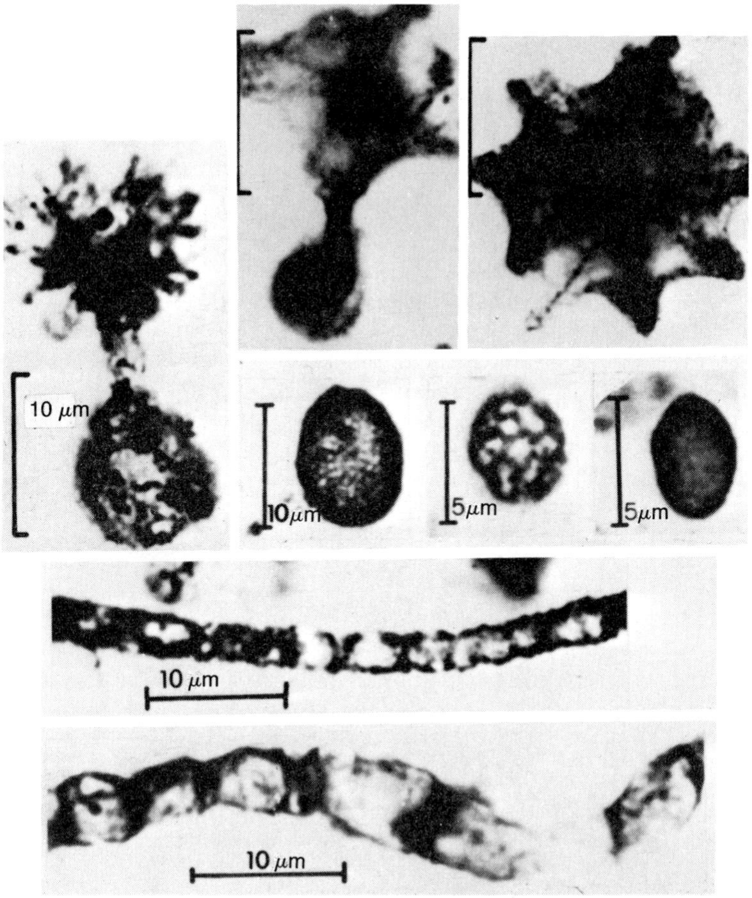

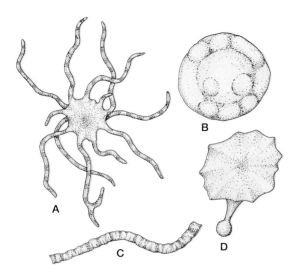

FIGURE 7–22 Fossil remains of microorganisms from the Gunflint Chert. The three specimens across the top with umbrella-like crowns are *Kakabekia umbellata*. The three subspherical fossils are species of *Huroniospora*. The filamentous microorganisms with cells separated by septa are species of *Gunflintia*.
(Photographs provided by J. W. Schopf, courtesy of J. W. Schopf and Elso S. Barghoorn.)

FIGURE 7–23 (A) *Eoastrion*, (B) *Eosphaera*, (C) *Animikiea*, and (D) *Kakabekia* from the Gunflint Chert. All specimens are drawn to the same scale. *Eosphaera* is about 30 microns in diameter.

The Rise of Eukaryotes

The evolution of eukaryotes surely qualifies as one of the major events in the history of life. Eukaryotes possess the potential for sexual reproduction, and this provides enormously greater possibilities for evolutionary change. Unfortunately, the fossil record for the earliest eukaryotes is somewhat ambiguous. This is not surprising when one considers that the first eukaryotes were microscopic unicells whose identifying characteristics (enclosed nucleus, organelles, and so on) are rarely preserved. There is, however, one important clue to the identification of a fossil as a eukaryote. It is based on the size of cells. Living spherical prokaryotic cells rarely exceed 20 microns in diameter, whereas eukaryotic cells are nearly always larger than 60 microns. Such larger (and hence probable eukaryotic) cells begin to appear in the fossil record by about 1.4 to 1.6 billion years ago.

A group of organisms particularly useful in correlating Proterozoic strata are known as acritarchs. **Acritarchs** are unicellular, spherical microfossils with resistant single-layered walls. The walls may be smooth or variously ornamented with spines, ridges, or papillae (Fig. 7–24). Although their precise nature is uncertain, acritarchs appear to be phytoplankton that grew thick coverings during a resting stage in their life cycle. Some resemble the resting stage of unicellular, biflagellate, marine algae known as dinoflagellates. (Dinoflagellates today cause the so-called red tide that periodically poisons fish and other marine animals.) Membrane-bounded nuclei can be detected within some acritarchs, and their size is comparable to that of living eukaryotes.

Acritarchs first appear in rocks about 1.6 billion years old. They reached their maximum diversity and abundance 850 million years ago and then suffered a steady decline. By about 675 million years ago, few remained. Their decline coincided with the major episode of glaciation underway near the end of the Proterozoic. Reduction of carbon dioxide and an increase in atmospheric oxygen accompanying glacial conditions may have been responsible for the extinction of all but a few species that managed to survive until Ordovician time.

In addition to acritarchs, protozoan eukaryotes were probably present in the Proterozoic. As is true today, these protozoans derived their energy by ingesting other cells. Amoebas and ciliates are living examples of protozoans. Protozoan fossils are common in many Phanerozoic rocks, largely because they had evolved readily preservable shells. Proterozoic protozoans were naked, shell-less creatures with little chance of preservation. The exception are tiny vase-shaped microfossils about 800 million years old found in both Arizona and Spitsbergen. The scarcity of Proterozoic protozoans, however, does not mean that they were not present in great abundance. It is likely that the algal mats and masses of decaying organic matter in late Proterozoic seas teemed with protozoans exploiting these rich sources of food.

Advent of the Metazoans

The Neoproterozoic fossil record for larger, multicellular animals is considerably better than that for protozoans. Indeed, since the 1960s, important discoveries of large, multicellular late Proterozoic animals have been made in Australia, Africa, South America, Russia, China, Europe, Iran, and North America

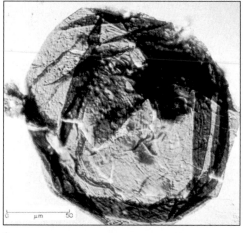

A

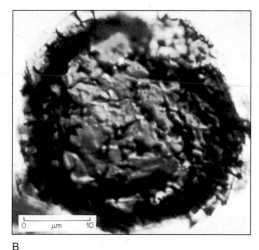

B

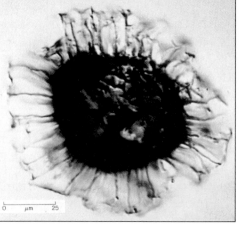

C

FIGURE 7–24   Acritarchs. (A) The striated acritarch *Kildinosphaera lophostriata* from the Chuar Group of Arizona (see Figure 7–13). (B) A spiny acritarch named *Vandalosphaeridium walcotti,* also from the Chuar Group. (C) The spiny acritarch *Skiagia scottica* from the early Cambrian (about 560 million years old) of northern Greenland, depicted here for comparison with the late Proterozoic forms.   *(Courtesy of G. Vidal.)*

FIGURE 7–25 Fossil metazoans from the Conception Group, Avalon Peninsula, Newfoundland. These early metazoans are similar to those of the Ediacaran fauna of Australia. The rocks in which they occur are thought to be late Proterozoic in age. The elongate forms have been interpreted as a form of colonial soft coral, and the circular fossil has been called a jellyfish. There are, however, other interpretations. *(Courtesy of B. Stinchcomb.)*

(Fig. 7–25). The fossils consist of impressions in sedimentary rocks of animals that are clearly metazoans. **Metazoans** are multicellular animals that possess more than one kind of cell and have their cells organized into tissues and organs.

The best-known fossils of Neoproterozoic metazoans were first found in 1946 as well-preserved impressions in what were once the sands of an ancient beach. These sands now comprise the Rawnsley Quartzite of the Pound Subgroup. Exposures of the fossil-bearing rock occur in Ediacara Hills in the Flinders Range of southern Australia (see Fig. 7–19). Thousands of impressions of large, soft-bodied animals, referred to as the **Ediacaran fauna,** have been recovered from these quartzites (Fig. 7–26). According to their shape, they fall into three groups, discoi-

FIGURE 7–26 Three members of the Ediacaran fauna. (A) *Pseudohizostomites,* a wormlike form of uncertain affinity. (B) *Tribrachidium,* an unusual discoidal form that appears to have no living relatives. (C) *Parvancorina,* possibly an arthropod. *(Specimen A is an imprint in the Rawnsley Quartzite; B and C are plaster molds made from the original fossils. All specimens courtesy of M. F. Glaessner.)*

FIGURE 7–27 Impression of a soft-bodied discoidal fossil in the Ediacaran Rawnsley Quartzite of southern Australia. This organism has been interpreted as a jellyfish and named *Cyclomedusa.* Some paleontologists, however, believe it is unrelated to any living organism. *(Courtesy of B. N. Runnegar.)*

FIGURE 7–29 An exceptionally well-preserved specimen of *Dickinsonia costata* in the Ediacaran Rawnsley Quartzite of southern Australia. This fossil has been interpreted as a segmented worm. Divisions on the scale are in centimeters. *(Courtesy of B. N. Runnegar.)*

dal, frond-like, and ovate to elongate. Discoidal forms such as *Cyclomedusa* (Fig. 7–27) were initially thought to be jellyfish. Another circular form, *Tribrachidium* (Fig. 7–26B), appears to have no modern counterpart and may be a member of an extinct phylum.

FIGURE 7–28 Diorama of the sea floor in which lived Ediacaran metazoans. The large, frondlike organisms are interpreted here as soft corals known today as sea pens. Silvery jellyfish are seen swimming about. On the floor of the sea, one can find *Parvancorina* and elongate, wormlike creatures. *(U.S. Natural History Museum, Smithsonian Institution.)*

The second group of Ediacaran specimens can be called frond fossils. They resemble the living soft corals informally called sea pens (Fig. 7–28). Sea pens look rather like fronds of ferns, except that tiny coral polyps are aligned along the branchlets. The polyps capture and consume tiny organisms that float by. Frond fossils similar to those in the Rawnsley Quartzite are also known from Africa, Russia, and England. In those from England *(Charniodiscus),* the frond is attached to a basal, concentrically ringed disk that apparently served to hold the organisms to the sea floor. The disks are frequently found separated from the fronds, which suggests that many of the discoidal fossils common in Ediacaran faunas around the world may be anchoring structures of frond fossils and not jellyfish.

The third group of Ediacaran fossils are ovate to elongate in form and were originally regarded as impressions made by flatworms and annelid worms. Typical of these fossils is *Dickinsonia* (Fig. 7–29), which attained lengths of up to a meter, and *Spriggina* (Fig. 7–30), a more slender animal with a distinctive crescent-shaped structure at its anterior end.

There is an interesting controversy currently taking place with regard to Ediacaran fossils. Most have long been interpreted as Proterozoic members of such existing phyla as the Cnidaria (which includes jellyfish and corals) and Annelida. Indeed, many paleontologists still retain this opinion. However, that view of the affinities of Ediacaran organisms has been challenged by the innovative German paleontologist Adolf Seilacher. Seilacher argues that the resemblance between living sea pens and frond fossils is only superficial. In support of his arguement, he

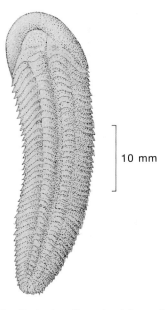

10 mm

FIGURE 7–30 *Spriggina floundersi* from the Ediacaran Rawnsley Quartzite of southern Australia.

notes that the branchlets in the frond fossils are fused together and do not have passages through which water currents might pass. Living sea pens have such openings, and this permits the polyps on the branchlets to gather food particles efficiently in the passing flow of water. With regard to the discoidal fossils thought to be jellyfish, Seilacher notes that living jellyfish have radial structures at their centers and concentric structures around the periphery. This arrangement is opposite to that found in the discoidal Ediacaran fossils. Finally, the rather superficial resemblance of *Spriggina* and *Dickinsonia* to worms may have little significance, for there is no evidence in the fossils of organs that are essential to modern-day worms, such a mouth, gut, or anus.

The Ediacaran animals may be fundamentally different from cnidarians and annelids in yet another way. Cnidarians and worms are rarely preserved in sandy deposits. Preservation of soft, delicate tissue is unlikely in such coarse sediment. Yet the Ediacaran animals have left distinct impressions in the enclosing rock. For this to occur, they must have had firm and tough outer coverings. The ribbed and grooved appearance of many Ediacaran impressions suggests to Seilacher that these animals had an exterior construction suggestive of an air mattress and that this construction provided the firmness needed to make an impression in sand.

Seilacher and his American colleague Mark McMenamin believe that Ediacaran animals lived with symbiotic algae in their tissues, as do modern corals. This relationship would have provided the animals with a way to derive nutrition from the photosynthetic activity of the algae. Perhaps also the broad, thin shapes of many of these animals gave

them sufficient surface area to allow for diffusion respiration, which would be essential to animals that had not yet evolved the complex circulatory, digestive, and respiratory systems of thicker animals.

The dissimilarities of Ediacaran creatures to animals that exist today and their own unique characteristics support a view that they should not be placed in existing phyla. Seilacher suggests placing them in a separate taxonomic category for which he proposed the name **Vendoza** (after **Vendian,** the final period of the Neoproterozoic in Russia). There are, however, many paleontologists who still believe that most of the Ediacaran animals are early members of existing phyla. The resolution of this controversy may have to await the discovery of fossils that provide more anatomic information.

The Ediacaran fauna survived for about 50 million years following their appearance about 630 million years ago. The fauna is important as a record of the Earth's first evolutionary radiation of multicellular animals. Although their record is still too patchy to trace phylogenetic relationships, certain Ediacaran animals were probably ancestral to Paleozoic invertebrates.

The Ediacaran fauna indicates that seas during Vendian time were populated largely by soft-bodied organisms. Shelly fossils are rarely found. One genus, first discovered in Vendian rocks of Namibia, Africa, is named *Cloudina* after the American geologist Preston Cloud. *Cloudina* secreted a tubular, calcium carbonate shell only a few centimenters long (Fig. 7–31). It has been interpreted as a tube-dwelling annelid worm.

Not all Precambrian fossils are body fossils (petrifactions, impressions, molds, or casts). Trails, burrows, and other **trace fossils** also provide important information about ancient life. Trace fossils of burrowing metazoans have been found in Vendian rocks

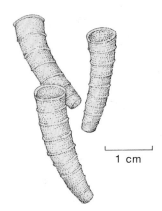

1 cm

FIGURE 7–31 *Cloudina,* the earliest known calcium carbonate shell-bearing fossils. *Cloudina* was first described from the Proterozoic Nama Group of Namibia. It is believed to be a tube-dwelling annelid worm.

FIGURE 7–32 Trace fossils made by a possible molluscan animal as it crawled across soft sediment on the sea floor. The host rock occurs at the Proterozoic-Cambrian boundary in British Columbia, Canada. Originally, the traces had the form of elongate depressions. Sediment deposited on top of the original layer filled the depressions, so that the crawling traces are in convex relief. *(Courtesy of H. J. Hofmann.)*

of Australia (Fig. 7–32), Russia, England, and North America. In every locality, they occur in rocks deposited after the late Proterozoic episode of glaciation (named the Varangian glaciation). It is interesting that the Proterozoic traces consist mostly of relatively simple, shallow burrows, whereas those of the overlying Cambrian are more complex, diverse, and numerous. One sees a similar increase in complexity, diversity, and abundance of metazoan body fossils as one passes from the Proterozoic into the Cambrian.

The sudden appearance of Ediacaran fossils near the end of the Proterozoic and the absence of such metazoans in older rocks may be related to the accumulation of sufficient free oxygen in the atmosphere to permit oxidative metabolism in organisms. On the other hand, the Ediacaran life may have evolved more gradually from earlier small and naked forms that were essentially incapable of leaving a fossil record. Perhaps the ancestral metazoans lived in "oxygen oases" in which marine plants were concentrated. After the atmospheric oxygen content had reached about 1 percent of the present level, these as yet undiscovered animals would have been free to leave their oases and spread widely in the seas. According to this view, the evolutionary development of metazoans may not have been abrupt, but their dispersal may have been rapid after suitable conditions became prevalent.

The Changing Proterozoic Environment

The Earth's early atmosphere and hydrosphere were largely devoid of free oxygen. Then about 2 billion years ago, the atmospheric oxygen began to accumulate because of increased plant photosynthetic activity (Fig. 7–33). One result of the oxygen buildup was the accumulation on land of considerable amounts of ferric iron oxide, which stained terrestrial sediments a rust-red color. Such sedimentary rocks are known as **red beds** and are considered a valid indication of the advent of an oxygenic environment. Of course, the oxygen level probably rose slowly and very likely did not approach 10 percent of present atmospheric levels of free O_2 until the Cambrian Period.

In general, Proterozoic rocks provide evidence for a wide range of climatic conditions, but there is no indication that these climates were especially unique in comparison with those of the Phanerozoic eras. Thick limestones and dolomites with reeflike algal colonies were deposited along the equator, where warm, tropical conditions prevailed much as they do today. During the Early Proterozoic, the equator lay close to the northern border of North America. Precambrian evaporite deposits in eastern Canada and in Australia suggest that conditions were periodically rather than arid during the Middle Proterozoic. In the middle and low latitudes, climates were more severe, as indicated by consolidated deposits of glacial debris (tillites) and glacially striated basement rocks. As previously discussed, the best known of these poorly sorted, thick, boulder-like deposits is the Gowganda Formation.

Mineral Wealth of the Proterozoic

Like the Archean, Proterozoic rocks have yielded important quantities of metals, including iron, nickel, silver, chromium, and uranium. Iron is the most notable of these ores in terms of tonnages that have been mined. The world's largest group of iron mines is in the Canadian shield. Of these, the most famous are the mines that process sedimentary ores in the Lake Superior region. Ore deposits were formed in local areas where iron-bearing sedimentary rocks were altered and enriched by removal of nonmetallic constituents. Other major sources of iron occur in Sweden, the Ukraine, South America, and South Africa.

Enormous amounts of copper were once recovered from Proterozoic rocks along the Keweenaw Peninsula of Lake Superior; however, today the easily accessible ores are almost exhausted. The abandoned mines are a testimony of the ultimate fate of even the richest ore deposits. Today, huge tonnages of copper are mined from Proterozoic rocks in Zambia, Africa.

Important deposits of lead and zinc occur in the Proterozoic rocks of Broken Hills, New South Wales (Australia) and Sullivan, British Columbia (Canada). A Proterozoic gabbro intrusion near Sudbury, Ontario has provided nearly 70 percent of the world's nickel.

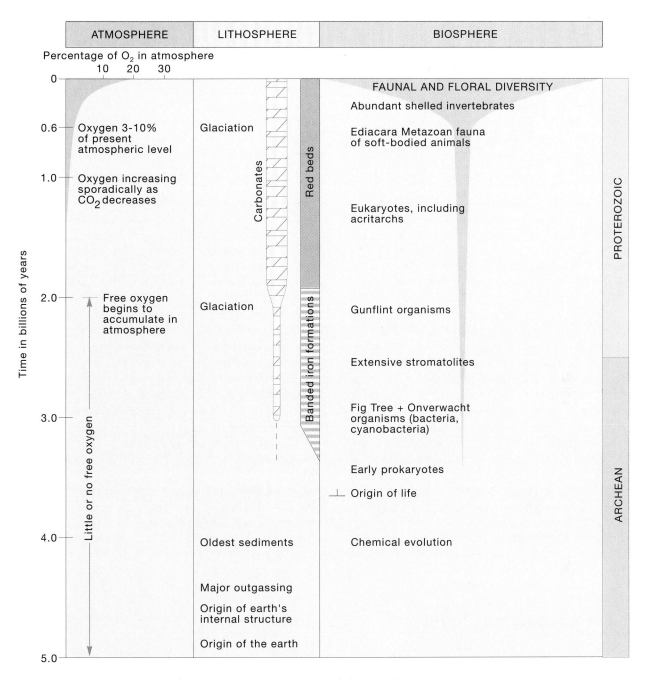

FIGURE 7–33 Correlation of major events in the history of the biosphere, lithosphere, and atmosphere.

Summary

By the beginning of the Proterozoic about 2.5 billion years ago, many of the smaller cratonic elements of the Archean had come together to form a large craton. Lines of contact between the once separate cratonic elements were thus marked by orogens. In North America, the large continent known as Laurentia was formed by this welding of smaller crustal segments. Laurentia and other large cratons of the Proterozoic experienced further growth largely as a result of accretion of sedimentary prisms along continental margins or additions of microplates. Proterozoic rocks in several parts of the Earth provide evidence that they were involved in a modern style of plate tectonics.

The Wopmay orogen of Canada's Northwest Territories is a well-studied example of Proterozoic sedimentation and tectonics. The Wopmay contains a sequence of sedimentary deposits reflecting the opening and closing of an ocean basin. Similar events occurred in the Trans-Hudson and Grenville orogens of Canada. Aulacogens, the failed rifts of three-armed rift systems, are also evident in Proterozoic depositional basins.

Rocks of the Proterozoic contain the oldest evidence of continental glaciers. Two episodes of glaciation occurred, each documented by ice-deposited conglomeratic beds (tillites), varved mudstones of meltwater lakes, and rocks that have been scratched and faceted by glacial action. The Gowganda Formation contains glacial deposits of a Paleoproterozoic ice age. The second episode of glaciation (the Varangian glaciation) occurred near the end of the late Proterozoic.

Banded iron formations (BIF) are notable for two reasons. They are an important mineral resource, and they reflect the buildup of sufficient oxygen in the atmosphere to oxidize iron at the Earth's surface. The development of banded iron formations is followed by deposition of red beds beginning about 2.0 billion years ago. The Keweenawan lavas of the Mesoproterozoic reflect an episode of severe rifting in the North American midcontinental region. The extensive systems of faults associated with rifting provided conduits for the extrusion of Keweenawan lavas. The rifts, however, did not extend to the edge of the continent and did not evolve into an opening oceanic tract. Near the end of the Mesoproterozoic, sedimentary rocks that had accumulated along the eastern margin of North America were compressed and intruded in a mountain-building event known as the Grenville orogeny. South of the Canadian shield, Proterozoic rocks are exposed in several of the Cordilleran ranges. Among these rocks, the Belt Supergroup was deposited along the rifted western margin of North America.

Life at the beginning of the Proterozoic resembled that of the Late Archean. It consisted of prokaryotes such as cyanobacteria. Stromatolites became increasingly abundant during the Proterozoic. By the Middle Proterozoic, the first eukaryotes appeared and subsequently diversified. Among the most common Proterozoic fossils of eukaryotic organisms are acritarchs: tiny spherical structures thought to represent the resting phase of planktonic, eukaryotic microorganisms.

The final and most dramatic evolutionary event of the Proterozoic was the appearance of metazoans (multicellular animals) near the end of the eon. The fossils of these animals are known from various localities around the world but were first discovered in Ediacara Hills in southern Australia. The fossils are impressions in a former sandstone. They are generally large, discoidal, frondlike, or elongate animals that have traditionally been considered early members of such existing phyla as the Cnidaria or Annelida. Another view is that they are so dissimilar to existing phyla that they should be given a separate taxonomic designation, for which the name Vendoza has been proposed.

The Proterozoic seems to have had climates generally similar to those of the Phanerozoic. They ranged from warm tropical or subtropical conditions suggested by the presence of thick, stromatolitic carbonate sequences to the cooler conditions that accompanied the two episodes of continental glaciation.

Review Questions

1. Describe the sequence of events recorded by rocks of the Wopmay orogen. Where in the Canadian shield is there evidence of a generally similar sequence of events?

2. What are aulocogens? How do they develop? Do aulocogens exist today and, if so, where?

3. What is the environmental significance of BIFs? What is the significance of the occurrence of red beds above sequences of BIFs?

4. During the Neoproterozoic, Precambrian continents were collected into a supercontinent. What is the name given to that Neoproterozoic supercontinent? What name is applied to the superocean surrounding that supercontinent?

5. What kind of tectonic activity was the underlying cause of the massive extrusions of basaltic Keweenawan lavas in the Lake Superior region?

6. What metallic ores are extracted from Proterozoic rocks in Zambia, Africa; Ontario (Sudbury); and the region just north of Lake Superior?

7. When do eukaryotes appear in the fossil record? How do eukaryotes differ from prokaryotes? (See Chapter 6.)

8. What are metazoans? What is the earliest known occurrence of abundant metazoans? With regard to their general morphology, what are the three major groups of Ediacaran metazoans?

9. What are the three major geochronologic divisions of the Proterozoic? What is the duration of each? When were Vendian rocks deposited?

10. When rafting through the Inner Gorge of the Grand Canyon of the Colorado River, what Proterozoic rock would you see exposed in the walls of the gorge?

Discussion Questions

1. A geologist engaged in the exploration for petroleum examines the geologic map of a large region that has Archean, Proterozoic, and Phanerozoic rocks. In rocks of which of these geochronologic divisions will he have the best prospect for finding petroleum? Why?

2. What is the evidence for episodes of continental glaciation during the Proterozoic? When did ice ages occur? Why is it unlikely that continental glaciation would have occurred in the Archean?

3. Discuss the nature and evolutionary level of organisms represented by the fossils of the Gunflint Chert as compared to those of the Rawnsley Quartzite.

4. What characteristics of the Ediacaran "jellyfish" suggest that they may not really have been jellyfish, that the Ediacaran frond fossils may not have been sea pens (soft corals), and that the Ediacaran annelids were not really annelids?

5. What are acritarchs? What feature of acritarchs has enhanced their preservation? When did acritarchs reach their maximum diversity and when did they nearly suffer extinction? What environmental conditions may have contributed to their decline?

6. Stromatolites were exceptionally widespread during most of the Proterozoic but became relatively rare thereafter. What may have been the cause of their decline?

7. Where is the lithostratigraphic unit known as the Belt Supergroup located? What evidence indicates that these rocks were deposited in shallow coastal areas?

8. Examine Figure 7–13. What kind of unconformity separates the Grand Canyon Group from Cambrian strata (left side of cross-section)? What type of unconformity separates the Vishnu Schist and Zoroaster Granite from the Grand Canyon Series? Is the conglomerate at the base of the Grand Canyon Group the expected lithology for the initial strata above the unconformity? Explain.

9. Figure 7–11 shows the Chief Mountain exposure of Proterozoic Belt Supergroup rocks resting on *younger* Cretaceous strata. Sketch a cross-section showing how this occurred as a result of a thrust fault.

Supplemental Readings and References

Conway Morris, S. 1990. Late Precambrian and Cambrian soft-bodied faunas. *Ann. Rev. Earth and Planetary Sci.* 18:101–122.

Conway Morris, S. 1987. The search for the Precambrian–Cambrian boundary. *Sci. Am.* 75:157–167.

Diver, W. L., and Peat, C. L. 1979. On the interpretation and classification of Precambrian organic-walled microfossils. *Geology* 7:401–404.

Glaessner, M. F. 1984. *The Dawn of Animal Life.* Cambridge, England: Cambridge University Press.

Hoffman, P. 1989. Precambrian geology and tectonic history of North America. In *Geology of North America,* Vol. A, pp. 447–511. Boulder, CO: Geological Society of America.

Margulis, L. 1982. *Early Life.* New York: Van Nostrand Reinhold.

McMenamin, M. A. S., and McMenamin, D. L. S. 1989. *The Emergence of Animals.* New York: Columbia University Press.

Plumb, K. A. 1991. New Precambrian time scale. *Episodes* 14(2):139–140.

Schopf, J. W. 1992. *Major Events in the History of Life.* Boston: Jones and Bartlett.

Seilacher, A. 1989. Vendoza: Organismic construction in the Proterozoic biosphere. *Lethaia* 22:229–239.

Vidal, G. 1973. The oldest eukaryotic cells. *Sci. Am.* 250:48–57.

The Lower Cambrian Tapeats Sandstone exposed along Bright Angel Trail, Grand Canyon of the Colorado River. The sandstone represents the initial shoreline deposit of a transgressing Cambrian Sea. *(Photo by Peter Kresan.)*

If in late Cambrian time you had followed the present route of Interstate 80, you would have crossed the equator at Kearney, Nebraska.

JOHN McPHEE, *In Suspect Terrain*, Noonday Press, 1983

CHAPTER 8

Early Paleozoic Events

OUTLINE

- **Lands, Seas, and Orogenic Belts**
  - Clues to Paleogeography
  - The Continental Framework
- **The Base of the Cambrian**
- **Early Paleozoic History**
  - Cratonic Sequences
  - Sauk and Tippecanoe
  - The Cordilleran Region
  - Deposition in the Far North
  - The Appalachian Belt
  - The Caledonides
  - The Uralian Seaway
  - Asia Begins to Take Form
  - Way Down South
- **Early Paleozoic Global Geography**
- **Aspects of Early Paleozoic Climates**
- **Mineral Resources of the Lower Paleozoic**

In many parts of the world, Phanerozoic rocks have yielded their secrets more readily than those of the older Archean and Proterozoic. They are often more accessible, less altered, and more fossiliferous. In Chapter 1, we noted that the Phanerozoic contains three eras: the Paleozoic, Mesozoic, and Cenozoic. We will first consider the geologic history of the Paleozoic Era, dividing it for convenience into informal early and late segments. The Cambrian, Ordovician, and Silurian periods constitute the early Paleozoic, which lasted about 190 million years. In the subsequent 180 million years, the Devonian, Mississippian, Pennsylvanian, and Permian periods ran their course. Each of the two approximate halves of the Paleozoic were characterized by general similarities in the historical sequence of events. For example, the early Paleozoic began on most continents with gradual marine invasions of low-lying interior regions. These wide epeiric seas moderated climate and provided habitats for a multitude of marine creatures. Near the end of the early Paleozoic, conditions on the continents became more unstable. Inland seas began an oscillating retreat back into the major ocean basins. Increasing thicknesses of terrestrial sediments and volcanic rocks appear in the stratigraphic sequence, and contorted strata attest to the growth of mountain ranges. It was a time when lands predominated, just as the preceding interval had been a time when seas covered much of the Earth. Gradually, however, the lands were reduced, and the seas returned as time passed from early Paleozoic into late Paleozoic.

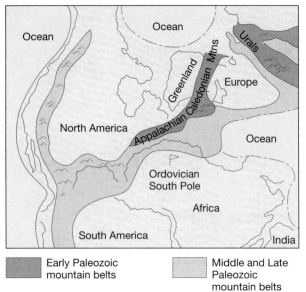

Early Paleozoic mountain belts

Middle and Late Paleozoic mountain belts

FIGURE 8–1 Pangea with mountain belts formed during the Paleozoic.

After we have examined the total physical history of the Phanerozoic, it will be apparent that long periods of gradual sedimentation were punctuated by intervals of severe change involving Earth movements and mountain building. The late Proterozoic supercontinent Rodinia (Proto-Pangea) had begun to break apart, and the return of the errant lands to form Pangea was associated with orogenies (mountain-building episodes) that affected most continents that now surround the Atlantic Ocean (Fig. 8–1). In North America, these events include the Taconic, Acadian, and Allegheny orogenies, whereas in Europe, the names Caledonian orogeny and Hercynian orogeny are applied to these same Paleozoic mountain-building events.

■ *Lands, Seas, and Orogenic Belts*

Clues to Paleogeography

Because of the many processes that have ceaselessly worn away or altered the older parts of the geologic record, it is difficult to reconstruct details of early Paleozoic geography. From paleomagnetic studies we have learned that during the Cambrian most of North America lay along the paleoequator, which was aligned nearly perpendicular to present parallels of latitude from north-central Mexico to Ellesmere Island in the Arctic (Fig. 8–2). The Ordovician equator extended slightly more northeastward, from what

is now Baja California to Greenland and thence across the British Isles and central Europe. These paleoequatorial positions are substantiated by paleoclimatic indicators. Salt and gypsum accumulations, for example, are found in Ordovician rocks of Arctic Canada at a paleolatitude of about 10° from the paleoequator. Great thicknesses of carbonate rocks bearing fossils of marine invertebrates are located within 30° of the paleomagnetically determined equator.

Although Rodinia had begun to break apart near the end of the Proterozoic, paleomagnetic and paleontologic evidence indicates that the continents were still relatively close together at the beginning of the Paleozoic. They appear to have been clustered in the Southern Hemisphere. The Gondwanaland region was turned about 180° from its present orientation, whereas what is now North America, Greenland, and western Europe lay somewhat east of South America. By late in the Cambrian, the ocean tracts separating the continents had widened (see Fig. 8–36), and the rifted passive margins of the separating land masses became collecting sites for continental shelf and rise sediments. Subsequent closure deformed these sediments and raised them into mountain ranges.

Figure 5–66 (in Chapter 5) illustrates in simplified diagrams the sequence of events as they are postulated by the Earth scientists James Valentine and Eldridge Moores. These configurations show Proto-Pangea divided into four continents by Late Cambrian and Early Ordovician time. Then, during Late Ordovician and Silurian time, Europe west of the Uralian trough pushed into North America, forming a chain of ancient mountains called the Caledonides and welding the two continental segments together. The next major collision occurred about 300 million years ago when the Northern Hemisphere continents joined Gondwanaland, forming the ancient Hercynian Mountains. These mountains have since been largely eroded away and covered with younger strata over much of Europe. Finally, near the end of the Paleozoic, Asia was joined to Europe. The sediments caught and crushed in the closing ocean formed the Ural Mountains. Pangea was now fully constructed.

The Continental Framework

As we consider the geologic history of the early Paleozoic, it becomes apparent that events of the more stable interiors of land masses differed from those of the often more dynamic continental margins.

The Stable Interior

In Chapter 6 we noted that continents can be described in terms of cratons and mobile belts or

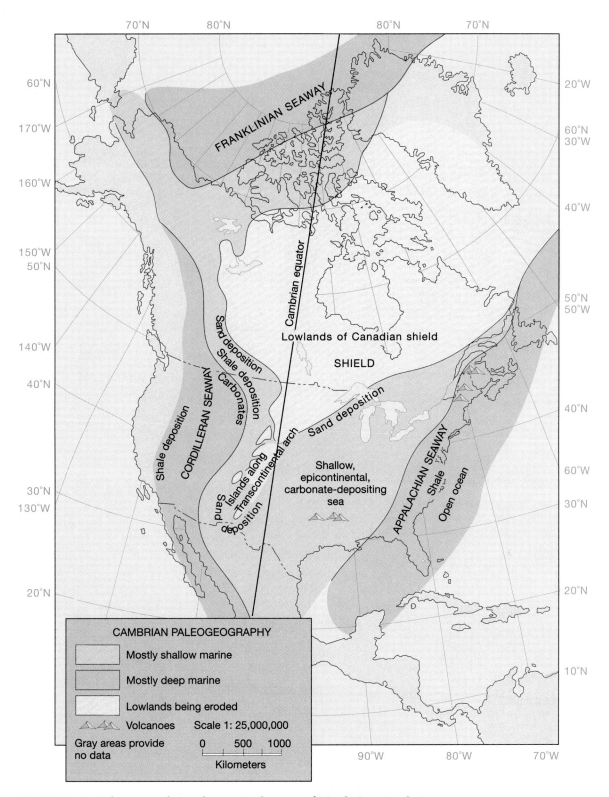

FIGURE 8–2 Paleogeographic and tectonic elements of North America during the Cambrian Period, showing position of the Cambrian paleoequator.

orogens. A craton is the relatively stable part of the continent consisting of a Precambrian shield and the extension of the shield that is covered by flat-lying or only gently deformed Phanerozoic strata. These cra-

tonic strata were originally wave-washed sands, muds, and carbonates deposited in shallow seas that periodically flooded continental regions of low relief. Here and there, these strata are warped into broad

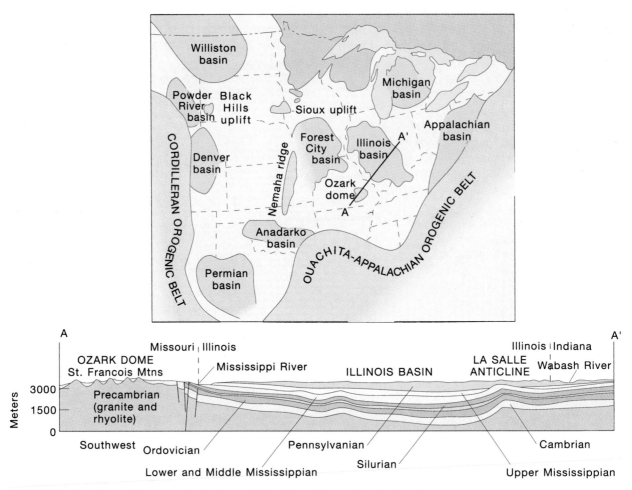

FIGURE 8–3 Map of the central platform of the United States showing major basins and domes. Structural section below the map crosses part of the Ozark dome and the Illinois basin. The basins and domes developed at different times during the Phanerozoic.

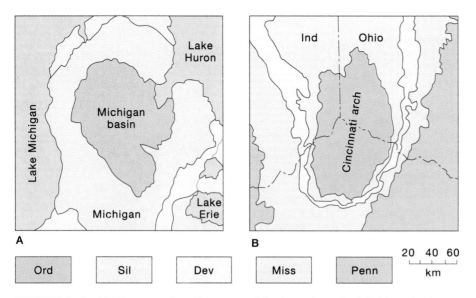

FIGURE 8–4 (A) In an erosionally truncated basin such as the Michigan basin, the youngest beds are centrally located. (B) In a domelike structure such as the Cincinnati arch, the oldest beds are located in the center.

synclines, basins, domes, and arches (Fig. 8–3). The resulting tilt to the strata is slight and is usually expressed in ft/mile rather than degrees. In the course of geologic history, the arches and domes stood as low islands or barely awash submarine banks.

Domes and arches seem to have developed in response to vertically directed forces unlike those that formed the compressional folds of mountain belts. Whether domes or basins, structures of the platform can be recognized by their pattern of outcropping rocks. Erosional truncation of domes exposes older rocks near the centers and younger rocks around the peripheries. Sequences of strata over arches and domes tend to be thinner. Also, because these structures were periodically above sea level, they characteristically have many erosional unconformities. Basins, in contrast, were more persistently covered by inland seas, have fewer unconformities, and developed greater thickness of sedimentary rocks. In erosionally truncated basins, younger rocks are centrally located and older strata occur successively farther toward the peripheries (Fig. 8–4).

Orogenic Belts

The North American craton is bounded on four sides by great elongate tracts that have been the sites of intense deformation, igneous activity, and earthquakes. One or more of such orogenic belts are present on all continents (Fig. 8–5). Most are located along present or past margins of continents, such as the North American Cordilleran belt. As described in Chapter 6, the passive margin of a continent may experience an early stage in which thick sequences of sediments accumulate along the continental shelf and rise. This stage may be followed by subduction of oceanic lithosphere along continental margins and may terminate in continent-to-continent collisions, as in the classic Wilson cycle. As we review the history of eastern North America later in this chapter, we will describe the geologic evidence for this sequence of events along the Appalachian orogen.

■ *The Base of the Cambrian*

At one time, recognition of the boundary between the Cambrian and the underlying Precambrian seemed relatively simple. The base of the Cambrian System could be identified by the first occurrence of shell-bearing multicellular animals. Among these, marine arthropods known as trilobites served as the principal marker fossils. A problem became evident in the 1970s, however, when a distinctive group of shelly fossils was found *beneath* the lowest occurrence of strata containing the first trilobites. Although these animals had mineralized skeletons, they were very small, which is why they had not been discovered earlier. They include early members of such existing phyla as Porifera, Brachiopoda, Mollusca, and possibly Annelida. Although originally discovered in Siberia, similar pretrilobite fossils have also been found recently in other continents. These discoveries, and the known occurrences of a variety of trace fossils, have resulted in a new classification of Lower Cambrian chronostratigraphic units (Fig. 8–6). Based on uranium-lead isotope dates obtained from rocks in northeastern Siberia, it can be said that the Cambrian began 544 million years ago. The boundary is marked by the lowest (oldest) occurrence of feeding burrows of the trace fossil *Phycodes pedum* (Fig. 8–

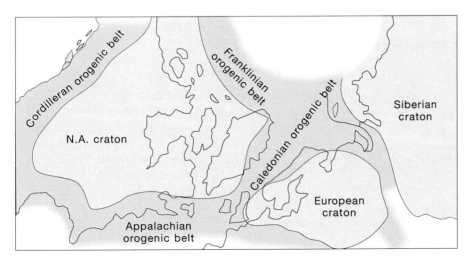

FIGURE 8–5 Cratons and mobile belts of North America and Europe.

7). The initial stage of the Lower Cambrian has been designated the Nemakit-Daldynian. It is followed by the Tommotian, Atdabanian, Butonian, and Toyonian stages.

■ *Early Paleozoic History*

The beginning of the Paleozoic Era was a geologically calm time for North America, as if the continent rested from the rigors of orogeny and severe climate

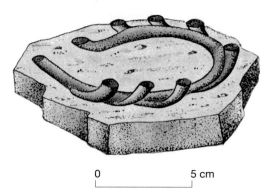

0   5 cm

FIGURE 8–7 The trace fossil *Phycodes*. The fossil represents the feeding behavior of a benthic organism that produced horizontal cylindrical burrows with sequential upward shafts as it explored for food. The base of the *Phycodes pedum* biozone is a marker for the base of the Cambrian system.   *(After Crimes, T. P. 1989. Trace Fossils, in The Cambrian–Precambrian Boundary. Oxford Monographs on Geology and Geophysics 12. Oxford, England: Clarendon Press.)*

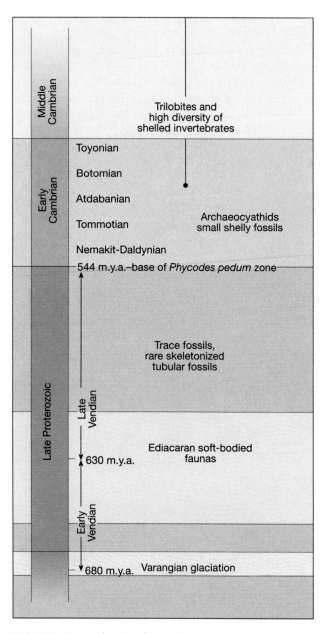

FIGURE 8–6 The Cambrian–Precambrian boundary is placed at the base of the *Phycodes pedum* zone, corresponding to a uranium–lead age determination of 544 million years.

that had occurred near the end of the preceding era. We know little of the Paleozoic ocean basins, for those old sea floors have long since been consumed at subduction zones. However, oceans did exist and gradually spilled out of their basins onto low regions of the continents. Although a large part of the sedimentary record has been removed by erosion, enough remains to indicate that practically all of the Canadian shield was inundated at times. Initially, the dominant deposits were sands and clays derived from the weathering and erosion of igneous and metamorphic rocks of the shield. Later, as these quartz- and clay-producing terranes were reduced, limestones and dolomites became increasingly more prevalent. These rocks contain abundant remains of lime-secreting marine organisms and indicate that shallow seas were common throughout much of the equatorial region of the Earth during the early Paleozoic. Indeed, the advances and retreats of these shallow seas were the most apparent events in the early Paleozoic history of the continental interiors.

Cratonic Sequences

It is convenient to discuss the geologic history of the continental mass that now constitutes North America in terms of its cratonic and orogenic regions. It is also advantageous to subdivide the early Paleozoic of the craton on the basis of advances (transgressions) and retreats (regressions) of epeiric seas. Regressions exposed the old sea floors to erosion and produced extensive unconformities that are used as stratigraphic markers.

Chapter 1 included a discussion of the largely European basis for developing the geologic time scale and defining its geologic periods. Although the cratonic rocks of North America can be correlated with those of Europe on the basis of fossils, most of the stratigraphic breaks in deposition do not correspond to those that form European period boundaries. Indeed, if the relative geologic time scale had been worked out in North America instead of Europe, there might have been only two periods in the early Paleozoic rather than three. The American stratigrapher Laurence Sloss has suggested a workable remedy for this disparity. He proposed that the Paleozoic

rocks of the North American craton be divided into unconformity-bounded *sequences* of deposition. Each sequence consisted of the sedimentary rocks deposited as the sea transgressed an old erosional surface, reached its maximum inundation, and then retreated. For the Paleozoic, he named these the **Sauk, Tippecanoe, Kaskaskia,** and **Absaroka** sequences (Table 8–1).

Because the cratonic sequences seen in North America can also be found in other continents, it appears that worldwide changes in sea level were re-

TABLE 8.1 Cratonic Sequences of North America*

Geologic Time		Cratonic Sequences		Orogenic Events	Biological Events	Ice Ages
		Center of craton	Margin of craton			
	Cenozoic		Tejas	Himalayan Alpine Laramide	Age of mammals	
			65 mya		Massive extinctions	
Mesozoic	Cretaceous		Zuni	Sevier Nevadan	First flowering plants Climax dinosaurs and ammonites	
	Jurassic				First birds Abundant dinosaurs and ammonites	
	Triassic	248 mya		Sonoma	First dinosaurs First mammals Abundant cycads Massive extinctions (including trilobites) Mammal-like reptiles	
Late Paleozoic	Permian		Absaroka			
	Pennsylvanian			Alleghenian	Great coal forests Conifers First reptiles	
	Mississippian		Kaskaskia	Antler	Abundant amphibians and sharks Scale trees Seed ferns	
	Devonian			Acadian-Caledonian	Extinctions First insects First amphibians First forests First sharks	
		408 mya				
Early Paleozoic	Silurian		Tippecanoe		First jawed fishes First air-breathing arthropods	
	Ordovician			Taconic	Extinctions First land plants Expansion of marine shelled invertebrates	
	Cambrian		Sauk		First fishes Abundant shell-bearing marine invertebrates Trilobites	
		544 mya				
	Late Proterozoic				Rise of the metazoans	

*The green areas represent sequences of strata. They are separated by major unconformities, indicated in yellow. Note that the rock record is most complete near cratonic margins just as the time spans represented by unconformities are greatest near the center of the craton. Major biologic, orogenic, and glacial events are added for reference. (Cratonic sequence model after Sloss, L.L. 1965. *Bull. Geol. Soc. Amer.* 74:93-114.)

sponsible for the repeated major transgressions and regressions. Widespread continental glaciation may also have been a cause for the eustatic changes. It is also reasonable that minor variations in the major cycle resulted from tectonic movements on the craton. Many geologists believe that the long-term changes in sea level were probably related to sea-floor spreading. For example, rapid spreading results in high midoceanic ridges. The ridges in turn displace water, causing a global rise in sea level.

There is an interesting correspondence between Sloss's cratonic sequences and the Vail curves discussed in Chapter 3. The Vail curves are derived from high-precision seismic profiles and permit one to "see" deep beneath the surface They indicate that unconformity-bounded sequences can be traced across the craton and into the thick, deep sections along the continental margins.

Sauk and Tippecanoe

During the earliest years of Sauk deposition, seas were largely confined to the continental margins (continental shelves and rises), so most of the craton was exposed and undergoing erosion. No doubt it was a bleak and barren scene, for vascular land plants had not yet evolved. Uninhibited by protective vegetative growth, erosional forces gullied and dissected the surface of the land. Over a span of at least 50 million years, the Precambrian crystalline rocks were deeply weathered and must have formed a thick, sandy "soil." Eventually, marine waters spilled out of the marginal basins and began a slow encroachment over the eroded and weathered surface of the central craton.

The craton was not a level, monotonous plain during Sauk time but had distinct upland areas composed of Precambrian igneous and metamorphic rocks. During times of marine transgression, these uplands became islands in early Paleozoic seas and provided detrital sediments to surrounding areas. The absence of marine sediments over these upland tracts provides evidence of their existence and extent. One of the largest of the highland regions was the transcontinental arch (see Fig. 8–2), which extended southwestward across the craton from Ontario to Mexico.

By Late Cambrian time, seas extended across the southern half of the craton from Montana to New York. A vast apron of clean sand (Fig. 8–8) was spread across the sea floor for many miles behind the advancing shoreline (Fig. 8–9). This sandy aspect, or facies, of Cambrian deposition was replaced toward the south by carbonates. Here the waters were warm, clear, and largely uncontaminated by clays and silts from the distant shield. Marine algae flourished and contributed to the precipitation of calcium carbonate. Invertebrates, although present, did not contribute to the volume of sediment to the degree they did in later periods.

In the Cordilleran region, the earliest deposits were sands, which graded westward into finer clastics and carbonates. An excellent place to study this Sauk transgression is along the walls of the Grand Canyon of the Colorado River. In this region one finds the Lower Cambrian Tapeats Sandstone (see chapter opening photograph) as an initial strandline deposit above the old Precambrian surface. The Tapeats can be traced both laterally and upward into the Bright Angel Shale, which was deposited in a more offshore environment (Fig. 8–10). Next is the Muav Lime-

FIGURE 8–8   The Upper Cambrian Galesville Sandstone exposed in a quarry near Portage, Wisconsin. The mature sandstone is being quarried for use in the manufacture of glass and in premixed cement.

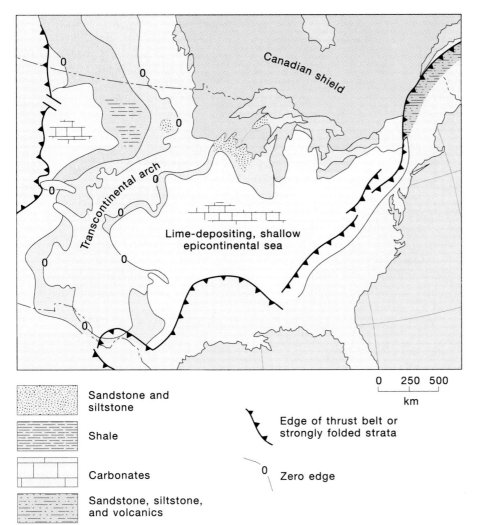

FIGURE 8–9 Upper Cambrian lithofacies map. *(Simplified and adapted from Stratigraphic Atlas of North America and Central America. Shell Oil Company, Exploration Department.)*

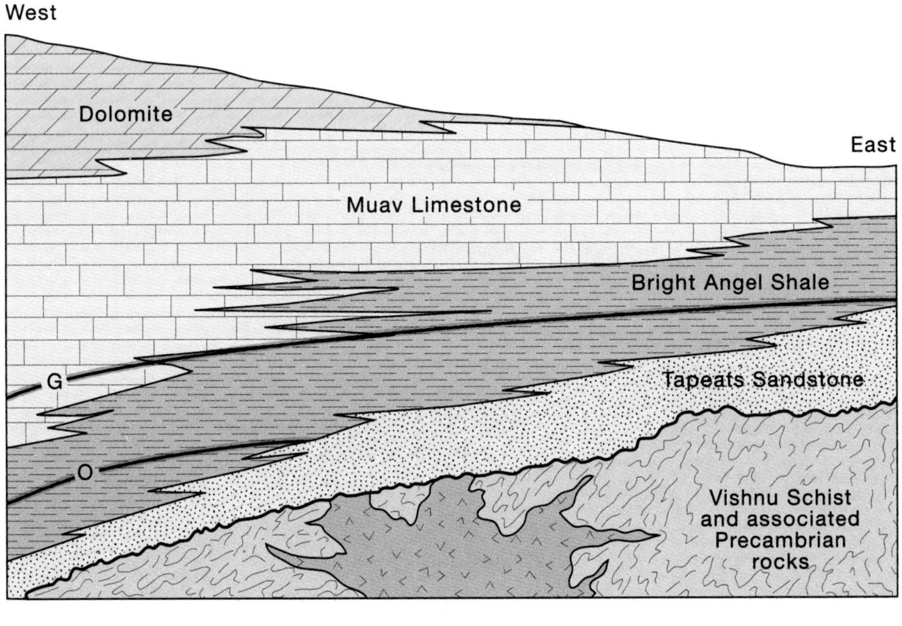

Canadian shield

Transcontinental arch

Lime-depositing, shallow epicontinental sea

0   250   500

km

Sandstone and siltstone

Shale

Carbonates

Sandstone, siltstone, and volcanics

Edge of thrust belt or strongly folded strata

0   Zero edge

West

East

Dolomite

Muav Limestone

Bright Angel Shale

Tapeats Sandstone

G

O

Vishnu Schist and associated Precambrian rocks

FIGURE 8–10 East–west section of Cambrian strata exposed along the Grand Canyon. The line labeled "O" indicates the top of the Lower Cambrian *Olenellus* trilobite zone. The line labeled "G" indicates the location of the Middle Cambrian *Glossopleura* trilobite zone. The section is approximately 200 km long, and the section along the western margin is about 600 meters thick. *(Adapted from McKee, E. D. 1945. Cambrian Stratigraphy of the Grand Canyon Region. Washington, Carnegie Institute, Publ. 563.)*

stone, which originated in a still more seaward environment. As the sea continued its eastward transgression, the early deposits of the Tapeats were covered by clays of the Bright Angel Formation, and the Bright Angel was in turn covered by limy deposits of the Muav Limestone. Together these formations form a typical transgressive sequence, recognized by coarse deposits near the base of the section and increasingly finer (and more offshore) sediments near the top.

Cambrian rocks of the Grand Canyon region not only provide a glimpse of the areal variation in depositional environments as deduced from changing lithologic patterns, but they also illustrate that particular formations are not usually the same age everywhere they occur. Detailed mapping of the Tapeats, Bright Angel, and Muav units combined with careful tracing and correlation of trilobite occurrences indicate that deposition of the highest facies (Muav) had already

begun in the west before deposition of the lowest facies (Tapeats) had stopped in the east. Correlations based on the guide fossils provided reliable evidence that the Bright Angel sediments were largely Early Cambrian in age in California and mostly Middle Cambrian in age in the Grand Canyon National Park area. This example illustrates the **principle of temporal transgression,** which stipulates that sediments deposited by advancing or retreating seas are not necessarily of correlative geologic age throughout their areal extent.

The episode of carbonate deposition that was so characteristic of the southern craton continued into the early Ordovician almost without interruption. Then, near the end of the early Ordovician, the seas regressed, leaving behind a landscape underlain by limestones that was subjected to deep subaerial erosion. That erosion produced a widespread unconfor-

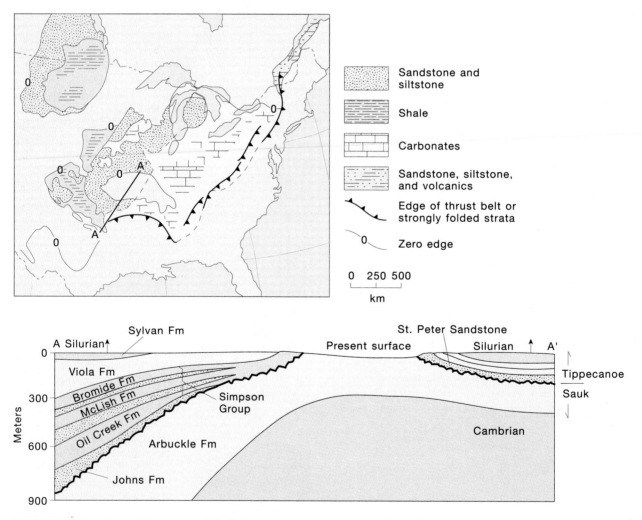

FIGURE 8–11   Lower Tippecanoe lithofacies map and cross-section along line A to A'. In the centrally located white area, Tippecanoe strata have been removed by erosion. Note the extensive unconformity between the Sauk and the Tippecanoe that is depicted on the cross-section.

mity that geologists use as the boundary between the Sauk and the Tippecanoe sequences (Fig. 8–11).

The second major transgression to affect the platform occurred when the Tippecanoe sea flooded the region vacated as the Sauk sea regressed. Again the initial deposits were great blankets of clean quartz sands. Perhaps the most famous of these Tippecanoe sandstones is the St. Peter Sandstone (Fig. 8–12), which is nearly pure quartz and is thus prized for use in glass manufacturing. Such exceptionally pure sandstones are usually not developed in a single cycle of erosion, transportation, and deposition. They are the products of chemical and mechanical processes acting on still older sandstones. In the St. Peter Formation, waves and currents of the transgressing Tippecanoe sea reworked Upper Cambrian and Lower Ordovician sandstones and spread the resulting blanket of clean sand over an area of nearly 7500 sq km. As described in Chapter 3, sandstones can be described by their texture and composition as either mature or immature. These textural and compositional traits are, of course, related to the processes that produced the sandstone. Sandstones with a high proportion of chemically unstable minerals and angular, poorly sorted grains are considered immature, whereas aggregates of well-rounded, well-sorted, highly stable minerals (such as quartz) are considered mature. The St. Peter Sandstone is so pure that it is geologically unusual and perhaps should be termed an ultramature sandstone.

The sandstone depositional phase of the Tippecanoe was followed by the development of extensive limestones, often containing calcareous remains of brachiopods, bryozoans, echinoderms, mollusks, corals, and algae (Fig. 8–13). Some of these carbonates were chemical precipitates, many were fossil frag-

FIGURE 8–13 Fossiliferous limestone deposited in the Tippecanoe sea. This specimen, from near Danville, Missouri, is about 20 cm in longest dimension. The most abundant fossils are brachiopods. On close examination, one can also find many twiglike branches of colonial bryozoans and echinoderm parts.

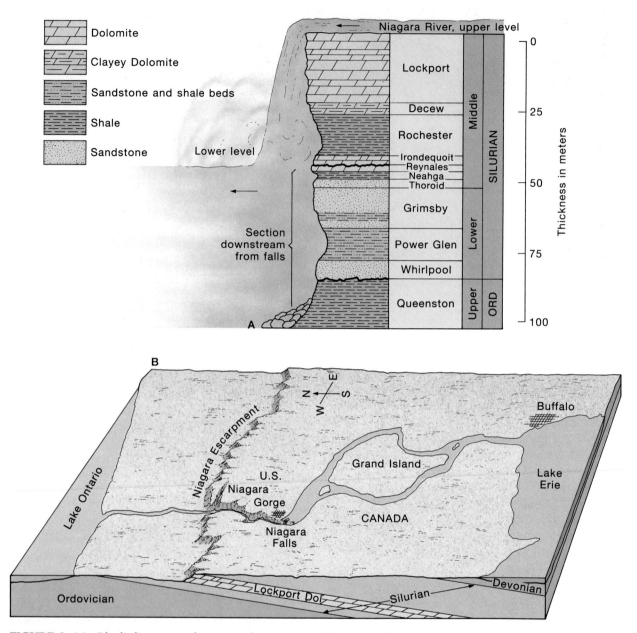

FIGURE 8–14 Block diagram and stratigraphic section of Niagara Falls. The Lockport Dolomite forms the resistant lip of the falls. The rocks dip gently to the south in this area, and where harder dolomite layers such as the Lockport Dolomite intersect the surface, they form a line of bluffs known as the Niagaran Escarpment. *(Map after E. T. Raisz.)*

ment limestones (bioclastic limestones), and some were great organic reefs. Frequently, the deposited calcite underwent chemical substitution of some of the calcium by magnesium and in the process was converted to the carbonate rock dolomite.

In the region east of the Mississippi River, the dolomites and limestones were gradually supplanted by shales. As we shall see, these clays are the peripheral sediments of the Queenston clastic wedge which lay

far to the east. The geologic section exposed at Niagara Falls is a classic locality in which to examine these rocks (Figs. 8–14 and 8–15).

Near the close of the Tippecanoe sequence, landlocked, reef-fringed basins developed in the Great Lakes region (Fig. 8–16). During the Silurian, evaporation of these basins caused the precipitation of salt and gypsum on an immense scale. These deposits clearly indicate the existence of excessively arid con-

FIGURE 8–15 Niagara Falls, formed where the Niagara River flows from Lake Erie into Lake Ontario. The classic section of the American Silurian System is exposed along the walls of the gorge below the falls. *(Guido Alberto Rossi/The Image Bank.)*

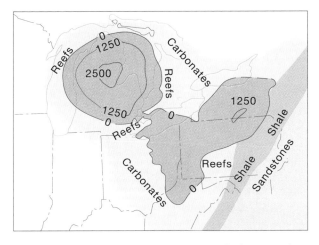

FIGURE 8–16 Isopach map showing thickness and area of evaporite basins during the late Silurian. Areas of evaporite precipitation were surrounded by carbonate banks and reefs. Gaps in the reefs and banks provided channelways for replenishment of the basins with normal sea water, thus replacing water lost by evaporation. *(Adapted from Alling, H. L., and Biggs, L. I. 1961. Bull. Amer. Assoc. Petrol. Geol. 45:515–547.)*

ditions. In the salt-bearing Salina Group, for example, evaporites have an aggregate thickness of 750 meters. The precipitation of this amount of salt and gypsum, if theoretically confined to a single continuous episode of evaporation, would require a column of sea water 1000 kilometers tall. Because such a column of water has never existed on Earth, it is likely that the evaporating basins had connections to the ocean through which replenishment could occur. One setting in which this might have occurred would be in a basin that had its opening to the sea restricted by a raised sill (or possibly also by a submerged bar). Evaporation within the basin would have produced heavy brines, which would have sunk to the bottom and been prevented from escaping because of the sill or bar (Fig. 8–17). This type of basin is called a **barred basin.**

The Cordilleran Region

During much of early Paleozoic time, large areas of the Cordilleran region of western North America lay

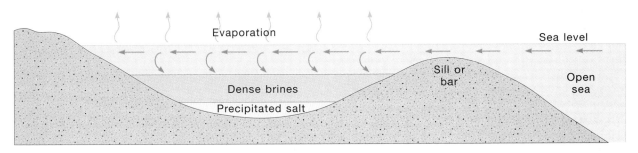

FIGURE 8–17 Cross-section illustrating a model for the deposition of evaporites in a barred basin. Sea water is able to flow more or less continuously into the basin over the partially submerged barrier. It is evaporated to form dense brines. Because of their density, the brines sink and are thus unable to return to the open sea because of the barrier. When the brine has become sufficiently concentrated, salts are precipitated.

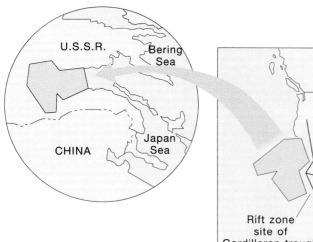
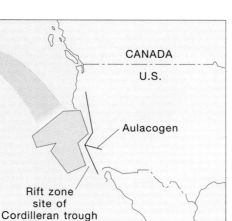

FIGURE 8–18 Conceptual drawing indicating the development of a passive margin trough along the western edge of North America by detachment of a land mass that ultimately was incorporated into the Siberian Platform of Russia. *(Concept from Sears, J. W., and Price, R. A. 1978.* Geology 6:227–270.)

beneath sea level. A coastal plain and shallow marine shelf extended westward from the Transcontinental arch. This relatively level expanse was bordered on the west by deep marine basins adjacent to the western continental margin. Many of the marine sedimentary basins of the Cordillera appear to have originated during the Proterozoic, when North America separated from another continental block that drifted westward, rotated counterclockwise, and was ultimately incorporated into the Siberian platform (Fig. 8–18). As a result of the separation, a passive margin was developed along the western border of North America. Evidence for this hypothesis consists of similarities in Proterozoic rocks of the two now widely separated segments as well as correspondence in the alignment of structural trends when the two blocks are placed in proper juxtaposition. In addition, there is geophysical evidence for the presence of a failed rift or aulocogen in Southern California. As is characteristic of aulocogens, this one is perpendicular to the possible line of separation between the two land masses. One notes further that the Belt Supergroup of Montana, Idaho, and British Columbia as well as the Uinta Series of Utah and the Pahrump Series of California were all deposited in fault-controlled basins formed during an episode of Proterozoic rifting.

Some of the grandest sections of Cambrian rocks in the world are exposed in the Canadian Rockies of British Columbia and Alberta. The formations have been erosionally sculpted into some of Canada's most magnificent scenery (Fig. 8–19). Lower Cambrian rocks include ripple-marked quartz sandstones that were probably derived from the Canadian shield. By Middle Cambrian time, the seas had transgressed farther eastward, and shales and carbonates became more prevalent. An interesting section of these Middle Cambrian rocks is exposed along the slopes of

FIGURE 8–19 Mount Burgess in Yoho National Park, British Columbia. The mountain is on the east side of Burgess Pass about 3 km southeast of the Burgess Shale fossil quarry, where a diverse fauna of Cambrian metazoans was discovered in 1909 by Charles Walcott. The rocks exposed along the flanks of Mount Burgess consist primarily of Cambrian shales. *(Courtesy of the Image Bank/Wm. A. Logan.)*

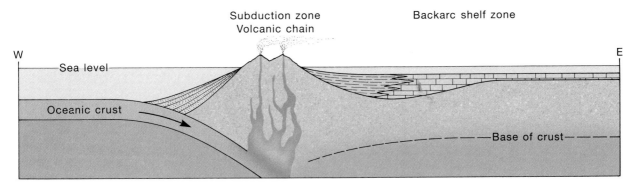

FIGURE 8–20 Interpretive cross-section of conditions across the Cordilleran region during early Paleozoic time.

Kicking Horse Pass in British Columbia near the border with Alberta. One of the units in this section, the Burgess Shale, has excited the interest of paleontologists around the world because it contains abundant remains of Middle Cambrian soft-bodied animals, most of which have never been discovered elsewhere. We will discuss these fossils in Chapter 10, along with other organisms of the early Paleozoic.

Patterns of sedimentation in the Cordilleran region during the Ordovician and Silurian indicate that the western edge of the continent was no longer a passive margin, as was the case in the late Proterozoic and earliest Cambrian. The Pacific plate was now moving against North America, and a subduction zone with an associated volcanic chain had formed (Fig. 8–20). A thick subduction zone complex, con-

sisting of graywackes and volcanics, was laid down in the trench above the subduction zone, whereas east of the volcanoes, siliceous black shales and bedded cherts accumulated. The shales are noted for their rich content of fossil colonial organisms called graptolites (Fig. 8–21). Graptolites are so prevalent in lower Paleozoic rocks that the sediments have been referred to as the "graptolite facies." The associated bedded cherts may have derived their silica by dissolution of particles of volcanic ash and the tiny shells of siliceous marine microorganisms.

Eastward of the subduction zone complex in the Cordillera, one finds a thick sequence of fossiliferous carbonates, shales, and sandstones (Fig. 8–22). These are the rocks of the continental shelf. Because many of the carbonates contain abundant fossils of marine

FIGURE 8–21 Branches (stipes) of the graptolite *Diplograptus. Diplograptus* is also common in dark shales of Ordovician age in both Europe and North America.

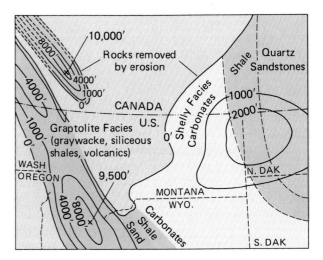

FIGURE 8–22 Ordovician sedimentation in the Montana Williston Basin region. Isopach lines indicate thickness of the Ordovician time–rock unit. Deep-water sediments reach thicknesses of 9500 ft (2900 meters) in Idaho and 10,000 ft (3080 meters) in British Columbia. *(After Sloss, L. L. 1950. Bull. Am. Assoc. Petr. Geol. 34:423–445.)*

invertebrates, they are sometimes dubbed the "shelly facies."

Silurian deposits of the Cordillera are generally similar to those of the Ordovician. Sediments of the subduction complex can be recognized in southeastern Alaska, California, and Nevada, where they occur in great thrust sheets. Nevada also contains exposures of Silurian carbonates deposited in shallower continental shelf areas.

Deposition in the Far North

The cold and sparsely inhabited northern part of North America is a difficult region in which to carry on geologic field studies. Thus, this region has received less attention than more hospitable parts of the continent. To add to the difficulty, lower Paleozoic outcrops are often sparse, covered with younger rocks, and deformed or metamorphosed. Nevertheless, there is evidence that the northern edge of the continent formed late in the late Proterozoic when a continental mass split off from the North American craton. The rifting and displacement were probably part of the general dismemberment of Rodinia occurring at this time. By Ordovician time, conditions had stabilized enough to permit the development of a carbonate platform on the continental shelf. To the north, the shelf dropped off abruptly, providing a deep-water environment in which turbidities were trapped. The Silurian section in this region is one of the thickest in the world. It contains extensive reef deposits, some of which stood more than 200 meters above the sea floor. There are thick deposits of evaporites as well. These rocks imply warm conditions, an interpretation strengthened by paleomagnetic data indicating that this presently frigid region lay within 15° of the early Paleozoic equator.

The Appalachian Belt

Bordering the interior platform of North America on the east from Newfoundland to Georgia is the Appalachian orogenic belt. (Its southwestward extension is called the Ouachita orogenic belt.) The Appalachian belt contains the record of three major orogenic events and many minor disturbances. Study of the rocks deformed during these events has been the basis not only for classic theories of mountain building but also for new ideas that view the Appalachian belt as a collage of microcontinents and other suspect terranes accreted to the eastern edge of North America by plate convergence.

The Appalachian depositional basin (see Fig. 8–2) can be neatly divided into a western belt composed of shales, limestones, and sandstones and an eastern belt composed largely of graywackes, volcanics, and sili-

ceous shales, which are often highly metamorphosed and intruded by granite masses of batholithic proportions. Today these rocks underlie the Blue Ridge and Piedmont Appalachian provinces, whereas rocks of the western belt occur beneath the Valley and Ridge province and Appalachian plateaus (Figs. 8–23 and 8–24).

At the beginning of the Paleozoic, the eastern margin of North America was a passive margin, much like it is today. Relatively shallow-water deposits of the shelly facies were being spread across the continental shelf, while deeper water sediments accumulated farther seaward along the continental rise.

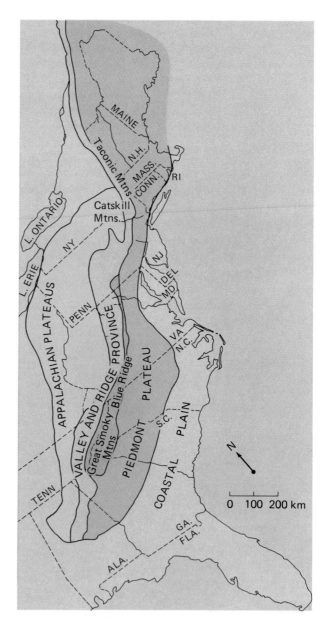

FIGURE 8–23 Physiographic provinces of the eastern United States.

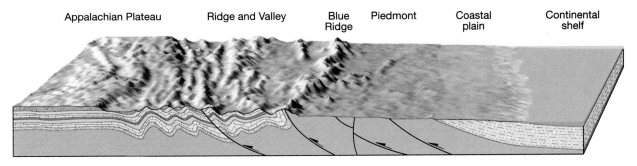

FIGURE 8–24 Simplified cross-section of the Appalachians showing relation of surface topography to structure.

The initial deposits along the shelf were largely quartz sandstones and shales derived from the stable interior, which was still largely emergent in earliest Cambrian time. Gradually, however, the seas spread inland and sandy strandlines migrated westward. The early sandstone deposits of the shelf were covered by a thick sequence of carbonate rocks that formed a shallow bank or platform extending from Newfoundland to Alabama. Mudcracks and stromatolites attest to the deposition of these carbonates (now dolomites) at or near sea level.

This rather quiet depositional scenario of the Cambrian and Early Ordovician changed dramatically during the Middle Ordovician (Fig. 8–25). At

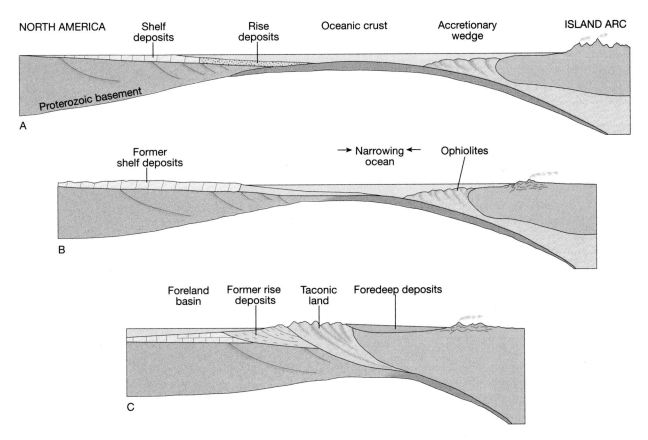

FIGURE 8–25 Plate tectonic forces that resulted in the Taconic orogeny. Following the Neoproterozoic break-up of Rodinia, a passive margin characterized the eastern border of North America (A). Subsequently, a large island arc converged on the passive margin and converted it to an orogenic belt with growing mountain ranges (B & C). *(Adapted from Rowley, D. B., and Kidd, S. F. 1981. Jour. Geol. 89:199–218.)*

that time, carbonate sedimentation ceased, the carbonate platform was downwarped, and huge volumes of graptolitic black shales and immature sands spread westward over the dolomites. The Ordovician section in New England contains such coarse detritus as well as volcanic pyroclastics and interbedded lava flows, which indicate that a subduction zone with an accompanying volcanic chain had formed along the eastern margin of North America. Apparently, the Proto-Atlantic (or Iapetus) Ocean that had opened during the Proterozoic and Cambrian was beginning to close again. That closure resulted in the subduction–volcanic arc complex of the Paleozoic and ultimately carried the continents that were to compose Pangea into contact with each other.

The pulses of orogenic activity that had begun in the Early Ordovician were followed by several more intense deformational events in Middle and Late Ordovician that constitute the **Taconic orogeny.** The Taconic orogeny was caused by the partial closure of the Iapetus Ocean. During that closure, an island arc, and possibly other terranes, collided with the formerly passive margin of North America. Later orogenic episodes resulted from further compression of opposing plates. In this scenario, Europe and Africa lay opposite and east of North America. An alternate hypothesis suggests that South America lay adjacent to the eastern margin of North America and that a southern segment of the Andes was contiguous with the Taconic mountains.

The effects of the Taconic orogeny are most apparent in the northern part of the Appalachians. Caught in the vise between closing lithospheric plates, sediments were crushed (Fig. 8–26), metamorphosed, and thrust northwestward along a great fault. In this fault, named **Logan's thrust,** continental rise sediments have been shoved over and across some 48 km of shelf and shield rocks. Today we are able to view the remnants of this activity in the Taconic Mountains of New York. Ash beds, now weathered to a clay called bentonite, attest to the violence of volcanism. Masses of granite now exposed in the Piedmont help record the great pressures and heat to which the sediments of the subduction zone complex were subjected. But even if the igneous rocks had never been found, geologists would know mountains had formed, for the great apron of sandstones and shales that outcrop across Pennsylvania, Ohio, New York, and West Virginia must have had their source in the rising Taconic ranges (Fig. 8–27).

COMMENTARY
A Colossal Ordovician Ash Fall

The Ordovician succession of marine sedimentary rocks in eastern North America and northwestern Europe contains many widespread ash beds that have been altered to bentonites (see *Commentary,* Glass from Sand). Because these ash beds were deposited in what can be termed a geologic instant, they constitute excellent time planes for correlation and provide the basis for a variety of geologic studies that require comparison of sedimentary rocks that are of identical age.

The most extraordinary of the Ordovician ash falls, called the Millburg bed in North America and the Big Bentonite in Europe, is estimated to have thrown an amount of ash into the atmosphere that is equivalent to 1140 km^3 of dense rock if it were pulverized to ash-size particles. The bentonite layer, now compacted, is 1 to 2 meters thick over an area of several million square kilometers. The mineralogy of the ash indicates a single eruption, and its thickness distribution suggests that the parent volcano was centered somewhere between eastern North America and Sweden. So large a volume of ash would have required a sustained eruption over a period of 10 to 15 days. It was clearly the largest ash-producing eruption recorded in all of the past 590 million years, and yet seems to have had little effect on Ordovician marine organisms. This observation casts doubt on theories that invoke atmospheric dust and ash as a primary cause of global extinctions.

Uranium-lead ages obtained from zircon crystals in the bentonite of the giant ash fall confirm that the beds now on opposite sides of the Atlantic are of the same age and that the eruption occurred 454 million years ago.

Reference
Huff, W. D., Bergstrom, S. M., and Kolata, D. R. 1992. Gigantic Ordovician volcanic ash fall in North America and Europe: Biological, tectonomagmatic, and event-stratigraphy significance. *Geology* 20:875–878.

FIGURE 8–26   These Ordovician shale beds exposed at Black Point, Port au Port, Newfoundland, are intensely folded as a result of the Taconic orogeny of Late Ordovician time. *(Courtesy of the Geological Society of Canada, Ottawa.)*

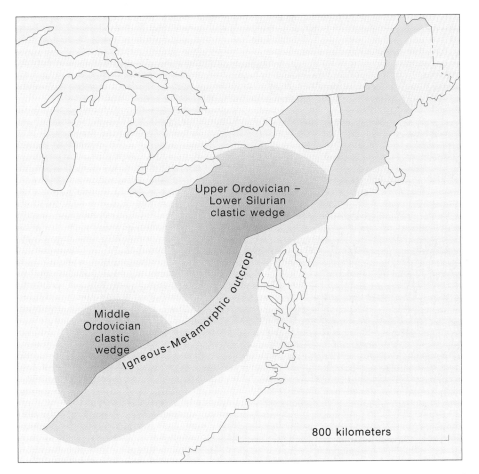

FIGURE 8–27   Great wedges of clastic sediments spread westward as a result of erosion of mountain belts developed during the early Paleozoic.   *(After Hadley, J. B. 1964. In Tectonics of the Southern Appalachians. V.P.I., Dept. Geol. Sci. Memoir 1.)*

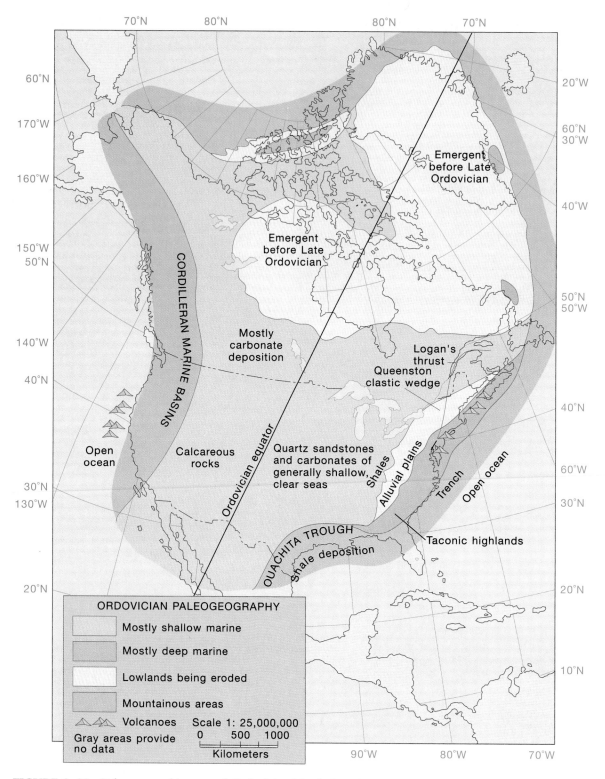

FIGURE 8–28   Paleogeographic map of Ordovician North America.

From a feather edge in Ohio, this barren wedge of rust-red terrestrial clastics called the **Queenston clastic wedge** (Fig. 8–28) becomes increasingly coarser and thicker toward the ancient source areas to the east (Figs. 8–29 and 8–30). Streams flowing from those mountains were laden with sand, silt, and clay. As they emerged from the ranges, they spread their load of detritus in the form of huge deltaic systems

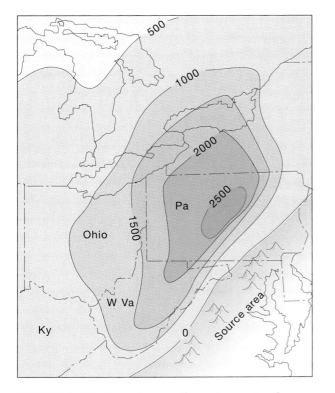

FIGURE 8–29 Isopach map illustrating regional variation in thickness of Upper Ordovician sedimentary rocks in Pennsylvania and adjoining states. *(After Kay, M. 1951. Geol. Soc. Am. Mem. 48.)*

that grew and prograded westward into the basin, often covering earlier shallow marine deposits. The deltaic aspect of many of these deposits accounts for the alternate name of Queenston delta for the Queenston clastic wedge. It has been estimated that over 600,000 km^3 of rock were eroded to produce the enormous volume of sediment in the Queenston clastic wedge. This indicates that the Taconic ranges may have exceeded 4000 meters (13,100 feet) in elevation.

During the Silurian, the locus of most intense orogenic activity seems to have shifted northeastward into the Caledonian belt. Meanwhile, erosion of the Taconic highlands continued. Early Silurian beds are often coarsely clastic, as evidenced by the Shawangunk Conglomerate of New York (Fig. 8–31). These pebbly strata give way upward and laterally to sandstones such as those found in the Cataract Group of Niagara Falls. Silurian iron-bearing sedimentary deposits accumulated in the southern part of the Appalachian region. The greatest development of this sedimentary iron ore is in central Alabama, where there are also coal deposits of Pennsylvanian age. The coal is used to manufacture coke, which is needed in the process of smelting. Limestone, used as blast furnace flux, is also available nearby. The fortunate occurrence of iron ore, coal, and limestone in the same area accounts for the formerly important steel industry of Birmingham, Alabama (Fig. 8–32).

Extending across the southern margin of the North American craton is the Ouachita–Marathon depositional trough (see Fig. 8–31). Although over 1500 km long, only about 300 km of its folded strata are exposed. Additional information about the distribution of early Paleozoic rocks within the belt is de-

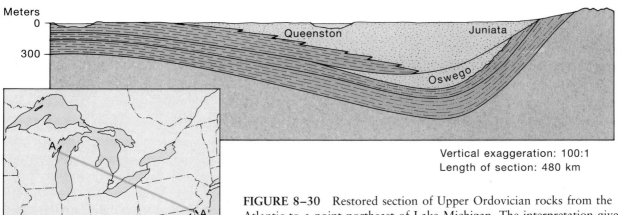

Vertical exaggeration: 100:1
Length of section: 480 km

FIGURE 8–30 Restored section of Upper Ordovician rocks from the Atlantic to a point northeast of Lake Michigan. The interpretation given to the cross-section is that a rising highland area to the east supplied clastic sediments to the basin until it filled. Continued sedimentation forced the retreat of the sea westward and extended a clastic wedge toward the west.

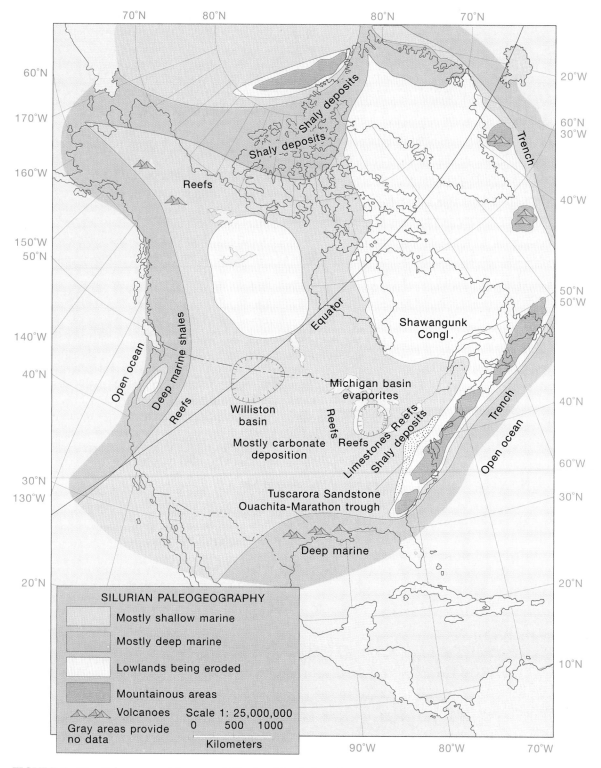

FIGURE 8–31 Paleogeographic map of Silurian North America.

rived mostly from oil well drilling and geophysical surveys.

Overall, nearly 10,000 meters of Paleozoic sediment fill the Ouachita–Marathon trough. However, of this thickness, only about 1600 meters were deposited in the Early Silurian. This indicates more rapid

subsidence and deposition during the late Paleozoic. Unfossiliferous metamorphosed graywackes, quartzites, and cherty beds of probable Early Cambrian age are the oldest Paleozoic sediments known from the Ouachita belt and indicate deposition along the continental rise. The next oldest rocks are Middle Cam-

COMMENTARY

An Appalachian–South American Connection?

In the conventional view of the dismembering of Rodinia, Laurentia separated from Europe and Africa during the Neoproterozoic. Iapetus separated the land masses, and subsequent closure of Iapetus caused island arcs and microcontinents to collide with the eastern margin of Laurentia. The result was the Taconic orogeny. There is, however, an alternate scenario for Ordovician Appalachian mountain building that involves collision between eastern North America and the western margin of South America. In South America, the collision produced the Famatinian orogeny. Several lines of evidence support this surprising, but as yet unproven, hypothesis. Rocks of the two regions have considerable stratigraphic similarity. Of particular significance is the observation that a Cambrian carbonate platform lies to the west of the Famatinian orogen and that the limestones of the platform contain a Cambrian Olenellid trilobite fauna otherwise known only from Laurentia (see Fig. 10–47C, *Olenellus*). It is likely that the carbonates bearing the trilobite fauna represent a segment of Laurentia that adhered to South America when the two continents broke apart. In addition, isotope dates have been obtained that indicate that both the Taconic and Famatinian orogenies experienced their optimum tectonic activity between 480 and 460 million years ago.

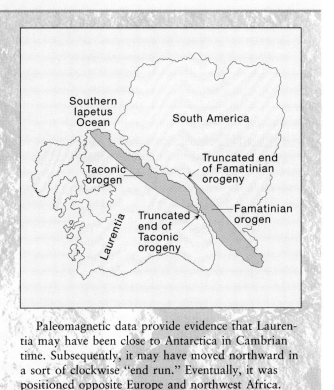

Paleomagnetic data provide evidence that Laurentia may have been close to Antarctica in Cambrian time. Subsequently, it may have moved northward in a sort of clockwise "end run." Eventually, it was positioned opposite Europe and northwest Africa. Collisions with these continents provided the Acadian and Alleghenian orogenies of the late Paleozoic.

FIGURE 8–32 Clinton iron ore from the Silurian Clinton Group near Birmingham, Alabama. The ore is an oölite of the iron oxide mineral hematite. It includes shells of marine fossils that have been replaced by iron oxide. Nearby coal beds provide the fuel needed to produce steel from the iron ore.

brian igneous bodies that are 525 to 535 million years old. These are overlain unconformably by basal sandstones and fossiliferous limestones of platform facies. Ordovician and Silurian deposits in the Ouachita region can be differentiated into distinct graptolitic facies and shelly facies of the continental shelf. The deeper water deposits lay to the south, whereas the shallower shelf sediments were deposited along the edge of the craton. Northward, the early Paleozoic section thins as it becomes part of the interior platform. Paleozoic rocks of the Ouachita belt are noted for their unusually siliceous and cherty character. Very likely, the abundant silica was derived from the submarine weathering of volcanic ash ejected by volcanoes that lay to the south of the Ouachita trough.

The Caledonides

Scotland's ancient name, still used in poetry, is Caledonia. Eduard Suess used the name for the Scottish

continued on p. 298

Shenandoah National Park

Shenandoah National Park in northwest Virginia is an excellent place to view the effects of the many orogenic events that have punctuated the geologic history of the Appalachian region. Much of the viewing can be done from overlooks along Skyline Drive, which runs the entire length of the park along the crest of the Blue Ridge Mountains. Seventy-one overlooks along the drive provide spectacular views of the Shenandoah Valley to the west and the Piedmont to the east. For those who prefer to explore on foot, there are 455 miles of trails, including a 95-mile segment of the Appalachian trail that extends from Maine to Georgia.

The oldest rocks in Shenandoah National Park originated during the Proterozoic Eon. These ancient metamorphics consist of gray–green gneisses of the Pedlar Formation. Exposures of the Pedlar rocks can be seen along Skyline Drive at the Tunnel Parking and Buck Hollow overlooks. About 1100 million years ago, an attractive, light-colored, coarsely crystalline granite intruded the Pedlar Formation. The intrusive body, named the Old Rag Granite after its exposure on Old Rag Mountain, can be seen from the Thorofare Mountain overlook and nearby trails. As is typical of many other granite bodies, the rock has developed a complex system of joints. Water flowing through the joints has eroded and chemically weathered the granite, slowly widening it and ultimately sculpting many of the separated blocks into huge, rounded boulders. (Rounding is a consequence of corners having greater surface area for their volume than edges or faces, so they weather more rapidly.)

Erosion following the intrusion of the Old Rag Granite stripped away much of the cover of older, less resistant rocks. The relatively quiet episode of erosion ended

Dark Hollow Falls. There are more than 200 waterfalls in Shenandoah National Park. They develop where streams flow across outcrops of resistant rock that are underlain by layers of more easily eroded rock. The less resistant layers are more rapidly eroded, resulting in an abrupt drop of the elevation of the stream bed. Eventually, waterfalls degenerate into rapids and then may disappear. *(Photograph by Jeff Gnass.)*

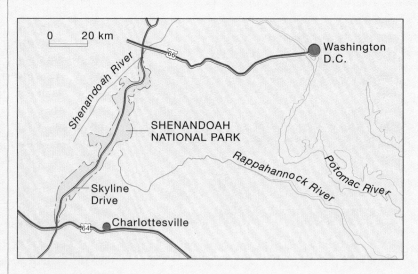

Location map, Shenandoah National Park, Virginia.

near the end of the Proterozoic, when rifting occurred and basaltic lavas rose to the surface along tensional faults. The lavas accumulated to thicknesses in excess of 600 meters. Some of them exhibit pillow structure, indicating that they were extruded beneath the sea. In subsequent orogenic events, these basalts were metamorphosed to become the greenstones that cap many of the mountains in the park. Rifting and volcanic activity continued into the Cambrian Period, as a reflection of the breakup of Proto-Pangea that had begun near the end of the Proterozoic. Gravels, sands, and clays, carried to the rift zone by streams, were deposited along the western margin of the widening sea. Subsequently, when this passive margin became a convergent zone, these sediments were altered to metaconglomerates, phyllites, and quartzites. The rocks constitute formations of the Cambrian Chilhowee Group. Chilhowee quartzites, because of their resistance to erosion, cap many of the ridges in Shenandoah National Park. In addition, the earliest evidence of life in the park consists of tubes and burrows of wormlike organisms discovered in Chilhowee metasediments.

For the remainder of the Paleozoic, Shenandoah National Forest and the Appalachian orogen were subjected to periodic and often severe tectonism. As noted in Chapters 9 and 10, these mountain-making events began with the Ordovician Taconic orogeny and culminated in the great Allegheny orogeny at the end of the Paleozoic. There are a few patches of Mesozoic rocks in the park. These consist primarily of basaltic dikes. One can view these dikes, which have well-developed columnar jointing, from Skyline Drive at Franklin Cliffs and Crescent

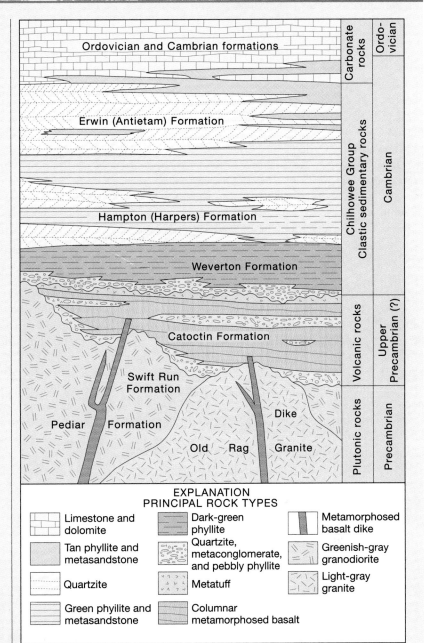

Generalized columnar section of rock formations, Shenandoah National Park. *(From Gathright, T. M., II, 1976. Geology of Shenandoah National Park, Virginia. Virginia Divisions of Mineral Resources Bulletin 86.)*

Rocks. Cenozoic geologic history in Shenandoah National Park has been dominated by erosion and periodic isostatic adjustments.

Reference
Gathright, T. M. II 1976. *Geology of Shenandoah National Park, Virginia.* Virginia Division of Mineral Resources, Bull. 86, Charlottesville, VA.

remnants of an early Paleozoic mountain range, the **Caledonides.** It is also the basis for the name of the Caledonian orogenic belt, which extends along the northwestern border of Europe (see Fig. 8–5).

The Caledonian and Appalachian belts have had a generally similar history. This is not surprising, for both are really part of one greater Appalachian–Caledonian system. Both elongate depositional sites evolved as a result of a cycle of ocean expansion and contraction, such as is schematically illustrated in Figure 8–25. The cycle began with a Late Precambrian to Middle Ordovician episode of spreading, as the Iapetus Ocean widened to admit new oceanic crust along a spreading center. On the continental shelves and rises of the separating blocks, thick lenses of sediment accumulated. This depositional phase is marked by the development of two distinct facies. The graptolite facies consists of more than 6000 meters of volcanics, graywackes, and shales. Graptolites are prevalent in this facies and are used for subdividing the Ordovician and Silurian into biostratigraphic zones. A shelly facies with clean sandstones and fossiliferous limestones identifies the shallower or continental shelf facies. Here and there along the margins of the Caledonian trough there are deposits of Upper Silurian freshwater shales that are noteworthy for their fossil remains of early fishes and strange arthropods known as eurypterids (Fig. 8–33).

The closure of the Iapetus Ocean and crumpling of the Caledonian marine basins began in Middle Ordovician time, when subduction zones developed along the margins of the formerly separating continents. This event is recognized from the volcanic rocks that occur in the Canadian Maritime Provinces, northwestern England, northeastern Greenland, and Norway. Little by little, the Iapetus Ocean closed until the opposing continental margins converted in a culminating mountain-building event termed the **Caledonian orogeny.** The orogeny reached its climax in Late Silurian to earliest Devonian time. It was most intense in Norway, where Precambrian and lower Paleozoic sedimentary rocks are drastically metamorphosed and thrust-faulted. Southwestward, in the British Isles, the effects were not as severe, although mountainous terrains also developed there.

South of the unruly Caledonian belt there is evidence of a land mass that during Late Silurian and Devonian received sandy detritus from the growing Caledonides. This region, on which was deposited clays and sands of the Old Red Sandstone, is often referred to as the Old Red Continent (Fig. 8–34). The deposits here are in the form of a thick, clastic wedge very similar to the Queenston clastic wedge on the opposite side of the Iapetus Ocean. In fact, the two clastic wedges are approximately mirror images of one another (see Fig. 8–25).

Rocks of the Caledonian belt are well exposed not only in the British Isles but also across northeastern Greenland and Spitzbergen. The geologic succession in the Caledonides of eastern Greenland begins with a thick (10,000-meter) sequence of Precambrian rocks. The uppermost unit in the Precambrian sequence contains tillites, indicating an episode of glaciation just prior to the beginning of the Paleozoic. In eastern Greenland, Cambrian deposition began with sandstones followed by shales, dolomites, and limestones. Carbonates continued to be prevalent in the Early Ordovician, but thereafter sedimentary patterns began to change in response to the closing of the Iapetus Ocean as it was manifested in the Caledonian orogeny. Erosion of the metamorphosed and folded rocks led to the accumulation of the continental Old

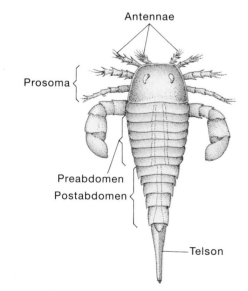

B

FIGURE 8–33 Two genera of eurypterids. *Eurypterus* (A) is noted for its broad, flipper-like paddles and blunt frontal margin. *Pterygotus* (B) is distinguished by a pair of formidable-looking frontal pincers. *(Drawing and model of* Eurypterus, ×1/3. *Reconstruction of* Pterygotus *courtesy of the National Natural History Museum, Smithsonian Institution.)*

A

A

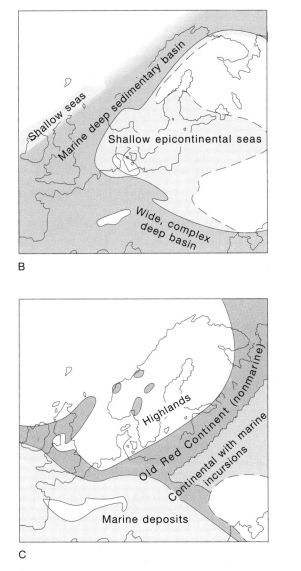

B

C

FIGURE 8–34 Paleographic maps of Europe during the (A) Cambrian, (B) Ordovician–Silurian, and (C) Devonian. *(After Brinkman, R. 1969. Geologic Evolution of Europe. Stuttgart: Enke; New York: Hafner Publishing Co.)*

Red Sandstone facies in basins adjacent to the growing mountains.

The Uralian Seaway

The elongate seaway in which Lower Paleozoic rocks of the Urals were deposited extended southward from the Russian Arctic along the course of the present Ural Mountains. The seaway separated the Russian platform on the west from the Angaran shield on the east. The western side of the Uralian belt contains early Paleozoic shelf carbonates, sandstones, and shales that thin toward the west and pinch out against the Russian platform (Fig. 8–35). Deep ma-

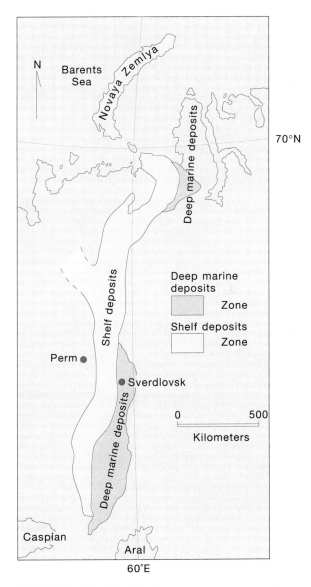

FIGURE 8–35 The Uralian seaway. *(Adapted from Read, H. H., and Watson, J. 1975. Introduction to Geology. New York: Halstead Press.)*

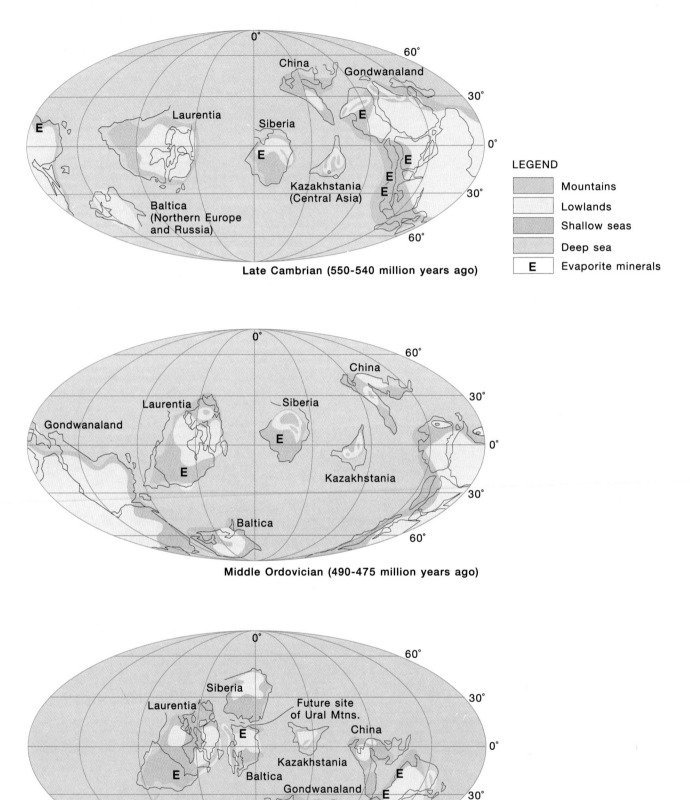

FIGURE 8–36 Global paleogeography of the early Paleozoic. *(From Scotese, C. R., Bambach, R. K., and Barton, C., et al. 1979. J. Geol. 87(3):217–277.)*

rine conditions characterized the eastern side of the Uralian tract. Here we find turbidites, shales, and volcanics.

Although the Uralian orogenic phase was not to come until the late Paleozoic, folded strata, unconformities, and mafic intrusions attest to plate convergence and subduction in the region. The history of the Urals seems to have followed an ocean expansion and contraction cycle (that is, a Wilson cycle) in some ways similar to that of the Caledonian belt. Paleomagnetic studies indicate that the Russian and Siberian platforms were separated by a wide oceanic tract during the early Paleozoic. During the Silurian and Devonian, subduction zones developed between the two opposing continental masses. As the oceanic crust was being subducted, the intervening ocean narrowed, and the continents began to converge on one another (see Middle Silurian in Fig. 8–36). By the end of the Paleozoic, and possibly continuing into the Triassic, the former depositional belt experienced its orogenic climax as the two continental blocks collided (see early Late Permian in Fig. 9–1).

Asia Begins to Take Form

Asia east of the Urals is a very ancient and complex terrane formed from the convergence and accretion of several crustal blocks. The sutures along which the continents collided are marked by ophiolite belts, narrow zones of high-pressure metamorphism, and volcanic rocks that characterize crustal tracts above subduction zones that descend at continental margins. Some geologists believe that as many as nine separate crustal blocks or microcontinents may have combined to form Asia. In the early Paleozoic, faunal and paleomagnetic evidence suggests that most of these blocks were separated by considerable widths of oceanic crust. By the Middle Ordovician, narrowing of these oceans and convergence of the continental crustal blocks was in progress, although major episodes of collisional orogenesis did not occur in most regions until late in the Paleozoic and in the Triassic and Jurassic.

Way Down South

Unlike the stratigraphic record of the early Paleozoic for North America and Europe, the record for the Southern Hemisphere land masses is relatively poor. This may be the result of the continents being prevailingly above sea level, which would favor erosion rather than deposition. However, there is a fairly complete record of sedimentation along the north–south trending Tasman belt of Australia (Figs. 7–19

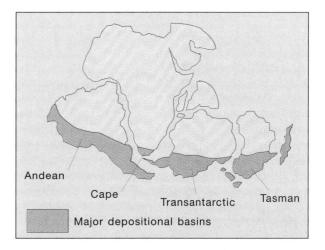

FIGURE 8–37 Principal depositional basins of Southern Hemisphere continents during early Paleozoic time.

and 8–37). Sedimentation began there just before the Paleozoic. These uppermost Proterozoic rocks are mostly sandstones that grade into overlying Lower Cambrian formations without a significant erosional unconformity. The relatively quiet period represented by these sediments was followed by a full-scale orogeny in the Late Cambrian. Thick graywackes and volcanics record the orogenic disturbances. During the Ordovician, a broad-shelf sea spread across central Australia and received deposits of well-sorted sandstones and limestones.

The African record of the early Paleozoic is far more obscure than that of Australia. The most complete sequences of strata are located north of the equator, where one finds Cambrian through Silurian rocks that were laid down in shallow embayments of a seaway that was the precursor of the Tethys basin. Africa south of the equator was persistently above sea level and was being eroded during most of the early Paleozoic. Strata of this age are indeed uncommon.

One feature in the geologic record of the northern part of Africa is rather astonishing. There is remarkable evidence of an Ordovician episode of glaciation in the area that is now the Sahara Desert. One can only surmise that this part of Africa was close to the Ordovician South Pole (as indicated by paleomagnetic data).

Lower Paleozoic rocks in South America are also poorly known and difficult to interpret. Strata deposited along the Andean tract have been greatly deformed and frequently obliterated by Mesozoic orogenic activity. Elsewhere, the Paleozoic sequence lies buried beneath an immense pile of sediment spread eastward from the eroding Andes Mountains. We do, however, have scattered bits of the record. We know that Cambrian and Ordovician seas occupied the western margin of South America and that graptolitic shales and shelly limestones were deposited there. Platform deposits were also laid down in shallow basins within the present Amazon region. In Ordovician rocks of Argentina, we find boulder beds and faceted cobbles produced by glaciers.

■ *Early Paleozoic Global Geography*

We are so accustomed to the shapes and locations of today's continents that the Earth's early Paleozoic geography seems alien to us. Geologists have attempted to reconstruct the positions of continents during this ancient time by combining all relevant paleomagnetic, paleoclimatologic, biogeographic, and tectonic information. This synthesis of data has resulted in global paleogeographic maps such as those provided in Figure 8–36. These maps indicate

that by late Cambrian time, the major land areas were positioned at low latitudes as rather isolated continents in a well-interconnected world ocean. The main continental blocks at this time were Kazakhstania (a piece of crust now located in central Asia north of Iran), Siberia, Laurentia, Baltica, and Gondwanaland. There was no land north or south of 60° latitude, and the poles were therefore in the middle of oceanic areas. During the Ordovician and Silurian, the most dramatic change in global geography was the movement of Gondwanaland from its equatorial position southward to a polar location. Baltica (northern Europe and Russia), meanwhile, gradually circled and set a collision course toward eastern Laurentia. The actual collision took place in the Devonian as part of the Acadian orogeny and the assembly of Pangea.

■ *Aspects of Early Paleozoic Climates*

During the early Paleozoic, the Earth had a north and south pole, was encircled by an equator, and must

C O M M E N T A R Y
The Big Freeze in North Africa

The news media inform us of critical environmental problems almost every day. We are ourselves to blame for many of these problems, including those caused by pollution, acid rain, deforestation, and modification of natural water systems. Yet environmental and ecological crises have occurred many times in the geologic past, long before there were humans to generate them. For example, consider the effects on life of the Ordovician Ice Age described in this chapter. We have evidence that the South Pole at this time was located in what is now a barren area between Algeria and Mauritania. Evidence for an enormous ice sheet in this region occurs as glacial grooves and striations, outwash plains, moraines, and meltwater channels that extend across thousands of square kilometers. There must have been a truly enormous buildup of continental ice, and that accumulation would surely have resulted in cooler temperatures in ocean regions far to the north. Eustatic

and climatic changes would have had a pervasive effect on life. Indeed, paleontologists have long recognized that the late Ordovician was marked by extinctions of large numbers of marine invertebrates, including entire families of bryozoans, corals, brachiopods, sponges, nautiloid cephalopods, trilobites, crinoids, and graptolites. Twenty-two percent of all families of known animals were wiped out. It was one of the greatest episodes of mass extinction in the geologic record.

It seems likely that the demise of these families of organisms resulted from general cooling of tropical seas as well as the draining of epeiric and marginal seas as sea level was lowered. The loss of the shallow-water environments in which early Paleozoic invertebrates thrived, and the crowding of species into remaining habitats, would explain this Ordovician ecological disaster.

have been characterized by latitudinal and topographic variations in climate, just as such variations exist today. As we have seen, during episodes of great marine transgressions and low-lying terrains, climates were probably somewhat milder; during those periods when continents stood high and great ranges diverted atmospheric circulation patterns, climates were more diverse and extreme, much as is the case today. However, there were also factors that would have made those climates of long ago rather unique. During the early Paleozoic, the Earth turned faster and the days were shorter. Tidal effects were stronger, and until late Ordovician, there were no vascular plants to absorb the sun's radiation. There may even have been changes in solar radiation or atmospheric composition that left no unambiguous clues. The total climatic effects of these possibilities can only be surmised.

At the very beginning of the Cambrian, the climate was probably somewhat cooler than the average for the entire early Paleozoic. The Earth was still recovering from a great ice age that had culminated near the close of the Precambrian. Soon, however, climates became warmer. Broad, shallow seas (epeiric seas) slowly began to encroach on the continental interiors. Paleogeographic reconstructions, such as those shown in Figure 8–36, suggest that North America, Europe, and even Antarctica may have lain astride or near the equator during the early Paleozoic. These geophysical data are compatible with the depositional record of thick limestones and extensive reefs, which today form only in the warmer regions of the oceans.

However, some evaporites and fossiliferous limestones have been found in rather high paleolatitudes, indicating that worldwide climates may also have been somewhat warmer than average for a time. In the Ordovician stratigraphic sequences of northern Canada, for example, thick deposits of gypsum and salt are found at paleolatitudes (latitudes determined by paleomagnetic studies) of about 10°S. The North American reefs of Ordovician age generally fall within 30° of the paleoequator, whereas the richly fossiliferous carbonate rocks of the craton would have been within about 40° of the paleoequator. The study of cross-bedding in sandstones suggests that eastern North America lay in a zone of trade winds that blew from northeast to southwest with respect to the present directional grid. These winds thus moved in a direction opposite to those prevalent in eastern North America today.

That there were severe as well as equable climates during the early Paleozoic is indicated by extensive Late Ordovician glacial deposits in the region of the present Sahara Desert. These deposits are sufficiently widespread that they must have been emplaced by continental glaciers and certainly must have been accompanied by a lowering of mean annual temperatures at middle and high latitudes. Study of the distribution of the glacial deposits, directions of glacial striations, and paleomagnetic data indicates that in Ordovician time the north African part of Gondwanaland was positioned astride the South Pole.

During the Silurian, there were latitudinal variations in climate somewhat similar to those of today. Glacial deposits of this age are found only in higher locations, usually in excess of 65° from the paleoequator. Coral reefs, evaporites, and desert sandstones are found within 40° of the Silurian equator. Certainly there were regions of marked aridity during the Silurian. Epeiric seas became relatively less extensive as the Silurian drew to a close. We can speculate that the Iapetus Ocean formed by the division of that hypothetical continent Proto-Pangea had already begun closing and that the Caledonian crunch was about to take place. The curtain was ready to rise on the world of the late Paleozoic.

■ *Mineral Resources of the Lower Paleozoic*

The lower Paleozoic systems contain several kinds of Earth materials that have economic importance. Among these are building stones, such as slates, which were once extensively quarried in New Jersey and Pennsylvania, and marbles derived from Ordovician limestones, which are still quarried in New England. The St. Peter Sandstone (see Fig. 8–12) has been and continues to be an important source of clean quartz for use in the manufacture of glass. As noted earlier, salt is also an important mineral resource of the lower Paleozoic (Silurian). In New York State, salt recovery and marketing is a major industry. The method used in recovery involves pumping water down wells into salt-bearing formations. The water dissolves the salt, which is then pumped to the surface to be reprecipitated in evaporating pans and refined.

The most notable ore deposits of the lower Paleozoic are sedimentary iron ores that occur in Newfoundland and in a band extending southward from New York into Alabama. At Birmingham, these iron ore beds were once the basis for a thriving steel industry. However, most of the mines are now closed. Nevertheless, a large amount of ore remains, and it may become economically feasible to reopen mines at some future date. Petroleum has been recovered from Ordovician and Silurian rocks in Ohio, Oklahoma, and Texas, as well as in western Europe.

COMMENTARY
Glass from Sand

Glass, a product formed by melting and cooling quartz sand and minor amounts of certain other common materials, is so familiar to us that we sometimes forget its importance. In fact, it would be difficult to imagine a world without glass. There would be no goblets, bottles, mirrors, or windows; no glass laboratory retorts, tubes, or beakers; no electric light bulbs or neon signs; no glass lenses for cameras, microscopes, telescopes, or motion picture projectors; and no fiber optics for use in viewing the inside of the human body. We see glass used everywhere in our buildings, electronic equipment, and medical devices. We not only see it, we see through it.

Glass has a long and colorful history. Natural glass, such as obsidian, was used by prehistoric humans to make arrowheads and knife blades. Human-made glass was made by Egyptians as long ago as 3000 B.C.E. At Ur in Mesopotamia, glass beads nearly 4500 years old have been found in archaeological excavations. The Greeks and Romans learned the art of making glass ornaments, bottles, vases, and trinkets that were prized throughout the ancient world. During the Middle Ages, Europeans produced beautiful stained-glass windows for cathedrals. Blood-red stained glass was made by adding copper compounds to the molten silica. The Europeans knew that cobalt gave the glass a rich, deep-blue color; manganese turned it purple; and antimony provided golden yellows. Iron oxides were added to color glass green or brown.

Today, the sands used in making glass have a silica content as high as 95 to 99.8 percent. As mentioned in this chapter, two sources of such exceptionally pure sands are the Ordovician St. Peter Sandstone of Missouri and Illinois and the sandstone members of the Devonian Oriskany Formation of West Virginia and Pennsylvania. These clean, pure sandstones were deposited in shallow marine environments where wave action could remove clay impurities and concentrate grains of sand-size quartz. Both sandstones are composed of grains that have been eroded from older sandstones, and this reworking has also contributed to their extraordinary purity.

In the manufacture of window glass, the mixture prepared for melting consists of about 72 percent silica from quartz grains. Certain metallic oxides, such as soda (Na_2O) and lime (CaO), serve to lower the temperature required for melting. That temperature for soda–lime–silica glass is 600 to 700°C. Once molten, the "batch" (as it is called) is cooled, poured, rolled, or blown into the shapes desired. Compounds of boron and aluminum can be added to provide heat-resistant glass, as in the Pyrex® brand of cookware. Fine cut glass and so-called crystal are a silica–lead–soda glass known not only for its brilliance but musical tone as well. By varying the kinds and amounts of metallic oxides, glass can be produced for a great variety of special uses and products. There is little doubt that it will always be an essential part of life in this age of technology.

Summary

The Paleozoic Era began about 544 million years ago. It began at a time when the supercontinent Rodinia had begun to pull apart, allowing the incursion of oceanic tracts between separating land masses. The Iapetus Ocean was an early Paleozoic proto-Atlantic ocean that was widening at the beginning of the Cambrian but was to close during later Paleozoic time as a result of conversion of tectonic plates bearing North America, western Europe, and Gondwanaland.

Most of North America at the beginning of the Paleozoic was above sea level and experiencing erosion. The initial sites of marine deposition were along the rifted continental margins. Gradually, however, the seas spilled out of these marginal basins onto the continental platform. The sediments deposited on the floors of epeiric seas that covered the platform record cycles of transgression and regression. The thicker sections are preserved in basins, and the thinner sequences with many unconformities developed over domes. The Sauk and Tippecanoe cratonic sequences record the transgressive–regressive cycles of the early Paleozoic. For much of the early Paleozoic, the transcontinental arch was a positive area providing sediment to surrounding seas.

By Ordovician time, the relative calm that characterized the Cambrian was interrupted by a plate convergence and mountain-building episode called the

Taconic orogeny. Periodic crustal disturbances associated with this orogeny raised mountains along the eastern margin of North America. Rapid erosion of these uplands produced huge volumes of detritus, which were spread westward by streams and deposited in a great apron of detrital sediment called the Queenston clastic wedge. Similar mountain-building activity was also under way in northwestern Europe, where the Caledonian depositional basin was compressed, intruded, and uplifted in the early stages of the Caledonian orogeny. Both the Taconic and Acadian orogenies were the result of plate convergence associated with the narrowing of the Iapetus Ocean.

The initial deposits of the early Paleozoic in western North America were continental shelf and rise deposits of the rifted continental margin. This early episode of relatively quiet sedimentation was interrupted during the Ordovician and Silurian by movement of the Pacific plate against the western edge of

North America. The result of this movement was the development of a subduction zone and chain of volcanic islands along the western margin of the continent.

In general, the range of climatic conditions during the early Paleozoic was not significantly different from that of the recent geologic past. Cold climates are indicated by the presence of Ordovician glacial deposits in Southern Hemisphere continents. Warmer conditions prevailed in regions where reefs were extensive and richly fossiliferous limestones were deposited. Aridity is indicated by the immensely thick sections of evaporites in the region around the Great Lakes. Although the range of climates may not have been greatly different from that of today, there were many nonclimatic differences. In the early Paleozoic the days were shorter and tides were stronger, and until Late Ordovician the lands had no cover of green plants.

Review Questions

1. What are the names and locations of the major orogenic events of the early Paleozoic, and when did each occur?
2. What is the name of the lowermost stage of the Cambrian System? What is the name of the Proterozoic stage on which it rests?
3. What geologic evidence indicates that the transcontinental arch was above sea level during most of the early Paleozoic?
4. What are evaporites? What is a barred basin? Why is a barred basin a particularly effective place for precipitation of evaporites? Where and when did such basins develop during the early Paleozoic?
5. What is a clastic wedge? Are clastic wedges primarily marine or nonmarine? Under what circumstances do they develop? What clastic wedge is associated with the Taconic orogeny?
6. Why are the unconformities that form the boundaries of cratonic sequences considered the result of eustatic lowering of sea level rather than tectonic uplift of the craton?
7. The St. Peter and Oriskany sandstones are considered mature sandstones. What are the characteristics of mature sandstones? How does the maturity of these sandstones contribute to their value as source materials for the making of glass?
8. How does the history of the Cordilleran region differ from that of the Appalachian region during the early Paleozoic?
9. What methods have geologists used to determine the locations of continents during the early Paleozoic so that they are able to construct maps such as those in Figure 8–36?

Discussion Questions

1. Why is the unconformity that separates the Sauk from the Tippecanoe not the same age at widely separated locations?
2. Domes and basins are characteristic structural features of the central stable regions of North America. Define each and explain how they differ in thickness of strata and unconformities. How might one distinguish between these two structures on a geologic map?
3. The guide fossil for the base of the Cambrian is a trace fossil (*Phycodes*). What are trace fossils? Discuss the possible advantages and disadvantages of trace fossils in the correlation and dating of chronostratigraphic units.
4. Why are rock units like the Tapeats Sandstone and Mauv Limestone not precisely the same age at different localities?

5. Why is it useful and logical to divide rocks of the craton into sequences rather than on the basis of the standard geologic time scale?
6. What evidence might you seek in the field to confirm the following?

a. the location of the paleoequator
b. former conditions of extreme aridity
c. former extensive episodes of glaciation
d. mountain building associated with tectonic plate collision

Supplemental Readings and References

Bambach, R. K., Scotese, C. R., and Ziegler, A. M. 1980. Before Pangaea: The geographies of the Paleozoic world. *Amer. Scientist* 68(1):26–38.

Cowie, J. W., and Brasier, M. D. (eds.). 1989. *The Precambrian–Cambrian Boundary.* Oxford, England: Clarendon Press.

Fairbridge, R. W. 1970. Ice age in the Sahara. *Geotimes* 15(6):18.

Frazier, W. J., and Schwimmer, D. R. 1987. *Regional Stratigraphy of North America.* New York: Plenum Press.

Kummel, B. 1970. *History of the Earth* (2nd ed.). San Francisco: W. H. Freeman Co.

Landing, E. 1994. Precambrian–Cambrian boundary global stratotype ratified and a new perspective on Cambrian time. *Geology* 22(2):179–182.

Sloss, S. L. 1963. Sequences in the cratonic interior of North America. *Geol. Soc. Amer. Bull.* 74:93–111.

Stearn, C. W., Carroll, R. L., and Clark, T. H. 1979. *The Geological Evolution of North America* (3rd ed.). New York: Ronald Press.

Red sandstone of the Permian Lyons Formation, Garden of the Gods, just west of Colorado Springs, Colorado. Erosion has sculpted these nearly vertical beds into dramatic pinnacles and spires, some of which stand 15 meters above their surroundings. Erosion of the Ancestral Rocky Mountains that lay to the west provided the sands that now compose the Lyons. The beds were tilted upward during the orogeny that produced the modern Rocky Mountains. *(Photograph by Djuna Bewley.)*

A continuous ocean spreads over the space now occupied by the British islands: in the tract covered by the green fields and brown moors of our own country, the bottom, for a hundred yards downwards, is composed of the debris of rolled pebbles and coarse sand intermingled, long since consolidated into the lower member of the Old Red Sandstone; the upper surface is composed of banks of sand, mud, and clay; and the sea, swarming with animal life, flows over all.

HUGH MILLER, *The Old Red Sandstone,* 1858

CHAPTER 9

Late Paleozoic Events

The Devonian, Mississippian, Pennsylvanian, and Permian are the geologic periods of the late Paleozoic in North America. Elsewhere, the Mississippian and Pennsylvanian are combined into the Carboniferous Period. The late Paleozoic was a time when most of the separate land masses of earlier periods were assembled into the great supercontinent of Pangea. During the pre-Pangean clustering of land masses, the margins of larger continents, like eastern North America, grew by accretion of island arcs and microcontinents. Ultimately, the larger continents collided. The **Caledonian** and **Acadian** orogenies brought North America and Europe together as a combined land mass named Laurussia (Fig. 9–1). Subsequently, the plate bearing Gondwanaland began to close on Laurussia. By late in the Carboniferous, it had come into contact with Laurussia, generating the **Hercynian orogeny** of central Europe and the **Allegheny orogeny** of eastern North America. It may seem startling that mountains in the southern Appalachian region were the result of a collision with northwestern Africa. Similarly, the now deeply eroded and largely buried mountains of the Ouachita orogenic belt were produced by collision with the South American segment of Gondwanaland. The part of South America that is now Venezuela apparently formed the southern margin of that late Paleozoic Ouachita mountain system.

The clustering of once separate continents to form Pangea was nearly complete by the Late Permian. Great mountain systems identified the suture zones where continents had come together. An enormous

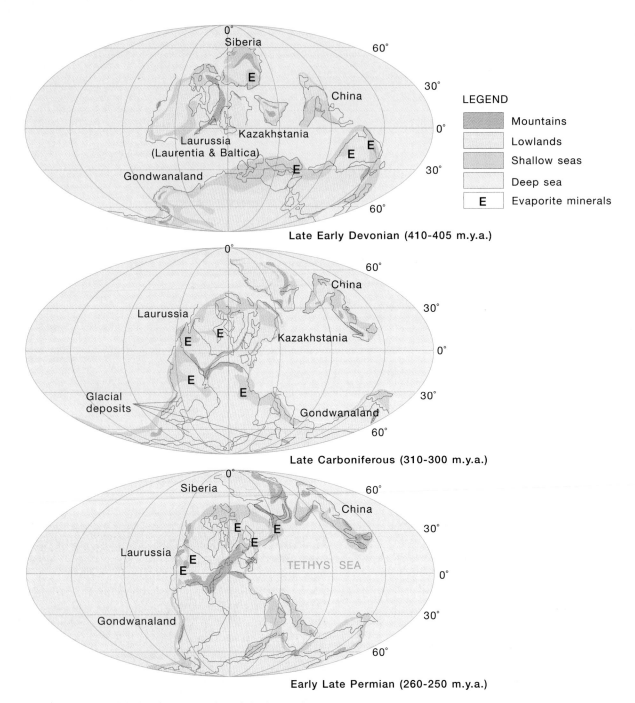

Late Early Devonian (410-405 m.y.a.)

LEGEND

Mountains	
Lowlands	
Shallow seas	
Deep sea	
E	Evaporite minerals

Late Carboniferous (310-300 m.y.a.)

Early Late Permian (260-250 m.y.a.)

FIGURE 9–1 Global paleogeography of the late Paleozoic. *(From Scotese, C. R., Bambach, R. K., Barton, C., Van Der Voo, R., and Ziegler, A. M. 1979. J. Geol. 87(3):217–277.)*

ocean called Panthalassa spanned the globe across nearly 300° of longitude.

In addition to continental collisions and recurrent orogenic activity, the late Paleozoic was a time of diverse sedimentation, progress in organic evolution, and diverse climatic conditions. During this time, there was widespread colonization of the land by both large land plants and vertebrates. Amphibians and reptiles, along with spore-bearing trees and seed ferns, were especially evident in the landscapes of the late Paleozoic. Near the end of the era, conifers and reptiles more tolerant of drier, cooler climates became abundant. Marine invertebrates thrived in shallow epicontinental seas that were especially extensive in North America during the Mississippian. In the central interior of the United States, limestones de-

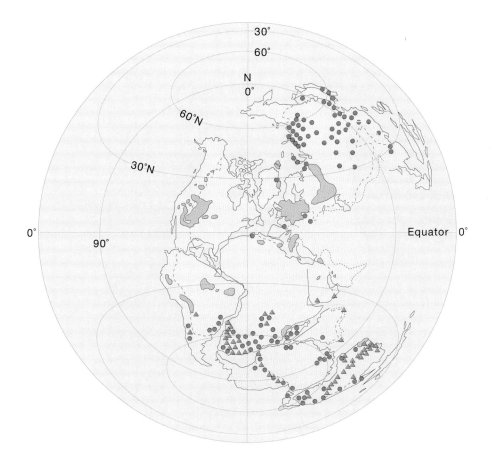

FIGURE 9–2 The distribution of glacial tillites (blue triangles), coal (red circles), and evaporites (irregular green areas) during the Permian, about 250 million years ago. *(From Dewey, G. E., Ramsey, T. S., and Smith, A. G. 1974. J. Geol. 82(5):539.)*

posited in these inland seas are over 700 meters thick and are among the most extensive sheets of limestone anywhere in the world. Adjacent to the orogenic belts and in regions where the epicontinental seas were drained as a result of uplift of the land or lowering of sea level, sediment-laden streams deposited their loads of sand and mud.

These nonmarine sediments contain the fossil plants and animals that aid in inferring climatic conditions on the continents. The accumulation of abundant vegetation in swampy areas provided the raw materials from which the famous coal deposits of Pennsylvanian time were formed. Now consolidated sands of desert dunes and layers of evaporites attest to arid conditions at many localities around the globe (Fig. 9–2). Tillites and glacial striations on bedrock provide evidence of glaciation. This is not unexpected, for the more southerly parts of Gondwanaland were located near the pole. Evidence of Late Carboniferous or Permian ice sheets can be found on the now separated land masses of South America, Africa, Australia (Fig. 9–3), Antarctica, and India. Decades before geologists had an understanding of plate tectonics, the ancient tillite deposits on these continents were cited as evidence of continental drift.

FIGURE 9–3 Glacial striations formed in Early Permian time on Proterozoic quartzite in Australia. The ice flowed to the right, parallel to the pen. Note the downflow smoothing and crescent-shaped "chatter" marks with their concave sides facing upflow (upper right). *(Courtesy of Warren Hamilton, U.S. Geological Survey.)*

■ *The Craton of North America*

In Chapter 8, we noted that the last event of the early Paleozoic cratonic region had been the withdrawal of inland seas. Many early Paleozoic formations were subjected to erosion following the regression of seas, and this exposed older and deeper formations in those areas that had experienced uparching. Gradual flooding of that early Paleozoic erosional surface was the first event of the late Paleozoic. The new inland sea is termed the **Kaskaskia sea** (Fig. 9–4), and in it were deposited marine rocks of the Kaskaskia se-

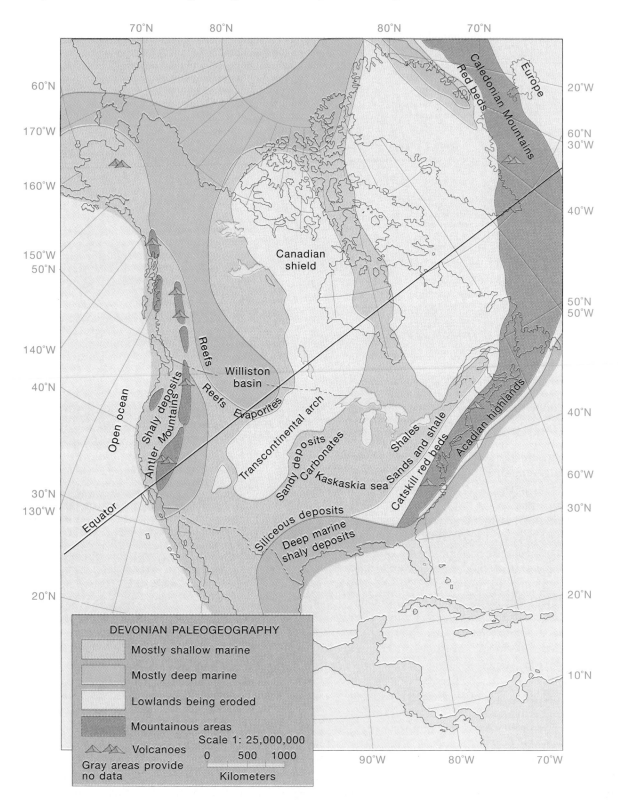

FIGURE 9–4 Paleogeography of North America during the Devonian Period.

quence (see Table 8–1). Except for minor regressions, the sea persisted over extensive areas of the craton until the end of the Mississippian Period. Erosion of the layers of sediment deposited on the Kaskaskia sea floor began, only to be interrupted by the advance of a second sea, the **Absaroka sea.**

The Kaskaskia Sequence

Along the eastern side of the North American craton, Kaskaskia sedimentation consists initially of clean quartz sands. The most famous of these sandy formations is the Oriskany Sandstone of New York and Pennsylvania (Fig. 9–5). Because of its purity, Oriskany Sandstone is extensively used in making glass.

Although quartz is by far the dominant mineral in the Oriskany, the formation also contains a very small percentage of other silicate materials, which, like quartz, are resistant to weathering and erosion but are heavier than quartz. These so-called *heavy minerals* can be used to determine the kinds of source rocks and source areas from which clastic sediments were derived. After study of the heavy minerals from scattered exposures of Oriskany sandstones, it became apparent that there are two distinctly different assemblages of heavy minerals. In the area south of New York, the formation contains exceedingly small amounts of only the most stable heavy minerals: tourmaline, zircon, and rutile. The grains in these sandstones are exceptionally well rounded and worn, and less stable heavy minerals are lacking. This part of the Oriskany is considered to have derived from older clastic sedimentary units to the east and north. In contrast, the Oriskany in New York State contains a heavy mineral assemblage of relatively unstable pyroxene, amphibole, biotite, and garnet. Garnet is ordinarily derived from metamorphic source rocks, whereas the others are usual components of a variety of igneous and metamorphic rocks. The grains are unaltered. One may infer, therefore, that the sands composing the Oriskany of the New York area may have had an igneous or metamorphic source area and were not recycled from older sandstones.

The Oriskany Sandstone is the initial deposit of a transgressing sea, as is indicated by the way the sands overlap older strata. As this marine advance continued, limy sediments were deposited over the Oriskany, and corals began to build reef structures. In areas where water circulation was restricted, salt and gypsum were deposited. During the Middle and Late Devonian of the region east of the Mississippi Valley, carbonate sedimentation gave way to shales. The change to clastic deposition was a consequence of mountain building associated with the Acadian orogeny in the Appalachians. As will be described shortly, highlands formed during this orogeny were

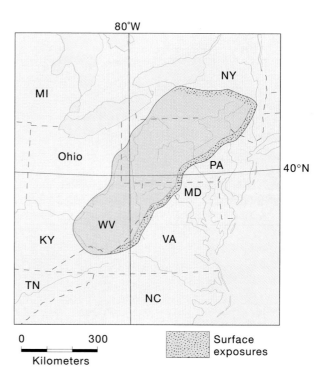

FIGURE 9–5 Areal extent of the Oriskany Sandstone. Surface exposures occur along the northern and eastern borders of the area shown in green.

rapidly eroded, and clastics were transported westward to form an extensive apron of sediments that are coarser and thicker near their eastern source area and give way to thin but widespread black shales in the east-central region of the United States.

In the far western part of the craton, Middle and Upper Devonian rocks are largely limestones, although there are shales as well. In a depressed area known as the Williston basin (from South Dakota and Montana northward into Canada), extensive reefs were developed. Arid conditions and restricted circulation in reef-enclosed basins resulted in the deposition of impressive thicknesses of gypsum and salt. The reefs provided permeable structures into which petroleum migrated, resulting in some of Canada's richest oil fields.

During the Late Devonian, mountains existed along the eastern margin of North America. A clastic wedge was built along the western side of the mountains. Conglomerates and sandstones were the characteristic deposits of the wedge, but farther to the west, suspended muds were carried into the shallow sea that covered the platform. These muds were deposited as a thin but extraordinarily widespread sheet of dark shale. There are many local names for the shale, but the formal name most commonly used is **Chattanooga Shale.** Because the shale is both wide-

A

B

FIGURE 9–6 (A) Specimen of crinoidal limestone (Burlington Formation) in which crinoid fragments (mostly stem plates) stand out in relief because of weathering. (B) Reconstruction of crinoids growing on the floor of the Kaskaskia (Mississippian) sea. *(Latter photo courtesy of the National Museum of Natural History, Smithsonian Institution.)*

spread and easy to recognize, it has served as an excellent marker in regional correlation. It also contains uranium and has been found to be 350 million years old.

In addition to its uranium content, the Chattanooga is rich in finely disseminated pyrite and organic matter. It is black or dark gray in color and contains few, if any, fossils of bottom-dwelling invertebrates. These characteristics indicate that the Chattanooga accumulated in stagnant, oxygen-deficient waters. Today, such environments are much more limited in size. It is difficult to conceive how a sea occupying an area as vast as the depositional area of the Chattanooga could sustain such oxygen-deficient conditions. One interesting suggestion is that phytoplankton proliferated catastrophically, consuming and thereby depleting the oxygen supply. Without the oxygen required for decay of organic matter, car-

bon would accumulate and contribute to the black color characteristic of the Chattanooga Shale.

During the passage from Devonian to Mississippian time, highland source areas that provided the Chattanooga Shale were reduced, and the quantity of muddy sediment decreased. Carbonates then became the most abundant and widespread kind of sediment in the epeiric seas of the platform. Cherty limestones (see Fig. 2–29), shelly limestones, limestones composed of the remains of billions of crinoids (Fig. 9–6) and other invertebrates, and limestones containing myriads of oöids (Fig. 9–7) formed extensive beds across the central and western parts of the craton (Fig. 9–8). You may recall from Chapter 2 that oöids are spherical grains formed by the precipitation of calcium carbonate around a nucleus such as a tiny particle of calcium carbonate. They form in clear, shallow sea water where currents exist to move parti-

A

B

cles about on the sea floor. Salinity somewhat greater than normal appears to enhance the formation of oöids.

The sea in which these crinoidal, oölitic, and other types of limestones were deposited was the most extensive that North America had experienced since Ordovician time (Fig. 9–9). The widespread blanket of carbonates deposited in this sea is often referred to as the great Mississippian lime bank. They record the last great Paleozoic flooding of the North American craton.

In Late Mississippian time, as the Kaskaskia seas began their final withdrawal, large quantities of detrital sediments and thin, scattered limestones were deposited. Today, these rocks are reservoirs for petroleum in Illinois and therefore have been studied extensively. Detailed maps of thicknesses of some of the sandstone units show that they are thickest along branching, sinuous trends that suggest old stream valleys developed on the former sea floor. Studies of

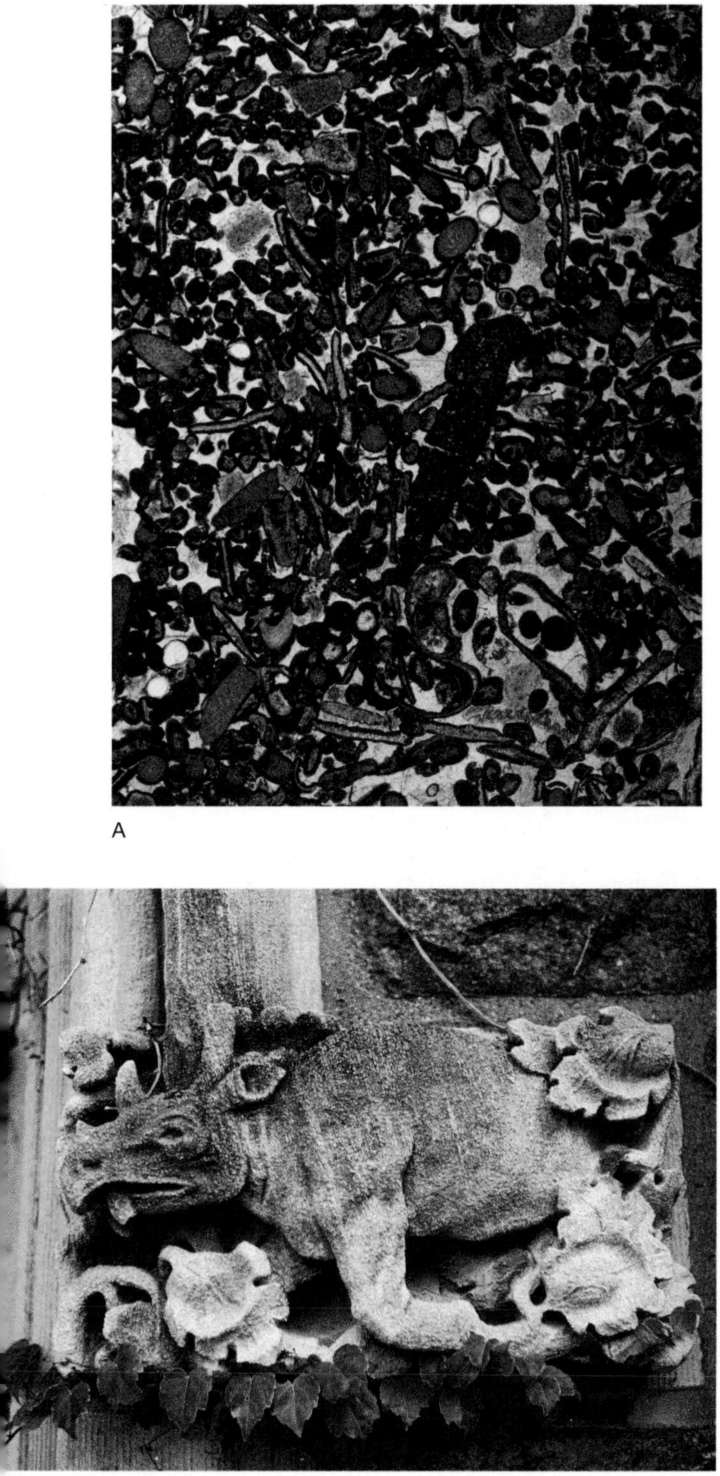

FIGURE 9–7 (A) Photomicrograph of Salem Limestone (Mississippian). The Salem consists primarily of fossil debris and oöids cemented with clear calcite. It is extensively quarried in southern Indiana and used as trim and building stone in public buildings throughout the United States. Width of section 2.5 mm. (B) *Uintatherium,* an Eocene mammal, carved in Salem Limestone. The sculpture is one of many carvings of fossils that decorate the geology building at Washington University.

FIGURE 9–8 Mississippian limestones (Chouteau and Burlington formations) overlying gray shale (Hannibal Shale) exposed during excavation for the Clarence Cannon Dam in northeastern Missouri. The Burlington Formation is noted for its rich content of crinoid remains. It also contains abundant chert nodules.

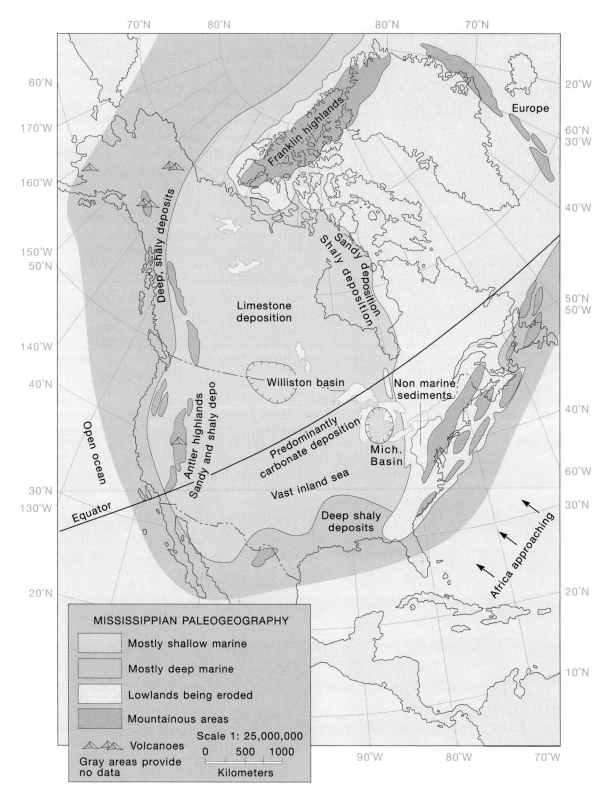

FIGURE 9–9   Paleogeography of North America during the Mississippian Period.

FIGURE 9–10 Cross-bedding in sandstones of the Upper Mississippian Tar Springs Formation, southern Illinois. *(Photograph courtesy of J. C. Brice.)*

grain size, cross-bedding (Fig. 9–10), and current-produced sedimentary structures strongly indicate that the detritus was derived from the northern Appalachians and was transported southwestward across the central interior.

The Absaroka Sequence

When the Kaskaskia seas had finally left the craton at the end of Mississippian time, the exposed terrain was subjected to erosion that resulted in one of the most widespread regional unconformities in the world. Not only was erosion extensive in areas, but it also eroded away entire systems of older rocks over arches and domes. The unconformity provides a criterion for separating those strata equivalent to the Carboniferous of Europe into the Mississippian and Pennsylvanian systems. It is appropriate to use these two names in North America, not only because of the unconformity but also because the rocks above the erosional hiatus differ markedly from those below. Indeed, the overlying Pennsylvanian strata are a consequence of different tectonic circumstances.

It was not until near the beginning of Middle Pennsylvanian time that the seas were able to encroach onto the long-exposed central region of the craton. The deposits of this seaway are those of the Absaroka sequence. In general, the Pennsylvanian rocks near the eastern highlands are thicker, and virtually all are continental sandstones, shales, and coal beds (Fig. 9–11). This eastern section of Pennsylva-

nian rocks gradually thins away from the Appalachian belt and changes from predominantly terrestrial to about half marine rocks and half nonmarine rocks. Still farther west, the Pennsylvanian outcrops are largely marine limestones, sandstones, and shales.

One of the most notable aspects of Pennsylvanian sedimentation in the middle and eastern states is the repetitive alternation of marine and nonmarine strata. A group of strata that records the variations in depositional environments caused by a single advance and retreat of the sea is called a **cyclothem**. A typical cyclothem in the Pennsylvanian of Illinois contains ten units (Fig. 9–12). Units one through five are continental deposits, the uppermost of which is a coal bed (Fig. 9–13). The strata deposited above the coal bed represent an advance of the sea over an old vegetated area.

Cyclothems are notable not only for their cyclicity but because they can be correlated over great distances. Those of the Appalachian region, for example, can be correlated to cyclothems in western Kansas. In Missouri and Kansas, at least 50 cyclothems are recognized within a section only 750 meters thick, and some of these extend across thousands of square kilometers.

It is apparent that cyclothems are the result of repetitive and widespread advance and retreat of seas. But what was the cause of such oscillations? One explanation involves periodic regional subsidence of the land to a level slightly below sea level, so that marginal seas could spill onto swampy lowlands. A relatively short time later, subsidence might cease and the shallow sea might be filled with sediment, forcing the sea out and reestablishing nonmarine conditions. Alternatively, temporary regional uplifts could have caused withdrawal of the sea. Finally, worldwide (eustatic) changes in sea level related to continental glaciation might have produced repetitive marine invasions and regressions. We have ample evidence that large regions of Gondwanaland were covered with glacial ice from Late Mississippian through Permian time. When the ice sheet grew in size, sea level would be lowered because of water removed from the ocean and precipitated as snow onto the continent. During warmer episodes, meltwater returning to the ocean would raise sea level. These sea-level oscillations could result in the shifting back and forth of shorelines that produced cyclothems. Additional support for this idea comes from recognition of cyclothems of the same age on continents other than North America; this indicates that the cause of the cyclicity was global.

Ordinarily, cratonic areas of continents are characterized by stability. The southwestern part of the North American craton provides an exception to this

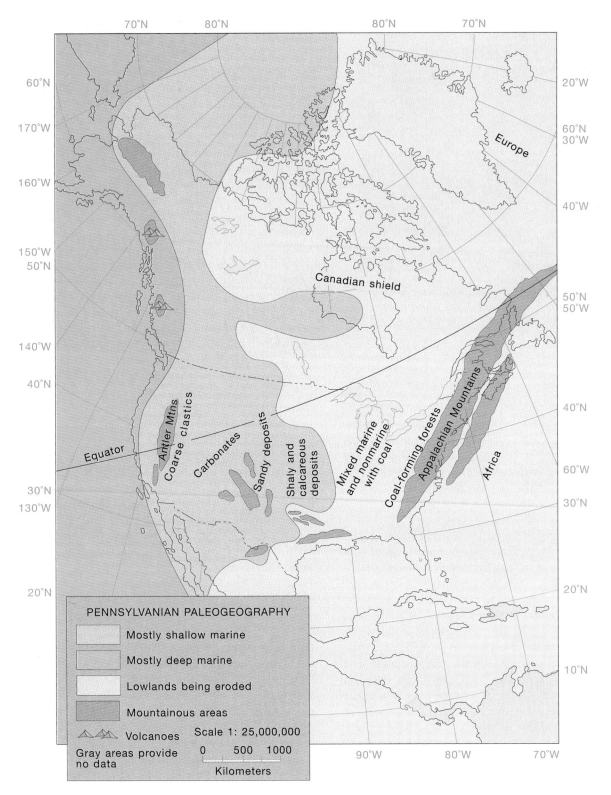

FIGURE 9–11 Paleogeography of North America during the Pennsylvanian Period.

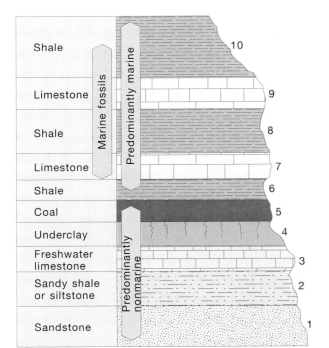

Shale		10
Limestone		9
Shale		8
Limestone		7
Shale		6
Coal		5
Underclay		4
Freshwater limestone		3
Sandy shale or siltstone		2
Sandstone		1

Marine fossils — Predominantly marine — Predominantly nonmarine

FIGURE 9–12 An ideal coal-bearing cyclothem showing the typical sequence of layers. Many cyclothems do not contain all ten units, as in this illustration of an idealized sequence. Some units may not have been deposited because changes from marine to nonmarine conditions may have been abrupt and/or units may have been removed by erosion following marine regressions. The number 8 bed usually represents maximum inundation and, correlated with the same bed elsewhere, provides an important correlative stratigraphic horizon.

FIGURE 9–13 Part of an Illinois cyclothem. The lowermost layer is the coal seam (cyclothem bed 5), followed upward by shale (bed 6) near the geologist's hand, limestone (bed 7), shale (bed 8), another limestone (bed 9), and the upper shale (bed 10). Part of another sequence caps the exposure. This cyclothem is part of the Carbondale Formation. *(Photograph courtesy of D. L. Reinertsen and the Illinois Geological Survey.)*

general rule, for during the Pennsylvanian this was a region of mountain building. The resulting highlands are generally called the Colorado mountains (also called the ancestral Rockies) and the Oklahoma mountains. These mountains and related uplifts appear to have resulted from movements of crustal blocks along large, nearly vertical reverse faults. The Colorado mountains included a range that extended north–south across central Colorado (the Front Range–Pedernal uplifts) and a segment curving from Colorado into eastern Utah (the Uncompahgre uplift). The central Colorado basin lay between these highland areas (Fig. 9–14). A separate range, the Zuni–Fort Defiance uplift, extended across northeastern Arizona on the southwestern perimeter of the Paradox basin. To the east of the Colorado moun-

tains lay the southeastward-trending Oklahoma mountains. Eroded stumps of this once rugged range form today's greatly reduced Arbuckle and Wichita mountains. Remains of the Amarillo mountains are now buried beneath younger rocks and are known principally from exploratory drilling for oil.

Judging from the tremendous volume of sediments eroded from the Colorado mountains, it is likely that they attained heights in excess of 1000 meters. They also were probably subjected to repeated episodes of uplift, continuing in some places into the Permian. Erosion of these highlands eventually exposed their Precambrian igneous and metamorphic cores. As erosion and weathering continued, great wedge-shaped deposits of red arkosic sediments were spread onto adjacent and intervening basins

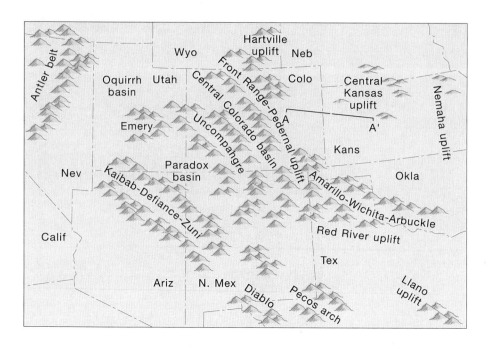

FIGURE 9–14 Location of the principal highland areas of the southwestern part of the craton during Pennsylvanian time.

(Fig. 9–15). A small part of this massive accumulation of clastic sediment is dramatically exposed in the Red Rocks Amphitheatre just a few miles west of Denver and the "flatirons" near Boulder, Colorado (Fig. 9–16).

Geologists have long puzzled over this episode of deformation of the craton, but plate tectonics has recently provided a hypothesis for this unusual event. It seems likely that the collision of Gondwanaland with North America along the site of the Ouachita oro-

genic belt generated stress in the bordering area of the craton to the north. Crustal adjustments to relieve these stresses resulted in the deformation that produced the highlands and associated basins.

The basins that lay adjacent to the ancient Colorado mountains are important in working out the geologic history of the southwestern craton because of the excellent stratigraphic record they contain. The Paradox basin is especially interesting. It was inundated by the Absaroka sea in Early Pennsylvanian

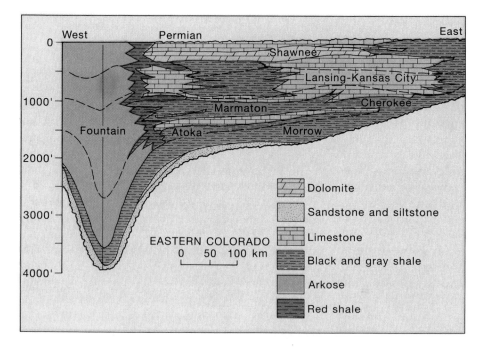

FIGURE 9–15
Pennsylvanian cross-section across eastern Colorado and western Kansas showing great accumulation of coarse arkosic sandstones east of the Colorado mountains. Line of section is along A—A' in Figure 9–14. *(From Stratigraphic Atlas of North and Central America. Shell Oil Company, Exploration Department.)*

FIGURE 9–16 Steeply dipping beds of the Fountain Formation form the "flatirons" just west of Boulder, Colorado. The red arkosic sandstones, conglomerates, and mudstones of the Fountain Formation were deposited during the late Pennsylvanian and early Permian. The source of Fountain formation sediment was the ancestral Rocky Mountains that lay to the west. The formation was tilted upward to form the flatirons, as well as dramatic ridges and hogbacks of the Colorado Front range, during the orogeny that produced the modern Rocky Mountains.

time. The initial Pennsylvanian deposits, largely shales, were deposited over a karst topography developed on Mississippian limestones. By Middle Pennsylvanian time, the western access of the Absaroka sea to the Paradox basin had become restricted, and thick beds of salt, gypsum, and anhydrite were deposited (Fig. 9–17). Fossiliferous and oölitic limestones developed around the periphery of the basin, and

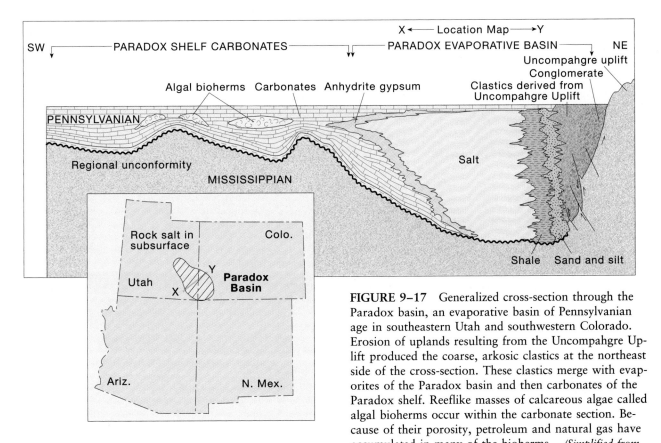

FIGURE 9–17 Generalized cross-section through the Paradox basin, an evaporative basin of Pennsylvanian age in southeastern Utah and southwestern Colorado. Erosion of uplands resulting from the Uncompahgre Uplift produced the coarse, arkosic clastics at the northeast side of the cross-section. These clastics merge with evaporites of the Paradox basin and then carbonates of the Paradox shelf. Reeflike masses of calcareous algae called algal bioherms occur within the carbonate section. Because of their porosity, petroleum and natural gas have accumulated in many of the bioherms. *(Simplified from Baars, D. L. et al. 1988. Basins of the Rocky Mountain region, Sedimentary Cover—North American Craton: U.S. DNAG (Decade of North American Geology) D-2:109–220.)*

patch reefs grew abundantly along the western side. The association of porous reefs and lagoonal deposits resulted in several sites suitable for the later entrapment of petroleum.

Near the end of the Pennsylvanian, the Paradox basin was filled to above sea level by arkosic sediments shed from the recently uplifted Uncompahgre highlands. Also at this time, the Absaroka sea, which had begun its transgression at the beginning of the Pennsylvanian, began a slow and irregular regression near the end of the same period. The withdrawal was still incomplete in Early Permian time, so marine sediments continued to be deposited in a narrow zone from Nebraska to western Texas. Fossiliferous limestones characterized these inland seas, although near highlands in Colorado, Texas, and Oklahoma, coarse clastics accumulated to sufficient thicknesses to bury surrounding uplands.

The deposits along what was the eastern edge of the seaway have been eroded away, but at several places along the western side one can observe the change from richly fossiliferous beds below to barren shales, red beds, and evaporites above. The thick and extensive salt beds of Kansas provide testimony to the gradual restriction and evaporation of Permian seas in the central United States. The last Permian marine conditions occurred in the western part of Texas and southeastern New Mexico, where a remarkable sequence of interrelated lagoon, reef, and open-basin sediments was deposited (Figs. 9–18 and 9–19). In this region, several irregularly subsiding basins developed between shallowly submerged platforms. Dark-colored limestones, shales, and sandstones were deposited in the deep basins, whereas massive reefs formed along the basin edges. Behind the reefs, in what must have been the shallow waters of extensive lagoons, the deposits are thin limestones, evaporites, and red beds. Late in the Permian, the connections of these basins to the south became so severely restricted that the waters gradually evaporated, leaving behind great thicknesses of gypsum and salt.

Much paleoenvironmental information has been obtained from study of the West Texas Permian

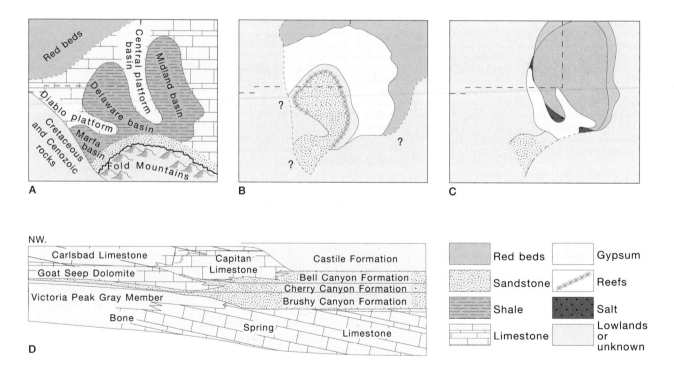

FIGURE 9–18 The Permian of North America can be divided into four stages. The oldest is the Wolfcampian, which is followed by Leonardian, Guadalupian, and Ochoan. A, B, and C show the paleogeography and nature of sedimentation in West Texas during the (A) Wolfcampian, (B) Guadalupian, and (C) Ochoan. (D) A simplified cross-section of Leonardian and Guadalupian sediments of the Guadalupe Mountains indicates the relationship of the reef to the other facies. *(After many sources, but primarily King, P. B. 1948. U.S. Geological Survey Professional Paper 215.)*

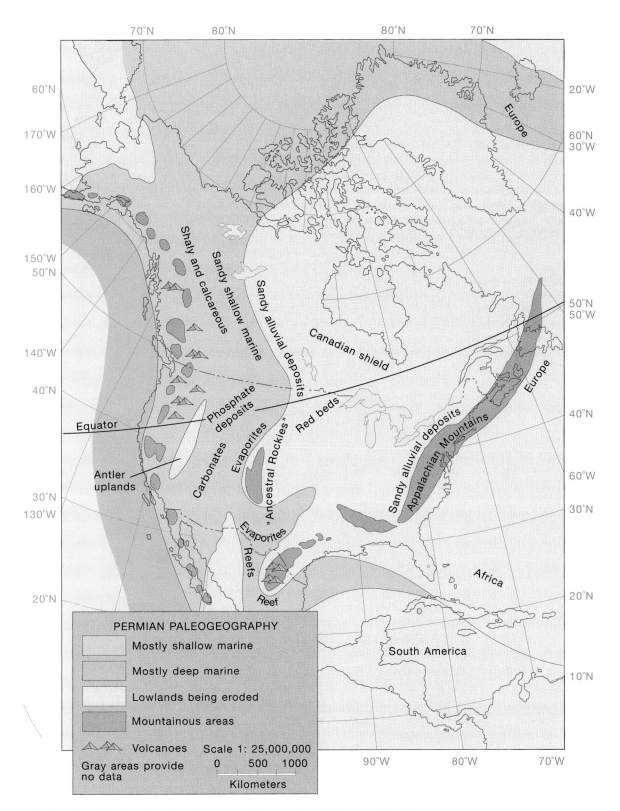

FIGURE 9–19 Generalized paleogeographic map for the Permian Period.

C O M M E N T A R Y

Reefs

Reefs are of great interest to geologists for many reasons. They comprise immense masses of carbonate material generated entirely by marine invertebrates and algae. Reefs are often well preserved and therefore provide intact representations of complex ancient and recent biological communities from which a wealth of ecological information can be obtained. Petroleum geologists are aware that ancient reefs serve as excellent traps for petroleum, and therefore they carefully map their locations and trends.

Reefs are shallow-water, wave-resistant buildups of calcium carbonate formed around a rigid framework of skeletal material. For most of the Phanerozoic, that framework has been built by corals, although the families of corals that construct reefs have varied through time as various groups evolved or became extinct. Before there were abundant reef-building corals, calcareous algae and bryozoa were the principal frame builders. The calcareous skeletons of other reef-dwelling organisms such as crinoids, brachiopods, mollusks, sponges, and algae add mass to the reefs and help to trap interstitial accumulations of sand and lime mud.

In most reefs, such as the spectacular Permian Capitan Reef of western Texas (see Figs. 9–18 and 9–20), one can recognize three major reef components or facies. The first is the **reef core** itself, which is formed of organisms that have built the reef upward from the shallow sea floor to sea level. Waves constantly crash along the front or windward side of the reef core. In response to this kind of vigorous wave action, organisms living along the reef front form low-growing or encrusting skeletal structures. Waves continuously break off pieces of this part of the reef, and the resulting debris forms a steeply dipping apron of debris that resembles a submarine talus deposit (see accompanying figure). This part of the reef is referred to as the **fore-reef facies**. A more sheltered area known as the **back-reef facies** is located landward of the reef core. Here carbonate sands, muds, and oöids accumulate among more erect growing frame builders. The back-reef is also known for the extraordinary diversity of its biological community.

rocks. From the lack of medium- and coarse-grained clastics, one may assume that surrounding regions were low lying. The gypsum and salt suggest a warm, dry climate in which basins were periodically replenished with sea water and experienced relatively rapid evaporation. Careful mapping of the rock units has helped to establish an estimated depth of about 500 meters for the deeper basins and only a few meters for the intrabasinal platforms. The basin deposits are dark in color and rich in organic carbon as a likely consequence of accumulation under stagnant, oxygen-poor conditions. Perhaps upwelling of these deeper waters may have provided a bonanza of nutrients on which phytoplankton and reef-forming algae thrived. Along with the algae, the reefs contain the skeletal remains of over 250 species of marine invertebrates. Today, because of their relatively greater resistance to erosion, these ancient reefs form the steep El Capitan promontory in the Guadalupe Mountains of West Texas and New Mexico (Fig. 9–20). Here one can examine the forereef composed of broken reef debris that formed a sort of talus de-

posit caused by the pounding of waves along the southeastern side.

■ *The Eastern Margin of North America*

During the latter half of the Paleozoic Era, the great Appalachian and Ouachita belts experienced their culminating and most intense episodes of orogenesis. The crumpling of these former depositional tracts was a consequence of the reassembly of the formerly separated continents into Pangea. By Devonian time, the Appalachians had taken the shock of a collision with a displaced continental fragment named the **Avalon terrane** (Fig. 9–21). Yet an additional collision occurred in late Carboniferous when northwest Africa moved against the southern part of the Appalachian belt. That encounter was the cause of the **Allegheny orogeny**. This great smashup not only raised old basins of sedimentation but also transmitted compressional forces into the interior, causing deep-

FIGURE 9–20 Example of a facies change in rocks at the southern end of the Guadalupe Mountains about 100 miles east of El Paso, Texas. The prominent cliff is called El Capitan. It is composed of a Permian limestone reef deposit, rich in sponges and associated marine invertebrate fossils. Behind the reef there once existed lagoons with abnormally high salinity, as suggested by evaporites, dolomites, and virtually unfossiliferous limestones. These lagoonal beds are exposed along the ridge on the right side of the photograph. Thus, a massive reef limestone facies lies adjacent to a lagoonal facies. To the south of the reef lay a marine basin in which normal marine sediments were deposited. *(Photograph courtesy of the National Park Service.)*

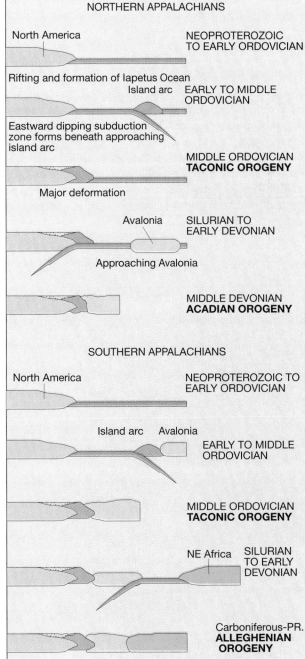

FIGURE 9–21 Simplified diagramatic plate tectonic sequence involved in the evolution of the northern and southern Appalachians. *(Adapted from Taylor, S. R. 1989. GSA Memoir 172.)*

seated deformations such as those that raised the Colorado and Oklahoma mountains.

The effects of the Devonian deformation, called the Acadian orogeny, are clearly seen in a belt extending from Newfoundland to West Virginia. Here one finds thick, folded sequences of turbidites interspersed with rhyolitic volcanic rocks and granitic intrusions. The intensity of the compression that affected these rocks is reflected in their metamorphic minerals, which indicate that mineralization occurred at temperatures exceeding 500°C and pressures equivalent to burial under 15,000 meters of rock. The overall result of the Acadian orogeny was to demolish forever the marine depositional basin and establish in its place mountainous areas in which erosion was the prevalent geologic process. Here and there in isolated basins among the mountains, Devonian nonmarine sediments were deposited. However, the greatest volume of erosional detritus was spread outward from the highlands as a great wedge of terrigenous sediment called the Catskill clastic wedge (Figs. 9–22 and 9–23).

The Catskill Clastic Wedge

For many reasons, the Devonian rocks of the **Catskill clastic wedge** (also called the Catskill delta) have excited the interest of geologists for well over a century. This apron of sediments provides an ideal area for the examination of facies from a varied assortment of both marine and nonmarine depositional environments (Fig. 9–24). Catskill facies characteristically exhibit rapid lateral changes from sandstones to shales. Such relationships form traps for petroleum

and are the reason for the thousands of wells drilled into Catskill strata. The flat slabs of red sandstones have also provided vast tonnages of flagstones for buildings in eastern cities. But for historical geologists, the most pervasive reason for studying Catskill rocks is the clues they provide to the locations and time of occurrence of phases of the Acadian orogeny. Most of these clues come from the study of subsidiary clastic wedges within the larger Catskill complex. The location of these wedges indicates that the Acadian orogeny was caused by the convergence of the Avalon terrane, and possible other terranes, against the irregular eastern margin of the North American craton. Additional pulses of orogenic activity fol-

continued on p. 330

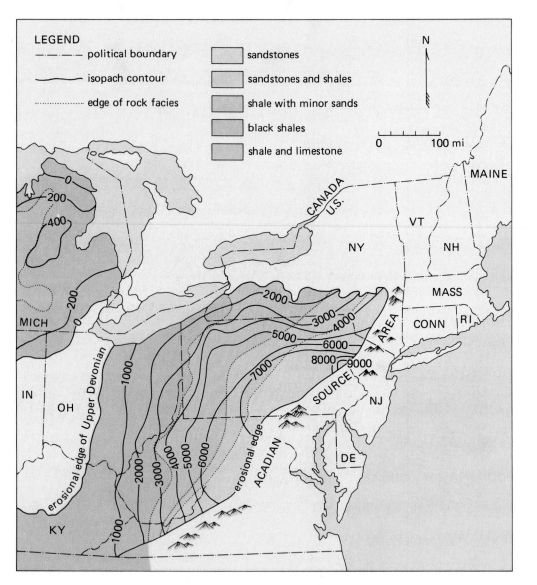

FIGURE 9–22 Isopach and lithofacies map of Upper Devonian sedimentary rocks in the northeastern United States. Thicknesses are given in feet. *(Modified from Sevon, W. D. 1985. Geol. Soc. Am. Special Paper 201:79–90; and Ayrton, W. G. 1963. Pa. Geol. Survey Rpt. 39(4):3–6.)*

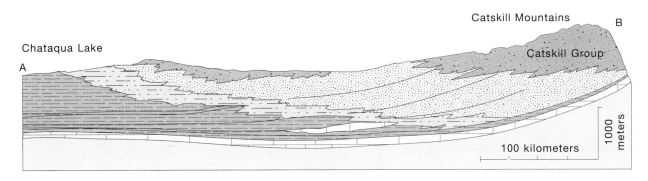

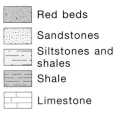
Red beds
Sandstones
Siltstones and
shales
Shale
Limestone

Line of section

FIGURE 9–23 East–west section across the Devonian Catskill clastic wedge, New York. Note that continental red beds interfinger with nearshore marine sandstones, and these in turn grade toward the west into offshore siltstones and shales. Continental deposits prograded upon the sea, pushing the shoreline progressively westward. *(Based on several classic studies by Chadwick, G. H., and Cooper, G. A. completed between 1924 and 1942.)*

FIGURE 9–24 Sedimentary rocks of the Catskill clastic wedge. (A) Catskill Formation sandstone filling an abandoned stream channel (scale divisions are 10 cm). (B) Stream deposited (fluvial) sandstones and mudstones exposed north of Trout Run, Pennsylvania (scale divisions are 50 cm). (C) Fluvial channel sandstones grading upward to floodplain mudstones, east of Haines Falls, New York. (D) Tidal flat deposits near Altoona, Pennsylvania. *(Courtesy of W. D. Sevon.)*

327

Acadia National Park

Acadia National Park in Maine is a delightful place in which to observe geologic features and processes. Thanks to the work of Pleistocene glaciers in stripping away soil and debris, one can walk directly on bedrock in many interior parts of the park. Along the shoreline, granite, diorite, and schist are revealed in wave-eroded cliffs and platforms. At Acadia one can witness the effects of igneous activity, explore glacial moraines, examine exotic rocks on cobble beaches, and observe the powerful action of waves in sculpting a rugged coast. All of this is easily reached by traveling southeast out of Bangor, through the town of Ellsworth to Bar Harbor.

Ice sheets covered Acadia and the rest of New England repeatedly during the Pleistocene Epoch. Each advance of the ice destroyed features formed by the previous advance, leaving features of the final glaciation clearly in evidence. The most apparent of these features are glacial scratches, glacial polish, and scars made in bedrock as rock debris carried along in the base of the glacier chipped out pieces of the underlying surface. There are also numerous erratics (see accompanying figure). Valleys in Acadia have the parabolic cross-section characteristic of valleys that have been glaciated. This distinctive shape results from the ability of glacial ice to erode not only at the base of the valley, as streams do, but well up onto the valley sides. As the glaciers moved down valleys, some areas were excavated deeper than others. Later, the deeper excavations filled with water to form so-called trough lakes (Sea Cove Pond, Eagle Lake, Jordan Pond). When the great Ice Age ended meltwater returned to the ocean caused sea level to rise. The ocean flooded into valleys and low areas, producing the remarkably irregular New England coast (appropriately, it is called a drowned coastline).

Granite cliffs of the rugged coast of Mount Desert Island, Acadia National Park. Waves breaking along the coast force water and air into fractures in the granite bedrock and thereby fragment huge masses of rock. The impact of waves may exert pressures of over 6000 pounds per square foot. *(Photograph by Jeff Gnass.)*

Along the drowned coastline of Acadia, one finds the most dramatic scenic attractions in the park (see accompanying figure). Here one achieves a vivid impression of the enormous energy released by waves as they batter sea cliffs and erupt in spray. Thunder Hole south of Bar Harbor provides a spectacular view of wave erosion in action. Thunder Hole is a deep gorge eroded along joints by pounding waves. When seas are heavy, water surges into the narrow chasm, compressing the trapped air; when the water and air are released, they resound like the clap of thunder. Surging waves along the coast lift and carry gravel, which is hurled against cliffs, undercutting them and causing collapse. These processes ultimately produce sea caves, sea arches, and wave-cut platforms. In quieter areas, the waves deposit their load, forming cobble and shingle beaches.

A particularly intriguing aspect of Acadia's geology relates to the derivation of its most ancient rocks. Acadia is a part of an exotic or suspect terrane, other components of which have been traced from Nova Scotia down into the southern Appalachians. As noted in Chapter 5, suspect terranes are pieces of continental crust or island arc moved by seafloor spreading from a distant place of origin to the margin of a continent and then sutured to the host continent at its subduction zone. They can be recognized as suspect terranes because they differ from their host continent in rock type, age, fossils, and paleomagnetic orientation.

The suspect terrane that in-

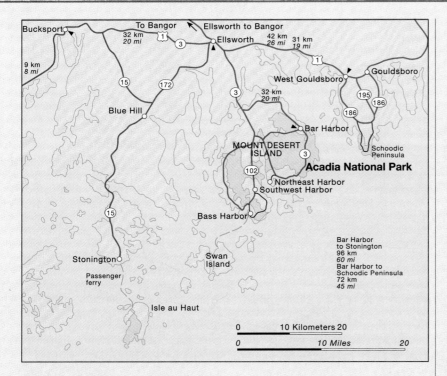

Location map for Acadia National Park *(Courtesy of U.S. National Park Service.)*

cludes Acadia National Park has been named Avalonia. It contains early Paleozoic rocks with trilobite and graptolite faunas distinctly different from those of adjacent areas of North America. In fact, these faunas more closely resemble those of Europe (Baltica).

When a suspect terrane arrives at a continent, it is said to have docked (rather like ships that arrive at a port one by one). Suspect terranes older than Avalonia had already docked against North America during the Ordovician Taconic orogeny. Avalonia came along somewhat later during the Devonian. Rocks of the terrane indicate that Avalonia is a vestige of an island arc that arrived at the subduction zone during the Acadian orogeny. After Avalonia had docked, at least three additional terranes were incorporated into the Appalachian orogen.

The Ordovician and Silurian rocks of Acadia are mostly meta-morphosed sediments and volcanic ash formed long before Avalonia had moved to North America. From oldest to youngest, these rocks consist of the Ellsworth Schist, Cranberry Series, and Bar Harbor Series. An erosional unconformity marks the upper boundary of these three units. Intruded into these units (and hence younger) are rocks of Devonian age. The earliest of these intrusives are diorites, which form dikes and massive sills. Some of the dioritic magma penetrated to the surface and erupted as fiery lava flows. After most of the dioritic magma had been emplaced and had crystallized, granitic magmas penetrated the diorites and metasediments. The initial fine-grained granites gave way to coarser grained and finally medium-grained granites, reflecting variations in the cooling history of the intruded bodies. All of this igneous activity was accompanied by compressional deformation of the eastern margin of North America. Ultimately, however, compression gave way to tensional forces as the break-up of Pangea began following the Permian Period. Erosion and intermittent uplifts were the principal post-Paleozoic events until the coming of the glaciers during the Pleistocene.

Glacial erratic boulder on Cadillac Mountain, Acadia National Park. An erratic is a rock fragment that has been carried by glacial ice and deposited at some distance from the outcrop from which it was derived. Usually, erratics rest on rock of different lithology and age. *(Photograph by Connie Toops.)*

329

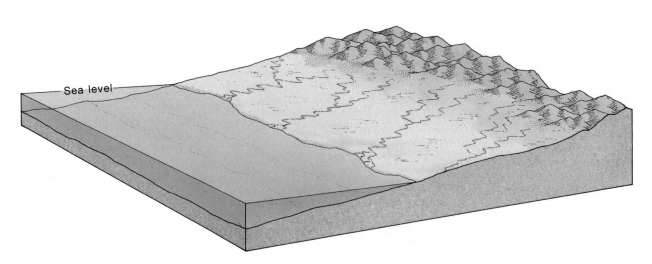

Sea level

FIGURE 9–25   A panoramic view of the Catskill clastic wedge as seen from a location above south-central Pennsylvania. Shoreline trends for about 300 km in a northeastern direction.   *(Modified from Woodrow, D. L. 1985. Geol. Soc. Am. Special Paper 201:51–63.)*

lowed as the westward-moving Avalon terrane encountered parts of the cratonic margin that projected eastward as promontories. The first orogenic pulse occurred during the early Middle Devonian, when the Avalon terrane collided with a promontory in the present vicinity of the St. Lawrence valley. Orogenic pulses followed in succession throughout the remainder of the Devonian as the Avalon terrane encountered the more southerly promontories of regions now occupied by the central and southern Appalachians.

In regard to Catskill facies, the older term, Catskill delta, is not appropriate, for these sediments were not deposited as large deltas of one or two major streams. Study of the directional properties of the fluvial sandstones indicates deposition from many small streams, all flowing westward out of the Acadian highlands (Fig. 9–25). In many areas, marine processes reworked the sediment supplied by streams so quickly that even small deltas did not develop.

Unlike the earlier Queenston rocks described in Chapter 8, the Catskill sediments were laid down at a time when land plants were abundant (Fig. 9–26) and provided a green mantle for the alluvial plains and hills. Fossil remains of these plants are most frequently encountered in the sediments deposited along former stream valleys. They indicate a tropical climate.

Nonmarine Catskill sedimentary rocks are dominated by sandstones and shales in which the contained iron is deeply oxidized. The oxide takes the form of the mineral hematite and causes the reddish-brown hues so characteristic of Catskill nonmarine

rocks. The majority of these so-called red beds represent deposits of braided or meandering streams.

The marine components of Catskill sediments consist mostly of fine sands and clays carried westward to an adjoining basin called the Catskill sea. Waves and currents shifted these clastics along the shoreline, forming lagoonal areas, bars, and beaches. More basinward deeper areas were characterized by deposition of dark clays interspersed with occasional sandy turbidite deposits.

Geologists in both North America and western Europe recognize the many similarities between the

FIGURE 9–26   Fossil plant root traces in mudstones of the Catskill clastic wedge (Oneonta Formation), near Unadilla, New York.   *(Courtesy of W. D. Sevon.)*

FIGURE 9–27 The Old Red Sandstone exposed in sea cliffs at St. Ann's Head, Pembrokshire, England. *(Courtesy of Institute of Geological Sciences, London.)*

nonmarine rocks of the Catskill clastic wedge and those that were spread across Europe south of the Caledonian orogeny. This vast region of largely alluvial deposition is named the **Old Red continent** after its most famous formation, the Old Red Sandstone (Fig. 9–27).

Post-Acadian Sedimentation

Mississippian strata crop out in the Appalachian region from Pennsylvania to Alabama. Nonmarine shales and sandstones predominate, indicating that erosion of the mountainous tracts developed during the Acadian orogeny was still in progress. Some of the finer clastics were spread westward onto the craton to form the vast deposits of Lower Mississippian black shales mentioned earlier. Particularly coarse sediments, including conglomerates, were deposited as part of the **Pocono Group.** Pocono sandstones (Fig. 9–28) form some of the resistant ridges of the Appalachian Mountains in Pennsylvania. Westward, the Pocono section thins and changes imperceptibly

FIGURE 9–28 Exposure of the Pocono Group (Mississippian) in a highway cut in western Pennsylvania. The section includes thick, cross-bedded fluvial sandstones as well as a few dark, coaly shales. The cut is across the axis of a syncline that was formed during the Allegheny orogeny. *(Courtesy of J. D. Glaser.)*

into marine siltstones and shales. Evidently, the depositional framework consisted of another great clastic wedge complex of alluvial plains that sloped westward and merged into deltas, which were being built outward into the epicontinental seas. The plains and deltas, standing only slightly above sea level, were backed by the rising mountains of the Appalachian fold belt. A large part of the coarser clastics had settled before reaching the southern portions of the epicontinental seas, where marine limestones are the most prevalent rocks.

Pennsylvanian rocks of the Appalachians are characterized by cross-bedded sandstones and gray shales that were deposited by rivers or within lakes and swamps. Coal seams are, of course, prevalent in the Pennsylvanian System of the eastern United States and reflect the luxuriant growths of mangrove-like forests that clothed the lands. This was an ideal environment for coal formation. Vegetation that accumulated in the poorly drained swampy areas was frequently inundated and killed off. Immersed in water or covered with muck, the dead plant material was not destroyed by oxidation to carbon dioxide and water. It was, however, attacked by anaerobic bacteria. These organisms broke down the plant tissues, extracted the oxygen, and released hydrogen. What remained was a fibrous sludge with a high content of carbon. Later, such peatlike layers were covered with additional sediments—usually siltstones and shales—and then compressed and slowly converted to coal.

The Allegheny Orogeny

The culminating deformational events affecting the eastern margin of North America occurred as Gondwanaland (Africa and South America) converged on North America and Europe (Fig. 9–29). In North America, the orogenic activity resulting from these collisions is termed the **Allegheny orogeny.** This great episode of mountain building began during the Mississippian Period and continued throughout the remainder of the Paleozoic. It affected a belt that extended across 1600 km, from southern New York to central Alabama. Like the earlier Acadian orogeny, the Allegheny deformations were the result of closure of the Iapetus Ocean and the ultimate consolidation of continents to form Pangea.

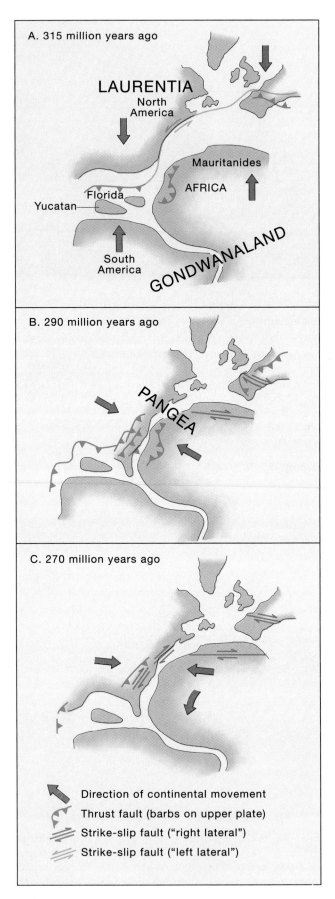

FIGURE 9–29 Plate tectonic model for late Paleozoic continental collisions proposed by P. E. Sacks and D. T. Secor, Jr. (A) Early Pennsylvanian, (B) late Pennsylvanian, (C) Permian. *(Adapted from Sacks, P. E., and Secor, D. T., Jr. 1990.* Science *250:1702–1705.)*

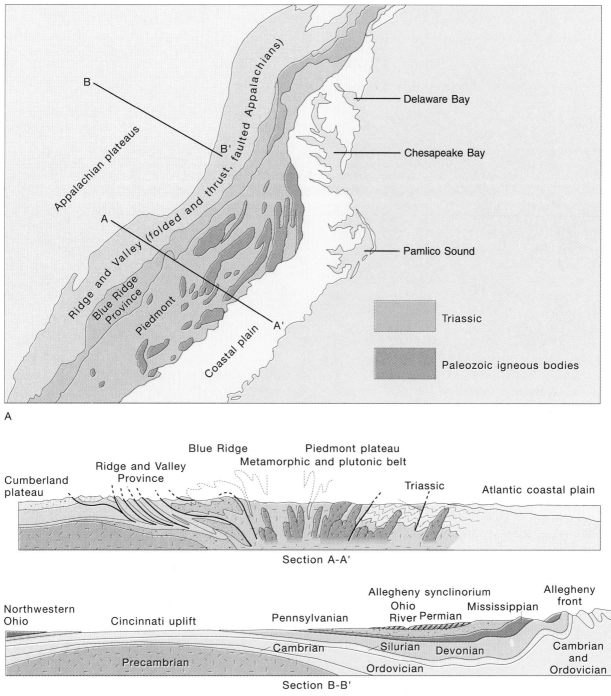

FIGURE 9-30 (A) Map of physiographic provinces of the Appalachian region. (B) Two cross-sections showing the structural relationships of the Appalachian Mountains. *(Map A from* Tectonic Map of the United States, *Geological Society of America, 1944; B from King, P. B. 1950. Bull. Am. Assoc. Petr. Geol.)*

The effects of the Allegheny orogeny were profound and included Permian compression of early continental shelf and rise sediments as well as strata deposited along the bordering tract of the craton. The great folds now visible in the Ridge and Valley Province were developed during this orogeny (Fig. 9–30). Less visible at the surface but no less impressive are enormous thrust faults formed along the east side of

FIGURE 9-31 The Waynesburg Sandstone of the Dunkard Series exposed along the Ohio River in Meigs County, Ohio. *(Photograph courtesy of the Ohio Division of Geological Survey.)*

the southern Appalachians. The folds are asymmetrically overturned toward the northwest, and the fault surfaces are inclined southeastward, suggesting that the entire region was moved forcibly against the central craton. Not unexpectedly, erosion of the rising mountains produced another great clastic blanket of nonmarine sediments. These mostly continental reddish sandstones and gritty shales compose the **Dunkard Group** (Fig. 9-31) and **Monongahela Group** of Permian–Pennsylvanian age.

The Ouachita Deformation

The deformation of the Ouachita orogenic belt was caused by the collision of the northern margin of Gondwanaland (northern South America and perhaps part of northwestern Africa) with the southeastern margin of the North American craton (see Fig. 9–29). That deformation began rather late in the Paleozoic, for the rock record indicates that from Early Devonian to Late Mississippian time the region experienced slow deposition interrupted by only minor disturbances. Carbonates predominated in the more northerly shelf zone, whereas cherty rocks known as **novaculites** accumulated in the deeper marine areas (Fig. 9–32).

A

B

FIGURE 9–32 (A) Tightly folded beds of Arkansas novaculite. (B) Close-up view of complex small folds in the novaculite beds. *(Photograph courtesy of C. G. Stone and the Arkansas Geological Commission.)*

Novaculites are hard, even-textured, siliceous rocks composed mostly of microcrystalline quartz. They are formed from bedded cherts that have been subjected to heat and pressure. Arkansas novaculite is used as a whetstone in sharpening steel tools. The

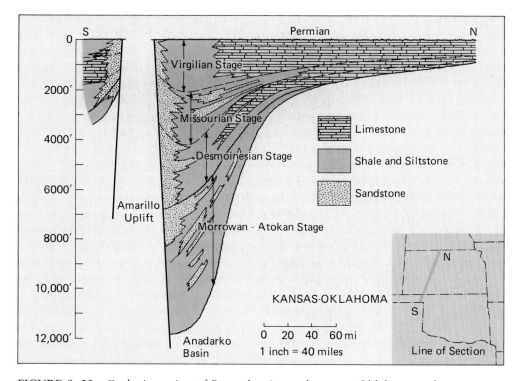

FIGURE 9–33   Geologic section of Pennsylvanian rocks across Oklahoma and Texas, showing the thick wedge of sediment shed from mountain ranges to the south.   *(From Stratigraphic Atlas of North and Central America. Shell Oil Company, Exploration Department.)*

pace of sedimentation increased dramatically toward the end of the Mississippian, when a thickness of over 8000 meters of graywackes and shales was spread into the depositional basin. The flood of clastic deposition continued into the Pennsylvanian, forming a great wedge of sediment that thickened and became coarser toward the south, where mountain ranges were being formed (Fig. 9–33). Radiometric dating of now deeply buried basement rocks from the Gulf Coastal states indicates that these rocks were metamorphosed during the late Paleozoic and were the likely source for the Pennsylvanian clastics spread into the belt. These coarse sediments document several pulses of orogenesis that ultimately produced mountains along the entire southern border of North America. In the faulting that accompanied the intense folding, strata of the continental rise were thrust northward onto the rocks of the shelf. By Permian time, stability returned, and strata of this final Paleozoic period are relatively undisturbed.

Since the time of active mountain building, erosion has leveled most of the highlands, leaving only the Ouachita Mountains of Arkansas (Fig. 9–34) and Oklahoma and the Marathon Mountains of southwest Texas as remnants of once lofty ranges. Although the Ouachita belt has been traced over a distance of nearly 2000 km, only about 400 km are

FIGURE 9–34   The Ouachita Mountains in Arkansas. *(Photo courtesy of Graham Thompson and Jon Turk.)*

exposed. Thus, the actual configuration of the belt has been determined by the examination of millions of well samples and other data obtained during drilling activities associated with petroleum exploration.

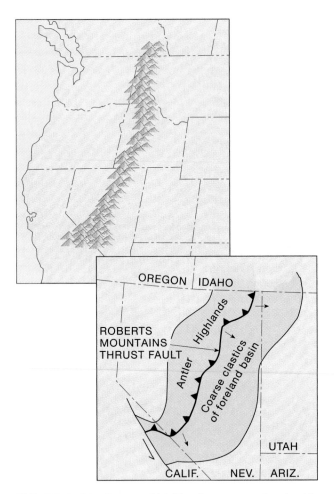

FIGURE 9–35   Extent of highland areas associated with the Antler orogeny, and location of the Roberts Mountains Thrust Fault.

■ *Sedimentation and Orogeny in the West*

The late Paleozoic history of the western or Cordilleran belt was almost as lively as that of the Appalachian. You may recall from Chapter 8 that during the early Paleozoic, a passive margin existed along the western side of the North American craton. The familiar belts of continental shelf, slope, and rise sediments were deposited along that relatively quiet margin. More dynamic conditions began in the Devonian, when subduction of oceanic lithosphere beneath the western margin of the continent was initiated. This was the beginning of a disturbance known as the **Antler orogeny.** During the Antler orogeny, a volcanic island arc converged on the continent, crushing the sediment in the intervening basin. The convergence was accompanied by thrust faulting on a massive scale. The effects are clearly seen in the Roberts Mountains Thrust Fault of Nevada and at other locations northward into Idaho (Fig. 9–35). One can recognize continental rise and slope deposits that have been thrust as much as 80 km over shallow-water sediments of the former continental shelf.

The Antler orogeny, which had begun in the Late Devonian, continued actively into the Mississippian and Pennsylvanian. Highlands created by the orogeny provided detrital sediment that was transported into adjacent basins (Fig. 9–36). Especially thick sequences of Pennsylvanian and Permian shelf sediments accumulated in the area now occupied by the Wasatch and Oquirrh mountains (Fig. 9–37). In the latter area, the Oquirrh Group is over 9000 meters thick.

Mississippian and Pennsylvanian deposits west of the Antler highlands include a great volume of coarse detritus and volcanic rocks. These materials were swept eastward from a volcanic arc source area that lay along the western side of North America (Fig. 9–38). Over 2000 meters of sandstones, shales, lavas,

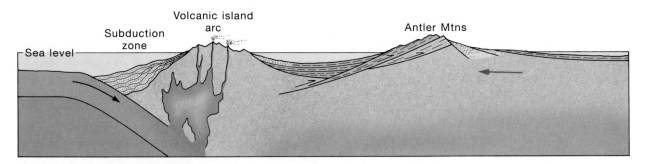

FIGURE 9–36   An interpretation of conditions in the Cordilleran orogenic belt in Early Mississippian time, shortly after the Antler orogeny.   *(Based on diagrams by Poole, G. F. 1974.* Soc. Economic Paleontologists and Mineralogists Special Publication *22:58–82.)*

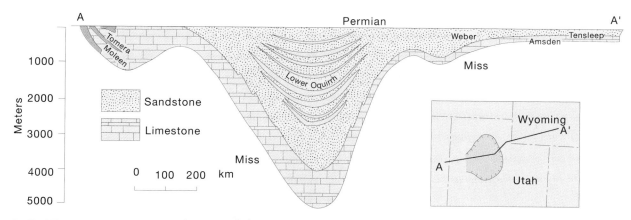

FIGURE 9–37 Section across the Oquirrh basin. *(From* Stratigraphic Atlas of North and Central America. *Shell Oil Company, Exploration Department.)*

and ash beds are found in the Klamath Mountains of northern California. Volcanic rocks in western Idaho and British Columbia attest to a continuation of vigorous volcanism from the Mississippian through the Permian. Crustal deformation along the west side of the Cordilleran belt is indicated by an extensive area of angular unconformities between Permian and Triassic sequences. These Permian–Triassic disturbances of the Cordilleran region have been named the **Cassiar orogeny** in British Columbia and the **Sonoma orogeny** in the southwestern United States. Like the earlier Antler orogeny, the Sonoma event was probably caused by the collision of an eastward-moving island arc against the North American continental margin in west-central Nevada. Oceanic rocks and remnants of the arc were thrust into the edge of the continent and became part of North America.

Permian conditions in the shelf area east of the Antler uplift were quieter than those to the west. The region was occupied by a shallow sea during much of the Permian Period. Within that sea, several large platform deposits accumulated. One of these was the Kaibab Limestone, an imposing formation that forms the vertical cliffs along the rim of the Grand Canyon. Beneath and eastward of the Kaibab Limestone, there are red beds that reflect deposition on coastal mudflats and floodplains. Sand dunes were prevalent nearby, as indicated by the massive, extensively cross-bedded Coconino Sandstone.

While the Kaibab was being deposited in the southwest, a relatively deep marine basin was developing to the north in the general area now occupied by Wyoming, Montana, and Idaho. In this basin, sediments of the Phosphoria Formation were deposited. Although the Phosphoria includes beds of cherts, sandstones, and mudstones, it takes its name from its many layers of dark phosphatic shales, phosphatic limestones, and phosphorites. **Phosphorite,** a dark

FIGURE 9–38 The Mississippian Bragdon Formation of the eastern Klamath Mountains of California. The Bragdon is composed of siltstones, sandstones, and felsic ash shed from a volcanic island arc that formed next to North America during late Paleozoic time. *(Courtesy of M. Miller.)*

gray concretionary variety of calcium phosphate, is mined from the formation and used in the manufacture of fertilizers and other chemical products. The unusual concentration of phosphates may have been produced by upwelling of phosphorus-rich sea water from deep parts of the basin. The metabolic activities of microorganisms may have assisted in the precipitation of phosphate salts.

■ *The Northern Seaway*

The late Paleozoic record of the Northern or Franklinian seaway in northern Canada is initiated by Devonian shales and limestones. Coral reefs are abundantly preserved in the Devonian System. The existence of these reefs is not surprising, for the region lay at about 15° north latitude during the Devonian and temperatures were much warmer than today. Following deposition of the Devonian marine rocks, an extensive mountain system was developed across northern Canada during an event called the **Ellesmere orogeny.** Erosion of the mountains provided immense volumes of detrital sediments to adjacent lowlands. One such depositional site, named the Sverdrup basin, formed late in the Mississippian and served as a trap for a nearly continuous sequence of strata ranging in age from Late Mississippian to Early Cenozoic. The Late Mississippian beds include plant-bearing deltaic shales and sandstones, red beds, and coal seams. By Pennsylvanian time, the mountains had been reduced, and hence the supply of detritus was diminished. The basin, however, continued to subside and became a relatively deep-marine environment of deposition. In this marine basin, thick sequences of Pennsylvanian and Permian sandstones, limestones, and evaporites were deposited.

■ *Europe During the Late Paleozoic*

During the late Paleozoic, Europe was bordered by the Uralian seaway on the east and the Hercynian on the south. Along the northern margin of Europe, the Caledonian orogeny had created the vast land area of the Old Red continent. Uplands provided a source for clastic sediments, which were swept out into numerous basins to accumulate to thicknesses exceeding 10,000 meters. Judging from the many interlayers of ash and lava, volcanic activity was frequent on the Old Red continent. Fossil fishes and sedimentary evidence indicate that Europe's climate was tropical and possibly semiarid at the time.

South of the Caledonian mountains, the continental deposits gradually thin and grade into marine shales and limestones (see Fig. 8–34C). For a time, quiet prevailed in western Europe. However, in the Late Devonian and Early Mississippian, the depositional sites were intensely folded, metamorphosed, and intruded by granites as continents began or continued the collisions that were to produce Pangea. The late Paleozoic orogenic event has been named the **Hercynian orogeny,** and its result was a great range of mountains across southern Europe. The Hercynian orogeny occurred at approximately the same time as the Allegheny orogeny in North America. For the most part, the eroded stumps of the Hercynian mountains are now buried, but here and there patches of the younger covering rocks have been eroded away, revealing the intensely folded, faulted, and intruded older rocks that lie beneath.

From the Hercynian uplands, gravel, sand, and mud were carried down into basins and coastal environments. The clastic deposits were quickly clothed in dense tropical forests. Burial and slow alteration of vegetative debris from these forests provided the material for coal formation in the great European coal basins. Plant fossils found in these coal seams are of the tropical kind and differ from the more temperate Gondwana flora of the Southern Hemisphere.

Although the greatest amount of Hercynian deformation occurred toward the end of the Early Carboniferous (Mississippian) in Europe, spasms of unrest continued in both the Late Carboniferous and the Permian. These later episodes of folding correlate with similar deformations in the southern Appalachian and Ouachita orogenic belts of North America.

No less important than the Hercynian basins in Europe was the lengthy Uralian tract, which extended along a belt now occupied by the Ural Mountains. The Uralian basins were already in existence at the beginning of the Paleozoic Era. Interestingly, there are discrepancies between the polar wandering curves for Paleozoic rocks from the cratons on either side of the Urals. The discrepancies indicate that the Russian and Siberian platforms were widely separated during early Paleozoic time and that they began to converge in the middle part of the era and ultimately collided by the end of the Paleozoic (see Fig. 9–1). Indeed, the western Urals consist largely of oceanic material scraped off against the edges of the converging plates. The collision resulted in the formation of a great mountain system along the entire Uralian orogenic belt and unified ancestral Siberia with eastern Europe.

In central Europe and parts of Russia, there were temporary Late Permian marine incursions that precipitated evaporites. The famous German potassium salts are a product of one of these inland seas, which has been named the **Zechstein sea** (Fig. 9–39). Apparently, aridity was as characteristic of western Eu-

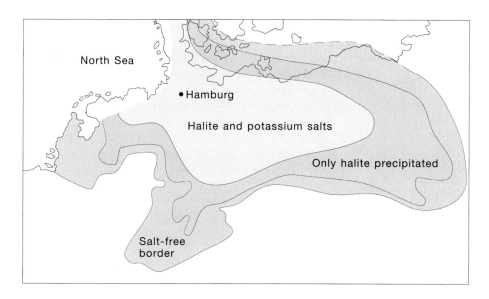

North Sea

• Hamburg

Halite and potassium salts

Only halite precipitated

Salt-free
border

FIGURE 9–39 Outcrop area of sedimentary rocks deposited in the Zechstein sea. *(After Brinkman, R. 1969. Geological Evolution of Europe. Translation from German by Sanders, J. E. and Enke, F. Stuttgart: Verlag; and New York: Hafner Publishing Co.)*

rope during the Permian as it was of Texas and New Mexico.

■ *Gondwanaland During the Late Paleozoic*

During the late Paleozoic, the great land mass of Gondwanaland remained fairly intact. It moved across the South Pole and more fully entered the side of the Earth on which Laurussia was located. In its northward migration, Gondwanaland caused the closure of the ocean that separated it from Laurussia, causing the Hercynian orogeny of Europe and the Allegheny orogeny of North America. Orogenic activity associated with subduction zones was also evident in the late Paleozoic history of Gondwanaland, particularly along the Andean belt of South America and the Tasman belt of eastern Australia.

The most dramatic paleoclimatologic event of Gondwanaland's late Paleozoic history was the growth of vast continental glaciers (see Fig. 9–2). Extensive layers of tillite (Fig. 9–40) and the scour marks of glaciers (see Fig. 9–3) have been found at hundreds of locations in South America, South Africa, Antarctica, and India. There are indications of at least four and possibly more glacial advances, suggesting a pattern of cyclic glaciation not unlike that experienced by North America and Europe less than 100,000 years ago. The orientation of striations chiseled into bedrock by the moving ice suggests that the glaciers moved northward from centers of accumulation in southwestern Africa and eastern Antarctica. During the warmer interglacial stages and in outlying less frigid areas, *Glossopteris* and other plants toler-

ant of the cool, damp climates grew in profusion and provided the materials for thick seams of coal.

In time the ice receded, and Gondwanaland's vast cratonic areas became sites for deposition of Permian nonmarine red beds and shales. As we shall see in the next chapter, some of these sediments contain the fossil remains of the ancestors of the Earth's first mammals.

FIGURE 9–40 Upper Carboniferous tillite at base of slope (beneath hammer) composed of Precambrian basalt. Tillite is consolidated and unsorted glacial debris. Kimberley, South Africa. *(Courtesy of Warren Hamilton, U.S. Geological Survey.)*

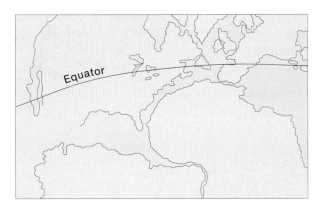

FIGURE 9–41 Approximate relationship of continents to the equator during the Carboniferous (Mississippian–Pennsylvanian).

■ *Climates of the Late Paleozoic*

The main climatic zones of the late Paleozoic tended to parallel latitudinal lines, just as zones do today. Of course, the continents were located differently, so the South Pole was in South Africa and the North Pole was over open ocean. The late Paleozoic equator trended northeastward across Canada and southward across Europe (Fig. 9–41). Coal beds, evaporites, coral reefs, and dune-deposited red beds developed within 40° of the paleoequator. Today, we cannot help but be startled to find the fossils of tropical amphibians and plants at locations within the Arctic circle. As already noted, cooler climates prevailed in Gondwanaland because of its late Paleozoic proximity to the South Pole, and by Permian time the southern continents were in the grip of a major ice

age. To the north, however, warmer conditions prevailed. Fossil plants in Pennsylvanian coal seams and shales reflect tropical conditions, as do coral reefs developed in areas of marine deposition. Evaporites associated with red beds suggest arid conditions in some areas. It is likely that the northern regions of Pangea lay within the zone of northeasterly trade winds. Much as today, as these winds traveled westward and encountered mountains, they would have been forced to ascend, form clouds, and produce precipitation on the windward side of the ranges. On reaching the far or lee side of the mountains, much of the moisture in the air would have been lost, and arid conditions would be the expected result, at least until moisture was again replenished as winds continued westward over epeiric seas.

■ *Mineral Products of the Late Paleozoic*

Although a great variety of mineral deposits were formed during the late Paleozoic, the fossil fuels (coal, oil, and gas) are particularly significant. Coal occurs in all post-Devonian systems. In Northern Hemisphere continents, it is particularly characteristic of the Late Carboniferous (Pennsylvanian). Thick deposits of Pennsylvanian coal occur in the Appalachians, the Illinois basin (Fig. 9–42), and the industrial heartland of Europe. Some sequences of Pennsylvanian strata may include many coal beds, as in West Virginia, where 117 different layers have been named. In the anthracite district of western Pennsylvania, orogenic compression has partially metamorphosed coal into an exceptionally high-carbon, low-

A   B

FIGURE 9–42 (A) A remote-controlled continuous mining machine with a flooded bed scrubber for dust control is removing coal in a West Virginia underground mine. (B) Large stripping shovel in the pit of an Illinois surface mine. *(Courtesy of Peabody Coal Co.)*

FIGURE 9–43 Drilling for oil in upper Paleozoic rocks of eastern Kansas.

which is exposed in Montana, Idaho, Utah, and Wyoming. The Phosphoria shales are extensively quarried for phosphate, which is sold as an important plant food.

Mountain building, with its attendant intrusive and volcanic igneous activity, is nearly always accompanied by the emplacement of metallic ores. The Hercynian and Allegheny orogenies of the late Paleozoic generated ores of tin, copper, silver, gold, zinc, lead, and platinum. Deposits of all the precious metals, as well as copper, zinc, and lead, are found in the Urals of Russia and in China, Japan, Burma, and Malaya. Tin, tungsten, bismuth, and gold are mined in Australia and New Zealand. It is self-evident that late Paleozoic rock sequences, with their stores of both metallic and nonmetallic economic minerals and their content of fossil fuels, are vitally important to the welfare of modern civilizations.

volatile variety that is prized for its industrial uses. Permian coal seams are found in China, Russia, India, South Africa, Antarctica, and Australia. The coal industry in North America, which had deteriorated for the last few decades, is currently experiencing an upsurge as a consequence of decreasing supplies of petroleum and natural gas.

Commercial quantities of oil and gas are frequently found in Upper Paleozoic strata. Devonian reefs within the Williston basin of Alberta and Montana have been exceptionally productive reservoir rocks for petroleum. Devonian petroleum has also been produced in the Appalachians. Indeed, in 1859, the first U.S. oil well was drilled into a Devonian sandstone. (Oil was struck at a depth of only 20 meters.) Carboniferous formations of the Rocky Mountains, midcontinent (Fig. 9–43), and Appalachians also contain oil reservoirs. However, wells drilled into reefs and sandstones of the Permian basin of West Texas have yielded the greatest amount of oil from Upper Paleozoic formations of the western United States. Oil trapped in Upper Paleozoic strata beneath the North Sea is now being actively utilized to supply the needs of Europe.

Arid, warm climatic conditions, particularly in northern continents during the late Paleozoic, provided a suitable environment for deposition of sodium and potassium salts. In Late Permian time, enormous amounts of phosphates (Fig. 9–44) were also deposited as part of the Phosphoria Formation,

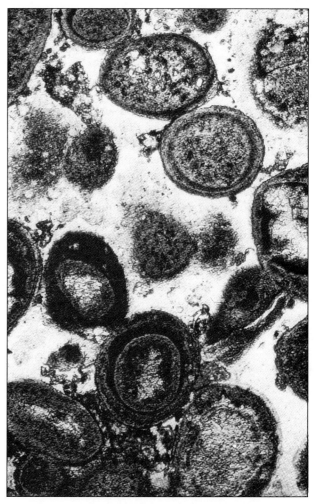

FIGURE 9–44 Pelletal, oölitic phosphate rock. The phosphatic pellets and oöids are cemented by microcrystalline quartz. This sample was obtained from an outcrop of phosphate rock in Madison County, Montana. (Average diameter of oöids, 0.8 mm.)

Summary

The closure of the Iapetus Ocean that had begun in the early Paleozoic continued throughout the remainder of the era. As the ocean narrowed, it carried island arcs and microcontinents to the margins of Europe and eastern North America, and these land areas were welded to their larger host continents. The collisions produced the early Paleozoic Taconic and Acadian orogenies. As oceanic closure continued into the late Paleozoic, North America and Baltica collided, resulting in the Acadian orogeny. The subsequent Allegheny orogeny occurred near the end of the era as northwest Africa moved against North America. To the south, the plate bearing South America converged on the underside of our continent and resulted in the Ouachita orogenic activity.

Whereas the process of bringing continents into contact resulted in mountain building along their converging margins, the interiors were affected primarily by broad epeirogenic movements and associated advances and withdrawals of epicontinental seas. In North America, these shifts in shorelines are reflected in the marine and nonmarine sediments that compose the Kaskaskia and Absaroka cratonic sequences. Kaskaskia rocks are predominantly fossiliferous limestones, with sandstones and shales increasing in volume toward the eastern and southern depositional areas. As the Kaskaskia seas withdrew, deltaic and fluvial deposits spread across the old sea floor. The regression resulted in a great regional unconformity that marks the boundary between Mississippian and Pennsylvanian systems in North America. The most distinctive feature of the Pennsylvanian sediments is their cyclic nature. In the midcontinent, they consist of alternate marine and nonmarine groupings. To the east, the Pennsylvanian sediments were largely terrestrial, whereas marine deposition prevailed in the western part of the craton. Near the end of the late Paleozoic, the epicontinental seas gradually withdrew, and evaporite and red bed sequences developed in the Permian basins of New Mexico and West Texas.

Mountain building was not confined to eastern North America during the late Paleozoic. In the west, crustal disturbances raised a range of mountains that extended from Arizona to Idaho. Thick deposits of sand and gravel attest to the former existence of these ranges. There was further activity during the Pennsylvanian, when uplands still visible were formed in Texas, Arkansas, and Oklahoma. A chain of mountains (the Colorado mountains) also developed across Colorado. These Colorado mountains were the source of sandstones that today form the famous red rocks of the Garden of the Gods along the Front Range of the Rockies.

The Old Red continent was the dominant feature of Europe at the beginning of the late Paleozoic. For a time, relative stability prevailed in Europe, and then much of the continent was thrown into an episode of folding, volcanism, and intrusion as it collided with Gondwanaland. These disturbances were part of the Hercynian orogeny. The collision of Gondwanaland with Laurussia was not only the cause of the Hercynian orogeny in central Europe but also the cause of the Allegheny orogeny along the eastern and southern margin of North America.

Gondwanaland was ringed by orogenic belts during the late Paleozoic, and the Andean and Tasman segments experienced late Paleozoic orogenesis. The central regions were occasionally covered with inland seas in which were deposited fossiliferous limestones and shales. During the Permian, when the northern land masses were relatively warm and sometimes arid, Gondwanaland appears to have been cooler. The South Pole for the Permian was located in what is now South Africa, and all of the now separated Gondwanaland continents exhibit the scars of widespread glaciations.

Review Questions

1. Describe the relation between the orogenies that produced the Appalachian and Ouachita mountains and the movement and position of tectonic plates.
2. What is the plate tectonic reason for the approximate time equivalence between the Allegheny and Hercynian orogenies?
3. Why is it logical to divide the late Paleozoic into two cratonic sequences? What are the names and durations of those sequences?
4. When did the Old Red continent develop and where was it located? What kind of sediment was being deposited across the Old Red continent? How do the deposits of the Old Red continent compare with those of the Catskill clastic wedge?
5. What is the geologic evidence for the occurrence of a geographically extensive episode of continental glaciation in the Southern Hemisphere during the late Paleozoic?
6. When and where were the Colorado mountains

("ancestral Rockies") developed? What is the nature of the sedimentary rocks that had their source in the Colorado mountains? Where can one travel to see these rocks?

7. Name three metallic and three nonmetallic mineral resources extracted from Upper Paleozoic rocks.

8. Explain the occurrence of oil and gas in the calcareous algal bioherms found in the shelf carbonates of the Paradox basin.

Discussion Questions

1. In the study of an ancient mountain range, how might a geologist recognize a displaced terrane? Having recognized a terrane as a displaced (alien or exotic), how might he or she determine that the terrane was from an island arc? From a microcontinent?

2. What is the paleoenvironmental significance of an association of red sandstone, salt, and gypsum deposits? Where and when during the late Paleozoic did deposition of this association of rocks form?

3. What is a cyclothem? Prepare a list and discuss each of the eustatic and tectonic conditions that might account for the cyclicity for which cyclothems are named.

4. Describe the environmental conditions under which the Chattanooga Shale was deposited. What problems are associated with hypotheses for the origin of this extensive blanket of dark shales?

5. Approximately where was the South Pole located during the Permian Period? How have geologists ascertained that position? What confirmation for the pole position can be obtained from the global distribution of corals?

Supplemental Readings and References

Bally, W., and Palmer, A. R. (eds.). 1989. The geology of North America; an overview. *Decade of North American Geology,* vol. A. Boulder, CO: Geological Society of America.

Bambach, R. K., Scotese, C. R., and Ziegler, A. M. 1980. Before Pangaea: The geographies of the Paleozoic world. *Amer. Scientist* 68(1):26–28.

Bird, J. M., and Dewey, J. F. 1970. Lithosphere plate–continental margin tectonics and the evolution of the Appalachian orogeny. *Bull. Geol. Soc. Am.* 81:1031–1060.

Burchfiel, B. C., Cowan, D. S., and Davis, G. A. (eds.). 1992. Tectonic overview of the Cordilleran orogen in the western U.S. *Decade of North American Geology,* vol. G-3. Boulder, CO: Geological Society of America.

Frazier, W. J., and Schwimmer, D. R. 1987. *Regional Stratigraphy of North America.* New York: Plenum Press.

Hatcher, R. D., Thompson, W. A., and Viele, G. W. (eds.). 1989. The Appalachian–Ouachita orogen in the United States. *Decade of North American Geology,* vol. F-2. Boulder CO: Geological Society of America.

Rodgers, J. J. W. 1993. *A History of the Earth.* Cambridge, England: Cambridge University Press.

Woodrow, D. L., and Sevon, W. D. (eds.). 1985. The Catskill Delta. Geological Society of America Special Paper 201. Boulder, CO: Geological Society of America.

Reconstructions of some of the arthropods preserved in the Cambrian Burgess Shale of British Columbia. The organisms provide a view of marine life as it existed 530 million years ago.

When you were a tadpole and I was a fish
In the Paleozoic Time,
And side by side on the ebbing tide
We sprawled through the ooze and slime,
Or skittered with many a caudal flip
Through the depths of the Cambrian fen,
My heart was rife with the joy of life,
For I loved you even then.

LANGDON SMITH, *Evolution*

CHAPTER 10

Life of the Paleozoic

The pace of evolution appears to have gone into double quick time at the start of the Cambrian Period. The rich fossil record stands in startling contrast to that of the earlier Precambrian. We have noted that most of the Precambrian was a world of bacteria, algae, and fungi. Multicellular animals appear only in uppermost Proterozoic strata, such as those at Ediacara Hills in Australia. Except for a few tiny tubular creatures, Precambrian animals had not acquired the ability to secrete shells.

Certainly the improvement in the fossil record so evident in Cambrian strata can be attributed to the spread of shell-building abilities. During the Cambrian, shell-bearing trilobites and brachiopods were particularly abundant, but an even greater expansion of shelly animals followed in the early Ordovician. Epeiric seas that spread across the cratons of the world provided a multitude of habitats and opportunities for diversification and expansion of phyla. Most of the phyla on hand by Ordovician time are still with us today. The proliferation of animals with shells was probably related to the advantages afforded by shells in providing protection and support for soft tissues.

Later in the Paleozoic, as invertebrates continued their remarkable success in populating the seas, plant evolution progressed steadily to the point at which the transition from water to land was accomplished. Vascular land plants spread across the continents, ultimately producing densely forested regions. Shortly after the appearance of the first true land plants, vertebrates with legs instead of fins came ashore.

■ *Plants of the Paleozoic*

The history of plants begins among the bacteria and algae of the Precambrian. That bacterial growths were abundant is indicated by stromatolites and the great Precambrian iron ore deposits, believed to have been produced by bacterial activity. Stromatolites were already present in the late Archean, expanded during the Proterozoic, and continued to be found in limestones of all younger periods. During the Cambrian, stromatolitic reefs were widespread (Fig. 10–1) but became more restricted during and after the Ordovician. Then as now, they may only have been able to form in areas lacking in marine invertebrates that grazed on stromatolite organisms. Many such grazing invertebrates had appeared by late Cambrian time.

Another relatively common group of marine algal fossils found in lower Paleozoic rocks are the **receptaculids** (Fig. 10–2). Receptaculids are lime-secreting green algae of the family Dasycladaea. In general appearance, they remind one of the seed-bearing central area of a sunflower; this has caused amateur fossil collectors to call them "sunflower corals." They are, however, neither sunflowers nor corals. Nor are they sponges, a designation once considered because of their spicule-like rods and circulatory passages. Although most frequently found in Ordovician rocks, members of this group also occur sparsely in Silurian and Devonian strata.

FIGURE 10–1   Hummocky stromatolites of Cambrian age exposed along the banks of the Black River in Missouri.

FIGURE 10–2   *Fisherites*, a receptaculid from the Ordovician Kimmswick Formation of Missouri. Sometimes called sunflower corals, Receptaculids are the fossil remains of a type of green algae and not related to corals. (About 10 × 16 cm.)

The evolution of cyanobacteria, such as those responsible for the stromatolitic growth referred to earlier, was an important first step in the colonization of continental environments. Although most Precambrian stromatolites were formed in shallow marine environments, some grew in freshwater bodies as well. Thus, they are formed by hardy organisms that were able to migrate into brackish bays and estuaries and eventually into streams and lakes.

The appearance of green algae or **chlorophytes** was the probable next step in the evolutionary journey toward land plants. The close relationship between chlorophytes and land plants is suggested by their adaptation not only to freshwater bodies but to moist soil as well. In addition, both green algae and land plants possess the same kind of green pigment and produce the same kind of carbohydrate during photosynthesis. Early in the Paleozoic, chlorophytes were probably symbiotically joined by fungi to form lichens.

Today's land plants include bryophytes (mosses, liverworts, and hornworts) and tracheophytes, or vascular plants. Tracheophytes, such as trees, ferns, and flowering plants, have vascular tissues that provide for transport of water and nutrients. Tubes and vessels convey fluid from one part of the plant to another. The importance of the vascular system is apparent when we remember that above-ground moisture is undependable in most land areas. There is nearly always a greater persistence of moisture and plant nutrients within the pore spaces of soil, but beneath the surface there is no light for photosynthesis. The vascular system of tracheophytes permits part of the plant to function underground, where there is water but no light, and another part to grow

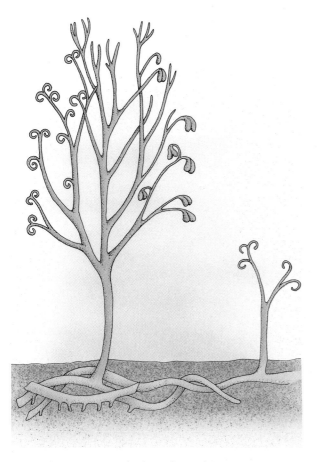

ductive and dispersal elements of the progenitors of both bryophytes and tracheophytes. By Silurian time (about 425 million years ago), spores with a distinctive three-rayed scar appeared. They are called trilete spores and are the characteristic reproductive bodies of primitive tracheophytes.

The spread of early Paleozoic tracheophytes profoundly altered the environment. Their roots slowed the erosion of land. Moisture lost from leaves during transpiration raised local humidity. But of even greater importance, the verdant cover of plants provided food and shelter for early land animals. Even after death, the litter of decaying vegetation promoted the formation of soils and provided for the survival of a host of microorganisms, worms, and early land-dwelling arthropods.

The transition from aquatic plants to land plants was apparently difficult, for it was late in coming. The first *unquestioned* remains of vascular plants are found in rocks of Middle Silurian age. These are **psilophytes.** They were small plants (less than 30 cm tall) consisting of horizontal stalks or rhizomes that grew just beneath the surface of the ground in moist soil. Extending vertically from the rhizomes (Fig. 10–3) were short, slender stems bearing smaller branches and spore sacs. Delicate strands of wood cells and a rudimentary vascular system have been recognized in psilophyte fossils.

The psilophytes paved the way for the evolution of large trees. Because of the evolution of wood, plants were able to stand tall against the pull of gravity and the force of strong winds. By late in the Devonian, there were forests of lofty, well-rooted, leafy trees (Fig. 10–4). One of these forests, located near the small town of Gilboa in upstate New York, has left many fossils as ample evidence of its existence. Some of the Gilboa trees were over 7 meters tall. Impressive as these trees were, they were to be dwarfed by some of their Carboniferous descendants.

FIGURE 10–3 A psilophyte from the Lower Devonian of the Gaspé Peninsula (plant at left about 16 cm tall).

where water supplies are uncertain but sunlight for photosynthesis is ensured.

The earliest fossil evidence for the invasion of continents by plants consists of groupings of four spore bodies in tetrahedral arrangement. These are called tetrads and have been obtained from rocks as old as Ordovician. They are thought to be the repro-

FIGURE 10–4 Restoration of a Middle Devonian forest in the eastern area of the United States. (A) An early lycopod, *Protolepidodendron.* (B) *Calamophyton,* an early form of the horsetail rush. (C) Early tree fern, *Eospermatopteris. (Courtesy of The Field Museum; painting by C. R. Knight, photo by R. Testa. Negative No. CK49T.)*

FIGURE 10–5   Fossilized bark of the Carboniferous tree *Lepidodendron.* Scale rows are about 1 cm wide.   *(Photograph courtesy of Wards Natural Science Establishment, Inc., Rochester, NY.)*

When one surveys the entire history of vascular plants, three major advances become apparent. Each involved the development of increasingly more effective reproductive systems. The first advance led to seedless, spore-bearing plants, such as those that were ubiquitous in the great coal-forming swamps of the Carboniferous. The second saw the evolution of seed-producing, pollinating, but nonflowering plants (gymnosperms). This was a late Paleozoic event. The evolution of plants with both seeds and flowers (angiosperms) came late in the Mesozoic Era.

Among the moisture-loving plants of the Carboniferous were the so-called scale trees, or **lycopsids.** Today, they are represented by their smaller survivors, the club mosses, of which the ground pine *Lycopodium* is a member. Small size was not particularly characteristic of the late Paleozoic lycopsids. The forked branches of *Lepidodendron* (Fig. 10–5) reached 30 meters into the sky. The elongate leaves of the scale trees emerged directly from the trunks and branches. After being released, they left a regular pattern of leaf scars. In *Lepidodendron,* the scars are arranged in diagonal spirals, whereas species of *Sigillaria* have leaf scars in vertical rows.

Another dominant group of plants that grew side by side with the Carboniferous lycopsids were the **sphenopsids.** Living sphenopsids include the scouring rushes and horsetails. Fossil sphenopsids, such as *Calamites* (Fig. 10–6) and *Annularia* (Fig. 10–7), possessed slender, unbranching, longitudinally ribbed stems with a thick core of pith and rings of leaves at each transverse joint. At the top, a cone bore the spores that would be scattered in the wind.

True ferns (Fig. 10–8) were also present in the coal forests. Many were tall enough to be classified as trees. Like the lycopsids and sphenopsids, they reproduced by means of spores carried in regular patterns on the undersides of the leaves.

Seed plants made their debut during the late Paleozoic. They probably arose from Devonian fernlike plants. These plants, appropriately dubbed "seed ferns," had fernlike leaves, but unlike true ferns they reproduced by means of seeds.

One of the most widely known seed fern groups is *Glossopteris,* which was restricted to the Southern

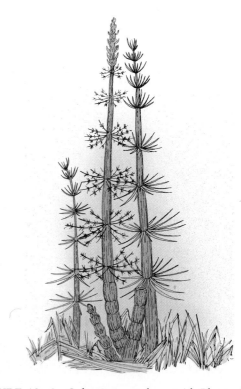

FIGURE 10–6   *Calamites,* a sphenopsid. Plants shown are about 3 to 5 meters tall.

FIGURE 10–7 *Annularia,* an abundant sphenopsid of Pennsylvania age.

Hemisphere during the Carboniferous and Permian (see Fig. 5–29). Most species of *Glossopteris* were plants with thick, tongue-shaped leaves. Because of certain anatomic traits and because of their association with glacial deposits, *Glossopteris* and associated plants are thought to have been adapted to cool climates.

Fossils of seed plants are present in the Northern Hemisphere also. Cordaites (primitive members of the conifer lineage) were abundant, some towering to 50 meters (Fig. 10–9). Their branching limbs were crowned with clusters of large, straplike leaves. These and somewhat more modern cone-bearing conifer plants spread widely during the Permian, perhaps as a consequence of drier climatic conditions. The first ginkgoes made their appearance during the Permian. Today, only a single species, *Ginkgo biloba,* remains as a survivor of this once flourishing group.

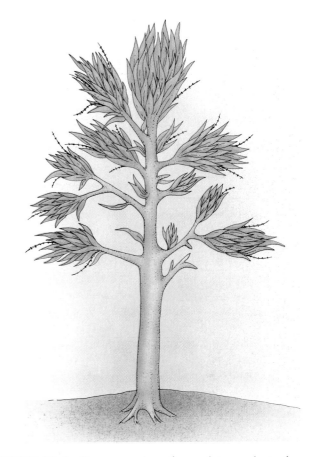

FIGURE 10–9 Reconstruction of a cordaitean plant of the Carboniferous.

■ *Invertebrate Animals of the Paleozoic*

People are accustomed to thinking of the ocean as the birthplace of life. Thus, it is no surprise to learn that the Earth's earliest animals lived in the sea. The only fossils ever found of animals that lived during the end of the late Proterozoic, the Cambrian, and the Ordovician have been those of ocean dwellers. As is the case today, these animals varied in their choices of habitat as well as in the kinds of food they required. In the geologic past, as in the present, there were basically three modes of life. Organisms that could live on the sea floor or burrow in sea-floor sediment are termed **benthic,** those that floated are termed **planktic,** and those that swam are termed **nektic.** In addition, many animals in the sea spend a phase of their lives in one of these modes and then change to accommodate themselves to another. In fact, the mobile larvae of a multitude of otherwise sedentary (stationary) "critters" are dispersed in this way.

As indicated by the Ediacaran fossils described in Chapter 7, multicellular marine organisms were al-

FIGURE 10–8 *Pecopteris,* a true fern from the Pennsylvanian of Illinois (note the penny for scale).

ready present near the end of the Precambrian. Those organisms, however, were soft-bodied and are only infrequently preserved as imprints in once soft sediment. When animals had achieved the ability to secrete hard parts, their chances of being preserved improved immensely. The oldest fossils to accomplish this feat occur in rocks of the latest Precambrian (see *Cloudina,* Fig. 7–31). A more diverse assemblage occurs in the Cambrian Tommotian Stage. The Tommotian fauna was widespread, as indicated by fossil sites in Siberia, Sweden, North America, Antarctica, and England. Represented are the shells and elements of tiny (rarely more than a few millimeters long) mollusks, sponges (known from spicules), and a variety of creatures of uncertain classification that secreted cap-shaped and tubular shells (Fig. 10–10). Shells are composed of either calcium carbonate or calcium phosphate. The phosphatic fossils can be recovered from carbonates by dissolving pieces of the rock in weak acid. Among the tubular fossils are a rather distinctive group called **anabaritids** (Fig. 10–11). The shell resembles three tubes open to one another where they are joined together. Each of these partial tubes is surmounted by a narrow keel. The keel may have provided support for the animal on soft mud of

FIGURE 10–11 *Anabarites,* an early Cambrian shelled fossil. The shell is only 5 mm in length. *(After a drawing by Dianna L. Schulte McMenamin in McMenamin, M.A.S., and McMenamin, D.L.S. 1990. The Emergence of Animals. New York: Columbia University Press.)*

the sea floor. Another form, *Lapworthella,* has a cap shape ornamented with grooves and ridges. It was probably not the single shell of an animal but rather one of many similar skeletal elements that covered all or part of the body. Such an interpretation is favored by the discovery of *Lapworthella* elements fused together in a side-by-side arrangement.

Following the Tommotian stage of the early Cambrian, animals with shells became abundant and diverse. Familiar fossils such as trilobites and brachiopods dominate the fossil record. But there were also many soft-bodied creatures as well. Some of these are known from extraordinary preservation at only a few localities, such as the Chengjian fossil site in China and the Burgess Shale site in British Columbia.

Window into the Past: Fossils of the Burgess Shale

High on a ridge near Mount Wapta, British Columbia, there is an exposure of Burgess Shale that contains one of the most important faunas in the fossil record (Fig. 10–12). The fossils in the Burgess Shale are reduced to shiny, jetlike impressions on the bedding planes of the black rock (Fig. 10–13). Most of them are the remains of animals that lacked shells. Altogether, they form an extraordinary assemblage (Fig. 10–14) that includes four major groups of arthropods (trilobites, crustaceans, and members of the taxonomic groups that include scorpions and insects) as well as sponges, onycophorans (see Fig. 10–12), crinoids, sea cucumbers, chordates, and many species that defy placement in any known phyla.

The Burgess Shale site was not found as the result of a purposeful search. It was stumbled upon acci-

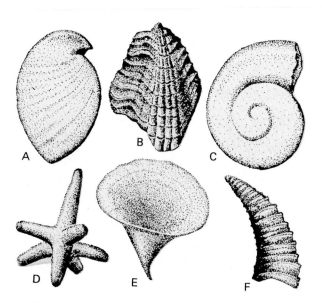

FIGURE 10–10 Late Precambrian and Early Cambrian shell-bearing fossils from Siberia. (A) *Anabarella,* ×20, a gastropod; (B) *Camenella,* ×18, affinity uncertain; (C) *Aldanella,* ×20, a gastropod; (D) sponge spicule, ×30; (E) *Fomitchella,* ×45, affinity uncertain; and (F) *Lapworthella,* ×20. *(After Matthews, S. L. and Missarzhevsky, V. V. J. 1975. Geol. Soc. London 131:289–304.)*

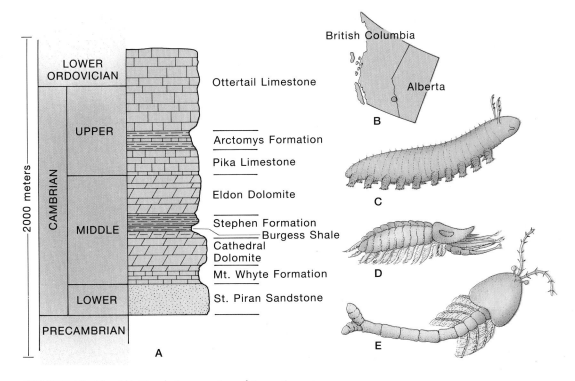

FIGURE 10–12 (A) Cambrian stratigraphic section at (B) Kicking Horse Pass, British Columbia. According to a now famous story, this is where a slab of fossiliferous shale was kicked over by a pack horse and caught the eye of paleontologist Charles Walcott, who followed the trail of rock debris up the side of the mountain to discover the Burgess Shale beds with their rich fauna of Middle Cambrian fossils. Among these were (C) *Aysheaia*, an onycophoran believed to be intermediate in evolutionary position between segmented worms and arthropods. Among the thousands of specimens were trilobite-like arthropods such as (D) *Leanchoilia superlata* and (E) *Waptia fieldensis*.

dentally. The initial discovery of a single fossil was made in August of 1909 by Charles D. Walcott. Walcott's field crew included his wife, Marrella, who was an avid fossil collector, and his two sons. As they were riding along a trail south of Mount Wapta, Mrs. Walcott's horse dislodged a slab of black shale with a silvery impression on its surface. The slab caught Walcott's eye, and he dismounted to examine it closely. It contained the remains of a crustacean that Walcott later named *Marrella*. Unfortunately, the discovery came at the end of the field season, and exposures at higher elevations from which the rock had come were snow covered. As a result, the Walcotts were forced to leave the site and return the following year, when they traced the fossil to its source in the Burgess Shale. Quarrying was begun and continued during the 1912, 1913, and 1917 field seasons.

FIGURE 10–13 *Haplophrentis*. This photograph illustrates the nature of preservation of Burgess Shale fossils. *Haplophrentis* had a tapering shell surmounted by a lid or operculum, which could be closed for protection. The lateral blades on either side may have served as props. The length of the shell is 2 cm. *(Courtesy of The National Museum of Natural History, Smithsonian Institution; photograph by Chip Clark.)*

FIGURE 10–14 The Burgess Shale diorama at the U.S. National Museum of Natural History. The reconstruction is based on actual fossil remains of organisms. It depicts a benthic marine community of the Middle Cambrian. A steep submarine escarpment forms the background. Slumping along this wall contributed to the preservation of the Burgess fauna by burying the organisms. Along the wall one can see green and pink vertical growths of two types of algae. The large purplish creatures that resemble stacks of automobile tires are sponges *(Vauxia)*. The blue-colored animals are trilobites *(Olenoides)*. The brown arthropods with distinct lateral eyes are named *Sidneyia*. Climbing out of the hollow on the sea floor are crustaceans called *Canadaspis*. The yellow animals swimming toward the right above the sea floor are *Waptia*. *Opabinia* has crawled out of the left side of the hollow. Burrowing worms are visible in the vertical cut at the bottom. *(Courtesy of the National Museum of Natural History, Smithsonian Institution; photograph by Chip Clark.)*

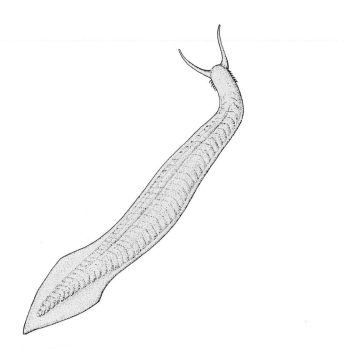

FIGURE 10–15 Reconstruction of *Pikaia*, the earliest known member of our own phylum, the Chordata. Note the rod along the animal's back that appears to be a notochord. (Length about 4.0 cm.)

Altogether, about 60,000 specimens were collected and stored in the U.S. National Museum of Natural History, where Walcott served as secretary of the Smithsonian Institution. In the 1960s, these fossils were re-examined by Harry B. Whittington, who then joined paleontologists of the Geological Survey of Canada in reopening the quarry and assembling a second collection. Whittington devoted the next 15 years to an incisive and critical scrutiny of the Burgess Shale fauna. He was able to describe and depict the anatomy of the Burgess animals with great clarity. During the study, it became apparent that earlier interpretations of many Burgess animals as ancestors to existing taxonomic groups were incorrect and that Burgess fauna included creatures so different as to warrant placement in new phyla.

Among the sometimes bizarre and unique creatures of the Burgess fauna were small, elongate animals that are currently believed to be the earliest known chordates. Chordates are animals that have, at some stage in their development or throughout their lives, a notochord (internally supportive rod) and a dorsally situated nerve cord. Chordates in which the notochord is supplanted by a series of vertebrae are termed vertebrates. These earliest known chordates have been assigned the generic name *Pikaia* in honor of Mount Pika near the Burgess fauna discovery site. Fossils of *Pikaia* (Fig. 10–15) exhibit two

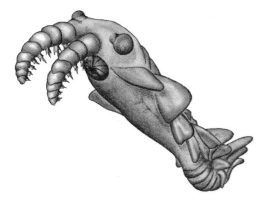

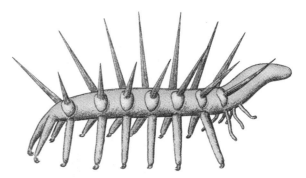

FIGURE 10–18 The early Cambrian Burgess Shale fossil *Hallucigenia*.

FIGURE 10–16 *Anomalocaris.* This giant predator (often 60 cm in length) captured its prey with its huge frontal appendages and passed the victims back to the circular mouth with its outer and inner circles of teeth. The side flaps were used in swimming, like underwater wings.

features upon which their designation as chordates is based. They possessed a notochord and a series of V-shaped muscle bands along the sides of the body, as is characteristic of the musculature in fish. These muscles, working with the rigid but flexible notochord, provided a sinuous, fishlike body motion required for swimming. Like *Pikaia*, humans are chordates. For this reason, we are keenly interested in this Cambrian animal.

Other particularly interesting members of the Burgess Shale community were *Anomalocaris*, *Opabinia*, and *Hallucigenia*. *Anomalocaris* (Fig. 10–16) was a fierce predator and reached lengths of over a half meter. On its long oval head were two stalked eyes and a pair of feeding appendages used to capture and carry prey to its ventrally placed circular mouth. The mouth was ringed with teeth, which may have been responsible for the wounds observed in fossils of other members of the community.

If *Anomalocaris* was the largest animal of its time, *Opabinia* (Fig. 10–17) might well qualify as the strangest. The creature had five eyes and a flexible nozzle equipped with nippers used in capturing its victims. Perhaps predators like *Opabinia* and *Anomalocaris* were responsible for selection pressures favoring the spread of protective skeletons in other early Paleozoic animals.

Hallucigenia (Fig. 10–18) was a particularly perplexing Burgess Shale fossil. Its original interpretation suffered from the limited amount of information that could be gleaned from its compressed and carbonized remains. In early reconstructions, *Hallucigenia* was depicted as walking on pairs of stiltlike legs, rather like some sea urchins are able to walk on movable spines. A single row of tentacles was thought to extend along the back of the animal. Subsequently, the Swedish paleontologist Lars Ramskold was able to flake away some of the covering shale

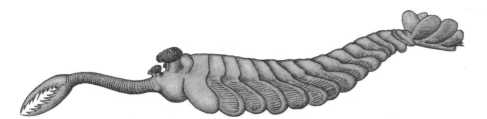

FIGURE 10–17 *Opabinia.* This strange Burgess Shale animal had a frontal "nozzle" with a jawlike structure at its end used for gathering food. There were five eyes on its head. Each side was covered with overlapping lobes bearing narrow strips interpreted as gills. This specimen was 7 cm long. *(Adapted from Whittington, H.B. 1975. Philosophical Transactions of the Royal Society of London, vol. B.)*

from a specimen and reveal a claw at the tip of one of the supposed tentacles. Claws normally are found on feet, so reconstructions were revised to show the former legs as paired dorsal spines and the former tentacles as paired legs. Although only one row of legs had been detected, the second row was assumed to lie in the shale matrix beneath the body. That interpretation was validated in 1991 when Ramskold re-examined the best specimen at the Smithsonian Institution and discovered parts of the second leg row. The head region of *Hallucigenia*, however, remains obscure. Examination of specimens recently uncovered in China may shed light on this aspect of *Hallucigenia*'s anatomy. For the present, *Hallucigenia* is seen as an animal with a cylindrical trunk about 2.5 cm. long. The trunk does not extend beyond the terminal leg pair. There are seven pairs of legs and seven pairs of dorsal spines.

Hallucigenia appears to be an onycophoran. Onycophorans are wormlike animals. Living forms are terrestrial and live in moist habitats beneath logs, rocks, and leaves. The group is considered to be the connecting link between annelids and arthropods.

The special significance of the Burgess Shale fauna lies in the perspective that it provides on life at the beginning of the Paleozoic. It reveals that the Cambrian seas teemed with diverse and complex animals and that marine life at the beginning of the Paleozoic had already reached an advanced stage of evolution. The discovery of the Burgess fossils further demonstrates that reconstructions of ancient communities based on the fossils of shelled animals (often all that are preserved) are likely to be seriously incomplete. Much like marine communities today, there were more animals with soft parts in the Burgess fauna than those with hard parts, and these soft-bodied animals were more varied than the shelled ones. Thus, the Burgess fauna is a more valid representation of life as it really was than is usually found for other fossil faunas.

Early Paleozoic Diversification

As indicated by the highly structured and diversified Burgess Shale community, as well as other communities known mostly from shelled animals, marine life during the early Paleozoic had evolved a variety of life styles and adapted to a multitude of habitats. There were **epifaunal** animals that lived on the surface of the sea floor, as well as **infaunal** creatures that burrowed beneath the surface. There were borers as well as burrowers, attached forms and mobile crawlers, swimmers, and floaters. Some were *filter-feeders* that strained tiny bits of organic matter or microorganisms from the water; others were *sediment-*

feeders that passed the mud of the sea floor through their digestive tracts to extract the nutrients within. There were animals that grazed on algae that covered parts of the ocean bottom, and carnivores that consumed these grazers. A host of scavengers processed organic debris and aided in keeping the seas suitable for life. Some classes of invertebrates maintained a single mode of life, whereas groups within other classes adapted to several different life styles. Snails, for example, included scavengers, herbivores, and carnivores. With the passage of geologic time, many evolutionary changes and extinctions were to occur among the invertebrates, but these changes tended to occur *within* the taxonomic groups that had been established during the early Paleozoic.

Unicellular Animals

The principal groups of early Paleozoic unicellular animals with a significant fossil record are the **foraminifers** and **radiolarians**. Foraminifers build their tiny shells by adding chambers singly, in rows, in coils, or in spirals. Some species construct the shell, or *test*, of tiny particles of silt; others secrete tests composed of calcium carbonate. The test is characteristically provided with holes through which extend "tentacles" of protoplasm for feeding. It is from these holes, or *foramina*, in the test that foraminifers take their name. Foraminifers range from the Cambrian to the present, but they are rare and poorly preserved in lower Paleozoic strata. During these early stages of their history, they were mostly simple, saclike, tubular, or loosely coiled benthic creatures with tests composed of sedimentary particles.

FIGURE 10–19 Fusulinid limestone of Permian age from the Sierra Madre of western Texas. *(Courtesy of W. D. Hamilton.)*

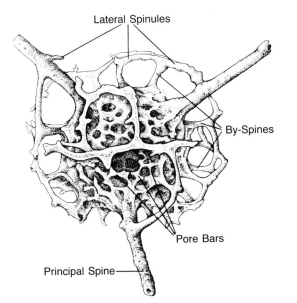

FIGURE 10-20 A Late Ordovician radiolarian (×150) from the Hanson Creek Formation of Nevada. It is unusual to find well-preserved radiolarians in rocks of the early Paleozoic. *(From Dunham, J. B., and Murphy, M. A. 1976. J. Paleo. 50(5):883.)*

Foraminifers become more numerous and varied by Carboniferous time. The increase is strikingly evident among the fusulinids, a group whose shells superficially resemble grains of wheat (Fig. 10–19). Although many genera of fusulinids are similar in external appearance, they have complex internal features that are distinctive and dissimilar at the species level.

Fusulinids had global distribution during the Pennsylvanian and Permian and were sometimes so abundant that they constituted a high percentage of the bulk volume of entire strata. Because they evolved rapidly and because they are so widespread and abundant, fusulinids are among the most important invertebrate guide fossils of the late Paleozoic.

In modern oceans, the tests of calcareous foraminifers rain down continuously on the sea floor, much like an unending limy snow. The accumulated debris forms a deep-sea sediment called globigerina ooze because of the prevalence of the remains of species of *Globigerina* and related planktonic foraminifers.

Radiolarians are also single-celled planktonic organisms that have been present on Earth at least since the Paleozoic began. Like the foraminifers, they have threadlike pseudopodia that project from an ornate, lattice-like skeleton of opaline silica or a proteinaceous substance. In some regions of the oceans today, radiolarian skeletons accumulate to form deposits of radiolarian or siliceous ooze. Although radiolarians do occur in lower Paleozoic rocks (Fig. 10–20), they are rare and not yet useful for stratigraphic correlation. They are more abundant in Mesozoic and Cenozoic rocks and have been shown to be particularly useful in correlating Pleistocene deep-sea deposits. Of special interest is the observation that Pleistocene radiolarian extinctions appear to be associated with dates of reversal in the Earth's magnetic field.

Cup Animals: Archaeocyathids

Archaeocyatha means "ancient cups." It is an appropriate name for a somewhat enigmatic group of Cambrian organisms that constructed conical or vase-shaped skeletons out of calcium carbonate (Fig. 10–21). **Archaeocyathids** hold two paleontologic records. They are the earliest abundant reef-building animals on Earth, and they are members of a separate phylum that suffered extinction. Although abundant during the Early Cambrian, they had entirely died out before the end of that period. Gregarious in habit, they carpeted the floors of sea adjacent to land areas. Low reefs formed primarily of archaeocyathids can be studied today in North America, Siberia, Antarctica, and Australia. Australian archaeocyathid reefs are often over 60 meters thick and extend horizontally in narrow bands for over 200 km.

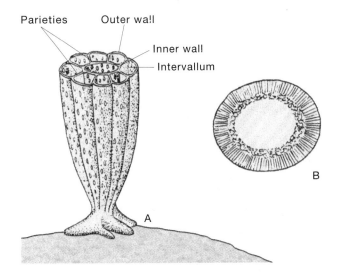

FIGURE 10-21 (A) Longitudinally fluted cup of an archaeocyathan, about 6 cm in height. (B) Transverse section of a nonfluted archaeocyathan having closely spaced parieties and a vesicular inner wall. (Maximum diameter 4 cm.)

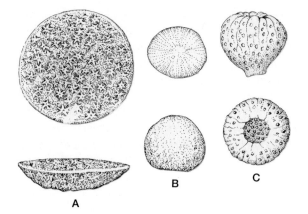

FIGURE 10–22   Early Paleozoic (Silurian) sponges. (A) *Astraeospongium* (it takes its name from the starlike spicules.) (B) *Microspongia.* (C) *Astylospongia* (½ natural size).

FIGURE 10–23   The Devonian siliceous sponge *Hydnoceras*, 15 cm in height.   *(Photograph courtesy of J. Keith Rigby, Brigham Young University.)*

Pore-Bearers

Among the many stationary animals to colonize the early Paleozoic sea floor were the sponges. Sponges are members of the phylum **Porifera.** They appear to have evolved from colonial flagellated unicellular creatures and thus provide insight into how the transition from unicellular to multicellular animals may have occurred.

Sponges are a relatively conservative branch of invertebrates that have a long history. Cambrian representatives of all but one modern class of Porifera are known as fossils. Some fossil sponges, such as *Protospongia* from the Cambrian; *Astraeospongium, Microspongia,* and *Astylospongia* (Fig. 10–22) from the Silurian; and the siliceous sponge *Hydnoceras* from the Devonian are all well-known guide fossils (Fig. 10–23).

Sponges have always been predominantly marine creatures, although a few modern species live in fresh water. Spicules formed in the walls and cell layers of sponges are distinguishing characteristics of the phylum and provide both protection and support. The spicules may be composed of silica, calcium carbonate, or a proteinaceous material termed *spongin.* Naturally, the mineralized spicules are more commonly preserved and are frequently found by geologists examining rocks that have been disaggregated for study. Spicules are also important in the classification of sponges. For example, the Desmospongea consist entirely or in part of spongin, which may be reinforced with siliceous spicules. Hexactinellida develop siliceous spicules of distinctive shape, and Calcarea are characterized by spicules made of calcium carbonate. The Sclerospongea are a minor class of tropical marine sponges with skeletons composed of siliceous spicules, spongin fibers, and a basal layer of calcium carbonate.

Although sponges vary greatly in size and shape, their basic structure (Fig. 10–24) is that of a highly perforated vase modified by folds and canals. The body is attached to the sea floor at the base, and there is an excurrent opening, or *osculum,* at the top. The wall consists of two layers of cells. Facing the internal space is a layer of collar cells (choanocytes), and on the outside is a protective wall of flat cells that resemble the bricks of a worn masonry pavement. Between these two layers one finds a gelatinous substance called *mesenchyme.* Here amoeboid cells go about the work of secreting the spicules. Sponges lack true organs. Water currents moving through the sponge are created by the beat of *flagellae.* These currents bring in suspended food particles, which are ingested by the collar cells. In a simple sponge, water enters through the pores, flows across sheets of choanocytes in the central cavity, and passes out through the osculum.

A group of early Paleozoic sclerosponges of particular interest because of their reef-building capabilities is the **stromatoporoids** (Fig. 10–25). These organisms constructed fibrous, calcareous skeletons of pillars and thin laminae that can nearly always be found in reef-associated carbonate rocks of the Silurian and Devonian. Apparently, stromatoporoids grew profusely and in close association with corals, brachiopods, and other invertebrate reef dwellers. Silurian coral–stromatoporoid reefs are well known to geologists primarily because of extensive study of reef exposures on Gotland Island off the Baltic coast of Sweden and in the region around and southwest of the Michigan basin. Reef development during the Devonian was similar to that in the Silurian, with blanket-like, massive, and cylindric stromatoporoids forming a considerable part of the total mass of the reef structures.

Corals and Other Cnidaria

Sea anemones, sea fans, jellyfish, the tiny *Hydra*, and reef-forming corals are all representatives of the phylum **Cnidaria**, which is known for the great diversity and beauty of its members (Fig. 10–26). The cnidarian body wall is composed of an outer layer of cells, the *ectoderm;* an inner layer, the *endoderm;* and a thin, noncellular intermediate layer, the *mesoglea.* In the endoderm are found primitive sensory cells, gland cells that secrete digestive enzymes, flagellated cells, and nutritive cells to absorb nutrients. A dis-

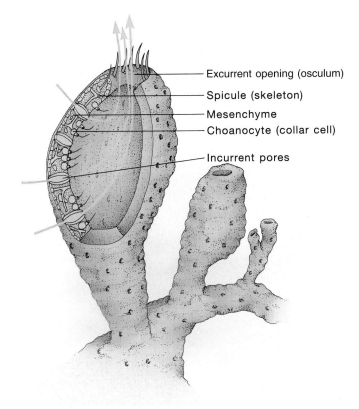

Excurrent opening (osculum)
Spicule (skeleton)
Mesenchyme
Choanocyte (collar cell)
Incurrent pores

FIGURE 10–24 Schematic diagram of a sponge having the simplest type of canal system. The path of water currents is indicated by arrows.

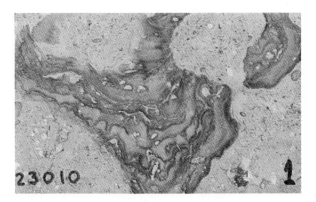

FIGURE 10–25 Polished slab of limestone containing the stromatoporoid named *Stromatoporella*, from Devonian rocks of Ohio.

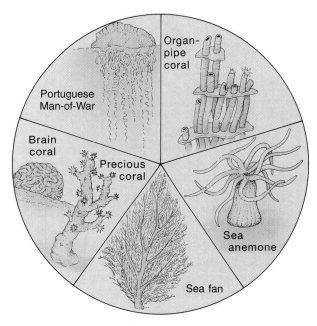

Organ-pipe coral
Portuguese Man-of-War
Brain coral
Precious coral
Sea anemone
Sea fan

FIGURE 10–26 Diversity among some common cnidarians. *(From Levin, H. L. 1975. Life Through Time. Dubuque, IA: William C. Brown Co.)*

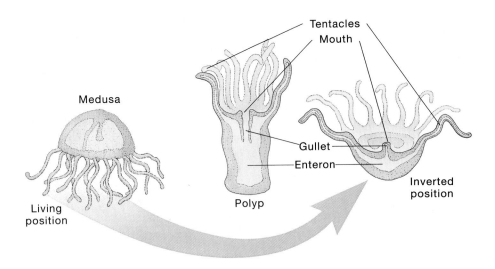

FIGURE 10–27 Comparison of polyp and medusa forms in cnidarians.

tinctive feature of cnidarians is the presence of sting-ing cells, which, when activated, can inject a paralyz-ing poison. These specialized cells, called cnidocytes, are a unique characteristic of all cnidaria.

Body form in cnidaria may be either polyp or medusa (Fig. 10–27). The medusoid form is seen in the jellyfish, which resembles an umbrella in shape. Jellyfish have a concave undersurface that contains a centrally located mouth; for this reason, it is desig-nated the oral surface. In jellyfish such as the living *Aurelia,* which is common along the eastern shore of the United States, the mouth is surrounded by four oral arms. All jellyfish have tentacles, and in *Aurelia* these are located around the margin of the umbrella. Most jellyfish swim by rhythmic contractions of the umbrella.

In the polyp form, as exemplified by *Hydra,* there is a circle of tentacles above the saclike body. Corals and sea anemones are among the frequently encoun-tered cnidaria with this polyp form. In stony corals (class Anthozoa), the polyp secretes a calcareous cup, in which it lives. The animal may live alone or may combine with other individuals to form large colo-nies. The cup, or **theca,** may be divided by vertical plates called **septa,** which serve to separate layers of tissue and provide support. As the animal grows, it also secretes horizontal plates termed **tabulae.** Corals are identified according to the nature of their septa, tabulae, and other skeletal features. For example, after the development on an initial embryonic set of six protosepta, the Paleozoic **rugose corals,** or Rugosa, insert new septa at only four locations dur-ing growth. In other Paleozoic corals, septa are ab-sent or poorly developed, so that tabulae are the most important features in classification. These are the Tabulata. **Tabulate corals** include many interesting colonial forms, such as the honeycomb and chain corals. The rugose corals and tabulate corals became

extinct at the end of the Paleozoic and were followed in the Mesozoic by anthozoans of the order Sclerac-tinia. In all scleractinian corals, septa are inserted be-tween the mesenteries in multiples of six.

The fossil record for Cnidaria begins with fossil jellyfish impressions discovered in upper Precam-brian rocks. The phylum is still poorly represented in Cambrian rocks. In the succeeding Ordovician the record improves dramatically, for the lime-secreting anthozoans begin to expand and diversify. The first stony corals were the tabulates, recognized by their simple, often clustered or aligned tubes divided hori-

FIGURE 10–28 The common tabulate coral *Favosites.* An individual polyp resided in each of the tubes. Tiny pores visible in the walls of the tubes allowed for inter-connections of tissues between adjacent polyps.

A

B

C

D

FIGURE 10–29 Devonian rugose corals. (A) The solitary horn coral *Zaphrenthis* with clearly visible radiating septa in the hornlike theca. (B) The compound (colonial) rugose coral *Lithostrotionella*. (C) A polished slab of the compound coral *Hexagonaria*. Water-worn fragments of this coral are found along the shore of Lake Michigan at Petoskey, Michigan, and this accounts for its being called Petoskey stone. Although not a rock, Petoskey stone is the designated state rock of Michigan. (D) Reconstruction of compound and solitary rugose corals on the floor of a Devonian epeiric sea. *(Diorama photograph courtesy of the National Museum of Natural History, Smithsonian Institution.)*

zontally by transverse tabula. These include the honeycomb coral *Favosites* (Fig. 10–28). Tabulate corals were the principal reef formers during the Silurian. They declined after the Silurian, but their role in reef building was assumed by the rugose corals. Both solitary and compound colonies of rugosids (Fig. 10–29) are abundant in the fossil record of the Devonian and Carboniferous, but this group declined and became extinct during the Late Permian.

Reefs built on the North American platform by Paleozoic corals were extensive and often exceeded 250 meters in total thickness. Most of these reefs lie beneath younger rocks and were discovered during oil explorations. Because of their porosity, buried Paleozoic reefs became sites for the accumulation of petroleum.

Moss Animals: Bryozoa

Bryozoans are minute, bilaterally symmetric animals that grow in colonies, which frequently appear twiglike when viewed without the aid of a magnifier.

The individuals, called *zooids,* are housed in a capsule, or *zooecium* (Fig. 10–30), which is often preserved as a result of its being calcified. The zooecia appear as pinpoint depressions on the outside of the colony, or *zoarium.* The zooid has a complete, U-shaped digestive tract with mouth surrounded by a tentacled feeding organ called the **lophophore.**

Today, there are more than 4000 living species of bryozoans, but nearly four times that number are known as fossils. Their earliest unquestioned occurrence is from Lower Ordovician strata of the Baltic, but they did not become abundant until Middle Ordovician and Silurian time. In rocks of these ages, their remains sometimes make up much of the bulk of entire formations. Like the corals, bryozoa contributed to the framework of reefs. One of the commonest of all the early Paleozoic bryozoans was *Fistulipora,* different species of which develop massive, incrusting, or arborescent colonies. The genus *Hallopora* (Fig. 10–31A) is one of many branching twig

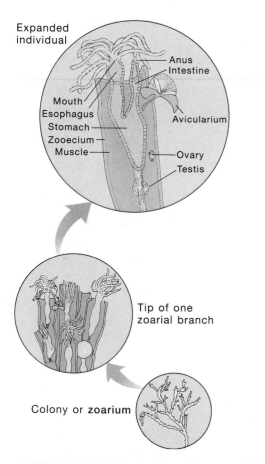

FIGURE 10–30 Relationship of bryozoan individuals to the zoarium. *(From Levin, H. L. 1975.* Life Through Time. *Dubuque, IA: William C. Brown Co.)*

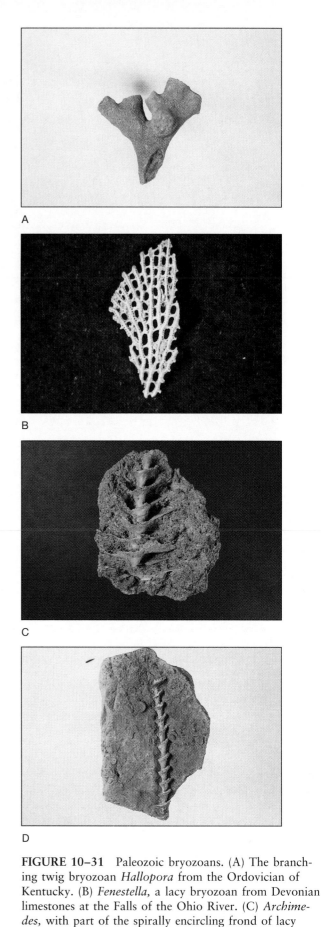

FIGURE 10–31 Paleozoic bryozoans. (A) The branching twig bryozoan *Hallopora* from the Ordovician of Kentucky. (B) *Fenestella,* a lacy bryozoan from Devonian limestones at the Falls of the Ohio River. (C) *Archimedes,* with part of the spirally encircling frond of lacy bryozoan colony attached and visible. (D) The central axis of *Archimedes.*

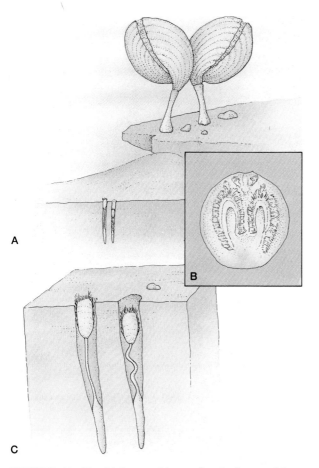

A

B

C

FIGURE 10–32 Living positions of articulate and inarticulate brachiopods. (A) The articulate living brachiopod *Magellania* attached to the sea floor by pedicle and with lophophore barely visible through the gape in the valves. (B) Interior of brachial valve showing ciliated lophophore. (C) The inarticulate brachiopod *Lingula,* which excavates a tube in bottom sediment and lives within it. The pedicle secretes a mucus that glues the animals to the tube.

FIGURE 10–33 The inarticulate brachiopod *Obolus* from the Cambrian Conasauga Formation near Coosa, Georgia. The specimens are about 0.6 cm in diameter.

bryozoans. Star-shaped patterns on the surface of *Constellaria* provided the inspiration for its generic name. Late Paleozoic bryozoans include varieties that constructed lacy, delicate, fanlike colonies (Fig. 10–31B). Associated with these so-called **fenestellid** colonies were the bizarre Mississippian corkscrew bryozoans known by the generic name *Archimedes* (Fig. 10–31C and D).

Brachiopods

Brachiopods are probably the most abundant, diverse, and useful fossils readily found in Paleozoic rocks. They are characterized by a pair of enclosing valves, which together constitute the shell of the ani-

mal (Fig. 10–32A). They resemble clams in this regard, but in symmetry of the valves and soft-part anatomy, brachiopods are different from bivalves. Brachiopod valves are almost always symmetric on either side of the midline, and the two valves differ from each other in size and shape. The valves of a clam are right and left, whereas those of a brachiopod are dorsal and ventral. The valves may be variously ornamented with radial ridges or grooves, spines, nodes, and growth lines. Calcium carbonate is the usual hard tissue of brachiopods, although the valves of some families are composed of mixtures of chitin and calcium phosphate. This is particularly true of the group called **inarticulates** (Figs. 10–32C and 10–33).

In the **articulate** brachiopods, the valves are hinged along the posterior margin and are prevented from slipping sideways by teeth and sockets. The less common inarticulates lack this definite hinge, are held together by muscles, and characteristically have simple spoon-shaped or circular valves. Although brachiopod larvae swim about freely, the adults are frequently anchored or cemented to objects on the sea floor by a fleshy stalk *(pedicle)* or by spines. Some simply rest on the sea floor. One of the more conspicuous soft organs of brachiopods is the *lophophore*, a structure consisting of two ciliated, coiled tentacles whose function is to circulate the water between the valves, distribute oxygen, and remove carbon dioxide (see Fig. 10–32). Water currents generated by *cilia* on the lophophore move food particles toward the mouth and short digestive tract.

FIGURE 10–34 Some early Paleozoic brachiopods. Along the top are Ordovician strophomenid brachiopods: (A) *Rafinesquina*, (B) *Strophomena*, and (C) *Leptaena*. Along the bottom are (D) *Hebertella*, an Orodovician orthid brachiopod; (E) *Hiscobeccus*, an Ordovician rhynchonellid; and (F) *Pentamerus*, a Silurian pentamerid ($\times\frac{1}{2}$).

Brachiopods still live in the seas today, although in far fewer numbers than during the Paleozoic. Earliest brachiopods were almost entirely the chitinous inarticulate types. These increased in diversity during the Ordovician but then declined. A few species of only three families remain today.

The articulates also first appeared in the Cambrian but became truly abundant during the succeeding Ordovician Period when there was a great expansion of all sorts of shelled invertebrates. Across the floors of early Paleozoic epeiric seas, large and small aggregates of these filter-feeders could be found. Today, their skeletons compose much of the volume of thick formations and provide the stratigrapher with essential markers for correlation. Among the articulate brachiopods that were particularly abundant during the early Paleozoic were the strophomenids, orthids, pentamerids, and rhynchonellids (Fig. 10–34). During the Devonian Period of the late Paleozoic, spiriferid brachiopods became particularly abundant. Spiriferids take their name from the calcified internal spirals that supported the lophophore (Fig. 10–35). The most characteristic brachiopods of the Carboniferous and Permian were the productids (Fig. 10–36). Productids were distinc-

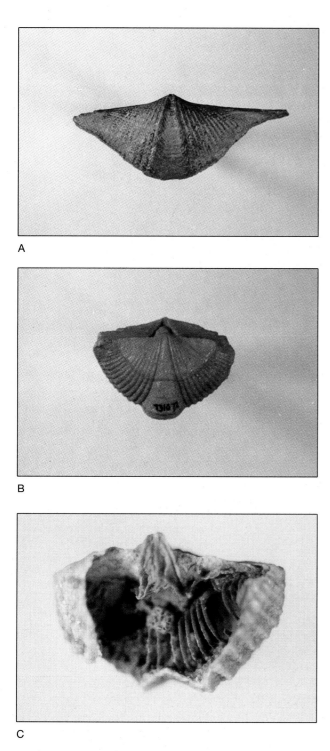

FIGURE 10–35   Devonian spiriferid brachiopods. (A) *Mucrospirifer.* (B) *Platyrachella.* (C) A spiriferid brachiopod with shell broken to reveal the internal spiral supports for the lophophore. (All ×0.75.)

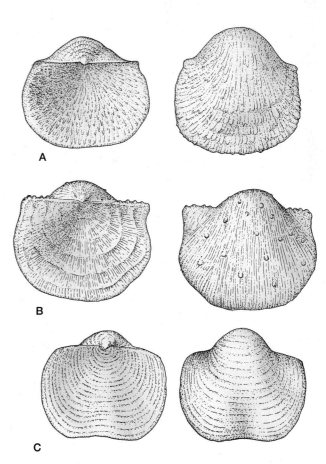

FIGURE 10–36   The Pennsylvanian productid brachiopods (A) *Juresania,* (B) *Linoproductus,* and (C) *Echinoconchus.* *(From Guidebook Series 8, Illinois Geological Survey, 1970.)*

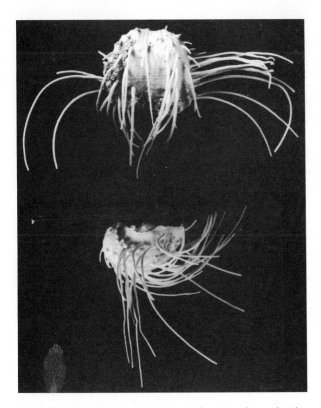

FIGURE 10–37 Ventral (upper photograph) and side (lower photograph) views of the Permian spinose productid brachiopod *Marginifera ornata* from the Salt Range of West Pakistan. Valves (not including spines) are about 2 cm wide. *(Courtesy of R. E. Grant, U.S. Geological Survey.)*

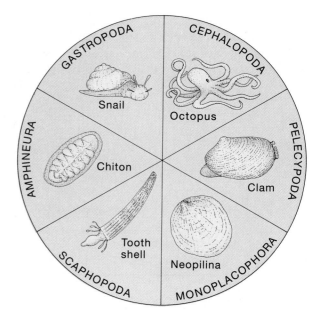

FIGURE 10–38 Some common members of the phylum Mollusca. *(From Levin, H. L. 1975. Life Through Time. Dubuque, IA: William C. Brown Co.)*

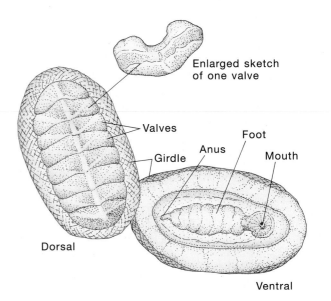

FIGURE 10–39 The common Atlantic Coast chiton.

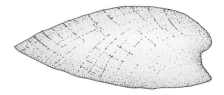

FIGURE 10–40 The "living fossil" *Neopilina ewingi.* (magnification ×$\frac{1}{2}$).

tively spinose (Fig. 10–37) with large, inflated ventral valves. They were so numerous during the Carboniferous that the period might well be dubbed the age of productids.

Mollusks

A stroll along almost any seashore will provide evidence that mollusks are today's most familiar marine invertebrates. The shells of most members of this phylum have been readily preserved and provide enjoyment to collectors and conchologists today. Such well-known animals as snails, clams, chitons, tooth shells, and squid are included within the phylum **Mollusca** (Fig. 10–38). Although most mollusks possess shells, some, such as the slugs and octopods, do not. The various classes of Mollusca differ considerably in external appearance, yet they have fundamental similarities in their internal structure. There is a muscular portion of the body called the *foot* that functions primarily in locomotion. In cephalopods, the foot is modified into tentacles. A fleshy fold called the *mantle* secretes the shell. In aquatic mollusks, respiration is accomplished by means of gills. Well-developed organs for digestion, sensation, and circu-

lation attest to the advanced stage of evolution that mollusks have attained.

Included within the phylum Mollusca are a group of marine animals that may be collectively referred to

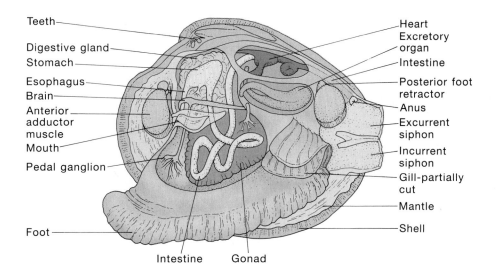

Teeth
Digestive gland
Stomach
Esophagus
Brain
Anterior adductor muscle
Mouth
Pedal ganglion
Foot

Heart
Excretory organ
Intestine
Posterior foot retractor
Anus
Excurrent siphon
Incurrent siphon
Gill-partially cut
Mantle
Shell

Intestine   Gonad

FIGURE 10–41 Internal anatomy of a clam.

as placophorans. **Placophorans** are relatively primitive mollusks that have multiple paired gills and, in shelled forms, a creeping foot much like that seen in snails. The most familiar placophorans are the polyplacophorans, represented by chitons (Fig. 10–39), which possess eight overlapping plates covering an ovoid, flattened body. Chitons are highly adapted for adhering to and grazing on algae that covers rocks and shells. Their fossil record begins in the Cambrian and extends to the present day.

As indicated by their name, **monoplacophorans** are placophorans that have a single shell resembling a flattened or short cone. Their fossil record extends back to the Cambrian, but they disappeared from the fossil record in the Devonian. In 1952, however, ten living specimens of a monoplacophoran were dredged from a deep ocean trench off the Pacific coast of Costa Rica. The specimens, later named *Neopilina* (Fig. 10–40), were of great interest to paleontologists, for they displayed a segmental arrangement of gills, muscles, and other organs indicating that mollusks branched off the invertebrate family tree at some point below the branch represented by segmented annelid worms and that the unsegmented condition seen in most mollusks is not primitive but rather a secondary development. Many paleontologists regard the monoplacophorans as ancestral to more familiar mollusks such as bivalves, gastropods, and cephalopods.

The **Bivalvia** (also termed the Pelecypoda) are a class of mollusks that includes clams, oysters, and mussels. They are characterized by layered gills, a muscular foot (Fig. 10–41), a bilobed mantle, and the absence of a definite "head." Although members of the class made their first appearance in the Cambrian, they did not become notably abundant until Pennsylvanian and Permian, when they populated the shallow seas of the time.

Gastropods first appear in Lower Cambrian strata. Earliest forms constructed small, depressed conical shells. During later Cambrian and Ordovician time, gastropods with the more familiar coiled conchs became commonplace (Fig. 10–42). It appears that coiling in a plane was not unusual initially

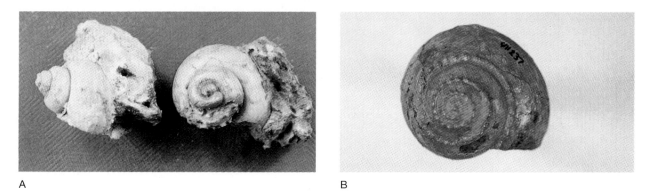

A   B

FIGURE 10–42 Two representative early Paleozoic gastropods. (A) *Taeniospira* from the Upper Cambrian Eminence Formation of Missouri. (B) *Clathospira* from the Ordovician Platteville Formation of Minnesota (×0.85).

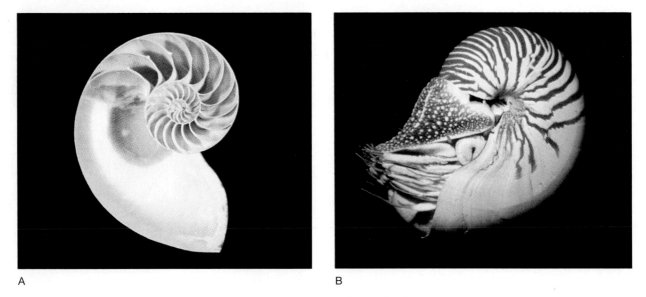

A   B

FIGURE 10–43 *Nautilus,* a modern nautiloid cephalopod. (A) Conch of a nautiloid sawed in half to show large living chamber, septa, and septal necks through which the siphuncle passed. (B) Living animal photographed at a depth of about 300 meters.   *(Photograph courtesy of W. Bruce Saunders.)*

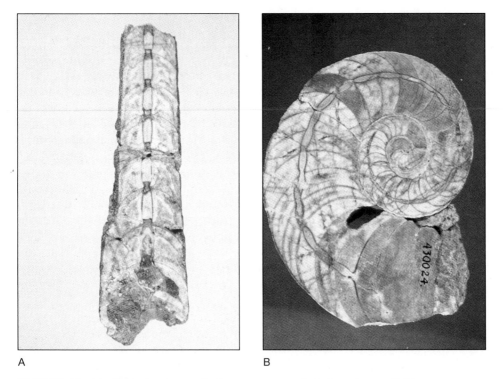

A   B

FIGURE 10–44 Variation in conch shape among early Paleozoic nautiloid cephalopods. Both of these specimens are from the Silurian of Bohemia. (A) A sawed and polished section of the straight conch of *Orthoceras potens* showing septa and siphuncle. (B) Sawed and polished section of *Barrandeoceras,* exhibiting a coiled form. Specimen (A) is 22.5 cm in length; (B) has a diameter of 18 cm.

but was supplanted by spiral coiling. Gastropods of the Ordovician and Silurian had shapes similar to those of living species. One presumes that their soft parts were also similar, with distinct head, mouth, eyes, tentacles, and a vertically flattened foot to provide for gliding movement. By Pennsylvanian time, gastropods had become abundant and diverse. During this period, the oldest known air-breathing (pulmonate) gastropods appeared. The succeeding Permian Period ended with widespread extinctions among marine invertebrates, and many families of gastropods were decimated.

The **cephalopods** may be the most complex of all the mollusks. Today, this marine group is represented by the squid, cuttlefish, octopods, and the attractive chambered nautilus. The nautilus in particular provides us with important information about the soft anatomy and habits of a vast array of fossils known only by their preserved conchs. In the genus *Nautilus* (Fig. 10–43), one finds a bilaterally symmetric body, a prominent head with paired image-forming eyes, and tentacles developed on the forward portion of the foot. Water is forcefully ejected through the tubular "funnel" to provide swift, jet-propelled movement.

At first glance, the conchs of cephalopods resemble snail shells, but the resemblance is only superficial. Although we have mentioned an exception to this rule, the gastropod shell generally coils in a spiral, whereas the cephalopod conch characteristically coils in a plane. More importantly, the planispirally coiled conch is divided into a series of chambers by transverse partitions, or septa. The bulk of the soft organs reside in the final chamber. Where the septa join the inner wall of the conch, *suture* lines are formed. These lines are enormously useful in the identification and classification of cephalopods. For example, cephalopods placed in the subclass **Nautiloidea** have straight or gently undulating sutures, whereas the **Ammonoidea** have more complex sutures.

The oldest fossils to be classified as cephalopods were small, conical conchs discovered in Lower and Middle Cambrian rocks of Europe. The class gradually increased in number and diversity and became ubiquitous inhabitants of Ordovician and Silurian seas. Indeed, by Silurian time, a great variety of conch forms—from straight to tightly coiled (Fig. 10–44)—had developed. Some of the elongate forms exceeded 4 meters in length. The first signs of a decline in nautiloid populations can be detected in Silurian strata. After Silurian time, the group continued to dwindle until today only a single genus, *Nautilus,* survives.

·During the Devonian, the first ammonoid cephalopods appeared. These were the **goniatites,** charac-

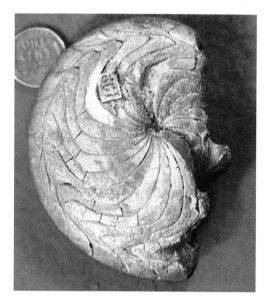

FIGURE 10–45 Goniatite ammonoid cephalopod exhibiting zig-zag sutures. Diameter is 4.6 cm.

terized by angular and generally zig-zag sutures without any additional crenulations (Fig. 10–45). The goniatites persisted throughout the late Paleozoic and gave rise to the ceratites and ammonites of the Mesozoic.

Arthropods

The **arthropod** phylum is enormous. It includes such living animals as lobsters, spiders, insects, and a host of other animals that possess chitinous exterior skeletons, segmented bodies, paired and jointed appendages, and highly developed nervous system and sensory organs. Members of Arthropoda that have left a particularly significant fossil record are the **trilobites, ostracods,** and **eurypterids.**

Trilobites (Fig. 10–46) were swimming or crawling arthropods that take their name from division of the dorsal surface into three longitudinal segments, or *lobes.* There are, for example, a central axial lobe and two lateral (pleural) lobes (Fig. 10–47). There was also a transverse differentiation of the shield into an anterior *cephalon,* a segmented *thorax,* and a posterior *pygidium.* The skeleton was composed of chitin strengthened by calcium carbonate in parts not requiring flexibility. As in many other arthropods, growth was accomplished by molting. Although many trilobites were sightless, the majority had either single-lens eyes or compound eyes composed of a large number of discrete visual bodies. The earliest

A

B

C

D

FIGURE 10–46 A gallery of trilobites. (A) The Middle Cambrian guide fossil *Ogygopsis klotzi* from Mount Stephens, British Columbia, near the site of the famous Burgess Shale quarry (length 6.0 cm). (B) *Ellipsocephalus* from the Middle Cambrian of Bohemia (average length 2.8 cm). (C) *Isotelus* from Ordovician limestones in New York state (length 3.8 cm). (D) *Dalmanites* from Silurian beds in Indiana (length 3.2 cm).

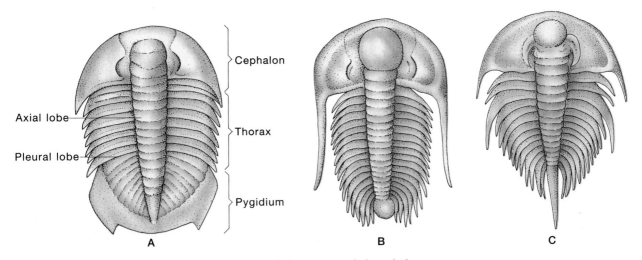

Cephalon

Axial lobe

Thorax

Pleural lobe

Pygidium

A

B

C

FIGURE 10–47 Three well-known Cambrian trilobites. (A) *Dikelocephalus minnesotensis* (Upper Cambrian), (B) *Paradoxides harlani* (Middle Cambrian), and (C) *Olenellus thompsoni* (Lower Cambrian).

C O M M E N T A R Y
The Eyes of Trilobites

"The eyes of trilobites eternal be in stone, and seem to stare about in mild surprise at changes greater than they have yet known," wrote T. A. Conrad. Indeed, trilobites may have been the first animals to look out upon the world, for they possessed the most ancient visual systems known. The eyes of trilobites include both simple and compound types. A simple eye appears as a single, tiny lens, resembling a small node. A few receptor cells probably lay beneath each simple lens. Simple eyes are rare among trilobites. Far more abundant are compound eyes composed of a large number of individual visual bodies, each with its own biconvex lens. Each lens is composed of a single crystal of calcite. One of the properties of a clear crystal of calcite is that in certain orientations, objects viewed through the crystal produce a double image. Trilobites, however, were not troubled by double vision because the calcite lenses were oriented with the principal optic axis (the light path along which the double image does not occur) normal to the surface of the eye.

In some compound trilobite eyes, termed holochroal, the many tiny lenses are covered by a continuous, thin, transparent cornea. Beneath this smooth cover there may be as many as 15,000 lenses. In other trilobites, such as *Phacops* and *Calliops* (see Fig. 10–44 and the figure in this Box), there are discrete individual lenses each covered by its own separate cornea and separated from its neighbors by a cribwork of exoskeletal tissue. Such compound eyes are termed schizochroal.

Trilobite eyes can sometimes provide clues useful in determining the habits of certain species. Eyes

The large schizochroal right eye of a well-preserved, silicified cephalon of the trilobite *Calliops*. Light directed toward the back of the specimen is emerging through some of the lensal pits.

along the anterior margin of the cephalon, for example, may indicate active swimmers. Eyes located on the ventral side of the cephalon may indicate that the trilobite was a surface-dweller. In the majority of trilobites, the eyes are located about midway on the cephalon in a position appropriate for an animal that crawled on the sea bottom or occasionally swam above it. Trilobites that either lacked eyes or were secondarily blind appear to have been adapted for burrowing in the soft sediment on the ocean floor.

---

trilobites had small pygidia and a large number of thoracic segments. These characteristics suggest that trilobites evolved from annelid worms.

If one considers the entire fossil record for only the Cambrian, the trilobites are clearly the most abundant and diverse. More than 600 genera are known from the Cambrian. They first appear in rocks above those of the Tommotian Stage of the Cambrian.

The earliest trilobites apparently were bottom-dwelling, crawling scavengers and mud processors. Some preferred limy bottoms of cratonic and shelf areas, and others, such as *Paradoxides* of the Middle Cambrian, inhabited the muddy or silty floors of deep-water tracts. As a result of such environmental preferences, it has been possible to delineate trilobite faunal provinces in North America and Europe, which suggest that these two continents were once close together. (For example, the Lower Cambrian *Olenellus* assemblage is found in exposures in Pennsylvania, New York, Vermont, Newfoundland, Greenland, and Scotland.)

The optimum time for trilobites was reached in the Late Cambrian. After that they began to decline,

FIGURE 10–48   Side view of the trilobite *Phacops rana* showing the discrete visual bodies of the large compound eye. Length is 3.2 cm.

perhaps in response to predation from cephalopods and fishes; however, they remained fairly abundant throughout the Ordovician and Silurian. Extinctions during the Middle and Late Devonian decimated the majority of trilobite groups, but the large-eyed group known as proparians (Fig. 10–48) persisted and now serve as guide fossils. The largest trilobites ever found lived during the Devonian. One giant reached a length of over 70 cm. After the temporary increase in diversity during the Devonian, trilobites continued to wane until they became extinct near the close of the Paleozoic . Nevertheless, they were not at all biologic failures, for they had been important animals in the oceans for over 300 million years.

Arthropod companions to the trilobites in the Paleozoic seas were the small, bean-shaped ostracods (Fig. 10–49). At first glance, ostracods appear so different from trilobites as to cause one to question their classification within the same phylum. The ostracods have a bivalved shell vaguely suggestive of some sort of tiny clam. However, this bivalved carapace encloses a segmented body from which extend seven pairs of jointed appendages. Adult animals are about 0.5 to 4 mm in length. The valves are composed of both chitin and calcium carbonate and are hinged along the dorsal margin.

Ostracods first appeared early in the Ordovician, and some limestones of this age are almost completely composed of their discarded carapaces. *Leperditia* and *Euprimitia* are common Ordovician genera. Ostracods continue in relative abundance to the present day. They occur in both marine and freshwater sediments. Because of their small size, they are brought to the surface in wells drilled for oil, and along with other microfossils, such as foraminifers and radiolaria, are used by exploration geologists in correlating the strata of oil fields.

FIGURE 10–49   Ostracods *(Leperditia fabulites)* in the Ordovician Plattin Limestone, Jefferson County, Missouri. Ostracods are the smooth, bean-shaped fossils. These average about 1 cm in length and are thus large ostracods. Two strophomenid brachiopods are also present on the surface of the limestone.

FIGURE 10–50   The fossil eurypterid *Eurypterus lacustris* from the Bertie Waterlime of Silurian age, Erie County, Pennsylvania. This well-preserved specimen is about 28 cm long.   *(Courtesy of Ward's Natural Science Establishment.)*

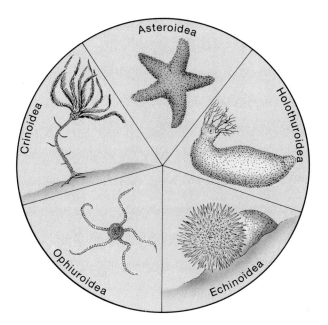

FIGURE 10–51 Representative living echinoderms. *(From Levin, H. L. 1975. Life Through Time. Dubuque, IA: William C. Brown Co.)*

Eurypterids (Fig. 10–50) are a group of arthropods that, because of their rarity, are less useful in stratigraphic studies than are either trilobites or ostracods. Nevertheless, they are among the most impressive of early Paleozoic marine invertebrates. Although many were of modest size, some were nearly 3 meters long and, had they survived, would have been suitable subjects for a Hollywood monster film. On their scorpion-like bodies were five pairs of appendages and a fearful-looking pair of pincers. Some were also equipped with a terminal spine. Eurypter-

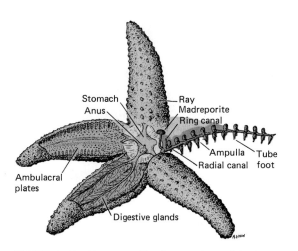

FIGURE 10–52 Partially dissected starfish showing elements of the water vascular system and other organs.

ids ranged across portions of the sea floor and brackish estuaries from Ordovician until Permian time but were especially abundant during the Silurian and Devonian.

Spiny-Skinned Animals: Echinoderms

Just as modern seas abound with starfish, sea urchins, and sea lilies, so were the oceans of the early Paleozoic populated with members of the phylum Echinodermata (Fig. 10–51). **Echinoderms** are animals with mostly five-way symmetry that masks an underlying primitive bilateral symmetry. They are well named "the spiny skinned," for spines are indeed present in many species. A unique characteristic of the phylum is the presence of a system of tubes—the water vascular system—which functions in respiration and locomotion (Fig. 10–52). Members of the phylum are exclusively marine, typically bottom-dwelling, and either attached to the sea floor or able to move about slowly.

Of considerable interest is the evolutionary relationship between echinoderms and vertebrates. There are so many allied embryologic parallelisms among primitive chordates and echinoderms that some zoologists speculate that both may have arisen from similar ancestral forms. For example, the larvae of echinoderms closely resemble those of the living protochordates called acorn worms. Also, in embryologic development, the mesodermal layer of cells, as well as certain other elements of the body, arises in the same way. Biochemistry has provided additional evidence for a relationship between echinoderms and chordates by revealing chemical similarities associated with muscle activity and oxygen-carrying pigments in the blood.

Among the many classes of the phylum Echinodermata, the Asteroidea (starfish), Ophiuroidea (brittle stars), Echinoidea (sea urchins), Edrioasteroidea, Crinoidea (sea lilies), Blastoidea, and Cystoidea are the most abundant and useful in geologic studies. The mostly attached and sessile crinoids, blastoids, and cystoids were particularly characteristic of the Paleozoic Era, whereas the more mobile asteroids and echinoids were more frequent in later eras.

Echinoderms appear to have evolved late in the Proterozoic, although fossil remains of Proterozoic forms are few and often enigmatic. The Ediacaran fauna, for example, includes a globular fossil named *Arkarua* that has five rays on its surface and may be related to a group of echinoderms called edrioasteroids. Edrioasteroids (Fig. 10–53) are considered by many paleontologists to be ancestral to starfish and sea urchins.

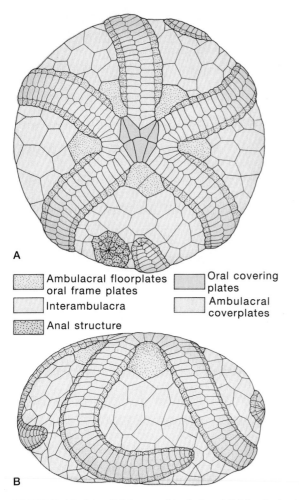

A

Ambulacral floorplates oral frame plates	Oral covering plates
Interambulacra	Ambulacral coverplates
Anal structure	

B

FIGURE 10–53 *Edrioaster bigsbyi*, a Middle Ordovician edrioasteroid. Specimen is 45 mm in diameter. (A) Oral surface. (B) Lateral view of the globoid fossil. *(From Bell, B. M. 1977. J. Paleo. 51(3):620.)*

FIGURE 10–54 The spiraled, spindle-shaped Early Cambrian echinoderm *Helicoplacus*. The specimen is about 2.6 cm long. *(After Durham, J. W., and Caster, K. E. 1963. Science 140:820–822.)*

FIGURE 10–55 A well-preserved specimen of the Silurian cystoid *Caryocrinites ornatus* from the Lockport Shale of New York. *(From Sprinkle, J. 1975. J. Paleo. 49(6):1062–1073.)*

By Early and Middle Cambrian time, several classes of rather peculiar echinoderms appeared, but this initial radiation was apparently not successful. None of these became abundant, and three became extinct before the end of the Cambrian. One of the most bizarre of these early echinoderms was *Helicoplacus* (Fig. 10–54), a form that has plates and food grooves arranged in a spiral around its spindle-shaped body.

The stemmed or stalked echinoderms first occur in Middle Cambrian strata but do not become abundant until Ordovician and Silurian time. Stalked forms called **cystoids** (Fig. 10–55) are the most primitive among this group. The striking pentamerous (five-way) symmetry that is evident in most echinoderms is often less well developed in cystoids. Beginning students of paleontology often recognize cystoids by the characteristic pores that occur in rhomboid patterns on the plates of the calyx. Although cystoids range from Cambrian to Late Devo-nian, they are chiefly found in Ordovician and Silurian rocks.

Unlike some of the cystoids, stalked echinoderms known as **blastoids** (Fig. 10–56) have a beautifully symmetric arrangement of plates. The five radial

FIGURE 10–56 Calyx of the blastoid *Pentremites symmetricus* from the Upper Mississippian near Floraville, Illinois. Height of calyx is 2.6 cm.

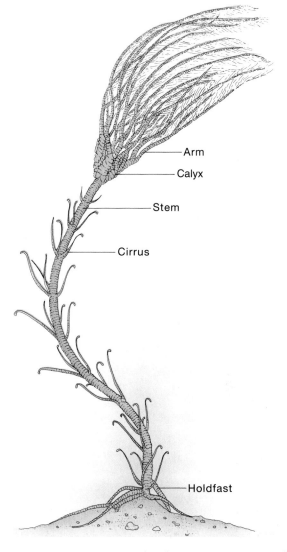

FIGURE 10–57 Crinoid in living position on sea floor.

areas (ambulacral areas) are prominent, bear slender branches or brachioles along their margins, and have a well-developed and unique water transport system. The blastoids first appeared in Silurian time, expanded in the Mississippian, and declined to extinction in the Permian.

Crinoids, like most cystoids and blastoids, are composed of three main parts: the *calyx* (which contains the vital organs), the *arms*, and the *stem*, with its rootlike *holdfast* (Fig. 10–57). The arms bear ciliated food grooves and, like the brachioles of blastoids, serve to move food particles toward the mouth. Crinoids are found in rocks that range in age from Ordovician to Holocene. In some areas, Ordovician, Silurian, and Carboniferous rocks contain such great quantities of disaggregated plates of crinoids that they are named crinoidal limestones. So abundant are crinoid remains in Mississippian rocks that the period has been dubbed the age of crinoids (Fig. 10–58).

A

B

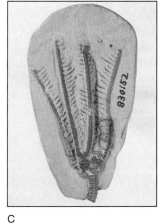

C

FIGURE 10–58 (A) Diorama depicting crinoids living on the floor of a Mississippian epeiric sea. (B) Calyx and arms of the Mississippian crinoid *Taxocrinus* (height 5.1 cm). (C) *Scytalocrinus* (height 4.5 cm). Both fossil crinoids are from the Keokuk Formation of Mississippian age, near Crawfordsville, Indiana.

(Diorama photograph courtesy of the National Museum of Natural History, Smithsonian Institution.)

FIGURE 10–59   The large Mississippian echinoid *Melonechinus* from the St. Louis Limestone, St. Louis, Missouri. Average diameter of specimens is 12 cm. *(Photograph by John Simon.)*

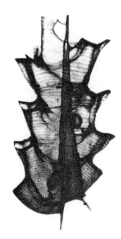

FIGURE 10–60   Part of a stipe of the Ordovician graptolite *Orthograptus quadrimucronatus* (×15). The specimen has been bleached for better visualization of internal features. The cuplike thecae are clearly visible on either side. One can also see the inverted cone of the sicula. The nema would rise from the pointed end of the sicula.   *(From Herr, S. R. 1971. J. Paleo. 45(4):628–632.)*

Although one group of crinoids survived the hard times at the end of the Paleozoic and gave rise to those still living today, most died out before the end of the Permian. The blastoids also died out before the end of the Paleozoic. Free-living echinoderms such as echinoids (Fig. 10–59) and starfish were locally abundant during the late Paleozoic but not as numerous or diverse as they would become in post-Paleozoic time.

Graptolites

While myriads of benthic invertebrates populated the sea floors of the early Paleozoic, there were also great numbers of floating and swimming organisms living above the ocean floor. Many of these planktic (or planktonic) creatures lacked skeletons and may never be revealed to us. One group, the **graptolites** (Fig. 10–60), had preservable chitinous skeletons that housed tiny individual organisms.

Graptolites made their appearance at the end of the Cambrian but did not become abundant until the Ordovician. Until recently, they were thought to have become extinct by the end of the Mississippian. In 1989, however, a specimen was recovered from South Pacific waters that may be a surviving species of the graptolite group. Marine biologist Noel Dilly named this apparent survivor *Cephalodiscus graptoloides*. The protein found in the skeletal material and the specimen's general morphology closely resemble that of fossil graptolites of the early Paleozoic. After the initial discovery of *Cephalodiscus graptoloides*, additional specimens were found by Noel Dilly while vacationing in Bermuda.

The structure of the graptolite skeleton is distinctive. The graptolite animals live in tiny cups or *thecae*, that are aligned along a stem or *stipe*. The stipes may be solitary or formed into a system of two or more branches. In general, evolution progressed from multibranched forms to those having only a single stipe. The entire colony is referred to as a rhabdosome. At the lower end of the rhabdosome there is a single thin filament—the *nema*. Some graptolites were attached to floating objects by the threadlike nema. Where the lower end of the nema reaches the base of a stipe, there is conical *sicula*, which may have served as the theca for the first individual. From the sicula, subsequent individuals and their thecae were added by budding.

Fossil graptolites characteristically occur as flattened and carbonized impressions in dark shales. Rarely, uncompressed specimens are found. These reveal an unexpected relationship to primitive living chordates called *pterobranchs*. Both groups secrete tiny enclosed tubes, and in both the unique structure of the thecae is so similar that a close relationship is virtually certain.

During the Ordovician and Silurian, graptolites were so abundant that they may have formed sargasso-like floating masses. They were apparently carried about by ocean currents and thus achieved the worldwide distribution required of guide fossils. Their use in stratigraphy, however, is limited in that they are seldom found preserved except in fine-grained, usually carbonaceous, shalelike sediments. Perhaps these occurrences came about as masses of graptolite colonies floated into toxic waters, died, and settled to the sea floor.

Continental Invertebrates

Because of the greater hazards of postmortem destruction of continental (terrestrial) invertebrates, their fossil record is not as complete as that for marine invertebrates. Nevertheless, as plants invaded the continents, animals were able to follow. Arthropods, with their sturdy legs and protective exoskeletons, were probably the marine animals best preadapted for coming ashore. Early invaders may have lived within the wet, nutritious plant debris washed up along the beaches of early Paleozoic seas. Their fossil record begins with possible millipede trace fossils in Ordovician rocks. More direct evidence consists of arthropod cuticles and bristles recovered along with spores from early Silurian strata. The same samples also contain fecal pellets believed to have been left by arthropods feeding on fungi and algae. Body fossils of centipedes and millipedes occur in late Silurian deposits of Britain. Near Gilboa, New York, Devonian rocks have yielded the fossil remains of mites, primitive wingless arthropods called archaeognaths, centipedes, the earliest known spiders, and spider-like predators known as trigonotarbids. Except for the primitive bristletails, insects do not become common until about 315 million years ago (late Mississippian). Pennsylvanian rocks contain a better record of insects, among which were giant dragonflies with wingspans exceeding 70 cm (2 feet). Giant cockroaches (Fig. 10–61) found their food in forest litter and rotting vegetation. Although never abundant, eurypterids persisted throughout the late Paleozoic. In Nova Scotia, an interesting collection of land snails has been collected from hollows within the stumps of fossil trees. Scorpions and centipedes have been found in abundance in a concretionary Pennsylvania siltstone exposed along Mazon Creek in northern Illinois.

Of all the land invertebrates, it is apparent that the arthropods and gastropods have been persistently successful. Many paleontologists believe this may be because of attributes already evolved by their aquatic

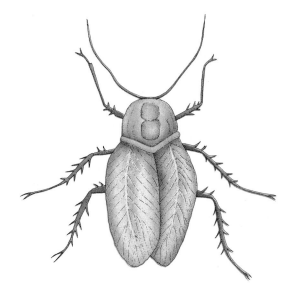

FIGURE 10–61 Reconstruction of a primitive Pennsylvanian cockroach. Length is 10 cm.

ancestors. As protection against desiccation, arthropods had evolved relatively impervious exoskeletons. Snails derived similar benefits from their shells. Both groups included very active animals with sufficient mobility to seek out food aggressively.

■ *Vertebrate Animals of the Paleozoic*

Conodonts

For more than a century, tiny fossils known as conodont elements have been known to geologists. Over this span of time, however, the nature of the animal that bore the elements and the function they served have been a persistent puzzle. Even without this knowledge, however, conodont elements have served admirably as guide fossils. They are found in a variety of marine sedimentary rocks that range from latest Paleozoic to Triassic in age.

Conodont elements consist of cones, bars and blades bearing tiny denticles or cusps, or variously shaped ridged structures. The elements are usually less than a millimeter in size and composed of a durable calcium phosphate variety of the mineral apatite. Fossil occurrences consist mostly of disassociated elements. In order to determine which elements occurred together in the parent organism, one must find conodont elements in natural association and not scattered after death. When, with good fortune, one finds such an assemblage, it is often possible to recon-

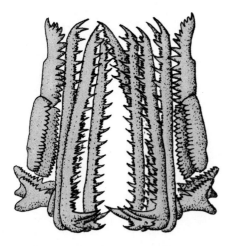

FIGURE 10–62   Conodont elements arranged in a natural assemblage.   *(After Scott, H. W. Conodont assemblages from the Heath Formation, Montana.* Jour. Paleo. *16(3):298–300.)*

struct the way differently shaped elements were arranged in the complete apparatus (Fig. 10–62; also see Fig. 4–52).

But what was the nature of the conodont animal? Since the discovery of conodont elements by Christian Pander in 1856, proposals have been advanced that they were either chewing or filtration or support structures for such diverse candidate animals as worms, fish, snails, cephalopods, or arthropods. Recently uncovered evidence, however, indicates that the conodont animal was a chordate. Several fossils of conodont animals have been reported over the past three decades, but the most informative fossil was found in 1995 in the Ordovician Soom Shale of South Africa. It consists of the remains of a creature with an elongate, eel-like body that in life was about 40 cm long. The creature had distinctive eye musculature, similar to that in primitive jawless vertebrates. Like the living cephalochordate *Branchiostoma* (see Fig. 10–63), muscles along either side of the creature were V-shaped units lying sidewise with the point forward. Scanning electron microscopy of the fossilized fibrous muscle tissue revealed similarities to the muscle tissue preserved in fossil fish. The Soom Shale specimen, named *Promissum,* has a longitudinal band that may mark the position of a notochord.

The above evidence strongly favors chordata status for the conodont animal. But what function was served by the conodont elements? Were they supports for food gathering organs? Did they act as sieves for filtering out fine food particles from water? Or were they used for chewing food? If conodont elements did function as teeth, they should show the effects of wear associated with chewing. Recent scanning electron microscope images of conodont elements reveal fine scratches, pitting, and other surface textures strongly suggesting the elements were used in chewing and shearing food. The conodont animal was probably not a docile deposit feeder, but an active predator, able to hold, crush, and cut the tissues of its prey.

The Rise of Fishes

A momentous biologic event occurred in early Paleozoic time. It was the birth of the chordate line. In our earlier discussion of the Burgess Shale fauna, we briefly described chordates as animals that have (at least at some stage in their life history) a stiff, elongate supporting structure. In addition, all chordates have a dorsal, central nerve cord, gill slits, and blood

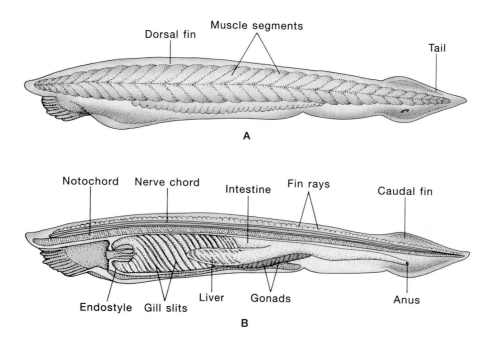

FIGURE 10–63
Branchiostoma, a member of the subphylum Cephalochordata. (A) External view; (B) longitudinal section. (Length 3 to 5 cm.)

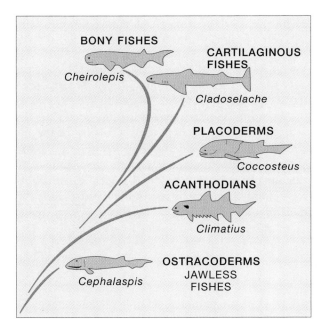

FIGURE 10–64 The basic evolutionary radiation of fishes that began during the early Paleozoic and produced the sharks and bony fishes during the Devonian. *(From Colbert, E. H. 1969. Evolution of the Vertebrates. New York: John Wiley. With permission of the author, artist Lois Darling, and the publisher.)*

that circulates forward in a main ventral vessel and backward in the dorsal. The supportive structure in primitive chordates is called a notochord and has been studied by generations of biology students in such animals as the lancelet *Branchiostoma* (Fig. 10–63). In taxonomic hierarchy, vertebrates are simply those chordates in which the notochord is supplemented or replaced by a series of cartilaginous or bony vertebrae.

As we have mentioned previously, the ancestors of the vertebrate lineage may lie somewhere among the echinoderms. The theoretic evolutionary progression that was to lead to vertebrates may have begun with sedentary, filter-feeding animals that had ex-

posed cilia located along their arms, somewhat in the fashion of crinoids. From such a beginning there may have evolved filter-feeders with cilia brought inside the body in the form of gills. In a subsequent stage, the organisms may have become free-swimming, gilled animals, in appearance not unlike small, simple fishes or frog tadpoles. *Pikaia* see Fig. 10–15), from the Middle Cambrian Burgess Shale, appears to have been such an animal. *Pikaia* or animals like it may be at the base of the evolutionary sequence leading to vertebrates. Presumably, we may call those earliest vertebrates fishes.

The oldest fossil remains of fishes consist of small pieces of external plates found in Upper Cambrian deposits of Wyoming. Their distinctive surface patterns and internal structure clearly indicate that these plates were part of the body covering of armored jawless fishes. Somewhat younger scales and plates of jawless fishes are found in Ordovician rocks of Colorado, Spitzbergen, and Greenland.

The vertebrates that we loosely call fishes are actually divided into at least five distinct taxonomic classes (Fig. 10–64). There are the jawless fishes, or Agnatha (which may include conodont-bearing animals); two groups of archaic jawed fishes, the Acanthodii and Placodermi; the cartilaginous fishes, or Chondrichthyes; and the familiar Osteichthyes with their highly developed bony skeletons. It is the first three of these categories—namely, the Agnatha, Acanthodii, and Placodermi—that frequented freshwater and saltwater bodies of the early Paleozoic.

The agnaths still exist today as the inelegant hagfish and lamprey. However, these specialized survivors are unlike the jawless fishes of the Ordovician and Silurian. Early Paleozoic agnaths are collectively termed **ostracoderms.** The name means "shell skin" and refers to the bony exterior that was a distinctive trait of many of these fishes. Ostracoderms comprised a rather diverse group that included the unarmored forms *Theolodus* and *Jamoytius* as well as such armored creatures as *Pteraspis* (Fig. 10–65).

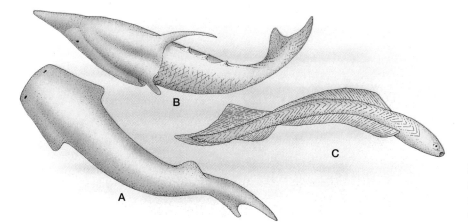

FIGURE 10–65 The early Paleozoic ostracoderms, (A) *Thelodus*, (B) *Pteraspis*, and (C) *Jamoytius*, drawn to the same scale.

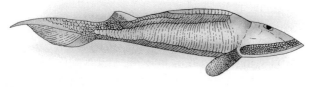

FIGURE 10–66 The ostracoderm *Hemicyclaspis*. The creature was about 15 cm long. *(After Stensio, A. E. 1932. Br. Mus., p. 3, fig. 15.)*

The purpose of the bony armor that is the hallmark of many of the ostracoderms is still being debated. A widely held view is that the armor provided protection against predators. Another theory stipulates that the dermal armor was not primarily for protection but rather was a device for storing seasonally available phosphorus. According to this idea, phosphates could be accumulated as calcium phosphate in the dermal layers during times of greater availability and then used during periods when supply was deficient. This cache of phosphorus may have been vital for early vertebrates to maintain a suitable level of muscular activity.

Hemicyclaspis (Fig. 10–66) is one of the most widely known of the ostracoderms. It is recognized by its large, semicircular head shield, on the top of which can be found four openings: two for the upward-looking eyes, a small pineal opening, and a single nostril. The pineal opening may have housed a light-sensitive third eye. The mouth was located ventrally in a position suitable for taking in food from the surface layer of soft sediment. Along the margin of the head shield were depressed areas that may have had a sensory function. Internally, the head contained a regular segmental arrangement of nerves, blood vessels, and gill pouches. These and other features of the soft internal anatomy have been discerned by dissection of exceptionally well-preserved specimens.

The ostracoderms continued into the Devonian but did not survive beyond that period. For the most part they were small, sluggish creatures restricted to mud-straining or filter-feeding modes of life. They were to be replaced gradually by fishes that had developed bone-supported, movable jaws. The evolution of the jaw was no small accomplishment, for it enormously expanded the adaptive range of the vertebrates. Fishes with jaws were able to bite and grasp. These new abilities led to more varied and active ways of life and to new sources of food not available to the agnaths.

The evolutionary development of the vertebrate jaw involved a remarkable transformation in which an older structure was modified to perform a role

that was entirely different from its original function. Those older structures were supports called *gill arches* that were located between the *gill slits* (Fig. 10–67). As the mouth extended posteriorly, the first two of these arches were sequentially eliminated. The third set of gill arches was gradually modified into jaws. Anatomic studies have shown the similarities in jaw architecture and neurology to elements of the gill arches and the nerves that serve these supports. In the earliest fishes, the upper jaw was attached to the skull by ligaments only. A full gill slit occurred immediately behind the jaw. In more advanced fishes, the upper jaw was anchored to the skull by the upper part of the next gill arch, the *hyomandibular*. The pair of gill slits that had once existed behind the jaw

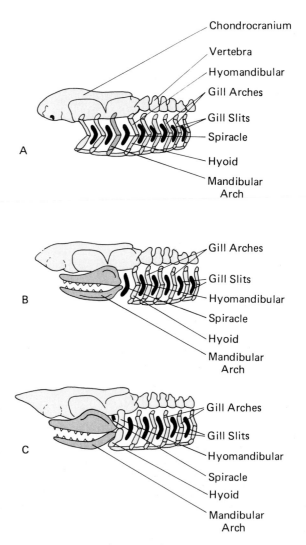

FIGURE 10–67 Origin of jaws. (A) Gill arches as they might exist in a primitive jawless fish. (B) Early jawed fish with a complete gill slit behind jaws. (C) A jawed fish (like a shark) with the first gill slit reduced to a spiracle. *(Modified from Cockrum, E. L., and McCauley, W. J. 1965. Zoology. Philadelphia: W. B. Saunders.)*

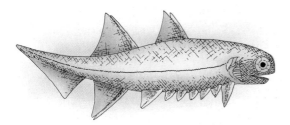

FIGURE 10–68 The Early Devonian acanthodian fish *Climatius.* *(After Romer, A. S. 1945.* Vertebrate Paleontology. *Chicago: University of Chicago Press.)*

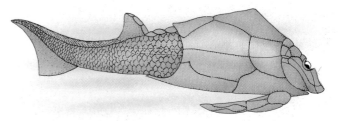

FIGURE 10–70 The Devonian antiarch fish *Pterichthyodes.* *(From Romer, A. S. 1945.* Vertebrate Paleontology. *Chicago: University of Chicago Press, p. 54, fig. 38.)*

was squeezed into a smaller space and ultimately developed into the *spiracle* openings found in members of the shark family today. The spiracle was a structure destined to become part of the ear apparatus in higher vertebrates.

The oldest fossil remains of jawed fishes are found in nonmarine rocks of the Late Silurian. These fishes, called **acanthodians** (Fig. 10–68), became most numerous during the Devonian and then declined to extinction in the Permian. Acanthodians were archaic jawed fishes and were distinct from the great orders of modern fishes. Another group of archaic fishes includes the **placoderms,** or "plate-skinned" fishes. They too arose in the late Silurian, expanded rapidly during the Devonian, and then in the latter part of that period began to decline and be replaced by the ascending sharks and bony fishes.

There was considerable variety among the placoderms. The most formidable of these plate-skinned fishes was a carnivorous group called arthrodires. *Dunkleosteus* (Fig. 10–69) was a Devonian arthrodire whose length exceeded 9 meters and whose huge jaws could be opened exceptionally wide to engulf even the largest of available prey. Other placoderms, called antiarchs (Fig. 10–70), had the heavily armored form and mud-grubbing habits of their predecessors, the ostracoderms.

Among the Late Silurian acanthodians and placoderms were the ancestors of the bony and cartilaginous fishes that were to dominate the marine realm in succeeding periods. The ascendancy of these more modern kinds of fishes began with a veritable evolutionary explosion during the Devonian. Two important categories of fishes, the cartilaginous chondrichthyians and the bony osteichthyians, made their debut in the Devonian. Today, the cartilaginous fishes are represented by sharks, rays, and skates. Among the better known of the late Paleozoic sharks

FIGURE 10–69 The gigantic armored skull and thoracic shield of the formidable late Devonian placoderm fish known as *Dunkleosteus. Dunkleosteus* was over 10 meters (about 30 feet) long. The skull shown here is about 1 meter tall. It is equipped with large bony cutting plates that functioned as teeth. Each eye socket was protected by a ring of four plates, and a special joint at the rear of the skull permitted the head to be raised and thereby provided for an extra large bite. *Dunkleosteus* ruled the seas 350 million years ago. *(Courtesy National Museum of Natural History, Smithsonian Institution; photograph by Chip Clark.)*

A

B

FIGURE 10–71   Models of (A) the Devonian marine shark *Cladoselache*, and (B) the Pennsylvanian freshwater shark *Xenacanthus*.

were species of *Cladoselache* (Fig. 10–71A). Remains of this shark are frequently encountered in the Devonian shales that crop out on the south shore of Lake Erie. During the Late Carboniferous, a group represented by *Xenacanthus* (Fig. 10–71B) managed to penetrate the freshwater environment. A third group of Paleozoic cartilaginous fishes were the bradyodonts. These had flattened bodies like modern rays and blunt, rounded teeth for crushing shellfish. Apparently, modern sharks arose from clado-selachian ancestors but retained their archaic traits until the Jurassic.

Because of the role of bony fishes in the evolution of tetrapods (four-legged animals) and because they are the most numerous, varied, and successful of all aquatic vertebrates, their evolution is of particular importance. Bony fishes may be divided into two categories: the familiar ray-fin, or Actinopterygii, and the lobe-fin, or Sarcopterygii.

As implied by their name, ray-fin fishes lack a muscular base to their paired fins, which are thin structures supported by radiating bony rays. Unlike the Sarcopterygii, they do not possess paired nasal passages that open into the throat. The ray-fins began their evolution in Devonian lakes and streams and

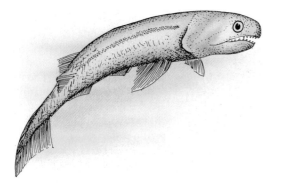

FIGURE 10–72   *Cheirolepis*, the ancestral bony fish that lived during the Devonian Period.

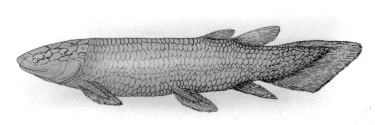

FIGURE 10–73   *Dipterus*, a Devonian lungfish.

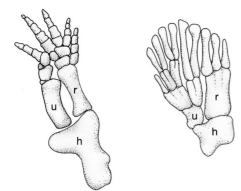

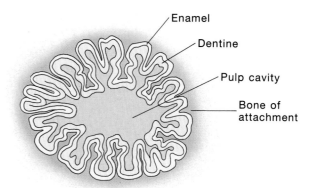

FIGURE 10–74 Comparison of the limb bones of a crossopterygian fish (upper right) and an early amphibian. *(From Levin, H. L. 1975.* Life Through Time. *Dubuque, IA: William C. Brown Co.)*

FIGURE 10–76 Cross-section of a crossopterygian tooth clearly exhibits the distinctive pattern of infolded enamel. This is also a characteristic of the teeth of the amphibian descendants of crossopterygian fishes. *(From Levin, H. L. 1975.* Life Through Time. *Dubuque, IA: William C. Brown Co.)*

quickly expanded into the marine realm. They became the dominant fishes of the modern world. The more primitive Devonian bony fishes are well represented by the genus *Cheirolepis* (Fig. 10–72). From such fishes as these evolved the more advanced bony fishes during the Mesozoic and Cenozoic.

The second category of bony fishes, the Sarcopterygii, are characterized by sturdy, fleshy lobe-fins and a pair of openings in the roof of the mouth that led to clearly visible external nostrils. Such fish were able to rise to the surface and take in air, which was passed on to functional lungs. Lungs and fins do not seem to occur together in modern fishes, but in late Paleozoic fishes the combination was not uncommon. Studies of living examples of Sarcopterygii indicate that lungs probably began their evolution as sac-like bodies developed on the ventral side of the esophagus and then became enlarged and improved for the extraction of oxygen. (In modern fishes, the lung has been converted to a swim bladder, which aids in hydrostatic balance.)

Two major groups of lungfishes lived during the Devonian. They are designated the **Dipnoi** and **Crossopterygii.** The Dipnoi, represented in the Devonian by *Dipterus* (Fig. 10–73), were not on the evolutionary track that was to lead to tetrapods. They are, nevertheless, an interesting group that includes the living freshwater lungfish of Australia, Africa, and South America. Their restricted presence south of the equator suggests that Gondwanaland was the probable center of dispersal for dipnoans. *Dipnoi* means "double breather." The name was suggested by the observation that living species are able to breathe by means of lungs during dry seasons. At such difficult times, they burrow into the mud before the water is gone. When the lake or stream is dry, they survive by using their accessory lungs, and when the waters return they switch to gill respiration.

Because of the arrangement of bones in their muscular fins (Fig. 10–74), the pattern of skull elements (Fig. 10–75), and the structure of their teeth (Fig. 10–76), fossil Crossopterygii are considered the an-

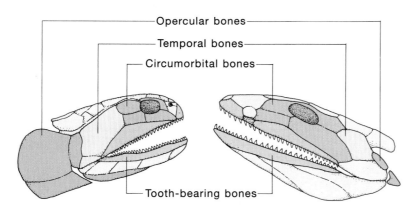

FIGURE 10–75 Comparison of skulls and lower jaws of a crossopterygian (left) and the Devonian amphibian *Ichthyostega.* *(From Levin, H. L. 1975.* Life Through Time. *Dubuque IA: William C. Brown Co.)*

FIGURE 10–77   The Devonian cros-sopterygian lungfish *Eusthenopteron* had sturdy fins whose structure fore-shadowed that of four-footed land animals that were to be their descen-dants.   *(St. Louis Science Center dio-rama.)*

cestors of the amphibians. A rather advanced fish that exemplifies Devonian crossopterygians is *Eus-thenopteron* (Fig. 10–77). In this genus, the paired fins were short and muscular. Internally, a single basal limb bone, the humerus (or femur for the pelvic structure), articulated with the girdle bones and was followed by two bones, the ulna and radius, for the pectoral fins and the tibia and fibula for the posterior fins.

Of course, the robust skeleton and sturdy limbs of the crossopterygians did not evolve because fishes had miraculously taken on a desire for life on land. These were adaptations for moving to and remaining in water during periods of drought. The air-gulping fishes found land a hostile environment and occa-sionally dragged themselves onto the land only as a means of reaching another pond or fresher body of water.

During the Devonian, two distinct branches of crossopterygians had evolved. One of these, the rhipidistians, led ultimately to amphibians, and the other led to saltwater fishes called coelacanths. Coe-lacanths were thought to have undergone extinction during the Late Cretaceous, but their survival to the present day has been documented by several catches of the coelacanth *Latimeria* (Fig. 10–78) near Mada-gascar.

Amphibians

It required tens of millions of generations to convert the crossopterygian fishes into animals that could live comfortably on solid ground (Fig. 10–79). Even so, the conversion was not complete, for amphibians continued to return to water to lay their fishlike, naked eggs. From these eggs came fishlike larvae, which, like fish, used gills for respiration.

A number of changes accompanied the shift to land dwelling. A three-chambered heart developed to route the blood more efficiently to and from the lungs. The limb and girdle bones were modified to overcome the constant drag of gravity and better hold the body above the ground. The spinal column, a simple structure in fishes, was transformed into a sturdy but flexible bridge of interlocking elements. To improve hearing in gaseous rather than fluid sur-roundings, the old hyomandibular bone, used in fishes to prop the braincase and upper jaw together, was transformed into an ear ossicle—the stapes. The fish spiracle (a vestigial gill slit) became the amphib-ian eustachian tube and middle ear. To complete the auditory apparatus, a tympanic membrane (ear drum) was developed across a prominent notch in the rear part of the amphibian skull.

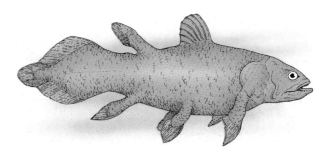

FIGURE 10–78   *Latimeria*, a surviving coelacanth liv-ing in the ocean near Madagascar and southeast Africa. *Latimeria* is a large fish, nearly 2 meters in length. *(From Levin, H. L. 1975. Life Through Time. Dubuque, IA: William C. Brown Co.)*

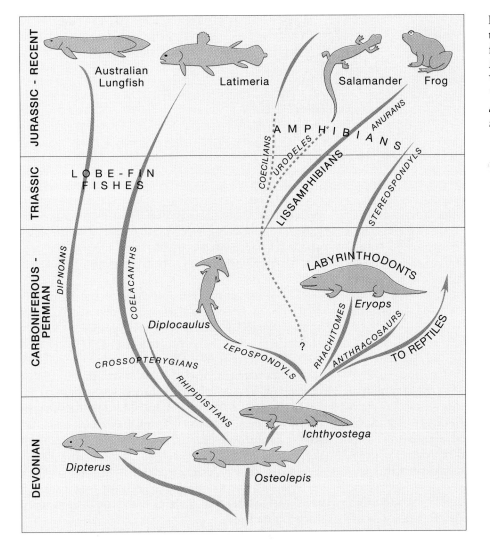

FIGURE 10–79 The evolution of amphibians and lobe-fin fishes. *(From Colbert, E. H. 1969. Evolution of the Vertebrates. New York: John Wiley. With permission of the author, artist Lois Darling, and the publisher.)*

The fossil record for amphibians begins with a group called **ichthyostegids** (Fig. 10–80). As suggested by their name, these creatures retained many features of their fish ancestors. The amphibians that followed the ichthyostegids fall mostly within a group collectively termed the **labyrinthodonts.** The labyrinthic wrinkling and folding of tooth enamel inherited from the rhipidistian Sarcopterygii provided the inspiration for the labyrinthodont name (see Fig. 10–76). During the Carboniferous, large numbers of labyrinthodonts wallowed in swamps and streams, eating insects, fish, and one another. A

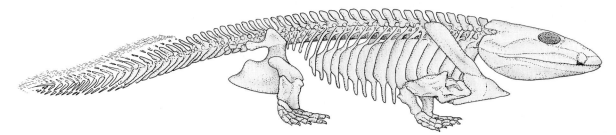

FIGURE 10–80 The skeleton of *Ichthyostega* still retains the fishlike form of its crossopterygian ancestors. *(From Levin, H. L. 1975. Life Through Time. Dubuque, IA: William C. Brown Co.)*

labyrinthodont that exemplifies the culmination of the lineage is *Eryops* (Fig. 10–81). Eryops was a bulky, inelegant creature with a flattish skull typical of labyrinthodonts and bony nodules in the skin for protection against some of its more vicious contemporaries. The labyrinthodonts declined during the Permian, and only a relatively few survived into the Triassic.

Reptiles

The evolutionary advance from fish to amphibians was no trivial biologic achievement, yet the late Paleozoic was the time of another equally significant event. Among the evolving land vertebrates were some that had developed a way to reproduce without returning to the water. They accomplished this feat by evolving enclosed eggs, in which the embryonic animal was allowed to pass through larval and other developmental stages before being hatched in an essentially adult form. This enclosed egg liberated the tetrapods from their reliance on water bodies and has been hailed as a major milestone in the history of

vertebrates. The animals that first evolved the so-called amniotic egg were amphibians. Evolutionary processes had provided them with a method for protecting developing young from predation and desiccation.

The oldest known remains of reptiles were discovered in Early Pennsylvanian swamp deposits of Nova Scotia. Named *Hylonomus*, these reptiles were only about 24 cm (9.5 inches) from head to tail and had a body form resembling that of a salamander. Their skeletal remains are found in holes within the stumps of Pennsylvanian scale trees. In slightly younger Late Pennsylvanian rocks of Kansas, additional early reptiles have been recovered from claystones and siltstones of former mudflats. The most common of these reptiles is called *Petrolacosaurus*. *Petrolacosaurus* had a longer neck and longer legs than *Hylonomus*. Details of its skull indicate that it was an early member of the lineage that would eventually lead to dinosaurs.

The small and relatively primitive reptilian groups of the Pennsylvanian provided the stock from which a succession of more specialized Permian reptiles

FIGURE 10–82   Permian reptiles. The sailback reptiles with the larger skulls and teeth are *Dimetrodon*. *Edaphosaurus*, a herbivorous form, is just right of center. The smaller lizard-like reptiles are *Casea*. *(Courtesy of The Field Museum of Natural History; painting by C. R. Knight; Negative No. CK45T.)*

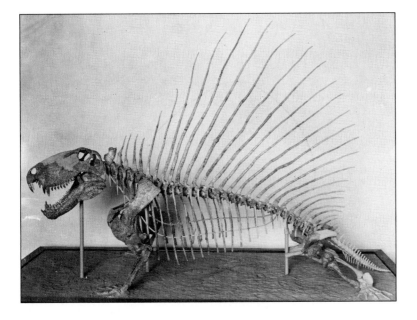

FIGURE 10–83 Mounted skeleton of the Permian "sail-reptile" *Dimetrodon gigas.* The tail was actually somewhat longer. *(Courtesy of the National Museum of Natural History, Smithsonian Institution.)*

evolved. The most spectacular of these were the **pelycosaurs,** several species of which sported erect "sails" supported by rodlike extensions of the vertebrae. The pelycosaurs were a varied group. Some, such as *Edaphosaurus,* were plant eaters, whereas others, as indicated by their great jaws and sharp teeth, ate flesh. *Dimetrodon* (Figs. 10–82 and 10–83) was one such predator. There have been many attempts to explain the function of the pelycosaurian sail. The most reasonable explanation is that the sail-fin acted in temperature regulation by serving sometimes as a collector of solar heat and at other times as a radiator giving off surplus heat. Because pelycosaurs had several mammalian skeletal characteristics, they are thought to be the early mammal-like reptilian group from which advanced mammal-like reptiles known as **therapsids** arose. In this regard, it is interesting that they were attempting a kind of body temperature control.

For paleontologists, therapsids are among the most fascinating of fossil vertebrates. They were widely dispersed during Permian and Triassic time. There is a fabulous record of these reptiles and their contemporaries in the Karoo basin of South Africa and more modest discoveries in Russia, South America, Australia, India, and Antarctica. Therapsids were predominantly small to moderate-sized animals that displayed at least the beginnings of several mammalian skeletal traits. There were fewer bones in the skull than generally found in reptile skulls, and there was a mammal-like enlargement of the lower jaw bone (dentary) at the expense of more posterior elements of the jaw. A double ball-and-socket articulation had evolved between the skull and neck. Teeth showed a primitive but distinct differentiation into

incisors, canines, and cheek teeth. The limbs were in more direct vertical alignment beneath the body, and the ribs were reduced in the neck and lumbar region for greater overall flexibility. These mammal-like features are well developed in the Permian therapsid *Cynognathus* (Fig. 10–84). Therapsids continued from the Permian through the Triassic but became extinct early in the Jurassic. However, before they died out completely, they gave origin to the early mammals.

FIGURE 10–84 Mammal-like reptiles. The scene depicts three carnivorous forms *(Cynognathus)* about to attack a plant-eating reptile *(Kannemeyeria).* *(Courtesy of The Field Museum of Natural History; painting by C. R. Knight; Negative No. CK22T.)*

■ *Mass Extinctions*

For most of the Paleozoic, the Earth was populated by a rich diversity of life. There were, however, times when the planet was less hospitable, and large groups of organisms suffered extinction (Fig. 10–85). Early geologists saw evidence of these mass extinctions in the fossil record and used this evidence to define the boundaries between geologic systems. Three episodes of extinction were particularly extensive: one at the end of the Ordovician, one in late Devonian, and one in the Permian that marks the end of the Paleozoic Era.

Late Ordovician Extinctions

Extinctions near the end of the Ordovician Period occurred in two phases. During the first phase, planktic organisms such as graptolites as well as such benthic creatures as trilobites and brachiopods were the principal victims. In the second phase, several trilobite groups that had survived the first wave of extinction perished. At the same time, corals, conodonts, and bryozoans were severely reduced in numbers and diversity. Both phases of extinction appear to be related to global cooling associated with the growth of Gondwana ice caps. As described in

Chapter 8 (Commentary, "The Big Freeze in North Africa"), an extensive ice sheet covered part of what is now northern Africa during the late Ordovician. The fossil record reveals that zones of many groups were shifted and compressed toward the equator in response to cooling at higher latitudes. As ice accumulated in the continental glaciers, sea level was lowered. This caused a loss of shallow epeiric seas and shallow marginal shelf environments. These seas had been optimum areas for the proliferation of many groups of marine invertebrates. Paleobiologist Steven Stanley has shown that when such a loss of shallow shelf environments is accompanied by cooling, mass extinctions are a likely consequence. Cooler conditions that began in the late Ordovician continued into the early Silurian, as can be inferred from the absence of significant limestone deposition. (Calcium carbonate is not readily precipitated in cold water.) Thus, the predominance of evidence seems to support the theory that late Ordovician mass extinctions resulted primarily from cooler conditions coupled with restriction of shallow-water environments.

Late Devonian Extinctions

Following the late Ordovician crisis, life expanded again, building slowly in the early Silurian but attain-

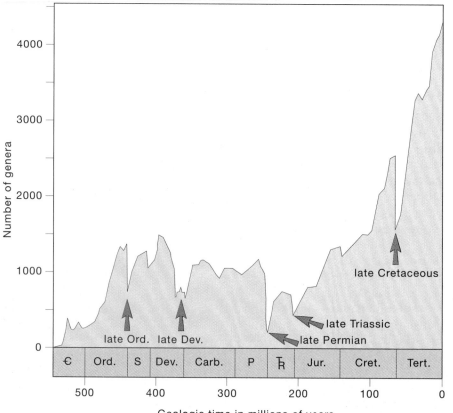

FIGURE 10–85 Diversity of marine animals compiled from a database recording first and last occurrences of more than 34,000 genera. The graph depicts five major episodes of mass extinction (global extinctions over a short span of geologic time). *(Adapted from Sepkoski, J. J., Jr. 1994. Geotimes 39(3):15–17.)*

ing rich levels of diversity during the Devonian. By the end of the Devonian, however, marine invertebrates were again confronted with adverse environmental change. That there was environmental stress is reflected in the decimation of once extensive Devonian reef communities. Reef-building tabulate corals and stromatoporoids are rarely seen in rocks deposited during the remainder of the Paleozoic. Of scores of rugose corals, only a few species survived. In addition, brachiopods, goniatites, trilobites, conodonts, and placoderms were severely reduced in numbers and variety.

Late Devonian extinctions occurred over an interval of several million years. Hence, a catastrophic sudden event, such as the impact of a meteor or asteroid, is unlikely to have been the cause. Nor is there evidence of fallout of heavy metals that normally results from such an impact. As is the case for late Ordovician extinctions, the basic cause was probably terrestrial and environmental. It appears to be related to continental glaciation in Gondwana. Glacial erratics, striations, and ancient tillites found in Brazil provide evidence that by late Devonian time, South America had drifted to a position over the South Pole. Global cooling and reduction of shallow-water environments as sea level declined may have been the basic reason for the die-off.

Late Permian Extinctions

The decimation of life at the end of the Permian Period has been described by Smithsonian paleontologist Douglas Erwin as "the mother of mass extinctions." The loss of biological diversity at this time exceeded that of all other extinctions, including the one that involved the demise of the dinosaurs 66 million years ago. By some estimates, more than 90 percent of all pre-existing marine species disappeared or were severely reduced. On land, spore-bearing ferns and other plants gave way to conifers, cycads, ginkgoes, and other gymnosperms. The newer groups probably evolved in response to cooler, drier climates. Many families of archaic amphibians, primitive reptiles, and mammal-like reptiles disappeared. Perhaps vegetative changes disrupted the food web for land animals, causing a detrimental ecologic chain reaction. Of all groups, however, tropical marine invertebrates experienced the most extensive losses. Fusulinids that had once blanketed large areas of the sea floor disappeared. Rugose corals, many families of crinoids, productid brachiopods, lacy bryozoa, and many groups of ammonoids became extinct. Trilobites were not involved in this episode of extinction, for they had disappeared earlier in the Permian.

Many factors may have contributed to the late Permian mass extinction. At the time of the extinctions, the supercontinent Pangea had completed its development. More rigorous climatic conditions existed across the great continent, as is characteristic of the interiors of large continents today. Epeiric seas had been drained or were very limited, reducing the number of favorable sites for the proliferation of shallow marine invertebrates. Frigid polar regions existed at both the north and south ends of Pangea. Organisms accustomed to warm waters shifted toward the lower latitudes, but many did not survive. The cooling prevented construction of organic reefs and subsequently impeded the formation of limestones. Pangea also blocked equatorial circulation in the ocean, disrupting the life zones of many organisms. The late Permian was also a time of extraordinary volcanic activity. One of the greatest episodes of flood basalt volcanism in the history of the Earth occurred in Siberia at the end of the Permian. Carbon dioxide released into the atmosphere during that volcanism may have triggered a greenhouse effect with consequent global warming. Together these events suggest that the extinctions at the end of the Permian may have been caused by a combination of several factors.

Summary

The fossil record of the Paleozoic is immensely better than that of the preceding Proterozoic. Shell building is one important reason for the improvement. The ability to secrete hard parts had already begun in earliest Cambrian time, and this adaptation led to a rapid radiation of marine invertebrates, with archaeocyathids, brachiopods, and trilobites becoming abundant. By Ordovician time, however, all major invertebrate phyla had become common. The best record of soft-bodied invertebrates is that of the Middle Cambrian Burgess Shale fauna, which includes many previously unknown arthropods, echinoderms, sponges, and cnidarians as well as an animal (*Pikaia*) believed to be the earliest known chordate. Among the other important invertebrate groups of the Paleozoic were foraminifers (fusulinids expanded during the Pennsylvanian and Permian), sponges, corals (tabulate and rugose), bryozoa, mollusks, echinoderms (especially blastoids and crinoids), and graptolites.

The first large fossils of vertebrate animals are found in rocks of Ordovician age. They are remains of a group of fishes known as ostracoderms. Archaic jawed fishes appear late in the Silurian but do not begin to dominate the marine realm until Devonian time. This was also the time when cartilaginous (chondrichthyian) and bony (osteichthyan) fishes began their rise to dominance. A special group of bony fishes called crossopterygian arose from osteichthyan stock. These fishes could breathe air by means of accessory lungs and possessed muscular fins that provided short-distance overland locomotion. The bones within the fins of crossopterygians resembled those of primitive amphibians. Other skeletal traits clearly suggest that these Devonian fishes were the ancestors of the first amphibians—the ichthyostegids. From a start provided by ichthyostegids, amphibians called labyrinthodonts underwent a successful adaptive radiation that lasted until the close of the Triassic. Long before their demise, however, they provided a lineage from which evolved the reptiles.

The transition from amphibian to reptile entailed a significant breakthrough in evolution. It involved the development of the amniotic egg, a biologic device that liberated land vertebrates from the need to return to water bodies to reproduce. The first known reptiles are Pennsylvanian in age and include such forms as *Hylonomus* and *Petrolacosaurus*. During very late Pennsylvanian and Permian, reptiles underwent an elaborate evolutionary radiation that produced the fin-back reptiles and mammal-like reptiles as well as several other major reptilian groups.

The earliest Paleozoic precursors to plants were mainly bacteria and algae, with the most obvious larger fossils being stromatolites and, by Ordovician time, the receptaculids. Fossil spores indicate that by late Ordovician, the first land plants, the psilophytes, made their appearance. It is probable that they evolved from the green algae (chlorophytes). Psilophytes were followed in the Devonian and Carboniferous by lycopsids, sphenopsids, and seed ferns. In temperate regions of Gondwana, a flora dominated by the "tongue fern" *Glossopteris* developed. Permian forests included plants more tolerant of drier and cooler conditions, such as conifers and ginkgoes.

Life did not expand steadily throughout the Paleozoic Era. There were difficult times when major groups of organisms became extinct. These mass extinctions occurred at the end of the Ordovician, Devonian, and Permian periods. The predominance of opinion favors environmental causes for the die-offs. In both the late Ordovician and Devonian, the Earth's South Pole was located over the Gondwanaland mass, and great ice caps formed. That global cooling ensued is indicated by the loss of many warm-water species, compression of living zones toward the equator, loss of reef communities, and reduction in carbonate deposition.

During the extinction at the end of the Permian, nearly half of the known families of animals disappeared. The decimation of marine animals was particularly dramatic and included the loss of fusulinids, spiny productid brachiopods, rugose corals, two orders of bryozoans, and many taxa of stemmed echinoderms, including blastoids. Among the vertebrates, over 70 percent of amphibian and reptilian families perished. Fortunately, a few groups survived the Permian crisis and were able to continue their evolution during the Triassic. The cause of the Permian biologic crisis is not definitely known. Many believe it was the result of the final consolidation of the great Pangean supercontinent, the high elevation of many regions on Pangea, reduction of warm epeiric seas, development of ice caps, and sharp climatic changes stemming from all of these factors.

Review Questions

1. What group of algae are considered the probable ancestors of the first land plants (tracheophytes)? For what reasons is this group considered ancestral to land plants?
2. In terms of evolution, climatic preference, and method of propagation, how did *Lepidodendron* differ from *Glossopteris*?
3. Describe the Burgess Shale animal named *Pikaia*. What is the evolutionary significance of this fossil?
4. To what geologic system(s) would rocks containing the following fossils be assigned?

   a. fusulinids
   b. archaeocyathids
   c. *Archimedes*
5. The phylum Cnidaria was formerly termed Coelenterata. Why is Cnidaria an appropriate name for these animals? What classes of Cnidaria lived only during the Paleozoic?
6. How do the sutures of nautiloids differ from those of ammonoid goniatites?
7. What are conodont elements? What other animal group has hard tissue (tooth or bone) of the same composition as conodont elements?

8. From what previous structure was the vertebrate jaw derived? What is the evidence for this evolutionary transformation?
9. What groups of Paleozoic invertebrates had become extinct by the end of the Paleozoic?
10. How are shallow shelf and epeiric sea environments affected by episodes of continental glaciation?

Discussion Questions

1. What was the probable function of the "sail" that characterized many pelycosaurs? How would you argue against the sail on pelycosaurs having importance as a secondary sexual characteristic? How would you argue against its importance for defense?
2. The Paleozoic episodes of extinction affected shallow marine invertebrates more than those that lived in deeper realms. Why?
3. What evidence suggests that global cooling was under way at the end of the Ordovician and Devonian periods?
4. Define and discuss the principal characteristics of the Echinodermata, Trilobita, Mollusca, Brachiopoda, Bryozoa, and Porifera.
5. Discuss the advantages that accrued to invertebrates that evolved shells.
6. Why was the evolution of a vascular system critical to the invasion of the lands by plants? What were the first plants to make the transition, and what were their characteristics?
7. Discuss those characteristics of therapsid reptiles that indicate that they were on the main line of evolution toward mammals.

Supplemental Readings and References

Bowring, S. A. et al. 1993. Calibrating rates of Early Cambrian evolution. *Science* 261:1293–1298.

Briggs, E. G., Erwin, D. H., and Collier, F. J. 1994. *The Fossils of the Burgess Shale.* Washington, DC: Smithsonian Institution Press.

Conway, M. S. 1987. The search for the Precambrian–Cambrian boundary. *Amer. Scientist* 75:157–167.

Cowen, R. 1995. *History of Life,* 2nd ed. Boston: Blackwell Scientific Publications.

Erwin, D. H. 1993. *The Great Paleozoic Crisis: Life and Death in the Permian.* New York: Columbia University Press.

Gould, S. J. 1989. *Wonderful Life.* New York: W. W. Norton Co.

Gray, J., and Shear, W. 1992. Early life on land. *Amer. Scientist* 80:444–453.

Stanley, S. M. 1987. *Extinction.* New York: W. H. Freeman.

Stewart, W. N., and Rothwell, G. W. 1993. *Paleobotany and the Evolution of Plants,* 2nd ed. Cambridge, England: Cambridge University Press.

Whittington, H. B. 1985. *The Burgess Shale.* New Haven: Yale University Press.

View of dipping Jurassic and Cretaceous sandstones along Waterpocket Monocline, Capitol Reef National Park, Utah. *(Scott Smith, photographer.)*

The earth, from the time of the chalk to the present day, has been the theater of a series of changes as vast in their amount as they were slow in their progress. The area on which we stand has been first sea and then land for at least four alternations and has remained in each of these conditions for a period of great length.

THOMAS HUXLEY, 1868, *On a Piece of Chalk*

CHAPTER 11

The Mesozoic Era

The extinction of Paleozoic animals and plants that occurred at the close of the Permian Period did not go unnoticed by the founders of geology. The fossil evidence for this time of crisis provided a natural boundary that was used to mark the end of the Paleozoic sequence. Overlying younger strata contained different and generally more progressive organisms, yet not as modern as those that live today. Thus, it seemed appropriate to formulate a middle chapter in the history of life. The founders of geology named that chapter the Mesozoic Era. In its turn, the Mesozoic also ended in a biologic crisis that marks its separation from the youngest era of all—the Cenozoic.

The Mesozoic Era lasted an estimated 160 million years, ending approximately 65 million years ago. The three periods of the Mesozoic are unequal in duration. The Triassic lasted about 30 million years, the Jurassic 55 million years, and the Cretaceous about 75 million years.

During the lengthy span of the Mesozoic, many new families of plants and animals evolved and experienced often spectacular radiations. It was the era in which two new vertebrate classes, the birds and the mammals, first appeared. It was also a time in which the supercontinent Pangea, which had formed in the previous era by the joining of ancestral continents, was gradually dismembered. As the continents moved slowly apart, the oceanic rifts between them deepened and broadened. A process of fragmentation and drift had begun that would ultimately lead to the present physical geography of the planet.

■ *The Break-up of Pangea*

Any discussion of the history of the Mesozoic Era must have Pangea as its starting point. The same forces that had drawn the plate segments together to form this great supercontinent were still in operation at the beginning of the Mesozoic; however, they now moved in different directions or had shifted their locations. In general, the dismemberment of Pangea seems to have occurred in four stages. The first stage began in the Triassic, with rifting and volcanism along normal fault systems. Events such as these result when a region of the lithosphere is subjected to tensional stresses. In this instance, the forces were associated with the separation of North America and Gondwanaland. As the rifting progressed, Mexico was decoupled from South America and the eastern border of North America parted from the Moroccan bulge of Africa (Fig. 11–1). Oceanic basalts were added to the sea floor of the newly formed and gradually widening Atlantic Ocean. The largely tensional geologic structures as well as the volcanic and clastic sedimentary rocks in the Triassic System of both the eastern United States and Morocco exhibit many striking similarities. These now widely separated regions were on opposite sides of the same axis of spreading and were affected by similar forces and events.

Recently, geologists have used a geophysical technique called *seismic reflection profiling* to locate the 300-million-year-old zone of convergence, or **suture zone,** formed by the convergence of Africa and North America. The zone, which trends roughly east–southeast across southern Georgia, was left behind when the two continents subsequently parted company about 190 million years ago. The discovery was particularly interesting to paleontologists, who had postulated the convergence and later separation of the two continents from fossil evidence.

Throughout Triassic time, Laurasia north of the Maritime Provinces remained intact. Sea-floor spreading was apparently more rapid in the central and northerly region of the newly opening Atlantic. It also appears that South America maintained its hold on Africa throughout most of the Triassic.

The second stage in the break-up of Pangea involved the rifting and opening of narrow oceanic tracts between southern Africa and Antarctica. This rift extended northeastward between Africa and India and developed a branch that separated northward-bound India from the yet unseparated Antarctica–Australia land mass. The rifting was accompanied by outpouring of great volumes of basaltic lavas.

In stage three of the break-up of Pangea, the Atlantic rift began to extend itself northward, and clockwise rotation of Eurasia tended to close the eastern end of the Tethys Sea, a forerunner of the Mediterranean. By the end of Jurassic time, an incipient breach began to split South America away from Africa. The cleft apparently worked its way up from the south, creating a long seaway somewhat reminiscent of the present Red Sea. Australia and Antarctica remained intact, but India had moved well along on its long voyage to Laurasia. By Late Cretaceous time, about 70 million years ago, South America had completely separated from Africa, and Greenland began to separate from Europe. However, northeastern-most North America was still attached to Greenland and northern Europe.

The fourth stage in the dismemberment of Pangea was not a Mesozoic event. It occurred early in the Cenozoic, during which time the North Atlantic rift slowly penetrated northward until eventually the North American and Eurasian continents were completely separated. Also during this final era of Earth history, Antarctica and Australia separated. This final separation of continents occurred only about 45 million years ago, with the total span of time needed for the fragmentation of Pangea being about 150 million years.

■ *The Mesozoic History of North America*

To the East and South

Triassic and Jurassic

At the beginning of the Mesozoic, conditions in the eastern part of North America were much the same as they had been near the end of the Paleozoic. The rugged Appalachian ranges that had been raised dur-

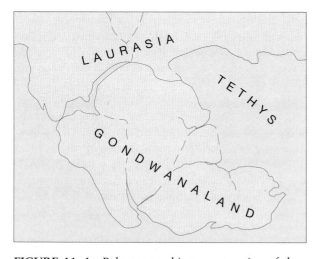

FIGURE 11–1 Paleogeographic reconstruction of the world about 180 million years ago when the break-up of Pangea was beginning.

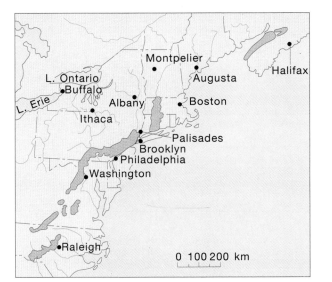

FIGURE 11–2 Outcrop areas of Triassic rocks in eastern North America. Green areas show troughlike deposits of Late Triassic age.

ing the Allegheny orogeny were undergoing vigorous erosion. The coarse clastics derived from the uplands filled intermontane basins and other low areas during the Early and Middle Triassic. Then, during the Late Triassic and Early Jurassic, North America began to experience the tensional forces that precede actual separation of continents. As a result, a series of fault-bounded troughs developed along the eastern coast from Nova Scotia to North Carolina (Figs. 11–2 and 11–3). Many of these faults represent reactivation of lateral strike–slip and reverse faults formed during late Paleozoic orogenic activity. The older faults provided zones of weakness that facilitated rifting during the subsequent Triassic. The downfaulted blocks of the Triassic faults provided traps for the accumulation of great thicknesses of poorly sorted arkosic red sandstones and shales. These sediments are nonmarine and indicate that insufficient separation had occurred during the Late Triassic to allow ingress of marine waters. Rocks deposited in the downfaulted basins constitute the **Newark Group** of Late Triassic and Early Jurassic age.

The poor sorting of the coarser Newark clastics and their high content of relatively unweathered feldspar grains indicate transportation and deposition by streams flowing swiftly down from the granitic highlands that bordered the fault basins (Fig. 11–4). In some places, drainage in the basins became impounded and lakes formed. These lake deposits contain the remains of freshwater crustaceans and fish. Ripple marks, mudcracks, raindrop impressions, and even the footprints of early members of the dinosaur line (Fig. 11–5) are frequently found on the lithified

surfaces of Newark sediments. Lavas, rising along steeply dipping fault surfaces, flowed out on the newly deposited sediment, and volcanoes spewed ash onto the surrounding hills and plains.

Three particularly extensive lava flows and an imposing sill are included within the Newark Group in the New Jersey–New York area. The exposed edge of a vertically jointed sill forms the well-known Palisades of the Hudson River (Fig. 11–6). Radiometric dates obtained for the Palisades basalt indicate that it solidified about 190 million years ago. Although most of the Newark beds are Triassic, recent spore and pollen studies suggest that Newark rocks in some areas accumulated during the Early Jurassic as well.

By Late Triassic time, the fault block mountains that had been produced as a result of rifting had been severely reduced by erosion. The topography was further reduced during the Jurassic and Early Cretaceous until only a broad, low-lying erosional surface remained. That old surface is called the Fall Line surface because its profile can best be seen along the Fall Line, which separates more resistant rocks of the Piedmont from the softer rocks of the Atlantic coastal plain. Beneath the Fall Line surface are Triassic faulted sediments and older crystalline rocks, and above it lie the more or less flat-lying Cretaceous sediments of the Atlantic coastal plain.

South of the old Appalachian–Ouachita tracts, a new depositional province, the Gulf of Mexico, began to take form. The initial deposits were mostly evaporites and do not lend themselves to radiometric dating. However, analyses of pollen from a sample of salt taken from this sequence suggest an age of Late Triassic. Evaporites continued to be deposited well into the Jurassic, indicating aridity in the Gulf region (Fig. 11–7, p. 396). Apparently, the Gulf of Mexico was rather like a great evaporating basin, concentrating the waters of the Atlantic Ocean and precipitating salt and gypsum to thicknesses exceeding 1000 meters (3280 feet) (Fig. 11–8, p. 397). The Jurassic evaporite beds are the source of the salt domes of the Gulf Coast. Salt domes are economically important structures associated with the entrapment of petroleum. When compressed by a heavy load of overlying strata, salt tends to flow plastically. As it moves upward in the direction of lesser pressures, the salt bends and faults overlying strata. The permeable, often faulted beds that slope downward away from the salt, or are arched over the dome, afford excellent sites for petroleum accumulation (see Figs. 11–8 and 2–30).

Evaporite conditions abated later in the Jurassic, and several hundred meters of normal marine limestones, limy muds, shales, and sandstones accumulated in the alternately transgressing and regressing seas of the Gulf embayment. Although there are surface exposures of these rocks where they extend into

(text continues on p. 396)

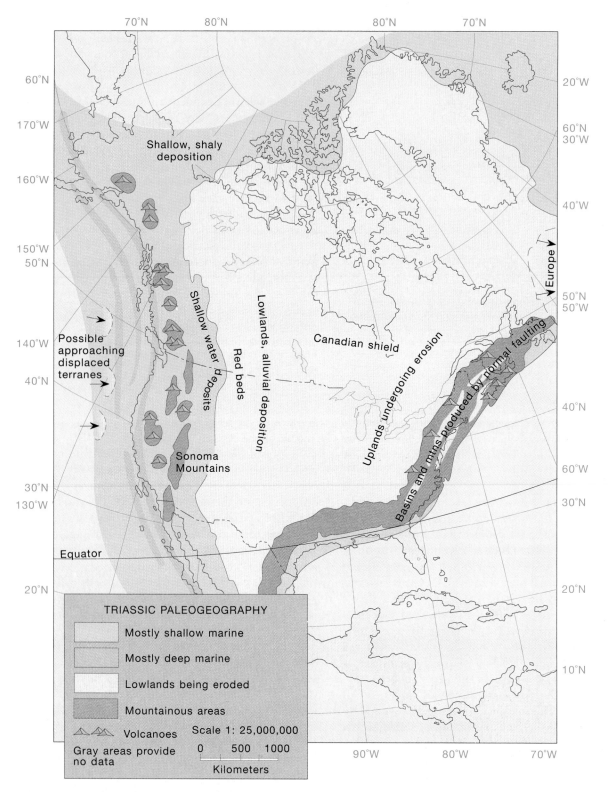

FIGURE 11–3 Generalized paleogeographic map for the Triassic of North America.

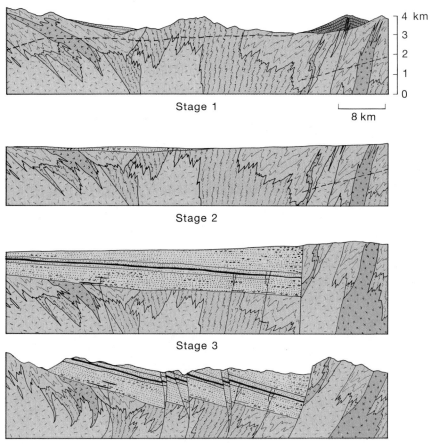

Stage 1

Stage 2

Stage 3

Stage 4

8 km

FIGURE 11–6 Palisades of the Hudson River. *(Photograph courtesy of Palisades Interstate Park Commission.)*

FIGURE 11–5 Slab of Newark Group sandstone containing the cast of a footprint made by a three-toed Triassic reptile. The footprint is about 20 cm long. The linear ridges in the slab are casts of mudcracks.

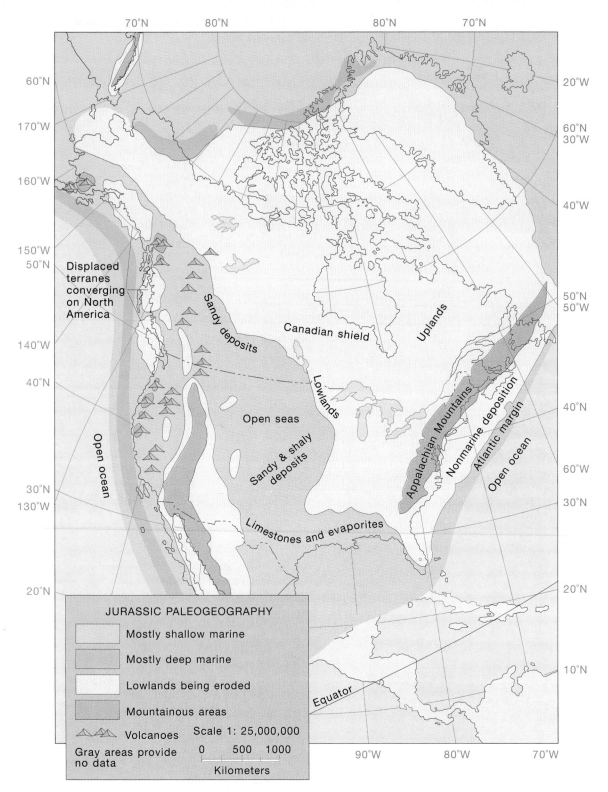

FIGURE 11–7 Generalized paleogeographic map for the Jurassic of North America.

northeastern Mexico, in the United States they are deeply buried beneath a thick cover of Cretaceous and Cenozoic sediments. Were it not for the drilling activities of oil companies, the nature of these rocks could only be inferred from geophysical data.

Cretaceous

The Cretaceous was a time of great marine inundations of land masses, and the Atlantic and Gulf coastal regions received their full share of flooding.

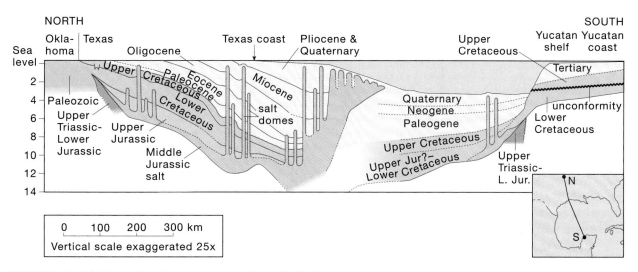

FIGURE 11–8 North–South cross-section of the Gulf of Mexico basin.
(Adapted from Salvadore, A. 1991. Geology of North America, pp. 1–12. Boulder, CO: Geological Society of America.)

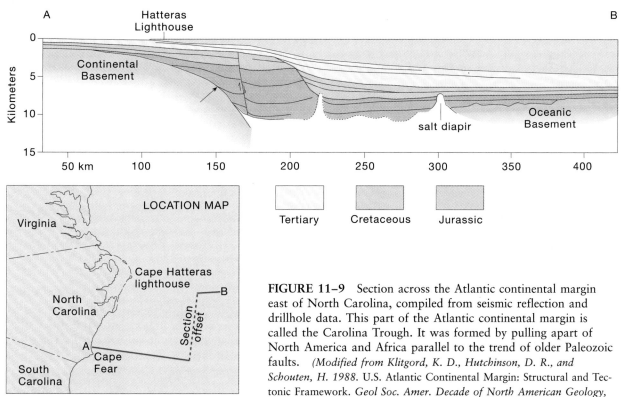

FIGURE 11–9 Section across the Atlantic continental margin east of North Carolina, compiled from seismic reflection and drillhole data. This part of the Atlantic continental margin is called the Carolina Trough. It was formed by pulling apart of North America and Africa parallel to the trend of older Paleozoic faults. *(Modified from Klitgord, K. D., Hutchinson, D. R., and Schouten, H. 1988. U.S. Atlantic Continental Margin: Structural and Tectonic Framework. Geol Soc. Amer. Decade of North American Geology, I-2:19–55. Boulder, CO: Geological Society of America.)*

Early in Cretaceous time, the Atlantic coastal plain, which had been experiencing erosion since the beginning of the Mesozoic, began to subside. At about the same time, the Appalachian belt that lay to the west was elevated. On the subsiding coastal plains, alternate layers of marine and deltaic terrestrial deposits accumulated and were gradually built into a great wedge of sediments that thickened seaward (Fig. 11–9). Today, the thinner eastern border of the wedge is narrowly exposed in New Jersey, Maryland, Virginia, and the Carolinas. But the greatest volume of Cretaceous sediments was deposited farther east, along the present continental shelf.

To the south, the Florida region was a shallow submarine bank during the Cretaceous. The oldest strata consist of limestones, but later in the period, terrigenous clastics were brought into the area by streams flowing from the southern Appalachians. These source areas were gradually worn down, and carbonate deposition prevailed.

The Cretaceous is noteworthy as a period of the Mesozoic in which carbonate reefs were particularly extensive in the warmer regions of both the Eastern and Western Hemispheres. Among the invertebrates that contributed their skeletal substance to these reefs, a group of bivalves called **rudists** were immensely important. In many Cretaceous reefs, the shells of these creatures (Fig. 11–10) form the basic framework of the reef. Because of their high porosity and permeability, rudistid reefs are reservoir rocks for some of the world's greatest Cretaceous oil accumulations.

Following the withdrawal of Jurassic seas, the region northwest of Florida consisted of low-lying coastal plains. However, early in the Cretaceous,

these relatively level lands were invaded by marine waters moving northward from the ancestral Gulf of Mexico. Nearshore deposits such as sandstones are overlain by finer sedimentary rocks that are characteristic of deeper water and provide clear evidence of the northward migration of the shoreline. The advance of the Cretaceous sea was not uniform. An extensive regression occurred near the end of the first half of the period; however, flooding resumed in Late Cretaceous time when a wide seaway occupied a tract from the Gulf of Mexico to the Arctic Ocean (Fig. 11–11). Among the Upper Cretaceous formations deposited in this inland sea, chalk was particularly prevalent. Chalk is a white, fine-grained, soft variety of limestone that is composed mainly of the microscopic calcareous platelets (called *coccoliths*) of golden-brown algae. It is common among beds of Cretaceous age in many other parts of the world (Fig. 11–12). Indeed, the Cretaceous System takes its name from *creta,* the Latin word for chalk.

To the West

As the eastern border of North America was experiencing the tensional effects caused by separation from Europe and Africa, compressional forces prevailed in the west. While the newly formed Atlantic Ocean was widening, North America moved westward, overriding the Pacific plate. Thus, deformation of western North America is related to events in the east. It has been shown, for example, that the pace of tectonic activity in the North American Cordillera was most intense when sea-floor spreading was most rapid in the Atlantic.

Accretionary Tectonics

During the Mesozoic, a steeply dipping subduction zone existed along the western margin of North America. This was not a simple Andean-type subduction zone, however, for the advancing Pacific plate carried considerably more than ocean basalts and sea-floor sediments. Entire sections of volcanic arcs, fragments of distant continents (the microcontinents described earlier), and pieces of oceanic plateaus were carried to North America's western margin as well. These fragments, now incorporated into the Cordilleran belt, constitute **displaced** or **allochthonous terranes** described briefly in Chapter 5. The better documented of these terranes can be identified by their distinctive age, rock assemblages, and mineral resources, which indicate the part of the world from which the displaced fragments came. Many of the fragments have one or more fault boundaries.

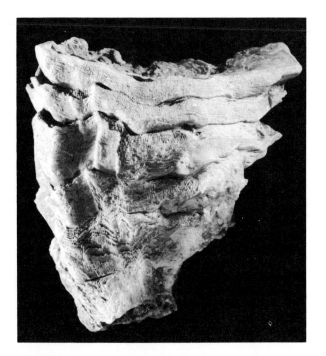

FIGURE 11–10 *Eoradiolites davidsoni,* a Cretaceous rudistid pelecypod from Texas. Rudistids appeared in the Jurassic and proliferated during the Cretaceous, when they were important reef-making organisms. They bear a close resemblance to some Paleozoic corals. This specimen is 8.6 cm tall.

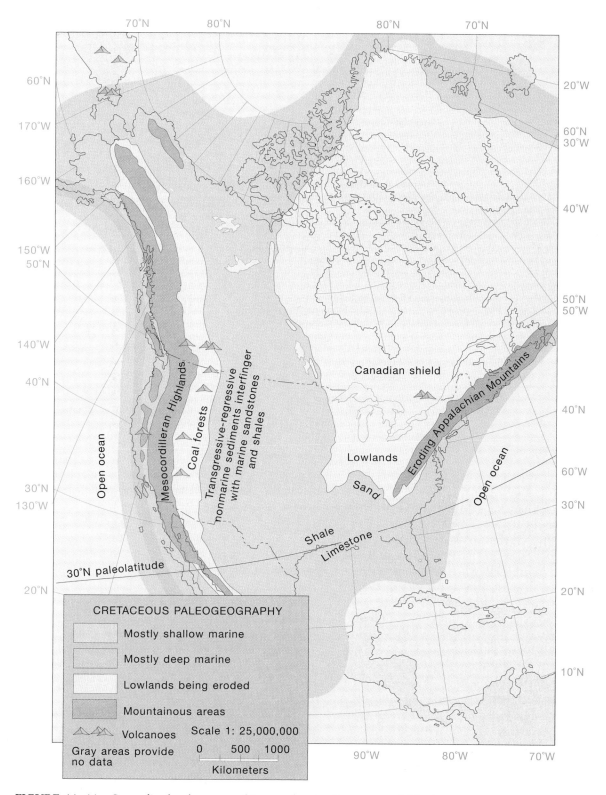

FIGURE 11–11 Generalized paleogeographic map for the Cretaceous of North America.

FIGURE 11–12   (A) Sea cliffs composed of chalk along the Dorset coast of England. (B) Close-up view of the chalk. The dark nodules are flint, a gray or black variety of chert.

Altogether, more than 50 displaced terranes have been recognized in the Cordillera. They may constitute as much as 70 percent of the total Cordilleran region. Thus, the growth of North America has occurred not only by accretion of materials along subduction zones but also by incorporation of huge crustal fragments formed elsewhere and subsequently conveyed to North America by sea-floor spreading. Some of these fragments were evidently so large that they caused changes in subduction zone configuration. Others composed of low-density rocks were too buoyant for subduction. These were scraped off the subducting plate, splintered, and thrust onto the continental margin. In contrast to subduction, the process whereby one rock mass rides up and over another is call **obduction.** Stated differently, in subduction, one plate stays at a constant level and the other plate is forced downward under it; whereas in obduction, one plate remains at a constant level, while the other plate is forced upward over it.

The growth of a continent by the progressive incorporation of crustal fragments in the aforementioned manner is termed **accretionary tectonics.** In an oversimplified analogy to accretionary tectonics, one can imagine icebergs in an ocean being driven onshore by powerful winds. Each crashes into the coast and is then struck from behind by subsequent arrivals until a jumbled region of displaced icebergs is con-

structed. In geology, an analogous region of displaced terranes is called a **tectonic collage.**

Triassic

The Cordilleran region as it existed during the Mesozoic can be divided into a western belt containing thick volcanic and siliceous deposits and a wide eastern tract adjacent to the more stable interior of the continent. Nearly 800 meters of clastic sediments and volcanics exposed in southwestern Nevada and southeastern California attest to the instability of the western zone. Geologists speculate that the belt may well have resembled the Indonesian island arc of today.

The initial orogenic event of the Cordilleran region is difficult to date precisely, although evidence for deformation at or near the Permian–Triassic boundary can be found from Alaska to Nevada. In the United States, the event has been named the **Sonoma orogeny.** Its effects can be studied in west-central Nevada. The Sonoma orogeny occurred when an eastward-moving volcanic arc, called the *Golconda arc,* collided with the Pacific margin of North America. At that time, North America was bordered by a westward-dipping subduction zone. As a result of the collision, the island arc and its adjacent accretionary wedge of sediment was welded onto the west-

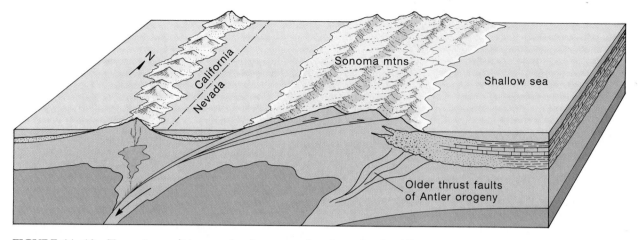

FIGURE 11–13 Tectonic conditions and paleogeography along the Cordilleran region during the Late Permian and Early Triassic.

ern edge of North America, adding as much as 300 km to the continent's east–west dimension. Oceanic rocks were thrust eastward onto the eroded structures of the old Antler orogen (Fig. 11–13). The larger of the thrust sheets moved scores of kilometers eastward and were themselves broken into many smaller internal faults, with each fault slice bringing a repeated stratigraphic sequence to the surface.

The Triassic rocks of the far western part of the Cordillera include great thicknesses of volcanics and graywackes presumably derived from the island arc. It is uncertain, however, whether these rocks were deposited where they are now found or whether they were part of a displaced terrane, called the **Sonoma terrane,** that actually originated in some unknown part of the Pacific.

Early Triassic rocks of the eastern Cordillera consist of sandstones and limestones of mostly shallow marine origin. The thickest section of these marine strata occurs in southeastern Idaho, where nearly 1000 meters of Lower Triassic sediments accumulated. Eastward from the Cordillera, these marine beds interfinger with continental red beds. Although seas remained in Canada during the Middle Triassic, in the United States they regressed westward, leaving vast areas of the former sea floor subject to erosion.

Upper Triassic formations rest on the unconformity resulting from this erosion. The Upper Triassic series consists mostly of continental deposits transported by rivers flowing westward across an immense alluvial plain. The lowermost strata consist of the sandy **Moenkopi Formation,** followed by the pebbly **Shinarump Formation** (Fig. 11–14) derived from uplifted neighboring areas in Arizona, western Colorado, and Idaho. Above the Shinarump lie the vividly

(text continues on p. 404)

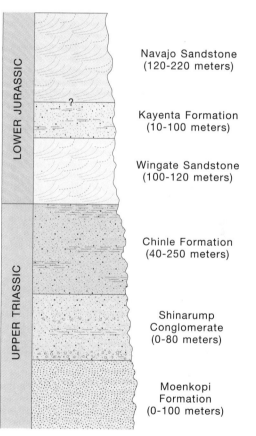

FIGURE 11–14 Generalized geologic section of Upper Triassic and Lower Jurassic sedimentary rocks of central Utah.

Zion National Park

Towering "temples" of Navajo Sandstone, colorful cliffs and arches, slot canyons, and the tumbling roar of streams greet the visitor to Zion National Park. For those who have had an introductory course in historical geology, the park is a place to see some of the most widely known rock units of the Mesozoic. Zion National Park is in the southwestern corner of Utah adjacent to Interstate 15. There is a scenic drive within the park that extends through spectacular Zion Canyon from the park's south entrance northward to an erosional feature called the Temple of Sinawava. The drive includes a mile-long tunnel in which windows have been carved so travelers can stop and view interesting features.

Zion Canyon has been (and continues to be) fashioned by mass wasting and running water. Mass wasting involves the movement of rock and soil downhill under the influence of gravity. In Zion it includes the sliding of large blocks of rock downslope, as resistant caprock is undermined by erosion and weathering. Rockslides and rockfalls are common. Mass wasting alone, however, cannot carve a canyon. The erosive and transportational power of streams is essential. The stream acts as a conveyor belt, carrying away what it has eroded from its channel as well as the debris supplied to it by mass wasting. Streams even cause some forms of mass wasting by removing material at the base of valleys to undermine cliffs and slopes and cause them to collapse. The North Fork of the Virgin River controlled the sculpting of Zion Canyon. It carries away about 3 million tons of mass-wasted debris each year. The ability of the stream to accomplish this extraordinary amount of work results from its steep gradient and the

Winter scene in Zion National Park, Utah. The narrow canyon in this view is appropriately dubbed the Subway. *(Photograph by Scott T. Smith.)*

high velocity this gradient produces. Increased gradients for streams in this region were initially developed about 10 to 12 million years ago, when much of the Colorado Plateau was uplifted. With uplift, streams were able to flow faster, entrench themselves, and, along with their tributaries, cut deeply into the plateau. In Zion, the major streams have cut their channels so rapidly that the floors of some tributary valleys are at a higher level than the floor of the main valley. Such hanging valleys are

normally a characteristic of glaciated areas. Where hanging valleys occur in Zion, waters of the tributaries cascade down to the main valley in waterfalls.

Most of the interesting erosional features in Zion have been carved from the Navajo Sandstone, which is part Triassic and part Jurassic in age. Among these features are towering monoliths referred to as temples. Zion takes its name from these cathedral-like features, reminiscent of the citadel in Palestine that was the nucleus of Jerusalem. Beneath resistant layers of sandstone, softer rock has often spalled away, leaving inset arches (blind arches). Free-standing arches have also developed, the largest of which in Zion is the Kolob Arch. The beauty of the arches, canyon walls, and temples is enhanced by an array of colors: shades of red, brown, and green from iron oxides; gray-green and bright orange from lichens; and shiny black from the iron and manganese oxides of desert varnish.

With clearly exposed rock units on all sides, Zion is a nearly ideal place to interpret geologic history. The lowermost (hence oldest) rock unit in the park is Permian in age. It is the Kaibab Limestone, a yellow–gray rock unit exposed in Hurricane Cliffs north of the Kolob Visitor Center. The Kaibab sediments were de-

posited in a shallow sea. That sea had withdrawn by the end of the Permian, and above the old sea bed shales and sandstone of the Triassic Moenkopi Formation were deposited. During this time, there were minor incursions of the sea, and gypsum was precipitated in the coastal lagoons associated with these incursions. After deposition of the Moenkopi, the region experienced uplift. Rejuvenated streams with steep gradients were able to carry gravels into the region, and these comprise the Shinarump Conglomerate at the base of the Chinle Formation. Most of the Chinle Formation consists of sands, shales, and beds of volcanic ash. Tree branches, logs, and stumps occur in the Chinle. These have been converted to petrified wood identical to that in the logs of Petrified National

Park. Apparently, much of the silica required for silicification of the wood was derived from dissolution of volcanic ash in the Chinle.

Above the Chinle beds in Zion National Park, one finds ripple-marked and cross-bedded silt-stones of the Moenave Formation. The Moenave in turn is overlain by siltstones of the Kayenta Formation. The footprints of dinosaurs have been discovered along the bedding surfaces of the Kayenta.

The Navajo Sandstone is next in the sequence of formations in Zion National Park. It is over 220 feet (670) meters thick in places and is the principal cliff former in the park. Lower beds of the Navajo show bedding features indicating that they were deposited in shallow water, but the greater part of the unit in Zion displays the kind of curved and wedge-shaped cross-bedding characteristic of wind-transported sediment. This interpretation is strengthened by the prevalence of frosted quartz grains in the Navajo. The frosted appearance is the result of grain-to-grain impact during wind transport. Cross-bedding in the Navajo sandstones is magnificently displayed in Zion at Checkerboard Mesa (see Fig. 11–26). One can envision the depositional area as a great coastal desert somewhat like the present Sahara, where dunes migrated along the coastal plain of a shallow sea.

Streams laden with red mud provided the material for the fluvial shales of the Temple Gap Formation, which rests on the sandstones of the Navajo Formation. These beds are covered by shallow-water limestones of the Jurassic Carmel Formation. The culminating events in Zion's history include extrusion of basaltic lavas from about 0.25 to 1.4 million years ago and intermittent episodes of further uplift.

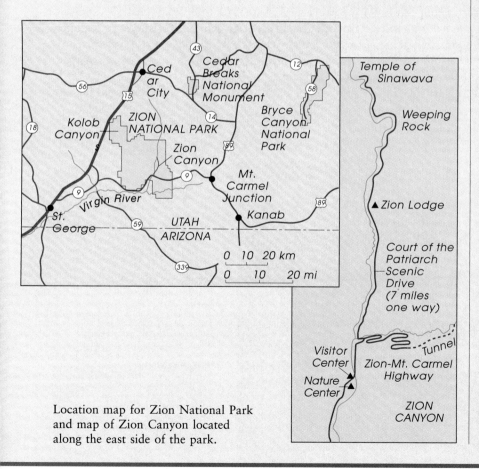

Location map for Zion National Park and map of Zion Canyon located along the east side of the park.

FIGURE 11–15 A typical scene in the Orange Cliffs area of Utah. The high cliffs along the horizon are Wingate Sandstone. They are capped by beds of the Kayenta Formation visible near the top center. The slope immediately below the cliffs is eroded in the Chinle Formation. The ridge in the foreground is capped by the Lower Triassic Moenkopi Formation. *(Photograph courtesy of the Utah Geological and Mineral Survey.)*

colored shales, siltstones, and sandstones of the **Chinle** and **Kayenta** formations (Fig. 11–15). The sediments of these two formations were deposited in stream valleys and lakes. They alternate with sand dune accumulations of the **Navajo Sandstone** (Fig. 11–16) and the Lower Jurassic **Wingate Sandstone** (Fig. 11–17). These formations are beautifully exposed in the walls of Zion Canyon in southern Utah.

They display sweeping cross-bedding, such as is formed in sandy deposits transported by winds. (However, such so-called festoon cross-bedding is not always caused by winds, for submarine dunes may be very similar in form.)

The Painted Desert of Arizona is developed mostly in Chinle rocks (Fig. 11–18). The formation is known throughout the world for the petrified logs of

FIGURE 11–16 Panoramic view of Triassic and Jurassic formations in Zion National Park, Utah. The lowermost slope is formed by shales of the Moenkopi Formation. Above that slope is a prominent ledge of Shinarump sandstones. The next slope above the Shinarump is the one formed in Chinle Formation. The towering cliffs at the top are Navajo and Carmel formations. *(Photograph courtesy of the U.S. National Park Service.)*

FIGURE 11–17 The erosional features in the foreground are sculpted from the Lower Jurassic Wingate Sandstone, beneath which lies the red Chinle Formation. The cliffs in the far background are also Wingate Sandstone. Colorado National Monument near Grand Junction, Colorado. *(Courtesy of R. J. Weimer.)*

conifers it contains. Each year, thousands of tourists examine these logs, now turned to colorful agate, in Petrified Forest National Monument (Fig. 11–19). Apparently, during times of Triassic floods, the trees were left on sandbars or trapped in log jams and covered by sediment. Percolating solutions of underground water subsequently replaced the wood with silica.

Jurassic to Early Tertiary Tectonics

Most of the orogenic activity in the Cordillera during the Mesozoic was a result of the continuing eastward subduction of oceanic lithosphere beneath the continental crust of the North American plate. That subduction varied in rate, in inclination, and, to a small degree, in direction. It resulted in eastward-shifting phases of deformation, which initially affected the far western part of the Cordillera and then proceeded eastward to reach the margin of the craton.

The deformational and magmatic activity associated with the western tract is sometimes termed the **Nevadan orogeny.** During the Triassic, and increasing in frequency during the Jurassic and Cretaceous, graywackes, mudstones, cherts, and volcanics that had been swept into the subduction zone were severely folded, faulted, and metamorphosed. The in-

FIGURE 11–18 Typical exposure of the Triassic Chinle Formation, Painted Desert, Arizona. *(Photograph by James Cowlin.)*

FIGURE 11–19 Petrified logs, Petrified National Park, Arizona. The petrified logs and wood fragments are *Araucarioxylon.* They have been weathered from the Chinle Formation of Triassic age. *(Courtesy of Lehi F. Hintze.)*

FIGURE 11–20 Strongly deformed bedded cherts of the Franciscan Formation exposed on the flank of Mt. Davidson near the center of San Francisco between Interstate 280 and Portola Drive. *(Courtesy of Brian C. Park-Li.)*

tensely crumpled and altered rock sequences that were affected by compression between the converging plates are appropriately termed *mélange*, which means a jumble. The Franciscan fold belt of California (Figs. 11–20 and 11–21) provides a good example of a mélange. In addition to the deformation of the sedimentary rock sequences, enormous volumes of granodiorite were generated above the subduction

FIGURE 11–21 Greenstones of the Franciscan Formation exposed at Lime Point, Marin Peninsula, California. The west side of the Golden Gate Bridge tower rests on rocks of the Franciscan Formation. *(Courtesy of U.S. Geological Survey.)*

FIGURE 11–22   Granodiorite of the Sierra Nevada batholith grandly exposed in Yosemite Valley, California.

zone and intruded overlying rocks repeatedly during the Jurassic and Cretaceous. The Sierra Nevada (Fig. 11–22), Idaho, and Coast Range batholiths are impressive evidence of the vast scale of this Nevadan magmatic activity (Fig. 11–23).

Somewhat before the Sierra Nevada batholith was emplaced during the Cretaceous, another series of deformations had begun east of the present-day Sierra Nevada. This second phase of the tectonic development of the Cordillera primarily affected shallow-water carbonates and terrigenous clastics deposited over a time span from Middle Jurassic to earliest Cenozoic. This new orogeny has been named the **Sevier.** During the Sevier orogeny, strata were sheared from underlying Precambrian rocks and broken along parallel planes of weakness to form multiple, imbricated, low-angle thrust faults (Fig. 11–24). The French word **décollement** (unsticking) has been used to describe this kind of structure, in which older rocks are thrust on younger in multiple, nearly parallel slabs of crust. It has been estimated that the compressional structures produced during the Sevier orogeny resulted in over 100 km of crustal shortening in the Nevada–Utah region. In addition to the major thrust faults, several large folds are known, as well as intricately folded strata that were deformed within the major thrust units.

Although the term *Sevier* is sometimes reserved for deformational episodes in the Nevada–Utah region, a similar style of deformation occurred to the north in Montana, British Columbia, and Alberta. Each of the major ranges in this region is a fault block

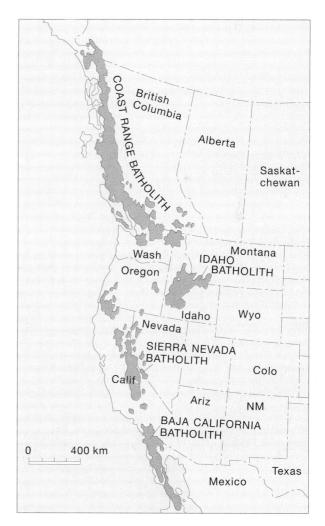

FIGURE 11–23   Mesozoic batholiths in west–central North America.

407

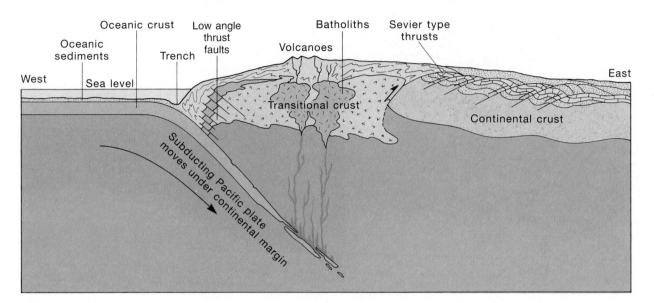

FIGURE 11–24 An advanced stage in the evolution of the North American Cordillera, with structures developing as a consequence of underthrusting of the continent by the Pacific oceanic plate. Note the multiple, imbricated, low-angle thrusts on the east side of the section. *(The diagram is simplified from Dewey, J. F., and Bird, J. M. 1970. J. Geophys. Res. 75(14):2638.)*

composed of Paleozoic strata of the shelf that have been thrust toward the east along westwardly dipping fault surfaces. The most famous of these faults is the Lewis thrust (Fig. 11–25), along which Proterozoic rocks of the Belt Supergroup were carried 65 km toward the east.

Magmatic activity along the far western edge of the North American plate had diminished near the end of the Cretaceous, and by early Tertiary time, much of the major thrusting of the marginal shelf region had subsided. Once again, deformational events shifted eastward to the cratonic region, where the Rocky Mountains of New Mexico, Colorado, and Wyoming are now located. These more eastward disturbances are called the **Laramide orogeny.** High angle reverse faults (which become thrust faults with

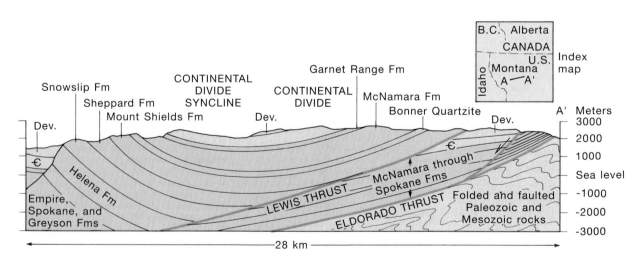

FIGURE 11–25 Northeast–southwest cross-section showing the Lewis and Eldorado thrust faults at a location about 66 km south of Glacier National Park. Formations shown in green are Proterozoic in age. Tan color indicates Paleozoic and Mesozoic rocks. *(After Mudge, M. R., and Earhart, R. L. 1980. U.S. Geological Survey Prof. Paper, No. 1174.)*

FIGURE 11–26
Cross-stratification in the Navajo Sandstone, Checkerboard Mesa, Zion National Park, Utah. *(Courtesy of Lehi F. Hintze.)*

depth) developed during the Laramide orogeny, but the more characteristic kind of structures are broadly arched domes, basins, monoclines, and anticlines. Many of the larger faults represent reactivation of older Precambrian faults. Strata composing the domes and anticlines are draped over central masses of Precambrian igneous and metamorphic rocks. In several instances, erosion has stripped away the cover of strata, exposing the central crystalline mass. Resistant layers of inclined beds that surround the central cores stand as mountains today. Many of these up-arched areas appear to have been produced by movements along underlying faults in the basement rocks. It has been suggested that these controlling structures were in turn developed by drag when the eastward-moving subducted oceanic lithosphere scraped along the sole of the cratonic margin of North America.

Most of the structures of the present Rocky Mountains are the result of Laramide phases of orogeny. The landscape we see today, however, is the result of repeated episodes of Cenozoic erosion and uplift. Erosion, acting on the geologic structures already present, sculpted the final scenic design.

Jurassic Sedimentation

During the Jurassic, when the mélange was being formed in the subduction zone complex, depositional sites inland experienced a less dramatic kind of sedimentation. Early Jurassic deposits consist of clean sandstones, such as the **Navajo Sandstone**. Large-scale cross-bedding is well developed in the Navajo (Fig. 11–26). However, thin beds of fossiliferous

limestone and evaporites occur locally and indicate that at least some parts of the formation were deposited in water. The Navajo and associated sand bodies were probably deposited in a nearshore environment, and it is likely that some of the deposits were part of a coastal dune environment (Fig. 11–27). They do not seem to have been laid down on the floors of a vast

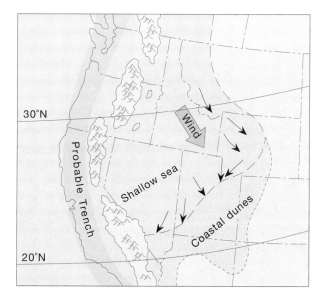

FIGURE 11–27 Paleogeographic map for the early Jurassic of the western United States, showing general extent of sea and land as well as paleolatitudes. *(From Stanley, K. O., Jordan, W. M., and Dott, R. H. 1971. Bull. Am. Assoc. Petrol. Geol. 55(1):13.)*

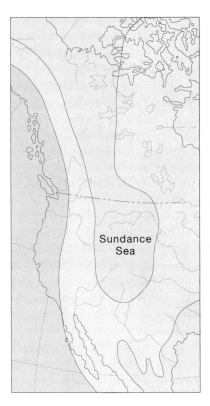

FIGURE 11–28 Region in the western North America inundated by the Middle Jurassic Sundance sea. (Land areas are shown in tan, marine areas in blue.)

interior desert, as has often been postulated. Judging from studies of cross-bedding orientations, the source area for these clean sandstones was probably the Montana–Alberta craton. Somewhat older but similar quartz sandstones were recycled and spread southward into Wyoming and Utah and then westward across Nevada.

Marine conditions became more widespread in the Middle Jurassic, when the entire west–central part of the continent was flooded by a wide seaway that extended well into central Utah (Fig. 11–28). This great embayment has been dubbed the **Sundance Sea.** Within the Sundance sea were deposited the sands and silts of the Sundance Formation, famous for its fossil content of Jurassic marine reptiles. Sediment for the Sundance Formation and overlying younger Jurassic rock units was derived from the Cordilleran highlands that lay to the west. These highlands continued to grow throughout the Jurassic, ultimately extending from Mexico to Alaska. Eventually the Sundance Sea regressed, leaving behind a vast, swampy plain across which meandering rivers built floodplains. These deposits compose the **Morrison Formation** (Fig. 11–29), which extends across millions of square kilometers of the American West. Enclosed within these floodplain deposits are the bones of more than 70 species of dinosaurs, including some of the largest land animals ever to have existed.

FIGURE 11–29 Morrison Formation exposed along the east flank of Circle Cliffs, Garfield County, Utah. *(Photograph courtesy of the Utah Geological and Mineral Survey.)*

COMMENTARY

Cretaceous Epeiric Seas Linked to Sea-Floor Spreading

During the Cretaceous, continents were extensively inundated by inland seas. Marine sedimentary rocks in North America, Europe, Australia, Africa, and South America indicate that about a third of the land area now emergent was under water. Because all the continents experienced marine transgressions at about the same time, the cause must have been a eustatic rise in sea level and not the subsidence of the land. However, there is little evidence of glaciation during the time that submergence was at its peak. It is therefore unlikely that meltwater from ice caps or continental glaciers was the cause of the eustatic rise in sea level. An alternative would be for crustal materials to take up space in the ocean and thereby cause a rise in sea level. This might occur if there were an increase in the rate at which oceanic crust is formed at midocean ridges.

Geologists have been able to estimate the volume of new ocean crust produced during 10-million-year intervals as far back as the beginning of the late Cretaceous. These studies indicate that the rate at which

ocean crust was being produced along midocean ridges was exceptionally rapid. Newly formed ocean crust is hotter than older crust and therefore occupies more volume. This is evident on the present ocean floor, where the surface of newly formed crust stands 2.8 kilometers below sea level, as contrasted to 5 kilometers for older, cooler oceanic crust. Also, when spreading rates are high (producing high-volume crust), the corresponding subduction of cooler, low-volume crust is also greater. Taken together, these sea-floor spreading processes decrease the space for water in the ocean basins, causing sea level to rise and resulting in marine transgressions across formerly emergent regions of the continents. In North America, the rise in sea level was responsible for epeiric seas that advanced southward from Alberta, Canada, and joined an embayment that spread northward from the Gulf of Mexico. At its peak, that seaway was over 1600 kilometers wide and divided our continent into two separate land masses.

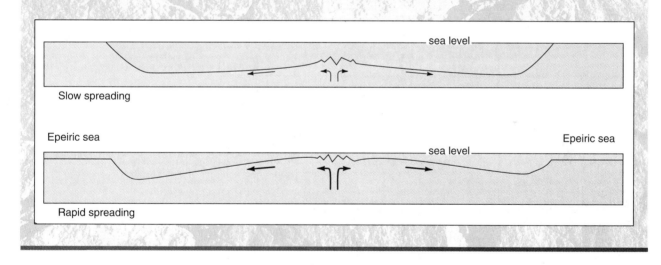

Cretaceous Sedimentation

During the Cretaceous, the Pacific border region of North America was a land of lofty mountains formed during the orogenesis that had begun in the Jurassic. Erosion of these ranges brought sediments downward into adjoining rapidly subsiding basins, many of which were open to the Pacific Ocean. In some places, over 15,000 meters of volcanics and clastics accumulated. Cycles of folding and intrusion occurred repeatedly, with exceptional deformations

taking place during Middle and then Late Cretaceous time.

In the more eastward regions of the Cordillera, the Cretaceous began with the advance of marine waters both northward from the ancestral Gulf of Mexico and southward from arctic Canada. These Early Cretaceous invasions did not meet, so an area of dry land existed in Utah and Colorado. Their advance was reversed by a general regression that resulted in the unconformity used to separate the Cretaceous into an Early and a Late division. The

FIGURE 11–30   Area of outcrop of Cretaceous limestone and marl rocks in the Atlantic and Gulf coastal plains, and generalized columnar sections of the Texas (left) and Alabama (right) Cretaceous.

flooding that followed in the Late Cretaceous was the greatest of the entire Mesozoic Era. This time, the embayment from the north joined with the southern seaway and effectively separated North America into two large islands.

Sedimentation in Cretaceous epicontinental seas was largely controlled by local conditions. Along the Gulf Coast, limestones and marls (clayey limestones) accumulated (Fig. 11–30). North of Texas, however, the great bulk of sediments consisted of terrigenous

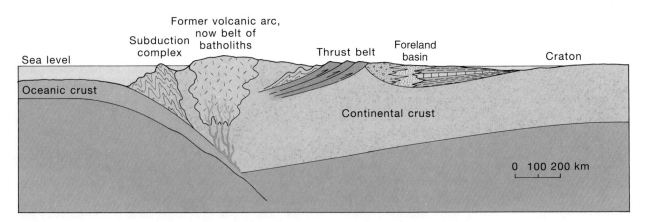

FIGURE 11–31   Conceptual cross-section indicating major tectonic features present during the Cretaceous across the western United States.

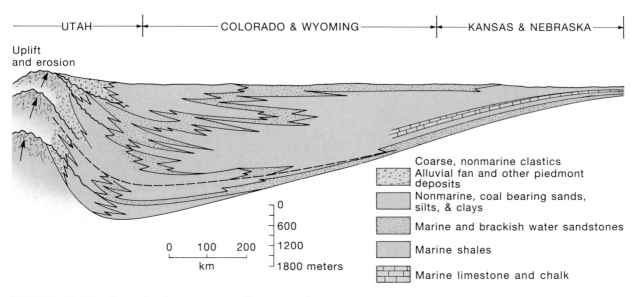

UTAH | COLORADO & WYOMING | KANSAS & NEBRASKA

Uplift
and erosion

Coarse, nonmarine clastics

Alluvial fan and other piedmont deposits

Nonmarine, coal bearing sands, silts, & clays

Marine and brackish water sandstones

Marine shales

Marine limestone and chalk

0

600

1200

1800 meters

0   100   200

km

FIGURE 11–32 Generalized cross-section illustrating the pattern of deposition of Cretaceous rocks in the back-arc basin of the western interior of the United States. *(Simplified from King, P. B. 1977.* Evolution of North America. *Princeton, NJ: Princeton University Press.)*

FIGURE 11–33 The north side of the impressive Interstate-70 road cut west of Denver, Colorado. The rocks visible in this view are part of the Dakota Group. They are dipping steeply toward the east (right). Sandstones of the Dakota Group serve as an important source of artesian groundwater beneath the Great Plains. They have also yielded oil and gas in eastern Colorado, Nebraska, and Wyoming.

clastics supplied by streams flowing from the Cordilleran highlands.

In terms of plate tectonics, the depositional basin that received these sediments (as well as smaller amounts from the craton) can be termed a **Foreland basin** (Fig. 11–31). The **foreland** basin of the western interior of North America was immense. It extended from the Arctic to the Gulf of California, and in places it was over 1600 km wide. It was bounded on the west mostly by fold and thrust belts, such as the Sevier, and on the east by the craton. Many of the rock formations in the basin yield commercial quantities of oil, natural gas, and coal. Exploration for these resources has provided hundreds of well logs and detailed surface studies and has permitted geologists to recognize and map Cretaceous rocks of the Piedmont, coastal plains, shorelines, marine shelf areas, and deep-water tracts of the central basin. The picture that has emerged shows coarse, terrigenous clastics immediately adjacent to the Cordilleran highlands. These interfinger with coal-bearing continental deposits that similarly interfinger with marine rocks farther to the east (Fig. 11–32). The marine and nonmarine facies shift repeatedly with the alternate advances and regressions of the Cretaceous sea. As is normally true, the transgressive phase is marked by sandstone beds, such as those of the Late Cretaceous **Dakota Group** (Fig. 11–33). The red, yellow, and brown layers of the Dakota are exposed in many places along the eastern front of the Rocky Moun-

FIGURE 11–34 Prominent north–south trending hogback formed by the eastward-dipping Dakota Sandstone of Cretaceous age, a few miles south of Golden, Colorado. The Dakota Sandstone at this location is a beach deposit of the Cretaceous sea. Well-developed ripple marks and occasional dinosaur footprints are preserved in the sandstone. *(Courtesy of U.S. Geological Survey.)*

FIGURE 11–35 An exposure of Niobrara chalk in Kansas sculpted by processes of erosion into a pinnacle. The pinnacle is about 7.5 meters tall. *(Photograph courtesy of the Kansas Geological Survey, photograph by Jim Enyeart.)*

tains, where the inclined beds form prominent ridges called *hogbacks* (Fig. 11–34). In parts of the Great Plains, the Dakota sandstones are an important source of underground water.

Cretaceous rocks of Wyoming and Colorado also include extensive beds of a soft, plastic, light-colored clayey rock called *bentonite*. Bentonite is composed of clay minerals that are formed by the alteration of volcanic ash. Volcanoes in the Idaho region during the Late Jurassic and Cretaceous explosively ejected tremendous volumes of ash that were carried into adjacent states by westerly winds. The resulting ash beds, subsequently converted to bentonites, represent single geologic events of short duration and can be traced for great distances, even across changing facies. Hence, bentonite beds are of great value in time-stratigraphic correlation. Volcanic ash also provided a source for the silica that occurs in some Cretaceous siliceous shales and bedded cherts.

Cretaceous carbonate formations include soft, clayey limestones and chalky shales of the **Niobrara Formation** (Fig. 11–35). The Niobrara has yielded the remains of a variety of marine creatures, including enormous numbers of oysters, a large Cretaceous diving bird (Fig. 11–36), marine reptiles, and the large flying reptile *Pteranodon.* Toward the end of the Cretaceous, the seas that supported these creatures began a slow withdrawal. This regression of the Cretaceous epicontinental sea from central North America was contemporaneous with the Laramide deformation that produced the Rocky Mountains. As

FIGURE 11–36 Skeleton of the Cretaceous diving bird *Hesperornis.* This bird was over a meter tall when standing. Propulsion was provided by its large webbed feet. A notable primitive trait of this bird was the retention of teeth. *(Courtesy of The National Museum of Natural History, Smithsonian Institution.)*

the seas withdrew, coal-bearing deltaic and other continental sediments were gradually spread across the old sea floor.

Geologic field work in frigid polar climates is difficult, and this has limited our knowledge of the history of the northern margin of North America. Within the last two decades, however, exploration for petroleum has provided new information about Mesozoic basins of deposition in the far north. Mesozoic sediments accumulated mainly in two basins along the northern margin of the continent belt (Fig. 11–37): the Brooks basin of northern Alaska and the Sverdrup basin farther east. The Mesozoic record in the Brooks basin begins with Triassic clastics and limestones, some of which are fossiliferous. Jurassic and Cretaceous beds consist of volcanic rocks, sandstones, and shales.

Mesozoic formations of the Sverdrup basin include both marine and nonmarine sandstones and shales. Some of the formations contain petroleum. Sandstones and siltstones predominate within the Triassic system. Similar rocks, interspersed with coal beds, were deposited during the Jurassic. Volcanic rocks, continental clastics, red beds, and additional coal beds characterize the Cretaceous rock record.

◼ *Eurasia and the Tethys Seaway*

Extending eastward from Gibraltar and across southern Europe and Asia to the Pacific is the great Alpine–Himalayan mountain belt. Most of the rocks composing the ranges within the belt were laid down in an important depositional trough known as the

FIGURE 11–37 Location of the Brooks and Sverdrup depositional basins.

COMMENTARY

Chunneling in the Cretaceous: The Channel Tunnel

For nearly three centuries, visionaries had dreamed of constructing a tunnel between England and France. That dream became reality in October of 1990 when English and French tunnelers that had been working toward one another met deep beneath the floor of the English Channel. They had excavated and installed the lining of the first of a three-part tunnel system that was completed by the summer of 1993. The Channel Tunnel is the longest undersea tunnel on Earth. It has two "running tunnels" flanking a central "service tunnel." High-speed trains carry passengers through the running tunnels from Folkestone, England (southwest of Dover) to France in 30 minutes.

About 85 percent of the "chunnel," as it has been dubbed, was excavated in a Lower Cretaceous for-

mation known as the Chalk Marl. The term *marl* indicates that the chalk contains clay, and the mixture of chalk and clay results in a rock that is relatively strong, stable, and impermeable. It is an excellent natural material in which to excavate a tunnel. In addition, throughout most of the tunnel route, the Chalk Marl is over 25 meters thick and thus easily accommodates the 8-meter diameter of the running tunnels. The Chalk Marl is overlain by the Upper Cretaceous Gray and White Chalk formations. Beneath the Chalk Marl is the Gault Clay, a unit unsuitable for tunneling because it is plastic and inherently weak.

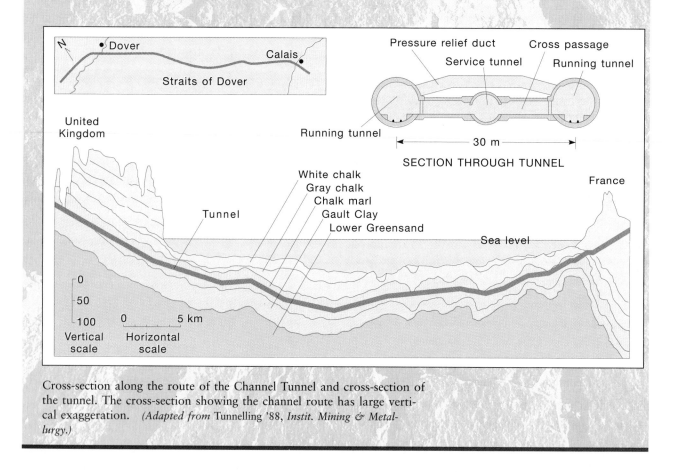

Cross-section along the route of the Channel Tunnel and cross-section of the tunnel. The cross-section showing the channel route has large vertical exaggeration. *(Adapted from* Tunnelling '88, *Instit. Mining & Metallurgy.)*

Tethys seaway. Long after their deposition, these rocks were deformed into mountain ranges as northward-moving segments of Gondwanaland collided with Eurasia.

The geologic history of the Tethys seaway can be traced back to the Paleozoic. But unlike the Appala-

chians, for which most deposition and tectonic activity occurred during the Paleozoic, the Tethys experienced most of its eventful geologic history in the Mesozoic and Cenozoic.

During the Triassic Period, limestone deposition predominated within the Tethys (Fig. 11–38). Within

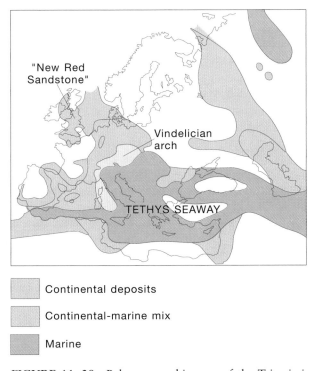

Continental deposits

Continental-marine mix

Marine

FIGURE 11–38 Paleogeographic map of the Triassic in Europe. *(Modified from Brinkmann, R. 1969. Geologic Evolution of Europe. New York: Hafner Publishing Co.)*

this lime-depositing sea, bivalves, cephalopods, crinoids, reef corals, and calcareous algae thrived and left a rich fossil record. Of particular importance for biostratigraphic correlation of Triassic strata are the abundant ammonoid cephalopods (see Chapter 12).

North of the Tethys, the development of a highland tract known as the Vindelician arch separated marine sedimentation to the south from different sedimentation in north-central Europe. In that northern region, the Triassic record begins with red and brown nonmarine clastic rocks that resemble those deposited under arid conditions during the preceding Permian Period. These continental deposits are overlain by shoreline sandstones, marls, evaporites, and limestones deposited during a temporary marine invasion of central Europe. The sea did not remain for long. Nor was it able to reach as far north as Great Britain, where a nonmarine sequence known as the New Red Sandstone was being deposited. As the sea withdrew, continental sedimentation much like that of the earliest Triassic resumed.

Marine conditions were far more widespread during the Jurassic than they had been during the Triassic. Shallow seas advanced from both the Tethys and the Atlantic and spread across Europe, leaving a rich sedimentary record in the basins that lay between the old Hercynian uplands. Although the predominant Jurassic sediment is limestone, there is often a grada-

tion to fine clastics adjacent to the highlands. Eventually, marine conditions extended from the Tethys across Russia and into the Arctic Ocean. The invasion was short-lived, however, and the shallow seas gradually drained from the continent by the end of the Jurassic. Marine conditions persisted in the Tethys, and a complicated pattern of facies changes developed in response to block-faulting on the sea floor.

The Cretaceous is the most eventful geologic period of European Mesozoic history. During this time, the African plate was moving northward, narrowing the region occupied by the Tethys seaway. Along the southern margin of Europe, formerly quiet depositional tracts of the Tethys were subjected to powerful compressional forces. The record of this event is seen in complex overturned folds and thrust faults and in a zone of ophiolites that trend east–west across southern Europe. As described in Chapter 5, ophiolites are greenish igneous rocks that were once part of the crust of the ocean that existed between Gondwanaland and Laurasia. The ophiolites occur in association with radiolarian cherts and lithified deep-sea oozes, which were also part of that sea floor. Subduction of the oceanic crust beneath Eurasia as the African plate approached created the ophiolite zone as well as the deformation and volcanism that characterized the southern margin of Eurasia during the Cretaceous.

The Late Cretaceous is noteworthy not only for its mountain-building events but also as a time of extensive marine transgressions. More of Europe was inundated during the Late Cretaceous than at any other time since the Cambrian (Fig. 11–39). A great embayment from the Tethys worked its way north-

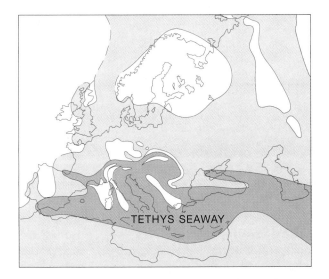

FIGURE 11–39 Areas covered by seas during the Cretaceous. *(Simplified from Brinkmann, R. 1969. Geologic Evolution of Europe. New York: Hafner Publishing Co.)*

ward until it ultimately joined with a similar south-ward encroachment from the region of the North Atlantic. One result of the linking of the two seaways was a rapid interchange of marine organisms from the two formerly separated regions. In the Tethys–Alpine region, block-faulting that had begun during Late Jurassic time continued until about the Middle Cretaceous, when further compression from the south resulted in folding of the thick sequence of marine Tethys sedimentary rocks. The axes of these folds paralleled the east–west trend of the Tethys. Most of these great anticlines and synclines remained submerged, but the tops of some of the folds rose above the level of the waves, forming elongate islands along the north side of the narrowing Tethys seaway. Erosion of the anticlinal islands produced terrigenous clastics that were periodically swept downward by turbidity currents into the adjoining basins, to be deposited as thick sequences of interbedded sandstones and shales.

An event that determined the geographic configuration of the northern coastlines of Spain and France occurred in Europe between Late Jurassic and Late Cretaceous time. As indicated by paleomagnetic data, there was no Bay of Biscay separating western France from Spain prior to the Late Jurassic. The Bay of Biscay opened as the Iberian Peninsula (Spain and Portugal) rotated about 35° from its former line of contact with France. The upheaval of the Pyrenees was another consequence of this rotation. The dating of the rotational event is confirmed by faults of Late Jurassic age on the northern border of the Bay of Biscay and by the fact that the oldest rocks in the bay are Late Cretaceous in age.

East of the Alpine region, the Tethys extended as a great loop past Burma and southeastward into Indonesia. Triassic and Jurassic deposits of the eastern Tethys consist of shales, fossiliferous limestones, and dolomites. During the Cretaceous, volcanic activity occurred throughout what is now the Himalayan region. Some areas seem to have subsided, and terrigenous rocks resembling those moved by submarine landslides or turbidity currents merged with marine volcanic tuffs, breccias, and radiolarian cherts.

■ *The Gondwana Continents*

Africa

Near the beginning of the Mesozoic, separations had already developed between the eastern coast of America and northwest Africa. However, as yet no rift existed between Africa and South America (see Fig. 11–1), and thus there was no South Atlantic Ocean. Most of Africa was relatively stable through-

out the Mesozoic, with only relatively minor transgressions along the northern and eastern borders. During the Cretaceous, the Tethyan waters spread across what is today the northern tier of countries from Algeria to Egypt. Terrigenous muds and carbonates were the most prevalent deposits.

Although the Mesozoic rock record in the more interior parts of Africa is often poor, some regions yield particularly interesting geologic information. One such region is the Karoo basin at the southern end of Africa (Fig. 11–40). The Karoo was a continental basin that was formed late in the Carboniferous and received swamp, lake, and river deposits until late in the Triassic Period. Rocks of the **Karoo sequence** are known to paleontologists around the world for their wealth of fossilized mammal-like reptiles. The fauna included large and small herbivores, diverse carnivores, and a range of insectivores and omnivores.

As the Triassic drew to a close, the terrestrial siltstones and sandstones of the Karoo sequence were covered by layer after layer of low-viscosity lava, which flowed onto the surface from fissures and, less commonly, from volcanoes to the southeast. The old Karoo landscape was buried beneath more than 1000 meters of basalt. Geologists speculate that these great outpourings of lava were associated with the pulling away of the Gondwanaland segments that once adjoined South Africa. Such fragmentation would have caused severe fracturing of the foundations of the continents and provided multiple avenues for the upwelling of molten rock. The extrusions continued well into the Jurassic Period. Contemporaneous lava floods and volcanism also occurred on the separating land masses of South America, Australia, and, somewhat later, Antarctica.

Africa experienced little tectonic activity throughout the remainder of the Jurassic and Cretaceous. Marginal seas extended along the eastern edges of the continent. Periodic advance and retreat of these seas brought an alternation of marine and nonmarine beds.

Australia and New Zealand

During the Triassic, Australia was still attached to Antarctica, and both continents were predominantly emergent. Terrestrial sandstones and shales resembling those of the Permian were deposited. Except for brief marginal embayments, deposition of lake and stream sediments continued to dominate until the middle of the Early Cretaceous (Fig. 11–41). At that time, a marine transgression brought sandstones and chalk beds into the interior.

New Zealand during the Mesozoic was geologically far more restless than Australia. It lay along the

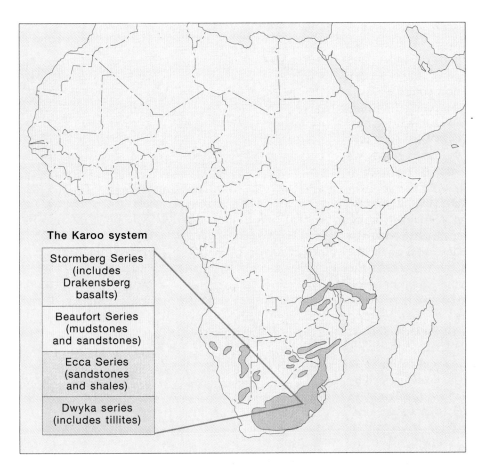

FIGURE 11–40 Outcrops of the Karoo System (green) of South Africa and the four principal units of the Karoo section.

The Karoo system

Stormberg Series (includes Drakensberg basalts)
Beaufort Series (mudstones and sandstones)
Ecca Series (sandstones and shales)
Dwyka series (includes tillites)

boundary between the southwestern part of the Pacific plate and the eastern edge of the Indian plate. The Mesozoic rock sequence includes turbidites, siliceous rocks, and abundant volcanics. Near the end of the Jurassic, a series of orogenic events occurred as a result of the collision and subduction of the Pacific plate beneath the island. In the course of the orogeny, mountains were erected and deep, unstable trenches developed. The activity continued into the Cenozoic and, intermittently, even to the present day.

India

During the Mesozoic, India moved steadily northward on its remarkable voyage to Laurasia. The Indian continent itself remained tectonically stable, even though it was moving at a rapid rate toward its ultimate collision with Laurasia in Cenozoic time. The land mass experienced only relatively minor marginal incursions of the sea. Across the interior, erosion of upland areas spread terrigenous clastic sediments into the lowlands and plains. Some of these continental formations have yielded a wealth of dinosaur bones and superb fossils of plants. By Cretaceous time, India was nearing the Tethyan region. Here and there, the floor of the Tethyan trough was

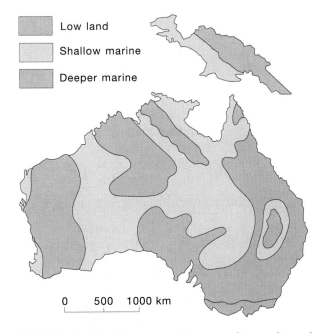

Low land

Shallow marine

Deeper marine

0   500   1000 km

FIGURE 11–41 Paleogeographic map of Australia and New Guinea during the Early Cretaceous, when most of Australia was inundated by shallow to slightly deeper seas.

buckled into elongate ridges that were indications of the coming tectonic collision. While these ridges were developing, the northwestern half of India was flooded with immense quantities of low-viscosity basaltic lava (Fig. 11–42). These now solidified lavas are flood basalts of the **Deccan Traps.** Radiometric dates obtained from the basalts indicate that the out-pourings continued from Cretaceous time well into the Cenozoic. It is likely that these basalt floods record the passage of India across a fixed hot spot in the mantle.

The Deccan Traps cover about 500,000 km^2 of western and central India with an aggregate volume in excess of 1,000,000 km^3. As such, they comprise the greatest volume of continental basalt on the Earth's surface. Erosion has removed much of the Deccan volcanics, and their original extent and volume were once far greater. The extensive fiery extrusions and huge volumes of carbon dioxide released along the web of fissures may have triggered climatic warming and chemical changes deleterious to life both on land and in the oceans. As will be noted in the next chapter, Deccan volcanism, accompanied by the eruption of huge volumes of lava in Brazil and on the ocean floor, may have contributed to the terminal Cretaceous episode of mass extinction.

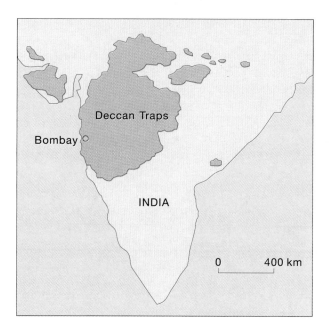

FIGURE 11–42   Present-day distribution of Deccan Trap basalts. Much of the original cover of lava has been lost to erosion, as indicated by the isolated remnants to the east of the large outcrop area.

South America

Much like Africa, South America stood well above sea level at the beginning of the Mesozoic. Along the western margin of the continent, two parallel zones of deposition were already established in Triassic time. The more western tract was characterized by turbidites, conglomerates, and siliceous sediments laid down in the deeper waters of the continental rise and slope. Eastward of these deep-water facies, carbonates and shales were deposited across shelf zones. The approximate boundary between the two suites of rocks can now be traced along the central ranges of the Andes. In addition to these marine tracts, broad basins in the interior accumulated continental deposits. Particularly remarkable are the eolian and fluvial sandstones and siltstones that spread across the southeastern region of the continent during the Triassic. The beds contain a rich fauna of Triassic vertebrates. Lava flows of the same age as the Karoo rocks of Africa are also prevalent in the upper part of this stratigraphic sequence.

By Jurassic time, the initial narrow split between South America and Africa had widened into a configuration resembling the present-day Red Sea. New sea floor formed along the spreading center of the developing South Atlantic. The western side of the continent was at the leading edge of the westward-moving tectonic plate. It was opposed by the Pacific plate, the forward margin of which plunged beneath South America as an eastward-inclined subduction zone.

Near the end of the Jurassic, the Andean belt experienced deformation and volcanism. This activity appears to have begun in the south and then sporadically shifted with time toward the north. The climax of orogenic deformation occurred in Late Cretaceous and early Tertiary time and was accompanied by regional uplift that led to widespread withdrawal of the seas. Evaporites were deposited in scattered isolated basins that for a time retained marine waters.

The frequency and intensity of deformational and volcanic activity along the Andean mobile belt increased during the Cretaceous. In the subduction zone created by the movement of the Pacific plate against South America, oceanic crust and its cover of sea-floor sediment were carried down to deep zones, where melting occurred. The melts in turn worked their way back to the surface as great intrusions and outpourings of andesite. Deformation and igneous activity continued well into the Cenozoic and formed many of the geologic structures now seen in the towering peaks of the Andes.

Antarctica

Antarctica was predominantly emergent throughout the Mesozoic. Eastern areas of the now frigid land mass were sites of continental deposition. Beds of volcanic ash and lava flows occur between these lake and stream deposits.

The nature of the sedimentation was markedly different in western Antarctica. Outcrops along the Antarctic Peninsula consist of rocks that are similar to those of equivalent age in the Andes. The strata of the Antarctic Peninsula most likely represent part of an orogenic belt that was continuous with the Andes to the north.

■ *Economic Resources*

Uranium Ores

Rocks of the Mesozoic System, like those of earlier eras, contain a wealth of important mineral resources. Notable among such resources are the nuclear fuels. In the United States, uranium ores are derived chiefly from continental Triassic and Jurassic rocks of New Mexico, Colorado, Utah, Wyoming, and Texas. The chief ore mineral is **carnotite** (Fig. 11–43), a yellow, earthy oxide of uranium that usually occurs within the pore spaces of sandstones, as encrustations, or as replacements of fossil wood. In some instances, petrified logs have provided amazing concentrations of not only uranium but also vanadium and radium. These are striking exceptions, however, and most of the ores are of very low concentration.

The uranium in Triassic and Jurassic beds of the western United States was originally dissolved from consolidated volcanic ash (tuff) and transported in groundwater. Precipitation of the ore occurred when the uranium-bearing solutions encountered rocks rich in organic matter in which chemically reducing (rather than oxidizing) conditions existed.

There is presently concern that the richer sources of fissionable uranium-235 in this country are likely to be exhausted by about the same time that our petroleum reserves are depleted. Among the alternate sources of nuclear energy now being studied are uranium-238 and thorium-232, which can be converted to fissionable isotopes by a process termed *breeding*. This will permit the use of the more abundant low-grade uranium ores. Of course, breeder reactors are not yet perfected and may not prove sufficiently safe or as enduring as other possible energy sources.

Fossil Fuels

Until either the breeder reactor or devices to utilize solar energy are perfected, the nations of the world will continue to rely on fossil fuels. Mesozoic rocks contain significant amounts of these critical re-

FIGURE 11–43 The bright yellow uranium mineral carnotite disseminated through and encrusting Shinarump sandstone. Specimen is 9 cm long.

sources. The Jurassic, for example, contains workable deposits of coal in Siberia, China, Australia, Tasmania, Spitzbergen, and North America. In North America, thick seams of Jurassic coal occur in British Columbia and Alberta. Cretaceous coal underlies more than 300,000 sq km of the Rocky Mountain region (Fig. 11–44). Much of this coal is now being mined, particularly because of its relatively low content of environmentally offensive sulfur.

In addition to coal, Mesozoic rocks supply large quantities of oil and gas to energy-hungry industrial nations. The oil provinces of the Middle East and North Africa contain more oil than the combined reserves of all other countries. Middle East petroleum comes primarily from thick sections of Jurassic and Cretaceous sedimentary rocks that accumulated in

FIGURE 11–44 Coal beds in the Mesaverde Formation (Upper Cretaceous) exposed in a road cut on Highway 6-50 at Castlegate, Utah.

FIGURE 11–45   Oil being pumped from Jurassic and Cretaceous formations in the Bighorn basin of Wyoming.   *(Photograph courtesy of L. E. Davis.)*

shelf and reef environments bordering the Tethys seaway. Other areas of petroleum production from Mesozoic rocks occur in the western United States (Fig. 11–45), Alaska, Arctic Canada, the Gulf Coastal states, western Venezuela, southeast Asia, beneath the North Sea, and beneath the eastern offshore area of Australia. Notwithstanding the size of some recent discoveries in rather forbidding parts of the world, it appears likely that the present rates of oil and gas consumption will cause exhaustion of the world's resources in less than a century. Therefore, oil and gas must be replaced by other energy sources in the near future, especially if we are to conserve these valuable materials for the chemical industry. The time is approaching when petroleum will be too valuable to burn.

Metallic Mineral Deposits

Metalliferous deposits were formed widely throughout the active orogenic belts of the Mesozoic world. A variety of metals now found in the Rocky Mountains, the Pacific coastal states, and British Columbia were emplaced during batholithic intrusions that accompanied Jurassic and Cretaceous orogenies. Among these ore deposits are the gold-bearing quartz veins known as the "Mother Lode." The California gold rush of 1849 was a consequence of the discovery

FIGURE 11–46   The Morenci open-pit copper mine in Arizona. The pit occupies the former location of the town of Morenci, Arizona. The Metcalf copper mine is in the background. Copper minerals at this location are disseminated throughout porphyritic igneous rock.   *(Courtesy of Erik Melchiorre.)*

of gold-bearing gravels eroded from the Mother Lode.

The copper, silver, and zinc veins of the Butte, Montana, and Coeur d'Alene, Idaho, mining district were emplaced as a result of Cretaceous igneous activity. One belt of copper-bearing rocks extends from Denver into northeastern Arizona. These deposits and similar ones in Utah occur in igneous rock that has a porphyritic texture (large crystals of feldspar and quartz in a finer matrix) and hence are called *porphyry copper deposits* (Fig. 11–46). The copper minerals, principally copper and copper–iron sulfides, are disseminated through rock that has been pervasively fractured during episodes of explosive volcanic activity. The many fractures provided passage for hot, mineralizing solutions from which the ore was precipitated. Porphyry copper deposits occur not only in western North America but also in South America, the Philippines, Iran, and eastern Europe.

Important nonmetallic deposits of the Mesozoic include sulfur and salt. Both are produced from the salt domes mentioned earlier (see Fig. 2–30). Diamonds are also obtained from Mesozoic igneous rocks. Siberia's diamond-bearing intrusions are believed to have penetrated the upper crust during the Triassic and Jurassic. Similar diamond pipes in Africa are probably of Cretaceous age.

Plate Tectonics and Ore Deposits

In recent years, geologists have begun to see numerous correlations between patterns of mineral distribution and locations of present and former tectonic plate boundaries. Plate boundaries are likely sites for movements of hot, aqueous fluids that bear important metals in solutions. With changes in temperature or pressures, or on contact with reactive rocks, such hydrothermal solutions will precipitate ore minerals. The process may operate at both convergent and divergent plate boundaries. Both the Cordilleran and Andean mobile belts are examples of convergent boundaries. An example of hydrothermal mineralization at a divergent boundary is provided by the Red Sea, which has pools of hot and exceptionally salty water along its bottom. The pools are rich in iron, manganese, zinc, and copper. The brines have percolated upward through the young oceanic crust, dissolving the metals en route. Metals brought to the surface in this way in the past may have been conveyed in thin layers across immense tracts of the ocean floor by sea-floor spreading. Sediments containing metallic ions extracted from the sea water itself often enriched the accumulation. Ultimately, sea-floor spreading moved the sediments and their contained metals into collision with another plate, providing an opportunity for their inclusion as ore bodies within the deformed belt.

Of course, the formation of ores in orogenic belts is a far more complicated process than this conceptual model suggests. Every ore body requires special local conditions and events for its emplacement. Many ore deposits have no relationship to plate boundaries. However, in the case of the metallic deposits of North and South America, a relationship is likely, for both of these regions have overriding convergent boundaries with the plates of the Pacific.

Summary

Within the approximate 160 million years of the Mesozoic Era, the pattern of lands and seas was extensively altered as large segments of Pangea were separated. The break-up appears to have begun during the Triassic, with the splitting off of North America from Gondwanaland. By Late Jurassic time, rifts had developed between all of the Gondwanaland segments except Australia and Antarctica. These two continents, like northwestern North America, Greenland, and northwestern Europe, retained a hold on one another until early in the Cenozoic.

At the beginning of the Mesozoic, most of North America was emergent. In the east, crustal tension was expressed by the development of large-scale block-faulting and volcanism. In downfaulted valleys, arkosic sandstones, conglomerates, and lacustrine shales accumulated. While this was occurring, the ancestral Gulf of Mexico began to take form far to the south. During the Jurassic and Cretaceous, thick sequences of limestones, evaporites, sandstones, and calcareous shales accumulated along what are now the Gulf and Atlantic coastal states.

Continental deposits, particularly red beds, characterize the Triassic of the Rocky Mountain regions of North America. Marine conditions persisted along the Pacific margin, where volcanic rocks and graywackes accumulated to great thicknesses. During the Jurassic, this western belt was strongly deformed as North America began to override the subduction zone at the leading edge of the eastward-moving Pacific plate. The advancing Pacific plate carried large fragments of volcanic arcs, oceanic plateaus, and microcontinents to the western margin of North America. These crustal fragments became incorporated into the Cordillera as a tectonic collage of displaced terranes. Growth of a continental margin in this way is referred to as accretionary tectonics.

In general, deformation of the North American Cordillera during the Mesozoic began along the Pacific coast and moved progressively eastward, until by Late Cretaceous time, inland seas that had occupied the western interior of North America were displaced. Majestic mountain ranges stood in their place. Crustal unrest continued to affect the western states even during the early epochs of the Cenozoic Era.

The focus of Eurasian Mesozoic events was the great east–west trending Tethys mobile belt. The initial Mesozoic deposits of the western Tethys were predominantly carbonates. Block-faulting developed along parts of the belt during the Jurassic. This middle part of the Mesozoic witnessed marine invasions that completely spanned the continent from the Tethys to the Arctic. During the Cretaceous, the Tethys was affected by powerful compressional forces as the northward-moving oceanic plate bearing peninsular India moved toward and under the Eurasian plate. One result of these movements was the development of a series of anticlinal welts along the northern side of the Tethyan seaway. Rapid erosion of these elongated, uplifted tracts produced thick sequences of coarse clastics in the intervening troughs. Along the southern margins of the Tethys, conditions were more stable, and relatively undisturbed sequences of carbonates and sandstones were deposited.

Although Australia was predominantly emergent during the Mesozoic, an active tectonic zone existed across a tract now occupied by New Zealand. This zone originated in the late Paleozoic and reached its orogenic climax during the Jurassic. New Zealand's Cretaceous rocks were for the most part deposited in shallow marginal seas adjacent to the ranges uplifted during the previous period.

The Mesozoic history of the interior regions of Africa is read primarily from continental deposits. Among these, none is more famous than the beds of the Karoo basin of South Africa. The Karoo sequence includes both Permian and Triassic strata, some of which have yielded splendid collections of the mammal-like reptiles that once inhabited southern Africa.

During the Mesozoic, a largely emergent peninsular India was progressing steadily toward its Eurasian destination. By Cretaceous time, it had approached sufficiently near to cause initial buckling in the eastern Tethys.

The Jurassic was an important geologic period for South America, which during this time began to separate from Africa. The separation continued during the remainder of the Mesozoic, and a new ocean, the South Atlantic, was born. The western margin of the continent formed the leading edge of the South American plate and was characterized by volcanic activity, folding, and intrusions.

The rocks of the Mesozoic have supplied the world with a variety of important economic resources. These include such fossil fuels as coal, petroleum, and natural gas as well as nuclear fuels. Orogenic activity during the Mesozoic resulted in the emplacement of such critical metals as copper, zinc, chromite, gold, silver, lead, mercury, and many others. Current theories of ore genesis suggest that metallic ores emplaced during mountain building along continental margins may represent concentrates of materials on the sea floor that have been conveyed to orogenic belts by sea-floor spreading. Once such materials are brought to the plate boundary, favorable physical and chemical conditions must exist locally to provide for concentration of the dispersed metals into ore deposits.

Review Questions

1. What is the total duration in years of the Mesozoic Era? Which period of the Mesozoic was the longest?
2. What was the probable source area for the sediments of the Morrison Formation of the Rocky Mountain region? What evidence indicates that the Morrison Formation was a continental rather than a marine formation?
3. At what time during the Mesozoic were epicontinental (epeiric) seas most extensive? During which period were such marine incursions most limited?
4. What was the Tethys seaway? What water bodies today are remnants of this seaway? Where might one go to examine the rocks deposited within the Tethys?
5. At what time in geologic history did the modern South Atlantic Ocean begin to form?
6. What nonmetallic mineral resources are derived from Mesozoic strata? What metallic ores occur?
7. What conditions favored the precipitation of uranium ores in Triassic and Jurassic rocks of the Colorado Plateau?
8. When did most of the ranges of the Rocky Mountains undergo their major deformation? In what areas was thrust faulting particularly prevalent?

Discussion Questions

1. What geologic event precedes the separation of a continent into two or more parts? Cite examples from Mesozoic geologic history.
2. Describe the general structural and environmental conditions associated with the deposition of the Newark Group. Why did normal faulting predominate rather than reverse faulting? Did older structures influence the location of Triassic faults? Explain.
3. Discuss the characteristics of structures produced by the Sevier orogeny.
4. What environmental conditions account for the presence of Jurassic evaporites in the Gulf Coast region? How are these evaporites related to petroleum traps in overlying Cretacous and Cenozoic strata?
5. Explain what is meant by the term *accretionary tectonics?* Provide examples of the occurrence of accretionary tectonics from the Mesozoic history of North America. How are allochthonous terranes related to accretionary tectonics?
6. Discuss the possible effects of Deccan and extensive ocean floor volcanism on global climate near the end of the Mesozoic Era.
7. Describe and contrast obduction with subduction and cite examples of the occurrence of each.
8. What is chalk? In what Mesozoic geologic period was it particularly abundant? How is the origin of chalk related to marine plankton?

Supplemental Readings and References

Ager, D. V. 1980. *The Geology of Europe.* New York: John Wiley.

Armstrong, R. L. 1968. The Sevier orogenic belt in Nevada and Utah. *Bull. Geol. Soc. Am.* 79:429–452.

Dietz, R., and Holden, J. C. 1971. The breakup of Pangaea. In *Continents Adrift: Readings from Scientific American.* San Francisco: W. H. Freeman and Co.

Greenwood, P. H. (ed.). 1981. *The Evolving Earth.* Cambridge, England: Cambridge Press.

Jones, D. L., Cox, A., Coney, P., and Beck, M. 1982. The growth of western North America. *Sci. Am.* 247(5):70–128.

Klitgord, K. D., Hutchinson, D. R., and Schouten, H. 1988. *U.S. Atlantic Continental Margin: Structural and Tectonic Framework.* Boulder, CO: *Geological Society of America.*

Manspeizer, W., Putter, J. H., and Cousminer, H. L. 1978. Separation of Morocco and eastern North America: A Triassic–Liassic stratigraphic record. *Bull. Geol. Soc. Am.* 89:901–920.

Reynolds, M. W., and Dolly, E. D. (eds.). 1983. *Mesozoic Paleogeography of the West-Central U.S.* Rocky Mountain Symposium. Society of Economic Paleontologists and Mineralogists, Denver, Colorado.

Rona, P. A. 1974. Plate tectonics and mineral resources. In *Planet Earth: Readings from Scientific American.* San Francisco: W. H. Freeman and Co., pp. 170–180.

Saleeby, J. B. 1983. Accretionary tectonics of the North American Cordillera. *Ann. Rev. Earth Planet Sci.* 15:45–73.

Salvadore, A. 1991. *Geology of North America,* pp. 1–12. Boulder, CO: *Geological Society of America.*

The theropod dinosaur *Yangchuanosaurus*, a large predator whose remains were recovered from late Jurassic rocks of China. *(Painting by Gregory S. Paul.)*

Tyrannosaurs, enormous bipedal caricatures of men,
would stalk mindlessly across the sites of future cities
and go their way down into the dark of geologic time.

LOREN EISELEY, *The Immense Journey, 1959*

CHAPTER 12

The Mesozoic Biosphere

The biosphere is the world of life. In this chapter, we direct our attention to the biosphere of the Mesozoic Era. The great nineteenth-century naturalist Louis Agassiz called the Mesozoic the *Age of Reptiles*. It was a clearly appropriate designation, for reptiles thrived on land, in the seas, and even in the air. Yet with all our fascination with the dinosaurs and other often bizarre reptiles of the Mesozoic, we should not forget that it was during this era that birds and mammals also began their evolution. In the world of plants, grasses and modern trees bearing seeds, nuts, and fruits expanded across the continents. As today, the distribution, evolution, and abundance of all these many forms of life were dependent on climate. Climate, in turn, was affected by the changing locations of continents, major marine transgressions and regressions, and the formation of great mountain systems.

■ *Mesozoic Climates*

Climate is influenced by a large number of factors, but the primary control of global climate is the balance between incoming and outgoing radiation from the sun. That balance is affected by a variety of factors, including the configuration and dimensions of the oceans and continents; the development and location of mountain systems and land bridges between continents; changes in snow, cloud, or vegetative cover (which affect the amount of solar radiation reflected back into space from the Earth); the carbon dioxide content of the atmosphere (which can trigger greenhouse warming); the location of the poles; the amount of radiation–blocking aerosols thrown into the atmosphere by volcanoes; and such astronomic factors as changes in the Earth's axis and orbit.

Cool climates seem to have characterized many continental areas during the final days of the Paleozoic Era. The vastness of the Pangea supercontinent, the upheaval of mountains, general uplifts, and withdrawal of inland seas in many regions were contributing causes of the generally cooler conditions. Gradually, however, the climate warmed. Glaciers in Africa, Australia, Argentina, and India began to melt as these continents drifted away from the South Pole. In general, during most of the Jurassic and Cretaceous, climates for most regions were warm and equable. Also during these periods, extensive regions were covered by epeiric seas, and this contributed to warmer conditions. Water bodies retain heat far better than do land areas. Furthermore, ocean waters are in constant circulation, distributing that warmth around the globe. Hence, when the proportion of ocean to land increases, warmer climates may result. The reverse holds true as well, and this was probably an important factor causing cooler climates at the end of the Paleozoic.

During the Triassic, the continents were still tightly clustered. The paleoequator extended from central Mexico across the northern bulge of Africa (Fig. 12–1). As noted in Chapter 11, the Triassic was a time of general emergence of the continents. Mountains, thrust upward at the end of the Paleozoic, inhibited the flow of moist air into the more centrally located regions, causing widespread aridity. Evaporites, dune sandstones, and red beds accumulated at both high and low latitudes and attest to relatively dry and warm conditions.

Reconstructions based in part on paleomagnetic studies suggest that during the Jurassic, the continents were at the approximate latitudinal positions they occupy today. Marine waters extended northward into a great trough formed by the opening of the Atlantic Ocean, and in many places shallow inland seas spilled out of the deep basins onto the continents. An arm of the **Protopacific** extended westward as the great Tethys seaway. It is not unreasonable to assume that warm, westward-flowing equatorial currents may have penetrated far into the Tethys. These same equatorial currents, deflected by the east coast of Pangea, were shunted to the north along coastal Asia and to the south along northeastern Africa and India. To complete the cycle, the cooled currents may have returned to the equator along the west side of the Americas. The presumed ocean and wind currents and the rather extensive coverage of seas, both upon

and adjacent to continents, brought mild climates to many regions during the Jurassic. There is also evidence, however, of aridity and climatic patterns characterized by monsoons. Glacial deposits of this age are not known, and coal beds occur in Antarctica, India, China, and Canada. Paleobotanical evidence suggests that tropical conditions prevailed over regions that are now temperate.

During most of the Cretaceous, climatic conditions were generally warm and stable as they had been during the Jurassic. A remarkably homogeneous flora spread around the world, with subtropical plant families thriving in latitudes 70° from the equator. Coal beds formed on nearly every continent and even at very high latitudes. However, such pleasant conditions did not persist, and toward the end of the Cretaceous, the climatic pendulum began a slow backward swing toward more rigorous conditions. The Late Cretaceous South Pole was centrally located in Antarctica, and the North Pole was located at the north edge of Ellesmere Island. Europe and North America had moved somewhat farther north, and the widespread inland seas had begun to recede. The worldwide regressions were accompanied in some regions by major orogenic disturbances.

In the Late Cretaceous seas, the golden-brown algae known as coccolithophorids spread in enormous profusion. Their calcium carbonate platelets (see Fig. 4–51) accumulated on the floors of the inland seas and were converted to thick layers of chalk. They are estimated to have been so abundant in the Late Cretaceous that their photosynthetic activity may have produced a carbon dioxide shortage in the atmosphere. Because carbon dioxide provides the Earth with a warming greenhouse effect it is at least theoretically possible that the great blooms of plankton may have contributed, along with other environmental conditions, to a temporary chilling of the Late Cretaceous world.

There are several lines of evidence that favor the interpretation of a terminal Cretaceous cooling. Terrestrial plants provide some clues. The tropical cycads underwent a sharp reduction, and ferns declined in both North America and Eurasia. Hardier plants such as conifers and angiosperms extended their realms. Evidence comes also from paleotemperature studies based on the oxygen isotope method described in Chapter 4. The oxygen isotope ratios obtained from open ocean planktonic calcareous organisms indicate a decline in ocean temperatures beginning about 80 million years ago. If there was indeed a dip in worldwide mean annual temperatures, it might have had a deleterious effect on plant and animal life. For this reason, some paleontologists speculate about its relationship to the extinctions that occurred at the end of the Mesozoic.

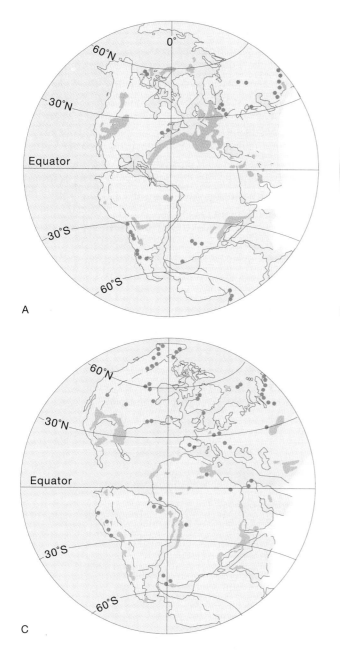

A

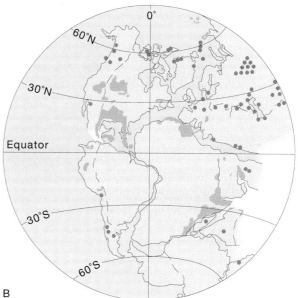

B

C

FIGURE 12–1 (A) Triassic, (B) Jurassic, and (C) Early Cretaceous paleogeographic maps showing positions of the equator, continents, and distribution of evaporites (green areas) and coal deposits (red circles). *(Adapted from Dewey, G. E., Ramsey, T. S., and Smith, A. G. J. Geol. 82(5):537, 1974.)*

■ *Mesozoic Flora*

The existence of animal life on Earth is ultimately dependent on plant life. This generalization was as valid during the Mesozoic as it is today. Then, as now, plants made up the broad base of the food pyramid. Their nutritious starches, oils, and sugars made possible the evolution and continuing existence of animals. Plants are a fundamental part of the Earth's essentially self-sustaining ecologic system. The operation of the system is dependent on oxygen and carbon dioxide. Animal respiration provides the carbon dioxide needed for plant photosynthesis, whereas plants supply—as a byproduct of photosynthesis—the oxygen needed by animals. In the geologic past, variations in plant productivity may have caused corresponding changes in the amount of carbon dioxide and oxygen in the atmosphere. Such variations may have favored the evolution of some organisms over others and may even have been responsible for the demise of particular groups. As an example, in 1984 geologist Dewey M. McLean suggested that the pronounced extinctions among planktonic foraminifera and coccolithophorids at the end of the Cretaceous Period may have been the result of rapid release of carbon dioxide derived from massive outpourings of lavas, such as occurred during the extrusion of the Deccan lavas in India. The surplus carbon dioxide may have then caused chemical changes in the upper 45 to 100 meters of the ocean that might have been lethal to microplankton. Without these organisms, extinctions of animals higher in the food chain would have been inevitable.

Marine Plants

Because plants that live in the oceans are suspended in water, they do not require the vascular and support systems that characterize land plants. Most marine plants are therefore unicellular, although they may grow in impressive colonies and aggregates. The major groups of marine plants are part of that vast realm of floating organisms called plankton; because they are plants, they are further designated as **phytoplankton.** The geologic record of the important fossil groups of phytoplankton is shown in Figure 12–2. As indicated on the chart, phytoplankton that did not secrete mineralized coverings predominated in the pre-Mesozoic eras. These include both cyanobacteria and green algae, as well as a group of cellulose-covered unicellular algae of uncertain affinity that are loosely called acritarchs. Of more immediate impor-

tance to us are the coccolithophorids, dinoflagellates, silicoflagellates, and diatoms, which were the more abundant phytoplankton groups of the Mesozoic.

Dinoflagellates are frequently encountered as fossils and are important aids in Mesozoic and Cenozoic stratigraphy. From the Jurassic on, they were among the primary producers in the marine food chain. Dinoflagellates (Fig. 12–3) are unicellular organisms that have a cell wall composed of a substance called sporopollenin like that in pollen. For propulsion, the organism is equipped with two flagella: One is longitudinal and whiplike, and the other is transverse and ribbon-like. During their life cycle, dinoflagellates develop a motile planktonic form and a cyst phase that is formed within the motile organism. Only dinoflagellate cysts, which have a covering extremely resistant to decay, are known as fossils.

The coccolithophorids also began their expansion during the Early Jurassic. Unlike the dinoflagellates, these calcium carbonate–secreting organisms have a splendid fossil record. Their abundant remains have formed many of the extensive coccolith limestones of the Mesozoic and early Cenozoic. Today, they are frequently present in the deep-sea sediment known as calcareous ooze. The coccolithophorid organism is one of several varieties of unicellular golden-brown

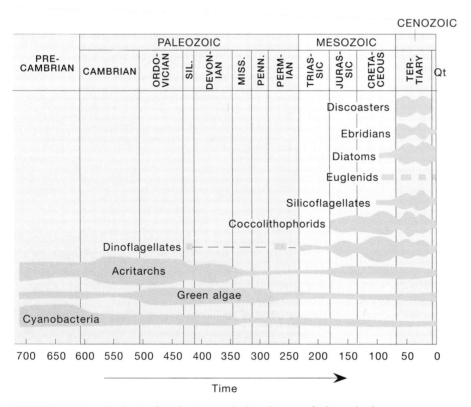

FIGURE 12–2 Geologic distribution and abundances of phytoplankton. *(From Tappan, H., and Loeblich, A. R., Jr. Geol. Soc. Am. Special Paper 127:257, 1970.)*

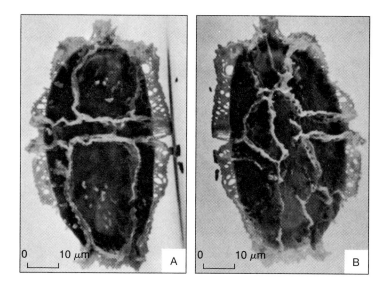

FIGURE 12–3 Fossil dinoflagellate cyst, *Prionodinium alveolatum,* from the Cretaceous of Alaska. *(From Leffingwell, H. A., and Morgan, R. P. J. Paleontol. 51(2):292, 1977.)*

algae. These algae deposit calcium carbonate internally on an organic matrix and construct tiny, shieldlike structures called **coccoliths.** Once formed, the coccoliths move to the surface of the cell and form a calcareous armor (Fig. 12–4). Coccoliths measure only from 0.002 to 0.01 mm in diameter. However, with the aid of the electron microscope, it is possible to discern their intricate construction. With such magnification, they are seen to consist of one, or sometimes two, superimposed elliptical or round plates that are concave on one side to fit snugly around the surface of a spherical cell. Each disc or plate is itself composed of smaller crystalloids that may be triangular, rhombic, or variously shaped by

the organism. The crystalloids are uniformly arranged, usually in a circular, radial, or spiral plan that is often astonishing in its precision (see Fig. 4–51). Because they are frequently fossilized, have experienced frequent evolutionary changes through time, and are readily dispersed by oceanic currents, coccoliths are extremely useful in stratigraphic correlation.

The earliest known silicoflagellates (Fig. 12–5) and diatoms appeared in the Cretaceous. Along with other phytoplankton, they experienced a decline at the end of the Cretaceous, and then all groups expanded again into the early epochs of the Cenozoic. The silicoflagellates and diatoms, along with the coccolithophorids, are members of the phylum Chrysophyta.

As suggested by their name, silicoflagellates are flagella-bearing organisms that secrete delicate siliceous skeletons in the form of simple, lattice-like

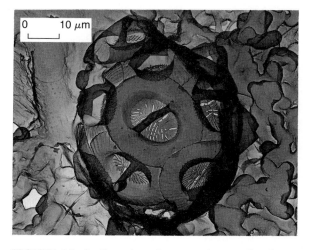

FIGURE 12–4 Scanning electron micrograph of a cocosphere covered with coccoliths. *(From McCormick, J. M., and Thiruvathukal, J. V., Elements of Oceanography, 2nd ed. Philadelphia: Saunders College Publishing, 1981. Photography courtesy of S. Honjo.)*

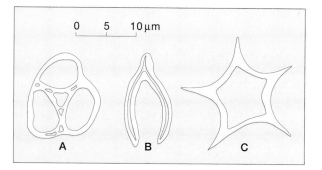

FIGURE 12–5 Three Mesozoic silicoflagellates. (A) *Corbisema* (Cretaceous–Eocene). (B) *Lyramula* (Cretaceous). (C) *Vallacerta* (Cretaceous).

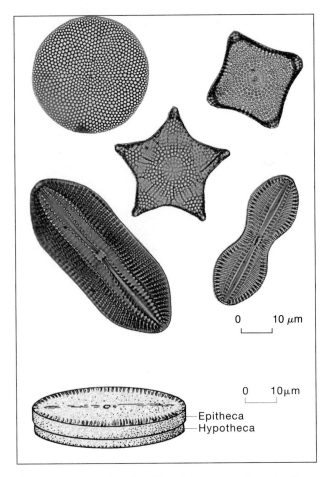

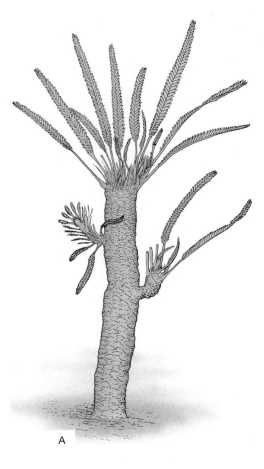

FIGURE 12–6   Photographs of diatoms collected off the coast of Crete. The drawing illustrates how the two parts of the frustule of a diatom fit together. *(Photographs courtesy of Naja Mikkelsen.)*

frameworks. Radiating spines characterize most genera, whereas others have stellate form. They range in size from 0.02 to 0.1 mm.

Like the silicoflagellates, **diatoms** also secrete siliceous coverings (Fig. 12–6). The covering is called a *frustule*. It is usually composed of an upper part (the *epitheca*) and a lower part (the *hypotheca*) that fit together like a lid on a box. The frustule may be circular, cylindric, triangular, or a variety of other shapes and is usually quite beautiful. Today, marine diatoms are prevalent in the cooler regions of the oceans. In the past, a proliferation of diatoms was often associated with volcanic activity. Apparently, the silica supplied to sea water as fine volcanic ash stimulated diatom productivity.

Terrestrial Plants

Although a vertebrate paleontologist might refer to the Mesozoic as the Age of Reptiles, paleobotanists

FIGURE 12–7   (A) Restoration of the cycadophyte *Williamsonia* from the Jurassic of India. This Jurassic plant grew to heights of about 2 meters. (B) A cycad growing in South Africa. The pineapple-like structures at the top are seed cones.   *(Part (A) from Sahni, B. 1932. Mem. Geol. Survey India, Paleontology Indica 20(3): 1–19; part (B) W. H. Hodge/Peter Arnold, Inc.)*

FIGURE 12–8 *Ginkgo biloba,* the ginkgo or maidenhair tree (so called because the leaves resemble those of the maidenhair fern). Note the naked, fleshy seeds that characterize female trees. The ginkgo represents the oldest genus of living trees. Fossils over 200 million years old are nearly identical to the living forms. *(W. H. Hodge/Peter Arnold, Inc.)*

might well argue that the term "Age of Cycads" would be equally appropriate. The **cycads,** more formally known as the class Cycadophyta (Fig. 12–7), are seed plants in which true flowers have not been developed. Jurassic cycads included tall trees with rough, columnar branches marked by the leaf bases of earlier growths and by crowns of leathery pinnate leaves. Actually, the Cycadophyta include two related groups: the cycadeoids (fossil cycads) and the Cycadales (true cycads). Although the adult plants in both of these groups were superficially similar, they differed in reproductive structures and in other anatomic details. Cycads experienced a marked decline in the Late Cretaceous, and only a few have survived to the present time. One such survivor is the sago palm, often used as a house plant.

In Chapter 4, it was noted that there were three important episodes in the evolution of land plants. The first stage led to the development of the spore-bearing leafy, treelike plants. The second involved the evolution of such nonflowering, pollinating seed plants as cycads, ginkgoes (Fig. 12–8), seed ferns, and conifers. As indicated in Figure 4–48, all but the

seed ferns have living representatives. Six families of conifers were present during the Jurassic and Cretaceous, including large numbers of pines. In 1994, the remains of over 39 huge pines were discovered in Wollemi National Park, Australia. Dubbed the "Wollemi pines," many of the trees towered 40 meters above the ground.

The third great episode of plant history is marked by the appearance of species having enclosed seeds and flowers. Such plants are known as **angiosperms.** Pollen grains resembling those produced by angiosperms provide the earliest evidence of these flowering plants. By Middle Cretaceous, angiosperms were widespread. Forested areas included stands of birch, maple, walnut, beech, sassafras (Fig. 12–9), poplar, and willow trees. Before the period came to a close, angiosperms had surpassed the nonflowering plants in both abundance and diversity. Flowering trees, shrubs, and vines expanded across the lands and, except for the absence of grasses, gave the landscape a modern appearance.

Either directly or indirectly, the evolution of Mesozoic insects, birds, reptiles, and mammals was influenced by this remarkable floral revolution. Angiosperms produced a variety of seeds, nuts, and fruits needed for the survival of the plant embryo. These plant products also provided nourishment to many groups of Mesozoic herbivores.

The relationship between insects and flowering plants is well known. The angiosperms, by encouraging insect visits, are able to use insects as delivery agents for pollen. This provides a far more efficient means of pollen dispersal than the more random wind pollination, which requires each plant to dis-

FIGURE 12–9 Fossil leaf of a Cretaceous sassafras tree from the Dakota Sandstone, Elsworth, Kansas. The slab is about 12.5 cm (5 inches) wide.

perse tens of millions of pollen grains rather than a few dozen. The insect pollinators are encouraged to visit particular flowers to obtain the nectar and pollen needed for food. Having had success with a flower of a certain form color, and scent, they move off to find others of the same kind to repeat the favorable experience. In this way, pollen is transported on the hair, legs, and bodies of insects from plant to plant of the same species. The selective competition for efficient pollinators has induced a constantly changing range in variations among both plants and insects. In the angiosperms, the need for each plant to be recognizably different results in a spectacular floral variety that has persisted from the Cretaceous down to the present.

■ *Mesozoic Invertebrates*

Terrestrial Invertebrates

Although paleontologists have assembled an enormous body of information about the marine invertebrates of the Mesozoic, relatively little is known about continental groups. The pulmonate, or air-breathing, snails are rarely found. Freshwater clams and snails are more commonly fossilized. Freshwater crustaceans, including nonmarine species of ostracodes, are frequently found in Mesozoic lake bed deposits. It is reasonable to assume that many varieties of worms existed, but they left few traces. The spiders, millipedes, scorpions, and centipedes that had been abundant in Carboniferous forests were undoubtedly also present in the Mesozoic, although their remains are rare. Among the insects, flies, mosquitoes, caddis flies, earwigs, wasps, bees, and ants are known from rocks as old as Jurassic. Many of the best fossils are collected from the Solnhofen Limestone, an unusual Jurassic formation in Bavaria. Even the Solnhofen collection is not truly representative, however, because most of the fossils are of larger species. Smaller insects are more difficult to find and are not as readily preserved. The fossil record of Cretaceous insects is also sparse. Insects preserved in amber (fossilized tree sap) of Cretaceous age are known from New Jersey, Arkansas, Canada, and Alaska. Specimens include bees, wasps, ants, beetles, flies, and mosquitoes. Butterflies, moths, termites, and fleas occur in rocks of the early Cenozoic but have not yet been found in rocks as old as Cretaceous.

Marine Invertebrates

At the end of the Paleozoic, many families of marine invertebrates either declined or became extinct. The

A

B

FIGURE 12–10   (A) The pelecypod *Gryphaea* is found in shallow marine Jurassic and Cretaceous strata around the world. It is a member of the oyster family and is characterized by a large left valve that is arched up and over the smaller right valve. (B) Fossils of *Gryphaea* exposed by weathering of the Tununk Shale Member of the Mancos Shale of Cretaceous age.   *(Tununk Shale photograph courtesy of Lehi Hintze.)*

Mesozoic resurgence of marine organisms was somewhat tardy, as indicated by the limited nature of Early Triassic faunas. However, after a few groups had become well established, marine life expanded dramatically. The bivalves became increasingly prolific from Middle Triassic on and eventually surpassed the brachiopods in colonization of the sea floor. Among the most successful pelecypods were the oysters, represented by such genera as *Gryphaea* (Fig. 12–10) and *Exogyra* (Fig. 12–11). Some members of the oyster group became giants of their kind. Other pelecypods grew conical shells that strikingly resembled the horn corals of the Paleozoic. In these forms, called **rudists** (Fig. 12–12), the left valve formed a small lid that closed the open end of the cone. Rudists formed reefs during the Jurassic and Cretaceous.

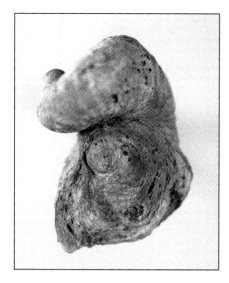

FIGURE 12–11 *Exogyra arietina* from the Cretaceous Del Rio Clay of Texas. *Exogyra* differs from *Gryphaea* in having the beak of the larger left valve twisted to the side so as not to overhang the right valve. Both genera occur in enormous numbers in shell banks of Cretaceous age along the Gulf and Atlantic coastal plains (actual size).

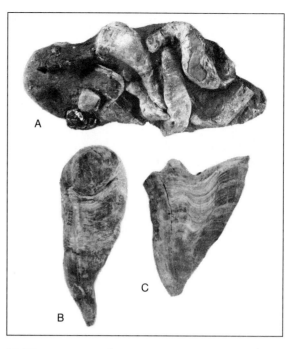

FIGURE 12–12 The Late Cretaceous rudist pelecypod *Coralliochama orcutti* from Baja California (0.5×). (A) Cluster of specimens from the reef deposit. (B and C) Two single specimens. *(From Marincovich, L., Jr. J. Paleontol. 49(1):212–223, 1975. Used with author's permission.)*

In those parts of the Mesozoic marine environment where the water was warm, relatively clear, and shallow, corals proliferated. During the Late Jurassic, for example, the Tethyan seaway was the site of major coralline evolution and reef building. Corals of the Mesozoic are called **scleractinians.** Scleractinians are divided into two important groups: the hermatypic scleractinians that built reefs and the ahermatypic or nonreef scleractinians. As is true today, hermatypic scleractinians had rather restrictive environmental requirements. Most could exist only in clear water of normal salinity no deeper than about 50 meters, with temperatures no lower than about 20°C. One reason hermatypic corals require shallow water is that they have a symbiotic relationship with an algae that lives within the coral polyp and is dependent on sunlight.

The great reefs of stony corals offered food and shelter to a host of other kinds of oceanic life. Persisting groups of brachiopods clung to the reef structures, as did bivalves, algae, bryozoans, and other sedentary creatures. Gastropods grazed ceaselessly along the reef structures, while crabs and shrimp scuttled about seeking food in the recesses and cavernous hollows of the reefs. In the quieter lagoonal areas behind the reefs and on the floors of the epicontinental seas, starfish, sea urchins, and crinoids thrived (Fig. 12–13).

Those cousins of the crinoids, the echinoids, became far more diverse and abundant in the Mesozoic

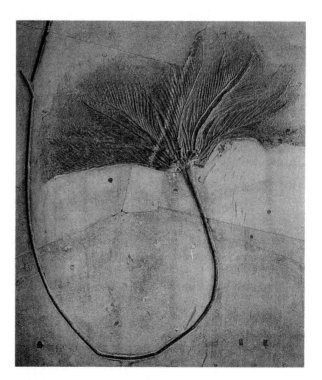

FIGURE 12–13 Large, well-preserved specimen of the crinoid *Pentacrinus subangularis* from Lower Jurassic strata near Holzmaden, Germany. The extended arms span about 1.2 meters. *(Washington University collection, photograph by John Simon.)*

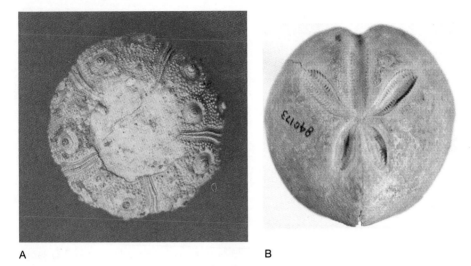

A   B

FIGURE 12–14 (A) The regular fossil echinoid *Cidaris* from Jurassic marine strata of Germany. (B) *Hemiaster,* a fossil irregular echinoid from the Cretaceous of Mississippi. Specimens are about 6 cm in diameter.

than they had ever been in the preceding era. Some Lower Cretaceous formations in the Gulf Coast region contain prodigious remains of these spiny creatures. Collectors in Europe prize the silicified echinoids obtained from Cretaceous chalk beds. The regular sea urchins were especially numerous. In these forms, the symmetry is fivefold, and the shell, or **test,** is nearly spherical (Fig. 12–14A). The regular forms were overtaken by the irregular echinoids during the Cretaceous (Fig. 12–14B). These are mostly flattened bilateral sea urchins that live as burrowers in the sediment of the sea floor. Starfish and ophiuroids (Fig. 12–15) were also abundant in Mesozoic seas.

For a paleontologist specializing in marine invertebrates, the Mesozoic might appropriately be designated "the age of ammonoids." Not only were these mollusks abundant, but they were so varied that they are exceptionally useful in worldwide correlation of Mesozoic rocks. Zones developed on the

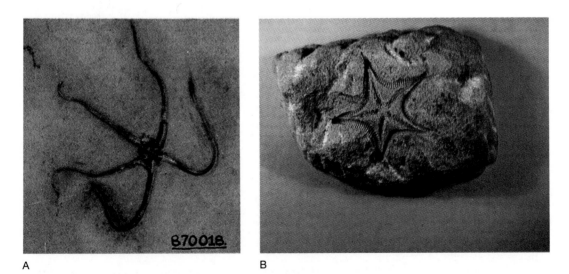

A   B

FIGURE 12–15 Two members of the echinoderm class Stelleroidea. (A) An ophiuroid (brittle star or serpent star) from the Triassic of England. Slab is about 10 cm wide. (B) An imprint of an asteroid (starfish) in a California Cretaceous sandstone. The starfish is about 15 cm across.

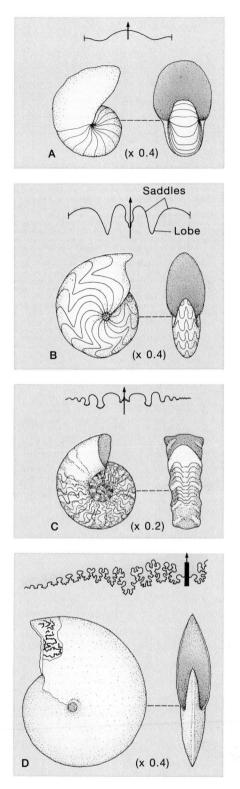

FIGURE 12–16 Sutures of cephalopods. (A) nautiloid cephalopod (arrow at midventral line pointing toward conch opening). (B) Ammonoid cephalopod with goniatitic sutures. (C) Ammonoid with ceratitic sutures. (D) Ammonoid cephalopod with ammonitic sutures. *(From Twenhofel, W. H., and Shrock, R. R. Invertebrate Paleontology. New York: McGraw-Hill Book Co., 1935. Copyright © 1935 McGraw-Hill Book Co.)*

basis of ammonoid guide fossils permit correlation of Mesozoic time–rock units with a level of precision surpassing even that of isotopic techniques. You may recall that two orders of cephalopods arose during the Paleozoic Era: the Nautiloidea, having relatively smooth sutures, and the Ammonoidea, having wrinkled sutures. **Sutures** of cephalopods are lines formed on the inside of the conch where the edge of each chamber's partition, or **septum,** meets the inner wall. Wrinkled sutures are simply a reflection of septa that, like the edges of a pie crust, are fluted. Based on the complexity of the suture patterns (Fig. 12–16), **ammonoids** can be subdivided into three groups: goniatites (see Fig. 10–45), which lived from Silurian to Permian time; ceratites, which were abundant in Permian and Triassic marine areas (Fig. 12–17); and ammonites (Fig. 12–18). Ammonites, although represented in all three Mesozoic periods, were most prolific during the Jurassic and Cretaceous.

Knowledge of the exact suture pattern of ammonoid cephalopods is necessary for their identification and hence their use in correlation. The function of the septal fluting that produced the suture patterns provides an interesting subject for speculation. Most

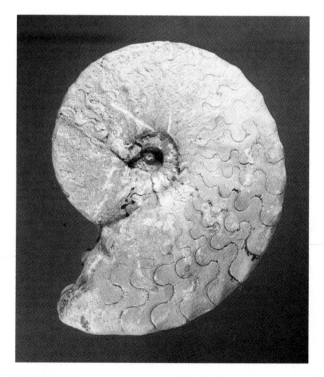

FIGURE 12–17 An ammonoid cephalopod displaying ceratitic sutures (note the tiny serrations on the lobes of the sutures). This Triassic specimen was recovered from the Upper Muschelkalk strata of Germany. *(Photography by E. Holdener.)*

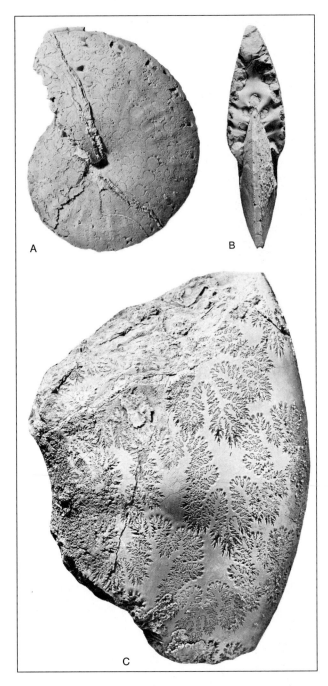

FIGURE 12–18 Cretaceous ammonoid cephalopods from the Mancos Shale of New Mexico. (A) *Hoplitoides sandovalensis* (vertical diameter of 9.3 cm). (B) Apertural view of another individual of same species as in part (A). (C) Part of outer whorl of *Tragodesmoceras socorroense* (maximum height of 16 cm). *(Photograph courtesy of W. A. Cobban, U.S. Geological Survey.)*

paleontologists favor the theory that, like the corrugated steel panels used in buildings, fluted septa provided greater strength. Comparison studies based on the living cephalopod *Nautilus* have revealed that the

gas-filled chambers exert only a slight outward pressure, whereas the water pressure on the outside of the conch wall is considerable. The septal fluting may have helped the animal to withstand the differences in pressure. Proponents of this concept call attention to the fact that ammonoid conchs (unlike nautiloid conchs) tend to thin toward the adult chambers. Apparently, to compensate for that thinner and presumably weaker conch wall, the septa in many species became more closely spaced and more intricately fluted.

The greater variety of Mesozoic ammonoids is an indication of their success in adapting to a variety of marine environments. They seem to have expanded not only within the shallow epicontinental seas but also in the open oceans. During the Cretaceous, many ammonites became aberrant in shape, so that the normally planispiral forms were joined by species with open spirals, straightened conchs, and even some that coiled in a helicoid fashion, like that of a snail. However, near the end of the Cretaceous, the entire diverse assemblage began to decline and, rather mysteriously, became extinct by the end of the era. Only their close relatives the nautiloids survived.

Another group of cephalopods that became particularly common during the Mesozoic were the squidlike **belemnites** (see Fig. 4–42). The belemnite conch was inside the animal. Its pointed end was at the rear and the forward part was chambered. A few remarkable specimens from Germany are preserved as thin films of carbon and clearly show the 10 tentacles and body form. X-ray study of these fossils has revealed details of internal anatomy as well. Much like the modern-day squid, the belemnites were probably able to make rapid reverse dashes by jetting water out of a funnel located at the anterior end.

The belemnites were highly successful during the Jurassic and Cretaceous. Triassic belemnites may well have been the ancestors to the squids, which were also numerous during the Jurassic and Cretaceous. Octopods, because they lack a conch, are inadequately recorded in the Mesozoic. However, their presence is affirmed by an imprint of an octopus found in strata of Late Cretaceous age from Lebanon.

Marine gastropods were also abundant during the Mesozoic. Many are found in sediments that represent old beach deposits. Then, as now, cap-shaped limpets grazed slowly across wave-washed boulders while a variety of snails with tall, helicoid conchs crawled about on the surfaces of shallow reefs. For the most part, the gastropod fauna had a modern appearance and included many colorful and often beautiful forms that have present-day relatives.

Modern types of marine crustaceans, such as crayfish, lobsters, crabs, shrimps, and ostracodes,

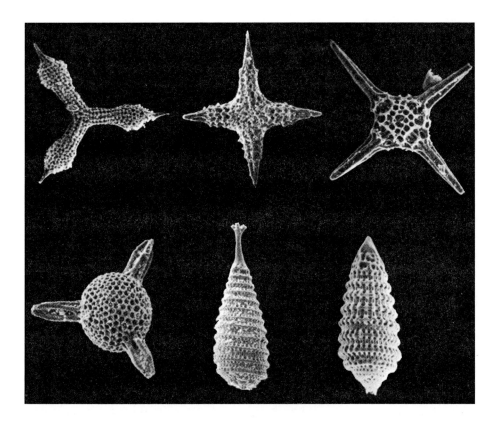

FIGURE 12–19 Scanning electron photomicrographs of Jurassic radiolaria from the Coast Ranges of California. Top row, left to right: *Paronaella elegans, Crucella sanfilippoae,* and *Emiluvia antiqua.* Bottom row, left to right: *Tripocyclia blakei, Parvicingula santabarbarensis,* and *Parvicingula hsui.* (From Pessagno, E. A., Jr. Micropaleontology 23(1):56–113, 1977, selected from Plates 1–12.)

were abundant by Jurassic time. At some localities, barnacles grew in profusion on reefs and wave-washed rocks.

Among the single-celled protozoans that crawled or floated in Mesozoic seas were the radiolarians (Fig. 12–19) and foraminifers (Figs. 12–20 and 12–21). As noted in Chapter 10, both of these groups appeared in the Paleozoic. Radiolarians make their lattice-like skeletons from opaline silica. In some regions today, radiolarian and diatom skeletal remains

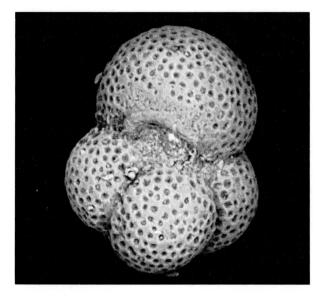

FIGURE 12–20 Electron micrograph of the planktonic foraminifer *Globigerinoides.* The height of the specimen is about 0.09 mm.

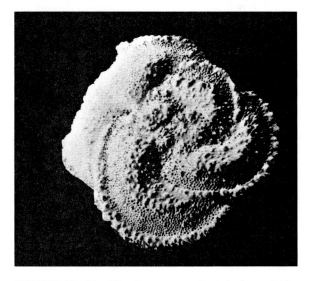

FIGURE 12–21 The Cretaceous planktic foraminifer *Globotruncana* (magnification ×100).

accumulate on the sea floor to form extensive deposits of siliceous ooze and in the past have contributed to the formation of chert beds.

The tests of foraminifers are generally more durable than those of radiolarians. The *forams,* as they are often called, left an imposing Mesozoic fossil record and one of great importance in stratigraphic correlation. They are especially important in petroleum exploration. Because of their small size and strong tests, large numbers of foraminifers can be obtained unbroken from the small pieces of rock recovered while drilling for oil. They are then used in tracing stratigraphic units from well to well. Forams are also sensitive indicators of water temperature and salinity and have provided data useful in reconstructing ancient environmental conditions.

Foraminifers were only meagerly represented in the Triassic but began to proliferate in the Jurassic and Cretaceous. Their expansion continued well into the Cenozoic. Nearly all were bottom-dwelling species until Cretaceous time, when plankic foraminifers colonized the upper levels of the ocean in prodigious numbers. Their empty tests blanketed the sea floor and became part of thick beds of chalk and marl that characterize marine Cretaceous sections in many parts of the world.

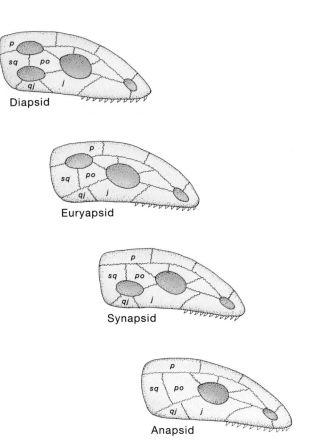

FIGURE 12–22   Reptile skull types. *(p,* parietal; *sq,* squamosal; *po,* postorbital; *j,* jugal; *qj,* quadratojugal.)

■ *Mesozoic Vertebrates*

Reptile Clans

The reptiles of the Mesozoic were a remarkably diverse group—so diverse that a simple classification will facilitate their study. One such classification is based on the position and number of openings on the sides of reptilian skulls, behind the eye orbits. Some reptiles had two such temporal openings, others only one, and still others lacked this feature altogether (Fig. 12–22). Those with two temporal openings can be placed in the subclass **Diapsida.** The dinosaurs, flying reptiles or pterosaurs, and all living reptiles except the turtles are diapsids (Fig. 12–23). Reptiles having only a single temporal opening are designated either **Synapsida,** if the opening is low on the skull (below the squamosal and postorbital bones), or **Euryapsida,** if it is high on the skull (above the squamosal and postorbital bones). Those reptiles such as turtles that lack a temporal opening constitute the **Anapsida.**

The Triassic Transition

The general unrest, broad uplifts, and upheavals that occurred during the Carboniferous and Permian periods caused regressions of epicontinental seas, resulted in a variety of continental environments, and generally provided the environmental stimulus needed to maintain the spread and diversification of land vertebrates. Although marine faunas changed rather abruptly in passing from the Paleozoic to the Mesozoic, there was considerable continuity among land animals. The labyrinthodont amphibians continued into the Triassic before becoming extinct, and the mammal-like reptiles were also able to cross the era boundary. The most progressive of the mammal-like reptiles, the ictidosaurs, succeeded their Permian precursors to become contemporaries of primitive Triassic mammals.

Many new reptile groups appeared in the Triassic. Among these were the ancestors of the first turtles. Well-preserved turtles have been recovered from Upper Triassic rocks in Germany. Although they appear similar to modern turtles, many of these early forms retained teeth in their jaws. The Triassic was also the time during which many lineages of marine reptiles appeared. The rhynchocephalians, repre-

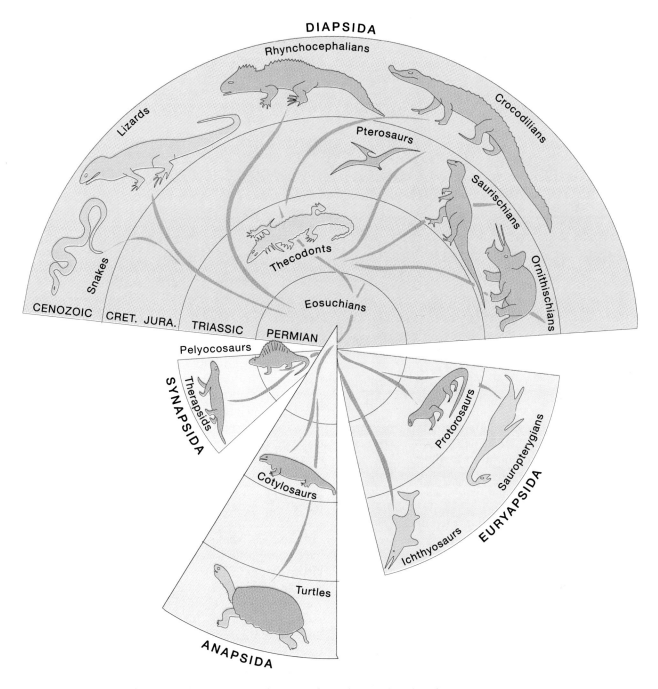

FIGURE 12–23 Evolution and general classification of reptiles. In this classification, reptiles are grouped according to the position and number of temporal openings in the skull (see Fig 12–22). Anapsida have no openings. Diapsida have two, Synapsida have one located low on the skull, and Euryapsida have one located higher on the skull. *(From Colbert, E. H. Evolution of the Vertebrates.* New York: John Wiley & Sons, 1969. Used with permission of the author and publisher.)

sented today by the tuatara of New Zealand, also were abundant. Most interesting of all, however, were reptiles known as **archosaurs**. The Archosauria are a large and important group of vertebrates that include crocodiles, the extinct flying reptiles, dinosaurs, and thecodonts. The thecodonts have a distinguished place in vertebrate history, for they are the ancestors of the dinosaurs.

FIGURE 12–24   *Hesperosuchus* from the Triassic of the southwestern United States. Adult *Hesperosuchus* was about 4 feet long.

The Thecodonts

Early **thecodonts,** as exemplified by *Hesperosuchus* (Fig. 12–24), were small, agile, lightly constructed reptiles with long tails and short forelimbs. They had already developed the unique habit of walking on their hind legs. This bipedal mode was an important innovation. Bipedalism permitted thecodonts to move about more speedily than their sprawling ancestors. Because their forelimbs were not used for support, they could be employed for catching prey; even more important, they could be modified for flight. Thus, the thecodonts were the ancestors not only of dinosaurs but of flying reptiles as well.

Not all thecodonts were nimble, bipedal sprinters; some reverted to a four-footed stance and evolved into either armored land carnivores or large crocodile-like aquatic reptiles called **phytosaurs** (Fig. 12–25). Occasionally in the history of life, initially unlike organisms from separate lineages gradu-

ally become more and more similar in form. The once distinctly different groups, in fact, change over many generations so that they are better adapted to a particular environment. The evolutionary process responsible for the trend toward similarity in form is called **convergence.** Phytosaurs and crocodiles are good examples of evolutionary convergence. Indeed, the most visible distinction between the two groups is the position of the nostrils, which are at the end of the snout in crocodiles but were just in front of the eyes in phytosaurs.

The Dinosaurs

Of all the reptiles that have ever lived on this planet, few are more fascinating than the dinosaurs (Fig. 12–26). Dinosaurs are the most awesome and familiar of prehistoric beasts. These headliners of the Age of Reptiles compose not one order, but two, each having evolved separately from the thecodonts. The two orders are the Saurischia (lizard-hipped) and the Ornithischia (bird-hipped). As suggested by these names, the arrangement of bones in the hip region provides the criterion for the twofold classification. The reptile pelvis is composed of three bones on each side. The uppermost bone is the ilium, which is firmly clamped to the spinal column. The bone extending downward and slightly backward is the ischium. Forward of the ischium is the pubis. In the saurischians, the arrangement of the three pelvic bones is triradiate, as it was in their thecodont ancestors (Fig. 12–27, and see Fig. 12–29). However, in the ornithischians the pubis is swung downward and backward so that it is parallel to the ischium, as in birds (Fig. 12–27).

The earliest dinosaurs were saurischians. They were small, lightly built, bipedal predators. The oldest known of these "first dinosaurs" include specimens of *Eoraptor* and *Herrerasaurus* discovered in Late Triassic beds of Argentina by paleontologist Paul Sereno. As indicated by potassium-argon dating of adjacent volcanic ash beds, the fossils are about 225 million years old, and thus Late Triassic in age.

FIGURE 12–25   *Rutiodon,* a Triassic phytosaur. Like many other phytosaurs, *Rutiodon* grew to lengths of 10 or more feet.

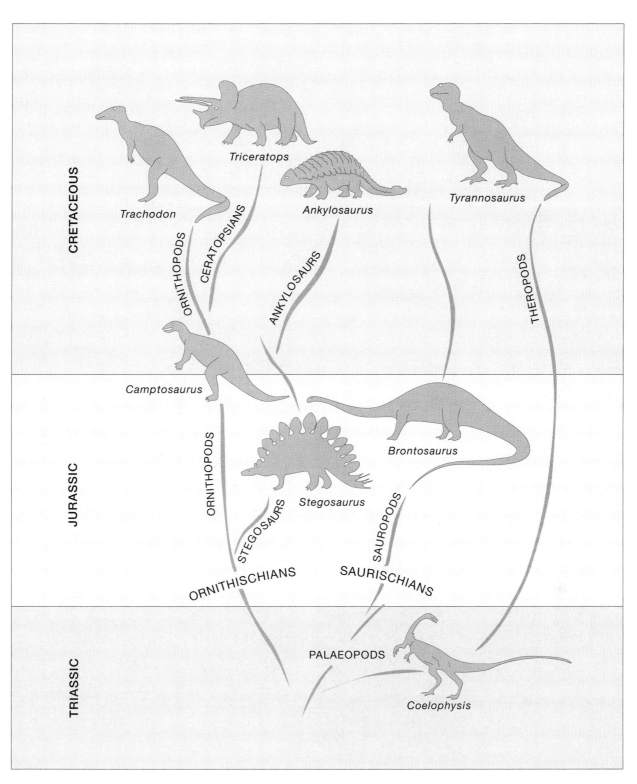

FIGURE 12–26   Evolution of the dinosaurs.   *(After Colbert, E. H. Evolution of the Vertebrates. New York: John Wiley & Sons, 1969. Used with permission of the author and publisher.)*

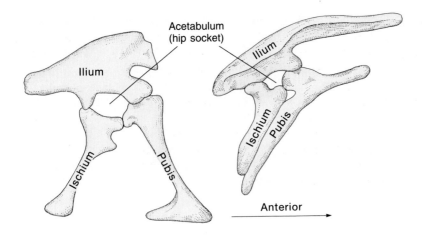

Acetabulum
(hip socket)

Ilium

Ilium

Ischium

Pubis

Ischium

Pubis

Anterior

FIGURE 12-27 Basis for the division of dinosaurs into two groups. On the left is the arrangement of pelvic bones in the *Saurischia;* on the right, the arrangement in *Ornithischia.* These views show one side only; the bones are duplicated on the other side.

FIGURE 12-28 The Late Triassic carnivorous dinosaur *Coelophysis.* This reptile was about 2.5 meters long. *(From* A Gallery of Dinosaurs and Other Early Reptiles *by David Peters. Alfred A. Knopf, Inc. Copyright © 1989, with permission.)*

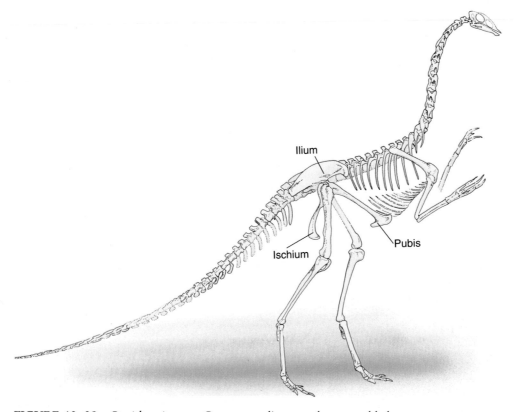

Ilium

Ischium

Pubis

FIGURE 12-29 *Ornithomimus,* a Cretaceous dinosaur that resembled an ostrich in size and form. *(From Osborn, H. F. Bull. Am. Mus. Nat. Hist. 35:733–777, 1917.)*

Eoraptor was only about a meter in length. His carnivorous life style is clearly indicated by a mouth full of curved serrated teeth and clawed, three-fingered hands suitable for slashing prey.

Coelophysis (Fig. 12–28) was a somewhat younger Late Triassic carnivorous saurischian whose bones have been recovered from the Chinle Formation of New Mexico. *Coelophysis* is one of a group of rather birdlike dinosaurs known as coelurosaurs that continued from the Late Triassic into the Jurassic and Cretaceous. *Ornithomimus* (Fig. 12–29) must have looked very much like an ostrich, with its long neck, toothless jaws, and small head. Because it lacked teeth, paleontologists speculate that it lived on the eggs laid by its contemporaries.

The larger carnivorous saurischians (including coelurosaurs) are called **theropods.** Of more dramatic dimensions than the coelurosaurs were the large carnosaurians such as *Allosaurus* (Fig. 12–30) of the Jurassic, and the Cretaceous dinosaurs *Deinonychus* (Fig. 12–31) and *Tyrannosaurus* (Fig. 12–32). The last-named beasts attained lengths of over 13 meters

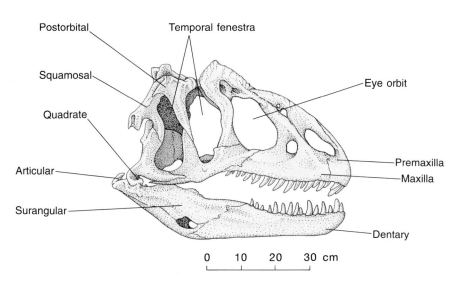

FIGURE 12–30 Skull of the carnivorous Late Jurassic theropod *Allosaurus.* Note the long jaws and dagger-like teeth, which *Allosaurus* used effectively in dealing with prey. The skull consisted largely of a framework of sturdy arches to which powerful muscles were attached. *(From Osborn, H. F. Bull. Am. Mus. Nat. Hist. 35:733–771, 1917.)*

FIGURE 12–31 The Cretaceous dinosaur *Deinonychus.* This reptile was about 8 feet long. It possessed a large skull, and the margins of the jaws were set with serrated, saber-like teeth.

FIGURE 12–32 *Tyrannosaurus,* a Late Cretaceous theropod.

and weighed in excess of 4 metric tons. Carnosaur hindlimbs were robust and muscular. Great curved claws for tearing flesh protruded from each of three toes, whereas a nearly functionless fourth toe bore a smaller claw. The forelimbs were small but powerful. As befits an animal that must kill with its jaws and teeth, the head of the carnosaur was large. Doubly serrated teeth, up to 6 inches long, lined the powerful jaws. These were truly spectacular predators.

The saurischian group included herbivorous sauropods as well as flesh-eating theropods. The

ancestry of the sauropods can be traced to the Late Triassic protosauropod known as *Plateosaurus* (Fig. 12–33). From this smaller, partially bipedal form, the more typical giant sauropods evolved at the beginning of the Jurassic and survived right up to the end of the Cretaceous. They are the animals that people first think of when they hear the word *dinosaur.* The best-known sauropods were enormous, long-necked, long-tailed beasts that had returned to the four-legged stance to support their tremendous bulk. A well-known Jurassic representative whose remains have been found in the Morrison Formation of Colorado is *Apatosaurus* (Fig. 12–34). This favorite of schoolchildren (formerly known as *Brontosaurus*) measured almost 20 meters in length and weighed over 3 metric tons. Yet *Apatosaurus* was a relative lightweight when compared to *Supersaurus* and *Ultrasaurus*, whose weights probably exceeded 80 tons. *Seismosaurus,* the Earth shaker, was over 35 meters (115 feet) long. *Diplodocus* (Fig. 12–35) was a smaller relative of *Seismosaurus.* This relative lightweight among sauropods weighed in at about 10 tons. The neck and tail were extremely long, the limbs slender, and the rib cage deep and narrow. In the early 1990s, *Diplodocus* remains in a Wyoming fossil quarry were found to include spines arranged along the dorsal midline of the animal's tail. It seems probable that the spines continued forward along the dorsal midline of the back and neck as well. Thus, future restorations of certain sauropods may require the addition of spines, at least along the tail.

For many years paleontologists speculated that, even with their pillar-like legs, these largest of land animals could not have supported their own weight continuously. It was therefore surmised that they

FIGURE 12–33 *Plateosaurus,* the Late Triassic ancestor of the giant sauropods.

FIGURE 12–34 The enormous Jurassic sauropod *Apatosaurus* (formerly called *Brontosaurus*).

dwelt in the buoyant waters of lakes and streams. However, there is little evidence to support this theory. Sauropod footprints and foot structure indicate that they walked on the tips of the toes on the front feet, with the heels of their hind feet resting on large pads, as seen in living elephants. They were clearly land-dwelling treetop browsers whose massive limbs provided adequate support on dry land.

Large size afforded certain advantages to the sauropods. Predators often avoid encounters with huge animals. In addition, great size in reptiles also serves to slow changes in body temperature. The ratio of surface area to mass for an animal decreases as size increases. Consequently, a large animal has a

proportionately smaller surface for heat loss, and just as a large pot of water loses its heat more slowly than a small pot, so does a large animal lose its heat more slowly than a small animal.

Sauropods made their first appearance in the Early Jurassic, and by the end of that period they were at the peak of their diversity, abundance, and size. After the Jurassic, they were less successful, but they did manage to survive until the end of the Cretaceous.

The other major dinosaur line, the Ornithischia, evolved near the end of the Triassic and thrived throughout the remaining Mesozoic. Ornithischians were plant eaters. The teeth in the forward part of the

FIGURE 12–35 Mounted skeleton of the lengthy Jurassic sauropod *Diplodocus*. From head to tail, *Diplodocus* measured 27 meters (88 feet) long. *(Courtesy of the National Museum of Natural History, Smithsonian Institution.)*

FIGURE 12–36 *Camptosaurus*, a small to medium-sized (2 to 3 meters long) herbivorous dinosaur. Although there were no teeth in the beaklike forward part of the jaw, the remainder of the jaw was equipped with a tight mosaic of row upon row of teeth suitable for grinding plant food.

jaws were replaced by a beak suitable for cropping vegetation. The group included both quadrupedal and bipedal varieties, with the bipedal condition considered more primitive. Even the most advanced quadruped ornithischians had shorter forelimbs, indicating their descent from bipedal forms.

The bipedal group of ornithischians is known as **ornithopods.** Their evolutionary history began in the Triassic, with relatively small species that lived primarily on dry land. A representative large Jurassic ornithopod is *Camptosaurus* (Fig. 12–36). This was a bipedal dinosaur of medium size with a heavy tail, short forelimbs, and long hind legs. The articulation of the jaw was arranged to bring the teeth together at the same time, an arrangement frequently seen in herbivores down through the ages. Leaves and stems

were cropped by the forward, beaklike part of the jaws and passed backward to the cheek teeth for chopping and chewing. From camptosaurid-like ancestors, the larger Cretaceous ornithopods developed. Among these was *Iguanodon* (Fig. 12–37), one of the first dinosaurs to be scientifically described. *Iguanodon*, sometimes called the thumbs-up dinosaur because of the horny spike that substituted for a thumb, is thought to have been a gregarious animal that may have moved about in herds. Evidence for this comes from a Belgian coal mine where 29 of these individuals were found together as a result of having fallen into an ancient fissure.

By Cretaceous time, ornithopods had moved into a variety of terrestrial environments. A particularly successful group were the trachodonts or hadrosaurs.

FIGURE 12–37 *Iguanodon* (in rear) was a herbivorous ornithischian dinosaur that lived during the Early Cretaceous. Approaching from the right is the carnivorous dinosaur *Baryonx*. The boy in the painting is for scale. *(From A Gallery of Dinosaurs and Other Early Reptiles, by David Peters. Alfred A. Knopf, Inc. Copyright © 1989, with permission.)*

FIGURE 12–38 Mounted skeletons of two Cretaceous duckbill dinosaurs (trachodonts) on display at the American Museum of Natural History in New York.

Some of the hadrosaurs appear to have been aquatic and inhabited lakes and streams. The evidence for this consists of skin impressions showing that the feet were webbed. The vertically flattened tail of these reptiles may have been used in swimming. Other hadrosaurs, however, seem to have lived in drier areas of coastal plains and floodplains. Discoveries of nesting sites in what were the foothills of the ancestral Rocky Mountains indicate that they may have moved to higher ground for nesting.

In hadrosaurs, the forward part of the face was broad, flat, and toothless, sometimes resembling the bill of a duck (Fig. 12–38); hence, the name duckbill dinosaurs for members of this group. Behind the toothless forward part of the jaw were lines of lozenge-shaped teeth that appear to have been well adapted for chewing coarse vegetation.

An interesting peculiarity of some hadrosaurs was the development of bony skull crests containing tubular extensions of the nasal passages (Figs. 12–39 and 12–40). The function of these bizarre structures is of special interest to paleontologists. It has been suggested recently that the cranial crests were used to catch the attention of sexual partners of the same species and could be further employed as vocal resonators for promoting an individual's success in obtaining a breeding partner.

Not all ornithopods had skull crests, and in some the crests lacked nasal tubes. One crestless ornithopod that certainly qualifies as a legitimate

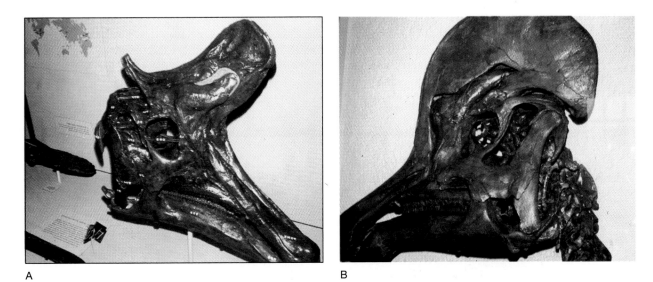

A   B

FIGURE 12–39 (A) Skull of the hadrosaur *Lambeosaurus* with its peculiar hatchet-shaped crest. (B) *Corythosaurus* had a helmet-shaped crest. These crests may have functioned as vocal resonators for dinosaur bellowing. The skulls are approximately 0.75 meter (30 inches) long. *(Courtesy of the National Museum of Natural History, Smithsonian Institution.)*

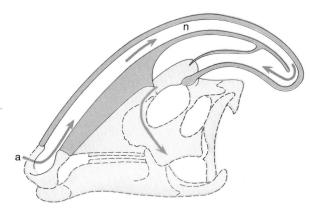

FIGURE 12–40 Internal structure of the skull crest of *Parasaurolophus cyrtocristatus*. The superficial bone of the left side has been removed to expose the left nasal passage *(n)*. Air enters nostrils at *a*, moves up and around partition in crest, and from there moves down and back to internal openings in the palate. *(From Hopson, J. A. Paleobiology 1:24, 1975.)*

FIGURE 12–41 Reconstruction of the head of *Pachycephalosaurus*, the "bone-head" dinosaur.

"bone head" was *Pachycephalosaurus* (Fig. 12–41). The skull in pachycephalosaurians consisted mostly of solid bone with only a small space to accommodate the unimpressive brain. One cannot help but wonder about the purpose of the thick skulls of the pachycephalosaurians. Perhaps these animals used their skulls as battering rams to butt heads against one another during competition for territory or females. Such behavior is seen today in bighorn sheep.

The best known of the quadrupedal ornithischians are the **stegosaurs** (Fig. 12–42). Stegosaurs had two pairs of heavy spikes mounted on the tail that were used for defense. However, their more identifying feature were the plates that stood upright along the back. Scientists have debated the purpose of these plates for many years. One theory is that the plates functioned in the regulation of body temperature by serving as body heat dissipaters. Indeed, the arrangement, size, and shape of the plates, and their probable rich supply of blood vessels, would favor the thermoregulatory suggestion, although the plates may have served as camouflage or protection as well.

During the Cretaceous, stegosaurs were succeeded by the heavily armored **ankylosaurs.** These bulky, squat ornithischians were completely covered by closely fitted bony plates that protected the entire length of their 6-meter backsides. The head was small and also armored. Small, weak teeth were present in the jaws of some ankylosaurs, but *Nodosaurus* (Fig. 12–43) was completely toothless.

FIGURE 12–42 *Stegosaurus.* The plates of *Stegosaurus* functioned as radiators and solar panels that helped the animal regulate its body temperature.

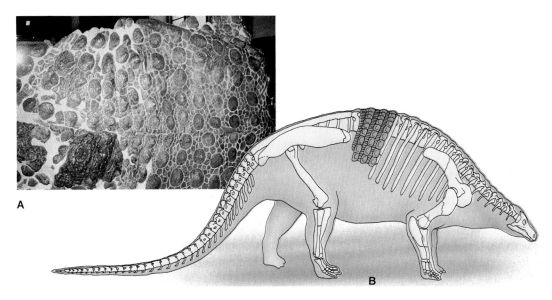

FIGURE 12–43 The Cretaceous ankylosaur *Nodosaurus*. This heavily armored dinosaur was about 5 meters long. Part of its armored carapace is shown in the photograph. *(Drawing from Lull, R. S. Amer. Jour. Sci. 5th ser. 1:97–126, 1921.)*

The fourth group of quadrupedal ornithischians are the **ceratopsians.** These beasts take their name from horns that grew on the face of all but the earliest forms. Typically, ceratopsians possessed a median horn just above the nostrils, and in some species an additional pair projected from the forehead. The head was quite large in proportion to the body and displayed a shieldlike bony frill at the back of the skull roof that served as protection for the neck region and as a place of attachment for powerful jaw and neck muscles. All ceratopsians possessed jaws that had the shape of a parrot's beak (Fig. 12–44). Judging from the scars that mark the shield bones of some ceratopsians, they were often attacked by the

FIGURE 12–44 The Cretaceous ceratopsian dinosaurs *Styracosaurus* (A) and *Triceratops* (B). To an advancing enemy, *Styracosaurus* presented a many-horned threat. *(Figure A from* A Gallery of Dinosaurs and Other Early Reptiles, *by David Peters. Alfred A. Knopf, Inc. Copyright © 1989, with permission.)*

great carnosaurs. No doubt they frequently emerged the victor. During the Late Cretaceous, ceratopsians moved eastward from Asia across the land connection to North America and inhabited the region that lay to the west of the epicontinental sea.

Were Dinosaurs Warm Blooded?

Since the late eighteenth century, when dinosaurs were first studied scientifically, they have been regarded as reptiles and hence ectothermic or cold blooded. Ectothermic animals have little or no ability to maintain a uniform body temperature by physiologic processes. Some ectotherms, however, may regulate their body temperature to a certain degree by seeking either sun or shade in response to temperature needs. In living reptiles, the pineal gland may play a role in directing this type of behavior. In extinct reptiles, certain anatomical features, such as the sail in *Dimetrodon* or the plates on the back of *Stegosaurus,* may have served to catch the sun's rays or dissipate body heat.

In contrast to largely ectothermic animals, endothermic animals such as mammals and birds maintain a constant body temperature by physiologic production of heat internally and the radiation of excess heat away from the body. Like other animals, mammals produce heat by oxidizing food. However, when body temperature rises, special physiologic mechanisms regulated by the hypothalamus (part of the brain) help to dissipate the heat. These mechanisms include expansion of the blood vessels in the skin, perspiring, or (in furry animals) panting. When temperatures fall, other mechanisms—such as restriction of blood vessels in the skin and shivering—minimize heat loss.

In 1968, paleontologist Robert Bakker boldly proposed that dinosaurs were warm blooded, or endothermic, and thus could no longer be classified as reptiles. Bakker knew that birds and dinosaurs are similar in general anatomical design. If, like birds, dinosaurs were truly endothermic animals, then it would appear logical to restructure the classification of vertebrates by removing dinosaurs from the class Reptilia and erecting a new taxonomic class that would include both dinosaurs and birds. The class Aves would be eliminated. In the new classification, there would be four classes of terrestrial vertebrates: the class Amphibia (ectothermic amphibians), the class Reptilia (ectothermic reptiles such as lizards and snakes), the class Dinosauria (endothermic dinosaurs and birds); and the class Mammalia. In a broad interpretation of the proposed reclassification of vertebrates, dinosaurs still "live" today—as birds.

Bakker supported his theory of dinosaurian endothermy with several interesting lines of evidence. One relates to the way dinosaurs stood and walked. Today's lizards and salamanders are ectotherms and most have a sprawling stance, with their limbs directed more or less to the side. In contrast, the limbs of mammals and birds are held directly beneath the body. Dinosaur stance resembles that of mammals and birds, and hence, by correlation, dinosaurs were endothermic. But is this an entirely valid correlation? Dinosaur posture may be only an evolutionary solution for supporting their enormous weight.

A second observation considered by some to favor dinosaur endothermy relates to the microscopic structure of bones. Dinosaur bones are richly vascular, like the bones of mammals. Most reptiles have less vascular bones, indicating a poorer supply of blood to the bone tissue. The correlation, however, is not absolute. Some of the bones in living crocodiles have considerable vascularity, and crocodiles are primarily ectothermic. Perhaps the high amount of vascular bone in dinosaurs evolved in response to requirements related to growth rates or size and was not related to endothermy.

Isotope analysis of bone also provides a test of dinosaur endothermy. In ectothermic animals, there are large differences in the oxygen isotope content of bones of the extremities as compared to bones of the body core. Endothermic vertebrates do not exhibit this disparity. Analyses of bone from Cretaceous theropods, ceratopsians, and hadrosaurs show isotope variability similar to that in warm-blooded vertebrates.

Among the additional arguments for dinosaur endothermy is one that makes a correlation between predator to prey proportions in mammals (endotherms) as opposed to living reptiles (ectotherms). Today's endothermic communities consist of about 3 percent predators and 97 percent plant-eating prey. It clearly takes a lot of food to fuel the energy requirements of an endothermic predator such as a lion or wolf. In the ectothermic community, 33 percent of the animals are predators and 66 percent are prey. Determining the proportion of predators to prey among dinosaurs is rather tricky because the fossil record cannot be as precise as data obtained from living communities. At present, the evidence is contradictory. Early studies indicated that the proportion of predators to prey among dinosaurs was rather like that seen in endothermic communities, but more recent studies provide ratios more like that of ectothermic communities.

We can expect the debate about dinosaur warm bloodedness to continue into the coming decade. However, even if it is demonstrated that dinosaurs were not warm blooded and were truly reptiles, this does not mean that some did not achieve a measure of body temperature regulation. Their large size provided the means for achieving this feat. As noted earlier, size determines the rate at which an animal loses heat to the outside environment. Large animals have less surface area per unit volume, so they lose heat very slowly and can maintain a more or less consistent internal temperature for a long time, especially in an environment where cold weather is not a factor.

Dinosaur Parenting

Dinosaurs reproduced by laying eggs. Clutches of dinosaur eggs have been found at several localities around the world. But did dinosaurs care for the eggs after they were laid and subsequently nurture the hatchlings? Recent evidence that at least some dinosaurs cared for their young has come principally from localities in Montana and in Mongolia.

Dinosaur eggs from Montana occur in the Cretaceous Two Medicine Formation in the western part of the state. Dubbed "egg mountain" by its discoverer John (Jack) R. Horner, the fossil site includes an entire hatchery of hadrosaurian dinosaurs complete with nests (Fig. 12–45), clutches of eggs, hadrosaur embryos, and nestlings. There was evidence that hadrosaur babies were nurtured by their parents, that they lived within the social structure of large herds, that they were warm blooded, and that, in general, they behaved more like birds than living reptiles.

Excavations within the Two Medicine Formation revealed that hadrosaurs had hollowed out bowl-shaped nests in soft fluvial sediments and that within each nest they had laid about 20 eggs in neatly ar-

ranged circles. Plant impressions in the sediment suggest that the eggs were covered with decaying vegetation so that fermentation would provide warmth. Nests were about 7.5 meters (24.6 feet) apart from one another, a distance approximating the length of the adult parent. Horner found two lines of evidence suggesting that the parent hadrosaurs, named *Maiasaura* (from the Greek word for good mother lizard), nurtured their young. He observed that many of the nests contained the bones of juveniles that were about a meter long. Thus, babies, which were only about 30 cm long at the time they hatched, stayed in their nests, where food was brought to them until they had grown sufficiently to fend for themselves. In addition, the teeth of these juveniles exhibited distinct signs of wear, suggesting that they had hatched earlier and had been feeding for some time while still in the nest. The interpretation that newly hatched babies stayed in the nest until they tripled their size also supported the interpretation that *Maiasaura* was warm blooded. A baby crocodile would require three years to triple its size. It is inconceivable that *Maiasaura* babies would have stayed in their nests three years. Like today's ostriches, they probably attained their meter-long size in just a few months. Such rapid growth occurs only in endothermic vertebrates.

In 1984, the hunt for dinosaur remains in western Montana provided Horner and his field party with their best evidence that *Maiasaura* lived and traveled together in enormous herds. The evidence consisted of a bed of rock containing nearly 30 million bones that represent about 10,000 individual *Maiasaura*. The bone bed is about 1¼ miles long be ¼ mile wide. A layer of volcanic ash occurs above the bone bed. Its presence suggests that the *Maiasaura* herd was killed by suffocating ash, lethal gases, or possibly mud flows associated with a nearby volcanic eruption. Such volcanic events were commonplace in the Rockies during the Cretaceous.

The relatively gentle plant-eating hadrosaurs were not the only dinosaurs to hover, like birds, over their nests. Predators appear to have had the parenting instinct as well. Evidence for this interpretation was discovered in 1993 by Mark A. Norell and his team while searching for dinosaur remains in Cretaceous rocks of the Gobi Desert of Mongolia. The evidence consisted of a nest of dinosaur eggs arranged in a semicircle and including the nearly complete skeleton of an embryonic theropod. The tiny dinosaur was about to hatch, and thus resembled a miniature adult. It was readily identified as the embryo of the predatory dinosaur *Oviraptor*.

The fossil eggs found by Mark Norell are identical to those found in 1923 in the Gobi Desert by the

FIGURE 12–45   Reconstruction of a nest of hadrosaur eggs recovered from the Cretaceous Medicine Bow Formation of western Montana.   *(Courtesy of the Museum of the Rockies, Montana State University, with permission.)*

FIGURE 12–46   Fossil dinosaur eggs from the Upper Cretaceous of Mongolia. Recent discovery of eggs containing fully formed embryo skeletons indicates that the eggs belonged to the theropod dinosaur *Oviraptor*.
(Courtesy of the National Museum of Natural History, Smithsonian Institution.)

famous fossil hunter Roy Chapman Andrews (Fig. 12–46). Those eggs were thought to belong to the ceratopsian dinosaur *Protoceratops*. On top of the nest were bones of *Oviraptor*, so named because that theropod was assumed to have died while attempting to steal or eat the eggs in the nest. Actually, the eggs belonged to *Oviraptor*, who died while either protecting or incubating them.

Aerial Reptiles of the Mesozoic

Again and again in the history of life, the descendants of a small group of animals that were initially adapted to a narrow range of ecologic conditions have, by means of evolutionary processes, radiated into peripheral environments. As the new groups diverged from their ancestral lineage, they changed in ways that made them better suited to their new surroundings. This process, known as **adaptive radiation,** is well demonstrated by the Mesozoic reptiles. The adaptive radiation originated with the stem reptiles of the Late Carboniferous and ultimately produced the enormous variety of large and small, herbivorous and carnivorous, dry land and aquatic animals of the Mesozoic. The radiation did not end with terrestrial vertebrates, however, for during the Mesozoic, reptiles invaded the marine environment and even managed to overcome the pull of gravity for short periods of time.

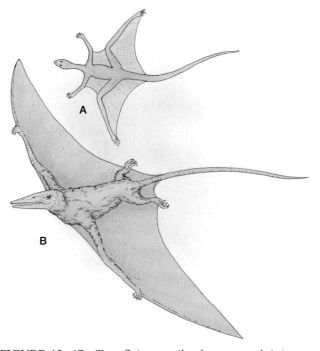

FIGURE 12–47   Two flying reptiles from central Asia. (A) *Sharovipteryx* is a Triassic form that is particularly interesting because it appears to represent an intermediate stage between primitive forms and more highly adapted flyers of the Jurassic and Cretaceous (B). The larger form is *Sordes pilosus* of Late Jurassic age. The hairy covering of *Sordes pilosus,* clearly visible in the fine limestone that enclosed the fossil, provides evidence that these creatures were warm blooded.

FIGURE 12–48 The Jurassic flying reptile *Rhamphorhynchus*. *Rhamphorhynchus* was a small reptile with a wingspan of about 50 cm. Its most distinctive trait was its long tail, which had a rudder-shaped membrane at its end.

The first reptiles to attempt to conquer the air were not active flyers, but rather were gliders similar to present-day flying lizards and so-called flying squirrels. For such animals gliding provided an easy way to move from tree tops to the ground or from branch to branch. The earliest of such gliders were discovered in rocks of Permian age. They have been given the name *Coelurosauravus*. The skin membranes that served as wings for *Coelurosauravus* were supported along the sides of the body by greatly elongated ribs. Rib-supported wings are not unusual in gliding reptiles. The Triassic lizard *Icarosaurus* (named after Icarus, the ill-fated son of Daedalus) also evolved a similar support for its wings. It is also a characteristic of the present-day *Draco* lizard of the Orient.

A different structure for flying appeared in the Triassic reptile *Longisquama*. The scales in this bizarre creature were as long as its body, and along the forward edge of its forelimbs they were nearly seven times as long. The long scales could be spread out to either side to make an effective wing. Paleontologists believe that feathers could have evolved from such scales.

The Triassic beds that contained the remains of *Longisquama* also yielded a reptile that was not a passive glider but an active flyer. Its name is *Sharovipteryx*. Because this reptile could truly fly and maneuver in the air, it is considered the first known **pterosaur**. In *Sharovipteryx*, the skin membranes ex-tended between the elbows and the knees and from the rear legs to the tail (Fig. 12–47A).

A discovery from Jurassic beds in the former Soviet Union confirmed the generally accepted theory that active, wing-flapping reptilian flyers must have been able to maintain a constant high internal temperature. Without that kind of metabolism, cold air would have reduced their power of exertion and they would have stalled. The actual evidence was found in a well-preserved pterosaur that had a covering of soft hair. Appropriately, it was named *Sordes pilosus* (Fig. 12–46B), meaning the hairy devil. The fur appears to have been longest on the animal's underside, prompting speculation that it was an adaptation used in incubating eggs or insulating hatchlings.

The pterosaurs may appear to us as ugly, graceless creatures, but their existence from Late Triassic until Late Cretaceous attests to their adaptive success. The Jurassic and Cretaceous pterosaurs that are most familiar typically had rather large heads and eyes and long jaws, which in most forms were lined with thin, slanted teeth. The bones of the fourth finger were lengthened to help support the wing, whereas the next three fingers were of ordinary length and terminated in claws. The wing was a sail made of skin stretched between the elongate digit, the sides of the body, and the rear limbs. There were two general groups of pterosaurs: the more primitive were the rhamphorhynchoids (Fig. 12–48), which had long tails, and the more advanced were tail-less

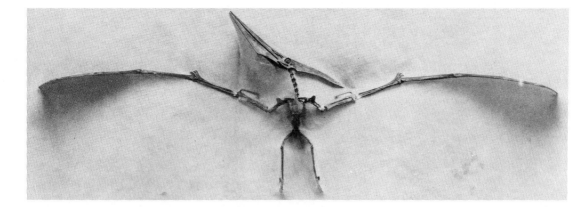

A

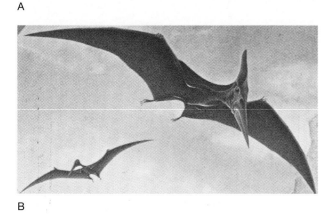

B

FIGURE 12–49 (A) Restored and mounted skeleton of the great flying reptile *Pteranodon ingens* from Cretaceous chalk deposits of Kansas. *Pteranodon* had a wingspan of over 7 meters. (B) *Pteranodon* in flight.

pterodactyloids. The latter group is exemplified by *Pteranodon* (Fig. 12–49), species of which had an astonishing wingspan of over 7 meters. The body of *Pteranodon* was about the size of that of a goose. The skeleton was lightly constructed, as is fitting for an aerial vertebrate. The animals probably soared along much like oversized sea birds, snapping up various sea creatures in their toothless jaws. Relative to body size, pterosaurs had somewhat larger brains than some of their land-dwelling relatives. Perhaps this was a result of the higher level of nervous control and coordination needed for flight.

The first prize for large size among pterosaurs must certainly go to *Quetzalcoatlus northropi* (Fig. 12–50) from Upper Cretaceous beds of western Texas. This giant, named after an Aztec god that took the form of a feathered serpent, had an estimated wingspan of 15.5 meters (50.8 feet). *Quetzalcoatlus northropi* ranks as the largest known flying vertebrate ever to soar over the Earth's surface.

A Return to the Sea

The marine habitat is one in which the archosaurs were not notably successful. Only one archosaurian group—the sea crocodiles—was able to invade the

FIGURE 12–50 The giant flying reptile *Quetzalcoatlus* discovered in Upper Cretaceous strata of West Texas. This reptile had a wingspan of 15.5 meters. The flying reptile *Pterodaustro* stands in front of *Quetzalcoatlus*. *(From* A Gallery of Dinosaurs and Other Early Reptiles, *by David Peters. Alfred A. Knopf, Inc. Copyright © 1989, with permission.)*

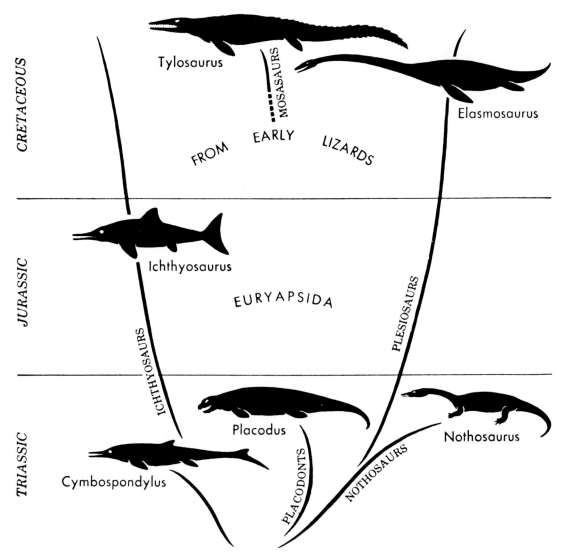

FIGURE 12–51 Evolutionary relationships of major groups of Mesozoic marine reptiles. *(From Colbert, E. H.* Evolution of the Vertebrates. *New York: John Wiley & Sons, 1969. Used with permission of the author, artist Lois Darling, and publisher.)*

oceanic environment. Other reptilian groups, such as the ichthyosaurs, plesiosaurs, mosasaurs, and sea turtles, however, were very successful in adapting to the oceanic environments (Fig. 12–51). Many fed on ammonoids, sharks, and bony fishes that had preceded them in populating the seas. During the Cretaceous, the most modern of the bony fish, the teleosts, made their appearance.

Not unexpectedly, the invasion of the marine environment required many modifications in form and function. Paddle-shaped limbs and streamlined bodies evolved to allow efficient movement through the

water. Because these reptiles were unable to abandon air breathing and reconvert to gills, their lungs were modified for greater efficiency. In those sea-going reptiles that were unable to lay their eggs ashore, reproductive adaptations provided for birth at sea.

Marine reptiles that had paddle-shaped limbs for locomotion were already present during the Triassic Period. One group, the **nothosaurs**, were just beginning to take on adaptations that would be perfected in their descendants, the plesiosaurs. The nothosaurs were joined in the Triassic by a group of mollusk-eating, flippered reptiles known as **placodonts**

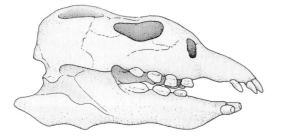

FIGURE 12–52 The skull of *Placodus,* a bulky, mollusk-eating, paddle-limbed marine reptile of Triassic age. The forward teeth were modified for plucking shellfish from the sea floor, and the cheek teeth for crushing their shells.

(Fig. 12–52). These bulky animals had distinctive pavement-type teeth in the jaws and palate, which they used for crushing the shells of the marine invertebrates upon which they fed.

By far the best known of the paddle swimmers were the **plesiosaurs** (Fig. 12–53). Their earliest remains are found in Jurassic strata. Sometimes nicknamed "swan lizards," they had short, broad bodies and large, many-boned flippers. In some species, the neck was extraordinarily long and was terminated by a smallish head. Slender, curved teeth, suitable for ensnaring fish, lined the jaws. *Elasmosaurus,* a well-known Cretaceous plesiosaur, attained an overall length in excess of 12 meters.

In addition to the long-necked types, there were many plesiosaurs characterized by short necks and large heads. It is likely that the short-necked plesiosaurs were aggressive divers. *Kronosaurus,* a giant, short-necked form from the Lower Cretaceous of Australia, had a skull size that probably exceeds that of any known reptile. It was 3 meters long.

The most fishlike in form and habit of all the marine reptiles were the Triassic and Jurassic **ichthyosaurs** (Fig. 12–54). In many ways, they were the reptilian counterparts of the toothed whales of present-day oceans. Ichthyosaurs had fishlike tails in which vertebrae extended downward into the lower lobe, boneless dorsal fins to help prevent sideslip and roll, and paddle limbs for steering and braking. The head was a suitably pointed entering wedge for cutting rapidly through the water. A ring of bony plates surrounded the large eyes and may have helped protect them against water pressure. Clearly, these were active predators with good vision and the ability to move swiftly through the water.

The **mosasaurs,** a group of giant monster lizards, were a highly successful group of Cretaceous sea dwellers. A typical mosasaur had an elongate body and porpoise-like flippers. The creatures propelled

FIGURE 12–53 Cretaceous long-necked plesiosaurs.   *(From a painting by David Peters, with permission.)*

FIGURE 12–54 Restoration of an ichthyosaur, a marine reptile similar to a modern porpoise in form and habits.

themselves through the water by the sculling action of their long, vertically flattened tails and the rhythmic undulations of their long bodies. The lower jaw had an extra hinge at midlength (Fig. 12–55), which greatly increased its flexibility and gape. Mosasaurs were primarily fish eaters, but some frequently dined on large mollusks. Conchs of cephalopods have been found with puncture marks that precisely match the dental pattern of their mosasaurian foes.

Perhaps less spectacular than the mosasaurs, but far more persevering, were the sea turtles. In this group we find a trend toward increased size. The Cretaceous turtle *Archelon*, for example, reached a length of nearly 4 meters. As an adaptation to its aquatic habitat, the carapace of the marine turtle was greatly reduced and the limbs were modified into broad paddles.

As briefly indicated earlier, the marine crocodiles (Fig. 12–56) were the only archosaurian group that entered the sea. They became relatively common during the Jurassic, but only a few remained by Early Cretaceous. It is likely that they did not fare well in competition with the mosasaurs.

The Birds

From the time of Darwin, naturalists have been aware of the structural similarities between birds and reptiles. These similarities prompted the statement by Thomas Huxley that birds are only glorified reptiles that have gained wings and feathers and lost their teeth. However, such comparisons depreciate the marvelous attainments of birds, not the least of which are their superior powers of flight and high level of endothermy. Both of these attributes are related to the evolution of feathers. Developmentally, feathers are homologous with reptilian scales. In their

FIGURE 12–56 Toothy skull of the Jurassic marine crocodile *Geosaurus*. Length of skull is about 45 cm.

FIGURE 12–55 Mounted skeleton of a Cretaceous mosasaur. These giant marine lizards attained lengths of over 9 meters (30 feet). *(Courtesy of the National Museum of Natural History, Smithsonian Institution.)*

A   B

FIGURE 12–57   The Jurassic bird *Archaeopteryx*. (A) The skeleton of *Archae-opteryx* in the Solnhofen Limestone of Germany. The Solnhofen beds formed from lime mud deposited on the floor of a tropical lagoon. The specimen resides in the Berlin Museum of Natural History.   *(John D. Cunningham/Visuals Unlimited).*   (B) Restoration of *Archaeopteryx*.   *(From a painting by Rudolph Freund, Courtesy of the Carnegie Museum of Natural History.)*

earliest development, they may not have functioned in flight, but instead for insulation, camouflage, or display.

There is little doubt that birds evolved from small Triassic theropods that were already birdlike in their bipedal stance and in the structure of their forelimbs and hindlimbs, shoulder girdle, and skull. Earlier in this chapter we noted a proposed new classification of vertebrates in which the close relationship between dinosaurs and birds is indicated by the inclusion of both groups within the same taxonomic class (the Dinosauria).

The first, undisputed fossil bird to be discovered, *Archaeopteryx*, is the perfect evolutionary link between small, bipedal theropods and modern birds. *Archaeopteryx* (Fig. 12–57) was about the size of a crow. With the exception of its distinctly fossilized feathers (arranged on the forelimbs as in modern flying birds), its skeletal features were largely reptilian. The jaws bore teeth, and the creature had a long,

FIGURE 12–58   Restoration of *Morganucodon*, an early mammal from the Late Triassic of Wales.

lizard-like tail that bore feathers. Unlike the wings of modern birds, in which the bones of the digits coalesce for greater strength, the primitive wings of *Archaeopteryx* retained claw-bearing free fingers for climbing and grasping. The light-weight sternum lacked a keel, indicating that the heavy muscles needed for sustained flight were lacking. Because of this, some paleontologists believe *Archaeopteryx* was a structurally primitive evolutionary side branch and that it did not give rise to more advanced fliers. Support for this view came in 1986 with the discovery of two crow-sized skeletons that not only were 75 million years older than *Archaeopteryx* but also possessed keel-like breast bones. Although feather impressions were not found with the skeletons, the tiny bones of the forearms possessed a series of nodes that may have been the sites for the attachment of feathers. This seemingly more advanced, yet older, vertebrate has been given the name *Protoavis*. Not all paleontologists agree that *Protoavis* is a bird, and some consider it merely a small theropod. Theropods, after all, have many birdlike traits and are the ancestors to birds.

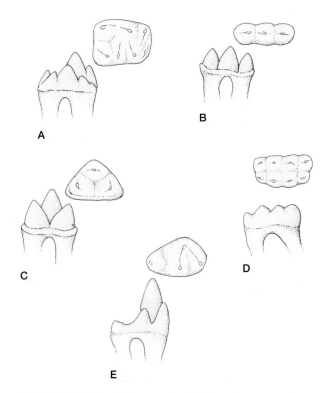

FIGURE 12–59 Molar teeth of Mesozoic mammals. Side views of lower molars (as viewed from inside the mouth) and top views of the oral surfaces.
(A) Docodont. (B) Triconodont. (C) Symmetrodont, (D) Multituberculate. (E) Eupantothere.

Small, delicate, hollow-boned animals are not readily preserved, and the fossil record for birds—especially Mesozoic birds—is not good. The Cretaceous provides only a partial glimpse of bird evolution. Toothed birds otherwise resembling terns and gulls are occasionally found in the Cretaceous deposits of the inland chalky seas. Some Cretaceous birds, such as *Hesperornis,* became excellent swimmers. They retained feathers and other birdlike characteristics but lost their wings and relied on their webbed feet for swimming (see Fig. 11–36). Marine sediments provided the rapid burial needed to preserve these aquatic birds. On land, preservation of Mesozoic birds was a rare occasion.

The Mammalian Vanguard

While the Mesozoic reptiles were having their heyday, small, furry animals were scurrying about in the undergrowth and unwittingly awaiting their day of supremacy. These shrewlike creatures were the primitive mammals. On the basis of rare and often minuscule remains, they are known from all three systems of the Mesozoic. Among the earliest of the mammals were the **morganucodonts** (Fig. 12–58), jaws, teeth, and skull fragments of which have been recovered in Upper Triassic rocks of south Wales. It is evident from these rather scrappy remains that morganucodonts still retained many vestiges of reptilian structures. However, the articulation of the jaw to the skull was mammalian, and the lower jaw was functionally a single bone, as in mammals. New discoveries of early mammals in regions other than England are currently being studied. One of these from the Triassic of Texas may be about 10 million years older then *Morganucodon.*

There were at least six additional groups of Mesozoic mammals. Each is recognized primarily on the basis of tooth morphology. Among those known from the Jurassic and Cretaceous are the docodonts, triconodonts, symmetrodonts, multituberculates, and eupantotherians (Fig. 12–59). The **docodonts** had multicusped molar teeth, which suggests that they may have been the stock from which present-day monotremes evolved. Very likely, they fed on insects, as did many of these primitive mammals. The **triconodonts** can be recognized by cheek teeth in which three cusps are aligned in a row. Some were as large as cats and may have preyed on smaller vertebrates. **Symmetrodonts** had molars constructed on a more or less triangular plan. As suggested by their name, **multituberculates** had teeth with many cusps. They may have been the first entirely herbivorous mammals. Their chisel-like incisors and the gap between the incisors and the cheek teeth gave them a

FIGURE 12–60 The rodent-like multituberculate *Taeniolabis*.

decidedly rodent-like appearance (Fig. 12–60) and indicate that they probably had rodent-like habits as well.

From an evolutionary point of view, the **eupantotheres** are the most important of the mammalian vanguard, for in the Late Cretaceous they gave rise to the marsupials and placentals. Marsupials, for example, have an asymmetrically triangular pattern of cusps that was evidently inherited from the pantotheres. In general, pantotheres were small, ratlike animals with long, slender, toothy jaws. It is likely that they preyed not only on insects but also on small lizards and mammals.

For the mammals, the Mesozoic was a time of evolutionary experimentation. For 100 million years, they effectively and unobtrusively lived among the great reptiles while simultaneously improving their nervous, circulatory, and reproductive anatomy. Equipped with an exceptionally reliable system for control of body temperature, they were able to thrive in cold as well as warm climates. As the reptile population declined near the end of the era, mammals quickly expanded into the many habitats vacated by the saurians.

■ *The Terminal Cretaceous Crisis*

Just as the end of the Paleozoic was a time of crisis for animal life, so also was the conclusion of the Mesozoic. Primarily on land but also at sea, extinction overtook many seemingly secure groups of vertebrates and invertebrates. In the seas, the ichthyosaurs, plesiosaurs, and mosasaurs perished. The ammonoid cephalopods and their close relatives, the belemnites, as well as the rudistid pelecypods, disappeared. Entire families of echinoids, bryozoans,

planktonic foraminifers, and calcareous phytoplankton became extinct. On land, the most noticeable losses were among the great clans of reptiles. Gone forever were the magnificent dinosaurs and soaring pterosaurs. Turtles, snakes, lizards, crocodiles, and the New Zealand reptile *Sphenodon* (the tuatara) are the only reptiles that survived the great biologic crash (see Fig. 12–23). Altogether, in the Late Cretaceous, extinctions eliminated about one-fourth of all known families of animals.

The question of what caused the decimation in animal life at the end of the Mesozoic continues to intrigue paleontologists. Scores of theories, some scientific and many preposterous, have been offered. Those that have the most credibility attempt to explain simultaneous extinctions of both marine and terrestrial animals and seek a single or related sequence of events as a cause. In general, the theories tend to fall into two broad categories. The first of these categories relies on some sort of extraterrestrial interference, such as an encounter with an asteroid or comet. Such events are considered catastrophic, in that their effects are concentrated within a relatively short span of time. The second group of hypotheses proposes that the cause of extinctions was events that occurred right here on Earth.

Extraterrestrial Causes of Extinctions

Asteroid Impact

Since the time that geologists first became aware of the extinctions that mark the Cretaceous–Tertiary boundary, there have been attempts to place the blame for those extinctions on meteorites, comets, or lethal cosmic radiation. Tangible evidence to support those ideas, however, was lacking. This situation changed in 1977 when geologist Walter Alvarez discovered a thin layer of clay at the Cretaceous–Tertiary boundary outside of the town of Gubbio, Italy (Fig. 12–61). Alvarez sent samples of the clay to his father Luis, a physicist, who had the clay analyzed. The result of the analysis was startling. The samples contained approximately 30 times more of the metallic element iridium than is normal for the Earth's crustal rocks. Where could this high concentration of iridium have come from? Iridium is probably present in the Earth's core and perhaps the mantle, but how could the metal from so deep a source find its way into a clay layer at the Cretaceous–Tertiary boundary? While volcanism is a possibility, iridium also occurs in extraterrestrial objects such as asteroids (or their fragments, meteorites). The father and son Alvarez team favored this extraterrestrial origin for the iridium in the clay layer and proposed

FIGURE 12–61   The coin marks the location of the iridium-rich layer of clay that separates Cretaceous and Tertiary rocks near Gubbio, Italy. The gray Cretaceous limestone below the coin contains abundant fossil coccolithophores, but few of these phytoplankton remain in the Tertiary beds above the iridium-rich clay layer. *(Lawrence Berkeley Laboratory, University of California.)*

that an iridium-bearing asteroid had crashed into the Earth at the end of the Cretaceous. The explosive, shattering blow from the huge body (presumed to be over 10 km in diameter) would have thrown dense clouds of iridium-bearing dust and other impact ejecta into the atmosphere. Mixed and transported by atmospheric circulation, the dust might have formed a lethal shroud around the planet, blocking the suns' rays and thereby causing the demise of marine and land plants, on which all other forms of life ultimately depend. As the dust settled, it would have formed the iridium-rich clay layers found at Gubbio and subsequently in Denmark, Spain, New Zealand, North America, Austria, Haiti, the former Soviet Union, and in sediment layers beneath the Atlantic and Pacific oceans (Fig. 12–62).

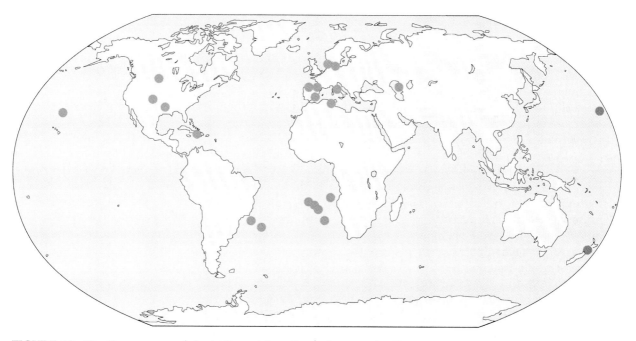

FIGURE 12–62   Occurrences of the iridium-rich sediment layer at the Cretaceous–Tertiary boundary. *(From Alvarez, W., et al. Geol. Soc. Amer. Special Paper 190:305–315, 1990.)*

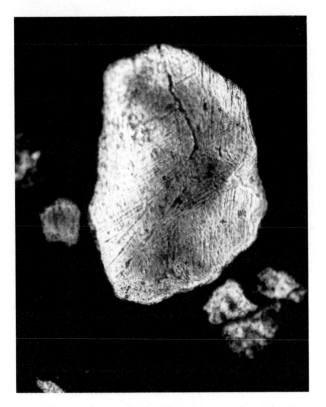

FIGURE 12–63 A shocked quartz grain from the Cretaceous–Tertiary boundary layer. Note the parallel intersecting sets of microscopic planes called shock lamellae. These intersecting lamellae are similar to those observed in quartz from rock fragments clearly known to have been subjected to meteorite impact. Maximum diameter of grain is 0.3 mm. *(Courtesy of Jim S. Alexopoulos.)*

In addition to the iridium-rich clay layer found at various sites around the world, there are other kinds of evidence for the impact of a large extraterrestrial body at the end of the Cretaceous. One of these is the widespread occurrence of **shocked quartz** (Fig. 12–63) at the Cretaceous–Tertiary boundary. These mineral grains are recognized by distinctive parallel sets of microscopic planes (called *shock lamellae*) that were produced when high-pressure shock waves, such as those emanating from the impact of a large meteorite, travel through quartz-bearing rocks. Often in the same stratum containing grains of shocked quartz one also finds tiny glassy spherules thought to represent droplets of molten rock thrown into the atmosphere during the impact event. These are termed **tektites**. A rare silicate mineral known as stishovite, produced only at high temperatures and pressures, is also found in the boundary clay. It is

taken as evidence of sudden extremely high pressures, such as those that would be associated with the impact of an asteroid. Finally, sediments at the Cretaceous–Tertiary boundary often include a layer of soot that may be the residue of vegetation burned during widespread fires caused by extraterrestrial impact.

Any large extraterrestrial object that explodes on striking the Earth is called a **bolide.** Bolides have collided with the Earth many times in the geologic past, but their scars left on continents have mostly been obliterated by the relentless action of weathering and erosion. Nevertheless a few can be discerned on photographs taken from spacecraft or during geologic investigations. Where ancient bolides have struck in the ocean, their craters have been carried by sea-floor spreading to zones of subduction or collision, where they are destroyed. In spite of these difficulties, however, if one or more bolides struck our planet 65 million years ago, one would hope to find traces of their craters. Among several possible candidates is a buried 180-km-wide apparent crater in the Yucatan Peninsula of Mexico (see Commentary Box, "Crater Linked to Demise of Dinosaurs").

On first examination, the bolide-impact hypothesis seems a tidy way to account for the extinction of dinosaurs and many of their animal and plant contemporaries. Like all hypotheses, however, it had to stand the test of scrutiny by the scientific community. Were the Alvarezes correct in assuming that the iridium was derived from an impacting asteroid?

Geologists have found evidence that iridium in clays like that at Gubbio may have its source in the Earth's mantle, from which it can move to the surface by way of conduits and blast into the atmosphere as iridium-rich volcanic ash and dust. Volcanism was indeed very prevalent during the late stages of the Cretaceous. Especially significant were the tremendous outpourings of lava in India. Known as the Deccan traps, these lava flows cover a large part of the Indian Peninsula. They were extruded about 65 million years ago, at the same time that intensive volcanic activity was under way in many other parts of the world, including the western United States, Greenland, Great Britain, Hawaii, and the western Pacific. Proponents of a volcanic source for the iridium note that the element is often distributed across sediment thicknesses of 30 to 40 centimeters, suggesting that it was deposited over the time span of several thousand years during which volcanism was most prevalent. If of bolide origin, one would suspect the iridium to be confined to a thin layer of sediment. Another aspect of this interesting scientific controversy is the presence of abundant antimony and arsenic in some of the beds containing the iridium. Although common in volcanic ash and lava, these elements are exceedingly rare in meteorites.

Planet X

Stimulated by the idea of an extraterrestrial cause for extinctions, scientists have provided several interesting alternatives to the asteroid-impact hypothesis. Blasts generated by comets encountering the Earth, for example, could also be disastrous for life. The arrival of one or more large comets would heat the atmosphere to intolerable levels and possibly provide a rain of toxic cyanide to doom life in the seas. Scientists promoting this hypothesis have been encouraged by evidence that extinctions, less drastic in effect than the ones at the Cretaceous–Tertiary boundary, seem to have occurred on Earth every 26 million years or so. This periodicity would seem to favor some sort of regular astronomic interference, in contrast to the randomness of asteroid impact. One hypothesis to account for the periodicity proposes the existence of a tenth planet, dubbed Planet X, in the outer reaches of our solar system. Astronomers have been predicting the existence of such a planet since the late nineteenth century to explain perturbations in the movements of Neptune and Uranus. Calculations suggest that Planet X would require about 56 million years to complete its huge inclined and elliptical orbit around the sun. Twice during each orbital trip (every 28 million years), it would pass through a cloud of comets (called the Oort cloud) that lies beyond the orbit of Neptune. As it traveled through the field of comets, Planet X would probably dislodge some of those comets and send them hurtling to the Earth, with devastating effect on the biosphere. This is an interesting hypothesis that awaits evidence, which can only come from continuing astronomic observation.

Cosmic Rays

If there were no actual impacts of comets, asteroids, or large meteorites to cause extinctions at the end of the Cretaceous, might there still have been other more subtle but equally effective extraterrestrial causes? One culprit might have been an influx of abnormally high amounts of cosmic radiation. Such radiation might have destroyed organisms directly or severely impaired their reproductive capabilities. Most proponents of this idea suggest that reversals of the Earth's magnetic field (a phenomenon known to occur) may have temporarily eliminated the planet's protection from harmful rays. The hypothesis, however, has some aspects that are difficult to defend. It does not explain why marine organisms, which could descend a few meters and escape radiation damage,

were decimated at levels far in excess of those among unprotected land plants. Also, several known periods of polar reversals do not correlate to episodes of mass extinction. In addition, the Mesozoic extinctions spanned a time interval far greater than that encompassed by documented magnetic reversals.

Another possibility for radiation damage unrelated to magnetic reversals might have been the arrival of a blast of lethal rays from a nearby supernova. Supernovas are colossal stellar explosions that radiate about as much energy as 10 billion suns and can affect the surface of planets located as far as 100 light-years away (1 light-year is approximately 8,898,000,000,000 km). Astronomers estimate that, on the average, a supernova may occur within 50 light-years of the Earth every 70 million years. If such an event did take place, there seems little doubt that plant and animal communities would have been severely affected. For the present, however, we must await firm evidence that such a stellar explosion did occur about 65 million years ago.

Terrestrial Causes of Extinctions

Several hypotheses that seek to explain the demise of plants and animals near the end of the Cretaceous propose that events on Earth upset the ecologic balance between organisms and the environment to which they had become adapted. Recall from Chapter 11 that continents during the Cretaceous were extensively covered by shallow, warm, epicontinental seas. Many geologists believe these inundations were caused by displacement of ocean water when midoceanic ridges were raised as a consequence of accelerated rates of sea-floor spreading. Whatever the cause, the warm inland seas that resulted were very favorable for marine life and helped to moderate and stabilize climates on the continents as well. Conditions favored the growth of living things, and plants and animals experienced remarkable increases in variety and abundance. but these favorable conditions were soon to change. Studies of the stratigraphic sequence across the Cretaceous–Tertiary boundary in Denmark indicate a global lowering of sea level at the end of the Mesozoic. Perhaps this change in sea level was caused by a slowing of rates of sea-floor spreading. Whatever its cause, the change in sea level spelled disaster to animals and plants of formerly extensive shallow coastal areas, and especially to the phytoplankton on which the marine food web depends. Without the moderating effects of vast tropical seas, continental regions would also have experienced harsher climatic conditions and more extreme seasonality. Many families of organisms that had become adjusted to the previous environment might not

C O M M E N T A R Y
Crater Linked to Demise of Dinosaurs

If we accept the hypothesis that a huge bolide smashed into our planet at the end of the Cretaceous (with catastrophic effects on life), where then is the crater produced by that lethal event? A possible candidate for the crater has been located near the town of Chicxulub, Mexico in the northern part of the Yucatán Peninsula (see accompanying map). At this location, magnetic and gravity surveys, as well as cores and logs of oil wells, reveal a buried, circular, crater-like structure about 180 km in diameter. Andesitic rock exists in the central core of the structure. The andesite has an isotopic and chemical composition similar to that of small particles of glass called tektites that are abundant in the Cretaceous–Tertiary (K/T) boundary layer at many locations in the Caribbean region. Tektites are thought to form by melting and rapid cooling during meteorite impact. Thrown into the air during impact, they can be transported worldwide by wind systems.

Further evidence of bolide impact is found in core samples of rocks penetrated during the drilling of oil wells in and around the Chicxulub structure. Prominent among these rocks are coarse breccias that occur both above and interbedded with andesite. The breccias appear to be part of a blanket of severely fragmented rock produced by a massive impact. Abundant grains of shocked quartz (see Fig. 12–63) and feldspar occur in clasts of the breccia.

Assuming the Chicxulub structure is indeed the crater formed by bolide impact, does its age correlate with the extinctions that were widespread at the end of the Cretaceous? Isotopic age determinations indicate an affirmative answer to this question. Melt rocks subjected to $^{40}Ar/^{39}Ar$ analyses provide an age

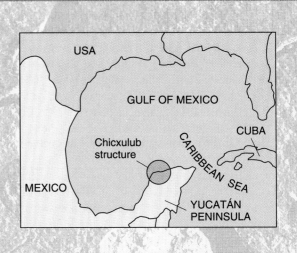

of 65.2 ± 0.4 million years. In addition, these same melt rocks acquired remanent magnetism, indicating that they had solidified during an episode of reverse geomagnetic polarity known to exist at the time of deposition of the K/T boundary layer.

If the Chicxulub structure is indeed an impact crater, as some evidence suggests, it is the largest ever blasted out of the Earth. The bolide that produced the crater would have an estimated diameter of more than 10 km. As it splashed down, it would have produced a wave over 34 km high that would flood land areas adjacent to the Caribbean and Gulf of Mexico. Rock would be vaporized, and the atmosphere would fill with dust, water vapor, and carbon dioxide from melted limestones. These conditions would have caused great stress on life of the Late Cretaceous.

have been able to adapt to the changes and, one by one, would have met their end. As is characteristic within the complex ecologic web, their demise would affect associated organisms, resulting in waves of extinctions among directly or indirectly dependent species.

There are many other variations on the general theme of environmental factors as a cause of extinctions. Some geologists dispute the evidence for a fall

in sea level and invoke other events to put stress on the global ecosystem. For example, there is evidence that the Arctic Ocean was landlocked during the Cretaceous, with the result that it became a freshwater body. Near the end of the Cretaceous, tectonic plate movements broke the land barrier containing the Arctic water, and the light, nonsaline water flowed southward over the denser, salty Atlantic. Marine plankton living in the vital upper zones of the ocean could probably not tolerate the change in water chemistry and marine circulation. A chain of ecologic disasters would have been a possible result.

As described earlier, the major hypothesis for a terrestrial cause of Cretaceous extinctions proposes that the die-off resulted from widespread volcanism. Volcanoes produce dust and aerosols, such as sulfuric acid, that block solar radiation and cause a decline in temperatures. The sulfuric acid would result in acid rain, and such precipitation might have changed the alkalinity of the oceans, placing lethal stress on plankton, other invertebrates, and other animals higher in the food chain. Thus volcanism can be considered harmful both to animals on land and in the sea.

There are still other hypotheses to account for the Late Cretaceous extinctions, but the examples provided here illustrate the complexity of the problem. The debate between those favoring sudden extraterrestrial causes and those supporting purely terrestrial hypotheses will surely continue for decades. What-ever the final outcome, it is a fact that hard times near the end of the Mesozoic doomed the dinosaurs, pterosaurs, ammonites, and over three-fourths of the known species of marine plankton. The extinctions, however, did not happen at the same time for all families, nor were they sudden occurrences. Indeed, families or groups of families died out sporadically over an interval of 0.5 to 5 million years. This fact alone is the most persuasive in the argument against a sudden extraterrestrial catastrophe. Perhaps the real cause of extinctions will be found in a combination of the two categories of hypotheses. The paleontologic evidence indicates that extinctions were under way in the Late Cretaceous. The debilitated and diminished survivors of environmental hard times may have been dealt a coup de grace by an extraterrestrial event. Fortunately, small mammals, birds, lizards, crocodiles, turtles, many fish groups, many deciduous plants, and certain mollusks survived the hard times and proliferated during the following Cenozoic Era.

Summary

Climates of the Mesozoic were in general mild and equable except for occasional intervals of aridity and an episode of cooler conditions near the end of the era. In the widespread Mesozoic seas, coccoliths and diatoms flourished, as did such invertebrate groups as ammonoids, belemnites, oysters, and other pelecypods, echinoderms, corals, and foraminifers. On land, seed ferns and conifers were common in Triassic and Jurassic forests. In the Cretaceous Period, the flowering plants expanded and with them a multitude of modern-looking insects. The changing composition of plant populations was matched by innovations among terrestrial vertebrates. The most dramatic of these changes involved the evolution of dinosaurs, pterosaurs, and birds from the small bipedal thecodonts of the Triassic.

The dinosaurs were the ruling reptiles of the Mesozoic. Both carnivorous and herbivorous varieties occupied a variety of habitats. Based on difference in pelvic structure, two orders of dinosaurs, the Saurischia and Ornithischia, are recognized. Certain of the dinosaurs as well as the pterosaurs may have been endothermic, as indicated by bone structure, posture, and predatory–prey relationships. The reptilian dynasty extended to the oceans as well, where ichthyosaurs, plesiosaurs, and mosasaurs competed successfully with the most modern of fishes, the teleosts. The Mesozoic is also noteworthy as the era during which two new vertebrate classes appeared: mammals and birds. The birds, with true feathers for insulation and flight, evolved from small carnivorous dinosaurs (theropods). The oldest remains that are unquestionably those of a true bird are of *Archaeopteryx* from the Jurassic. *Protoavis,* from Triassic strata, is currently being evaluated as a possibly older bird.

Mammals made their debut during the Triassic. In general, most of these primitive mammals remained small and inconspicuous. One group, the eupantotheres, gave rise to marsupial and placental mammals during the Cretaceous.

Like the Paleozoic, the Mesozoic Era closed with an episode of extinctions. Many groups of both marine and terrestrial reptiles succumbed, as did the ammonoids and belemnites. Several of the major taxonomic classes survived, but entire families within those classes were exterminated. Still other groups of Mesozoic organisms experienced rapid evolutionary development and thereby were able to keep pace with environmental changes.

Geologists are not agreed on the cause of terminal Cretaceous extinctions. There is, however, an impressive body of evidence that a large meteorite or asteroid struck the Earth about 66 million years ago, and the effects of the impact may have caused or contributed to the mass extinction. Others argue that the extinctions can be attributed to extensive volcanic activity, withdrawal of epeiric seas, and changes in climate.

Review Questions

1. During what geologic period of the Mesozoic is chalk particularly abundant? What group of organisms provide skeletal remains for chalk deposition? Why are chalk formations rare in Paleozoic rocks?

2. What are diatoms? How do they differ in composition and morphology from coccolithophorids? How might the distribution of diatoms and coccolithophorids be related to the acidity or alkalinity of ocean water?

3. How did the terrestrial plant flora of the Jurassic differ from that of the Cretaceous? What environmental conditions might have driven the change in floras?

4. What are ammonoid cephalopods? What attributes of ammonoids result in their having special value as guide or index fossils?

5. What are foraminifers? Why are they of particular value to petroleum geologists involved in the correlation of subsurface strata?

6. What two classes of vertebrates appear for the first time during the Mesozoic Era?

7. Cite two examples of evolutionary convergence among animals living during the Mesozoic Era.

8. Formulate single-sentence descriptions of each of the following Mesozoic animals or animal groups:

anapsid	ornithischian	theropod
euryapsid	ankylosaur	plesiosaur
rudist	ornithopod	ceratopsians
teleost	stegosaur	pterosaur
diapsid	saurischian	thecodont
mosasaur	belemnite	placodont
scleractinian	phytosaur	

9. What is the particular evolutionary importance of the following: (a) thecodonts, (b) *Archaeopteryx,* (c) eupantotheres?

10. Prepare a list of the reptilian groups that survived the mass extinction at the end of the Cretaceous Period.

Discussion Questions

1. Discuss the differences between the marine invertebrate faunas of the Mesozoic and those of the Paleozoic. What Paleozoic groups are not seen in the Mesozoic?

2. Discuss several lines of evidence that indicate that certain Mesozoic reptile groups were endothermic. Why would endothermy be less important for the giant saurischians as compared to small dinosaurs?

3. What attributes of Cretaceous mammals may have contributed to their survival during the biological crisis at the end of the Mesozoic?

4. Discuss the differences between world geography at the end of the Permian as compared to the end of the Cretaceous.

5. Oceans cover about 71 percent of the Earth's surface, yet evidence of impact craters on the ocean floor is rarely seen. Why?

6. What evidence at the boundary between the Cretaceous and Tertiary systems at many localities favors the bolide impact hypothesis for the extinction of the dinosaurs and many other animal groups? What arguments can be advanced against this popular hypothesis?

7. When were the Deccan traps extruded? Discuss the possible role of this and other synchronous volcanism as a cause for mass extinctions.

8. Discuss possible natural events on Earth that might cause global warming. What conditions might result in global cooling?

Supplemental Readings and References

Alvarez, W., and Asaro, F. 1990. What caused the mass extinction? An extraterrestrial impact. *Scientific American* 246:78–84.

Bakker, R. T. 1986. *The Dinosaur Heresies*. New York: William Morrow & Co.

Benton, M. J. 1988. The origins of the dinosaurs. *Modern Geology* 13:41–56.

Colbert, E. H., and Morales, M. 1991. *Evolution of the Vertebrates*. New York: John Wiley & Sons.

Courtillot, V. E. What caused the mass extinction? A volcanic eruption. *Scientific American* 246:85–92.

Desmon, A. J. 1976. *The Hot-Blooded Dinosaurs*. New York: Dial Press.

Horner, J. R. 1984. The nesting behavior of dinosaurs. *Scientific American* 250:130–137.

Lucas, S. G. 1994. *Dinosaurs, The Textbook*. Dubuque, IA: William C. Brown.

Paul, G. S. 1988. *Predatory Dinosaurs of the World*. New York: Simon & Schuster.

Weishample D. B., Dodson P., and Osmolska, H. 1990. *The Dinosauria*. Berkeley, CA: University of California Press.

The landscapes of the modern world acquired their present form during the Cenozoic Era. Even the spectacular scenery of the Colorado Plateau has been sculpted since about middle Tertiary time. This view of the Colorado River was taken from Dead Horse Pass, Utah. *(Courtesy of Lehi Hintze.)*

*Many an aeon moulded earth before
her highest, man, was born,
Many an aeon too may pass when
earth is manless and forlorn.
Earth so huge and yet so bounded—
pools of salt and plots of land—
Shallow skin of green and azure—
chains of mountains, grains of sand!*

A. TENNYSON, *Locksley Hall: Sixty Years After*, 1866

CHAPTER 13

The Cenozoic Era

The Cenozoic is the final era of geologic time. It is the era during which the continents and their landscapes acquired their present form. Many of the animals and plants of the present world have been shaped and modified by Cenozoic events.

There are two sets of subdivisions used for the Cenozoic (Table 13–1). In a scheme long used by geologists, the era is separated into a Tertiary and a Quaternary Period. Many European geologists, however, prefer a time table that divides the era into **Paleogene, Neogene,** and **Quaternary** periods, arguing that this represents a more equal division of the epochs and also a more natural way of dividing Cenozoic rocks of Europe. The Paris basin is the type area for most of the Cenozoic epochs. In that area, a major unconformity representing a marine regression does indeed serve nicely as a boundary between the Paleogene and Neogene systems. However, because predominant usage still seems to favor the old scheme, it will be employed here.

The Cenozoic was and is a time of considerable tectonic plate motion and sea-floor spreading. It has been estimated that approximately 50 percent of the present ocean floor has been renewed along mid-ocean ridges during the past 66 million years. Much of this new ocean floor was emplaced in the expanding Atlantic and Indian oceans. As this widening progressed, the Americas moved westward. The area that is now California came into contact with the northward-moving Pacific plate and thereby produced the San Andreas fault system. South America moved firmly against the Andean trench and actually

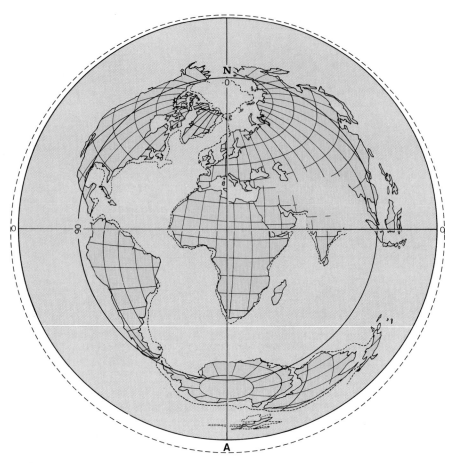

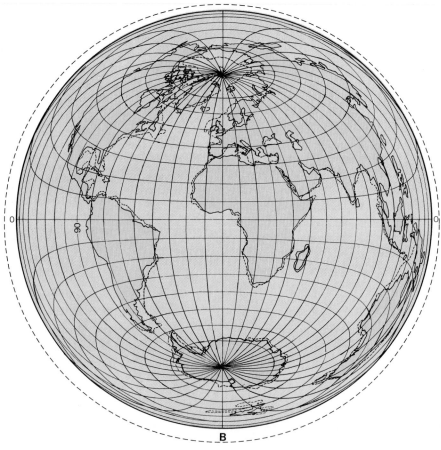

FIGURE 13–1 (A) Location of major land masses during the Eocene about 50 ± 5 million years ago. Note that Antarctica and Australia are still connected. (B) The world as it is today plotted on the same (Lambert Equal Area) projection. *(From Briden, J. C., Drewey, G. E., and Smith, A. G. J. Geol. 82:556, 558, 1974.)*

TABLE 13–1 Geochronologic Terminology Used for Divisions of the Cenozoic Era

ERA	PERIOD		EPOCH
CENOZOIC	QUATERNARY		RECENT
			PLEISTOCENE
	TERTIARY	NEOGENE	PLIOCENE
			MIOCENE
		PALEOGENE	OLIGOCENE
			EOCENE
			PALEOCENE

bent and displaced it. Orogenic and volcanic activity was vigorous along the western backbone of the Americas and resulted in the formation of the Isthmus of Panama, which today links North and South America. The North Atlantic rift extended itself to the north, separating Greenland from Scandinavia and destroying the land connection between Europe and North America. Far to the south, Australia separated from Antarctica (Fig. 13–1) and moved northward to its present location. During the Cenozoic, a branch of the Indian Ocean rift split Arabia away from Africa and in the process created the Gulf of Aden and the Red Sea (Fig. 13–2). The most dramatic crustal events of the era, however, were the collisions of Africa and India with Eurasia. These magnificent smashups transformed the Tethys seaway into such lofty mountain ranges as the Alps and the Himalayas.

The interior regions of continents stood relatively high during the Cenozoic, and as a result, marine transgressions were limited. Climatic zones were more sharply defined than in earlier eras. A cooling trend that was to culminate in the Pleistocene Ice Age is clearly indicated by paleobotanical studies.

■ *Before the Ice Age*

Eastern United States

The eastern margin of North America experienced little orogenic activity during the Cenozoic. Erosion continued in the Appalachians. Periodically, as the uplands were beveled by erosion, broad, gentle uplifts occurred. Streams, revitalized by the uplifts,

FIGURE 13–2 The Red Sea, shown here in a photograph taken from the *Gemini* spacecraft, formed about 30 million years ago when the Arabian Peninsula broke away from Africa. The seaway continues to widen today. The view is toward the south. The wedge of land in the lower left is the Sinai Peninsula, bordered on the right by the Gulf of Suez and Egypt, and on the left by the Gulf of Aqaba and Saudi Arabia. *(Courtesy of NASA.)*

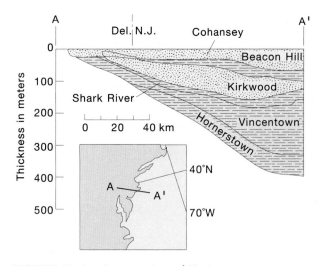

FIGURE 13–3 Cross-section of Tertiary strata across the New Jersey coastal plain. *(From* Stratigraphic Atlas of North and Central America, *courtesy of Shell Oil Company.)*

shore (Fig. 13–3). Southward, in the vicinity of Florida, terrigenous clastics were less available. Carbonate sediments accumulated to thicknesses of more than 2500 meters along a series of subsiding, elongate coralline platforms that were probably much like the Bahama Banks (Fig. 13–4). In the final ages of the Tertiary, uplift along the northern end of this tract raised the land area of Florida above the waves.

Gulf Coast

The best record of Cenozoic strata in North America is found in the Gulf coastal plain (Fig. 13–5). Altogether, eight major transgressions and regressions are recorded in this region. The Paleocene transgression brought marine waters as far north as southern Illinois. Frequently, during marine regressions, near-shore deltaic sands were deposited above offshore shales. The resulting interfingering of permeable sands and impermeable clays provided ideal conditions for the eventual entrapment of oil and gas.

"A wedge of sediments that thickens seaward" is a particularly suitable description for the Cenozoic formations of the Gulf Coast. Geophysical measurements suggest that the thickness of Tertiary sediments beneath the northern border of the Gulf may exceed 10,000 meters. The area must have been subsiding rapidly as it received this great mass of sediment.

Rocky Mountains and High Plains

While marine sedimentation was under way along the eastern coastal regions, terrestrial deposition prevailed in the Rocky Mountain region. The exception

sculpted a new generation of well-defined ridges and valleys. The most recent of the uplifts in the Appalachian belt was accompanied by gentle tilting of the Atlantic coastal plain and adjacent parts of the continental shelf. Terrigenous clastic sediments were brought out of the highlands by streams and deposited on the plains and, on occasion, were reworked by waves and currents of periodic marine transgressions. The Cenozoic section of marine rocks is thinner near the Appalachian source areas and becomes thicker and less clastic toward the region that lies off-

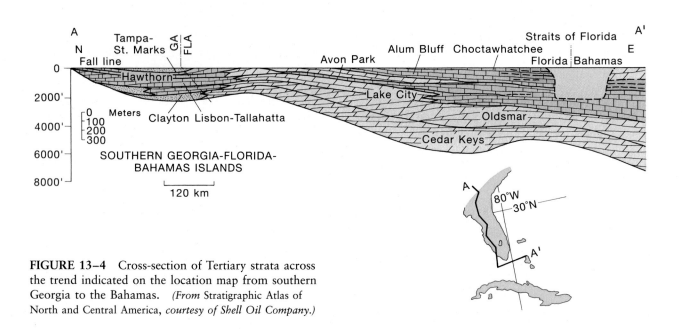

FIGURE 13–4 Cross-section of Tertiary strata across the trend indicated on the location map from southern Georgia to the Bahamas. *(From* Stratigraphic Atlas of North and Central America, *courtesy of Shell Oil Company.)*

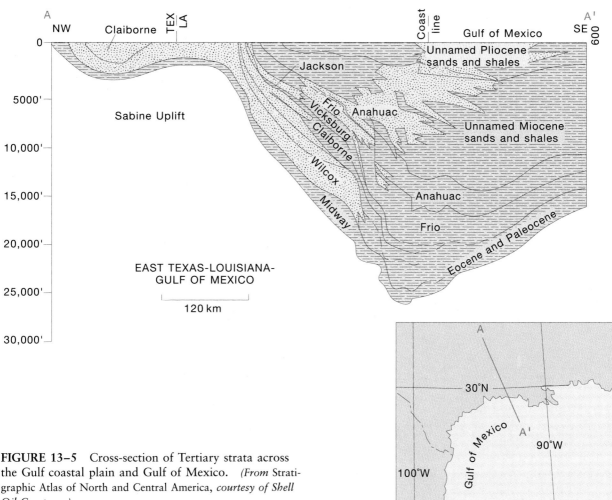

FIGURE 13–5 Cross-section of Tertiary strata across the Gulf coastal plain and Gulf of Mexico. *(From* Stratigraphic Atlas of North and Central America, *courtesy of Shell Oil Company.)*

was a single area of marine sedimentation during the Paleocene (Fig. 13–6). The strata recording the presence of this seaway are found in western North Dakota and consist of dark shales containing over 150 species of invertebrates. There is no evidence of a connection between this sea and either the Gulf of Mexico or the Arctic, and it is therefore believed to be a vestige of the more extensive Late Cretaceous epicontinental sea.

The Late Cretaceous and Early Tertiary phases of deformation described in Chapter 11 were largely responsible for the major structural features of the Western Cordillera. However, the present topography of this region is due primarily to erosion following uplifts that began during the Miocene. As always, erosion acting on the tilted and folded hard and soft layers was the final factor in shaping the landscape. Following Late Cretaceous orogenesis, erosional debris was trapped in lowlands and intermontane basins, and these terrestrial sediments provide the rec-

ord of the earlier Tertiary epochs. Later uplifts and erosion resulted in the spectacular relief of the Rockies and caused the detritus of the older basins and newer uplands to be spread over the plains that lay to the east. The result was the creation of a vast apron of nonmarine Oligocene through Pliocene sands, shales, and lignites that underlies the western high plains. Beds of volcanic ash interspersed within the Tertiary sections of the Western Cordillera attest to periodic episodes of igneous pyrotechnics.

The lower Tertiary sedimentary rocks formed from the continental sediments deposited in intermontane basins consist of gray siltstones and sandstones, as well as carbonaceous shales, lignite, and coal. These rocks are well represented in the Fort Union Formation, a unit that can be recognized in several intermontane basins in the Rocky Mountain region. The Fort Union is approximately 1800 meters (5900 ft) thick and holds immense tonnages of low-sulfur coal in its lower levels. The coal beds are indi-

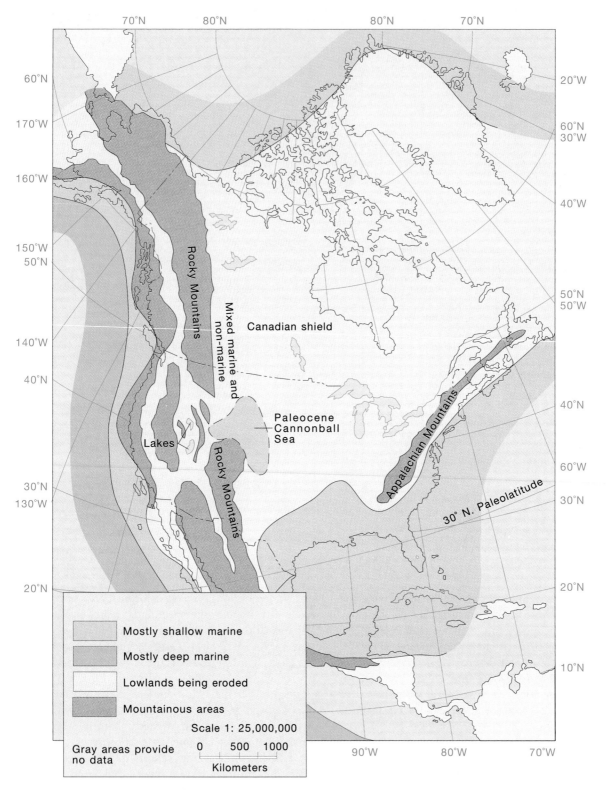

FIGURE 13–6 Paleogeographic map of North America during the early Tertiary.

FIGURE 13-8 Cedar Breaks National Monument in Utah. Here the Eocene Cedar Breaks Formation has been eroded into steep ravines, pinnacles, and razor-sharp divides.

FIGURE 13-7 Pinnacles, often called hoodoos, eroded in the Cedar Breaks Formation, Bryce Canyon National Park, Utah. *(Photograph by James Cowling, with permission.)*

cators of widespread swampy conditions in this region during Paleocene time.

Eocene deposits of the Rocky Mountains include such well-known formations as the Wasatch and Green River formations. The sediments of the Wasatch Formation were deposited by streams, whereas the Green River is a lake deposit. Wasatch sedimentary rocks are often coarse and gravelly near the base of the formation, but at higher levels consist of finer clastics. These are red mudstones interbedded with drab mudstones and lenticular sandstones. The bright colors contrast markedly with the underlying drab gray shales of the Fort Union. At Bryce Canyon (Fig. 13-7) and Cedar Breaks in Utah, the attractively colored Cedar Breaks Formation is eroded into the pinnacles of a magnificent badlands topography (Fig. 13-8).

Basins between the Wind River and Owl Creek mountains received not only the erosional detritus from those mountains, but also the discharge of their streams. During the Eocene, that volume of stream water filled many of the basins and formed extensive shallow-water lakes. One lake, notable for its lacustrine deposits, formed in the Green River basin of southwestern Wyoming (Fig. 13-9), creating the

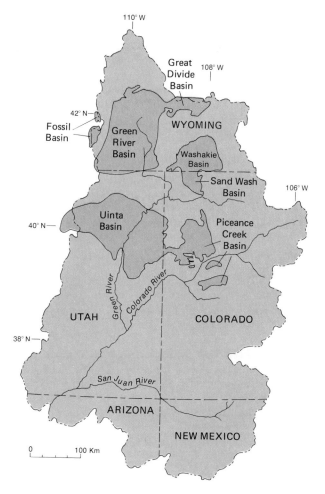

FIGURE 13-9 Cenozoic basins of Colorado, Utah, and Wyoming containing important oil shale deposits. Map boundary is the Upper Colorado River drainage basin. *(Simplified from Rickert, D. A., Ulman, W. J., and Hampton, E. R. [eds.] 1979. Synthetic Fuels Development, U.S. Geologic Survey publication.)*

GEOLOGY OF NATIONAL PARKS

Badlands National Park, South Dakota

Travelers driving west on Interstate 90 toward Rapid City, South Dakota, will miss an extraordinary vista if they fail to take exit 131 and travel only 8.5 km southward on highway 240 to Badlands National Park. Immediately on entering the park, visitors are in a strange and alien land of steep ravines and ethereal spires. They are now in the world's best example of badlands topography (see Fig. A).

Badlands are areas nearly devoid of vegetation where erosion, rather than forming ordinary hills and valleys, dissects the land into a labyrinth of chasms, razor-sharp ridges, and pinnacles. Bedrock of impermeable clays and shales, the absence of a protective cover of plants, and infrequent heavy rains are the primary conditions under which badlands form. All three of these conditions exist in Badlands National Park.

The rocks in which Badlands National Park have been fashioned consist of horizontal layers of clay, shale, and volcanic ash. Much of the ash has been weathered to a clayey sedimentary rock called bentonite. In addition, there are beds of river sandstones. The sandstone layers are more difficult to erode. They often form a caprock at the top of buttes and provide an overhanging ledge for erosional features called mushroom rocks.

Running water has been the most important agent in forming the badlands. Rain falling suddenly during cloudbursts cannot

FIGURE A   Badland topography of Badlands National Park, South Dakota. Badlands are areas of little vegetative cover and easily eroded sediment that have developed an intricate maze of narrow ravines and divides. The term was first applied by early French fur traders, whose wagons encountered this troublesome *mauvais terres* (French for "badlands"). *(Photograph by Connie Toops.)*

Green River Formation, noted earlier (Fig. 13–10). Over 600 meters of freshwater limestones and fine, evenly laminated shales were deposited in this basin as its slowly subsided. The laminations are varves, each of which consists of a thin, dark winter layer and a lighter-colored, summer layer. By counting the varves, it has been determined that more than 6.5 million years were required to deposit the Green River sediments. Within the fine shales are well-preserved fishes (Fig. 13–11), insects, and plant frag-

infiltrate the impermeable clays and shales. It runs off and becomes channeled into numerous small rivulets that erode a dense system of ever-enlarging, coalescing gullies and ravines. The impact of raindrops as they pelt the soft surfaces dislodges particles of rock, speeding the erosional processes. Bentonite in the beds also facilitates erosion. The clay in bentonite swells enormously when it becomes wet. On drying, bentonite crumbles and disintegrates, forming masses of easily eroded sediment. The White River has taken its name from the whitish color imparted to the water by submicroscopic particles of bentonitic clays.

The actual formation of the topography seen in Badlands National Park began with an episode of regional uplift late in the Pliocene Epoch. As a result of the uplift, major streams were rejuvenated and entrenched themselves into the underlying poorly resistant beds. White River became one of these entrenched streams. The process provided steeper gradients to tributary streams that flowed southward into White River. The southern perimeter of the uplifted area that paralleled the White River Valley to the north became intricately gullied and dissected, forming an early badlands topography that was subsequently eroded away. As tributary streams continued to extend their channels northward into the upland area, the badlands

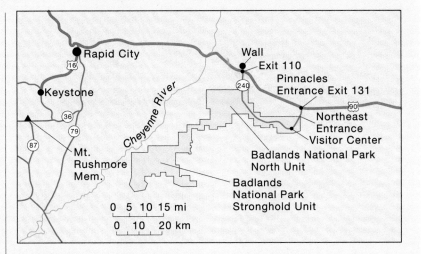

FIGURE B Location map for Badlands National Park.

we view today began their development.

Badlands National Park has an interesting geologic history. The oldest formation in the park is the Pierre Shale. The beds of the Pierre Shale were deposited in the shallow but extensive sea that covered much of western North America during the Late Cretaceous. The Cretaceous ended with regional uplift, and as the sea receded, erosion produced the unconformity on which the principal beds of the badlands were deposited. These are sediments of the Oligocene White River Group. They consist primarily of the deposits of slow-moving streams

that flowed across a landscape characterized by wide floodplains, marshlands, lakes, and ponds. At this time, there was abundant plant food for a rich fauna that included turtles, lizards, alligators, huge titanotheres, aquatic rhinoceroses, three-toed horses, early camels, entelodonts, oreodonts, and tapirs. Many of these herbivores were prey for predatory members of the dog and cat families. Figure 14–40 provides a panoramic view of the environment and its inhabitants. The rich fossil discoveries in Badlands National Park have prompted some geologists to refer to the Oligocene as "The Golden Age of Mammals."

ments. In addition, the shales are rich in waxy hydrocarbons. For this reason, they are called **oil shales** and can be processed to yield up to 200 liters of petroleum per ton. The Wasatch Formation, also of Eocene age, underlies and interfingers laterally with the

Green River beds. Several oil fields in Wyoming pump oil from the Wasatch Formation. The source rock for the oil was probably the Green River Shale.

The Uinta basin south of the Green River basin is notable because it is structurally the deepest of those

FIGURE 13–10 The Green River Formation exposed in a road cut about 10 miles west of Soldier Summit, Utah. The exposure here consists of dark shales and thin, lighter colored limestones.

FIGURE 13–11 Fossil remains of an Eocene freshwater fish *(Diplomystis)* from the Green River Formation, Wyoming. The fish is 8 cm long.

in the Colorado Plateau. It received a particularly thick succession of Tertiary sediments (Fig. 13–12). These formations can still be observed in exposures in the interior of the basin, where they have been only slightly affected by post-Eocene erosion. Strata along the northern edge of the basin are sharply upturned, forming hogbacks.

During the late Eocene and Oligocene, explosive volcanic activity blanketed the region that today includes Yellowstone National Park (Fig. 13–13) and the San Juan Mountains with layers of volcanic ash. For paleontologists, however, the more interesting rocks of this epoch are floodplain deposits of the White River Formation. This famous formation contains entire skeletons of Tertiary mammals in extraordinary number, variety, and excellence of preservation. If one recalls newsreels of recent floods, it is not difficult to understand how Oligocene mammals became buried in the sediment of overflowing streams. The clays, silts, and ash beds of the White River Formation are the sediments from which the Badlands of South Dakota have been sculpted (Fig. 13–14).

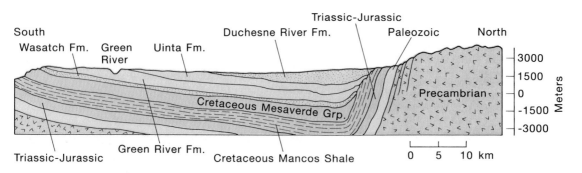

FIGURE 13–12 North–south cross-section through the Uinta basin. Tertiary units are above the Mesaverde Group.

FIGURE 13–13 Yellowstone Falls and canyon in Yellowstone National Park. Rocks exposed in the canyon walls consist of Lower Tertiary lava flows and volcanic ash, often altered to bright colors by hydrothermal activity. *(Courtesy of L. E. Davis.)*

Another particularly interesting occurrence of Oligocene beds is exposed near Florissant, Colorado. Explosive volcanic activity in this area produced a great deal of ash, which settled to the bottom of a neighboring lake, burying thousands of insects (Fig. 13–15), leaves, fish, and even a few birds. The Florissant flora, known not only from leaves but also from spores and pollen, indicates subhumid conditions for the region and elevations of between 300 and 900 meters.

The Miocene was a time of continued fluvial and lacustrine sedimentation in intermontane basins as well as on the plains east of the mountains. By Miocene time, climates had become somewhat cooler. There were expanding areas of grasslands populated by Miocene camels, horses, rhinoceroses, deer, and other grazing mammals. In the central and southern Rockies, Miocene formations include beds of volcanic ash as well as interspersed lava flows attesting to vigorous volcanic activity. The well-known gold deposits at Cripple Creek, Colorado, are mined from veins associated with a Miocene volcano. Regional uplift of the Rockies also began in the Miocene and was accompanied by increased rates of erosion. Great volumes of terrigenous detritus eroded from uplifted areas, filled intermontane basins, and spread eastward, where the detritus contributed to the construction of the Great Plains. Some of these crustal move-

FIGURE 13–14 Poorly consolidated, horizontal claystone and shale beds of the Oligocene White River Formation form the picturesque pinnacles of the South Dakota badlands. Lacking the protection afforded by vegetation, the soft sediments are susceptible to severe erosion during infrequent but torrential rains. *(Courtesy of R. Arvidson.)*

FIGURE 13–15 Fossil spider (upper left) and fossil insects from the Oligocene tuff beds near Florissant, Colorado. (Tuff is consolidated volcanic ash.)

FIGURE 13–16 The jagged Teton Range has been cut by water and ice erosion from a crustal block uplifted along a nearly vertical fault. Movement along the fault began in the Pliocene and continued sporadically through the Pleistocene. *(Courtesy of Eva Moldovanyi.)*

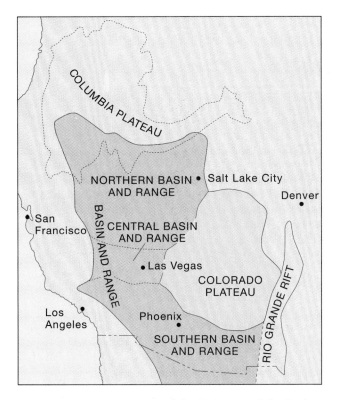

FIGURE 13–17 Location of the U.S. part of the Basin and Range Province, the Colorado Plateau, and the Columbia Plateau. *(Modified from Wernicke, B. 1992. Geol. of North America G-3:553–581. Boulder, Colorado, Geological Society of America.)*

FIGURE 13–18 The California and Nevada Nopah Range of the Basin and Range Province.

Basin and Range Province

A larger scale physiographic feature that is the result of Late Tertiary crustal extension (stretching) is the Basin and Range Province, which occupies a broad zone from Nevada and western Utah southward into central Mexico (Fig. 13–17). This region had been folded and overthrust during the Mesozoic and existed as a structural arch during the early Tertiary. Then, beginning in the Miocene, the arch subsided between great normal faults that developed on both the west and the east sides. Similar faults with general north–south alignment developed in the interior of the region. The uplifted blocks formed linear mountain ranges that became sources of sediment for the adjacent downdropped basins (Figs. 13–18 and 13–19).

ments continued throughout the remaining epochs of the Cenozoic and brought some of the highest peaks of the Rockies to spectacular elevations.

As indicated earlier, the present Rocky Mountain topography is largely the result of the uplift and erosion that began in the Miocene. Much of the erosional detritus was spread on top of the White River deposits and equivalent beds over extensive areas of South Dakota and Nebraska. The general character of the sedimentary layers does not change in the overlying Pliocene. Remains of plants and animals in some of these sediments indicate that the Pliocene was somewhat cooler and drier than the Miocene had been. This cooling was an early indication that an ice age was about to begin.

At particular localities west of the Great Plains, Tertiary geologic events produced some of the most striking scenic attractions in the world. Normal faulting and volcanism accompanying late Tertiary uplifts were responsible for some of these features. In northwestern Wyoming, the lofty Teton Range (Fig. 13–16) was elevated along great normal faults, with displacements of up to 6000 meters. The magnificent east face of the Tetons is a fault scarp that rises nearly 2.5 km to an elevation of over 4000 meters.

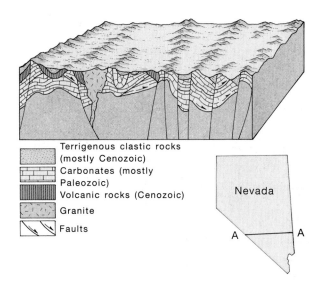

Terrigenous clastic rocks (mostly Cenozoic)
Carbonates (mostly Paleozoic)
Volcanic rocks (Cenozoic)
Granite
Faults

Nevada

A — A

FIGURE 13–19 Geologic section across the Basin and Range Province along line A–A' in southern Nevada.

C O M M E N T A R Y
Hellish Conditions in the Great Basin

If the poet Dante (1265–1321) had been strolling in the Great Basin during the Oligocene and Early Miocene, his vision of the lower world might have included turbulent clouds of white-hot ash, jets of searing gases, and fiery explosions of incandescent molten rock. During the Mid-Cenozoic, the Great Basin was indeed the scene of such volcanic activity. As one hikes the trails of Great Basin mountain ranges, the evidence is nearly always underfoot in the form of hardened ash falls and solidified lava beds. Volcanic rocks also lie between the mountain ranges but are blanketed by layers of sediment.

The volcanic rocks of the Great Basin are derived from high-silica and hence highly viscous magmas. They have the same general composition as the granodiorities of the Sierra Nevada. Such magmas produce volcanism of the explosive Peléean type, in which hot gases, steam, ash, and pyroclastics of all sizes are discharged with enormous force. Nuée ardentes characteristically accompany such eruptions. As these fiery turbulent clouds rushed swiftly down slopes, they incinerated all life and ultimately deposited their load of ash on the surrounding lowlands. In the Great Basin, they buried the landscape. Glassy particles in these fiery clouds were still so hot and plastic when they came to rest that they fused together to form hard, welded rocks called ignimbrites. The volume of ash expelled was truly incredible. A single ash fall covered over 10,000 km to a depth of over 180 meters.

Commonly, high-silica magmas cool slowly at depth and form great batholiths composed of granite and granodiorite. The magma of the Great Basin, however, breached the crust and produced the intense episode of volcanism. What may have been the cause? The answer may lie in a hypothesis that relates events in the Great Basin to plate tectonics. During the Mid-Cenozoic, the North American plate was moving westward at the rapid rate of about 15 cm per year. It overrode the Pacific plate so quickly that the plate was unable to descend into the mantle. Rather, it slid nearly horizontally under North America. The sea-floor plate was close to its origin. It had only recently moved from its spreading center, and consequently it was still very hot. As the hot sea-floor plate moved along beneath what is now the Great Basin, it might have provided much of the thermal energy responsible for the wave of volcanism that engulfed this region of North America during the Mid-Cenozoic.

Subsequent geologic history in the region is recorded in coarse terrigenous clastics washed out of the mountains. These Miocene to Holocene sediments blocked the passes and caused lakes to develop. Miocene sediments often include gypsum and salt layers formed when these lakes evaporated.

The cause of the large-scale tensional faulting is still being debated by geologists. On the basis of paleomagnetic evidence, some geologists believe that uparching crustal extension and faulting in this region occurred when westward-moving North America overrode part of the oceanic plate and spreading center (East Pacific rise) that was being subducted along the coast of California. When the subducted spreading center reached the region beneath eastern Nevada and western Utah, it caused uplift and stretching of the crust in east and west directions. Recently, however, this idea has fallen into disfavor because of a lack of evidence of progressive eastward deformation that should accompany the passage of a spreading center beneath a continent.

As an alternative theory, many geologists now believe that the normal faulting in the Basin and Range is simply the way the crust adjusted to the change along the California coast when oblique shearing of the edge of the continent during the early Miocene replaced the earlier subduction zone. Yet another model proposes extension and uplift of the crust from the remnants of an oceanic plate that had been carried beneath the Basin and Range region by an earlier episode of subduction. When subduction ceased, the oceanic slab may have formed a partially molten buoyant mass that pressed upward against the overlying crust and caused tensional faulting and escape of lava along fault and fracture zones. Finally, there is the possibility that the crustal extension and tensional faulting is related to convectional movements beneath the continental plate similar to those that cause the breaking apart of continents.

Colorado Plateau

One of the most magnificent regions of uplift in the American West is the Colorado Plateau (Figs. 13–20 and 13–21). Somehow this block of crust had remained undeformed during the Mesozoic orogenies,

FIGURE 13–20 Canyon of the Dolores River, Colorado Plateau south of Grand Junction, Colorado.

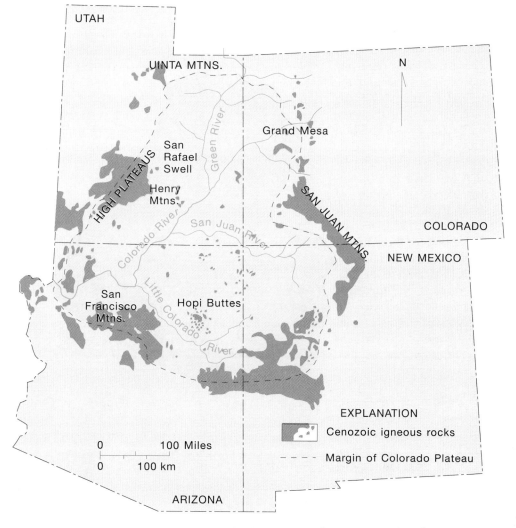

FIGURE 13–21 Location of volcanic and other centers of igneous activity of Cenozoic age in the Colorado Plateau and its periphery. *(From Hunt, C. B., Cenozoic Geology of the Colorado Plateau, U.S. Geological Survey Professional Paper 279, 1956.)*

FIGURE 13–22 Vertical aerial photograph of a large cinder cone in the San Francisco volcanic field of northern Arizona. The solidified flow issuing from the cone is 7 km long and more than 30 meters thick.   *(Courtesy of the U.S. Geological Survey.)*

for its Paleozoic and Mesozoic rocks are relatively flatlying. The region formed a buttress around the perimeter of which folding and faulting produced highlands. The plateau was repeatedly raised during early to middle Pliocene time (about 5 to 10 million years ago). Steep faults developed on the plateau during its rise and provided avenues for the upward escape of volcanic materials. The San Francisco Mountains near Flagstaff, Arizona, are a group of impressive recent volcanoes and cinder cones built above the level of the plateau (Fig. 13–22).

The best-known feature resulting from the linked processes of uplift and erosion on the Colorado Plateau is the Grand Canyon of the Colorado River (Fig. 13–23). This awesome monument to the forces of erosion is more than 2600 meters (8500 feet) deep in some places. The river has penetrated through Phanerozoic strata into crystalline Precambrian basement rocks.

Columbia Plateau and Cascades

Unlike the Colorado Plateau, which is constructed of layered sedimentary rocks, the Columbia Plateau in the northwestern corner of the United States has been built by volcanic activity (Fig. 13–24). During the late Tertiary and Quaternary, basaltic lavas erupted along deep fissures in the region. The liquid rock spread out and buried more than 500,000 km^2 of existing topography under layer after layer of lava. In some places, these low-viscosity lavas flowed 170 km from their source. Their combined thickness exceeds 2800 meters.

West of the Columbia Plateau lies an uplifted belt that was also the site of extensive volcanic activity. Here, however, the outpourings of more viscous lavas resulted in the mountains of the Cascade Range. Volcanism in this region began about 4 million years ago.

The fact that volcanism has continued up to the present was made dramatically obvious to Americans by the eruption of Mount St. Helens on May 18, 1980 (Fig. 13–25). The volcano exploded with a force equivalent to 50 million tons of TNT and with a roar heard over 300 km away. A great turbulent cloud of hot gases, steam, pulverized rock, and ash burst laterally from the north side of the mountain, followed by a vertical column of gas and ash that rose 20,000 meters into the atmosphere and began to drift slowly toward the east. An estimated 1 km^3 of airborne ash and other rock debris from the explosion blocked out the sun's light and caused automatic street lamps to switch on in towns hundreds of kilometers downwind from the volcano. Turbulent hot masses of gas and ash surged down the sides of the mountain, flattening and scorching millions of trees. Hot gas and ash from the volcano melted part of the Mount St. Helens snow and ice cap. The meltwater mixed with ash and formed huge mud flows that surged down the mountain slopes at speeds as great as 80 km per hour. A steaming turbulent mass of debris crashed through the once beautiful valley of

FIGURE 13–23 The Grand Canyon of the Colorado River.

FIGURE 13–24   Lava flows exposed on either side of Summer Falls, Columbia Plateau.   *(Photograph courtesy of U.S. Department of the Interior, Bureau of Reclamation, Columbia Basin Project.)*

FIGURE 13–25   Aerial view of erupting Mount St. Helens volcano taken at about 11:30 A.M. on May 18, 1980. Clouds of steam and ash are being blown toward the northeast from the prominent plume, which reached about 20,000 meters in altitude. The white linear features at the base are logging roads, and the dark patches are stands of mature trees. The May 18 eruption produced an amount of ash roughly equivalent to that ejected during the A.D. 79 eruption of Mount Vesuvius that buried Pompeii.   *(Photograph by Austin Post, courtesy of U.S. Geological Survey.)*

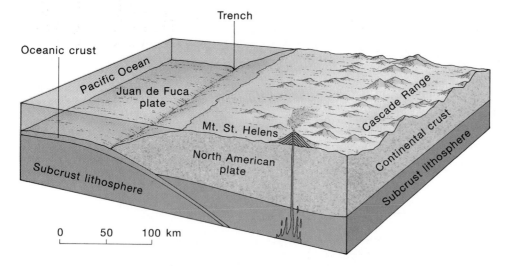

FIGURE 13–26 As the small Juan de Fuca plate plunges beneath Oregon and Washington, molten rock is generated and rises to supply the volcanoes of the Cascade Range.

the Toutle River and demolished 123 homes in the nearby town of Toutle. People camping or working near the volcano were killed by heat, gases, or burial under the downpour of ash. The May 18 eruption was a major volcanic event. The fact that it was actually rather puny when compared with the famous eruptions of Krakatoa or Santorini was no consolation to the citizens of Washington as they viewed the devastated gray landscape, the hundreds of square kilometers of valuable trees toppled like matchsticks, buried roads, harbors choked with ash-sludge, and smothered fields of wheat.

The recent activity of Mount St. Helens and the older eruptions that gave us the volcanic peaks of the Cascades are surface manifestations of an ongoing collision between the American plate and the small Juan de Fuca plate of the eastern Pacific, which moves eastward on a collision course toward the coasts of Oregon and Washington. The Juan de Fuca plate plunges beneath the coastline, and molten rock generated as the plate moves downward rises to supply the lava of the volcanoes (Fig. 13–26).

Mount St. Helens is not the only famous volcanic mountain in the Cascades, nor is it the only peak having periodic eruptions. As recently as 1914, Mount Lassen extruded lava and ash for a year and then exploded violently on May 19, 1915, producing a **nuée ardent** (fiery cloud) that roared down the mountainside, destroying everything in its path. Other well-known Cascade peaks include Mount Baker, Mount Hood, and Mount Rainier. The last major eruption of Mount Rainier was about 2000 years ago, although minor disturbances occurred frequently in the late 1800s. Crater Lake in Oregon (Fig.

13–27) was formed from the volcanic cone of Mt. Mazama after eruptions had drained a large portion of molten rock from the magma chamber beneath the volcano. Unsupported by underlying molten rock, the roof of the magma chamber collapsed (Fig. 13–28). The circular depression with a rela-

FIGURE 13–27 View of Crater Lake, Oregon, at dawn. Wizard Island, seen on the left, is a cinder cone formed on the floor of the crater about 1000 years ago, as estimated from the age of trees on the island. *(Photograph by Tom Till, with permission.)*

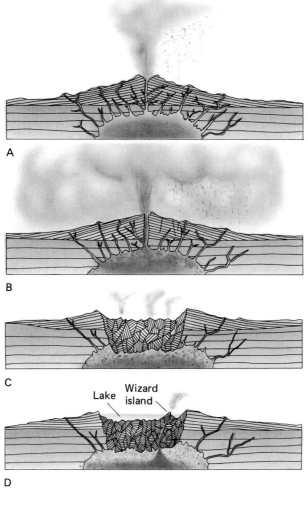

The building of ancient Mount Mazama. Magma chamber is filled and supplying magma to the volcano.

A

Often explosive eruptions begin to exhaust the reservoir of magma, leaving the upper part of the magma chamber empty.

B

Collapse of the summit into the vacated chamber below.

C

Minor eruptions build small volcanoes on caldera floor, and caldera gradually fills with water to form lake.

D

FIGURE 13–28 The origin of the caldera at Crater Lake, Oregon. *(After Williams, H., Crater Lake: The Story of Its Origin.* Bulletin, Dept. of Geological Sciences, vol. 25. University of California Publications, Berkeley, 1941.)*

tively flat floor and steep surrounding walls produced by the collapse is called a **caldera,** from the Spanish word for kettle. Other calderas form as a result of violent explosive activity.

Sierra Nevada and California

The ranges of the Sierra Nevada lie south of the Cascades. The rocks of these majestic mountains were folded and intruded during the Jurassic Nevadan orogeny. For most of the Tertiary time, the peaks were steadily reduced by erosion until their granitic basement lay exposed at the surface. Then, in Pliocene time and continuing into the Pleistocene, the entire Sierra Nevada block was raised along normal faults bounding its eastern side and tilted westward (see Fig. 5–44). Its high eastern front was lifted an astonishing 4000 meters. The depressed western side formed the California trough. As these movements were under way, rejuvenated streams and powerful valley glaciers eroded the magnificent landscapes of the present-day Sierras.

In the early epochs of the Cenozoic, the region west of the Sierras was most directly affected by subduction tectonics. As will be discussed shortly, however, this region underwent a change from subduction tectonics to strike–slip tectonics at about the beginning of Miocene time. Complex fault movements resulting from this shift in tectonic style created islands and intervening sedimentary basins in which fine-grained marine clastics, diatomites, and bedded cherts accumulated. Also during the Miocene, folding and uplift resulted in marine regressions in many areas, and by the end of the Tertiary, seas were restricted to a narrow tract along the western edge of California. Ultimately, Quaternary uplifts caused the regression of this tract as well.

The Tertiary stratigraphic succession of California is impressive, but also imposing are sections of Tertiary rocks far to the north in Alaska and the arctic islands of Canada. Coal found associated with these strata, along with fossil spores and pollen, indicates that a temperate or even warmer climate prevailed in these northern lands. Along the coastal

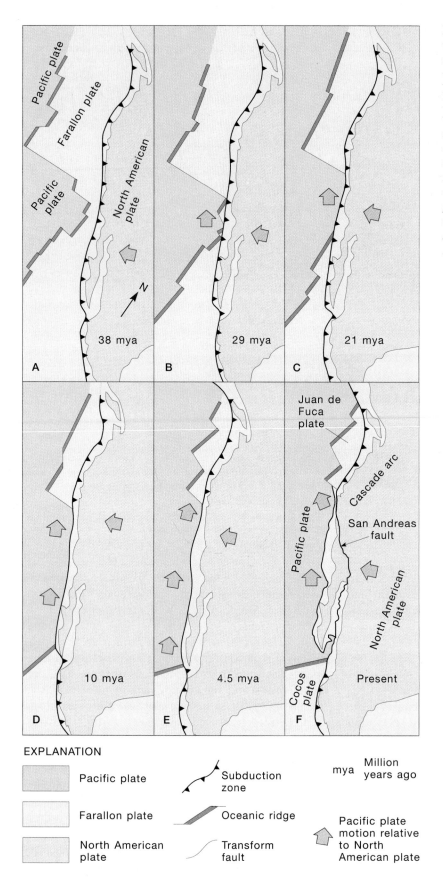

FIGURE 13–29 Schematic model of the interaction of the Pacific, Farallon, and North American plates for six time intervals during the Cenozoic. Note how the Farallon plate was largely subducted by Late Cenozoic time, leaving only remnants to the north (Juan de Fuca plate) and to the south (Cocos plate). The San Andreas and associated faults were caused by right-lateral movements beginning about 29 million years ago. *(Adapted from Nilsen, T. H. San Joaquin Geological Society Short Course No. 3, 1977. Data source from Atwater, T. Bull. Geol. Soc. Am. 81[12]:3513–3536, 1970.)*

EXPLANATION

Pacific plate

Farallon plate

North American plate

Subduction zone

Oceanic ridge

Transform fault

mya — Million years ago

Pacific plate motion relative to North American plate

mountains and Aleutian chain, Tertiary sands and shales are interspersed with layers of pyroclastics and lava flows. Then, as now, the area of present-day Alaska was tectonically active.

West Coast Tectonics

The western edge of North America during most of the Cenozoic was the site of an eastward-dipping subduction zone. This subduction was in one way or another responsible for the batholiths, compressional structures, volcanism, and metamorphism that accompanied Mesozoic and early Tertiary orogenies. The oceanic plate that was being fed into the subduction zone has been named the **Farallon plate.** During the Cenozoic, the Farallon plate was being consumed at the subduction zone faster than it was receiving additions at its spreading center. As a result, most of the Farallon plate and part of the East Pacific rise that generated it was gobbled up at the subduction zone along the western edge of California (Fig. 13–29).

Today, the small Juan de Fuca plate near Oregon and Washington and the Cocos plate off the coast of Mexico are the remnants of the once more extensive Farallon plate.

With the loss of the Farallon plate near California, the North American plate was brought into direct contact with the Pacific plate, and a new set of plate motions came into operation. Before making contact with the west coast, the Pacific plate had been moving toward the northeast. As a result, when contact was made, the Pacific plate did not plunge under the continental margin but rather slipped along laterally, giving rise to the San Andreas fault. No longer did the California sector of the west coast have an Andean type of subduction zone (Fig. 13–30). Strike-slip movements now characterized the southwestern margin of the United States.

An important result of the shearing and wrenching of the west coast was the tearing away of Baja California from the mainland of Mexico about 5 million years ago. Because it was located west of the San

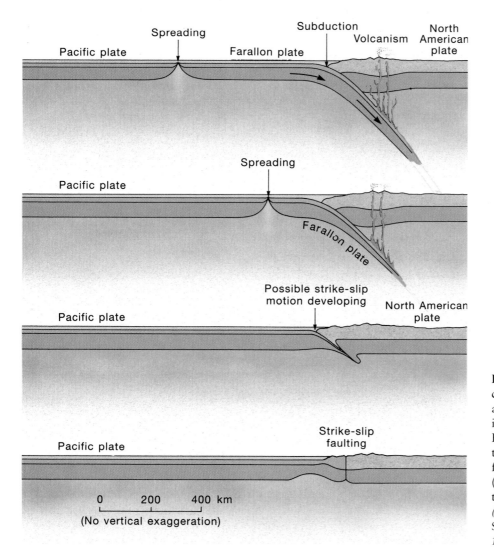

FIGURE 13–30 Sequence of cross-sections of California and its offshore area illustrating the subduction of the Farallon plate. In this model, the Pacific plate is considered fixed as the spreading center (East Pacific Rise) encounters the continental margin. *(After Atwater, T. Bull. Geol. Soc. Am. 81[12]:3513–3536, 1970.)*

COMMENTARY
When the Mediterranean Was a Desert

In 1970, geologists Kenneth J. Hsu and Brian B. F. Ryan were collecting research data while aboard the oceanographic research vessel *Glomar Challenger.* An objective of this particular cruise was to investigate the floor of the Mediterranean and to resolve several questions about its geologic history. The first question was related to evidence that the invertebrate fauna of the Mediterranean had changed abruptly about 6 million years ago. Most of the older organisms were nearly wiped out, although a few hardy species survived. A few managed to migrate into the Atlantic. Somewhat later, the migrants returned, bringing new species with them. Why did the near extinction and migrations occur?

A second question concerned the origin of an enormous buried gorge that lay beneath—and nearly parallel to—the present course of the Rhone River, which flows from Switzerland to France. The buried canyon is nearly 1000 meters deep. What had happened in the geologic past to give a stream sufficient erosive powers to cut so deeply?

The third task for the *Challenger*'s scientists was to try to determine the origin of the domelike masses buried deep beneath the Mediterranean sea floor. These structures had been detected years earlier by echo-sounding instruments, but they had never been penetrated in the course of drilling. Were they salt domes such as are common along the United States

Gulf Coast, and if so, why should there have been so much solid crystalline salt beneath the floor of the Mediterranean?

Finally, there was the presence—again detected on seismic profiles—of a hard layer of sedimentary rock 100 meters or so below the present sea floor. This hard layer was a strong acoustic reflector (layer of sediment or rock capable of reflecting sound waves) and was easily recognized nearly everywhere in the Mediterranean. Most deep-sea deposits are soft, poorly reflective oozes. What was the composition and origin of the hard layer?

With such questions as these clearly before them, the scientists aboard the *Glomar Challenger* proceeded to the Mediterranean to search for the answers. On August 23, 1970, they recovered a sample from the surface of the hard layer. The sample consisted of pebbles of hardened sediment that had once been soft, deep-sea mud, as well as granules of gypsum and fragments of volcanic rock. Not a single pebble was found that might have indicated that the pebbles came from the nearby continent. In the days following, samples of solid gypsum were repeatedly brought on deck, as drilling operations penetrated the hard layer at its predicted depth. Clearly, the strong acoustic reflector was a layer of gypsum. Furthermore, the gypsum was found to possess peculiarities of composition and structure that suggested it

---

Andreas fault system, Baja California was sheared from the American plate and now accompanies the northward movement of the Pacific plate.

South America

Extensive deformation, metamorphism, and emplacement of granitic masses had characterized the Andean belt during the Cretaceous. Folding and volcanism continued to be widespread during the Cenozoic as well and was especially intense during the Miocene. Subsequently, the Andean highlands were eroded to low relief and then were uplifted in late Pliocene. Erosion of the uplifted surface is responsible for the present relief and topography of the Andes. Cenozoic instability in the Andes has been related to the continued movement of the floor of the South Pacific Ocean beneath South America. An oceanic trench off the Pacific coast and an eastwardly inclined seismic zone characterized by

deep-focus earthquakes (a Benioff zone) attest to this interpretation.

As a consequence of the frequent upheavals that beset the west side of South America, epicontinental seas were either marginal or limited. Sedimentation was primarily terrestrial, as silt, sand, and gravel eroded from the Andes was washed down into the Amazon and Orinoco basins and across the Pampas. Riftlike basins created by block faulting served as collecting sites for prodigious thicknesses of clastic sediments.

The Tethyan Realm

The conversion of a major seaway separating Eurasia and Gondwanaland to a spectacular array of mountains and plateaus must be considered one of the greatest events of the Cenozoic Era. During most of the era, the Tethys seaway lay to the south of Europe. Its waters spilled onto the northern margin of Africa

had formed on desert flats. Sediment above and below the gypsum layer contained tiny marine fossils, indicating open-ocean conditions. As they drilled into the central and deepest part of the Mediterranean basin, the scientists took solid, shiny, crystalline salt from the core barrel. Interbedded with the salt were thin layers of what appeared to be windblown silt.

The time had come to formulate a hypothesis. The investigators theorized that about 20 million years ago, the Mediterranean was a broad seaway linked to the Atlantic by two narrow straits. Crustal movements closed the straits, and the landlocked Mediterranean began to evaporate. Increasing salinity caused by the evaporation resulted in the extermination of scores of invertebrate species. Only a few organisms especially tolerant of very salty conditions remained. As evaporation continued, the remaining brine became so dense that the calcium sulfate of the hard layer was precipitated. In the central deeper part of the basin, the last of the brine evaporated to precipitate more soluble sodium chloride. Later, under the weight of overlying sediments, this salt flowed plastically upward to form salt domes. Before this happened, however, the Mediterranean was a vast "Death Valley" 3000 meters deep. Then about 5½ million years ago came the deluge. As a result of crustal adjustments and faulting, the Strait of Gibraltar opened, and water cascaded spectacularly back into the Mediterranean. Turbulent waters

tore into the hardened salt flats, broke them up, and ground them into the pebbles observed in the first sample taken by the *Challenger*. As the basin was refilled, normal marine organisms returned. Soon layers of oceanic ooze began to accumulate above the old hard layer.

The salt and gypsum, the faunal changes, and the unusual gravel provided abundant evidence that the Mediterranean was once a desert. But how did this knowledge relate to the buried gorge beneath the Rhone Valley? The answer was now clear. If the Mediterranean had been emptied, the surrounding lands would have stood high above the floor of the basin. The gradients of streams would have markedly increased, and their swiftly flowing waters would have incised rapidly, eroding deep canyons. Such canyons should also have been cut by other streams that once flowed into the Mediterranean. As news of the *Glomar Challenger*'s findings spread, geologists who had worked in the region alerted the investigators to the presence of gorges in North Africa. A Russian geologist recalled a 240-meter chasm that had been encountered at the time the foundations had been prepared for the Aswan Dam in Egypt. Oil company geologists in Libya sent word of many similar gorges that they had detected from seismograph records. A final piece of the puzzle seemed to fall into place.

as an epicontinental sea. Disturbances prophetic of impending tectonic upheavals began along the Tethys during the Eocene with large-scale folding and thrust-faulting. It was at approximately this time that the northward-moving African block first encountered the western underside of Europe and crumpled the strata that now compose the Pyrenees and Atlas mountains. Then, with a sort of scissor-like movement, the Alpine region began to be squeezed. At first, marine conditions persisted between the emerging folds. Dark siliceous shales, poorly sorted sandstones, and cherts accumulated between elongate submarine banks. In the Alps, these marine sediments are referred to as **flysch.**

By Oligocene time, compression from the south caused enormous recumbent folds to rise as mountain arcs out of the old seaway and to slide forward over the lands that once lay to the north of the Tethys. Great folds were cut along their undersides by thrust faults and pushed on top of one another as

spectacular monuments to the forces involved in plate collisions. North of these folded thrust structures lay a topographic depression that received the piedmont deposits eroded from the mountains. These terrestrial clastics, termed **molasse** by European geologists, resemble similar deposits swept eastward from the Rocky Mountains during the Cenozoic.

Even after the Oligocene, the compressions continued. During the Pliocene, further thrusts from the south carried the older folded belts northward over the molasse deposits and crumpled the Jura folds, which today form the northern front of the Alps. The thrusting was followed by spasmodic regional uplifts that continue even to the present day.

That part of the Tethys that extended far to the east of the Alps experienced its own paroxysm of mountain-making during the Cenozoic. Volcanism, folding, thrusting, and emplacement of massive granitic intrusions began early in the era and increased markedly during the Miocene. Great elongate tracts

FIGURE 13–31 Shivling Peak in the Himalayas. The mountain has been eroded by glaciers into a glacial horn. *(Courtesy of D. Bhattacharyya.)*

of the former sea floor were squeezed into folds and thrust southward. Much of the early deformation occurred at or near sea level, but ultimately the sea-floor section was forced up and above the level of the waves. A broad, subsiding trough formed along the northern edge of newly arrived peninsular India. On this lowland, more than 5000 meters of continental sediments were deposited. These sediments contain an important fossil record of Cenozoic mammals and plants. In the two final epochs of the era, regional (epeirogenic) uplifts brought the plateaus and ranges to lofty elevations and caused the retreat of those marginal seas that still remained. The uplift may very well have been caused by the continued movement of the Indian block northward beneath the southern edge of the Asian plate.

The history of the Tethys provides another example of a region of the Earth's crust that has undergone extensive geographic and physiographic change. The once quiet seaway that bordered Eurasia on the south was transformed into a structurally complex region of rugged highlands that includes the Alps, Appennines, Carpathians, Caucasus, and Himalayas (Figs. 13–31 and 13–32).

One event in the history of the Tethys was not directly related to orogeny in the region. As described

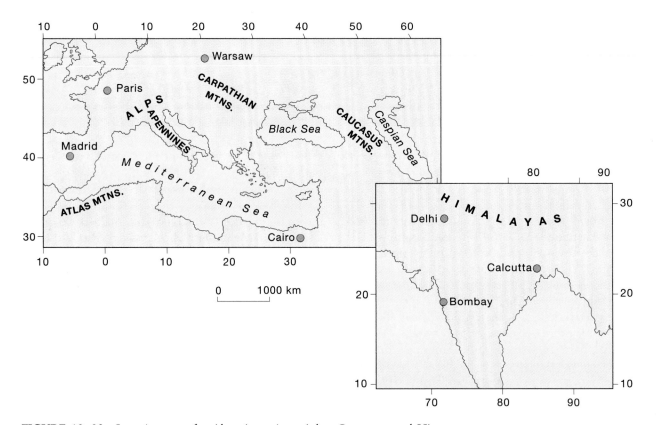

FIGURE 13–32 Location map for Alps, Apennines, Atlas, Caucasus, and Himalaya Mountains.

FIGURE 13-33 The Giant's Causeway, near Portrush, Antrim, Northern Ireland. This is one of the best-known locations in the world for viewing columnar jointing in basaltic rock. *(Photograph by Tom Till, with permission.)*

in the Commentary Box entitled "When the Mediterranean Was a Desert," there was an interval about 5.5 million years ago when the Mediterranean dried up and may actually have resembled the present Death Valley region of California. This extraordinary change occurred near the end of the Miocene and has been named the **Messinian event,** in that the Messinian is the final stage of the Miocene in Italy. The Messinian event was an indirect result of a sudden but temporary expansion of the Antarctic ice sheet and closure of the Straits of Gibralter by the counterclockwise rotation of the Iberian Peninsula. Water needed to form the ice came from the ocean, and the loss in volume of ocean water lowered mean global sea level by as much as 50 meters. This decline left the Mediterranean isolated from other oceans. Without a way to replenish water lost by evaporation, the Mediterranean dried up. Evidence for this event comes primarily from an extensive evaporite bed encountered in cores that penetrated deep into the Mediterranean sea floor.

North of the Tethys

Important geologic events were also occurring in regions north of the Tethys during the Cenozoic. In the early part of the era, lavas were extruded in Scotland, Ireland, Spitzbergen, Greenland, and Baffin Island. In Ireland, these lavas are vertically jointed and form the famous Giant's Causeway (Fig. 13-33). The volcanic activity appears to have accompanied the separation of Greenland and Europe.

Across the English Channel in northeastern Europe there is a record of repeated marine incursions during the Paleocene, Eocene, and Oligocene. The most extensive of the transgressions occurred during the Oligocene. Like the earlier incursions, waters flooded toward the southeast from the North Sea. The rocks deposited during the alternate marine transgressions and regressions have been thoroughly studied in the Paris basin. The Paris basin is not only a sedimentary basin, but a structural basin as well. This is quickly apparent on a geologic map of the area (Fig. 13–34), which shows younger Tertiary deposits surrounded by Mesozoic rocks. After the Oligocene, general uplift prevented further marine incursions. Thus, the late Tertiary record is read largely from lake and stream deposits.

East of Europe in Mongolia and China, epeirogenic uplift occurred during the time that the Himalayas were being deformed. Lakes and swamps were plentiful across Asia, and some extended over vast regions. Meandering streams built wide flood-plains. Important Cenozoic mammalian remains have been unearthed in these Asian fluvial deposits.

Africa

The southern margin of the Tethys seaway formed the upper boundary of Africa during most of the Cenozoic. The region had a much quieter tectonic his-

tory than did the northern Tethys. In Libya and Egypt, the formations are flat or only moderately folded carbonates; the only severely folded mountain ranges are in western North Africa, which experi-enced pulses of orogeny throughout the Cenozoic.

The larger part of Africa that lay south of the Tethys was generally emergent during the Cenozoic. It was a time in which erosion prevailed. The most conspicuous changes that occurred were those associ-ated with the rift valleys and uplands along the east side of the continent. During the late Cenozoic, much of eastern Africa was arched upward nearly 3000 meters above sea level. Fracturing and faulting oc-curred across the crest of the arch. The famous **rift valleys** of east Africa formed as narrow slivers of crust that had slipped downward between great fault blocks. As the faulting continued, volcanoes devel-oped along the fault trends. Mount Kenya and Mount Kilimanjaro are two of the most notable of

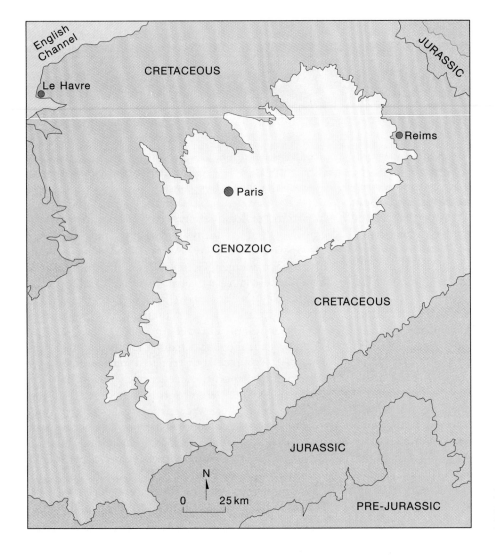

FIGURE 13–34 Geologic sketch map of the Paris Basin.

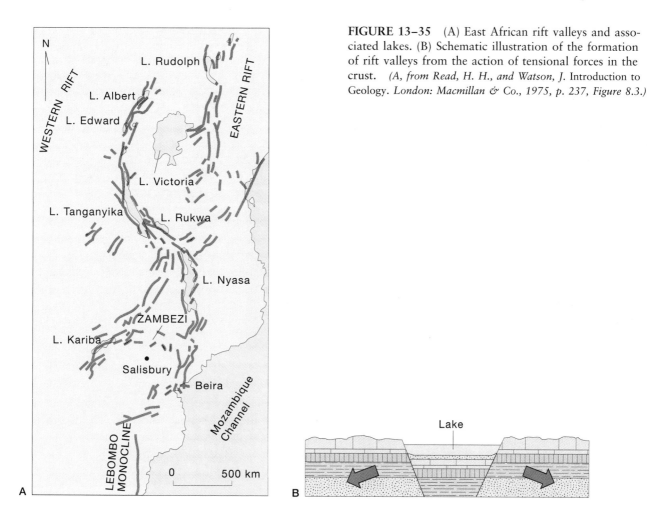

FIGURE 13–35 (A) East African rift valleys and associated lakes. (B) Schematic illustration of the formation
of rift valleys from the action of tensional forces in the
crust. *(A, from Read, H. H., and Watson, J.* Introduction to
Geology. *London: Macmillan & Co., 1975, p. 237, Figure 8.3.)*

these volcanoes, but there are many others as well. In addition to the volcanoes, a series of elongate lakes (Fig. 13–35) formed within the downfaulted blocks. Today, Lake Nyasa and Lake Tanganyika are splendid examples of these fault-controlled inland water bodies. This region of lakes, volcanoes, and fault-controlled ranges is well known for its exceptional scenic grandeur.

The Western Pacific

Most of Australia was comparatively stable during the Cenozoic Era. Marine deposits were restricted to peripheral basins. Near the center of the continent, coal-bearing terrestrial clastics are found. However, the general lack of tectonic activity that prevailed over most of the continent was not characteristic of the extreme southeastern border, where Paleocene and Eocene volcanism was intense. This activity was apparently associated with the separation of Antarctica from Australia an estimated 50 or 60 million years ago (see Fig. 13–1).

The Cenozoic history of New Zealand and the region north of Australia was also marked by exceptional instability. Throughout the islands that extend from the Aleutians down through Japan, the Phillipines, Indonesia, and into New Zealand, folding, volcanism, and spectacular upheavals were common occurrences. Tropical rains and storms caused rapid erosion and deposition of coarse sediments. Deep sedimentary basins received the erosional detritus and frequently spread it seaward as vast deltas.

The enormous volumes of Cenozoic andesites and basalts found in New Zealand clearly indicate its participation in the instability. Indeed, a belt of active volcanoes persists today on the north island. Late in the Cenozoic, vertical movements raised a great fault segment an incredible 18 km along the east side of the north island.

The crustal activity that characterized the western Pacific during the Cenozoic has continued to the present. If one were to speculate about the most active region of the globe during the near geologic future, the sinuous arcs from the Kamachatka Peninsula through Japan and down into Indonesia (Fig.

FIGURE 13–36 Lightning strikes during the eruption of Galunggung volcano, West Java, Indonesia, December 3, 1982. *(Photo by Ruska Hadian, volcanology survey of Indonesia. Courtesy of the U.S. Geological Survey.)*

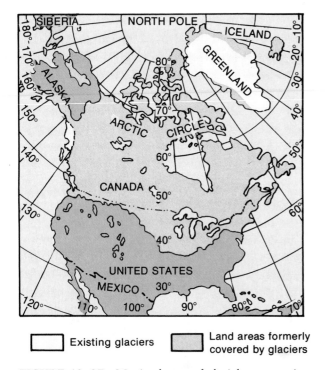

Existing glaciers

Land areas formerly covered by glaciers

FIGURE 13–37 Maximal area of glacial coverage in North America. *(From the U.S. Geological Survey Pamphlet, The Great Ice Age, No. 0-357-128. U.S. Government Printing Office, 1969.)*

13–36) would be likely candidates. (The Caribbean region might qualify as a second choice.)

Antarctica

One of the most interesting of the inferences that have resulted from the study of Cenozoic rocks of Antarctica is that this now frigid land had a genial or semitropical climate throughout the early part of the era. It was not until the Miocene that snow and ice began to accumulate and spread across the land mass. Deep-marine clastic sediments and volcanics accumulated along the Pacific side of Antarctica. The instability of west Antarctica can be inferred to be the result of the eastward movement of an oceanic plate against and beneath the continent.

■ *The Great Pleistocene Ice Age*

The final two epochs of the Cenozoic Era are the Pleistocene and Holocene (or Recent). They represent only about 2 million years of geologic time, but for humans they are an exceedingly significant interval. It was during the Pleistocene that primates of our own species evolved and rose to a dominant position. Second, during the Pleistocene, more than 40 million km^3 of snow and ice were dumped on about one-third of the land surface of the globe (Figs. 13–37 and 13–38). Such an extensive cover of ice and snow had profound effects not only on the glaciated terrains themselves (Fig. 13–39), but also on regions at

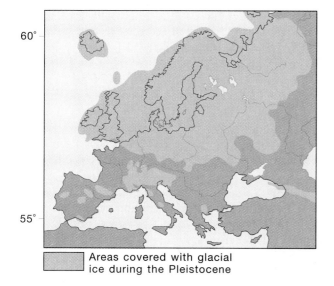

FIGURE 13-38 Areal extent of major glaciers in Europe during the Pleistocene Epoch.

FIGURE 13-39 Glaciated Canadian Shield north of Montreal, Canada. *(Courtesy of the Royal Canadian Air Force.)*

great distances from the ice fronts. Climatic zones in the Northern Hemisphere were shifted southward, and arctic conditions prevailed across northern Europe and the United States. Mountains and highlands in the Cordilleran and Eurasian ranges were sculpted by spectacular mountain glaciers, rather like those depicted in Figure 13-40. While the snow and ice accumulated and spread in higher latitudes, rainfall increased in lower latitudes, with generally beneficial effects on plant and animal life. Even as late as the beginning of the Holocene, presently arid regions in north and east Africa were well watered, fertile, and

populated by nomadic tribes. Peoples of middle and late Pleistocene time seemed to thrive by hunting along the fringes of the continental glaciers, where game was abundant and the meat kept longer with less danger of spoilage. Animal furs provided warm clothing, and following the discovery of fire, the caves were warm against the arctic winds.

In addition to the glaciations, the Pleistocene was a time of recurring crustal unrest. Volcanoes were active in New Mexico, Arizona, Idaho, Mexico, Iceland, Spitzbergen, and the Pacific borders of both North and South America.

FIGURE 13-40 Oblique aerial view of glaciers in the Fairweather Mountains of Alaska. Note the debris flowlines. *(Courtesy of M. E. Dalechek, U.S. Geological Survey.)*

Pleistocene crustal uplifts occurred in the Tetons, the Sierra Nevada, and the ranges of the central and northern Rocky Mountains. Pulses of uplift during the Pleistocene also characterized the Alps, the Himalayas, and the ranges that lay between. All of this crustal and climatic activity may suggest that the Pleistocene was a unique time in geologic history. However, this is not the case, for evidence suggests that widespread glaciations have also occurred in the Precambrian, late in the Ordovician, in the Permo-Carboniferous, and possibly during the Oligocene and Pliocene as well. Indeed, study of relatively accessible Pleistocene and Holocene glacial deposits has greatly improved our ability to interpret the vestiges of more ancient glaciations. Carbon and oxygen isotope analyses of foraminifers from deep-sea cores have provided evidence of several episodes of ocean water cooling that may not necessarily have been associated with glaciation on land.

Pleistocene and Holocene Chronology

It is a popular but biostratigraphically untenable conception that the Pleistocene Epoch began with a sudden worldwide onslaught of frigid climates and that the lower boundary of the Pleistocene Series is easily recognized by sedimentologic and paleontologic clues to that frigidity. However, as is the case with the other epochs of the Cenozoic, the Pleistocene was defined by Charles Lyell in 1839 according to the proportion of extinct to living species of mollusk shells in the layers of sediment. For example, strata containing 90 to 100 percent of present-day mollusk species were designated as belonging to the Pleistocene. Thus, the definition of what should be designated Pleistocene is straightforward, but it is no simple matter at all to find suitably fossiliferous sediments in various parts of the globe that can be confidently correlated to Lyell's type section in eastern Sicily.

At the present time, the most widely accepted time for the beginning of the Pleistocene is 1.8 million years ago. This date, however, does not coincide with the beginning of glaciation (as indicated by glacial sediments). The most extensive glaciation appears to have been initiated about 1 million years ago. Also, the low temperatures and abundant precipitation needed for extensive continental glaciation occurred at different times at different places.

For Pleistocene marine beds, fossils are indispensable for correlation. In some cases, a horizon correlative to the standard section can be recognized by the earliest appearance of certain cooler water mollusks and foraminifers. When studying cores of deep-sea sediments, one can frequently recognize the basal Pleistocene oozes by the extinction point of fossils called discoasters. **Discoasters** (Fig. 13–41) are calcareous, often star-shaped fossils believed to have

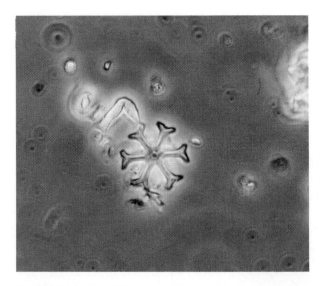

FIGURE 13–41 Discoaster *(Discoaster challengeri)* seen through a light microscope at a magnification of ×1200.

TABLE 13–2 Classic Nomenclature for Glacial and Interglacial Stages of the Pleistocene Epoch

NORTH AMERICA	ALPINE REGION	YEARS BEFORE PRESENT
		—10,000
WISCONSIN	Würm	
		—75,000
Sangamon	Riss-Würm	
		—125,000
ILLINOIAN	Riss	
		—265,000
Yarmouth	Mindel-Riss	
		—300,000
KANSAN	Mindel	
		—435,000
Aftonian	Günz-Mindel	
		—500,000
NEBRASKAN	Günz	
		—1,800,000
Pre-Nebraskan	Pre-Günz	

In North America, the glacial stages are Nebraskan, Kansan, Illinoian, and Wisconsinian. These terms correspond approximately to the Günz, Mindel, Riss, and Würm in Europe.

FIGURE 13–42 The Antarctic ice sheet near Victoria Land, Antarctica. *(Courtesy P. B. Larson and the U.S. Geological Survey.)*

been produced by golden-brown algae related to coccoliths. In continental deposits, the fossil remains of the modern horse *(Equus)*, the first true elephants, and particular species of other vertebrates are used to recognize the deposits of the lowermost Pleistocene.

Before the mid-1970s, geologists customarily thought of the Pleistocene as divisible into four distinct glacial stages, with intervening interglacial stages (Table 13–2). Recent investigations indicate, however, that there may have been as many as 30 periods of severely frigid conditions over the past 3 million years. Widespread cold climates thus existed well before the beginning of the Pleistocene, and indeed the first subtle evidence of the coming cold began to appear 40 million years ago. Antarctica (Fig. 13–42) has been frozen into an ice age for at least the past 15 million years.

The end of the Pleistocene (and beginning of the Holocene) is considered to be the time of melting of the ice sheets to approximately their present extent and the concomitant rise in sea level. This would mean that the Pleistocene ended about 8000 years ago. However, there is some argument about the date of this boundary. Some geologists prefer to set the Pleistocene–Holocene boundary at the midpoint in warming of the oceans, in which case the Ice Age would have ended between 11,000 and 12,000 years ago.

Whatever date is confirmed for the end of the Pleistocene Epoch, it is evident from historical records and carbon-14 dating of old terminal moraines that cold spells have recurred periodically into the

Holocene. Climates on our planet are in delicate balance with atmospheric, geographic, and astronomic variables. A change in any one of these variables is likely to affect climate. Meteorologists, for example, have found a very good correlation between periods of minimum sunspot activity and episodes of colder conditions on Earth. One well-documented period of cooler and drier conditions in the Northern Hemisphere occurred between A.D. 1540 and 1890, when temperatures were often 2° to 4°F cooler than today. These four centuries encompass the so-called **little ice age**. During the little ice age, frigid conditions periodically extended across most of Europe and in the United States as far south as the Carolinas. Sea ice intruded far to the south of its present margin. The cold caused loss of harvests, with resulting famine, food riots, and warfare in Europe. In the middle of the fourteenth century, harsh conditions of the little ice age were a contributing cause of the demise of a once flourishing Norse colony in Greenland.

Stratigraphy of Terrestrial Pleistocene Deposits

Glacial deposits are difficult to correlate. They often consist of chaotic mixtures of coarse detritus completely lacking in fossil remains. Moving glaciers may carry away even these materials, leaving only a bare surface of scoured and polished bedrock. However, if the ice does not advance for a time, a terminal moraine may be deposited in front of the glacier (Fig.

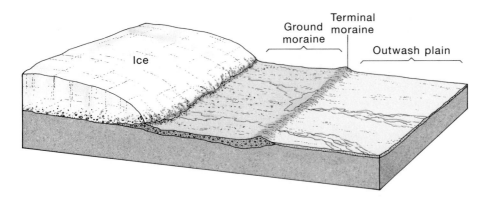

FIGURE 13–43 · Cross-section of a terminal moraine, ground moraine, and outwash plain.

13–43). As the ice melts, a widespread ground moraine is left behind. Moraines may consist of till, an unsorted mixture of particles from clay to boulder size dropped directly by the ice (Fig. 13–44). **Stratified drift,** on the other hand, has been washed by meltwater, is better sorted, and has many of the characteristics of stream deposits. In either case, however, the stratigrapher is faced with similar-looking masses of debris that are woefully deficient in distinguishing characteristics. As a result, several mutually substantiating criteria must be used in making correlations. The degree to which one sheet of glacial debris has been dissected by streams may indicate that it is older or younger than another sheet. The depth of oxidation and amount of chemical weathering may also provide some estimate of the relative age of interglacial soils. Fossil pollen grains in thick exposures of bog or lake deposits often clearly reflect fluctuations of climate and can be used to mark times of glacial advance and retreat. Varved clays deposited in lakes

near the glaciers can sometimes be correlated to similar sediments of other lakes and, in addition, may provide an estimate of the time required for deposition of the entire thickness of lake deposits. However, the most accurate means of dating and correlating Pleistocene sediments is to extract pieces of wood, bone, or peat and determine their age by radiocarbon dating techniques. Unfortunately, even this method has its limitations, the most significant of which is the relatively short half-life (5570 years) of carbon-14. This limits the method to materials less than about 100,000 years old and thus restricts the use of the carbon-14 technique to deposits of the last glacial stage.

Pleistocene Deep-Sea Sediments

Sediments deposited on the ocean floor during the Pleistocene are easier to date and correlate than terrestrial deposits. The ocean basins are sites of a rela-

FIGURE 13–44 Glacial till in the moraine of a valley glacier, Kenai Peninsula, Alaska. *(Courtesy of U.S. Geological Survey.)*

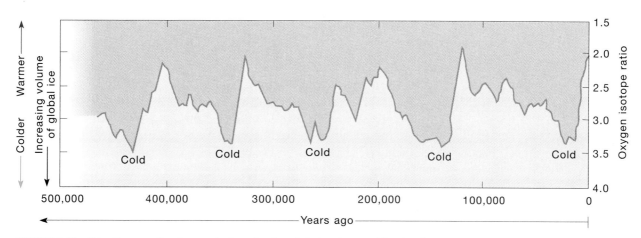

FIGURE 13–45 Curve reflecting variations in the global volume of ice (and, indirectly, paleotemperatures) during the past 500,000 years. Data are from radiometric dating and isotope measurements of cores from the Indian Ocean. *(Data from Hays, J. D., and Shackleton, N. J. Science 194:1121–1132, 1976.)*

tively continuous sedimentary record, the column of sediments contains abundant fossil remains, and the deposits can be dated by relating them to paleomagnetic data and to radiometric isotopes having short half-lives.

Continuous sections of deep-sea sediments can be obtained by means of piston coring devices. These tools provide cores more than 15 meters long. The cores can then be analyzed in various ways. One approach is to use the oxygen isotope ratios obtained from the shells of calcareous foraminifers within the cored sediment to determine the volume of ocean water that was stored at the time as glacial ice. Because of its greater mass, less oxygen-18 is evaporated and precipitated as snow. Hence, during an ice age when ice is accumulating globally, there will be relatively greater percentages of oxygen-18 relative to oxygen-16 in sea water and in the shells of marine invertebrates. Plotted against depth in a deep-sea core, these isotope ratios indicate variations in ice volume and temperature with time (Fig. 13–45). Indications of cooler conditions can then be correlated with declines in worldwide sea level.

Another method used in the correlation of Pleistocene marine sediments is to plot at each level in the core the relative abundance of foraminifers known to be especially sensitive to temperature. For example, the tropical species *Globorotalia menardii* (Fig. 13–46) is alternately present or absent within Pleistocene cores from the equatorial Atlantic. The absence of the species in part of a core is taken to indicate an episode of cooler climates and glaciation. Carbon-14 dates in the upper part of such cores permit one to determine the rates of sedimentation.

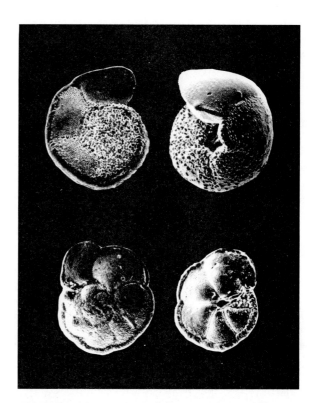

FIGURE 13–46 Two well-known species of planktonic foraminifers used in correlating deep-sea sediments. At the top are views of both sides of *Globorotalia truncatulinoides*. Below are views of opposite sides of *Globorotalia menardii*. Magnification of all specimens ×50.

Another foraminifer, *Globorotalia truncatuli-noides,* also permits one to recognize alternate cooling and warming of the oceans (Fig. 13–47). This species is trochospirally coiled, which means that it coils in a spiral with all the whorls visible on one side but only the last whorl visible on the other side. Individuals that coil to the right dominate in warmer water, whereas left-coiled individuals prefer colder water. During glacial advances, when ocean temperatures declined, the right-coiled populations of *G. truncatulinoides* in middle and low latitudes were replaced by populations in which left coiling predominated. The record of such changes is clearly apparent in the deep-sea cores.

Remanent magnetism in some deep-sea sediments is sometimes sufficiently strong and stable that a particular section of the core can be correlated to a magnetic reversal documented on land in volcanic rocks of known age. However, the analysis is likely to be in error if the original orientation of detrital magnetic grains has been disturbed by burrowing organisms (a process termed bioturbation).

Effects of Pleistocene Glaciations

Earlier it was noted that during maximum glacial coverage, over 40 million km^3 of ice and snow lay on the continents, equivalent to a tremendous amount of water; removal of that water from the oceans had a multitude of effects on the Pleistocene environment. It has been estimated that sea level may have dropped at least 75 meters during maximum ice coverage. Extensive tracts of the present continental shelves became dry land and were covered with forests and grasslands. The British Isles were joined to Europe, and a land bridge stretched from Siberia to Alaska. During interglacial stages, marine waters returned to the low coastal areas, drowning the flora and forcing terrestrial animals inland.

The glaciers had a direct impact on the erosion of lands and the creation of glacial land forms. The great weight of the ice depressed the crust of the Earth over large parts of the glaciated area, in some places to a level of 200 to 300 meters below the preglacial position. With the removal of the last ice sheet, the downwarped areas of the crust gradually began to return to their former positions. The rebound is dramatically apparent in parts of the Baltic, the Arctic, and the Great Lakes region of North America, where former coastal features are now elevated high above sea level (Fig. 13–48).

As the great continental glaciers advanced, they obliterated old drainage channels and caused streams to erode new channels. These dislocations are especially evident in the north–central United States. Prior to the Ice Age, the northern segment of the Missouri River drained northward into Hudson Bay, and the northern part of the Ohio River flowed northeastward into the Gulf of St. Lawrence. The lower Ohio drained into a preglacial stream named the Teays

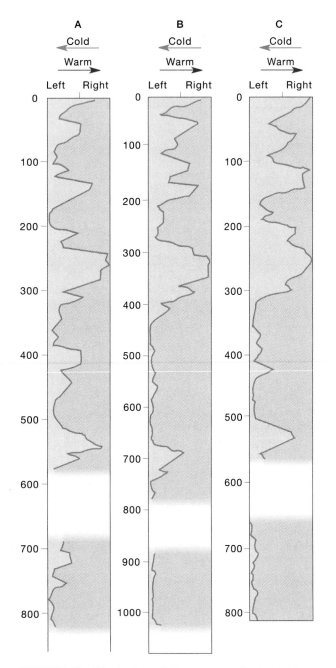

FIGURE 13–47 A, B, and C are curves depicting the percentages of right- and left-coiled *Globorotalia truncatulinoides.* The scales run from 100 percent right-coiling forms at the right margins of each column to 100 percent left-coiling forms at the left margins. Left-coiling forms prefer colder waters. Numbers to the left of the columns are deep-sea core depths in centimeters. Cores were not recovered in white areas. *(Adapted from Erickson, D. D., and Wollin, E. Science 162[3859]:233.)*

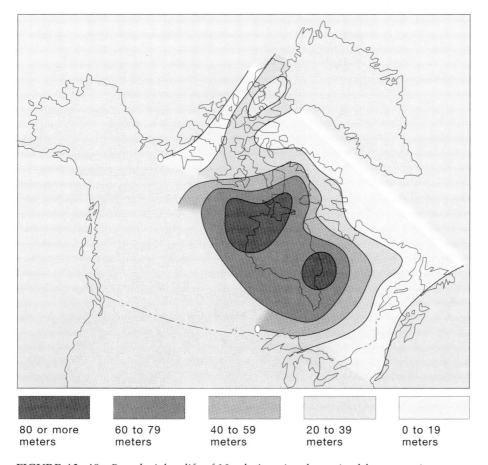

80 or more meters	60 to 79 meters	40 to 59 meters	20 to 39 meters	0 to 19 meters

FIGURE 13–48 Postglacial uplift of North America determined by measuring elevation of marine sediments 6000 years old. *(Simplified and adapted from Andrews, J. T., The pattern and interpretation of restrained, postglacial and residual rebound in the area of Hudson Bay,* in *Earth Science Symposium on Hudson Bay, Ottawa, 1968, Canadian Geological Survey Paper 68–53, p. 53, 1969.)*

River. Today, geologists recognize the former location of the Teays by a thick linear trend of sands and gravels. Those parts of the Missouri and Ohio rivers that once flowed toward the north were turned aside by the ice sheet and forced to flow along the fringe of the glacier until they found a southward outlet. The present trend of the Missouri and Ohio rivers approximates the margin of the most southerly advance of the ice. The equilibrium of streams was also affected, for with glacial advances there was a lowering of sea level and thus an increase in stream gradients near coastlines. A reverse effect occurred with wasting away of the ice sheets.

Prior to the Pleistocene, there were no Great Lakes in North America. The present floors of these water bodies were lowlands. Glaciers moved into these lowlands and scoured them deeper. As the glaciers retreated, their meltwaters collected in the vacated depressions (Fig. 13–49). Niagara Falls, between Lake Erie and Lake Ontario, came into

existence when the retreating ice of the last glacial stage uncovered an escarpment formed by southwardly tilted resistant strata. Water from the Niagara River tumbled over the edge of the escarpment, which is supported by the resistant Lockport Limestone. Weak shales beneath the limestone are continually undermined, causing southward retreat of the falls (see Fig. 8–14).

Another large system of ice-dammed lakes covered a vast area of North Dakota, Minnesota, Manitoba, and Saskatchewan. The largest of these lakes has been named Lake Agassiz in honor of Jean Louis Rodolphe Agassiz, the great French naturalist who initially insisted on the existence of the Ice Age. Today, rich wheatlands extend across what was once the floor of the lake.

Other lakes developed during the Pleistocene, occupying basins that were not near ice sheets. These were formed as a consequence of the greatly increased precipitation and runoff that characterized

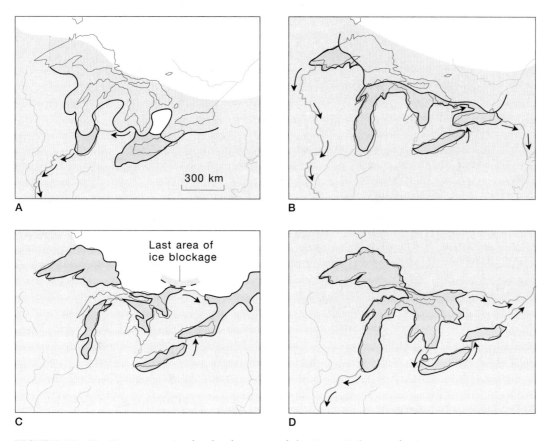

FIGURE 13–49 Four stages in the development of the Great Lakes as the ice of the last glacial advance moved away. *(After Hough, J. L., Geology of the Great Lakes. Urbana: Univ. of Illinois Press, 1958, figs. 56, 69, 73, and 74.)*

regions south of the glaciers. Such water bodies are called **pluvial lakes** (which comes from the Latin *pluvia*, meaning rain). Pluvial lakes were particularly numerous in the northern part of the Basin and Range Province of North America, where faulting produced more than 140 closed basins. So-called pluvial intervals, when lakes were most extensive, were generally synchronous with glacial stages, whereas during interglacial stages, many lakes shrank to small saline remnants or even dried out completely. Lake Bonneville in Utah was one such lake. It once covered more than 50,000 km² and was as deep as 300 meters in some places. Parts of Lake Bonneville persist today as Great Salt Lake, Utah lake, and Sevier Lake.

A particularly spectacular event associated with the formation of Pleistocene lakes occurred in the northwestern corner of the United States. Lobes of the southwardly advancing ice sheet repeatedly blocked the Clark Fork River, and the impounded water formed a long, narrow lake extending diagonally across part of western Montana. The freshwater body, called Lake Missoula, contained an estimated 2000 km³ of water. With the recession of the glacier,

FIGURE 13–50 Conglomeratic scabland deposits exposed near Lyons Ferry, Washington. *(Courtesy of U.S. Geological Survey.)*

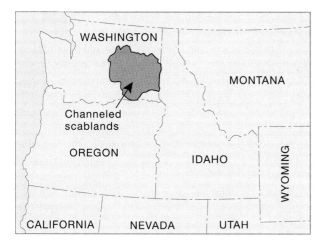

FIGURE 13–51 Location of the channeled scablands.

the ice dam broke, and tremendous floods of water rushed out catastrophically across eastern Washington, causing severe erosion and depositing huge volumes of gravel, boulders, and cobbles (Fig. 13–50). The dissected region is appropriately termed the **channeled scablands** (Fig. 13–51).

The glacial conditions of the Pleistocene also had an effect on soils. In many northern areas, fertile topsoil was stripped off the bedrock and transported to more southerly regions, which are now among the world's most productive farmlands. Because of the flow of dense, cold air coming off the glaciers, winds were strong and persistent. Fine-grained glacial sediments that had been spread across outwash plains and floodplains were picked up and transported by the wind and then deposited as thick layers of wind-blown silt called **loess**. Such deposits blanket large areas of the Missouri River valley, central Europe, and northern China.

Cause of Pleistocene Climatic Conditions

The results of oxygen isotope research indicate that world climates grew progressively cooler from the Middle Cenozoic to the Pleistocene. The culmination of this trend was not a single sudden plunge into frigidity but rather an oscillation of glacial and interglacial stages. Any theory that adequately explains glaciation must consider not only the long-term decline in worldwide temperatures but the oscillations as well. Further, the theory must include reasons for the ideal combination of temperature and precipitation required for the buildup of continental glaciers. Although geologists, physicists, and meteorologists have speculated about the cause of the Ice Age for more than a hundred years, no undisputed cause has been found. Indeed, it seems likely that Pleistocene climatic conditions came upon us as a result of several simultaneous factors.

A widely accepted theory for the Ice Age was developed by the Yugoslavian mathematician Milutin Milankovitch (1879–1958). After 30 years of careful study, Milankovitch convincingly proposed that irregularities in the Earth's movements and their influence on the amount of solar radiation received by the Earth could account for glacial and interglacial stages of the Pleistocene. His calculations were based on three variables: the Earth's axial tilt, precession, and orbital eccentricity. With regard to the first of these variables, Milankovitch recognized that the tilt of the Earth's axis varies between about 22° and 24° over a period of about 41,000 years. This results in a corresponding variation in the seasonal length of days and in the amount of solar radiation received at higher latitudes. Precession, the second variable, refers to the way the axis of rotation moves slowly in a circle that is completed about every 26,000 years. The effect is equivalent to tilting a rapidly spinning top, whose axis responds by describing a cone in space (Fig. 13–52). The third variable is the eccentricity of

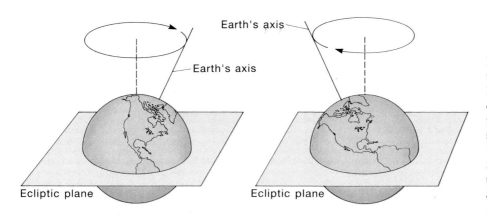

FIGURE 13–52 Two positions in the precession of the Earth's axis. The period of a complete precession is 26,000 years. Thus, the two positions shown are separated by 13,000 years. *(From Pasachoff, J. Contemporary Astronomy, 3rd ed. Philadelphia: Saunders College Publishing, 1985.)*

the Earth's orbital path around the sun, which over an interval of about 100,000 years varies about 2 percent. As a result of this variation, the Earth is at times closer or farther away from the sun.

According to Milankovitch's calculations, the combination of these variables periodically results in less solar radiation received at the top of the Earth's atmosphere, and this might suffice to cause cooling and recurrent glaciations. The Milankovitch cycles correspond rather well to episodes of glaciation over the past 100,000 years. That the Milankovitch cycles do have a role in causing the alternation of glacial and interglacial stages is supported by oxygen isotope analyses of the calcium carbonate shells of foraminifera recovered from deep-sea cores. These analyses confirm an interval of about 100,000 years between the times of coldest temperatures over the past 600,000 years.

If, however, the **Milankovitch effect** has been in operation through most of the Phanerozoic eon, why haven't there been Pleistocene-like glaciations continuously throughout geologic time? Apparently, other factors must be involved. One such factor might be a variation in the amount of solar energy reflected from the Earth back into space rather than being absorbed. The fraction of solar energy reflected back into space is termed the Earth's **albedo.** At the present time, it is about 33 percent. Theorists suggest that by the beginning of Pleistocene time, when continents were fully emergent and hence highly reflective, temperatures may have been lowered enough for the Milankovitch effect to begin to operate. This is not an unreasonable suggestion, for only a 1 percent lessening of retained solar energy could lead to as much as an 8°C drop in average surface temperatures. This would be sufficient to trigger a glacial advance if ample precipitation were available over continental areas. Still other geologists speculate that absorption of solar energy was hindered by cloud cover, volcanic ash, and dust in the atmosphere or fluctuations in carbon dioxide. A decrease in carbon dioxide content would cause a corresponding decrease in the warmth-gathering greenhouse effect, for example. On the other hand, it is also possible that a buildup of CO_2 might trigger glaciation, for as warming occurred, there might be more rapid evaporation and an increase in highly reflective cloud cover.

Other hypotheses for Ice Age origin stress the need for both ample snowfall and the preservation of that precipitation on suitably located continental areas. Ample snowfall may have been produced by the northward deflection of the Gulf Stream after its westward movement was blocked by the formation of the Isthmus of Panama about 3.5 million years ago. Precipitation resulting when the moist air was shunted northward fell on land areas as snow, and continental glaciers developed.

Although scientists are not yet ready to formulate a complete and unified theory for the cause of the Pleistocene's multiple glaciations, it appears that they may be very near the answer. Few deny that the Milankovitch effect, albedo, and pole positions have played a role in creating Pleistocene climatic conditions. The problem is to determine the relative importance of these factors and how they relate to one another. If Pleistocene-like climatic cycles are fundamentally the result of changes in the Earth's orbital geometry, one can make some interesting speculations about future climate. Calculations indicate, for example, that the long-term climatic trend over the next 20,000 years is toward extensive Northern Hemisphere glaciation and, of course, cooler conditions. What effect human activities, such as the burning of coal, oil, and gas, might have on this trend is uncertain.

■ *Cenozoic Climates*

The worldwide cooling that culminated in the great Ice Age actually had begun during the Early Cenozoic. However, the decrease in temperatures was not entirely uniform. There were warming trends during the Late Paleocene and Eocene, as indicated by fossils of palm trees and crocodiles found in Minnesota, in Germany, and near London. Trees that are today characteristic of moist temperature zones thrived in Alaska, Spitzbergen, and Greenland. Coral reefs grew in regions 10° to 20° north of their present optimum habitat. Cooling resumed during the Oligocene, and coral reefs began a slow retreat toward the equator. On land, temperate and tropical forests were also displaced toward lower latitudes. Another pulse of warmer conditions occurred during the Miocene, after which climates grew steadily cooler, as if in anticipation of the approaching Pleistocene glaciations.

Although the most extensive glaciations of the Cenozoic were those of the Pleistocene, glaciation occurred on a more limited scale at other times during the Cenozoic. Deep-sea cores taken near Antarctica indicate that an ice sheet had already begun to form during the Eocene. In addition, Miocene glacial conditions are indicated both by terrestrial glacial deposits in Antarctica and ice-rafted glacial deposits in Miocene deep-sea cores from the Bering Sea. The accumulation of Miocene glacial ice and resulting lowering of sea level contributed to the drying up of the Mediterranean Sea. Evidence for Pliocene glaciation has been recognized in the Sierra Nevada, Iceland, South America, and the former Soviet Union.

FIGURE 13–53 Drilling for offshore oil. Shown here is a huge semisubmersible drilling rig. *(Courtesy of Reading and Bates Drilling Company.)*

■ *Mineral Resources of the Cenozoic*

Most of the known accumulations of oil have been found in Tertiary reservoir rocks. In the United States, about half of all the oil ever discovered has been in Tertiary rocks. Paleocene petroleum reservoirs occur in Libya and beneath the North Sea. Petroleum trapped in Eocene permeable limestones and sandstones is being pumped in Texas, Louisiana, Iraq, the former Soviet Union, Pakistan, and Australia. Strata of Oligocene age yield oil in western Europe, Burma, California, and the Gulf Coast states of the United States. Miocene reservoir sandstones yield oil on every continent except Australia. They are extraordinarily productive in Saudi Arabia, Kuwait, California, Texas, and Louisiana. Offshore oil fields in the Gulf Coast (Fig. 13–53) and California tap oil held in rocks of Pliocene age. Finally, the largest reserves of oil shale in the world occur in the Eocene Green River Formation of the western United States (Fig. 13–54).

Coal also occurs in Cenozoic rocks. Most Cenozoic coal is of the lignitic or sub-bituminous variety, but because of its low content of sulfur, it is extensively used. The coal beds of the Paleocene Fort

FIGURE 13–54 Eocene oil shale from northeastern Colorado. *(Courtesy of U.S. Geological Survey.)*

C O M M E N T A R Y
Oil Shale

The Eocene oil shale (see Fig. 13–54) that occurs in parts of Wyoming, Utah, and Colorado is rich in an oil-yielding organic compound known as kerogen. When heated to 480°C, the kerogen in oil shale vaporizes. The vapor can then be condensed and will form a thick oil, which, when enriched with hydrogen, can be refined into gasoline and other products in much the same way as ordinary crude oil. The retort in which the shale is heated resembles a giant pressure cooker that fuels itself with the very gases generated during heating.

Many oil shales yield from one-half to three-quarters of a barrel (42 U.S. gallons) of oil per ton of rock. If one were to mine only the shale layers thicker than 10 meters, they would yield an impressive 540 billion barrels of oil. If this oil were to be produced during the next decade, it would appreciably reduce the United States' dependency on foreign crude oil. Production of oil from oil shale, however, requires the use of vast amounts of water in a region of North America where water is in short supply.

Two additional problems have prevented full-scale mining and retorting of oil shales. The first is the waste disposal problem. In the process of being crushed and retorted, the shale expands to occupy about 30 percent more volume than was present in the original rock. Geologists call this the popcorn effect. Where can this great volume of light, dusty material be placed? The second problem relates to air quality, for the processing of huge tonnages of oil shale is likely to release large amounts of dust into the atmosphere. Unfortunately, the shales occur in arid regions, and there is little water available for use in processing the rock to contain the dust and provide for revegetation of the land. Until these problems are solved, production of oil from oil shale will be slow. Perhaps this will work to our advantage, for we may need this oil for processed goods a century or two from now, when the internal combustion engine may be as rare as the horse and buggy is today.

Union Formation are mined in the Dakotas, Wyoming, and Montana. They represent the largest recoverable fossil fuel deposits in the United States. Scattered coal beds, mostly of Eocene age, are exploited along the Pacific coast of the United States.

Metals mined from Cenozoic rocks include placer gold gathered by dredging and hydraulic mining of gravel deposits in California. The gold in these deposits has been eroded from older Jurassic gold-bearing quartz veins in the Sierra Nevada. About 60 percent of the world's supply of tin is also supplied by Cenozoic stream deposits. Manganese, an essential additive in the manufacture of steel, is mined extensively from lower Tertiary rocks of the former Soviet Union. These deposits comprise 75 percent of the world's richest manganese ores. Another metal used in making certain types of steel is molybdenum. One deposit in Climax, Colorado, supplies 40 percent of the world's molybdenum. Other metals, such as copper, silver, lead, mercury, and zinc, are frequently found associated with Tertiary intrusive rocks of western North America and the Andes and with Tertiary orogenic belts along the western margins of the Pacific Ocean. Similar Cenozoic rocks were formed

FIGURE 13–55 Massive, jointed diatomite of the Bruneau Formation of middle Pleistocene age, Elmore County, Idaho. *(Courtesy of U.S. Geological Survey.)*

in southern Europe and Asia as a consequence of the deformation of the Tethys belt.

Nonmetallic materials from Cenozic rocks are also important resources. Diatomite (Fig. 13–55), the white, porous rock formed from the silica coverings (frustules) of diatoms, is mined in great quantities in California. Cenozoic building stone, clay, phosphates, sulfur, salt, and gypsum are quarried in North America, Europe, the Middle East, and Asia.

Summary

A compilation of the great historical events of the Cenozoic would certainly include the formation of the Alpine–Himalayan mountain system, the drift of continents to their present locations, the uplift of mountains around the perimeter of the Pacific, the development of an ice age, and the evolution of *Homo sapiens*. For most of the duration of the Cenozoic, the continents stood at relatively high elevations; epicontinental and marginal seas were of limited extent.

Along the east side of North America, the Appalachian chain was subjected to cycles of erosion and uplift that served to sculpt the surface to its present topography. Along the Atlantic and Gulf coastal plains, repeated transgressions and regressions reworked the clastics transported there from the interior. Florida was a shallow, limestone-forming, submarine bank until it was uplifted late in the era. In the Gulf of Mexico, Cenozoic deposits accumulated in thicknesses in excess of 9000 meters as the depositional basin was continuously downwarped.

In western North America, compressional forces that had produced many of the structures of the Cordillera ceased, and in their place vertical crustal adjustments occurred. The erosional debris worn from the highland areas filled the basins between ranges and was then spread eastward to form the clastic wedge of the Great Plains. Lakes formed in some basins and served as collecting sites for oil shales. Crustal movements later in the Tertiary caused the elevation of the Sierra Nevada along a great fault and produced the Basin and Range Province. Intense volcanic activity accompanied these crustal movements. Volcanism continues today along sectors of the west coast of North America that lie adjacent to subduction zones, such as in the Cascades, which are fed by lavas from the subducting Juan de Fuca plate. Also in the far west, the Farallon plate and its spreading center were overridden by the westward-moving continent. As a result, the formerly convergent plate boundary was changed to a shear boundary characterized by strike–slip, rather than subduction, tectonics. The development of the San Andreas fault system was a direct result of this change in the style of west coast tectonics.

In the Tethys region, there was intense deformation during the Cenozoic as major tectonic plates moved against one another. Orogenies that accompanied the compressions and uplifts along the southern margin of the Tethys region resulted in the Atlas Mountains, whereas along the northern border, the Alps, Carpathians, Apennines, Pyrenees, and Himalayas were constructed from the sediments that had accumulated in the former seaway. Orogenic activity was also prevalent over much of the Middle East and Far East and along the western borders of the Pacific. Africa witnessed Cenozoic marine deposition along its northern border. However, the most dramatic occurrences in Africa were the continuing development of the great rift valleys that extend along the eastern side of the continent.

It is apparent from fossil evidence and oxygen isotope studies that the Earth's climate was on a somewhat fluctuating cooling trend during the Cenozoic. The culmination of that trend was the beginning of extensive glaciation about 1 million years ago. Variations in climate during the Pleistocene caused the vast ice sheets that had formed on the northern continents to alternately advance and recede. There were at least four major advances, with minor fluctuations before and within each. The ice, at times attaining thicknesses of 3000 meters, depressed the Earth's crust, smoothed and rounded the landscape, disrupted drainage patterns, indirectly altered climates of adjacent regions, and was responsible for the development of numerous lakes, not the least of which are the Great Lakes. As the ice alternately accumulated and melted, sea level was caused to rise and fall with profound effects on low-lying coastal areas. It is a possibility that today we live in an interglacial stage and that some thousands of years from now the ice will again accumulate in Canada and northern Europe and advance southward.

Although a wealth of gold, tin, copper, silver, and other metals is found in deposits or associated with intrusions of Cenozoic rock, the greatest resource of the era has been petroleum. Indeed, most of the world's petroleum has been found in strata that range from 1 to 60 million years of age.

Review Questions

1. Describe the manner in which each of the following major physiographic features developed:
   a. Mountains of the Basin and Range
   b. Great Plains
   c. Columbia and Snake River Plateau
   d. Red Sea
   e. Teton Range
   f. Cascade Range
2. What Eurasian mountain ranges resulted from the compression and upheaval of large areas of the Tethys seaway?
3. What epochs of the Cenozoic Era are included in the Tertiary Period, the Quaternary Period, the Paleogene period, and the Neogene Period?
4. What is the economic importance of the Green River Formation? The Fort Union Formation?
5. What is the origin of Lake Nyasa and Lake Tanganyika in eastern Africa?
6. What is the origin of Great Salt Lake in Utah? Why is it so salty?
7. In a 30-meter-long deep-sea core that penetrated most of the Pleistocene section, how might you differentiate sediment that had been deposited during a glacial stage from that of an interglacial stage?
8. What land bridge, important in the migration of Ice Age mammals, resulted from the lowering of sea level during glacial stages?
9. What is the explanation for the gradual rise in land elevations that has occurred within historic time around Hudson Bay, the Great Lakes, and the Baltic Sea?
10. What changes in movement of tectonic plates along the western border of the United States were responsible for the development of the San Andreas Fault?

Discussion Questions

1. Discuss the evidence that the Mediterranean had dried up during the Neogene, and describe how this discovery illustrates the use of the scientific method.
2. Discuss the evidence that the Gulf of Mexico experienced profound subsidence during the Cenozoic Era.
3. Relate plate tectonics to the following:
   a. The origin of the Cascades
   b. The origin of the Basin and Range
   c. The origin of the Himalayas
4. Prepare an argument for a debate in which you defend the premise that the Milankovitch effect alone would not be sufficient to have caused an ice age.
5. Describe and prepare drawings that illustrate the three variables that account for Milankovitch cycles.
6. Discuss the advantages and limitations of the carbon-14 isotope dating method for determining the age of Pleistocene deposits of stratified drift.
7. Discuss the advantages and disadvantages or problems associated with the exploitation of Green River oil shales for their content of hydrocarbons.
8. Discuss conditions on Earth that might result in increased albedo. What are the possibilities that such conditions existed at the beginning of the Pleistocene?
9. What conditions during the Pleistocene favored the formation of extensive loess deposits? Where did this sediment come from? What accounts for the observation that the grains in loess are narrowly restricted in size, being mostly 1/16 or 1/32 mm in diameter?
10. The percentage of the isotope oxygen-18 relative to the percentage of oxygen-16 in the calcium carbonate shells of foraminifers increases during the colder glacial stages of the Pleistocene. Why?

Supplemental Readings and References

Ager, D. V. 1980. *The Geology of Europe*. New York: Halstead Press.

Atwater, T. 1970. Implications of plate tectonics for the Cenozoic evolution of western North America. *Bull. Geol. Soc. Am.* 81:3513–3536.

Covey, C. 1984. The earth's orbit and the ice ages. *Sci. Am.* 250(2):58–77.

Curtis, B. F. (ed.). 1975. Cenozoic history of the southern Rocky Mountains. *Geol. Soc. Am. Memoir* 144:1–279.

Fiero, B. 1986. *Geology of the Great Basin*. Reno: University of Nevada Press.

Hsu, K. J. 1983. *The Mediterranean was a Desert.* Princeton, NJ: Princeton University Press.

Levenson, T. 1989. *Ice Time: Climate Science and Life on Earth*. New York: Harper and Row.

Match, C. L. 1976. *North America and the Great Ice Age*. New York: McGraw-Hill.

Van Andel, T. H. 1985. *New Views of an Old Planet*. Cambridge, England: Cambridge University Press.

Cenozoic lizard from the Dominican Republic, La Toca Mine, preserved in amber. The specimen is 30 to 40 million years old.

(Photograph by G. O. Poinar, Jr.)

During the Cenozoic Era or the Age of Mammals, the vegetation was of modern cast and life conditions in the main were similar to those of today, although there is much evidence of a gradual elevation of nearly all lands with a consequent increase in aridity and diminution of moisture-loving vegetation.

RICHARD SWANN LULL, *Organic Evolution*, 1924

CHAPTER 14

Life of the Cenozoic

Because biological developments of the Cenozoic have occurred so recently, and because Cenozoic fossils are topmost and more assessable, we have more facts about life of this era than about the far lengthier ones that preceded it. Armed with this more adequate data, paleontologists are better able to compare biological evolution with environmental and geographic changes in the Cenozoic. Continental fragmentation clearly stimulated biological diversity and resulted in distinctive faunal radiations on separated land masses, as well as in isolated marine basins. In the seas, bivalves, gastropods, crustaceans, echinoids, and bony fishes flourished. The ammonites and such Mesozoic marine reptiles as ichthyosaurs, plesiosaurs, and mosasaurs were no longer present in Cenozoic seas. On land, major groups of mammals that have surviving species were present, including rodents, bats, whales, carnivores, elephants, bison, camels, horses and rabbits. South America, Australia, and Antarctica were separated from North America and Eurasia during most of the Cenozoic. As a result, a distinctive assemblage of mammals evolved on these southern continents showing convergent evolution with Northern Hemisphere species. When the Panamanian land bridge developed, many species of South American marsupials were driven to extinction by aggressive migrants from the north. Among the many interesting evolutionary developments of the Cenozoic, however, none seem more intriguing than the changes experienced by primates, which by late Tertiary had produced species believed to be the ancestors of humans. In the Pleistocene Epoch, our own species, *Homo sapiens*, appeared.

515

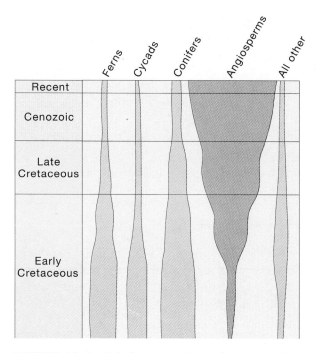

FIGURE 14–1   Relative proportions of genera in Cretaceous to Recent floras.

Plants and the Mammalian Response to the Spread of Prairies

Although they are the most recent plant group to evolve, the flowering plants are now the most widespread of all vascular land plants. Angiosperm floras did not explode upon the lands until mid-Cretaceous (Fig. 14–1). There were few spectacular floral innovations during the Cenozoic. Rather, this was a time of steady progress toward the development of today's complex plant populations. The Miocene is particularly noteworthy as the epoch during which grasses appeared and grassy plains and prairies spread widely over the lands. In response to the proliferation of this particular kind of forage, grazing mammals began their remarkable evolution.

Numerous evolutionary modifications among herbivorous mammals can be correlated to the development of extensive grasslands. Especially evident were changes in the dentition of grazing mammals. Grasses are abrasive materials. Many contain siliceous secretions, and because they grow close to the ground, they are often coated with fine particles of soil. To compensate for the wear that results from chewing grasses, the major groups of herbivores evolved high-crowned cheek teeth that continue to grow at the roots during part of the animals' lives. To provide space for these high-crowned teeth, the overall length of the face in front of the eyes increased. Enamel, the most resistant tooth material, became

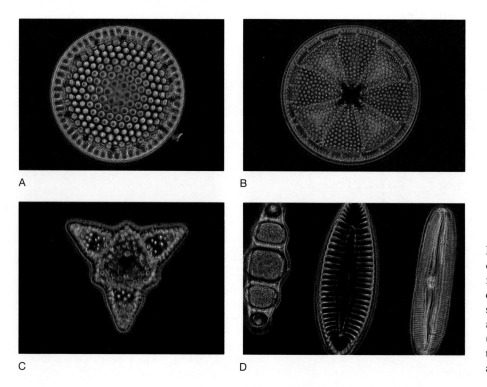

FIGURE 14–2   Cenozoic diatoms exhibiting various frustule shapes, including discoidal, triangular, and spindle-shaped. (A) *Arachnoidiscus*. (B) *Actinoptychus*. (C) *Triceratum*. (D) From left to right, *Biddulphia*, *Surirella*, and *Pinnulria* (×50).

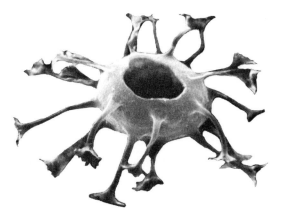

FIGURE 14–3 Dinoflagellate as seen with the aid of a scanning electron microscope. From Paleocene sedimentary rocks of Alabama. ×450 *(Courtesy of Standard Oil Company of California, photographer, W. Steinkraus.)*

folded, so that when the tooth was worn, a complex system of enamel ridges was formed on the grinding surface (see Fig. 4–19). In general, incisors gradually aligned into a curved arc for nipping and chopping grasses.

In the open plains environment, it was difficult to escape detection by predators. As a result, many herbivores evolved modifications that permitted speedy flight from their enemies. Limb and foot bones were lengthened, strengthened, and redesigned by the forces of selection to prevent strain-producing rotation and permit rapid fore-and-aft motion. To achieve greater speed, the ankle was elevated, and, like sprinters, animals ran on their toes. In many forms, side toes were gradually lost. Hoofs developed as a unique adaptation for protecting the toe bones while the animals ran across hard prairie sod. Herbivores called ungulates evolved the four-chambered stomach in response to selection pressures favoring improved digestive mechanisms for breaking down the tough grassy materials. The response of mammals to the spread of grasses provides a fine example of how the environment influences the course of evolution.

■ *Marine Phytoplankton*

As noted in Chapter 12, entire families of phytoplankton experienced extinction at the end of the Cretaceous. Only a few species in each major group survived and continued into the Tertiary. However, the survivors were able to take advantage of decreased competitive pressures and rapidly diversified. In general, peaks in diversity of species were reached in the Eocene and Miocene. A decrease in diversity has been recorded in the intervening Oligocene. Diatoms (Fig. 14–2), dinoflagellates (Fig. 14–3), and coccolithophores (Fig. 14–4) provided the most abundant populations of Cenozoic marine phytoplankton.

■ *Invertebrates*

The invertebrate animals of the Cenozoic seas included dense populations of foraminifers, radiolari-

FIGURE 14–4 A coccolith of the coccolithophorid *Coccolithus* as seen with the aid of the transmission electron microscope. Maximum diameter is 16 μm.

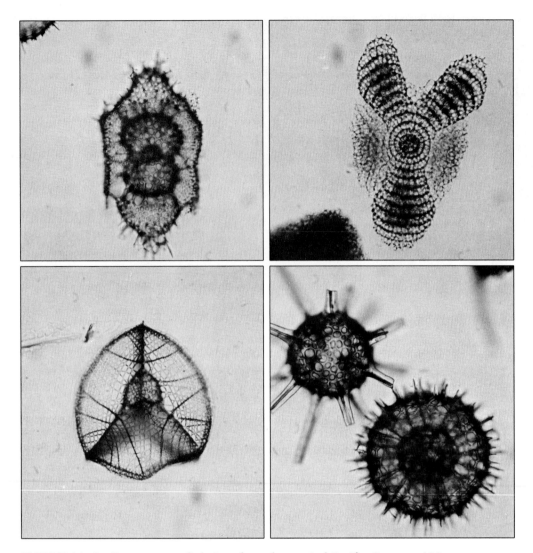

FIGURE 14–5   Quaternary radiolarians from the tropical Pacific Ocean, ×100. *(Courtesy of Annika Sanfilippo, Scripps Institute of Oceanography.)*

ans (Fig. 14–5), mollusks of all classes, corals, bryozoans, and echinoids. The combined fauna had a decidedly modern aspect. Once successful groups such as ammonites and rudistid bivalves were no longer present. No new major groups of invertebrates appeared during the Cenozoic.

Protozoans

Foraminifers were enormously prolific and diverse during the Cenozoic (Fig. 14–6) and included large numbers of benthic as well as planktic forms (Fig. 14–7). Among the larger genera, nummulitic foraminifers that were the size and shape of coins thrived in the clear Tethys seaway, as well as in warm water areas of the western Atlantic. The tests of these organisms have accumulated to form thick beds of nummulitic limestone. The ancient Egyptians quar-

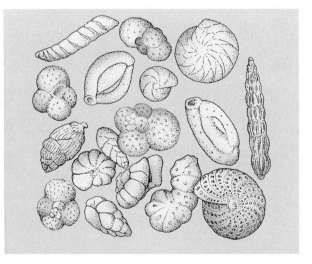

FIGURE 14–6   Diversity of form in Cenozoic foraminifers. Magnification ×50.

FIGURE 14–7   Present-day planktic foraminifers. (A) *Globorotalia tumida.*
(B) *Globigerinoides conglobatus.* (C) *Globigerinoides rubra.*   *(Courtesy of C. G.*
Adelseck, Jr., and W. H. Berger, Scripps Institute of Oceanography.)

ried this rock and used it to construct the Pyramids of
Gizeh. Even the famous Sphinx was carved from a
large residual block of nummulitic limestone. The
incredible numbers and variety of foraminifers have
permitted their extensive use in correlating Cenozoic
strata in oil fields of the Gulf Coast, California, Venezuela, the East Indies, and the Near East.

Corals

Corals (Figs. 14–8 and 14–9) grew extensively in the
warmer waters of the Cenozoic oceans. Fossil solitary
corals are commonly found in shallow-water deposits
of Early Tertiary age in Europe and the United States
Gulf Coastal region. Reef corals were most extensively developed in parts of the Tethyan belt, West
Indies, Caribbean, and Indo-Pacific regions. Careful comparison of coral species in Cenozoic rocks on
either side of the Isthmus of Panama has given geologists clues to when the Atlantic and Pacific
oceans were connected across this present-day barrier.
Continuous deposition of reef limestones occurred
throughout the Cenozoic in atolls of the open Pacific.
Atolls are ringlike coral reefs that grow around a central lagoon. In 1842, Charles Darwin brilliantly discerned the relationship between atolls and volcanic
islands. In his subsidence theory, he proposed that,
because of their great weight, volcanic islands subside

FIGURE 14–8   *Fungia,* a solitary coral that takes its
name from its resemblance to the underside of the cap
of a mushroom. It has a geologic range from the Miocene to the present and is common today in the seas of
the Indo-Pacific region. This specimen is about 9.0 cm in
diameter.

FIGURE 14–9   The living coral *Montastrea cavernosa,*
with polyps extended for feeding.   *(Courtesy of Charles*
Seaborn.)

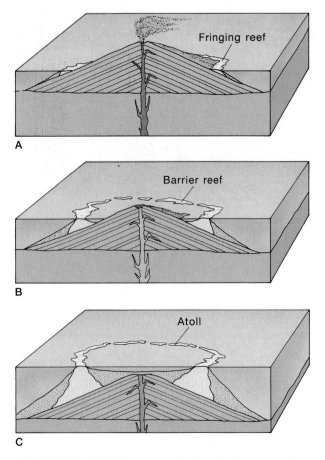

FIGURE 14–10 Three stages in the development of an atoll. (A) In initial stage, a fringing reef develops around the shoreline of a volcanic island. (B) The island begins to subside, and corals build upward to stay in their optimal shallow water life zone. The result is the development of a barrier reef backed by lagoons. (C) As subsidence continues, the original land area has become inundated, and a circle of reefs and coralline islands remains.

slowly. Furthermore, such subsidence is usually slow enough to permit the corals that grow around the fringe of the sinking island to grow upward and maintain their optimum habitat near the ocean's surface. Eventually, the central island would be submerged, but the encircling, living reef would persist (Fig. 14–10). Several deep borings into atolls have validated Darwin's theory. A boring drilled at Eniwetok island, an atoll in the Marshall Islands of the west Pacific, encountered the basaltic summit of the volcano at depths of about 1200 meters. Eocene reefs lie directly above these igneous rocks.

It is likely that the subsidence associated with the development of some atolls is related to sea-floor spreading rather than to the weight of the volcanic mass, as Darwin thought. Spreading begins along the crest of a midoceanic ridge, and as the plates move away from the ridge, they subside very slowly due to thermal contraction and isostatic adjustment. The subsidence is estimated at about 9 cm per year for the first million years. Thus, a volcano formed at the crest of the midoceanic ridge is carried down the slight incline of the ocean plate until it is completely submerged. Coral growth that is able to keep pace with the subsidence will result in the formation of barrier reefs and atolls (Fig. 14–11).

There is an alternate method for atoll formation. It is possible that corals might build **fringing reefs** around the perimeters of volcanic islands that had been erosionally truncated during a glacial period of low sea level. If sea level subsequently rose (as it did when the great Pleistocene ice sheets began to melt), then the corals would build the reef upward in order to stay at their optimum living depth. It seems not

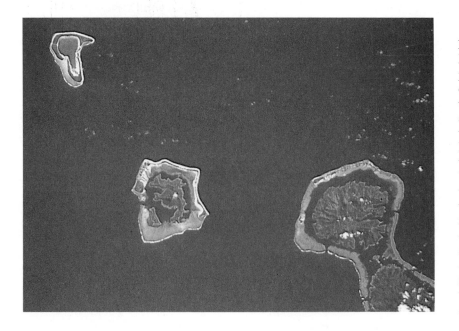

FIGURE 14–11 The Leeward Islands of the French Society Islands provide an illustration of the origin of coral reefs in deep oceans. At the bottom right, corals have constructed a fringing reef around the volcanic islands of Tahaa and Raiatea. Moving northward, the volcanoes are older and surrounded by barrier reefs. Bora Bora, near the center of the photograph, illustrates this phase of reef development. Tupai, visible in the top left of the photograph, is an atoll, and all that remains of the volcano around which the reef originally grew lies below the shallow floor of the central lagoon. *(Courtesy of L. E. Davis.)*

unreasonable that particular atolls may be primarily the result of subsidence of the island, whereas others may have been triggered by a rise in sea level.

Mollusks

Cenozoic shells of mollusks look very much like those likely to be found along coastlines today. Bivalves and gastropods were the dominant classes of Cenozoic mollusks (Fig. 14–12). Their range of adaptation was truly outstanding. Arcoids, mytiloids, pectinoids, cardioids, veneroids, and oysters were particularly abundant bivalves. Although the climax of cephalopod evolution had passed with the demise of the ammonites, nautiloids similar to the modern pearly nautilus lived in Tertiary seas, as they do today. Shell-less cephalopods, such as squid, octopi, and cuttlefish, were also well represented in the ma-

FIGURE 14–12 Common Cenozoic bivalves (A, B, C, D) and gastropods (E, F). (A) *Cardium.* (B) *Pecten.* (C) *Ostrea.* (D) *Mya.* (E) *Turritella.* (F) *Crepidula,* a small marine snail having a horizontal shelf that extends across the posterior part of the shell.

FIGURE 14–13   The freshwater shrimp *Bechleja rostrata* from the Eocene Green River Formation. *(Photograph courtesy of Rodney M. Feldmann. From Feldmann, M. et al. J. Paleontol. 55[4]:788–799, 1981.)*

rine environment, although their fossil record is understandably sparse.

Echinoderms, Bryozoans, Crustaceans, and Brachiopods

Many other invertebrates have continued successfully through the Cenozoic. Echinoderms, mainly free-moving types, were particularly prolific. Members of the phylum Bryozoa are common in Tertiary rocks and are still very abundant in many parts of the ocean today. It was during this final era that the modern crustaceans became firmly established both in fresh-water bodies (Fig. 14–13) and in the oceans. Brachio-pods declined in abundance and diversity during the Cenozoic. Fewer than 60 genera survive today. They consist mostly of terebratulids (Fig. 14–14A), rhynchonellids, and such inarticulate brachiopods as *Lingula* (Fig. 14–14B).

■ *Vertebrates*

Fishes and Amphibians

Bony fishes that evolved to the highest level of ossi-fication and skeletal perfection are the **teleosts.** Teleosts have achieved an enormous range of adaptive radiation during the Cenozoic. That radiation has produced such varied forms as perches, bass, snappers, seahorses, sailfishes, barracudas, sword-fishes, flounders, and others too numerous to mention. The Green River strata of Wyoming are well known for their content of beautifully preserved Eo-cene teleosts (see Fig. 13–11). Eocene strata at Monte Bolca in Italy also yield well-preserved teleost fossils (Fig. 14–15).

In addition to the bony fishes, the cartilaginous sharks were at least as common in the Tertiary as they are today. *Carcharodon,* an exceptionally large shark, had teeth as big as a man's hand and was more than 12 meters long.

Amphibians throughout the Cenozoic have re-sembled modern forms. All have been small-bodied,

A

B

FIGURE 14–14   (A) Cluster of present-day articulate terebratulid brachiopods, *Terebratulina septentrionalis.* (B) *Lingula,* a persistent primitive inarticulate brachiopod with a thin shell of protein-aceous material and a long, fleshy, muscular pedicle.

FIGURE 14–15 The Eocene flat-bodied fish *Gasteronomus* from the Mt. Bolla fossil locality near Verona, Italy. The specimen is 42 cm long. *(Photograph by John Simon, with permission.)*

smooth-skinned creatures not at all like their Paleozoic ancestors. Frogs, toads, and salamanders were relatively abundant. The first frogs appeared during the Triassic, and by Jurassic time they were already completely modern in appearance. Thus, they have continued almost unchanged for more than 200 million years.

Reptiles

By the beginning of the Cenozoic, the dinosaurs had disappeared from the lands, as had the flying reptiles from the air and several groups of marine reptiles from the seas. The reptiles that have managed to survive and continue to the present include the tuatara, which inhabits islands off the coast of New Zealand, as well as turtles, crocodilians, lizards, and snakes (Fig. 14–16). The tuatara, formally known as *Sphenadon,* resembles a large lizard. It is the sole survivor of a group of ancient reptiles known as rhynchocephalians that evolved and diversified during the Triassic.

FIGURE 14–16 Surviving reptiles. (A) monitor lizard, (B) hognose snakes emerging from their shells, (C) *Alligator mississippiensis,* (D) three-toed box turtle. *(B from Animals Animals © 1992 Zig Leszczynski; C from Ed Reschke; D from L. Stone/The Image Bank.)*

Cenozoic turtles are the descendants of a lineage that can be traced back into the Late Permian. Turtles are readily recognized by everyone because of their distinctive adaptations. The most apparent of these adaptations is the shell. In turtles, the ribs have expanded differentially to form a broad *carapace*. On the underside, a bony growth called the *plastron* provides a similar covering. Both carapace and plastron are covered with a horny sheath. The jaws in turtles are toothless and covered by a beak that is used effectively in slicing through plants or flesh.

Crocodilians also began their evolution during the Triassic and were very successful contemporaries of the dinosaurs throughout the Mesozoic. Modern crocodilians include the broad-snouted alligators, the narrow-snouted crocodiles, and the very narrow-snouted gavials.

Both lizards and snakes belong to an order of reptiles known as the Squamata. The Squamata are by far the most varied and numerous of living reptiles. Lizards are the ancestors of snakes. In fact, snakes are essentially modified lizards in which the limbs are lost, the skull bones modified into a highly flexible and mobile structure for engulfing prey, and the vertebrae and ribs greatly multiplied to provide the elongate form.

As evidence of their tetrapod ancestry, certain primitive snakes retain vestiges of rear limb and pelvic bones. Poisonous snakes evolved during the Miocene. They are characterized by specialized teeth or fangs for the injection of poison. One type of poison (neurotoxin) affects parts of the nervous system that control breathing and heart action, whereas a second type (hemotoxin) destroys red blood cells and causes disintegration of small blood vessels. It has been suggested that the evolution of poisonous snakes and the feeding behavior and skull structure of all snakes may be tied to the diversification of the Cenozoic mammals that served as prey.

Birds

The Cenozoic record for birds is generally poor because they are rarely preserved. The rather fragmentary record indicates, however, that birds have been essentially modern in basic skeletal structure since the beginning of the Cenozoic. Distinctive skeletal features of birds include the fusion of bones of the "hand" to help support the wing, development of a vertical plate or keel on the sternum for attachment of the large muscles leading from the breast to the wings, and fusion of the pelvic girdle and vertebrae to provide rigidity during flight. Other characteristics include a body covering of feathers, light and porous bones, jaws in the form of a toothless horny beak, a four-chambered heart, and constant body temperature. The avian fauna is composed of a rich variety of families, including song birds such as robins, up-

FIGURE 14–17 *Diatryma* from the Eocene (Wasatch Formation) of Wyoming. This giant killer bird stood nearly 3 meters high, and had a deep, powerful beak capable of delivering a bone-crushing bite to its prey.

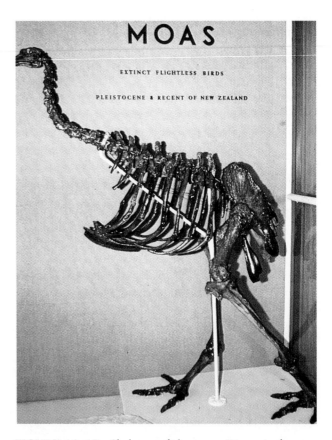

FIGURE 14–18 Skeleton of the moa, *Dinornis*, from the Pleistocene of New Zealand. This specimen is more than 2 meters tall.

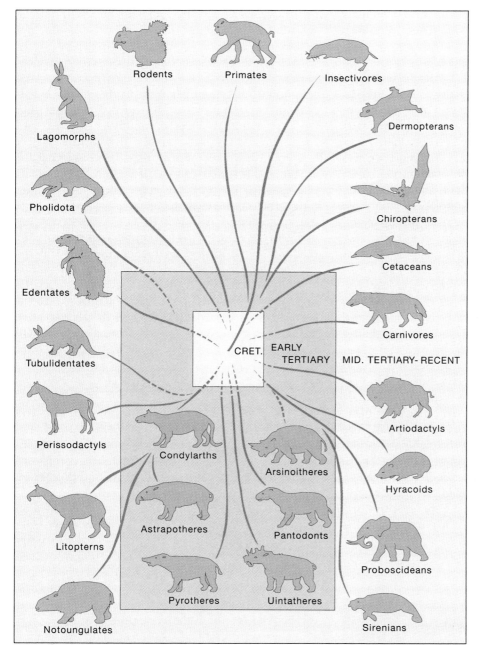

FIGURE 14–19 General relationships of the major orders of placental mammals. *(From Colbert, E. H. Evolution of the Vertebrates. New York: John Wiley, 1969. Used with permission of the author, artist Lois Darling, and publisher.)*

land birds (pheasants), forest birds (owls), oceanic birds (albatrosses), wading birds (plovers), flightless aquatic birds (penguins), and flightless land birds (ostriches).

The fossil record for large terrestrial flightless birds is somewhat better than for small flying varieties. *Diatryma* (Fig. 14–17), an Eocene representative of this group, stood 3 meters tall. In New Zealand, huge moas (Fig. 14–18) lived until relatively recent time. Some were over 3 meters tall and laid eggs that had a 2-gallon capacity. Perhaps the most famous of all Cenozoic ground birds was the dodo, which lived on the island of Mauritius east of Madagascar until about 1700, when it was exterminated by sailors searching for provisions. The moas of New Zealand were also exterminated by humans. Maori people killed them for food. The African ostrich, the South American rhea, and the emus and cassowaries of Australia are surviving flightless land birds.

Mammals

During the Cenozoic, mammals came to dominate the Earth in much the same way as reptiles had done during the preceding Mesozoic era (Fig. 14–19). As we noted in Chapter 12, however, the evolution of mammalian traits had already begun among the therapsids of the Permo-Triassic. The Karoo beds of Africa, for example, contain bones of near-mammals that had almost made the transition from reptile to mammal.

Just as birds are readily recognized by the possession of feathers, so are mammals by the possession of hair. Hair, however, is not an exclusive characteristic of mammals, for it was apparently also present in some mammal-like reptiles and such flying reptiles as *Psordes* (see Fig. 12–47). Like feathers, hair functions as an insulating body cover. Mammals are also recognized by the presence of mammary glands. However, neither body hair nor mammary glands are particularly helpful to the paleontologist, who must recognize mammalian remains on the basis of skeletal characteristics. Among these, the lower jaw is particularly useful. It consists of a single bone, the dentary, on either side (Fig. 14–20). In reptiles and birds, there are always several bones in the lower jaw. Another mammalian trait is the bony mechanism that conveys sound across the middle ear. In mammals, there is a chain of three little bones rather than a single one, as in lower tetrapods. Two of these ossicles, the incus and malleus, were derived from bones of the reptilian jaw. Unfortunately, these delicate ossicles are so small that they are rarely preserved.

Typically, mammals have seven cervical, or neck, vertebrae, regardless of the length of the neck. The skull is usually recognizable by the expanded braincase, and teeth are nearly always of different kinds, serving different functions in eating. As an aid to their endothermic metabolism, mammals have a well-developed secondary palate that separates the oral cavity from the nasal passages. This makes simulta-neous breathing and feeding possible. Without the secondary palate, an infant mammal could not suckle.

Paleontologists speculate that more rigorous climatic conditions during the Permo-Triassic favored selection of the mammalian traits of warm-bloodedness and postnatal care. All evidence indicates that the first mammals were diminutive creatures, and small animals lose heat rapidly. A high rate of heat loss could have been partly offset by insulating fur and an ample supply of food. If the earliest mammals laid eggs, as do primitive mammalian monotremes today, then the newly hatched young might also have had to cope with loss of body heat. They were very likely kept warm by snuggling next to the soft fur that may have formed an incubation patch on the female. Perhaps at the same time, they were nourished by secretions from glands that preceded the development of true mammary glands. Unfortunately, fossils provide few clues to the reproductive characteristics of early mammals, and theories will continue to rely heavily on inference.

Earliest Mammals

The mammalian fossil record begins with rare and difficult-to-find small jaw fragments, tiny teeth, and scraps of skull bone. The fossils, however, are sufficient to indicate that the earliest members of our own taxonomic class were very small, that they evolved from mammal-like reptiles such as those previously described from the late Paleozoic, and that, based on tooth patterns, they were insect eaters. Evidence from brain casts made of their inner cranial walls also suggests that the parts of their brains dealing with smell and hearing were particularly well developed, as in nocturnal animals. Perhaps the most significant and ironic fact about these tiny creatures is that they came on the scene at about the same time as the dinosaurs that they were destined to succeed. For 140 million years of dinosaur dominance, they quietly thrived as if waiting for just the right conditions for their own biological triumph.

Largely on the basis of tooth morphology (see Fig. 12–59), the early mammals are separated into categories bearing the names morganucodonts, docodonts, triconodonts, symmetrodonts, eupantotheres, and multituberculates. The oldest remains are of **morganucodonts** from Upper Triassic beds of Great Britain and from recent discoveries in South Africa (see Fig. 12–58). All of the groups just named lived during the Mesozoic, and the rodent-like **multituberculates** managed to cross the era boundary and live until Eocene time. The **docodonts** seem the possible ancestors of modern monotremes. However, of all the Mesozoic groups, the **eupantotheres** were of

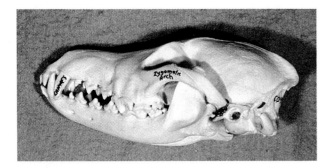

FIGURE 14–20 Skull of a coyote. One bone on either side, the *dentary*, forms the lower jaw. As is characteristic of most mammals, teeth are not identical in shape. Instead there are incisors at the front of the jaw, followed by dagger-like canines for stabbing, carnassial teeth that shear past one another like scissors for cutting meat into smaller pieces, and more robust back teeth for crushing tougher objects and bone. Note also the ample size of the braincase.

TABLE 14–1 Evolutionary Modifications of Mammals for a Variety of Habitats*

Habitat	Limbs	Teeth	Other Features
Primitive walking (ambulatory)	Limbs generalized; feet flat on ground	Incisors for grasping; canines for piercing; premolars for cutting; molars for grinding	Tongue, stomach, and intestine generalized; cecum small
Running herbivores (cursorial) Horse and cow	Limbs elongate, toes elongate and reduced in number; hoof of horse is one toenail; in cow, two toenails	Incisors for nipping grass; cheek teeth heavy and ridged to withstand wear of grass	Stomach (cow) or intestine (horse) large and complex; cecum long
Running carnivores (cursorial) Wolf, cheetah	Limbs elongate; often walk on toes; claws often long and sharp	Canines well-developed; premolars sharp for cutting flesh	Stomach large, intestine short
Heavy body Elephant	Bones in limbs massive; flat jointed; toes in circle around pad	Cheek teeth massive and ridged for grinding plant materials	Nose elongated into trunk to reach food on ground or in trees
Tree-climbing (arborial) Tree squirrel, monkey	Limbs elongate; wide range of motion; toes long, flexible	Cheek teeth usually smooth surfaced and flattened for crushing	Tail sometimes prehensile; diet usually fruit or vegetation
Diet of small, colonial insects Anteaters, spiny echidna	Claws enlarged for digging after insects	Reduced or absent	Tongue long, sticky, and extensile
Burrowing (fossorial) Moles, marsupial moles Various rodents	Limbs short and stout; forefeet often enlarged, shovel-like; claws long and sharp	Teeth various, depending on diet (vegetation or insects and worms)	Pinna of ears usually short or absent; fur short; tail short
Aquatic Whales, seals, sea cows	Limbs shortened; feet enlarged and paddle shaped or absent	Often reduced to simple, conelike structures, sometimes absent	Tail often modified into swimming organ
Volant Flying phalangers, flying squirrels	Limbs extendible to sides to hold gliding membrane outstretched	Various but usually show herbivorous adaptations	Tail often flattened for use in balance
Flying Bats	Forelimbs and especially fingers of forelimbs elongate to support wing	Various for diets of insects, fruits and nectar, or other specific foods	Tongue, stomach, and intestine various, correlated with diet

*From Cockrum, E. L., and McCauley, W. J. *Zoology.* Philadelphia: W. B. Saunders Co., 1965.

the greatest importance, for during the Cretaceous they gave rise to both the marsupial and the placental mammals.

The stage was thus set for the varied mammalian groups that evolved and spread across seas and continents during the Cenozoic. The absence of dinosaurs and other groups of previously numerous reptiles favored a vigorous mammalian radiation. Evolutionary experimentation among Mesozoic ancestors had provided Cenozoic mammals with more efficient nervous and reproductive systems, greater speed and agility, reliable systems of body temperature control, larger brains, and levels of animal intelligence that far

exceeded those of earlier classes of vertebrates. Armed with these attributes, mammals quickly expanded into the habitats vacated by the reptiles and found new pathways of adaptation as well (Table 14–1). By Paleocene time, 18 mammalian orders had appeared, and throughout the early Tertiary, there was an intricate interplay of appearances of new groups and extinctions of older groups that ultimately produced more than 30 taxonomic orders of mammals. The list is far too long to treat each group, but a brief review of some of the more interesting forms suggests the spectacular variety of the Cenozoic mammalian fauna.

A   B

FIGURE 14–21 Two present-day monotremes. (A) The duck-bill platypus of Australia and Tasmania. (B) The spiny anteater of Australia. *(A from Visuals Unlimited, © E. C. Williams.)*

Monotremes

The most primitive of all living mammals are called **monotremes.** These relics of an older time still lay eggs in the reptilian manner. However, unlike reptiles, monotremes have primitive mammary glands and provide their newly hatched offspring with nourishment, if only for a brief period. The oldest fossil remains of monotremes are from beds of the lower Cretaceous. The platypus of Australia and Tasmania and two species of spiny anteaters of New Guinea and Australia (Fig. 14–21) are the only three species of monotremes still in existence.

Marsupials

Marsupials are mammals that nurture their young in a special pouch, or marsupium (Fig. 14–22). Today, they are represented by such animals as kangaroos,

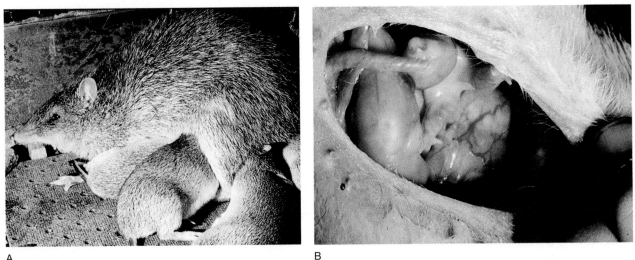

A   B

FIGURE 14–22 The short-nosed bandicoot, a common marsupial of Australia. (A) Advanced offspring that have left the pouch return for nursing. (B) A fetal bandicoot attached to a teat in the mother's pouch.

Marsupial "mouse"

Phalanger

Tasmanian "wolf"

Wombat

Koala

Bandicoot

Marsupial "mole"

Kangaroo

Opossum

FIGURE 14–23 Adaptations of marsupials for various habitats is evident in their diversity. (The Tasmanian wolf is now thought to be extinct.)

wallabies, wombats, phalangers, bandicoots, koalas of Australia, and opossums of the New World (Fig. 14–23). Compared with the placentals, it is apparent that they are a dwindling group.

Marsupials had their greatest success in Australia and South America. Both continents were more or less isolated during most of the Tertiary, and marsupials were able to evolve without excessive competition from placentals. Although a few middle Tertiary Australian mammal sites have recently been found, the fossil record of Australian marsupials is not good until Pleistocene time. An older and more complete record exists in South America, where marsupials evolved from opossum-like ancestors and produced an array of mostly flesh-eating types. *Thylacosmilus* (Fig. 14–24), a large Pliocene form, was similar in appearance to the North American saber-toothed cats. South American marsupials fared poorly in competition with placental immigrants from the north when North and South America became connected in the early Pleistocene.

Placental Mammals

Placental mammals appear during the Cretaceous as small, unspecialized insectivores. Modern moles are

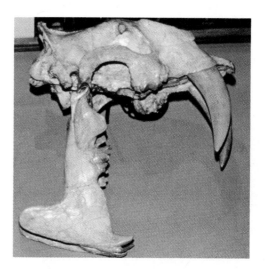

FIGURE 14–24 *Thylacosmilus,* a Pliocene South American carnivorous marsupial comparable to the placental saber-toothed cats. The bladelike upper canine teeth were about 18 cm long and were protected when the jaws were closed by a deep flange of bone in the lower jaw.

FIGURE 14–25 The common tree shrew most resembles the ancient insectivores that gave rise to the primates. *(Warren Garst/Tom Stack & Associates.)*

FIGURE 14–27 The great ground sloth *Megatherium*, which lived during the Pleistocene and was nearly as large as a present-day elephant. This specimen is on display at The Field Museum in Chicago.

members of the order **Insectivora,** but the tiny shrew (Fig. 14–25) is more representative of the kind of animal from which other orders of Cenozoic placentals evolved. The descendants of the insectivores include the edentates, bats, primates, rodents, flesh-eating mammals, a host of herbivores, and various marine mammals.

Armadillos, tree sloths, and South American anteaters are **edentates** that have not become extinct. Fossil species of armadillos have been found in rocks as old as Paleocene. Among the extinct edentates are the glyptodonts, which survived up until the Pleistocene. *Glyptodon* (Fig. 14–26) was a walking fortress with a spike-covered knob on its tail for bludgeoning pesky predators. Quite unlike the glyptodonts were the ungainly Cenozoic ground sloths. This group of

edentates includes truly colossal Pleistocene beasts such as *Megatherium* (Fig. 14–27).

The rodents have been exceptionally successful Cenozoic mammals. Today, they probably outnumber all other mammals and have invaded every habi-

FIGURE 14–26 Restoration of a scene during the Middle Pleistocene in Argentina. The heavily armored animals are glyptodonts. On the left is a giant ground sloth. *(Copyright © The Field Museum, painting by C. R. Knight, with permission.)*

FIGURE 14–28 The Paleocene rodent *Paramys,* which resembled a large squirrel in size and form.

tat of the Earth. Their diversification has produced such burrowers as the marmot, the partially aquatic muskrat and beaver, the desert-dwelling jerboas and kangaroo rats, and the arboreal squirrels. The fossil record of rodents begins in the Paleocene with squirrel-like forms, such as *Paramys* (Fig. 14–28), and the horned rodent *Ceratogaulus.*

Beavers are well represented among fossil rodents. During the Pleistocene, beavers, along with many other mammals, grew to exceptional sizes. Some were as large as modern large bears. The other gnawers and nibblers of the Cenozoic that are not classified as rodents are the rabbits, hares, and pikas. The fossil record for these **lagomorphs** begins with skeletal fragments of a primitive rabbit found in Paleocene beds of Mongolia.

Bats were also evolving during the Cenozoic. Their teeth have been discovered in Paleocene strata of France. A well-preserved skeleton was recovered from Lower Eocene strata in Wyoming and indicates that by that time, bats were already very similar to their living relatives. Bats are the only mammals to have achieved true flight. They have greatly elongated finger bones for the support of the membrane that forms the wing. Bats fly mostly at night when insect prey are abundant in the air. Although they occur in temperate zones, they are most abundant in the tropics. Vampire bats, which live in South America, feed on the blood of other mammals. Their front teeth are adapted to pierce the skin of their prey. When blood oozes from the wound, it is lapped up by the bat.

The history of flesh-eating placental mammals began in the Cretaceous with the advent of small, weasel-like animals called **creodonts** (Fig. 14–29). They were soon joined by members of the order **Carnivora,** which have common ancestry with the

FIGURE 14–29 The Eocene creodont, *Patriofelis. (Courtesy of the National Museum of Natural History, Smithsonian Institution.)*

Creodonta. It appears that toward the end of Eocene time, with the coming of speedier and more progressive plant-eaters, the creodonts gradually lost ground to carnivores. Most became extinct by the end of the Eocene, although one hyena-like group persisted until Pliocene time. The expansion of the Carnivora accelerated during the late Tertiary, producing a host of familiar flesh-eaters, including bears, raccoons, weasels, genets, hyenas, dogs, and cats (Fig. 14–30). Wild dogs were doing very well on Earth long before humans came along to make companions of them. The modern genus *Canis*, in the form of the dire wolf, was much in evidence during the Pleistocene. The most famous of the extinct cats are the so-called stabbing cats, exemplified by *Smilodon* (Fig. 14–31). This robust flesh-eater preyed on the larger herbivorous animals of the Pleistocene. A second line of cat evolution produced the biting cats, which resembled

modern leopards and pumas. The skeletons of these carnivores indicate that they were strong, speedy, and agile predators.

As is true of carnivores today, Cenozoic carnivores were an essential element in the evolutionary process. To survive, they had to equal or better the speed and cunning of the herbivores on which they fed. The herbivores, in turn, responded to the carnivore threat by evolving adaptations for greater speed and defense. Then, as now, predators were not villains but necessary constituents of the total biological scheme. They were able to cull out the weak, deformed, or sickly animals and thereby help to counteract the effects of degenerative mutation and to prevent overpopulation.

Not all the Carnivora of the Cenozoic were land dwellers. Some, such as seals, sea lions, and walruses, gathered their food in or at the edge of the sea. As

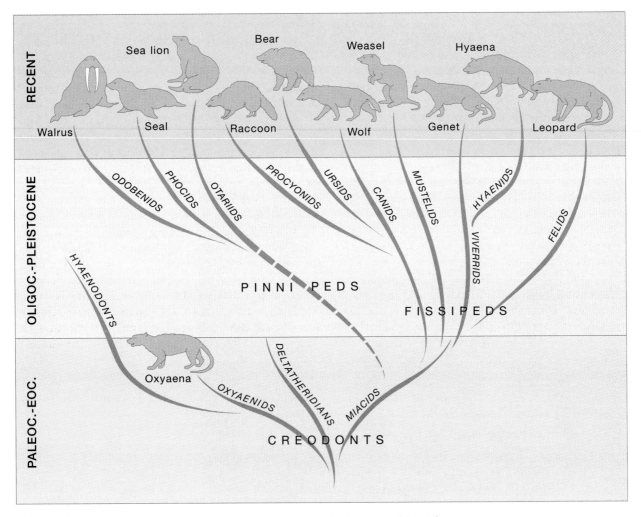

FIGURE 14–30 General relationships of major groups of flesh-eating placental mammals. *(From Colbert, E. H. Evolution of the Vertebrates. New York: John Wiley, 1969. Used with permission of the author, artist Lois Darling, and publisher.)*

FIGURE 14–31 The saber-toothed cat *Smilodon.*

indicated by their sharp, pointed teeth, early Cenozoic seals and sea lions ate fish as they do today. The walrus had (in addition to tusks) broad, flat teeth for crushing the shells of mollusks. Seals, sea lions, and walruses are known as **pinnepeds.** They probably descended from semiaquatic mammals somewhat similar to present-day otters, but the transitional forms are as yet undiscovered.

In adaptation to the marine environment, **cetaceans** (whales and porpoises) have made the most complete adjustment. Whales are descendants of carnivorous land animals, and the earliest whales known had legs and could both swim and walk. Among these earliest whales, *Pakicetus* lived in Eocene streams, lakes, and estuaries about 50 million years ago. *Pakicetus* was succeeded about 40 million years ago by another whale that had legs and was known as *Ambulocetus.* It was about the size and weight of a large sea lion. *Basilosaurus* (Fig. 14–32), from the Late Eocene of Egypt, continued the trend toward increased size by attaining a length of 60 feet. Tiny hind limbs were still present in *Basilosaurus.* They were so small, however, that they were useless in

swimming. Possibly the animal used them to guide and grasp its partner during copulation.

Modern whales arose from the group that included *Basilosaurus* and divided into two lineages: the toothed whales and the whale-bone whales. Toothed whales, which made their debut during the Oligocene, today include porpoises, killer whales, and sperm whales like Melville's famous Moby Dick. The titanic blue whale, right whale, and Greenland whale are all representatives of the second cetacean group. These plankton-feeding giants first appeared in the Miocene. Instead of teeth, they possess ridges of hardened skin that extend downward in rows from the roof of the mouth. The ridges are fringed with hair, which serves to entangle the tiny invertebrates on which these cetaceans feed—thus, the paradox of the largest of all animals feeding on some of the smallest. The great blue whales far exceed in size even the largest dinosaurs, for some have attained lengths of 30 meters and weights of over 135 metric tons.

The largest category of Cenozoic plant-eaters are the **ungulates.** Simply defined, ungulates are animals that walk on hoofs and feed on plants. The earliest

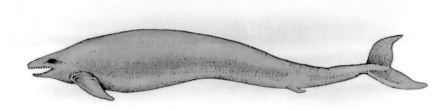

FIGURE 14–32 The Eocene whale *Basilosaurus.* The tendency toward increase in size was already evident in this whale, which was over 20 meters long.

FIGURE 14–33 The Early Tertiary archaic herbivorous mammal *Phenacodus* was a member of a group of primitive plant-eaters called condylarths. Like most of the Early Tertiary herbivores, *Phenacodus* walked on all five toes in a method termed plantigrade. Speed was not essential to this forest-dwelling browser. *(Courtesy of the National Museum of Natural History, Smithsonian Institution.)*

ungulates were members of a group called condylarths. *Phenacodus* (Fig. 14–33), a representative condylarth, had simple primitive teeth and five toes, each terminated with a small hoof. The animal walked mostly in a flat-footed fashion, for these un-

gulates had not yet evolved the ability to walk or run on their toes for greater speed. This type of stance, termed **plantigrade,** was characteristic of Early Tertiary ungulates, including the huge six-horned *Uintatherium* (Fig. 14–34). Primitive plantigrade and rel-

FIGURE 14–34 Eocene mammals. (A) *Uintatherium* is the large six-horned and tusked animal in the upper right. Other animals on this restoration are (B) the small, fleet rhinoceros *Hyrachus;* (C) *Trogosus,* a gnawing-toothed mammal; (D) *Mesonyx,* a hyena-like flesh-eater; (E) *Stylinodon,* a gnawing-toothed mammal; (F) three early members of the horse lineage *(Orohippus);* (G) a saber-toothed mammal, *Machaeroides;* (H) *Patriofelis,* an early carnivore; and (I) *Palaeosyops,* an early titanothere. Restorations are based on skeletal remains from the Middle Eocene Bridger Formation of Wyoming. *(Courtesy of the National Museum of Natural History, Smithsonian Institution, J. H. Matternes mural, with permission.)*

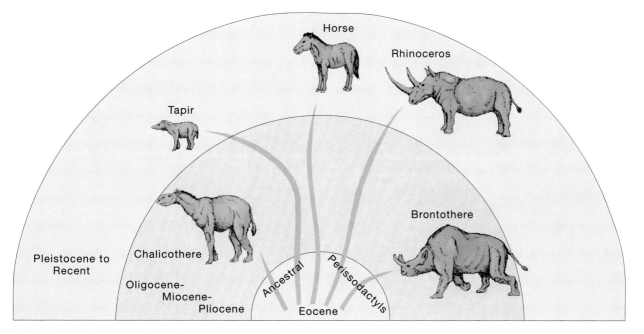

FIGURE 14–35   Generalized diagram of the evolution of perissodactyls.

atively small-brained plant-eaters are often termed **archaic ungulates.**

The modern and more familiar ungulates fall into two categories: the **perissodactyls** and the **artiodactyls.** The perissodactyls include the modern horses, tapirs, and rhinoceroses and the extinct chalicotheres and titanotheres (Fig. 14–35). They seem to have originated from condylarth ancestors and reached the peak of their evolutionary history during the Miocene Epoch. Since that time, they have declined steadily.

Perissodactyls have certain distinctive characteristics. For example, the number of toes on each foot is usually odd, and the axis of the foot, along which the weight of the body is primarily supported, lies through the third, or middle, toe. There is a tendency toward reduction of the lateral toes. In the modern horse, only the single, central toe remains. Perissodactyls are clearly **digitigrade,** meaning that they run or walk on their toes in order to attain a longer stride and greater speed.

The oldest known perissodactyl, *Hyracotherium* (Fig. 14–36), has been found in upper Paleocene and Eocene strata of both North America and England. Its ancestors were probably condylarths. Familiarly known as eohippus, *Hyracotherium* has the distinc-

FIGURE 14–36 *Hyracotherium,* the small Eocene horse more popularly known as eohippus, the "dawn horse." This tiny horse was only about a half meter in length. *(Courtesy of the American Museum of Natural History.)*

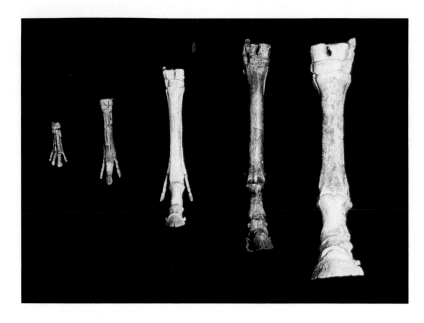

FIGURE 14–37   Evolution of the lower foreleg in horses, beginning at far left with *Hyracotherium* and ending with the modern horse *Equus* at the far right.   *(Courtesy of the National Museum of Natural History, Smithsonian Institution.)*

tion of being the earliest member of the horse family. This little horse was not much larger than a fox. It had four hoofed toes on the front feet (Fig. 14–37) and three on the hind feet. The back curved like that of a condylarth, and the dentition was primitive, with canines still present and the premolar teeth not similar to the molars, as they are in more recent perissodactyls. In addition, the molars were bluntly cusped for browsing and had not yet developed the ridged occlusion surface and high crowns that characterize modern grazers.

From ancestors such as *Hyracotherium,* the horse family evolved, at least until the Miocene, in a rather straightforward manner (see Fig. 4–18). The skeletal

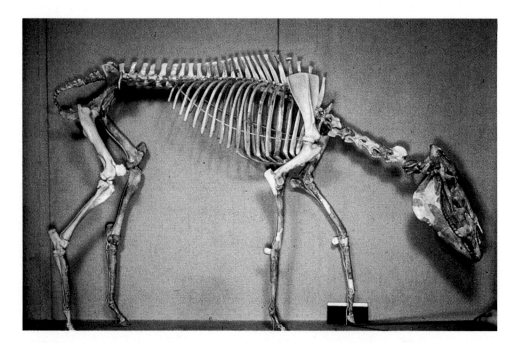

FIGURE 14–38   The Pliocene horse *Pliohippus.* Modern horses belonging to the genus *Equus* are descendants of *Pliohippus.*   *(Courtesy of the National Museum of Natural History, Smithsonian Institution.)*

FIGURE 14-39 The giant Oligocene to early Miocene hornless rhinoceros *Baluchitherium.*

remains clearly show progressive increases in size, length of legs, height of crowns of cheek teeth, and brain size. The curved back became straightened, and the middle toe was strengthened and emphasized at the expense of lateral toes (see Fig. 14-37). The premolars came to resemble the molars, and their grinding surfaces developed increasingly complicated patterns of resistant enamel ridges. *Mesohippus,* an Oligocene horse about twice the height of *Hyracotherium,* showed the beginnings of many of these trends but still retained relatively low-crowned teeth best adapted for browsing.

With the spread of grass-covered prairies during the Miocene, the family tree of horses became more complicated as more conservative species stayed in forested areas while others took to the more open environment of the prairies. *Merychippus,* a horse in the vanguard of this new group, ran on a single middle toe; the side toes were much reduced and of little use. The head was deeper and the face longer to accommodate longer cheek teeth. Later in the Miocene, a group of three-toed horses typified by *Hipparion* made their way from North America into Europe, where they survived until the beginning of the Pleistocene Epoch. A more progressive branch of the horse family tree produced *Pliohippus* (Fig. 14-38), a completely one-toed horse with lateral digits reduced to mere splints concealed beneath the skin. By Pliocene time, the descendants of *Merychippus* had divided into two distinct lineages. Horses of one branch, represented by *Hippidium,* migrated across the Panamanian land bridge into South America. The other

branch led to the *Equus,* the genus to which all modern horses belong.

During the Pleistocene, several different species of horses thrived. Some, such as *Equus occidentalis,* were the size of small ponies, but others, such as *Equus giganteus,* were larger than the great draft horses of today. They spread across the length of North America, Eurasia, and Africa. Only a few thousand years ago, horses suffered extinction in North America. (The continent was later restocked with the progeny of domestic horses that had escaped from the Spanish explorers in the sixteenth century.) The cause of the extinction of horses in North America is somewhat of a mystery. Some believe it was brought on by contagious disease, whereas others speculate that the cause was overkill by prehistoric human hunters.

Among the other surviving perissodactyls are the tapirs and rhinoceroses. Tapirs retain the primitive condition of four toes on the forefeet and three on the rear, as well as low-crowned teeth. They are forest-dwelling, leaf-eating animals whose fossil record begins in the Oligocene. More impressive to humans are the rhinoceroses, which began their evolution during the Eocene as small, swift-running creatures and eventually produced such giants as *Baluchitherium* (Fig. 14-39). *Baluchitherium* is the largest land mammal thus far discovered. It stood 5 meters tall at the shoulders.

The perissodactyls that did not survive into modern times were the **titanotheres** and **chalicotheres.** The most familiar titanothere of the Cenozoic was

FIGURE 14–40   A restoration of Oligocene mammals based primarily on skeletal remains from the White River Formation of South Dakota and Nebraska. The flora is based on nearly contemporaneous plant fossils from the Florissant beds of Colorado. (A) *Trigonias,* an early rhinoceros. (B) *Mesohippus,* a three-toed horse. (C) *Aepinocodon,* a remote relative of the hippopotamus. (D) *Archaeotherium,* an entelodont. (E) *Protoceras,* a horned ruminant. (F) *Hyracodon,* a small, fleet rhinoceros. (G) The giant titanothere *Brontotherium.* (H) Oreodonts named *Merycoidodon.* (I) *Hyaenodon,* an Oligocene carnivore. (J) *Poëbrotherium,* an ancestral camel.   *(Courtesy of the National Museum of Natural History, Smithsonian Institution, J. H. Matternes mural, with permission.)*

ponderous *Brontotherium* (Fig. 14–40G), readily remembered because of the pair of hornlike processes that grew over the snout. Chalicotheres differed from all other Cenozoic perissodactyls in having three claws rather than hoofs on their feet. These odd creatures (Fig. 14–41A) were rather similar to a horse in the appearance of the head and torso. The fore legs were longer than the hind legs, and the back sloped rearward. They lived from Eocene into the Pleistocene, at which time they became extinct.

The even-toed ungulates, or artiodactyls, have been far more successful than the perissodactyls in terms of survival, variety, and abundance (Fig. 14–42). Modern artiodactyls include pigs, deer, hippos, goats, sheep, cattle, and camels. The center of artiodactyl evolution was not North America, as had been the case with the odd-toed ungulates, but rather Eurasia and Africa. Artiodactyls were already present during the Eocene, and by late Tertiary time they had clearly achieved numerical and varietal superiority over other herbivores.

Most artiodactyls have an even number of toes on each foot, either four or two. Most of the weight of the animal is carried on the two middle toes, which form an identical pair and thus a cloven, or split, hoof. Unlike advanced perissodactyls, the molar and premolar teeth are dissimilar. Among the Tertiary artiodactyls that became extinct, the **oreodonts** and **entelodonts** are particularly interesting. Oreodonts (see Fig. 14–40H) were short, stocky grazers that roamed the grassy plains of North America in enormous numbers. Entelodonts, some of which were as large as buffalo, were repulsive-looking, hoglike beasts. Their trademarks were curious bony pro-

FIGURE 14–41 Mural depicting an assemblage of early Miocene mammals.
(A) The chalicothere *Moropus*. (B) The small artiodactyl *Merychyus*.
(C) *Daphaenodon*, a large wolflike dog. (D) *Parahippus*, a three-toed horse.
(E) *Syndyoceras*, an antelope-like animal. (F) *Dinohyus*, a giant, piglike en-
telodont. (G) *Oxydactylus*, a long-legged camel. (H) *Stenomylus*, a small camel.
*(Courtesy of the National Museum of Natural History, Smithsonian Institution, J. H.
Maternes mural, with permission.)*

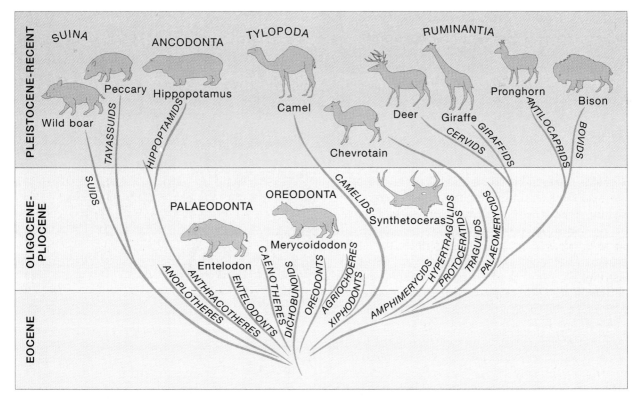

FIGURE 14–42 The evolutionary radiation of the even-toed ungulates, or artiodactyls. *(From Colbert, E. H. Evolution of the Vertebrates.* New York: John Wiley,
1969. Used with permission of the author and the publisher.)

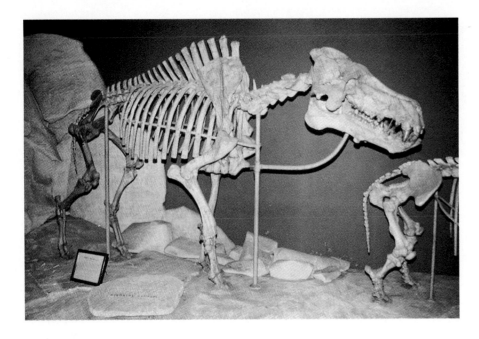

FIGURE 14–43 *Dinohyus,* a giant piglike entelodont from the Miocene of America. The animal was about 3 meters long. *(Courtesy of the Denver Museum of Natural History.)*

cesses that grew along the sides of the skull and jaws (Fig. 14–43; also see Fig. 14–41F).

The major radiation of artiodactyls began in the Eocene and ultimately produced the three major surviving categories: swine, camels, and ruminants. Of these, the swine family has probably remained the most primitive. They have kept their four toes, even though most of the weight is carried by the two middle digits. The swine group includes both pigs and the more lightly constructed and primarily South American peccaries. Hippopotami are the only modern amphibious artiodactyls. They are relatives of the pig family, having arisen from a group of Miocene piglike animals called anthracotheres.

People are often surprised to learn that much of the evolutionary development of the camel lineage—**Tylopoda**—occurred in North America. The geologic history of camels and llamas began in Eocene time with tiny creatures about the same size as a small goat. As they evolved through the Tertiary, they lost their side toes and increased the length of their legs and neck. The long-necked and long-limbed *Oxydactylus* (Fig. 14–44) of the Miocene probably browsed on leaves in much the same manner as the giraffe. By Pleistocene time, there were numerous modern-looking camels and llamas in North America. The llamas moved southward and took up residence in the highlands and plains of South America. Camels migrated across the Bering land bridge to Eurasia and Africa, where they established themselves in arid regions. Those that were left behind in North America mysteriously became extinct.

The **ruminants** are the most varied and abundant of modern-day artiodactyls. This group takes their name from the *rumen,* or the first of four compartments in their multichambered stomach. For the most part, they are cud chewers. The earliest ruminants were small, delicate, four-toed animals called **tragulids.** They are represented today by the skinny-

FIGURE 14–44 Skeleton of the Miocene camel, *Oxydactylus.* This small, graceful camel was only one of a varied group of camels that populated the grasslands of North America. *(Courtesy of the Denver Natural History Museum.)*

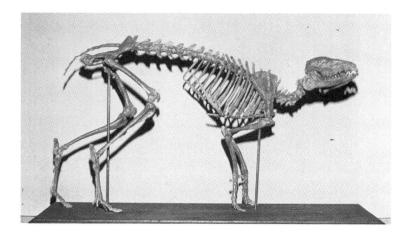

FIGURE 14–45   The primitive ruminant *Hypertragulus* from Lower Oligocene beds of North America. The tiny animal stood only about 30 cm high at the shoulder. *(Courtesy of the Denver Museum of Natural History.)*

legged and timid mouse deer of Africa and Asia. *Hypertragulus* (Fig. 14–45), whose remains are found in Lower Oligocene strata, was such an animal. Later lines of tragulids sometimes developed peculiar assortments of horns, exemplified by *Synthetoceras* (Fig. 14–46B). This Pliocene tragulid had not only a pair of horns over the eyes but also a Y-shaped horn on the snout.

Ruminants such as sheep, cattle, giraffes, and deer are called **pecorans**. Deer are primarily browsers that have made the forests their principal habitat ever since their first appearance in the Oligocene. The

FIGURE 14–46   A variety of early Pliocene mammals. (A) *Amebelodon,* the shovel-tusked mastodon. (B) *Synthetoceras.* (C) *Cranioceras.* (D) *Merycodus,* an extinct prong-horned antelope. (E) *Epigaulis,* a burrowing, horned rodent. (F) *Neohipparion,* a Pliocene horse. (G) The giant camel *Megatylopus* and smaller *Procamelus.* (H) *Prosthennops,* an extinct peccary. (I) The short-faced canid *Osteoborus.*   *(Courtesy of the National Museum of Natural History, Smithsonian Institution, J. H. Matternes mural, with permission.)*

FIGURE 14–47 *Megaloceros,* the giant Irish "elk," was actually a deer whose remains are frequently found in the peat bogs of Pleistocene age in Ireland. From tip to tip, the antlers were 3.6 meters in breadth. *(Specimen in Sedgwick Museum, Cambridge, England.)*

early stages of deer evolution are exemplified by *Blastomeryx,* a small, hornless creature that lived in North America during the Miocene. Subsequent evolution involved increases in size and development of antlers. The culmination of this trend is represented by the Pleistocene *Megaloceros,* whose antlers measured more than 3 meters from tip to tip (Fig. 14–47).

The giraffe family probably branched off from the deer lineage sometime during the Miocene and became specialized in browsing on leaves of trees that grew rather sparsely in areas where most other herbivores were grazers. *Palaeotragus,* the ancestral giraffe, looked much like the modern okapi of Africa. From this general form, the obvious evolutionary trends emphasized leg and neck elongation.

Of course, the bovids—cattle, sheep, and goats—are presently the most numerous of ruminants. Miocene strata provide the earliest fossil record of bovids. The bison are particularly interesting to Americans. Seven species of bison lived in North America during the Pleistocene, some of which were truly giants with horns that measured 2 meters from end to end.

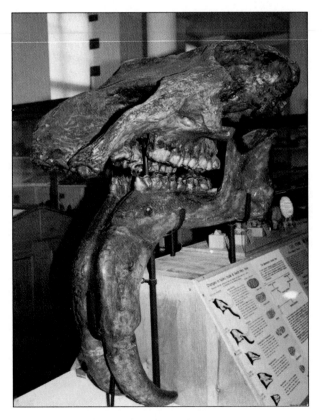

FIGURE 14–49 Skull of *Dinotherium,* a Miocene proboscidean. Length of skull is about 1.2 meters. This specimen is on display in the Sedgwick Museum of Cambridge University.

FIGURE 14–48 The first proboscideans (members of the group that includes elephants) were moeritheres, named after this typical genus *Moeritherium,* which lived in Egypt during the Late Eocene.

COMMENTARY
How the Elephant Got Its Trunk

Among the many mammals that evolved during the Cenozoic, few inspire more interest than the proboscideans—mastodonts, mammoths, and elephants. Our interest seems to relate not just to their great size, but to their most distinctive feature, the trunk. The elephant's trunk is a highly muscularized, much elongated organ containing nostrils that extend downward through its entire length to the tip. The Indian elephant uses a one-finger process to pick up small objects with the tip of the trunk, while the African elephant uses a two-finger process.

The question, "How did the elephant get its trunk?" might be readily answered by a child on reading Rudyard Kipling's *The Elephant Child.* According to this story, an overly inquisitive elephant child tried to find out what the crocodile had for dinner. When the elephant child lowered his head along the bank of the great, gray-green, greasy Limpopo River, the jaws of the crocodile closed on his nose. There ensued a fierce struggle in which the "bicolored-python-rock-snake" came to the aid of the poor elephant child, but in the tug-of-war, the elephant child's poor nose was stretched into the trunk we know today. Actually, the elephant achieved its wonderfully long and prehensile trunk by means of evolutionary processes that operated over many generations and resulted in an enormous elongation of the nose and upper lip. Scientists refer to such extraordinary growth of a body part as hypertrophy, and the elephant got its trunk by hypertrophy of the nose and upper lip. But why?

As one views the bones of proboscideans, beginning with the relatively small and primitive groups of the Eocene through the giant Pleistocene mastodonts and mammoths, several trends are apparent. One obvious trend was the increase in size, not only of the body, but of the huge proboscidean head. Supporting the weight of the ponderous head required a short, powerful neck. As the proboscideans grew taller, reaching food at ground level when equipped with a short neck would have been difficult. Selection paralleling that of the large head and short neck resulted in the development of the trunk. The elephant's trunk is a structural adaptation for feeding that could be used not only for raising food from the ground, but for stripping leaves from branches and stuffing them into the mouth. Accompanying the evolution of the trunk in proboscideans, the lower jaw became shorter, allowing the trunk to hang downward. Tusks in the lower jaw that were present in older proboscideans grew smaller and gradually disappeared.

A discussion of Cenozoic mammals would not be complete without a brief mention of elephants and their older relatives, the mammoths and mastodons. Because these animals have a trunk, they are collectively termed **proboscideans.** The trunk is the elephant's principal means of bringing food to its mouth. Other animals as tall as elephants reach food on the ground easily because of their long necks. However, proboscideans, with short, muscular necks to support their massive heads, have evolved their own unique anatomic solution to food gathering. Paleontologists are able to follow the development of the trunk in early proboscideans by noting the position of the external nasal openings at the front of the skull. Those openings recede toward the rear of the skull in sequential stages of trunk development. Another elephantine trademark is the tusks, which evolved by elongation of the second pair of incisors.

The early fossil record of proboscideans includes a trunkless, tapir-like animal named *Moeritherium* (Fig. 14–48). Bones of this ancestral proboscidean have been found in Eocene and Oligocene beds near Lake Moeris in Egypt. It is likely that the moeritheres arose from early Tertiary condylarths. From an ancestry represented by the moeritheres, proboscideans separated into two branches. One led toward a group of Miocene and Pliocene animals called **dinotheres** (Fig. 14–49). The tusks of dinotheres were distinctive in that they were present only in the lower jaws and curved downward and backward, an orientation presumably useful for uprooting plants and digging for roots and tubers.

The other branch of the proboscidean family tree produced mastodons and elephants. *Palaeomastodon,* in the Oligocene of North Africa, was probably representative of the group from which true mastodons evolved. The face was long, the trunk was short, and the second incisors in both the upper and the lower jaws had developed into tusks. By Miocene time, a larger proboscidean named *Gomphotherium* had found its way into North America by way of the Bering isthmus. Subsequent proboscidean evolution produced a variety of long-jawed mastodons. *Trilophodon,* from the Pliocene, had lower jaws al-

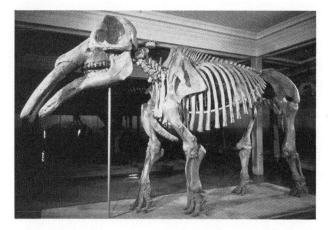

FIGURE 14–50 A four-tusked mastodon of the Late Miocene and Early Pliocene. The mounted skeleton is in the collection of the Denver Museum of Natural History.

most 2 meters long. The most bizarre proboscidean was the "shovel-tusked" *Amebelodon* (see Fig. 14–46A) of the Pliocene. The tusks of the lower jaws in these beasts were flattened and formed two sides of a broad, scooplike ivory shovel. Mastodons with shorter jaws and longer trunks are best represented by species of *Mammut* (Fig. 14–50), which were common throughout North America during the great Ice Age and survived until comparatively recent times. *Mammut* was heavier but not as tall as the African elephants. It had the sharply crested molars typical of mastodons, a stocky build, and great curving tusks nearly 2 meters long.

True elephants and mammoths evolved from gomphotheres in the Old World. Their line of evolution can be traced backward to a Miocene animal named *Stegolophodon*. Next on the evolutionary scene was *Stegodon*, a Pliocene form with a short, tuskless lower jaw and molar teeth already containing the numerous cross-ridges that characterize true elephants. Such teeth are more suitable for a diet of abrasive grasses. Further along in the evolution of the elephant, the trunk grew longer, the height of the skull increased enormously, and the jaws became so short that they could accommodate no more than eight grinding teeth at one time. As the molars of elephants gradually wear down, new teeth form in the posterior region of the jaws and push forward until the older, worn teeth break out. The last molars appear when the animals are 40 to 50 years old.

Mammoth is a term loosely applied to the Ice-Age elephants of North America, Europe, and Africa. They were a magnificent group of animals that included the famous woolly mammoths (Fig. 14–51) drawn by our own ancestors on the walls of their caves. The great imperial mammoth reached heights of 4.5 meters and ranged widely across California, Mexico, and Texas. The Columbian elephant had immense spiral tusks that in older individuals overlapped at the tips, thus becoming useless for digging purposes. Many of the late Pleistocene proboscideans were hunted by humans. Then, about 8000 years ago, all but two genera of elephants became extinct. Their continued survival, as is the case with many wild animals, will depend on our good judgment and that of the governments we support.

FIGURE 14–51 The woolly mammoth. In the Late Pleistocene, these magnificent animals lived along the borders of the continental glaciers. Their remains have been found frozen in the tundra of northern Siberia. *(Courtesy and copyright of The Field Museum; painting by C. R. Knight, photograph of painting by R. Testa, with permission. Negative No. CK30T.)*

FIGURE 14–52 An assemblage of animals that lived in Central Alaska about 12,000 years ago, during the Late Pleistocene. Fossil remains of this period are abundant and indicate a fauna in which grazing animals predominated, with the remaining fauna composed of browsers and predators. *(Courtesy of the National Museum of Natural History, Smithsonian Institution, J. H. Matternes mural, with permission.)*

■ Demise of the Pleistocene Giants

At the time of maximum continental glaciation (approximately 11,000 years ago), the Northern Hemisphere supported an abundant and varied fauna of large mammals, comparable to that which existed in Africa south of the Sahara several decades ago (Fig. 14–52). The fauna included giant beavers, mammoths, mastodonts, elk, most species of perissodactyls, many even-toed forms, and ground sloths. From all available evidence, most of these great beasts maintained their numbers quite well during the most severe episodes of glaciation, but they experienced rapid decline and extinction in the period around 8000 years ago, when the climate became milder. What might have been the cause for the extermination of these large Pleistocene mammals? Explanations tend to fall into two categories. In one, theories attempt to explain the extinctions as being related to human overkill; a second interpretation relates the losses to climatically controlled environmental changes.

The overkill concept is based on the inference that early humans had developed the ability to hunt in highly organized social groups and skillfully bring down large numbers of big animals on open prairies and tundras. Human predators may have killed then, as they do now, in excess of their needs. Unlike such predators as wolves, early human hunters probably did not seek out the weak or sick to kill but brought down the best animals in the herds. By decimating particular species of herbivores, the other, nonhuman predators would have necessarily also suffered. In support of this idea, paleontologists note that most

Pleistocene extinctions involved large terrestrial animals. Marine genera, protected by the sea from human predation, continued to thrive. Small mammals also survived; although frequently hunted, they were difficult to exterminate because of their more rapid breeding rate, their greater numbers, and the probability that they were killed only one at a time. The big, gregarious herbivores were easily accessible and reproduced more slowly. Entire herds could be driven off cliffs or into ravines, thus providing opportunities for slaughter of hundreds at the same time.

Another line of evidence favoring overkill stresses the observation that an early wave of extinction began in Africa and southern Asia, where predatory humans first became prevalent. The better known extinctions of the colder north did not occur until much later, when humans had moved into this region after having first devised the means to better clothe and shelter themselves.

The overkill concept has not, however, gone unchallenged. Many paleontologists feel that human populations were too small and that nomadic tribes shifted too frequently to account for the extermination of entire species of mammals. Some animals, such as the cave lion, were probably never hunted as food, and yet they too became extinct along with the mammoths and mastodons. Moreover, primitive tribes that today still live at the Stone Age level of development do not endanger the survival of the animals they hunt.

These arguments have led some paleontologists to favor theories that relate extinctions to climatic warming. The great beasts of the Pleistocene flourished during the cold 100,000-year glacial stages, even though those cooler episodes were interrupted by interglacial stages lasting only about 10,000 years. They were primarily adapted to cooler climates. With the onset of warmer conditions following the last glacial advance, their reproductive capacity may have been curtailed. Geologist D. M. McLean notes that this would be particularly true for the females of large mammals because of their relatively smaller surface area to volume ratio. (They have less surface area to give off stored body heat than in smaller animals.) McLean proposes that pregnant females may have experienced increased blood flow to peripheral tissues simply in order to cool down. This would lessen blood flow to the uterus, thereby reducing the supply of oxygen, nutrients, and water to the developing embryo. Heat generated by embryonic metabolic processes would also be difficult to dissipate. The final result would be elimination of many huge Pleistocene mammals, as well as evolutionary selection for smaller animals.

Summary

The Cenozoic was a time of gradual approach to the conditions of the present. The flowering plants continued to expand and diversify, with the most important single occurrence being the Miocene proliferation of grasses as extensive prairies and savannahs. Among the marine invertebrates, large discoidal foraminifera such as nummulites were characteristic of areas in the Tethys seaway and the tropical west Atlantic. Smaller planktonic foraminifers proliferated and underwent a marked diversification during the Cenozoic. Gastropods and pelecypods were the dominant higher marine invertebrates. Reef corals grew extensively across tracts of shallow warm water in the Tethys belt, the Caribbean, the West Indies, and around the many volcanic islands of the Pacific.

The reptile dynasty had collapsed at the end of the Mesozoic. Only turtles, crocodiles, lizards, snakes, and rhynchocephalians lived through the Cenozoic and survived to modern times. Although the fossil record for birds is less than adequate, it is likely that most modern orders were present by Eocene time. Nearly all the continents seem to have had one or more species of large, flightless, flesh-eating birds during the Cenozoic.

With considerable justification, the Cenozoic has been called the Age of Mammals. During the Cenozoic, mammals expanded rapidly into the multitude of potential habitats vacated by the reptiles. From ancestral shrewlike creatures, either directly or indirectly, the higher orders of placental mammals evolved. In the vanguard were small-brained archaic ungulates, such as amblypods and condylarths, as well as primitive flesh-eating creodonts.

Following this early Cenozoic radiation, a second wave of placentals began to evolve in the middle part of the era, and most of the modern mammalian orders became established. This adaptive radiation was as remarkable as had been that of the earlier Mesozoic reptiles. Mammals in the form of bats conquered the air, whereas others, such as whales, seals, and walruses, returned to the sea, where the evolution of vertebrates had begun long ago.

The Carnivora diversified and became dependent on the multitude of advanced ungulates that populated forests and grasslands. Ungulates included two great categories: the perissodactyls and artiodactyls. The former include horses, rhinoceroses, and tapirs, as well as the extinct titanotheres and chalicotheres. Among artiodactyls, or even-toed ungulates, the large, piglike entelodonts and smaller oreodonts failed to survive beyond the Oligocene. However, living artiodactyls include swine, camels, hippopotami, and a diverse host of ruminants. As all of these ungulates evolved during the Cenozoic, they underwent specialization of teeth for grazing and limbs for running across relatively open plains. Grass tends to wear down teeth rapidly. To compensate, grass-eaters evolved teeth with complicated ridges of hard enamel and exceptionally high crowns.

Similar changes also occurred in members of the proboscidean line. The earliest proboscideans are recorded as fossils in Eocene beds in Egypt. Their radiation produced an impressive array of variously tusked and specialized mastodonts and mammoths. The remains of those great beasts are frequently found in Pleistocene bog deposits in the eastern United States, Alaska, and Siberia.

The Pleistocene was a time of splendid diversity among mammals. The fauna was probably more varied and impressive than that of Africa a century ago. The Pleistocene was an age of giants, in which colossal ground sloths, beavers over 2 meters tall, giant dire wolves, bison, and tall, rangy mammoths and woolly rhinoceroses ranged far and wide across the continents. One might think an assemblage of beasts that had proved their stamina against the difficulties of an ice age would persist for at least as long as the dinosaurs. Yet widespread extermination and extinction was the dominant biological theme at the end of the Pleistocene. Overkill by human hunters may have contributed to the demise of many Pleistocene mammals, or the extinction of these animals may be related to an inability to adjust to climatic warming.

Review Questions

1. What is the depth of water preferred by reef corals? Why is this water depth necessary for coral growth? For the development of atolls?
2. Which groups of marine phytoplankton proliferated during the Cenozoic Era?
3. How did the cephalopod faunas of the Cenozoic differ from those of the preceding era?
4. Which group of bony fishes is particularly characteristic of the Cenozoic?
5. What is the importance of the order Insectivora in the evolutionary history of placental mammals? What is the importance of the Jurassic mammals known as eupantotheres?
6. How might a paleontologist determine that a fossil lower jaw containing teeth belonged to a mammal rather than a reptile?
7. What are ruminants? What advantages may be inherent in the ruminant type of digestion?
8. How is it possible to trace the development of the trunk or proboscis in the skeletal remains of proboscideans when the trunk itself is not preserved?
9. How do you account for the reductions and extinctions of marsupialian mammals in South America during the early Pleistocene?
10. If evolutionary success can be measured in terms of diversity of a particular group of animals, how would you rate perissodactyls relative to artiodactyls?

Discussion Questions

1. Prepare a list of the changes that have occurred in the evolution of horses from the appearance of *Hyracotherium* to *Equus*.
2. Prepare arguments for and against the premise that the extinction of large Pleistocene mammals was a consequence of overkill by early human hunters.
3. Discuss the theory that climatic warming following the last glacial stage selectively damaged the reproductive capacity of mammoths and mastodons and thereby led to their extinction.
4. Describe global geography during the Cenozoic and discuss its effect on the diversity of mammalian faunas.
5. Describe the adaptations seen in Cenozoic herbivores that are related to the spread of prairies.
6. Prepare an argument that the adaptive radiation of Cenozoic mammals was no less successful than the adaptive radiation of Mesozoic reptiles.
7. What groups of reptiles survived the great mass extinction at the end of the Cretaceous? Discuss possible factors or conditions that prevented these surviving reptiles from expanding to dominance during the Cenozoic.
8. Cenozoic whales demonstrate a remarkable example of evolutionary convergence with the marine reptiles known as ichthyosaurs of the Mesozoic. What characteristics of whales, however, are distinctly not reptilian?

Supplemental Readings and References

Carroll, R. L. 1988. *Vertebrate Paleontology and Evolution.* New York: W. H. Freeman and Co.

Colbert, E. H., and Morales, M. 1991. *Evolution of the Vertebrates,* 4th ed. New York: John Wiley.

MacFadden, B. J. 1992. *Fossil Horses.* Cambridge, England: Cambridge University Press.

Martin, P. S., and Wright, H. E. 1967. *Pleistocene Extinctions.* New Haven, CT: Yale University Press.

Match, C. L. 1976. *North America and the Great Ice Age.* New York: McGraw-Hill.

Savage, R. J. G., and Long, M. R. 1987. *Mammalian Evolution: An Illustrated Guide.* London: British Museum of Natural History.

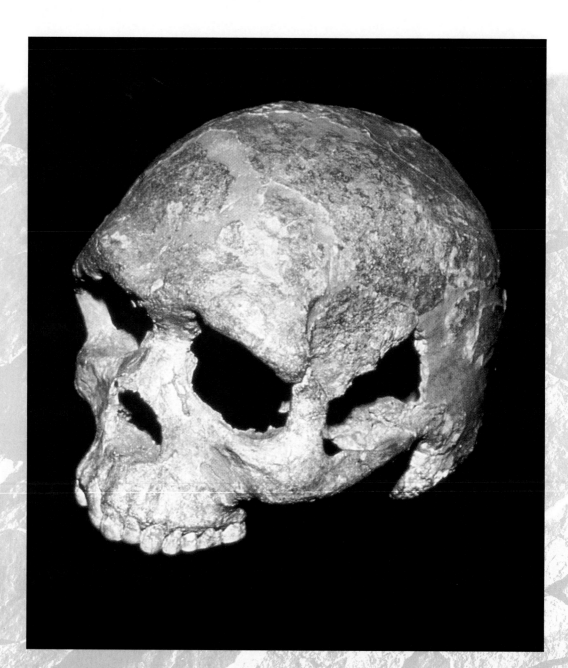

Skull of an early *Homo sapiens*. This early member of our own species had a very modern looking skull. The specimen is about 100,000 years old and pre-dates the classic European Neanderthals. *(Courtesy of Christopher Stringer, with permission.)*

It walked on its hind feet, like something out of the vanished Age of Reptiles.

The mark of the trees was in its body and hands.

It was venturing late into a world dominated by fleet runners and swift killers.

By all the biological laws this gangling, ill-armed beast should have perished, but

you who read these lines are its descendant.

LOREN EISELEY, 1971, *The Night Country*

Human Origins

Evolution during the late epochs of the Cenozoic produced a mammal capable of shaping and controlling its own environment. That mammal, of course, was our own species, *Homo sapiens*. Unlike previous vertebrates, this remarkable creature profoundly changed the surface of the planet, modified the environment in ways both beneficial and destructive, and so manipulated the populations of other creatures as to change the entire biosphere. For these reasons, it is fitting that the final chapter in our history of the Earth deal with the evolution of humankind.

In this chapter we will review the general characteristics of the order Primates. *Homo sapiens* share many characteristics with other primates. For example, it is apparent that we resemble such primates as the great apes in basic body structure. On the other hand, we have become quite distinct from other primates in many ways. We have a larger, more complex brain. We stand and walk erect and have structural modifications of our vertebral column, legs, and pelvic bones to make such erect posture possible. Our face is flatter, our teeth are less robust, and we have achieved extraordinary manual dexterity. This attribute has contributed to our ability to manufacture and use sophisticated tools. Moreover, the *Homo sapiens* we call human exceeds all other members of our order in a level of intelligence that has led to language and aesthetic sensibility. Looking closely at the bones, artifacts, and cave drawings of our Stone Age ancestors, we can discern the beginnings of these familiar attributes.

■ *Primates*

Primate Characteristics

What traits should an animal have to qualify as a primate? It seems that this question cannot be answered easily, for primates have remained structurally generalized compared with most other orders of placental mammals. They retain the primitive number of five digits, have teeth that are not specialized for dealing with either grain or flesh, and have never developed hoofs, horns, trunks, or antlers. In the course of their evolution from shrewlike insectivores, their principal changes have involved progressive enlargement of the brain and certain modifications of the hand, foot, and thorax that were related to their life in the trees and their manner of obtaining food. These adaptations were not insignificant, however, for they enabled the human primate to shape a mode of life qualitatively different from that of any other animal.

The fictional pig Snowball in George Orwell's book *Animal Farm* remarked that "the distinguishing

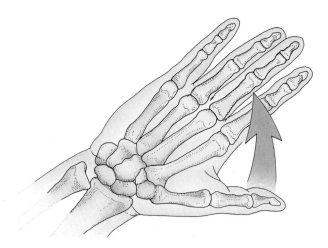

FIGURE 15–1 The right hand of a human (palm up). The human hand is not used in locomotion and can be used to manipulate small objects between the fingers and thumb.

TABLE 15–1   A Simplified Classification of the Order Primates

Order	Suborder	Superfamily	Common Names of Representative Forms	
Primates	Prosimii	*TUPAIOIDEA*	Tree shrew	
		LEMUROIDEA	Lemur	
		LORISOIDEA	Bush baby, Slender loris	
		TARSIOIDEA	Tarsier	
	Anthropoidea	*CEBOIDEA*	Howler monkey, Spider monkey Capuchin, Common marmoset, Pinche moneky	
		CERCOPITHECOIDEA	Macaque, Baboon, Wanderloo, Common langur, Proboscis monkey	
			Family	
		HOMINOIDEA	HYLOBATIDAE	Gibbon, Siamang
			PONGIDAE	Orangutan, Chimpanzee, Gorilla
			HOMINIDAE	Humans

mark of man is the hand, the instrument with which he does all his mischief." Although this is a biased point of view, it is correct in its assessment of the importance of the primate grasping hand with its opposable thumb (Fig. 15–1). This characteristic not only permitted primates a firm grip on their perches but also allowed them to grasp, release, and manipulate food and other objects. The forearms retained their primitive mobility, in which rotation of the ulna and radius upon one another permitted the hands to be reversed in position.

The development of the grasping, mobile hand was accompanied by improvement in visual attributes. The eyes of primates became positioned toward the front of the face so that there was considerable overlap of both fields of vision, resulting in an improved ability to judge distances. It would seem that grasping hands and feet and good binocular vision are obvious adaptations for an animal that leaps from branch to branch and seeks its food from precarious boughs. Yet many tree-dwelling animals such as squirrels, civets, and opossums lack the short face, close-set eyes, and opposable digits and get along very well. For this reason, anthropologists have recently suggested that the visual attributes of primates originated as predatory adaptations that allowed early insectivorous primates to gauge accurately the distance to insect prey without movement of the head. By being able to grasp narrow supports securely with its feet, the animal was able to use both of its mobile hands to catch the prey. Claws, which are used by animals like squirrels in moving about on relatively wide branches, would not be as advantageous for climbing about on thin boughs and vines

and might not provide a sufficiently secure hold for the quick catch of a moving insect. Thus, it seems that binocular vision and the grasping hand may not have been originally related to arboreal locomotion alone, but to the cautious, well-controlled, manual capture of visually located prey in the insect-rich canopy of tropical forests.

Other evolutionary modifications of primates were related primarily to changes in the eyes and limbs. On the outer margin of each eye orbit, a vertical ridge of bone developed (postorbital bar) that protected the eyes from bulging jaw muscles and accidental impact. As the eyes became positioned more closely together, the snout was reduced so that the face became flatter. In response to a brachiating habit (swinging from branches), forelimbs and hindlimbs diverged in form and function and an inadvertent predisposition toward upright posture developed.

Modern Primates

The primates are divided into the suborders **Prosimii** (which include tree shrews, lorises, and tarsiers) and the suborder **Anthropoidea** (which includes monkeys, apes, and humans) (Table 15–1). The Prosimii retain a greater number of primitive characters relative to the Anthropoidea. The tree shrews (see Fig. 14–25), for example, possess clawed feet and a long muzzle with eyes in the lateral positions. The lemurs (Fig. 15–2), which are confined largely to Madagascar, still have a long snout and lateral eyes, although these are positioned more to the front than those in the shrew. Some of the digits are clawed, whereas others are equipped with flattened nails. The East

FIGURE 15–2 The ring-tailed lemur of Madagascar. *(Frans Lanting/Minden Pictures.)*

FIGURE 15–3   *Tarsius,* a small East Indian nocturnal, arboreal, prosimian with large, forwardly directed eyes. *(Gary Milburn/Tom Stack & Associates.)*

Indian tarsier (Fig. 15–3) has a relatively flat face. As befits a nocturnal animal, the eyes are large and positioned toward the front to provide stereoscopic vision. The fingers and toes terminate in nails rather than claws.

In the more progressive Anthropoidea, the trends initiated by prosimians are further developed. Monkeys are the more primitive members of the Anthropoidea. They are grouped into New World and Old World forms (Fig. 15–4). The *Ceboidea,* or New World monkeys, are an early branch and were not involved in the eventual evolution of humans. Included among the ceboids are the spider monkey and marmoset, as well as the familiar little capuchin, or organ-grinder monkey. New World monkeys can be identified by their flattish faces, widely separated nostrils, and prehensile tails. Most are small in comparison of their Old World cousins.

The more advanced Old World monkeys, or *Cercopithecoidea,* are widely distributed in tropical regions of Africa and Asia. They include the familiar macaque or rhesus monkey of laboratories and zoos, the Barbary ape of Gibraltar, langurs, baboons, and mandrills. In this group of monkeys, the nostrils are

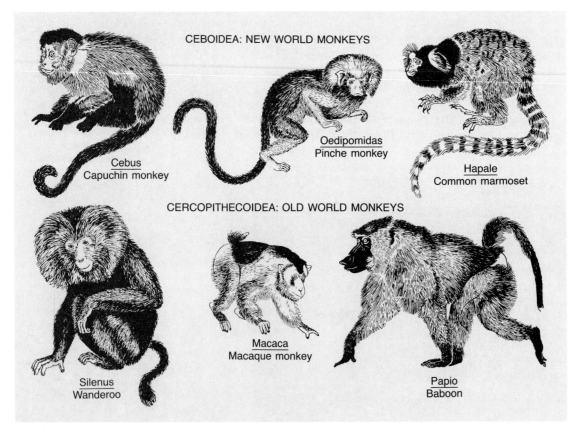

FIGURE 15–4   Representative New World and Old World monkeys. Most New World monkeys have prehensile tails that can function almost as effectively as another limb.   *(From Cockrum, E. L., et al. Zoology. Philadelphia: Saunders College Publishing, 1966.)*

A

B

C

D

FIGURE 15–5 The apes. (A) Gibbons are considered by many scientists as the most perfectly arboreal of the anthropoid apes. (B) Orangutans are solitary apes that seldom leave the protection of the trees. (C) A gorilla knuckle walking. (D) Chimpanzees live in groups and have a complex social behavior. *(A, © 1991 A. G. Nelson/Dembinsky Photo Associates. B, Tom McHugh/ Photo Researchers, Inc.; C, John D. Cunningham/Visuals Unlimited; D, Animals Animals © 1988, Zig Leszczynski.)*

close together and directed downward (as in humans) and the tail is not prehensile.

The anthropoid apes (Fig. 15–5) are tail-less primates. Modern species probably evolved from the same ancestral stock that produced humans. However, the divergence probably took place back in the middle Tertiary. The more primitive branch of the tail-less apes is composed of the *Hylobatidae,* best represented by those long-armed acrobats, the gib-

bons. Orangutans, chimpanzees, and gorillas are grouped within the *Pongidae.* The red-haired orangutan, or "wild man" of Borneo and Sumatra, is considerably larger than the gibbon and has a larger brain as well. Chimpanzees and gorillas are similar in many ways and tend to be less pronounced in their arboreal adaptations. Gorillas are essentially ground dwellers and spend only a small fraction of their time in the trees.

FIGURE 15–6   The Paleocene rodent-like prosimian *Plesiadapis*.

FIGURE 15–7   The Eocene lemur *Notharctus*.

■ *The Prosimian Vanguard*

The early fossil record for primates includes a creature named *Purgatorius*, known only from a few teeth discovered in the Hell Creek Formation at Purgatory Hill in Montana. These finds indicate that the earliest primates were contemporaries of the last of the dinosaurs, at least in the tropical latest Cretaceous environments of North America. The fossil record improves somewhat in the early epochs of the Cenozoic. *Plesiadapis* (Fig. 15–6), found in Paleocene beds of both the United States and France, has the distinction of being the only genus of primate (other than that of humans) that has inhabited both the Old World and the New World. The presence of this **prosimian** on the now widely separated continents is one of many clues indicating that the continents were not yet completely separated by the widening Atlantic in the Paleocene. *Plesiadapis* was a rather distinctive and specialized primate and represents a sterile offshoot of the primate family tree. The incisors were rodent-like and were separated from the cheek teeth by a toothless gap, or **diastema**. Fingers and toes terminated in claws rather than nails. The rodent-like characteristics and habits of the plesiadapiformes may have contributed to their extinction by late Eocene time. About that time, rodents were having their own radiation. Rodents are highly successful as gnawing, seed-eating animals and are able to reproduce at a rate with which no primate can compete.

General trends in prosimian evolution during the Eocene Epoch (37 to 55 million years ago) involved reduction in the length of the muzzle, increase in brain size, shifting of the eye orbits to a more forward position, and development of a grasping big toe.

These trends are evident in the fossil remains of *Cantius* from the Wind River basin of Wyoming. In the Wyoming stratigraphic sequence, one can trace the evolution of successive species of *Cantius* from lower to upper beds. Near the top of the section, a new genus, *Notharctus* (Fig. 15–7), occurs as the apparently direct descendant of *Cantius*. *Notharctus* has been known from other localities for over a century. In general form, it appears to be just the kind of lemur-like animal from which monkeys and apes are derived. Prosimian populations during the Late Paleocene and Eocene also included tarsiers. *Tetonius* was an Eocene form with tarsioid traits of closely spaced large eyes and a shortened muzzle.

Both tarsiers and lemurs were abundant and widely dispersed on Northern Hemisphere continents during the Eocene. However, with the advent of cooler Oligocene climates, they virtually deserted North America, whereas in the Eastern Hemisphere they were forced southward into the warmer latitudes of Asia, Africa, and the East Indies. Surviving prosimians are much reduced in variety and number. The prevailing opinion is that many have been replaced by the monkeys.

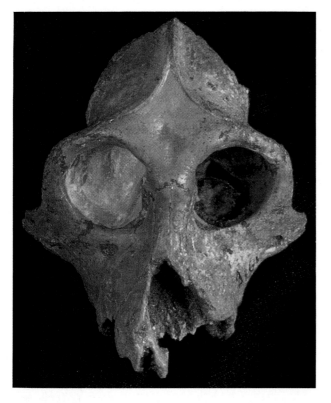

FIGURE 15–8 *Aegyptopithecus zeuxis* from Oligocene fossil beds in the Fayum district of Egypt, near Cairo. *Aegyptopithecus* is considered an early ape that preceded the Miocene apes exemplified by *Dryopithecus.* *(Courtesy of D. T. Rasmussen.)*

■ *The Early Anthropoids*

The next step in the primate evolutionary advance was the appearance of the first **anthropoids.** Discoveries in the Fayum region of Egypt have provided a wealth of information about the early anthropoids. The fossils there occur in several Oligocene horizons and include well over 100 specimens. Although many of the skull fragments and teeth retain subtle vestiges of prosimian ancestry, none of the remains are from prosimians. They are fossils of primates that have reached what may be called the monkey stage of organization. One primate discovered at Fayum is *Aegyptopithecus* (Fig. 15–8), a relatively robust arboreal primate with monkey-like limbs and tail, a brain larger than that of *Notharctus,* and eye orbits rotated to the front of stereoscopic vision. The fossil of the species *Aegyptopithecus zeuxis* is dated as 33 to 34 million years old, indicating that the prosimian–anthropoid transition had taken place by Oligocene time.

Fossils of the earliest known New World monkeys are known from Oligocene and Miocene strata of

South America. Although they appear superficially similar to some Old World forms, New World monkeys evolved independently from prosimian ancestors and without genetic contact with their Old World cousins.

Because evolution is a continuing process, with each animal truly a transitional link between older and younger species, it is a difficult and often arbitrary task to assign the fragment of a fossil jaw or tooth to either the monkey or the ape category. One of many clues used in attempting to make this distinction is the pattern of cusps on certain molar teeth. In Old World monkeys, specific molars have four cusps. Among apes (and humans), these same molars have five cusps, with an intervening Y-shaped trough. Molars belonging to *Aegyptopithecus* display the Y-5 pattern (Fig. 15–9), which is one reason anthropologists consider the genus to be the possible very early ancestor of Miocene apes. It seems safe to state that the Oligocene Epoch may have witnessed not only the transition from prosimians to anthropoids, but also the differentiation of apes from monkeys.

During the Miocene, plate tectonics had a significant influence on the evolution of primates. Africarabia (Africa and Arabia) was drifting north-

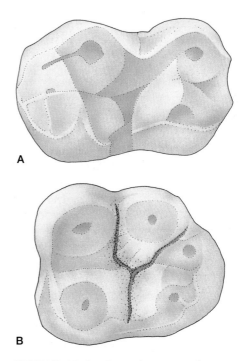

A

B

FIGURE 15–9 General pattern of cusps on the molars of Old World monkeys. (A) The lower molar of a baboon showing a cusp at each corner. (B) The lower molar of a chimpanzee showing the "lazy Y-5" pattern characterized by a Y-shaped depression that separates five cusps.

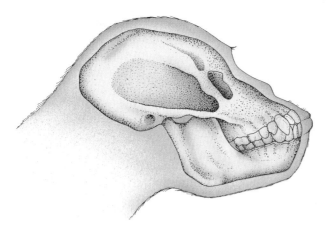

FIGURE 15–10   Skull of *Proconsul* (a dryomorph) from Lake Victoria, Kenya.

ward and ultimately collided with Eurasia. (The Arabian portion of the continent then rotated counterclockwise to eventually produce the Red Sea.) As a result of these movements, east–west circulation of tropical currents across the former Tethys seaway was prevented, and East Africa became cooler and dryer. Extensive grass-covered savannas replaced former areas of dense forests. The evolution of Miocene primates was strongly affected by these changes.

Among the new players that came on the scene during the Miocene were a group called **dryomorphs** (formerly called dryopithecines). The dryomorphs take their name from *Dryopithecus fontani,* a species

first discovered in France in 1856. The dryomorphs varied in size and appearance and include forms bearing different generic names from France, Spain, Greece, Hungary, Turkey, India, Pakistan, and Africa. Many anthropologists consider the important species *Proconsul africanus* to be either a very early dryomorph or the immediate dryomorph ancestor. *Proconsul africanus* was discovered by Mary and Louis Leakey in 1948 on an island in Lake Victoria, Kenya. The skull, jaws, and teeth of this small primate are decidedly apelike rather than monkey-like (Fig. 15–10). The slightly more advanced Middle Miocene dryomorph descendants of *Proconsul* are probably ancestral to both modern African apes and the first hominids, the australopithecines. This means that *Proconsul* is distantly ancestral to humans as well.

The study of Miocene primates clearly indicates an absence of orderly sequential change along a single trend. The family tree of primates is a complex of parallel and diverging branches. For this reason, the attempt to trace the ascent of humans up through the many bifurcations and dead ends is an exciting but difficult task. Paleontologists must be ready to reinterpret the story each time new fossil material is uncovered.

■ *The Australopithecine Stage and the Emergence of Hominids*

The story of the emergence of humans (hominids) begins with Raymond Dart's 1924 discovery of fossil

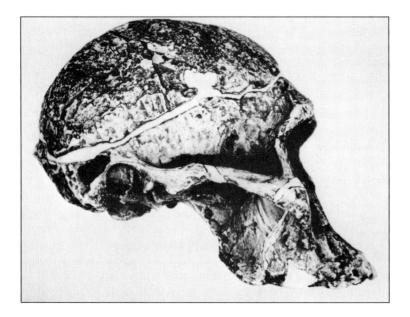

FIGURE 15–11   *Australopithecus africanus* from the Transvaal of South Africa. *(Photograph of Wenner–Gren Foundation replica by David G. Gantt.)*

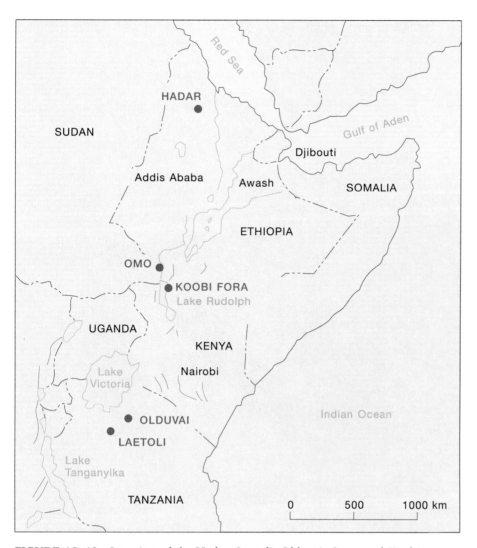

FIGURE 15–12   Location of the Hadar, Laetoli, Olduvai, Omo, and Koobi Fora (East Lake Rudolf) paleontologic sites.

remains of an immature primate in a limestone quarry in South Africa. Dart named the fossil *Australopithecus africanus* (Fig. 15–11). In succeeding years, many additional skeletal fragments of species of *Australopithecus* have been found in Upper Pliocene or Lower Pleistocene deposits of Africa, Java, and China. Fossil sites in East Africa have become increasingly important because of the veritable bonanza of hominid bones they have yielded. The new collections of fossils have given paleoanthropologists an unsurpassed record of human evolution over the past 4 million years. Some of the East African sites, such as Olduvai Gorge (Fig. 15–12), are now famous as a result of the lifelong research programs of Mary Leakey and the late Louis Leakey. Their efforts, and those of their son Richard, have provided fossils of

relatively heavy-bodied, robust **australopithecines** as well as lighter, so-called gracile types. Among the latter are remains of creatures that may well belong within our own genus.

East Africa during the Cenozoic experienced a lively history of volcanic activity. As a result, many of the fossil sites have interspersed layers of volcanic ash and lava flows. This fact has been of great benefit to the paleoanthropologists, for samples of ash can often be dated by the potassium–argon method. Dates from two succeeding ash beds would then provide a close estimate of the age of fossils found in the intervening layers of sediment.

Among the oldest fossils of hominids thus far discovered are skull fragments, teeth, arm bones, and part of a lower jaw from a fossil site near the village of Aramis in Ethiopia. The skeletal remains are neatly sandwiched between beds of volcanic ash that can be dated isotopically with great accuracy. The bones are 4.4 million years old. They are the remains of a hominid about 4 feet tall that has been named *Australopithecus ramidus* (recently renamed *Ardipithecus*

ramidus). This primate may be close to the common ancestor of both extinct species of australopithecines as well as modern humans (Fig. 15–13). It is possible, however, that a future discovery will find an even older candidate for that position. In any case, it is apparent that *Ardipithecus ramidus* walked the hills

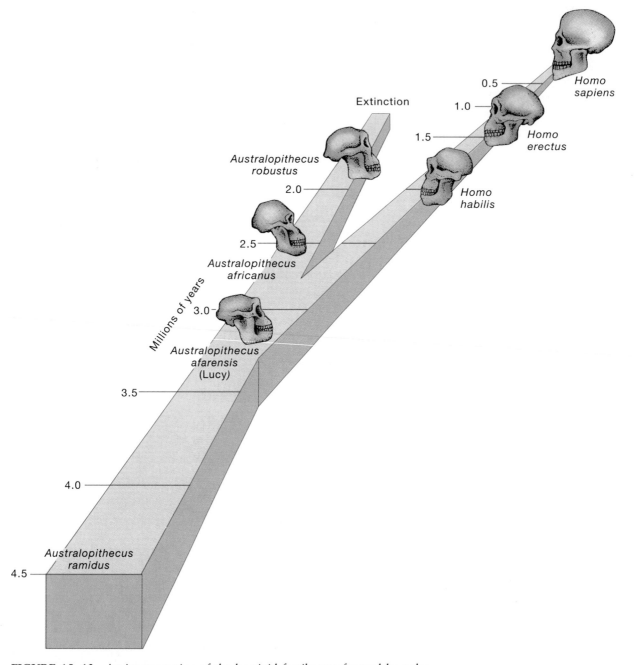

FIGURE 15–13 An interpretation of the hominid family tree favored by paleoanthropologist Donald C. Johanson, the discoverer of Lucy *(Australopithecus afarensis)*. In this interpretation, *A. afarensis* is considered the ancestor for species of the genus *Homo,* and later australopithecines are on a path leading ultimately to extinction. *(After Johanson, D. C., and White, T. D., Science 203[4378]:321–330, 1979.)*

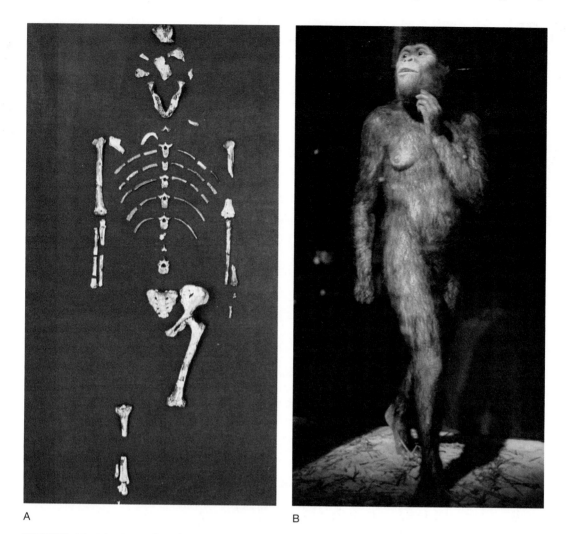

A   B

FIGURE 15–14 *Australopithecus afarensis,* informally known as "Lucy." Lucy was a young, erect walking hominid that lived in East Africa about 3.5 million years ago. (A) The specimen shown here is the oldest known, most complete skeleton of an erect walking human ancestor. (B) A reconstruction of Lucy on display at the St. Louis Zoo.   *(Photograph of skeletal remains courtesy of the Cleveland Museum of Natural History, with permission.)*

of Ethiopia over a million years before its famous descendant, *Australopithecus afarensis,* discovered in 1974 by Donald J. Johanson. Named "Lucy" after a popular Beatles song, "Lucy in the Sky with Diamonds," this diminutive young female is known from many skeletal parts (Fig. 15–14). She weighed about 80 pounds and had arms that were slightly longer relative to body size than our own. Her head was about the size of a softball and contained a brain comparable in size to that of a chimpanzee. The shape of the skull was more like that of an ape than a human, with the jaw thrust forward and no chin. Lucy and her kin lived in east Africa about 3.2 million years ago.

It is evident from the shape of the pelvic and leg bones in *Australopithecus afarensis* that this species walked erect. Further evidence of bipedalism came with the discovery of footprints in layers of volcanic ash (tuff) at Laetoli (Fig. 15–15). The footprints can be traced over a distance of 9 meters (about 30 feet) and were made by two contemporaries of Lucy as they walked side by side over a layer of soft, moist, volcanic ash. (Subsequently, another ashfall formed a protective seal over the footprints.)

The Omo and Koobi Fora sites in East Africa (see Fig. 15–12) have yielded not only the remains of *Australopithecus* but early members of our own genus *(Homo)* as well. The Omo digs are located along the Omo River in remote southwestern Ethiopia. Here one finds a nearly continuous section of sediments and volcanics that span 2.2 million years. The Koobi

A

B

FIGURE 15–15 (A) Footprints made by *Australopithecus afarensis* about 3.2 million years ago, as a pair of these hominids walked across wet volcanic ash. The footprints confirm skeletal evidence that the species had a fully erect posture. (B) Lucy and a male companion leave telltale footprints in wet volcanic ash as they stroll across the Late Pliocene landscape of eastern Africa. *(B, with permission of Peter Jones; A, American Museum of Natural History, with permission.)*

Fora locality is on the east side of Kenya's great alkaline Lake Turkana (formerly Lake Rudolf) (Fig. 15–16). It has been investigated by Richard Leakey and Glynn Isaac. Altogether, the Omo, Koobi Fora, Olduvai, Hadar, and Laetoli localities have given us sufficient skulls, jaws, teeth, and other bones to indicate that the australopithecines were not an entirely homogeneous group. However, certain shared characteristics appear in nearly all of the specimens unearthed. From the structure of their pelvic girdles, it is known that they stood upright in a fashion more human-like than apelike. Australopithecine dentition was essentially human, although the teeth were more robust than in modern humans. In contrast to these human-like characteristics, australopithecine cranial capacity was comparable to that of modern large

FIGURE 15–16 The Koobi Fora fossil site in East Africa where the skeletal remains of many australopithecines, as well as the skull designated KNM-ER-1470, were discovered. In addition to the hominid skeletal remains unearthed at Koobi Fora, paleontologists found pebble choppers and flake tools associated in deltaic deposits with the bones of a hippopotamus. The position of both the tools and hippo bones suggests that this was a butchering site. Apparently one day about 1.8 million years ago, some australopithecines came across the body of a hippopotamus that had recently died. They feasted on its carcass, using the sharp edges of flaked chert to get at the meat. *(Courtesy of Kay Behrensmeyer.)*

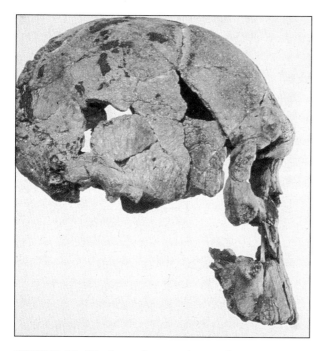

FIGURE 15–17 Lateral view of a nearly complete skull, designated KNM-ER-1470, from the Koobi Fora fossil site in Kenya. The creature may be an early representative of the genus *Homo*.

Recent investigations by anthropologists working in Kenya and Ethiopia have provided evidence that there may have been several human-like lineages co-existing in Africa between 1 and 3 million years ago. One of these lineages included *Australopithecus africanus* and the more recent *Australopithecus habilis* (some prefer *Homo habilis*) from Olduvai Gorge. The possible existence of a second lineage became apparent in 1972, when a team led by Richard Leakey unearthed a skull at the Koobi Fora site in beds that are about 2.0 million years old. The skull (Fig. 15–17), assigned the code name KNM-ER-1470, had a cranial capacity of about 800 cc, much larger than that of any *Australopithecus* and close to the cranial capacity of later hominids already assigned to the genus *Homo*. Although KNM-ER-1470 may not have quite reached the *Homo* stage of evolution, it was tending in that direction. The fossil indicates that large-brained, human-like creatures may have been present on Earth much earlier than previously suspected.

Yet another group of primates at the australopithecine stage had smaller brains but were larger than the *Australopithecus africanus* line. This group is exemplified by *Paranthropus* and may have evolved into a still heavier creature appropriately named *Gigantopithecus*. Such species as *Australopithecus boisei* (once called *Zinjanthropus boisei*) and *Australopithecus robustus* (formerly *Paranthropus robustus*) (Fig. 15–18) apparently represent sterile side branches that failed to give rise to any known descendant. Careful study of the teeth of some of these

apes and reached a volume of 600 or 700 cc (cubic centimeters). However, even with a brain only about half the size of *Homo sapiens*, australopithecines were able to make several kinds of crude implements from horn, teeth, and bone.

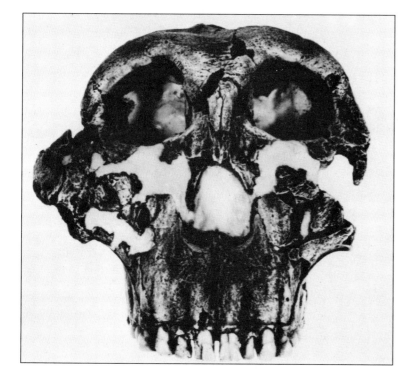

FIGURE 15–18 *Australopithecus robustus.* Because of the observation that forms such as this one had larger molars and premolars than did *Australopithecus africanus*, some paleoanthropologists believe that it was primarily a plant-eater and was taxonomically distinct from *A. africanus*. Recent interpretations favor inclusion of these creatures under the generic name *Australopithecus*. This specimen was derived from Bed I on the south side of the main ravine at Olduvai Gorge.
(Photograph of Wenner–Gren Foundation replica by David G. Gantt.)

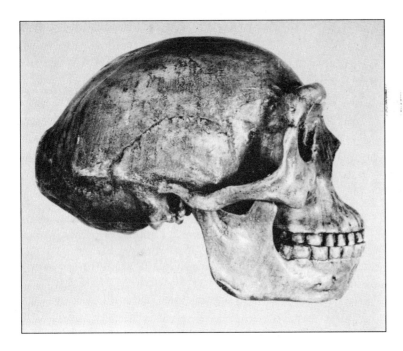

FIGURE 15–19 Replica of *Homo erectus*, previously known as *Sinanthropus pekinensis.* *(Photograph of Wenner–Gren Foundation replica by David G. Gantt.)*

larger australopithecines suggests that, unlike *A. africanus,* they may have been primarily plant-eaters.

From the preceding discussion, it is apparent that paleoanthropologists have established a rather large number of names for various australopithecine groups that display relatively minor differences. Some experts do not support the recognition of so many different species and believe that many forms now designated as separate species may prove to be local races or subspecies.

■ *The* Homo erectus *Stage*

The next stage in hominid evolution is represented by *Homo erectus,* formerly known as *Pithecanthropus erectus* from Java and *Sinanthropus pekinensis* (Fig. 15–19) from China. Other noteworthy discoveries include a skull cap of *Homo erectus* from Bed II of Oldovai Gorge (Fig. 15–20) that was found to be 750,000 years old, and a 1.8-million-year-old lower jaw with teeth from the former Soviet republic of

FIGURE 15–20 Upper portion of skull of a specimen of *Homo erectus* from Olduvai Gorge. This specimen was found associated with stone implements of the type known as Chellan. *(Photograph of Wenner–Gren Foundation replica by David G. Gantt.)*

C O M M E N T A R Y
Good and Bad News About Being Upright

Humans are unique among primates by having a body that is structured for standing fully erect and for walking on two legs. Our nonhuman ancestors, however, walked on all fours, and the evolutionary transition from those ancestors to hominid bipedalism has not been without problems. The human vertebral column, for example, is somewhat imperfect as a vertical support. That imperfection is one reason why so many people are plagued by lower back pain. Under the weight of the upper body, and provoked by lifting and unusual movements, intervertebral discs may become herniated and protrude, causing pressure against nerve structures. Severe pain and disability may follow, the treatment for which often requires delicate neurosurgery. In addition, evolutionary changes associated with our upright stance have increased the distance between the lowest ribs and the top of the pelvis. This has given us our distinctive waist, but it has also resulted in a weakening of the abdominal wall.

As a result, humans are particularly prone to hernias (ruptures).

Although these conditions affect many of us at various times during our lives, the advantages of erect posture nevertheless far outweigh the disadvantages. An upright stance and bipedalism in human evolution freed the hands for the kind of precise manipulation that led to the manufacture of tools and the advent of technology. For humans, hands are not merely motor organs. They are also sensory organs that can investigate by touch, "seeing" into places the eye may not be able to reach. Hands explore and react to what has been discovered. They can gesture, give instruction, indicate directions, and even express emotion. If we merely walked on our hands, what chance would we have had to evolve the complex human brain that provides for abstract thought, symbolic communication, and the development of culture?

Georgia. If the dating of the Georgian fossil is correct, then *Homo erectus* must have found his way from Africa into Europe far earlier than previously thought. Evidently, during the Middle Pleistocene, *Homo erectus* was widely dispersed in Africa and Eurasia. The species is considered the first true species of humans and is generally regarded as having evolved from species of *Australopithecus*.

The fossil evidence already available indicates a certain amount of variability between different groups of *Homo erectus*. However, the variability was probably no greater than that which exists today among members of our own species.

The bones of the axial skeleton and limbs of *Homo erectus* were quite similar to those of modern humans. In this regard, these hominids were more advanced than the australopithecines. For example, the pelvic bones of *Australopithecus* indicate that, although these creatures were fully bipedal, they walked with their feet turned outward in a sort of half-running, rolling gait. *Homo erectus,* on the other hand, was an excellent walker.

Although the postcranial skeleton of *Homo erectus* was modern, the skull was not. Cranial capacity ranged from about 775 cc in earlier forms to nearly

1300 cc in specimens of more recent age. Brain capacity in modern *Homo sapiens* ranges from 1200 to 1500 cc, and thus the brain size of *Homo erectus* overlaps at least the lowermost range of modern peoples. Pithecanthropines represent a stage in the evolution of hominids during which relatively rapid increases in brain size had begun. No doubt the expansion of the brain involved the reshaping not only of the cranium but also of the birth canal. Selection would very likely have favored large pelvic inlets to accommodate fetuses with larger heads and brains.

The skull of *Homo erectus* was massive and rather flat, with heavy supraorbital ridges over the eyes (see Fig. 15–19). The forehead sloped, and the jaw jutted forward at the tooth line in a condition termed **prognathous**. A definite jutting chin was lacking, and very likely the nose was broad and flat. These were rather primitive traits. However, except for their being somewhat more robust, the teeth and dental arcade in *Homo erectus* were essentially modern.

Evidence exists that *Homo erectus* made good use of his larger brain. From the bones of other animals found at living sites, it is clear that these hominids were good hunters. They were also skilled at making

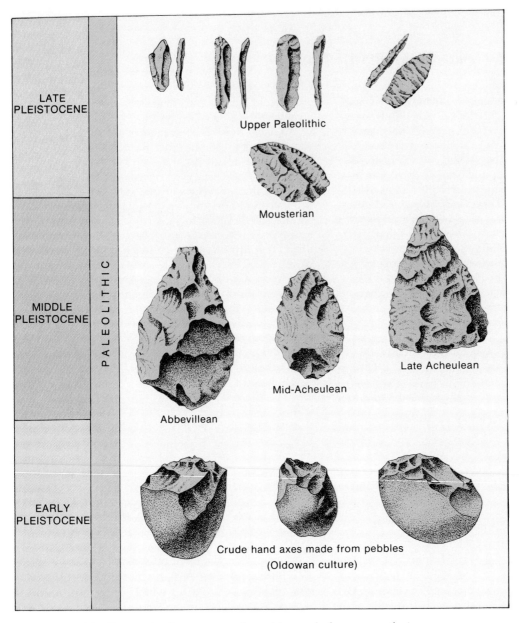

LATE PLEISTOCENE

Upper Paleolithic

Mousterian

MIDDLE PLEISTOCENE

PALEOLITHIC

Late Acheulean

Mid-Acheulean

Abbevillean

EARLY PLEISTOCENE

Crude hand axes made from pebbles
(Oldowan culture)

FIGURE 15–21   Progressive improvement in making tools from stone during
the Pleistocene. The crude stone tools of the Early Pleistocene were produced by
australopithecines. *Homo erectus* produced the better shaped tools of the Middle
Pleistocene. The Upper Paleolithic tools included carefully chipped blades and
points. The next stage, not shown here, would be the Neolithic, characterized by
refined and polished tools of many kinds.

simple implements out of flint and chert (Fig. 15–21), such as axes and scrapers, the former being equipped with wooden handles. Some also appear to have engaged in cannibalism, perhaps not for food but as part of a ritual system. Anthropologists do not know if *Homo erectus* spoke a distinctive language, wore clothes, or built dwellings. There is vague evidence that they hunted together in bands. What appear to be pithecanthropine slaughter sites have been found in Europe. Unfortunately, these sites did not yield a single bone of *Homo erectus*. A few *Homo erectus* fossil localities in Europe and China contain traces of carbon. However, this cannot be taken as indisputable evidence that *Homo erectus* had learned to use fire, for the carbon may have been the product of naturally occurring brush fires.

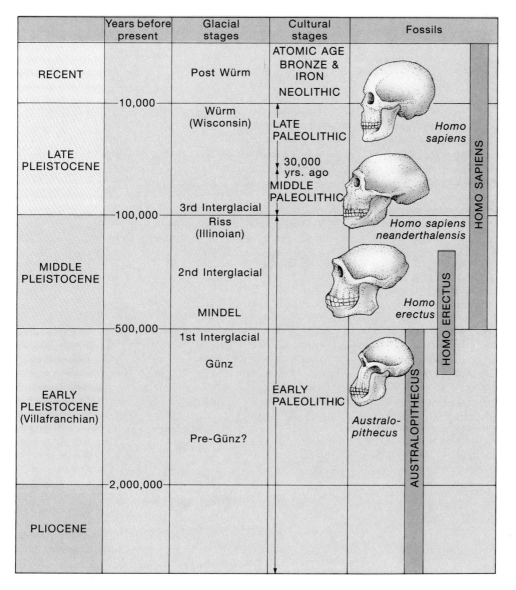

	Years before present	Glacial stages	Cultural stages	Fossils		
RECENT		Post Würm	ATOMIC AGE BRONZE & IRON NEOLITHIC			
	—10,000—	Würm (Wisconsin)	LATE PALEOLITHIC	*Homo sapiens*		HOMO SAPIENS
LATE PLEISTOCENE			↕ 30,000 yrs. ago MIDDLE PALEOLITHIC	*Homo sapiens neanderthalensis*		
	—100,000—	3rd Interglacial Riss (Illinoian)				
MIDDLE PLEISTOCENE		2nd Interglacial MINDEL		*Homo erectus*	HOMO ERECTUS	
	—500,000—	1st Interglacial Günz				
EARLY PLEISTOCENE (Villafranchian)			EARLY PALEOLITHIC	*Australo-pithecus*	AUSTRALOPITHECUS	
		Pre-Günz?				
	—2,000,000—					
PLIOCENE						

FIGURE 15–22   Chronologic chart of Pleistocene fossil humans.

■ Homo sapiens

The Neanderthals

From the *Homo erectus* stage of the Middle Pleistocene (Fig. 15–22), it is only a short step to Late Pleistocene *Homo sapiens neanderthalensis* (Fig. 15–23). The initial specimen of this early human was found in the Neander Valley near Dusseldorf, Germany. Subsequently, sufficient additional fossils have been found to indicate that the **Neanderthal** people ranged across the entire expanse of the Old World. With their heavy brow ridges and prognathous, chinless jaws, Neanderthals have become the very personification of the "cave man." Indeed, the face seems a brutish carryover from Middle Pleistocene. However,

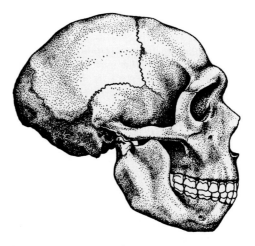

FIGURE 15–23   Right lateral view of the skull of *Homo sapiens neanderthalensis*.

C O M M E N T A R Y

Neanderthal Ritual

The Neanderthals lived from about 125,000 to about 40,000 years ago. Their culture included chipped flint tools, crude carvings, the use of fire, and evidence of deliberate burial of their dead in carefully prepared graves. Some of the burial sites contain indications that these ancient members of our own species may have had religious beliefs. The La Chapelle fossil site in southwestern France, for example, contains the body of a Neanderthal male placed in ritual position within a shallow grave with the leg of a bison on his chest. Flint tools were also placed in the grave, possibly in the belief that they could be used in an afterlife.

At the Ferrassie Neanderthal site in France, an apparent family cemetery was discovered within a rock shelter. Two adults, presumed to be a father and mother, are buried head to head. Nearby, three small children and a newborn infant are buried. Once again, flint flakes and bone splinters were placed in the grave of the adult male, and a heavy flat stone was placed over his head and shoulders.

Perhaps the stone was to protect the body, or possibly to restrain the deceased from returning to life.

That Neanderthals had a special attitude about death is also evident at a fossil site known as Teshik-Tash in former Soviet Uzbekistan. Anthropologists working at Teshik-Tash uncovered an array of goat horns surrounding the buried body of a nine-year-old child.

Yet another particularly interesting Neanderthal fossil site is located in Shanidar, Iraq. The remains of nine Neanderthals have been uncovered at Shanidar, and four of these were deliberately buried. One of the burials contains an assortment of pollen grains, indicating that flowers had been placed in the grave. Although not related to funerary ritual, the Shanidar site includes the remains of a 30-year-old male who had been born with a crippled arm. The arm had been amputated below the elbow, and this hardy Neanderthal survived the surgery. The excessive wear on his teeth suggests that he used them to compensate for the missing limb.

---

in other features, the neanderthaloids were modern. Below the neck, their skeletons matched our own, and their brain size equalled or exceeded that of present humans. Thus, the once popular depiction of Neanderthal as a bent-kneed, flat-footed, in-toed, bull-necked brute with a curved back is false. Reexamination of the skeleton on which the original restoration had been based revealed that it was that of an elderly individual, between 60 and 70 years old, who had been severely afflicted with osteoarthritis.

Most of the so-called classic Neanderthals were sturdy people of small stature that had adapted to life in the cold climates near the edge of the ice sheet. The relatively short limbs and bulky torso may have been an advantage in helping them to conserve body heat. Neanderthals often lived in caves (Fig. 15–24) and successfully hunted many contemporary cold-tolerant mammals, including cave bears, mammoths, woolly rhinoceroses, reindeer, bison, and fierce ancestors of modern cattle known as aurochs. They

FIGURE 15–24 Reconstruction of a Neanderthal family group. *(Courtesy of the National Museum of Natural History.)*

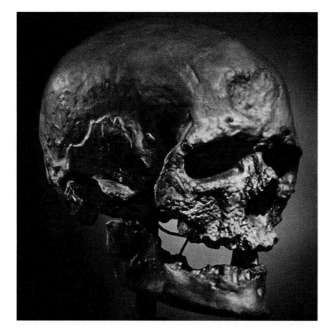

FIGURE 15–25 Skull of a Cro-Magnon human dated at about 30,000 years old. As in present-day humans, the face of Cro-Magnon was vertical rather than projecting. The heavy supraorbital ridges of the European Neanderthals are not present, and there is a prominent chin. *(Photograph of Wenner–Gren Foundation replica by David G. Gantt.)*

were not at all devoid of culture and manufactured a variety of stone spear points, scrapers, borers, knives, and saw-edged tools. In addition, Neanderthals made ample use of fire and apparently could ignite one at will in the hearths they excavated in the floors of caves.

The use and control of fire had many advantages for Pleistocene hominids. It provided light in caves that otherwise would have been avoided because of the near helplessness of humans in deep darkness. It gave warmth to creatures that had evolved in the tropics but had moved northward into colder realms, and it gave protection at night from predators. During the frigid winters of the Ice Age, fire provided the means to thaw meat that had become quickly frozen after a kill. Conceivably, cooking may have been a simultaneous result of the thawing process. Cooked foods provided the added advantages of promoting easier digestion and destroying harmful microorganisms by heat.

In addition to their use of fire, there is evidence that Neanderthals constructed shelters of skins, sticks, and bones in areas where caves may not have been available. Perhaps as a hunting ritual, these people killed cave bears and stacked the skulls carefully in chests constructed of stones. That they also pondered the nature of death and believed in an afterlife is indicated by their custom of burying artifacts along with the dead.

Cro-Magnon

About 34,000 years ago, during the fourth glacial stage, humans closely resembling modern Europeans moved into regions inhabited by the Neanderthals and in a short span of time completely replaced them, probably by tribal warfare and by competition for hunting grounds. The new breed, designated *Homo sapiens sapiens* but informally dubbed **Cro-Magnon** (Figs. 15–25 and 15–26), were mostly taller than their predecessors, had a more vertical brow, and had

FIGURE 15–26 Cro-Magnon hunters. *(Courtesy of the Field Museum, Chicago.)*

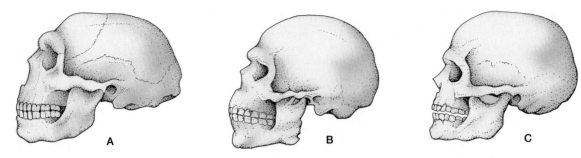

FIGURE 15–27 Comparison of the skulls of (A) *Homo neanderthalensis;* (B) a neanderthal/Cro-Magnon transitional form from a rock shelter on the slope of Mount Carmel; and (C) Cro-Magnon. The Mount Carmel skull is intermediate in both form and age between neanderthal and Cro-Magnon.

a decided projection to the chin. In short, Cro-Magnon's bones were modern, and anthropologists have recognized definite Cro-Magnon skull types among today's western and northern Europeans, as well as North Africans and native Canary Islanders.

The modern types of *Homo sapiens* did not appear suddenly on the evolutionary scene. Changes in dentition and supporting facial architecture were not an "overnight" occurrence. Transitional forms between Neanderthal and Cro-Magnon do exist (Fig.

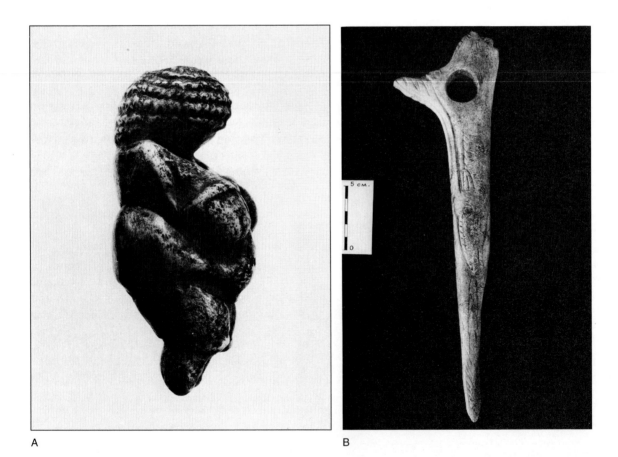

FIGURE 15–28 Prehistoric art of Late Pleistocene *Homo sapiens.* (A) Venus figure originally found in Austria ("The Venus of Willendorf"). (B) Thong-stropper used either to work hide thongs or to straighten arrow shafts. The tool is made from an antler and is intricately carved. *(Courtesy of Musée de l'Homme, Paris.)*

15–27) and include fossil skulls from Palestine, South Africa, Germany, and Czechoslovakia. These forms vary among themselves but generally show less robust brow ridges, the beginnings of the modern chin, and less pronounced muscle markings than are evident in typical Neanderthal specimens. The transitional forms seem to have evolved outside Europe and, after they had reached the Cro-Magnon stage, migrated into Europe and Asia during a temporary regression of the ice sheets about 35,000 years ago.

The cultural traditions of the Neanderthals appear to have been continued and further developed by Cro-Magnon. The variety and perfection of stone and bone tools were increased. Finely crafted spear points, awls, needles, scrapers, and other tools are found in caves once inhabited by these people. These caves also retain splendid paintings and drawings on their walls and ceilings. Carvings and sculptures of obese women (Fig. 15–28), probably used in fertility rites, were produced from fragments of bone or ivory, as were small, elegant engravings and statues of mammoths and horses. Cro-Magnon people enjoyed wearing body ornaments and frequently fashioned necklaces from pieces of ivory, shells, and teeth. Burial of the dead became a truly elaborate affair. Hunters were buried with their weapons and children with their ornaments. This apparent concern for an afterlife, and the sense of self-awareness that resulted in art and complex ritual, suggests that the beginning of the age of the philosopher had arrived.

Through most of their early history, *Homo sapiens* was a wandering hunter and gatherer of wild edible plants. However, about 10,000 to 15,000 years ago, near the beginning of the Holocene Epoch, tribes began to domesticate animals and cultivate plants. They learned to grind their tools to unprecedented perfection and to make utensils of fired clay. With more reliable sources of food, permanent settlements were developed, and individuals, spared the continuous demands of searching for food, were able to build and improve their cultures. Languages improved, and symbols developed into forms of writing. The era of recorded history had begun.

■ *Early Humans in the Americas*

The fossil record for hominids in the New World is discontinuous and often ambiguous. More often than is desirable, anthropologists must trace human presence on the basis of fragments of tools or weapons rather than actual skeletal remains. These clues, however, suggest that humans wandered into Alaska during the late phase of the Wisconsin glacial stage, when the Bering Straits provided a dry land bridge. It is likely that tribal groups seeking new hunting areas then drifted southeastward across the less mountain-

ous parts of Alaska, Canada, and western United States. Precisely when these migrations occurred is difficult to determine and is often hotly debated. Very crude flints, tentatively reported to be artifacts, have been found in alluvial fan sediments in California and may be around 30,000 years old. Charred bones of a dwarf mammoth believed to have been deliberately roasted in a campfire have been found on Santa Rosa Island off the coast of California. Carbon-14 dating indicates that the bones are about 28,000 years old. At Tule Springs, Nevada, fire pits determined to be 24,000 years old have been found in questionable association with flaked flints. In Mexico, an obsidian blade was found under a log that was dated at about 23,000 years. All of these dates, however, are questionable.

There are a number of more recent sites that yield indirect evidence of **paleoindians**. Projectile points are found in definite association with extinct Late Pleistocene elephants and bison in the western United States. Sites near Clovis and Sandia, New Mexico, have yielded the distinctive tools of two separate cultures that may have lasted from about 13,000 to 11,000 years ago. A somewhat younger group of paleoindians lived from about 11,000 to 9000 years ago. This group has been named the Folsom Culture. Folsom people manufactured short, finely flaked, and fluted points that were mounted on shafts. These flints have been found in kill sites in association with extinct species of bison.

The search for skeletal parts of early humans in America has provided fewer discoveries than has the search for their artifacts. However, there have been a few significant finds. A skull unearthed near Laguna Beach, California, has been dated by radiocarbon techniques as between 15,680 and 18,620 years old. Charred human bones from a cave on Marmes Ranch in southeastern Washington state have been dated at 11,500 to 13,000 years old. A projectile point, a bone needle, and the roasted bones of rabbits, deer, and elk have been found in association with the remains of "**Marmes Man.**" The oldest known human remains from South America consist of a portion of a jaw and teeth from Guitarrero Cave in northern Peru. These remains are believed to be about 12,600 years old. Other Pleistocene diggings in Florida, Mississippi, California, Texas, Venezuela, Chile, and Argentina clearly indicate the presence of humans in the Americas prior to 11,000 years ago.

■ *Population Growth*

One of the many lessons of historical geology is that the advent of *Homo sapiens* represents only one recent and momentary event along the sinuous, branching, 500-million-year evolution of vertebrates. We humans are linked by similarities of structure and body chemistry to lungfish struggling out of stagnating Devonian pools, to the small, shrewlike precursors of the primate order, and to the axe-carrying hunters that wandered along the margins of great ice sheets. The descendants of those and other Pleistocene hunters have emerged as the dominant species of higher life presently on this planet. Like other animals, *Homo sapiens* has been shaped by the combined powers of genetic change and environment. However, our species has quickly become a pervasive force in modifying the very physical and biological environment from which it was shaped. Humanity has come to rely heavily on science to improve its lot, but has had great difficulty in finding ways to manage the resulting technology and to control the burgeoning problems arising from too few resources for too many humans.

About 10,000 years ago (a mere moment to a geologist), there were more than 6 million humans on the Earth. By the year A.D. 1, the number had jumped 50-fold to 300 million. In 1970, the figure stood at 3.6 *billion*. If we project current rates of increase into the future, there will be 7.5 billion people on Earth by the year 2000 (Fig. 15–29). Only a decade into the twenty-first century, global population is likely to exceed 10 billion. Can the Earth sustain so enormous a population? The resources of our planetary home are limited, and environments are readily damaged. Already we have squandered about a fifth of the topsoil needed to grow food. Over the past five decades we have lost a third of the globe's forested areas, and about an eighth of all agricultural land has been altered to barren desert, partly as a result of overgrazing associated with the demand for meat. There has been loss of soil productivity stemming from accumulation of mineral salts in soils caused by the extensive practice of irrigation. Water shortages have become critical in many regions around the world. In the air, greenhouse gases have increased by more than a third, and there has been a 5 percent reduction of the ozone layer that protects all life from lethal ultraviolet radiation. Our mineral resources are also finite. As they are consumed and become rare or require more expensive extraction and refining, costs rise and standards of living decline in even the richest of nations.

With concerns such as these, few would doubt that the survival of *Homo sapiens* will depend on our ability not only to stabilize world population, but to protect agricultural lands, reforest the Earth, conserve and recycle mineral resources, and stop polluting our environment. The task is vast in scale and complex. Yet if we fail in the endeavor, other animals may replace our species and add their own distinctive chapters to the history of life on Earth.

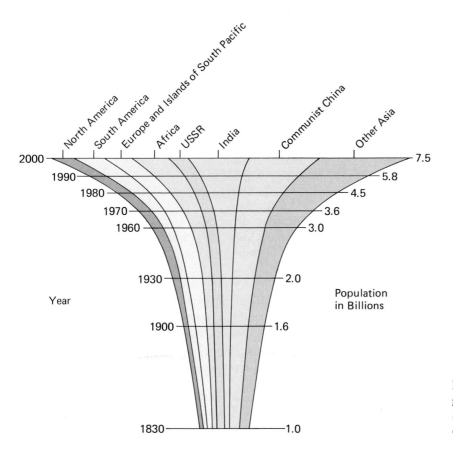

FIGURE 15–29   World population growth projected to the year 2000. *(From U.S. Dept. of State information circular.)*

Summary

During the early epochs of the Cenozoic, the main groups of placental mammals were differentiated. One group, the probable descendants of primitive insectivores, took to the trees of dense early Cenozoic forests, where food was plentiful and natural enemies few. The earliest of these primates have been named prosimians, literally "pre-monkeys." They were widespread during the Paleocene and Eocene but decreased during the Oligocene Epoch. One reason for the decline among prosimians was the appearance and differentiation of monkeys and apes. Looking back over early Cenozoic primate evolution, we can see the beginnings of the transformation of snouty, squirrel-like prosimians into progressive anthropoids. The head was reshaped and rounded, the muzzle was lost, and the face was flattened. The eyes were moved forward to provide good binocular vision, not only for quickly estimating the distance of nearby insect prey but also for accurately gauging the distance to the next bough. Manual dexterity became important for hunting, feeding, and gripping branches. Forelimbs with clawed digits became arms and hands equipped with an opposable thumb and an improved sense of touch. Swinging from branch to branch by the arms paved the way toward the erect posture that would be exploited at a later stage in primate history.

During the Miocene, a group of apes appeared that were the probable ancestors of both modern apes and the hominid line. They were the dryomorphs (the dryopithecines of older nomenclature). *Proconsul africanus* was a Miocene primate whose skull, face, and teeth make it a candidate for the subsequent apes, but whose limbs were not yet apelike. Other dryomorphs of the Miocene were more completely apelike and were widespread throughout Europe and Asia. In many ways, the Miocene dryomorphs were the evolutionary products of changing late Cenozoic environments. Dense, junglelike forests that had cloaked entire regions of Africa were being replaced by grassy, parklike habitats called savannas. The semierect dryomorphs and later the australopithecines were quick to respond with a restructuring of the axial skeleton, pelvic girdle, and hind feet. With this improved, erect stability, the hands were now completely freed for manipulative purposes. These creatures probably took advantage of the protection inherent in unified group action and may have had the ability to communicate, at least to the same degree as do modern apes. All of these functions led ultimately to greater development of the cerebral cortex, the part of the brain associated with dexterity, memory, learning, and thought. Great reproductive advantage must have accrued to those species that made the best use of their brains. Although growth in brain size was slow until the end of the Pliocene, it accelerated phenomenally during the Pleistocene.

The australopithecines appeared on Earth in the Pliocene about 4.4 million years ago. Their cranial capacity was about 600 cc, only about a third as large as that of humans. However, their dentition closely resembled that of modern people. Australopithecines were upright in stature and used crude tools. There is no evidence that they had discovered the use of fire. Structurally, the australopithecines were close to the grade of primate evolution that might warrant their being considered members of the genus *Homo*. In fact, they seem to have been either the ancestors or the contemporaries of larger brained creatures currently being designated as species of *Homo*.

The next stage of human evolution witnessed the appearance of *Homo erectus*, about 500,000 to 1 million years ago. These erect-walking primitive people had a brain capacity that ranged from 775 to 1300 cc. By Middle Pleistocene time, *Homo erectus* had spread widely throughout Europe and Africa and had progressively improved their tool-making abilities.

The Late Pleistocene was marked by the appearance of the very distinctive and famous Neanderthal people. Remains of the Neanderthals have been discovered in dozens of European sites, as well as in the Middle East and Africa. The Neanderthal skull was more massive than our own, had a less prominent forehead and chin, and had more prominent ridges above the eyes. The dozens of skulls already collected indicate that the Neanderthal peoples varied as widely in characteristics as do modern human populations. Neanderthals had brains as large as or larger than modern humans. Their tools were more finely constructed than those of *Homo erectus*, they fashioned objects for aesthetic purposes, they apparently had a religious concern about an afterlife, and they knew the use of fire.

Fire was of immense importance to Pleistocene humans. Because of the light that fire provides, the length of time during which they could effectively accomplish tasks was increased. Fire provided protection during the night and permitted *Homo erectus* to occupy cave shelters with less danger from great bears and other predators. (This alone has been a boon to anthropologists, in that it narrows down the number of places that are likely to have good fossils.) Fire also gave early humans the means to thaw and cook food. Its most important value, however, was that it provided warmth, thus enabling humans to extend their range into frigid latitudes.

Modern *Homo sapiens* appeared during the second interglacial stage of the Pleistocene. They continue down to the present day in the form of you and me. An early group, known as Cro-Magnon, replaced the Neanderthals in Europe and continued their cultural traditions. Even finer tools and weapons in bone and flint were manufactured by these peoples. They exhibited remarkable artistic skills, painting pictures of the animals vital to their survival on the walls of their caves and carving voluptuous Venus-type statues to encourage fertility. Humans came to the Americas at a rather late date. Artifacts and sparse skeletal remains indicate their presence somewhat over 20,000 years ago. Their spear heads are sometimes found among the remains of the extinct large mammals that they hunted.

Unlike our Ice Age ancestors, twentieth-century humans face difficult problems caused by the population explosion. Overpopulation is important because it is a major contributor to all other environmental and resource problems. More people use more oil and more metals, cut down more trees, deplete more farm land, and add more sewage to rivers, more garbage to landfills, and more pollutants to the air. For these reasons, finding ways to achieve population stability should have high priority among the nations of the world.

Review Questions

1. Distinguish between the following:
   a. Hominoidea and Hominidae
   b. Prosimii and Anthropoidea
   c. Hylobatidae and Pongidae
   d. Cercopithecoidea and Ceboidea
2. Early arboreal primates such as *Notharctus* had close-set eyes and grasping hands. Other than facilitating movement among tree branches, what other function might these attributes have provided?
3. During what geologic period do we find the earliest known remains of primates? When does the first member of the genus *Homo* appear?
4. What physical properties of chert and flint render them useful in the manufacture of spear points, axes, and scrapers by early humans? How do Oldowan, Acheulean, and Mousterian tools differ? When during the Pleistocene was each being manufactured?
5. What is the evolutionary importance of *Proconsul africanus*?
6. Why is it easier to find fossils and to date them in the East African fossil sites than in many other parts of the world?
7. What distinctly human-like traits were possessed by *Australopithecus*? What apelike characteristics were retained?
8. What features of Cro-Magnon people differentiate them from European Neanderthal people?

Discussion Questions

1. Discuss the general changes in the skulls of fossil primates as one progresses from an animal like *Plesiadapis* to modern humans.
2. Discuss the route by which *Homo sapiens* entered the New World. Would the migrations have been easier during a glacial or interglacial epoch? Why?
3. Discuss the relationship between the width of the pelvic girdle in female *Homo sapiens* and the enlargement of the brain seen in the evolution of Pleistocene hominids.
4. Prepare text for a debate in which you first argue for and then against the premise that life will be more difficult for humans during the year 2050 than it is today because of overpopulation.

Supplemental Readings and References

Brace, C. L. 1991. *The Stages of Human Evolution,* 4th ed. Englewood Cliffs, NJ: Prentice Hall.

Fleagle, J. G. 1988. *Primate Adaptation and Evolution.* San Diego: Academic Press.

Johanson, D. C., and Edey, M. A. 1981. *Lucy, the Beginnings of Humankind.* New York: Simon and Schuster.

Keyfitz, N. 1989. The growing human population. *Sci. Am.* 261(3).

Leakey, R. E. F. 1976. Hominids in Africa. *Am. Scientist* 64:174–178.

Lewin, R. 1984. *Human Evolution.* New York: W. H. Freeman and Co.

Lewin, R. 1993. *The Origin of Modern Humans.* New York: Scientific American Library.

Pilbeam, D. 1972. *The Ascent of Man.* New York: Macmillan.

Reader, J. 1981. *Missing Links: The Hunt for Earliest Man.* Boston: Little, Brown & Co.

Rosen, S. I. 1974. *Introduction to the Primates.* Englewood Cliffs, NJ: Prentice Hall.

Simons, E. L. 1972. *Primate Evolution.* New York: Macmillan.

Tattersall, Ian. 1993. *The Human Odyssey: Four Million Years of Human Evolution.* Englewood Cliffs, NJ: Prentice Hall.

False-color image of Venus produced by the *Magellan* mission using a special radar to see through the planet's clouds and map the surface below. The mosaic combines many consecutive image strips, each about 20 km wide. The bright areas usually denote rough surfaces and the dark areas are smooth or possibly covered with dust. *(Courtesy of NASA.)*

"You appear to be astonished," [Sherlock Holmes] said, smiling at my expression of surprize. "Now that I do know, I shall do my best to forget it . . ."

"But the Solar System!" I protested.

"What the deuce is it to me?" he interrupted impatiently. "You say that we go around the sun. If we went around the moon it would not make a pennyworth of difference to me."

SIR ARTHUR CONAN DOYLE, *A Study in Scarlet*

CHAPTER 16

Meteorites, Moons, and Planets

In this book, we have traced much of the history of the Earth since its origin about 4.6 billion years ago. We are aware that the Earth is only one of nine planets that circle the Sun and that each of these neighboring planets has an interesting history as well. It thus seems appropriate to consider briefly the characteristics of these planets and the other extraterrestrial bodies that are the Earth's nearby companions in space, including meteorites and asteroids.

Meteorites and asteroids have pock-marked the surface of our nearest celestial neighbor, the Earth's moon. Centuries ago, Moon watching provided an understanding of some planetary movements, a basis for the calendar, and the means to predict tides. In our laboratories, we have Moon rocks that contain clues to the satellite's origin.

Beyond the Moon lie the inner or terrestrial planets: Mercury, Venus, Earth, and Mars. All are relatively small and dense (Table 16–1). Mercury retains the scars of millions of meteoric impacts, has volcanic activity, and possesses a magnetic field. Beyond Mercury lies Venus. Our understanding of that planet has been extended by high-resolution radar images provided from NASA's *Magellan* spacecraft. Mars data collected decades ago still provide insight into the red planet's atmosphere, surface features, and both internal and surficial processes.

The giant planets, Jupiter, Saturn, Uranus, and Neptune, are much larger but less dense than the

TABLE 16–1   Planet Data

	Mercury	Venus[2]	Earth	Mars	Jupiter	Saturn	Uranus	Neptune	Pluto
Mean distance from the Sun (AU)[1]	0.387	0.723	1	1.524	5.202	9.555	19.218	30.109	39.439
Sidereal period of orbit (years)	0.24	0.62	1	1.88	11.86	29.46	84.01	164.79	247.68
Mean orbital velocity (km/s)	47.89	35.04	29.79	24.14	13.06	9.64	6.81	5.43	4.74
Equatorial radius (km)	2439	6052	6378	3397	71492	60268	25559	24764	1140
Polar radius (km)	same	same	6357	3380	66854	54360	24973	24340	same
Mass of planet (Earth = 1)[3]	0.06	0.82	1	0.11	317.89	95.18	14.54	17.15	0.002
Mean density (gm/cm[3])	5.44	5.25	5.52	3.94	1.33	0.69	1.27	1.64	2.0
Body rotation period (hours)	1408	5832.R	23.93	24.62	9.92	10.66	17.24	16.11	153.3
Tilt of equator to orbit (degrees)	0	2.12	23.45	23.98	3.08	26.73	97.92	28.8	96
Number of observed satellites	0	0	1	2	16	18	15	8	1

[1]AU indicates one astronomical unit, defined as the mean distance between the Earth and sun ($\approx 1.495 \times 10^8$ km).

[2]R indicates that planet rotation is retrograde (opposite to planet's orbit).

[3]Earth's mass is approximately 5.976×10^{26} grams.

Source: Boyce, J. M., and Maxwell, T. 1992. *Our Solar System.* Washington, DC: NASA, Publication NP 157.

inner planets. They have many moons, and these are studied for information about the origin of the planets they orbit. Jupiter, with its colored bands and Great Red Spot, is the largest planet in the solar system. Saturn, with its rings composed of thousands of ringlets, is no less interesting. Uranus has a thick blanket of methane clouds that obscure its surface. It is a planet whose axis lies in the plane of the solar system. Finally, there is the tiny deviant named Pluto. Together with its moon Charon, it travels the most eccentric orbit in the solar system.

FIGURE 16–1   A meteor crossing the field of view of the Palomar telescope while it was taking a 15-minute exposure of the Comet Kobayashi–Berger–Milon, August 1975.   *(Hale Observatory photography/John Huchra; from Pasachoff, J. M.* Contemporary Astronomy, *4th ed., Philadelphia: Saunders College Publishing, 1989.)*

FIGURE 16–2 Aerial view of Meteor Crater, formed by the impact of a large meteorite about 50,000 years ago. Conconino County, Arizona. *(Courtesy of W. B. Hamilton.)*

■ *Meteorites: Samples of the Solar System*

Because meteorites are masses of mineral or rock that have reached the Earth from space, they are of great importance to scientists interested in the origin and history of planets. They are the only objects from the universe beyond the Moon and the Earth that can currently be held in the hand and scrutinized in the laboratory. This research indicates that meteorites are fragments of asteroids formed when asteroids collided in space and were shattered. Such collisions would send chunks of rock into Earth-crossing orbits. Many meteorites bear compositional and textural evidence of the shock of such collisions. A few meteorites found in Antarctica are chemically similar to lunar basalts and are believed to have come from our Moon. Also of interest is at least one group of meteorites having similarities with the Martian materials analyzed by *Viking* landers. Tentatively, they may be considered to have come from that planet.

Some meteorites may have once been part of meteors (Fig. 16–1), those rapidly moving, luminous bodies we refer to as shooting stars. Most of these particles burn up when they enter the Earth's atmosphere, but if a portion survives its fall through the atmosphere and crashes into the Earth, it is then termed a **meteorite**. About 500 meteorites at least as large as a baseball reach the Earth's surface each year. Weathering and erosion on the Earth have erased most of the craters made by their impacts. Only about 70 partially preserved craters or clusters of craters remain.

The best preserved meteor crater in the world is Meteor Crater in Arizona (Fig. 16–2). The impact that produced Meteor Crater 50,000 years ago is estimated to have released energy equivalent to a 20-megaton nuclear bomb. In the past, much larger impacts have occurred. Meteor showers or the impacts of asteroids may even have contributed to episodes of mass extinctions on Earth. There is evidence, for example, that an asteroid with a diameter of 10 km or greater struck the Earth about 66 million years ago, lifting millions of tons of dust into the atmosphere. The dust blocked the Sun's rays so that darkness and cold prevailed for several weeks, or possibly even several months. These effects could certainly have been catastrophic for life on Earth and have been proposed as a major cause for the extinction of the dinosaurs (see Chapter 12).

Meteorites can be classified according to their composition as ordinary chondrites, carbonaceous chondrites, achondrites, irons, and stony-irons (Fig. 16–3). The most abundant of these types are the or-

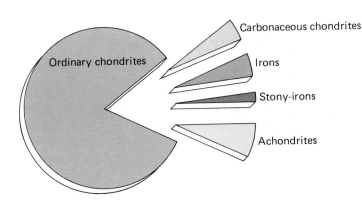

FIGURE 16–3 The major categories and proportions of meteorites.

FIGURE 16-4 Stony meteorite of the type termed a chondrite. Chondrites contain small, rounded particles called chondrules composed of ferromagnesian minerals. In most cases, chondrules are only about 2 or 3 mm in diameter, but the one visible in the lower right area of this specimen is approximately 2 cm in diameter. *(Washington University collection.)*

FIGURE 16-5 A carbonaceous chondrite. This is a piece of a larger meteorite that apparently fragmented on entering the atmosphere above Mexico in 1869. The outer crust has been fused by heat. The white inclusions are minerals (refractory minerals) that withstand high temperatures without melting. Carbon imparts some of the dark color to the meteorite. *(Courtesy of T. J. Bernatowicz.)*

dinary chondrites (Fig. 16-4), which are crystalline stony bodies composed of high-temperature ferromagnesian minerals. Ordinary chondrites can be dated by the uranium–lead, strontium–rubidium, and potassium–argon methods, and they are found to be about 4.6 billion years old. Many ordinary and carbonaceous chondrites contain spherical bodies called **chondrules** (see Fig. 16-4). Their spherical shape suggests that chondrules solidified from molten droplets splashed into space during collisions of objects swirling about in the solar nebula.

Less abundant than the ordinary chondrites are the **carbonaceous chondrites** (Fig. 16-5). In these meteorites, one finds the same abundance of metallic elements as in ordinary chondrites, with the addition of about 5 percent organic compounds, including inorganically produced amino acids. Suspended in the blackish earthy matrix of carbonaceous chondrites are both chondrules and irregular pieces of high-temperature minerals that may have condensed from a cooling vapor.

One of the most interesting findings about carbonaceous chondrites is that they contain nonvolatile elements in approximately the same proportion as those that occur in the Sun. Indeed, if it were possible to extract some solar material, cool it down, and condense it, the condensate would be chemically similar to carbonaceous chondrites. This similarity suggests that carbonaceous chondrites are samples of primitive planetary material that formed when the Sun formed and have never been remelted.

A small percentage of stony meteorites do not contain chondrules and therefore are termed **achon-**

FIGURE 16-6 Meteorite of the achondrite variety, found in Sioux County, Nebraska. *(Washington University collection.)*

FIGURE 16–7 An iron meteorite. The meteorite has been cut, polished, and etched with acid to show an interlocking mosaic of the nickel-poor mineral kamacite, and nickel-rich mineral taenite. The distinctive interlocking texture is called Widmanstätten pattern. *(From the Washington University collection.)*

drites (Fig. 16–6). Most of these have compositions similar to terrestrial basalts, and many exhibit angular broken grains. These fragmental textures indicate that achondrites may be the products of collisions of larger bodies.

The **iron meteorites** (Fig. 16–7) comprise about 6 percent of all observed falls of meteorites. Most iron

meteorites are intergrowths of two varieties of iron–nickel alloy. The large size of the crystals and metallurgic calculations indicate that some iron–nickel meteorites cooled as slowly as 1°C per million years. Such a slow rate of cooling would be possible only in objects at least as large as asteroids. Thus, the history of the iron meteorites probably involved a long episode of slow crystallization within a parent body sufficiently large to provide insulation for the hot interior, followed by violent disruption of that body by a collision.

The least abundant of all the categories of meteorites are the **stony-irons** (Fig. 16–8). As indicated by their name, they are composed of silicate minerals and iron–nickel metal in about equal proportions. The stony-iron meteorites are generally considered to represent fragments from the interfaces between the silicate and metal portions of asteroidal bodies. Some of them may have originated at the boundary between an iron core and the silicate mantle of a planetary body destroyed by collision.

■ *The Moon*

Moon Features

As satellites go, our Moon is large in relation to the size of its parent planet. It has a diameter of over one-fourth that of the Earth, and its density of about 3.3 g/cm³ is the same as that of the upper part of the Earth's mantle. Gravity on the Moon is insufficient to maintain an atmosphere. The Moon and the Earth revolve around their mutual center of gravity as a sort of "double planet." The Moon orbits the Earth and rotates on its axis at the same rate. As a result, we always see the same side of the moon and must depend on transmissions from space vehicles for images of the far side (Fig. 16–9). These photographs have revealed a densely cratered surface interrupted only by few and relatively small level areas.

Telescopic observation of the near side of the Moon was first made in 1609 by a rather irascible professor of mathematics at the University of Padua. The mathematician was Galileo Galilei. With the use of his primitive telescope, Galileo was able to see that the surface of the moon was not smooth but "uneven, rough, full of cavities and prominences." He recognized "small spots," which we now know are craters, and dark patches that children imagine to be the eyes and mouth of the "man in the moon." These dark patches lie within large basins. Galileo named them **maria** ("seas") and incorrectly suggested that they were filled with water. Today we know the dark

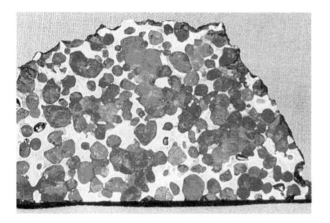

FIGURE 16–8 Stony-iron meteorite. The light-colored areas consist of iron–nickel, and the darker minerals are mostly olivine. *(Washington University collection.)*

FIGURE 16–9   The far side of the Moon as photographed by *Apollo 16*.
(Courtesy of NASA.)

areas have been flooded, not with water, but with basaltic lavas. In many maria, the lava has been contained within depressions that are hundreds of kilometers across and 10 to 20 km deep. These depressions are termed **mare basins** (Figs. 16–10 and 16–11). Mare basins had to have developed prior to the extrusion of the dark lavas that form the maria, for these lavas are contained within the basins.

Galileo also provided the name for the lighter hued rougher terrains of the Moon. He called these regions terrae, although modern space scientists refer to them as **lunar highlands.** The highlands are the oldest parts of the Moon, having already formed long before the earliest mare lavas appeared. The highlands are heavily cratered regions reminiscent of a battlefield following a heavy artillery barrage. They provide stark evidence of the Moon's early episodes of intense meteoritic bombardment.

On the Moon, as on the Earth, craters are roughly circular, steep-sided basins normally caused by either volcanic activity or meteoritic impact (Fig. 16–12). Although a very few lunar craters show questionable evidence of volcanic origin, the great majority are clearly the result of meteoritic impact. Unlike the Earth, the Moon has insufficient atmosphere to burn up approaching meteorites, and thus the frequency of impact is high. Lunar geologists are able to determine the relative ages of certain areas on the Moon's surface by the density of craters. Younger areas have fewer craters that do older ones. Another aid to recognizing differences in the age of lunar features is provided by the rays of material splashed radially outward from the crater at the time of impact. Younger rays and other impact features cross and partially cover older features. Light-colored impact rays composed of finely crushed rock are exception-

FIGURE 16–10 The near side of the Moon. To show the whole Moon but still reveal the detail that does not show up well at full Moon, the Lick Observatory has put together this composite of first and third quarters. Note the dark maria and the lighter, heavily cratered highlands. Two young craters, Copernicus and Kepler, can be seen to have rays of light (that is, lighter than the background) material emanating from them. This is a ground-based view of the Moon, and it is shown inverted as it would be when viewed through a telescope. *(Lick Observatory photograph; from Pasachoff, J. M.* Contemporary Astronomy, *4th ed. Philadelphia: Saunders College Publishing, 1989.)*

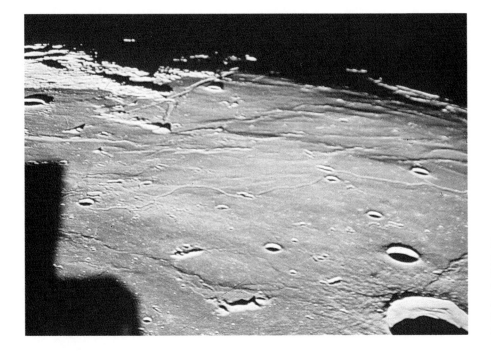

FIGURE 16–11 A view of part of the Mare Tranquillitatis (Sea of Tranquility), where astronauts aboard the *Apollo 11* spacecraft made their landing. *(Courtesy of NASA.)*

C O M M E N T A R Y

Is There a Bolide Impact in Our Future?

After reading about the probability that a huge extraterrestrial body smashed into the Earth about 66 million years ago, causing a biologic catastrophe (Chapter 13), one cannot but wonder about the possibility of a similar natural disaster for the future human population. Although often the subject of science fiction, a bolide impact capable of destroying civilization is not an unrealistic cause for concern. Meteorites the size of cobbles and boulders are frequent visitors to the Earth at the present time. Most do little harm, although globally about 16 buildings a year are damaged by meteorites. These are, however, only small chunks of extraterrestrial rock. What might be the result of an impact like the one that produced Meteor Crater in Arizona (see Fig. 16–2) about 50,000 years ago? That impacting body excavated a crater over 1000 meters in diameter while releasing energy equivalent to a 20-megaton atomic bomb. Meteor Crater is only one of about 130 such impact structures now recognized on Earth. Of more immediate concern is the knowledge that the orbits of thousands of large meteorites are known to cross the orbital path of the Earth.

Scientists at NASA estimate that a meteorite, in the size range from 50 to several hundred meters, strikes the Earth every 200 to 300 years. On impacting the Earth, the smallest of these would release energy equivalent to a 15-megaton nuclear bomb. Yet that impact would be trivial compared to the catastrophe accompanying a 1000-meter-wide impacting body. Such a smashup, releasing the energy equivalent of a million megatons of TNT, would have the potential to wipe out a quarter of the planet's human population.

As one indication that the possibility of future large meteorite impacts should be taken seriously, NASA scientists are currently examining ways to detect and intercept bodies headed our way. It has been suggested that a global network of telescopes be constructed for early detection of incoming large meteorites and asteroids. If the incoming body is spotted sufficiently early, it may then be possible to push it into a nonthreatening trajectory by exploding a nuclear bomb off to its side. On the side nearest the explosion, heat from radiation would vaporize the surface of the body, and the resulting jet of vapor might act as a rocket in nudging it off course. (It would be important that the bomb not strike the asteroid directly, for this would cause it to shatter and result in a rain of lethal fragments.)

Clearly, these are only preliminary suggestions for safeguarding civilization against a future cosmic catastrophe. The search for feasible solutions will require many years of research.

ally well developed in the large lunar crater known as Copernicus (Fig. 16–13). This familiar feature is about 90 km across and is rimmed by concentric ridges of hummocky material blasted out by the impact of the crater-forming body. Not all craters have rays as spectacular as those around Copernicus. Impact rays are sometimes obliterated by the rather continuous rain of tiny meteorites that strike the Moon. For this reason, younger craters may still be adorned with impact rays, whereas these features have been lost around older craters.

There is ample evidence of former igneous activity on the Moon. We have already noted the extensive lava floods of mare basins. In addition to these flood basalts, a few dome-shaped volcanoes measuring

FIGURE 16–12   A view of the heavily cratered far side of the Moon, photographed by the crew of the *Apollo 11* spacecraft.   *(Courtesy of NASA.)*

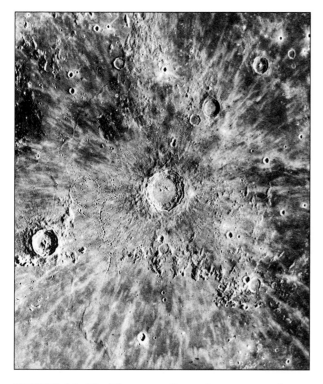

FIGURE 16–13 The crater Copernicus, as seen from an Earth-based telescope, has rays of lighter colored material emanating from it. This light material was ejected radially when the meteorite that formed the crater struck the Moon's surface. *(From Pasachoff, J. M. Contemporary Astronomy, 4th ed., Philadelphia: Saunders College Publishing, 1989.)*

about 10 km in basal diameter have been observed. About 1 percent of the lunar craters may have resulted from volcanic activity. Meandering channelways occur on the surface of our satellite. Most of these are thought to have been produced by the turbulent flow of lava or by the collapse of lava tunnels, although the evidence is not conclusive. These winding features are distinctly different from straight channels, which probably result from faulting. Tall volcanic peaks are not found on the Moon. Perhaps this is because lunar lavas, like lavas associated with shield volcanoes on the Earth, are highly fluid and spread laterally rather than piling up around a vent.

Are lunar volcanoes still active today? In 1958, Russian astronomers observed evidence of gas emission from the crater Alphonsus, and on two occasions in 1963, observers at the Lowell Observatory saw glowing red spots in and near the crater Aristarchus.

Moon Rocks

Direct examination of samples returned from the Moon (Fig. 16–14), as well as instrumental data

A

B

C

FIGURE 16–14 Moon rocks. (A) A photograph of a lunar basalt from the Sea of Tranquillity. This mare basalt was collected by *Apollo 15* astronauts. The lava solidified so rapidly that bubbles formed by escaping gases were trapped and formed vesicles. (B) A nonvesicular lunar basalt. (C) A photomicrograph of a thin section of lunar basalt composed of interlocking grains of augite (larger crystals), plagioclase (the zoned crystals), and olivine (center cluster of small crystals). The cubes are 2.5 cm square and indicate size. The area of the thin section is 3.7 mm. *(Courtesy of NASA.)*

FIGURE 16–15 Lunar rock from the lunar highlands. This coarsely crystalline rock is composed of approximately equal amounts of olivine (yellow grains) and plagioclase. *(Courtesy of NASA.)*

gathered by orbiting spacecraft, indicate that there are three main kinds of rocks on the Moon. One of these forms the floor of the smooth mare regions, whereas the other two are characteristic of the lunar highlands. The mare rocks are basalt, the finely crystalline, dark-colored igneous rock that was described in Chapter 2. Calcium plagioclase, pyroxene, olivine, and ilmenite are the principal minerals in basalt.

The rock types that occur in the lunar highlands differ from the mare basalts in both texture and composition. They are coarser (Fig. 16–15) and richer in calcium plagioclase. Most of them are aggregates of crushed rock mixed with material that appears to have been molten. Although several compositional varieties of highland rocks have been identified, in general they resemble the family of rocks on Earth known as gabbros.

The samples brought to Earth from the Moon were all obtained from the blanket of rock fragments and dust that covers the Moon's surface. Within this loose material, called **regolith**, it is possible to find fragments of rocks of all sizes, ranging from microscopic grains (Fig. 16–16) to huge boulders. Some of the boulders and cobbles consist of aggregates of smaller rock fragments welded together to form lunar breccias.

At Tranquility Base, site of the first Moon landing (1969), the regolith was estimated to be 3 to 4 meters thick. It is believed to have formed as a result of meteorite impacts, each of which could dislodge a mass of debris many hundred times greater than its own mass.

Radiometric dating of lunar rocks has been of great importance to the understanding of the se-

quence of events in the Moon's history. Most mare basalts are between 3.2 and 3.8 billion years old and were extruded during the great lava floods that produced the mare basins (Fig. 16–17). Most lunar breccias were formed during impact events that occurred prior to these floodings. The oldest rocks on the Moon were taken from the lunar highlands, and they have been found to be about 4.6 billion years old. This age represents the time that meteoritic impacts formed the rugged highland terrain and crushed and melted the rocks. Samples of Moon rocks now being studied in the United States have been found to retain weak magnetism imparted to the rocks at the time they solidified. This indicates that at one time the core of the Moon was liquid and that the Moon had a significant magnetic field between about 4.2 and 3.2 billion years ago. The Moon has no general magnetic field today.

Lunar Processes

On Earth there are processes capable of raising mountain ranges, as well as such opposing processes as erosion and weathering, which are constantly at work wearing away the lands. Geologic processes also operate on the Moon, but they are slower and of a different kind. The Moon is a desolate world in which there is no air or surface water. Whenever gases or liquids escape to the lunar surface, they are dissipated into space because the Moon lacks sufficient gravitational attraction to hold them. There can be no stream-cut canyons and valleys, no glacial deposits, and no accumulations of wave-sorted sand.

FIGURE 16–16 Lunar regolith (soil), which consists of impact fragments and tiny spheres of dark glass. Scale divisions at upper right are in millimeters. *(Courtesy of NASA.)*

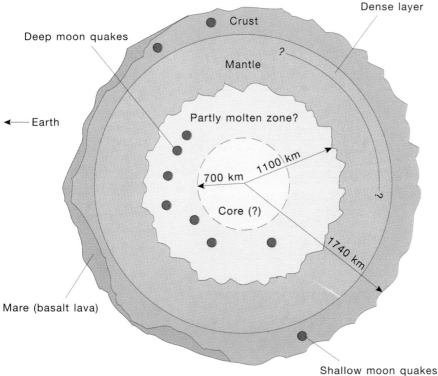

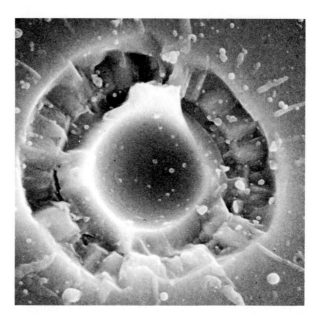

FIGURE 16–17 Cross-section through the Moon showing an outer crust, an inner mantle, and an innermost zone that may still be partly molten. Within the innermost zone there may be a small metallic core (dashed circle.) The Earth-facing side is relatively smooth and shows larger accumulations of the mare basin lavas. The far side is more rugged and has almost no lava. The drawing is not to scale, and the surface relief is exaggerated. Data for the drawing were provided by A. M. Dainty, N. R. Goins, and M. N Toksoz. *(Courtesy of NASA. From French, B. M., What's New on the Moon?)*

Rocks brought back to Earth from the Moon seem bright and newly formed and lack the discoloration common to weathered Earth rocks. Yet erosion does occur on the Moon, primarily through the continuous bombardment of the lunar surface by large and small meteorites. The larger meteorites produce craters and rough terrain, but it is the rain of small meteorites and **micrometeorites** (those with diameters less than 1 mm) that subtly alter the Moon's surface and are important in the formation of lunar soil. The upper or exposed surfaces of lunar rocks brought back to Earth from the Moon are riddled with tiny glass-lined pits produced by the impact of fast-moving micrometeorites. Geologists call these tiny depressions microcraters or **zap pits.** Most are less than 0.01 mm in diameter (Fig. 16–18). The micrometeorite barrage continues today, as was evident on examination of the *Apollo* spacecraft after its return to Earth. Ten tiny microcraters were chipped into its windows.

Although gravity on the Moon is only one-sixth that on Earth, it is nevertheless capable of moving lunar materials from high to low places. Whenever loose material on slopes is unable to resist the pull of gravity, it will break away and produce masses of slumped debris. Steplike slump masses are recognized as a common feature along the rims of some lunar craters. Elsewhere, photographs reveal tracks in lunar soils that record the rolling and sliding of large boulders down slopes.

FIGURE 16–18 Photograph of a *zap pit* or microcrater. The tiny pit is only 20 μm (0.02 mm) in diameter and was photographed with a scanning electron microscope. *(Photo by F. Horz, courtesy of G. J. Taylor.)*

A

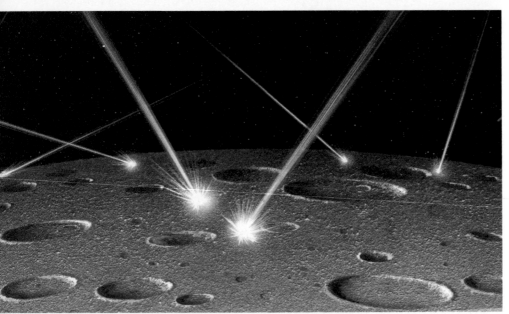

B

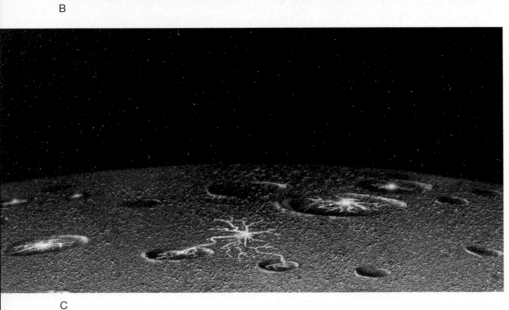

C

FIGURE 16–19 A brief history of the Moon. (A) According to one theory, the Moon was formed about 4.5 billion years ago when a Mars–size object struck the Earth, sending a cloud of vaporized rock into orbit. The vaporized rock rapidly coalesced to form the Moon. (B) During its first half-billion years, intense meteorite bombardment cratered the Moon's surface. (C) The dark, flat maria formed about 3.8 billion years ago as lava flows spread across portions of the surface. Today all volcanic activity has ceased.

Origin and Early History of the Moon

Data derived from the *Apollo* lunar missions have caused many cherished hypotheses about the origin of the Moon to be discarded and new ones formulated. The most popular of these newer hypotheses proposes that the Moon formed when an "impactor" somewhat larger than Mars smashed into the Earth about 4.4 billion years ago (Fig. 16–19). As indicated by the known composition of lunar materials, the impactor had already differentiated into a metallic core and silicate mantle. The collision spun out a huge cloud of vaporized debris derived mostly from the mantle of the impactor. Rather like the rings of Saturn, the debris cloud assumed disklike form. Shortly thereafter, its constituent particles coalesced to form the solid body we know as the Moon.

Coalescence to form the Moon involved meteor showers on a vast scale, and these infalls of impacting objects, coupled with gravitational compression, generated such huge amounts of heat that the entire outer part of the Moon became molten. While in this state, low-density alumino–silicates floated to the surface of the lunar "magma ocean" and formed the crust. High-density ferromagnesian minerals sank to deeper levels, where they ultimately crystallized to form the lunar mantle. By this time, heat from radioactive decay may have raised the temperature of the already hot interior to the point at which a small core may have formed.

For the next 300 million years or so, the Moon continued to be very hot. Magma invaded and assimilated already solidified crustal rocks, producing a complex and varied assemblage of rock types. Some magmas flowed onto the lunar surface as volcanic eruptions. The dark plains seen on the Moon's near side are composed of flows that spread across lunar landscapes between about 3.8 and 3.0 billion years ago. After the last of these eruptions, the Moon was less active. Meteorite impacts, however, continued to sculpt its surface (Fig. 16–20), sometimes forming spectacular craters such as Copernicus (see Fig. 16–13).

■ The Earth's Neighboring Inner Planets

As noted earlier, the first four planets in our solar system (Mercury, Venus, Earth, Mars) are sometimes called **terrestrial planets** because, like the Earth, they are small, dense, rocky, and rich in metals. Jupiter, Saturn, Uranus, and Neptune are large and dominated by hydrogen and helium. They are called **Jovian planets**.

Mercury

Mercury has the distinction of being the smallest and swiftest of the inner planets. It revolves rapidly, making a complete journey around the Sun every 88 Earth days. Mercury makes a complete rotation on its axis every 59.65 days, which means that the planet rotates three times while encircling the Sun twice.

FIGURE 16–20 The *Apollo 11* lander returning to the command module after the first humans landed on the Moon. In the background the Earth rises over the flat lunar plain. *(Courtesy of NASA.)*

FIGURE 16–21 Photomosaic of Mercury as photographed by the *Mariner* spacecraft. *(Courtesy of NASA.)*

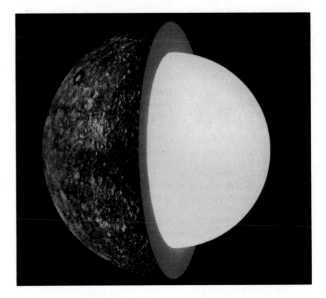

FIGURE 16–22 The interior of Mercury is dominated by a metallic core about the same size as our moon. *(University of Arizona, courtesy of Robert Strom.)*

Mercury has a weak magnetic field, a lightweight crust covered with fine dust, and an iron core that comprises nearly 75 percent of its radius (Fig. 16–22). The existence of the iron core is inferred from Mercury's high density, its magnetic field, and other evidence suggesting that the planet experienced melting and differentiation.

Venus

Although similar to the Earth in mass and volume (and therefore density and gravity), other characteristics of Venus are quite different from those of the Earth. Venus has no oceans, and its searing surface temperature of 470°C is hot enough to melt lead. Atmospheric pressure is 90 times greater than on Earth, equivalent to pressure 1000 meters beneath the Earth's ocean. In this dense atmosphere, even gentle winds are effective in transporting fine sediment. The Venusian atmosphere is 98 percent carbon dioxide, which serves as an insulating blanket that traps solar energy by the greenhouse effect. Without oceans in which carbon dioxide can be combined with calcium and magnesium to form carbonate rock, there is no process to remove carbon dioxide from the atmosphere. A thick layer of clouds composed of sulfuric acid droplets extends 40 km above the surface (Figure 16–23).

Venus, known to us as both the morning and evening star, makes one complete rotation each 243 days. Its direction of rotation is opposite that of the Earth and most other planets. A year on Venus is 225

Mariner 10, launched in 1973, provided our first close look at Mercury (Fig. 16–21). Following its launching, *Mariner 10* flew by Venus and, using the gravity of that planet to deflect its trajectory, went on to Mercury. Indeed, the spacecraft's orbit permitted visits over Mercury on three occasions. Photographs transmitted back to Earth revealed a planet with a moonlike surface. Densely cratered terrains as well as smooth areas resembling maria were apparent. Some of the craters exhibited rays of light-colored impact debris, just as do craters on the Moon. Long linear scarps were discerned, and these are thought to be fracture zones along which crustal adjustments occurred. Perhaps tidal forces or shrinkage of the planet as it cooled may have produced these elongate scarps.

FIGURE 16–23 The clouds of Venus as photographed by *Pioneer Venus Orbiter* spacecraft. In visible light, no structure at all can be discerned. *(Courtesy of NASA.)*

days. From Venus's size and density, one can infer an interior rather like that of the Earth.

The dense cloud cover of Venus had obscured its surface until the recent development of high-resolution radar imaging systems. A flood of new information began with NASA's 1978 *Pioneer* Venus mission and continued with the *Venera 15* and *Venera 16* missions of the former Soviet Union. The most spectacular images, however, have come from NASA's *Magellan* radar-mapping mission, which began in 1990. Analyses of *Magellan* and earlier images reveal that 65 percent of the planet consists of rolling plains, about 27 percent is lowlands, and the remaining 8 percent is highlands (Fig. 16–24). Among the highlands are mountains that tower 11 km above the adjacent plains. Several of the highland tracts remind one of continents (Fig. 16–25). The largest, Aphrodite Terra, is about the size of Africa and extends nearly halfway around the planet along the equator. Next in size is Ishtar Terra, a crustal segment as large

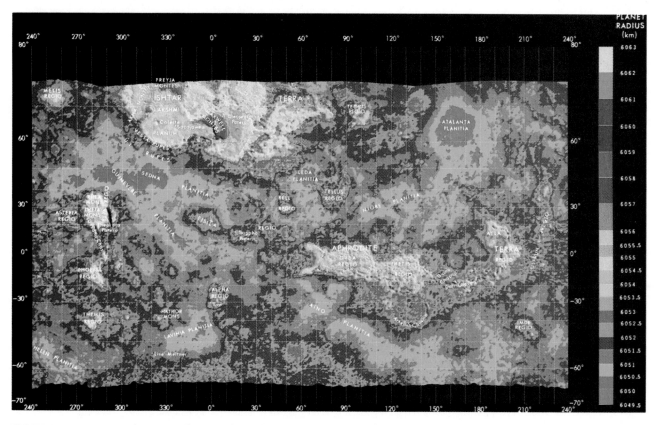

FIGURE 16–24 A radar map of Venus from the *Pioneer Venus Orbiter*. Most of Venus is covered by a rolling plain, shown in green and blue. The highlands (yellow and brown contours) sit atop the plain, like continents. Aphrodite Terra is half as big as Africa. Ishtar Terra is the size of the United States but looks relatively large in this Mercator projection. The lowlands, although resembling Earth's ocean basins, cover only 16 percent of the planet. *(Experiment and data from MIT; map from U.S. Geological Survey; NASA/Ames spacecraft.)*

FIGURE 16–25 An artist's conception of Ishtar Terra, a magnificent plateau about the size of the United States in area and higher than the Tibetan plateau. The reconstruction is based on measurements taken by the *Pioneer Venus Orbiter* spacecraft. *(Courtesy of NASA.)*

as the United States (Fig. 16–25). Tectonics operate on Venus in a different way than on Earth. There is little evidence of horizontal movements of the crust. One does not see transform faults or deep trenches. Instead, vertical movements appear to predominate, suggesting that areas of the crust were raised from below by rising plumes of viscous magma. In addition to this so-called "blob tectonics," volcanic eruptions and massive lava flows have been important in the evolution of the Venusian crust. Many of the vol-

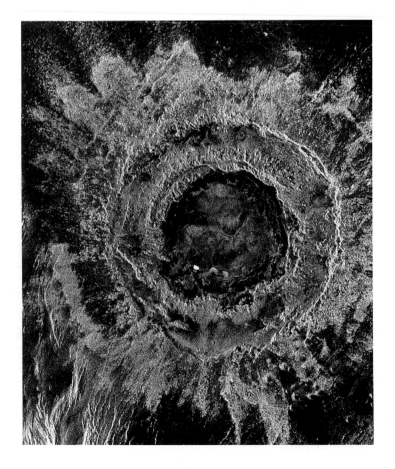

FIGURE 16–26 *Magellan* image of the impact crater Lise Meitner. This Venusian crater is about 150 km in diameter. The radar-bright rings are scarps similar to those seen around impact basins on the Moon. The lighter area surrounding the crater comprises the blanket of debris resulting from the impact. The scale bar is 100 km.
(Courtesy of NASA as seen in Alexopoulos, J. S., and McKinnon, Wm. K. 1994. Geological Society of America Special Paper 293.)

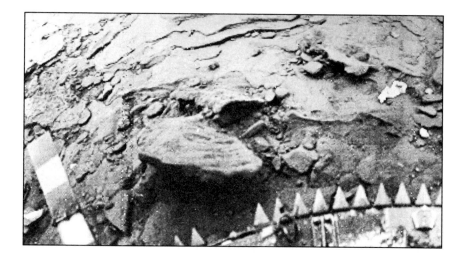

FIGURE 16–27 A view of the surface of Venus taken by *Venera 14*. The rocks are interpreted to be flow-layered basalts. The photograph was given by Russian scientists to the U.S. Geological Survey in appreciation of help in choosing the landing sites for *Venera 13* and *Venera 14*.

canoes are enormous, measuring nearly 400 km in diameter and rising 10 km above surrounding lowlands.

Along with volcanoes, the Venusian surface is pock-marked with impact craters (Fig. 16–26). On average, there are about two craters per million square miles. Long before the *Magellan* mission, scientists predicted that there would be few if any small-impact craters on Venus, because such small bodies would be destroyed on entering the atmosphere. The predictions were correct, and no craters smaller than 1.5 km in diameter have been found.

The first close look at the ground surface of Venus was provided in 1975 when the Russian robot spacecraft *Venera 9* and *Venera 10* succeeded in landing on the planet. In spite of the crushing atmospheric pressure and scorching temperatures (sufficient to melt tin), these spacecraft survived for about an hour. Even more successful, however, were the landings of *Venera 13* and *Venera 14* in March 1982. Like their predecessors, the two robots separated from their mother ships and drifted under parachute through the corrosive clouds to touch down on a mountainous region called Phoebe just below the Venusian equator. *Venera 13* survived two full hours, whereas its sister craft stopped functioning in less than an hour. The electronic eyes on both landers began working immediately and transmitted to Earth remarkable photographs of a landscape strewn with rust-colored slabby rocks and rubble (Fig. 16–27). In addition to accomplishing their photographic mission, the landers were able to drill out a few centimeters of Venusian soil and determine its composition. The analyses indicated that the samples were basalt. In fact, they bore a strong resemblance to oceanic basalts on Earth.

We still have much to learn about Venus. There is little doubt that scientists will have an increasingly

better understanding of the planet's geochemistry, tectonics, and both interior and exterior processes as the *Magellan* data continue to be analyzed.

Mars

The planet Mars (Fig. 16–28) has always been of special interest to humans because of speculation that some form of life, however humble and microscopic, may exist or at one time may have existed there. It is

FIGURE 16–28 Planet Mars as photographed from the approaching *Viking* spacecraft in 1976. Red coloration is the result of dust particles containing iron oxide. *(Courtesy of NASA.)*

a planet that is only a little farther from the Sun than is our own, it has an atmosphere that includes clouds, it has developed white polar caps, and it has seasons and a richly varied landscape. That landscape (Fig. 16–29) has been splendidly revealed to us by the *Mariner* and *Viking* missions. It includes magnificent craters, colossal volcanic peaks, deep gorges, sinuous channels, and an extensive fracture system. Evidence of wind erosion is clearly seen in photographs of the Martian surface, and there are indications of the

work of ice as well. The planet has a diameter of only about 50 percent that of the Earth and it has only 10 percent of the mass of the Earth.

A year on Mars is 687 Earth days long, and the planet's axial tilt is similar to that of the Earth. Carbon dioxide is the principal gas in the Martian atmosphere, although small amounts of nitrogen, oxygen, and carbon monoxide are also present. Atmospheric pressure, however, is less than 1 percent that of the Earth. The planet has virtually no greenhouse effect,

FIGURE 16–29   Heavily eroded canyonlands on Mars. This Viking photograph shows an area about 60 km across.   *(NASA/USGS, courtesy of Alfred McEwen.)*

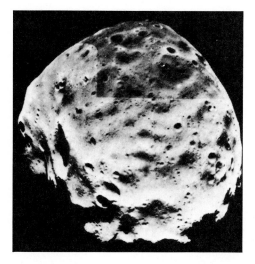

FIGURE 16–30 Photomosaic of Phobos, the inner satellite of Mars. The photographs incorporated into this mosaic were obtained when *Viking Orbiter 1* approached to within 480 km of Phobos in 1977. Phobos is about 21 km across and 19 km from top to bottom. North is at the top.   *(Courtesy of NASA.)*

and temperatures vary at the equator from about +21° to −85°C.

Mars has two moons. Their names are Phobos (Fig. 16–30), meaning fear, and Deimos, meaning panic. They are large chunks of rock about 21 and 15 km across, respectively.

The somewhat lower density of Mars (compared with that of other terrestrial planets) and its lack of a detectable magnetic field suggest that most of its iron may be scattered throughout the planet. This interpretation is strengthened by the color photographs transmitted from the *Viking* landers. Those photographs show the rusty red and orange hues typically developed on Earth by limonite and hematite. Indeed, the sky often takes on a reddish hue, probably because of the rust-colored dust. Apparently, iron at the Martian surface has reacted with water vapor and oxygen to provide the color that so intrigued early planetary observers.

Martian Landscape Features

As a result of the work accomplished by the *Mariner* and *Viking* orbiters, nearly the entire surface of Mars had been photographed by 1976. Crisp high-resolution photographs revealed craters, dunes, volcanoes, and canyons in unsurpassed clarity. On a grand scale, the planet's physiography is divisible into two rather different halves. The southern hemisphere is mostly heavily cratered rough terrain. This surface is inferred to be older than the smoother, lightly cratered surface of the northern hemisphere.

The two *Viking* spacecraft that reached Mars in the summer of 1976 each contained two parts—namely, a lander and an orbiter. The landers provided scientists with the first ground view of Mars. The photographs transmitted from the landers to the orbiters and thence to the Earth show a sandy-looking terrain littered with cobbles and boulders (Fig. 16–31). Here and there, undisturbed bedrock was

FIGURE 16–31 *Viking* lander photograph of the surface of Mars. The device on right is part of the lander that was purposely dropped for scale. It is about 20 cm long. Note the two trenches made by the lander's arm in taking samples of Martian regolith. *(Courtesy of NASA; computer color balancing by M. A. Dale–Bannister.)*

visible. Many of the rocks strewn about were clearly of volcanic origin. Some had spherical holes made by gas bubbles, whereas others were finely crystalline like many igneous rocks on Earth. The devices on the lander that analyzed the soil indicated that the fine, loose material was rich in iron, magnesium, and calcium. It was very likely derived from parent rocks rich in ferromagnesian minerals.

Craters and Dunes

The density of craters in the southern half of Mars is similar to the crater density in the lunar highlands.

Most of these craters were probably excavated between 4 and 4.5 billion years ago by the same torrential barrage of meteorites that pelted the lunar highlands.

Photographs clearly show row upon row of dunes on the floors of some of the larger craters. If the thinness of the Martian atmosphere is taken into account, then the size and spacing of these dunes can be used to estimate the velocity of the winds that formed them. Calculations indicate that the dune-forming winds blew at speeds of up to 320 km per hour.

Dust storms on Mars are prodigious events. As *Mariner 9* began its orbit of the planet, a dust storm

FIGURE 16–32 The gigantic Martian volcano Olympus Mons, photographed by *Viking Orbiter* in July 1976. The original black-and-white photograph has been colored in by an artist. *(Courtesy of NASA/JPL; from Abell, G. O., Realm of the Universe, 3rd ed. Philadelphia: Saunders College Publishing, 1983.)*

was in progress, and useful photographs of the Martian surface could not be transmitted until the dust subsided. *Viking* orbiters encountered similar problems. The dust storms are not driven by excessively violent winds, however. One of the *Viking* landers recorded a gust of 120 km per hour, but average wind velocities were much lower. Dust on Mars is sometimes swirled upward into towering tornado-like columns that, on a smaller scale, would be called "dust devils." Unlike those on Earth, Martian dust devils loom a spectacular 6 km above the parched terrain.

Volcanic Features

Although most of the craters on Mars have been formed by the impact of meteorites, a smaller number are certainly volcanic. Easily the most spectacular of the volcanic craters is Olympus Mons (Fig. 16–32). This gigantic feature is more than 500 km across at its base and rises 30 km above the surrounding terrain. It is equal in volume to the total mass of all the lava extruded in the entire Hawaiian Island chain. Around its summit are lava flows and the narrow rills interpreted as old lava channels or possibly collapsed lava tubes. Flood lavas are also evident, not only in the smoother northern hemisphere but also among the craters of the rugged southern half of Mars.

Channels and Canyons

So little water exists today in the Martian atmosphere that if all of it were condensed in one place, it would probably provide only enough water to fill a community swimming pool. Yet the surface of the planet is dissected with channels and canyons that grow wider and deeper in the downslope direction. Some of these channels have braided patterns, sinuous form, and tributary branches such as those characteristically developed by streams on Earth (Fig. 16–33).

These features suggest that flowing water was an important geologic agent at some earlier time in Martian history. There may once have been a considerable reserve of water frozen in the subsurface as a kind of permafrost. Permafrost is ground that remains below the freezing temperature and contains ice throughout the year. Many investigators theorize that the larger channels, some of which are over 5 km wide and 500 km long, were excavated by water gushing out onto the surface as the subsurface layer melted. Large tongues of debris closely resembling mud flows may also have received their water from such a process. Many channels on Mars dwarf our own Grand Canyon in size and, in order to form, would have required torrential floods so spectacular that they are hard to visualize by Earth standards.

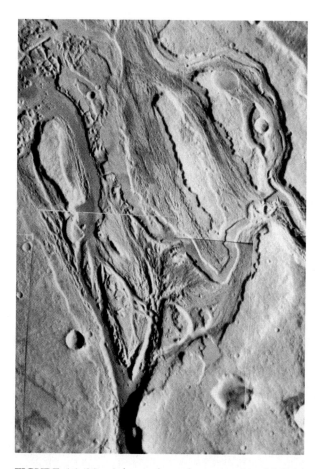

FIGURE 16–33 A large channel system named Mangala Vallis on Mars. The system, which resembles some dry river beds on Earth, flowed into the lower plains of the northern Martian hemisphere. Note the streamlined islands and intricate connections between channels. *(Courtesy of NASA.)*

Ice Caps

Long before spacecraft landed on Mars, observers using telescopes noticed that the planet had ice caps. There are actually two types of ice caps on Mars. Seasonal ice caps composed of frozen carbon dioxide develop only during Martian winters by direct condensation from the atmosphere. Beneath this seasonal ice, there are also permanent ice caps at both poles. The southern permanent ice cap has a diameter of 350 km and is composed of carbon dioxide. The northern permanent cap is 1000 km in diameter and is composed of ordinary water ice. It represents a large reservoir of water relative to the very small amount of water present as vapor in the Martian atmosphere.

Martian History

Like the Earth, Mars was transformed from a protoplanet by accretion of a multitude of smaller objects. This accretion was followed by an episode of differentiation during which the once homogeneous body became partitioned into masses that had different chemical compositions and physical properties. Very likely, the differentiation process involved melting, so that fluids could move from one region of the planet to another.

During the initial billion years or so of Martian history, the planet experienced heavy bombardment of meteorites and asteroids, with the result that the exposed crust became densely cratered. The craters have been modified by relatively recent wind erosion. The crust was disturbed not only by cratering, but also by the faulting and fracturing that accompanied volume changes in the mantle.

It is likely that a dense atmosphere formed on Mars as a result of volcanic outgassing (removal of embedded gas by heating) and that the production of water and carbon dioxide far exceeded its loss into space. The planet was warmer at that time, perhaps because of a greenhouse effect. For a time, running water was an important geologic agent, as evidenced by the sinuous furrows cut into the cratered terrain. In certain regions of the planet, a small decline in surface temperatures may have caused the water contained in surface materials to freeze, thus trapping still liquid subsurface water in pockets and channels. With local melting, the subsurface water may have gushed upward and eroded large channels while simultaneously causing entire areas of surficial material to collapse in mud flows.

Eventually, the supply of atmospheric water and carbon dioxide became so depleted that the planet began to cool. Remaining carbon dioxide and water migrated to polar caps and to subsurface reservoirs, where it has remained.

Because Mars was once warmer, had more water, and possessed the elemental raw materials thought to

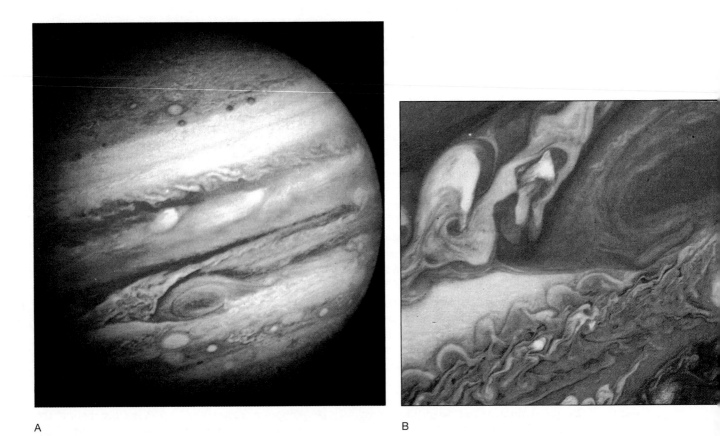

A   B

FIGURE 16–34 (A) Jupiter, photographed by *Voyager 1* spacecraft from a distance of 33,000,000 km. (B) A close-up of Jupiter's famous Great Red Spot. Note the turbulence to the west (left) of the Great Red Spot. *(Courtesy of NASA.)*

be required for the development of organisms, scientists have long speculated on the possibility of finding primitive forms of life on the planet. For this reason, a series of experiments aboard the *Viking* lander was designed to search for signs of present or former life. The results were disappointing. Not a single microorganism or even a trace of organic molecules could be detected in the Martian soil at the landing site. However, it is still too early to rule out the possibility that continuing research and exploration of other sites may provide evidence of organisms.

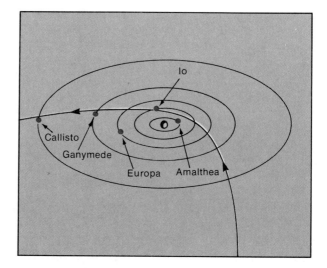

FIGURE 16–35 *Voyager 1*'s flight path allowed scientists to obtain close-range photographs of five of Jupiter's 13 satellites. Each is shown at its closest point to *Voyager 1*'s outbound flight away from Jupiter. The spacecraft approached within 280,000 km of Jupiter. *(Courtesy of NASA.)*

■ *The Outer Planets*

Beyond the orbit of Mars lies the ring of asteroids, and beyond the asteroids are the orbits of Jupiter, Saturn, Uranus, Neptune, and Pluto. The first four of these planets are also called the **Jovian** (for Jupiter) **planets.** They are similar to one another in having large size and relatively low densities. Pluto is not a Jovian planet. It is so far away and small that it is difficult to observe. As a result, there is considerable uncertainty about its origin.

Jupiter

Jupiter, named for the leader of the ancient Roman gods, is the largest planet in our solar system. The diameter of this huge planet is 11 times greater than the Earth's diameter, and its volume exceeds that of all the other planets combined. Its mass is 318 times that of the Earth, yet it is only one-fourth as dense. Jupiter spins rapidly on its axis, making one full rotation in slightly less than ten Earth hours. This rapid rotation results in the formation of the encircling colored atmospheric bands for which the planet is known (Fig. 16–34). Hydrogen, helium, and lesser quantities of ammonia and methane are the predominant gases in Jupiter's atmosphere. On the basis of experiments with these gases, some investigators believe that the reddish and orange colors in the atmosphere are derived from sulfur or phosphorus-based compounds. It is impossible to discern the solid surface of Jupiter with optical telescopes.

Jupiter's atmosphere, which is several hundred kilometers thick, passes gradually to liquid and eventually to solid matter. The interior of the giant planet may consist of highly compressed hydrogen, possibly surrounding a rocky core. Jupiter has at least 16 moons. The large planet with its many moons resembles a miniature solar system.

We have learned about Jupiter not only from Earth-based telescopes but from close encounters by unmanned spacecraft. The most successful of the robot explorations occurred in 1979 when the two *Voyager* spacecraft swept past Jupiter and transmitted splendid images of the planet's brightly colored atmosphere, the tempestuous Great Red Spot (see Fig. 16–34), five of Jupiter's moons (Fig. 16–35), and a ring of debris less than 0.6 km thick that circles the planet about 55,000 km above the tops of the clouds. The discovery of the ring makes Jupiter the fourth planet in the solar system known to possess such a feature (Saturn, Uranus, and Neptune are the other planets known to have rings).

Jupiter's Great Red Spot, which has twice the diameter of the Earth, is an intriguing feature known to astronomers for centuries. As clearly revealed by the *Voyager* mission, this many-hued disturbance is a gargantuan atmospheric storm that extends deep into the cloud cover and turns in a complete counterclockwise revolution every six Earth days.

In addition to the Great Red Spot, devices aboard the *Voyager* spacecraft radioed back the distinct images of great bolts of lightning, whose occurrence had only been suspected before the flyby. Another discovery was an auroral display far brighter than any northern lights ever seen on Earth. *Voyager* also confirmed the presence of an immense magnetic field sur-

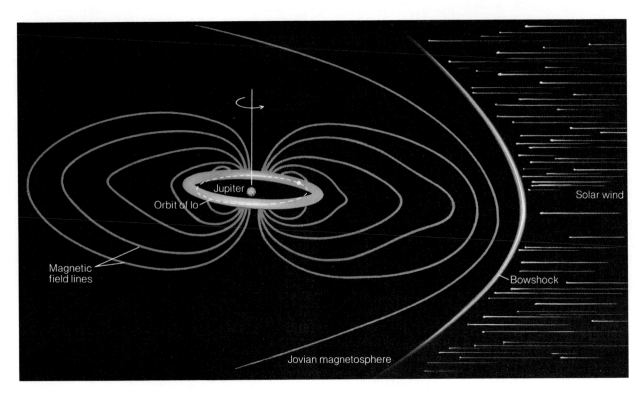

FIGURE 16–36   The magnetosphere of Jupiter as revealed by the *Pioneer* and *Voyager* spacecraft explorations. The blue zone consists of hot ionized gases (plasma).   *(From Abell, G. O., Morrison, D., and Wolff, S. C. Exploration of the Universe. Philadelphia: Saunders College Publishing, 1991.)*

rounding Jupiter (Fig. 16–36). First detected by the *Pioneer 10* spacecraft in 1973, the magnetic field extends outward from the planet for 16 million km and encompasses all of the larger satellites. The presence of the field indicates that Jupiter has a magnetic core resulting from fluid movements within its spinning interior.

The two *Voyager* spacecraft were not launched at the same time. *Voyager 2* reached Jupiter four months after its companion spacecraft and was thus able to fill in many of the gaps in information. *Voyager 2* also provided even better images of Jupiter's larger moons than had been transmitted by its sister ship. These larger satellites are called the Galilean moons to commemorate their discoverer. Individually, they are named Callisto, Ganymede, Europa, and Io. Each has characteristics quite different from those of its neighbors. **Callisto** (Fig. 16–37), the outermost of the Galilean moons, is riddled with craters resulting from meteorite impact. On Callisto's icy, dirt-laden surface *Voyager* cameras detected huge, concentric rings that outline extraordinarily broad impact basins. Callisto's nearest neighbor is **Ganymede** (Fig. 16–38), the largest of Jupiter's satellites. Ganymede also shows the effects of meteorite bombardment in its dark, cratered terrain. Individual cra-

FIGURE 16–37   Callisto, the outermost of the Galilean moons and the most heavily cratered body in our solar system.   *(Courtesy of NASA.)*

C O M M E N T A R Y
Io's Volcanoes

One of the most interesting results of the *Voyager* flybys of Jupiter was the discovery of active volcanoes on the Galilean satellite Io (see Fig. 16–41). Before the images had been received, most planetary scientists believed Io would be a cold, dead satellite, rather like our own Moon. On Earth, volcanoes are ultimately powered by heat from decaying radioactive materials in the interior. Io, however, has only about one-sixth of the mass of the Earth and, like our Moon, should have exhausted most of its original radioactivity billions of years ago. How then did the satellite acquire the heat needed to power the volcanoes seen by the *Voyager* spacecraft? It now appears that Io's heat is generated by alternate squeezing and unsqueezing in response to gravitational forces. We know that Io has a huge tidal bulge resulting from the gravitational attraction of Jupiter. This is to be expected, for Jupiter is hundreds of times more massive than the Earth and yet is about the same distance from Io as the Earth is from its moon. If Io were Jupiter's only satellite, it would keep the same face toward Jupiter, just as our moon does while orbiting the Earth, and the tidal bulge on Io would not generate heat. However, as the Galilean satellites Europa and Ganymede pass Io, they cause perturbations in Io's orbit. As a result, the huge tidal bulge rises and falls, alternately compressing and relaxing the satellite. The process generates a tremendous amount of frictional heat, and that heat powers Io's volcanoes.

Io's volcanoes are apparently of the explosive type. On Earth, the major driving force for explosive volcanoes is steam. Io, however, is a waterless, rocky body. It appears that sulfurous gases emanating from molten sulfur deep beneath the crust cause the explosive activity of Io's volcanoes.

FIGURE 16–38 Ganymede, Jupiter's largest satellite, as photographed by *Voyager 1* from 216 million km. *(Courtesy of NASA.)*

FIGURE 16–39 A view of the surface of Europa. The satellite's surface is composed of ice crossed by a complex system of cracks and low ridges. *(Courtesy of NASA.)*

ters have bright streaks or rays that fan out from the crater rims like huge splash marks. Sinuous ridges have been detected on Ganymede's surface as well as criss-crossing fractures, suggesting that fault movements have occurred on the satellite.

Europa (Fig. 16–39), the brightest of the Galilean moons, has a surface crust of ice but is otherwise a predominantly rocky satellite about the size of our own moon. It has few impact craters, and its surface is marked by huge fractures and ridges. These observations indicate that Europa's surface is being contin-

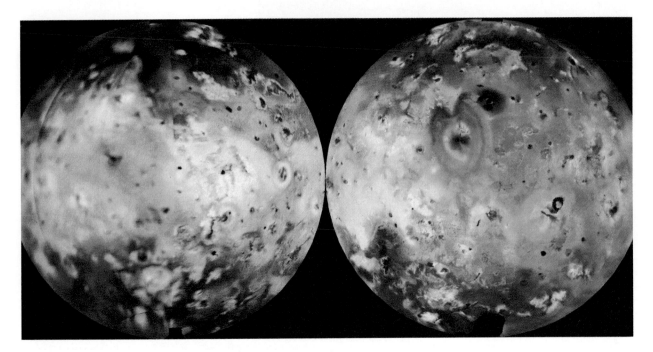

FIGURE 16–40 The satellite Io. The relatively smooth surface of Io and its active volcanoes suggest that it may be Jupiter's youngest satellite. Large amounts of sulfur and sulfur dioxide frost are thought to be present on the surface. *(Courtesy of NASA.)*

uously renewed, perhaps by the development and slow migration of glacier-like masses of ice.

Io (Fig. 16–40) is the innermost of Jupiter's large satellites. Like Europa, it is an essentially rocky body, but it differs in having a considerable amount of active volcanism (Fig. 16–41). Eight volcanoes were erupting during the *Voyager* flybys of Jupiter. Graceful plumes of sulfur and sulfur dioxide ejected from the volcanoes extended for hundreds of kilometers into space. On cooling, the sulfur and sulfur dioxide in the plumes condense into solid particles and fall steadily onto Io's surface. The blanket of sulfurous deposits has probably buried impact craters that may have once existed on the surface of the satellite. There is evidence of former volcanism as well, for one can detect lava flows and shield volcanoes on the surface of Io. Judging from their color, Io's lava flows are probably composed of sulfur.

Voyager 2 also passed near Amalthea, Jupiter's innermost satellite. The spacecraft's camera showed that Amalthea is a reddish-colored nonspherical satellite about 165 km long and 150 km wide. It is without doubt the most irregularly shaped satellite known in our solar system.

In July of 1994, Jupiter watchers were able to witness bolides smashing into a planet. The impactors

FIGURE 16–41 *Voyager 2* photograph of Io showing two active volcanoes. *(Courtesy of NASA.)*

were fragments of the comet Shoemaker–Levy 9, which one by one struck the far side of Jupiter's southern hemisphere, releasing energy equivalent to about 40 million tons of TNT. The impacts sent plumes of dark matter to heights more than 1000 km above Jupiter's clouds. One of these plumes covered an area equal to the diameter of the Earth.

Saturn

Saturn's density is only about 70 percent that of water. Its mass, however, is equivalent to 95 Earth masses. Measurements of Saturn's size and density suggest that the planet may have a central core of heavy elements (probably mostly iron) surrounded by an outer core of hot, compressed volatiles such as methane, ammonia, and water. These two core regions, however, constitute only a small part of the total volume of the planet, in which hydrogen and helium are overwhelmingly the predominant elements.

As was true for Jupiter, our knowledge of Saturn has been greatly improved as a result of data collected by unmanned spacecraft (Fig. 16–42). In particular, the electronic eyes of *Voyager 2* in August 1981 provided truly spectacular views of this second largest planet, its rings, and its moons. The data transmitted back to Earth helped planetary scientists confirm that Saturn has a magnetic field, trapped radiation belts, and an internal source of heat.

The general appearance of Saturn's atmosphere is similar to that of Jupiter. The alternating light and dark bands and swirls of gas, however, are partially veiled by a haze layer above the denser clouds. Temperatures near the cloud tops range from 86°K (−305°F) to 92°K (−294°F), with the coolest temperatures occurring near the center of the equatorial zone. The planet is well illuminated, for even its dark face receives a substantial amount of light reflected from the rings. On this dark side, *Voyager 2* detected radio emissions like those given off during lightning discharges.

The Rings of Saturn

There is perhaps no more beautiful sight in the solar system than the rings of Saturn (Fig. 16–43). With his primitive telescope, Galileo was the first to recognize these rings, but he was unable to discern them clearly. Fifty years later, in 1655, Huygens was able to differentiate three concentric rings, the brightest and broadest of which was in the center. There are, however, far more than three rings around Saturn. As a result of pictures taken by *Pioneer 11* and the more recent and truly dazzling photographs taken by *Voyager 1* and *Voyager 2*, we now know that the once neatly defined grouping of three rings is really a complex system of thousands of rings and rings within rings. Saturn's rings are composed of billions of orbiting particles of dust and ash, as well as many larger fragments measuring tens of meters in diameter. All of these revolve around Saturn in the approximate plane of the equator. Saturn's 17 or more moons also revolve in this plane. The entire ring sys-

FIGURE 16–42 Saturn and its rings as photographed by *Voyager 2* from a distance of 18 million km. *(Courtesy of NASA.)*

FIGURE 16–43   Computer-enhanced photograph of Saturn's rings as photographed by *Voyager 2*. The exaggerated color shows the reflection and transmission of different wavelengths of light by particles of different sizes in various parts of the ring system.   *(Courtesy of NASA.)*

tem has an outside diameter of about 275,000 km. Although the area of the rings is truly enormous, most of the material is concentrated within a thickness of only about 16 km. Thus, the entire system reminds one of a phonograph record, with the rings corresponding to the record's many tiny grooves. Scientists are uncertain about the origin of Saturn's rings. Some suggest that the particles in the rings are remnants of an exploded satellite. Others argue that the rings are a remnant of the particle cloud from which satellites form by accretion.

Saturn's Moons

Largely as a result of the *Voyager* missions, we know that Saturn ranks first among the planets in the number of its satellites (Fig. 16–44). Altogether, over 17 confirmed moons have been recognized. Some have been formally named, whereas others bear only numbers. The largest of the named satellites is **Titan.** Titan is about 5120 km in diameter and has a relatively low density about twice that of water ice. The satellite's surface is obscured by thick, dense haze. Titan has a nitrogen-rich atmosphere, and lakes of liquid nitrogen may exist at the poles, where surface temperatures are an estimated 90°K (−300°F). The more important inner satellites of Saturn are named Mimas, Enceladus, Tethys, Dione, and Rhea (Fig. 16–45). Their densities and surface brightness suggest that they too are composed mainly of water ice. The five satellites range in size from the 390-km diameter of Mimas to the 1530-km diameter of Rhea. All are densely cratered. The outer satellites of Saturn include tiny Phoebe, potato-shaped Hyperion, and Iapetus.

FIGURE 16–44 Saturn and some of its moons are shown in this montage compiled from *Voyager 1* and *Voyager 2* photographs. The satellites shown clockwise from upper left are Titan, Iapetus, Tethys, Mimas, Enceladus, Dione, and Rhea. *(Courtesy of NASA.)*

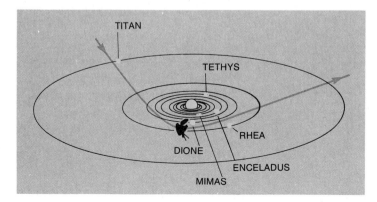

FIGURE 16–45 *Voyager 1* approached within 124,000 km (77,000 miles) of Saturn's cloud tops. Six of the satellites that were photographed are shown in their approximate positions at closest approach by the spacecraft. *(Courtesy of NASA.)*

603

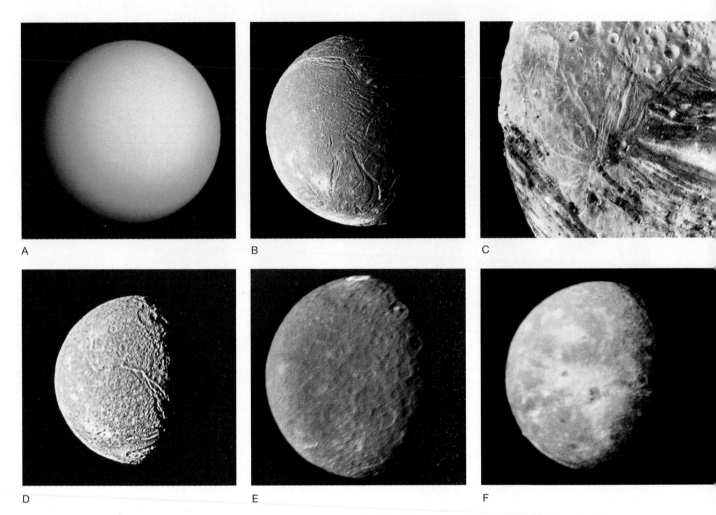

FIGURE 16-46 Uranus and its five larger moons as photographed by *Voyager 2* in January 1986. (A) Uranus. (B) Mosaic of Ariel, with its many faults and valleys. (C) Miranda, whose grooved and cratered surface indicates a complex geologic history. (D) Titania, with prominent fault valleys nearly 1600 km long. (E) Umbriel, with cratering, at a distance of 557,000 km. (F) Oberon, showing craters and a large peak along the lower left edge of the photograph. *(Courtesy of NASA.)*

Uranus

Beyond the orbit of Saturn lies the planet Uranus (Fig. 16–46). Like Saturn and Jupiter, Uranus also has rings. They have been photographed from Earth with infrared film. Uranus has a low density (1.2 g/cm^3) and a frigid surface temperature of about −185°C. Five large moons (and five recently detected small moons) circle Uranus. The larger moons are named Ariel, Miranda, Oberon, Titania, and Umbriel. Thus far, Earth-based instruments have only been able to detect hydrogen and methane on Uranus, although helium may also be present. The methane may be responsible for the planet's greenish hue. Perhaps the most distinctive characteristic of Uranus is the orientation of its axis of rotation, which lies nearly in the plane of its orbit. Because it is upside down, the planet's direction of rotation is the reverse of that of all other planets except Venus. This retrograde direction is also characteristic of the orbits of the five larger satellites around Uranus.

Neptune

Because they are similar in size and other physical properties, Neptune and Uranus are often called the twin planets. Through the telescope, Neptune, like Uranus, appears as a greenish orb. Neptune's density of 1.7 g/cm^3 is just slightly greater than the density of Uranus. Neptune's atmosphere may also resemble that of Uranus. Both hydrogen and helium have been detected spectroscopically, and methane is probably also present. Two moons, named Triton and Nereid, circle the planet. Triton is one of the four largest

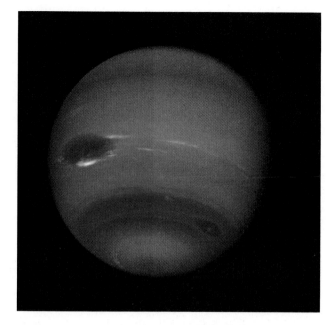

FIGURE 16–47 The planet Neptune as photographed in 1989 by *Voyager*. *(Courtesy of NASA/JPL.)*

moons in the solar system (the other three being Ganymede, Titan, and Callisto). Its surface may contain ice made of frozen methane, and it may even have a thin atmosphere of methane gas. Recently, astronomers have found evidence that Neptune may have a third moon and possibly even a ring system.

When *Voyager 2* approached Neptune in 1989, a huge, counterclockwise rotating area of atmosphere became noticeable. It was named the Great Dark Spot (Fig. 16–47) and seemed to be analogous to Jupiter's Great Red Spot in relative size and location in the southern hemisphere. Cirrus clouds composed of methane were detected around the margins of the Great Dark Spot. Other areas of atmospheric circulation on Neptune were also recognized, indicating that the planet has active weather systems, perhaps generated by internal heat. *Voyager 2* was also able to confirm the presence of two circular rings that lie very close to Neptune's equator.

Pluto

Far out on the outer limits of our solar system lies the orbit of Pluto. It is so small and so distant that less is known about it than about the other planets. We do observe that Pluto's orbit is tilted an unusual 17° to the ecliptic plane and that its orbit crosses that of Neptune. As a result, Pluto is actually closer to the Sun than Neptune during certain periods (as during the current period from 1979 to 1999). Because of the high inclination of Pluto's orbit, however, there is no danger of a collision with Neptune. Its small size

and orbital peculiarities suggest that Pluto may not be a true planet but rather a former satellite of Neptune that escaped the gravitational pull of that planet and now occupies a separate orbit. Evidence for this theory, however, is not conclusive.

In 1978, scientists of the U.S. Naval Observatory found that Pluto has a moon. They named it Charon after the mythologic boatman who ferried the souls of the dead across the river Styx to Hades. Some scientists believe the existence of Pluto's moon makes the "escaped satellite" theory for Pluto's origin unlikely. Others argue that Pluto and Charon may have been ejected from the Neptunian system as a satellite pair.

■ *The Future of the Solar System*

Anyone interested in the history of the Earth is naturally led to a consideration of its future. Clearly, all that has happened, is happening, and will happen to this planet is inevitably linked to the history and destiny of the Sun. The Sun is a star and, like other stars, will run its course from birth to death. By observing other stars that are at various stages in their histories, astronomers have pieced together a story of what happens when these bodies begin to deplete their store of nuclear fuel. Our star has been "burning" for about 5 billion years. In this span of time, it has consumed about half of its nuclear fuel and so has about another 5 billion years of life remaining.

What will "doomsday" for the Sun and Earth be like? Most astronomers believe that once the hydrogen in the Sun's core has been largely converted to helium by fusion, the core will begin to contract. As the core contracts, the outer layers of the Sun will expand and cool, causing the Sun to puff up into an enormous sphere and to turn blood-red. Our Sun will have become what astronomers call a red giant. As the great ball of rarefied gases continues to expand, it will engulf and vaporize Mercury and Venus.

What will happen to the Earth can be readily imagined. The searing heat will boil away the oceans and heat the crust even to the melting point of some surface rocks. This purgatorial state of affairs may continue for a billion years or so, until the Sun finally dissipates its energy. The dying Sun may then begin to shed its outer layers, leaving behind a dense, hot white dwarf star about the size of the Earth. Such white dwarf stars have exhausted their supply of nuclear fuels and remain incandescent only because of their residual internal heat. Eventually, even this light will fade as the Sun becomes a black dwarf, so called because it will be too cool to radiate visible light. The scorched Earth will continue as a barren, frigid sphere unable to free itself from the gravitational grasp of the dead Sun.

Summary

There are a great many chunks of matter in interplanetary space, ranging up to tens of meters in diameter. When one of these chunks survives its passage through the Earth's atmosphere or strikes the surface of a moon or planet, it is called a meteorite. Meteorites are important to planetary geologists because they are truly extraterrestrial samples of the solar system. They can be classified as ordinary chondrites, carbonaceous chondrites, achondrites, irons, or stony-irons and are believed to be fragments resulting mostly from collisions of one or more small planetary bodies.

The Moon, which has been studied by telescope for centuries, was first visited by humans in July 1969. Its general features include the relatively smooth, dark mare basins and the rugged lunar highlands. Samples of the loose regolith on the Moon have been brought back to Earth. They include basaltic rocks from the maria and specimens of anorthosite from the highlands. Study of the age and distribution of these rocks suggests that lunar history began about 4.6 billion years ago, when the Moon had developed its surficial skin of anorthositic igneous rock. This surface was then flooded in various regions by the basalts of the maria about 1 billion years later.

The planets of our solar system can be divided into two main groups. Those resembling the Earth are the inner (or terrestrial) planets. The outer planets resemble Jupiter and are termed Jovian planets. The terrestrial planets tend to be small, dense, and rocky, whereas the Jovian planets are large and have many satellites and dense, thick atmospheres.

The terrestrial planet closest to the Sun is Mercury. It is the smallest of the inner planets and is very moonlike in appearance. Next in the progression outward from the Sun is Venus. Venus is about the size of the Earth. It has a dense carbon dioxide atmosphere, which traps heat from the Sun. The surface temperature of Venus is higher than that of any other planet in the solar system. Its rocky surface is enshrouded by a cloud cover too dense to see through except by radar.

Mars is of interest because of its great variety of surface features. It has a cratered topography reminiscent of the Moon's and also has features that appear to be the result of processes similar to those operating on Earth. Dunes, stratified sediments, and sinuous rills suggest erosion by wind, water, and ice. Mars has a thin carbon dioxide atmosphere and polar caps composed of water ice and frozen carbon dioxide.

Among the outer planets, Jupiter is known for its tremendous size. It is larger than all other planets combined; has a massive hydrogen, helium, methane, and ammonia atmosphere; and rotates so rapidly as to form beautiful arrays of colored cloud bands. The planet has an internal heat supply. Its strong magnetic field suggests the presence of a metallic core.

Saturn is similar to Jupiter although smaller. Saturn's most distinctive feature is the thin, complex ring system that encircles its equator like the brim of a straw hat. At least 17 moons accompany Saturn on its revolution around the Sun.

Uranus and Neptune are similar in size and general characteristics. Both have atmospheres rich in hydrogen compounds, and both appear to have small ring systems. Beyond Neptune lies Pluto. Pluto is roughly the size of the Moon and is thus the smallest planet in the solar system.

Review Questions

1. In order of their increasing distance from the Sun, list the planets of the solar system. How do the inner four planets differ from the Jovian planets?
2. What are the principal kinds of meteorites? How do meteorites originate? What are chondrules?
3. What evidence suggests that the Moon does not have an iron–nickel core?
4. How do rocks of the lunar maria and lunar highlands differ in composition and age?
5. Why does one side of the Moon have more craters than the other side?
6. How do volcanoes on Io differ from those on Earth?
7. Account for the deficiency of small craters on the surface of Venus.
8. Account for the reddish coloration of the surface of Mars.

Discussion Questions

1. Discuss the differences between Mars and Earth with respect to density, internal structure, and average range of equatorial temperatures.
2. Describe how such ordinary geological precepts as superposition and cross-cutting relationships can be used to determine the relative ages of craters, lava flows, and other features on planets like Mercury, Venus, and Mars.
3. Compare the similarities and differences between Earth and its "twin" planet Venus. Do the same for Neptune and Uranus.
4. How is the absence of an ocean on Venus related to the persistence of carbon dioxide in that planet's atmosphere?
5. Antarctica is the best region on Earth for finding meteorites. Over 5000 have been found on that frigid continent. How do you account for the greater success in finding meteorites in Antarctica?

Supplemental Readings and References

Abell, G. O., Morrison, D., and Wolff, S. C., 1991. *Exploration of the Universe*, 6th ed. Philadelphia: Saunders College Publishing.

Cadogan, P., 1981. *The Moon: Our Sister Planet.* New York: Cambridge University Press.

Carr, M. H., 1981. *The Surface of Mars.* New Haven: Yale University Press.

Hartmann, W. K., 1993. *Moons and Planets: An Introduction to Planetary Science*, 3rd ed. Belmont, CA: Wadsworth Publishing Co., Inc.

Hodge, P. 1994. *Meteorite Craters and Impact Structures of the Earth.* Cambridge, England: Cambridge University Press.

Pasachoff, J. M., 1991. *Astronomy: From the Earth to the Universe*, 4th ed. Philadelphia: Saunders College Publishing.

Skinner, B. J. (ed.). 1981. *The Solar System and Its Strange Objects: Readings from American Scientist.* Los Altos, CA: Wm. Kaufmann, Inc.

Wilhelms, D. E. 1987. The geologic history of the Moon. USGS Professional Paper 1348. Washington, DC: U.S. Government Printing Office.

APPENDIX A

Classification of Living Things

Early students of biology found it convenient to divide all organisms into two great realms or kingdoms, designated the Animalia and the Plantae. However, by the late nineteenth century, many biologists suggested that perhaps a third kingdom—the Protista—should be established for certain single-celled organisms that seemed to be neither plant nor animal. Even this was considered inadequate by some biologists, for as this three-kingdom classification was gaining adherents, evidence was accumulating that clearly indicated the need for still further major groupings. Studies of unicellular organisms had revealed two different forms of cell structures. There were, for example, prokaryotic unicellular organisms, such as bacteria and blue-green algae, that lacked a cell nucleus and possessed other traits that set them clearly apart from eukaryotic unicellular organisms, which had a true nucleus, well-defined chromosomes, and organelles. Any classification that sought to categorize organisms according to similarity of origin and fundamental differences could not ignore the contrast between eukaryotes and prokaryotes. Finally, it was recognized that the fungi deserved special taxonomic consideration also, for they are dependent on a supply of organic molecules in their environment, as are animals, yet they absorb their food through cell membranes, as do plants. The fungi appear to have had an evolutionary radiation distinct from the other major groups.

In an effort to account for these differences and better represent the evolutionary relationships of organisms, R. H. Whittaker of Cornell University proposed a five-kingdom system of classification.* In this classification, plants, fungi, and animals are regarded as distinct in terms of being specialized for different modes of nutrition, photosynthesis, absorption, and ingestion. The Protista is composed of unicellular eukaryotic organisms, and the Monera include the simplest of organisms that are inferred to be similar to the primitive forms of life from which other kingdoms evolved. A modified version of Whittaker's classification is presented here. (Geologic ranges of major fossil groups are indicated.)

Biologists are currently considering a modification of Whittaker's classification, based on evolutionary relationships revealed by study of molecular structures and sequences. In these classifications, organisms are grouped into three so-called domains, the Bacteria, Archaea, and Eucarya.† The Eucarya, for example, would contain the kingdoms Animalia, Plantae, and Fungi.

---

*Whittaker, R. H. 1969. New concepts of kingdoms of organisms. *Science* 163:150–160.

†Woese, C. R., Kandler, O., and Wheelis, M. L. 1990. Towards a natural system of organisms: Proposal for the domains Archaea, Bacteria, and Eucarya. *Proc. Natl. Acad. Sci.* 87:4576–4579.

KINGDOM MONERA

Prokaryotes (*Archean to Holocene*)

PHYLUM METHANOCRETRICES — Methane-producing bacteria

PHYLUM OMNIBACTERIA — Rigid, rod-shaped heterotrophic bacteria, many pathogens (such as *E. coli*)

PHYLUM CYANOBACTERIA — Photosynthetic bacteria, formerly called blue-green algae (*Archean to Holocene*)

PHYLUM MYXOBACTERIA — Unicellular or filamentous gliding bacteria

PHYLUM APHRAGMABACTERIA — Small, nonmotile, heterotrophic bacteria that lack cell walls

PHYLUM SPIROCHAETAE — Spirochetes; tightly coiled forms with flagella

PHYLUM ACTINOBACTERIA — Filamentous bacteria. Source of many antibiotics today. Cause of leprosy, tuberculosis, and other diseases

OTHER MONERANS — Nitrogen-fixing aerobic bacteria, psudomonads, chemo-autotrophic bacteria

KINGDOM PROTOCTISTA
(formerly Protista)

Solitary or colonial unicellular eukaryotic organisms that do not form tissues

PHYLUM EUGLENOPHYTA — Euglenoid organisms

PHYLUM XANTHOPHYTA — Yellow-green algae

PHYLUM CHRYSOPHYTA — Golden-brown algae, diatoms, and coccolithophorids (*Paleozoic? Triassic to Recent*)

PHYLUM PYRROPHYTA — Dinoflagellates and cryptomonads (*Triassic to Recent*)

PHYLUM HYPHOCHYTRIDIOMYCOTA — Hypochytrids

PHYLUM PLASMODIOPHOROMYCOTA — Plasmodiophores

PHYLUM SPOROZOA — Sporozoans (parasitic protists)

PHYLUM CNIDOSPORIDIA — Cnidosporidians

PHYLUM ZOOMASTIGINA — Animal flagellates, protozoa that have whiplike cytoplasmic protrusions (flagellae)

PHYLUM SARCODINA — Rhizopods, protozoa with pseudopodia for locomotion (*Cambrian to Recent*)

PHYLUM CILIOPHORA — Ciliates and suctorians, movement accomplished by beating of cilia (adult suctorians are attached to objects)

KINGDOM PLANTAE

DIVISION* RODOPHYTA — Red algae, usually marine, multicellular

DIVISION PHAEOPHYTA — Brown algae, multicellular, often with large bodies, as in seaweeds and kelps

DIVISION CHLOROPHYTA — Green algae (*Proterozoic? Paleozoic to Recent*)

DIVISION CHAROPHYTA — Stoneworts

DIVISION BRYOPHYTA — Liverworts, mosses, and hornworts (*Late Paleozoic to Recent*)

DIVISION PSILOPHYTA — Extinct leafless, rootless, vascular plants (*Middle Paleozoic*)

DIVISION LYCOPODOPHYTA — Club mosses, with simple vascular systems and small leaves, including scale trees of Paleozoic (lycopsids) (*Silurian to Recent*)

DIVISION EQUISETOPHYTA (ARTHROPHYTA) — Horsetails, scouring rushes, including sphenopsids such as *Calamites* and *Annularia* of the Late Paleozoic (*Devonian to Recent*)

DIVISION POLYPODIOPHYTA — The true ferns or pteropsids (*Devonian to Recent*)

---

*In botany, it is conventional to use the terms *division, subdivision*, and so on, in place of *phylum* and *subphylum*.

DIVISION PINOPHYTA	The "gymnosperms," including conifers, cycads, and many evergreen plants; no true flowers (*Middle Paleozoic to Recent*)
Class Lyginopteriodopsida	The seed ferns, known from fossils of the Late Paleozoic and including such forms as *Neuropteris* and *Blossopteris* (*Late Paleozoic to Recent*)
Class Bennettitopsida	The extinct cycadeoids (*Triassic to Recent*)
Class Cycadopsida	Cycads (*Triassic to Recent*)
Class Ginkgoopsida	Ginkgoes (*Early Miocene to Recent*)
Class Pinopsida	Conifers, as well as the extinct *Cordaites* (*Late Paleozoic to Recent*)
Class Gnetopsida	Certain climbing shrubs and small tropical trees
DIVISION MAGNOLIOPHYTA	Flowering plants or "angiosperms"; seeds enclosed in an ovary (*Cretaceous to Recent*)
Class Magnoliopsida	Dicotyledonous plants: embryos with two cotyledons or seeds; leaves with netlike veins (*Cretaceous to Recent*)
Class Liliopsida	Monocotyledonous plants: embryos with only one seed leaf; leaves with parallel veins (*Cretaceous to Recent*)

KINGDOM FUNGI

DIVISION MYXOMYCOPHYTA	Slime molds
DIVISION EUMYCOPHYTA	True fungi

KINGDOM ANIMALIA

PHYLUM MESOZOA	Mesozoans
PHYLUM PORIFERA	Sponges; includes forms with calcareous spicules, siliceous spicules, and proteinaceous spicules; may also include the extinct stromatoporoids (*Cambrian to Recent*)
PHYLUM ARCHAEOCYATHA	Extinct spongelike organisms (*Cambrian*)
PHYLUM CNIDARIA (COELENTERATA)	Jellyfishes, corals; radially symmetric, aquatic, with body wall of two layers of cells, in the outer of which are stinging cells (*Proterozoic to Recent*)
Class Hydrozoa	Hydra-like animals
Class Scyphozoa	True jellyfishes
Class Anthozoa	Corals and sea anemones
Subclass Zoantharia	Hexacorals of modern seas (*Triassic to Recent*)
Subclass Rugosa	Paleozoic tetracorals (*Cambrian to Permian*)
Subclass Tabulata	Paleozoic tabulate coals (*Ordovician to Permian*)
PHYLUM CTENOPHORA	Modern comb jellies or "sea walnuts." Not known as fossils
PHYLUM PLATYHELMINTHES	Flatworms
PHYLUM NEMERTEA	Proboscis worms
PHYLUM NEMATODA	Roundworms
PHYLUM ACANTHOCEPHALA	Hook-headed worms
PHYLUM NEMATOMORPHA	Horsehair worms
PHYLUM ROTIFERA	Small, wormlike animals with a circle of cilia on the head
PHYLUM GASTROTRICHA	Small, wormlike animals resembling rotifers but lacking circle of cilia
PHYLUM BRYOZOA	The bryozoans, sometimes considered two phyla: *Entoprocta* and *Ectoprocta* (*Ordovician to Recent*)
PHYLUM BRACHIOPODA	Marine animals with two parts (valves) to their shell (dorsal and ventral) (*Cambrian to Recent*)
Class Inarticulata	Primitive brachiopods having phosphatic or chitinous valves, lacking hinge (*Cambrian to Recent*)
Class Articulata	Advanced calcareous brachiopods with valves that are hinged (*Cambrian to Recent*)

PHYLUM PHORONIDA	Wormlike marine animals that secrete and live within a leathery tube (*Early Mesozoic to Recent*)
PHYLUM ANNELIDA	The segmented worms (*Proterozoic to Recent*)
PHYLUM ONYCOPHORA	Rare tropical animals considered intermediate between annelids and arthropods (*Cambrian to Recent*)
PHYLUM ARTHROPODA	Segmented animals with jointed appendages
Subphylum Trilobita	Trilobites, common marine arthropods of the Paleozoic Era (*Cambrian to Permian*)
Subphylum chelicerata	
Class Xiphosura	Horseshoe crabs (*Silurian to Recent*)
Class Eurypterida	Eurypterids (*Ordovician to Permian*)
Class Pycnogonida	Sea spiders
Class Arachnida	Scorpions, spiders, ticks, and mites
Subphylum Crustacea	Lobsters, crabs, barnacles, and ostracodes (*Cambrian to Recent*)
Subphylum Labiata	
Class Chilopoda	Centipedes
Class Diplopoda	Millipedes
Class Insecta	The 24 orders of insects (*Devonian to Recent*)
PHYLUM MOLLUSCA	Unsegmented, soft-bodied animals, usually with shells
Class Monoplacophora	Primitive forms with cap-shaped shells (*Cambrian to Recent*)
Class Amphineura	Chitons, marine forms with shells composed of eight segments
Class Scaphopoda	Tusk shells, curved tubular shells open at both ends
Class Gastropoda	Snails, abalones, asymmetric animals with single spiral conch or no shell (*Cambrian to Recent*)
Class Bivalvia (Pelecypoda)	Shells of two valves (right and left); includes clams, mussels, oysters, and scallops (*Cambrian to Recent*)
Class Cephalopoda	Marine animals with tentacles around head and well-developed eyes and nervous system
ORDER NAUTILOIDEA	Nautiloids; cephalopods with simple suture lines (*Cambrian to Recent*)
ORDER AMMONOIDEA	Ammonoids; cephalopods with complexly folded sutural lines (*Devonian to Cretaceous*)
ORDER BELEMNOIDEA	Belemnites (*Late Mississippian to Early Tertiary*)
ORDER SEPIOIDEA	Cuttlefishes (*Jurassic to Recent*)
ORDER TEUTHOIDEA	Squids (*Jurassic to Recent*)
ORDER OCTOPODA	Octopi
PHYLUM HYOLITHA	Marine, bilaterally symmetrical, solitary metazoans with conical shells (*Cambrian to Permian*)
PHYLUM POGONOPHORA	Beard worms
PHYLUM CHAETOGNATHA	Arrow worms
PHYLUM ECHINODERMATA	Marine animals that are radially symmetric as adults (bilateral as larvae), have calcareous, spine-bearing plates and unique water vascular systems
Class Asteroidea	Starfishes (*Ordovician to Recent*)
Class Ophiuroidea	Brittle stars and serpent stars (*Ordovician to Recent*)
Class Echinoidea	Sea urchins and sand dollars (*Ordovician to Recent*)
Class Holothuroidea	Sea cucumbers
Class Crinoidea	Sea lilies and feather stars (*Cambrian to Recent*)
Class Blastoidea	The extinct blastoids of the Paleozoic (*Silurian to Permian*)
PHYLUM HEMICHORDATA	Includes the extinct Class Graptolithina (*Cambrian to Pennsylvanian*), as well as living acorn worms that have larval forms resembling those of echinoderms

PHYLUM CONODONTA	Extinct marine animals whose skeletal parts consist of microscopic mineralized elements arranged in apparatuses (*Proterozoic to Triassic*)
PHYLUM CHORDATA	Bilaterally symmetric animals with notochord, dorsal hollow neural tube, and gill clefts in the pharynx
Subphylum Urochordata	Sea squirts or tunicates; larval forms have notochord in tail region
Subphylum Cephalochordata	*Branchiostoma* ("*Amphioxus*"); small marine animals with fishlike bodies and notochord
Subphylum Vertebrata	Animals with a backbone of vertebrae, definite head, ventrally located heart, and well-developed sense organs
Class Agnatha	Living lampreys and hagfishes as well as extinct ostracoderms; agnatha lack jaws (*Cambrian to Recent*)
Class Acanthodii	Primitive, extinct, spiny fishes with jaws (*Middle to Late Paleozoic*)
Class Placodermi	Primitive, often armored, Paleozoic jawed fishes (*Middle Paleozoic*)
Class Chondrichthyes	Sharks, rays, skates, and chimaeras (*Middle Paleozoic to Recent*)
Class Osteichthyes	The bony fishes (*Devonian to Recent*)
Subclass Actinopterygii	Ray-finned fishes (*Devonian to Recent*)
Subclass Sarcopterygii	Lobe-finned, air-breathing fishes (*Devonian to Recent*)
ORDER CROSSOPTERYGII	Lobe-finned fishes, ancestors of amphibians
ORDER DIPNOI	Lungfishes
Class Amphibia	Amphibians, the Earth's earliest land-dwelling vertebrates; include extinct Labyrinthodontia of Late Paleozoic and Triassic (*Devonian to Recent*)
Class Reptilia	Reptiles, reproducing with use of amniotic eggs
Subclass Anapsida	Turtles, as well as the extinct acquatic mesosaurs and terrestrial stem reptiles called cotylosaurs (*Permian to Recent*)
Subclass Synapsida	Mammal-like reptiles, including sailback forms and therapsids (*Permian and Triassic*)
Subclass Euryapsida	Extinct, generally marine reptiles, including plesiosaurs, ichthyosaurs, and placodonts (*Permian to Cretaceous*)
Subclass Diapsida	Reptilian group that includes the extinct dinosaurs and crocodilians, lizards, snakes, and the modern tuatara (*Permian to Recent*)
Class Aves	Birds; warm-blooded, feathered, and typically winged animals; primitive forms with reptilian teeth, modern forms toothless (*Jurassic to Recent*)
Class Mammalia	Warm-blooded animals with hair covering; females with mammary glands that secrete milk for nourishing young (*Triassic to Recent*)
Subclass Eotheria	Primitive, extinct Triassic and Jurassic mammals
Subclass Prototheria	Monotremes such as the duck-billed platypus and spiny anteater. Egg-laying mammals (*Triassic to Recent*)
Subclass Allotheria	Extinct early mammals with multicusped teeth; multituberculates (*Triassic to Early Cenozoic*)
Subclass Metatheria	Pouched mammals or marsupials (*Cretaceous to Recent*)
Subclass Eutheria	The placental mammals; young develop within uterus of female, obtain moisture via the placenta (*Cretaceous to Recent*)
ORDER INSECTIVORA	Primitive insect-eating mammals, including moles and shrews (*Cretaceous to Recent*)
ORDER DERMOPTERA	The colugo

ORDER CHIROPTERA	Bats (*early Tertiary to Recent*)
ORDER PRIMATES	Lemurs, tarsiers, monkeys, apes, and humans (*early Tertiary to Recent*)
ORDER EDENTATA	Living armadillos, anteaters, and tree sloths; extinct glyptodonts and ground sloths
ORDER RODENTIA	Squirrels, mice, rats, beavers, and porcupines
ORDER LAGOMORPHA	Hares, rabbits, and pikas
ORDER CETACEA	Whales and porpoises (*Early Tertiary to Recent*)
ORDER CREODONTA	Extinct, ancient carnivorous placentals
ORDER CARNIVORA	Modern carnivorous placentals, including dogs, cats, bears, hyenas, seals, sea lions, and walruses
ORDER CONDYLARTHA	Extinct ancestral hoofed placentals (ancestral ungulates) (*Early Tertiary*)
ORDER AMBLYPODA	Extinct primitive ungulates (*Early Tertiary*)
ORDER TUBULIDENTATA	Aardvarks
ORDER PHOLIDOTA	Pangolins
ORDER PERISSODACTYLA	Odd-toed hoofed mammals, including living horses, rhinoceroses, and tapirs; and extinct titanotheres and chalicotheres (*Early Tertiary to Recent*)
ORDER ARTIODACTYLA	Even-toed hoofed animals, including living antelopes, cattle, deer, giraffes, camels, llamas, hippos, and pigs; and extinct entelodonts and oreodonts (*Early Tertiary to Recent*)
ORDER PROBOSCIDEA	Elephants and extinct mastodons and mammoths (*Early Tertiary to Recent*)
ORDER SIRENIA	Sea cows

Formation Correlation Charts for Representative Sections

The following charts may be used to ascertain the approximate equivalence of some of the better known rock formations deposited during the Phanerozoic Eon. The vertical scale does not indicate thickness but rather is an approximation of time. Time–stratigraphic terms are indicated on either side of the chart. The diagonal lines indicate formations missing by nondeposition of erosion.

Fm = formation (generally including two or more lithologic types); Ss = sandstone; Sh = shale; Ls = limestone; Dol = dolomite; Qtzite = quartzite; Congl = conglomerate; Grp = group (two or more formations); Mem = member. Solid horizontal lines between the formations indicate well-established boundaries, whereas broken lines indicate formation boundaries that are not firmly established.

CAMBRIAN

SERIES	Southern Appalachian	Wisconsin	Missouri, Ozark Region	Arbuckle, Wichita Mtns., Okla.	Sawback Range, British Columbia	House Range, Utah	Nopah Range, Calif.	Northern Wales	STAGES	SERIES
CROIXAN	Copper Ridge Dol	Trempealeau Grp.	Eminence Fm / Potosi Dol	Butterfly Dol / Signal Mtn Fm / Royer Dol	Lyell Fm	Notch Peak Fm		Dolgelly Beds (Lingula Flags)	Trempealeauan	UPPER
		Franconia Grp.	Elvins Grp — Doe Run Fm / Derby Fm / Davis Fm	Fort Sill Lms / Honey Creek Fm		Orr Fm	Nopah Fm	Festiniog Beds (Lingula Flags)	Franconian	
		Dresbach Grp.	Bonneterre Dol	Reagan Ss	Sullivan Fm	Weeks Fm		Maentwrog Beds	Dresbachian	
			Lamotte Ss							
ALBERTAN	Conasauga Fm				Eldon Dol	Marjum Fm		Clogan Shales (Menevian Beds)		MIDDLE
						Wheeler Sh	Cornfield Sprs. Fm			
					Stephen Fm	Swasey Ls		Cefn Goch Grit (Menevian Beds)		
						Whirlwind Fm	Bonanza King Fm			
						Dome Ls				
						Chisolm Sh				
					Cathedral Dol	Howell Ls	Cadiz Fm	Harlech Grits		
					Ptarmigan Fm	Tatow Fm				
WAUCOBAN	Rome Fm				Mount Whyte Fm	Pioche Fm	Wood Canyon Fm			LOWER
	Shady Dol				St. Piran Ss	Prospect Mtn. Qtzite	Stirling Qtzite			
	Weisner Qtzite				Fort Mtn Ss					

ORDOVICIAN

SERIES	STAGES		Taconic Area of Western Vt.	Northwestern New York	Virginia	Missouri	Oklahoma	Colorado	Utah	GREAT BRITAIN SERIES
CINCINNATIAN	Gamachian									ASHGILL
	Richmondian			Queenston Sh (Red beds)	Juniata Fm	Girardeau Ls Orchard Ck Sh Thebes Ss Maquoketa Sh Fernvale Ls	Polk Ck Sh	Fremont Ls	Fish Haven Dol	
	Mays-villian			Oswego Ss						
	Edenian			Lorraine Grp	Martinsburg Fm					
CHAMPLAINIAN	Mohawkian	Trentonian	Snake Hill Sh	Utica Sh						CARADOC
			Rysedorf Congl	Trenton Ls		Kimmswick Ls	Big Fork Chert			
			Normanskill Sh		Bays Fm	Decorah Grp		Harding Ss		
		Black River		Black River Ls	Edinburg Fm	Plattin Grp				
	Chazian				Lincolnshire Ls		Womble Sh		Swan Peak Qtzite	LLANDEILO
					Whistle Ck Sh	Joachim Ls				
	White-rockian				New Market Ls	St. Peter Ss				LLANVIRN
			Deepkill Sh			Everton Fm				
CANADIAN	Stages not yet established					Smithville Fm			Garden City Ls	
						Powell Fm	Blakely Ss			ARENIG
						Cotter Dol				
					Knox Dol (upper beds)	Jefferson City Grp	Mazarn Sh	Manitou Ls		
			Shaghticoke Sh			Roubidoux Fm				
						Gasconade Dol	Crystal Mtn Ss			TREMADOC

SILURIAN

Series in North America	Eastern Penn.	Western N.Y.	W. Ohio to E. Kentucky	Oklahoma	Northern Ill.	Central Nevada	Southern Wales	Series in Europe
CAYUGAN	Keyser Group: Manlius Ls / Rondout Ls / Decker Ls	CoblesKill Dol	Bass Island Group				Red Marls	PRIDOLIAN
	Bossardville Ls			Missouri Mtn. Slate			Temeside beds	
	Poxono Island Sh						Downton Castle Ss	
		Salina Grp		Henryhouse Sh			Upper Ludlow Sh	LUDLOVIAN
NIAGARAN	Bloomsburg red beds	Albemarle Group: Guelph Dol	Peebles Dol				Aymestry Ls	
		Albemarle Group: Lockport Dol	Durbin Grp		Racine Dol		Lower Ludlow Sh	
		Decew Dol					Wenlock Ls	WENLOCKIAN
			Lilley Fm			Lone Mtn. Ls		
		Rochester Sh			Waukesha Dol		Wenlock Sh	
	Shawangunk Congl	Irondequoit Lms	Bisher Fm	St. Clair Ls	Joliet Dol	Roberts Mtn. Fm		
			Ribolt Sh					
		Estill Sh		Blaylock Ss				LLANDOVERIAN
	Merriton Ls						Llandovery Sh	
ALEXANDRIAN	Neahga Sh							
	Thorold Ss	Neland Fm						
	Grimsby Ss						Llandovery Ss	
	Power Glen Fm	Brassfield Ls	Chimneyhill Ls	Kankakee Dol				
	Whirlpool Ss	Centerville Clay						
					Edgewood Dol			

DEVONIAN

SERIES NA	STAGES NA	New York Catskill Mtns.	New York Western	Virginia, Md.,Penn.	Western Tenn.	Nevada (Eureka Area)	Devon, England	Scotland	SERIES IN EUROPE
UPPER	BRADFORDIAN		Oswayo Grp				Pilton Beds		FAMENNIAN
			Cattaraugus Congl	Catskill Red beds	Chattanooga Shale	Pilot Shale	Baggy Beds	Upper Old Red Ss	
	CASSADAGIAN						Upcott Beds		
			Conneaut Grp				Pickwell Downs Ss		
			Canadaway Grp				Morte Slates		
			Java Fm						
	COHOC-TONIAN	Wittenburg Congl	West Falls Fm	Chemung Ss		Devils Gate Ls			FRASNIAN
	FINGER-LAKESIAN	Walton Sh	Sonyea Fm	Brallier Sh					
		Oneonta Fm	Genesee Fm	Harrell Fm			Ilfracombe Beds		
MIDDLE	TAGHAN-ICAN			Burket Sh					GIVETIAN
				Tully Fm					
	TIOUGH-NIOGAN	Hamilton Grp	Hamilton Grp	Mahantango Fm					
	CAZE-NOVIAN			Marcellus Sh			Hangman Grits		EIFELIAN
	ONESQUETH-AWAN	Onondaga Ls	Onondaga Ls	Onondago Ls		Nevada Fm	Lynton Beds		
LOWER	ESPUSIAN	Schohari Fm	Bois Blanc Fm	Needmore Sh	Camden Chert				EMSIAN
		Carlisle Fm							
		Esopus Sh							
	DEER-PARKIAN	Oriskany Ss	Oriskany Ss	Oriskany Ss	Harriman Fm				SIEGENIAN
				Shriver Chert	Flat Gap Ls			Lower Old Red Ss	
				Licking Ck Ls					
	HELDERBERGIAN	Helderberg Grp		New Scotland Ls	Ross Fm				GEDINNIAN
				Elbow Ridge Ss					
		Rondout Ls		Keyser Ls					

MISSISSIPPIAN

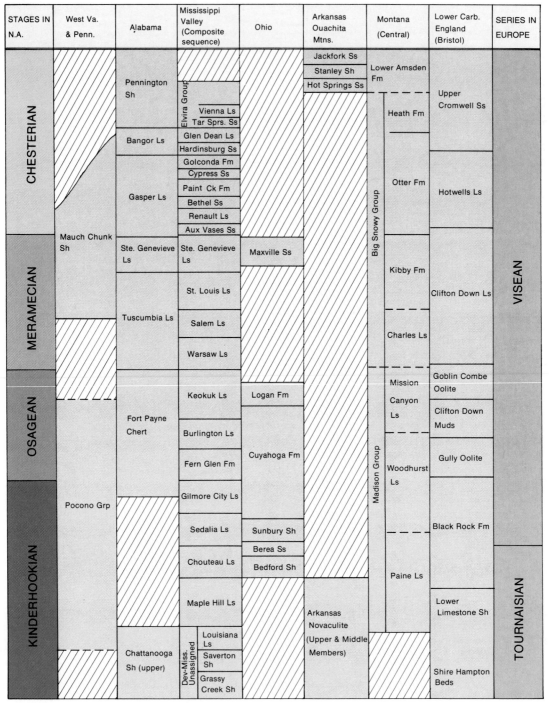

STAGES IN N.A.	West Va. & Penn.	Alabama	Mississippi Valley (Composite sequence)		Ohio	Arkansas Ouachita Mtns.	Montana (Central)		Lower Carb. England (Bristol)	SERIES IN EUROPE
CHESTERIAN	Mauch Chunk Sh	Pennington Sh	Elvira Group			Jackfork Ss	Lower Amsden Fm		Upper Cromwell Ss	**VISEAN**
				Vienna Ls		Stanley Sh				
				Tar Sprs. Ss		Hot Springs Ss	Heath Fm			
		Bangor Ls		Glen Dean Ls						
				Hardinsburg Ss					Hotwells Ls	
		Gasper Ls		Golconda Fm			Otter Fm	Big Snowy Group		
				Cypress Ss						
				Paint Ck Fm						
				Bethel Ss						
				Renault Ls						
				Aux Vases Ss						
MERAMECIAN		Ste. Genevieve Ls	Ste. Genevieve Ls		Maxville Ss		Kibby Fm		Clifton Down Ls	
		Tuscumbia Ls	St. Louis Ls							
			Salem Ls				Charles Ls			
			Warsaw Ls							
OSAGEAN	Pocono Grp	Fort Payne Chert	Keokuk Ls		Logan Fm		Mission Canyon Ls	Madison Group	Goblin Combe Oolite	
			Burlington Ls						Clifton Down Muds	
			Fern Glen Fm		Cuyahoga Fm		Woodhurst Ls		Gully Oolite	
			Gilmore City Ls						Black Rock Fm	
KINDERHOOKIAN			Sedalia Ls		Sunbury Sh					**TOURNAISIAN**
			Chouteau Ls		Berea Ss		Paine Ls			
					Bedford Sh				Lower Limestone Sh	
			Maple Hill Ls			Arkansas Novaculite (Upper & Middle Members)				
		Chattanooga Sh (upper)	Dev.-Miss. Unassigned	Louisiana Ls					Shire Hampton Beds	
				Saverton Sh						
				Grassy Creek Sh						

PENNSYLVANIAN

Stages in N.A.	Pennsylvania	West Virginia	Arkansas (Ouachita Mtns.)	Southern Illinois	Missouri	Colorado	Northeast England (Upper Carboniferous)	SERIES IN EUROPE
VIRGILIAN	////	Monongahela Grp.	////	////	Waubaunsee Grp	////	////	STEPHANIAN
					Shawnee Grp			
					Douglas Grp			
MISSOURIAN	Conemaugh Grp.	Conemaugh Grp.		McLeansboro Fm	Pedee Grp			
					Lansing Grp			
					Kansas City Grp			
					Pleasanton Grp			
DESMOINESIAN	Allegheny Grp.	Allegheny Grp.		Carbondale Fm	Marmaton Grp	Fountain Fm	Upper Coal Measures	WESTPHALIAN
					Cherokee Grp		Middle Coal Measures	
ATOKAN	Pottsville Grp.	Pottsville Grp.	Atoka Fm	Tradewater Fm	Riverton Fm			
					Burgner Fm			
					McLouth Fm		Lower Coal Measures	
					Cheltenham Fm	////		
MORROWAN	////		Johns Valley Sh	Caseyville Fm	Hale Fm	Glen Eyrie Sh		
			Jackfork Ss	////	////			
			Stanley Sh			////	Millstone Grit	NAMURIAN
			Hot Springs Ss	////				

PERMIAN

Stages in N.A.	Ohio and W. Va.	Western Texas	Arizona	Wyoming	South Africa	Germany	Russia (Ural Mtns.)	Stages in Europe
OCHOAN		Dewey Lake Fm					Tartarian	TARTARIAN (in part)
OCHOAN		Rustler Fm				Zechstein Fm	Tartarian	TARTARIAN (in part)
OCHOAN		Salado Fm				Zechstein Fm	Tartarian	TARTARIAN (in part)
OCHOAN		Castile Fm				Zechstein Fm	Kazanian	KAZANIAN
GUADALUPIAN		Bell Canyon Fm			Beaufort Series	Zechstein Fm	Kazanian	KAZANIAN
GUADALUPIAN		Cherry Canyon Fm		Phosphoria Fm		Upper Rotliegend	Kungurian	KUNGURIAN
GUADALUPIAN		Brushy Canyon Fm		Phosphoria Fm		Upper Rotliegend	Kungurian	KUNGURIAN
LEONARDIAN		Bone Spring Ls	Kaibab Ls		Ecca Series	Lower Rotliegend	Artinskian	ARTINSKIAN
LEONARDIAN		Bone Spring Ls	Toroweap Fm		Ecca Series	Lower Rotliegend	Artinskian	ARTINSKIAN
LEONARDIAN		Bone Spring Ls	Coconino Ss		Ecca Series	Lower Rotliegend	Artinskian	ARTINSKIAN
LEONARDIAN		Bone Spring Ls	Hermit Sh		Ecca Series	Lower Rotliegend	Artinskian	ARTINSKIAN
WOLFCAMPIAN	Dunkard Grp	Hueco Ls	Supai Fm		Dwyka Series	Lower Rotliegend	Sakmarian	SAKMARIAN
WOLFCAMPIAN	Dunkard Grp	Hueco Ls	Supai Fm		Dwyka Series	Lower Rotliegend	Sakmarian	ASSELIAN

TRIASSIC

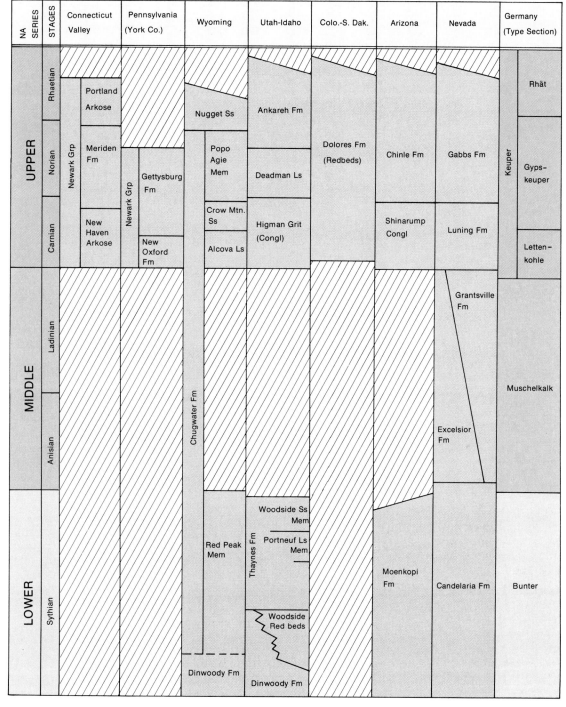

NA SERIES	STAGES	Connecticut Valley	Pennsylvania (York Co.)	Wyoming	Utah-Idaho	Colo.-S. Dak.	Arizona	Nevada	Germany (Type Section)
UPPER	Rhaetian	Portland Arkose (Newark Grp)	(Newark Grp)	Nugget Ss	Ankareh Fm	Dolores Fm (Redbeds)	Chinle Fm	Gabbs Fm	Rhät
UPPER	Norian	Meriden Fm (Newark Grp)	Gettysburg Fm (Newark Grp)	Popo Agie Mem	Deadman Ls	Dolores Fm (Redbeds)	Chinle Fm	Gabbs Fm	Gyps-keuper (Keuper)
UPPER	Carnian	New Haven Arkose (Newark Grp)	New Oxford Fm (Newark Grp)	Crow Mtn. Ss / Alcova Ls	Higman Grit (Congl)	Dolores Fm (Redbeds)	Shinarump Congl	Luning Fm	Letten-kohle (Keuper)
MIDDLE	Ladinian			Chugwater Fm				Grantsville Fm	Muschelkalk
MIDDLE	Anisian			Chugwater Fm				Excelsior Fm	Muschelkalk
LOWER	Sythian			Red Peak Mem / Dinwoody Fm	Woodside Ss Mem / Portneuf Ls Mem / Woodside Red beds / Dinwoody Fm (Thaynes Fm)		Moenkopi Fm	Candelaria Fm	Bunter

JURASSIC

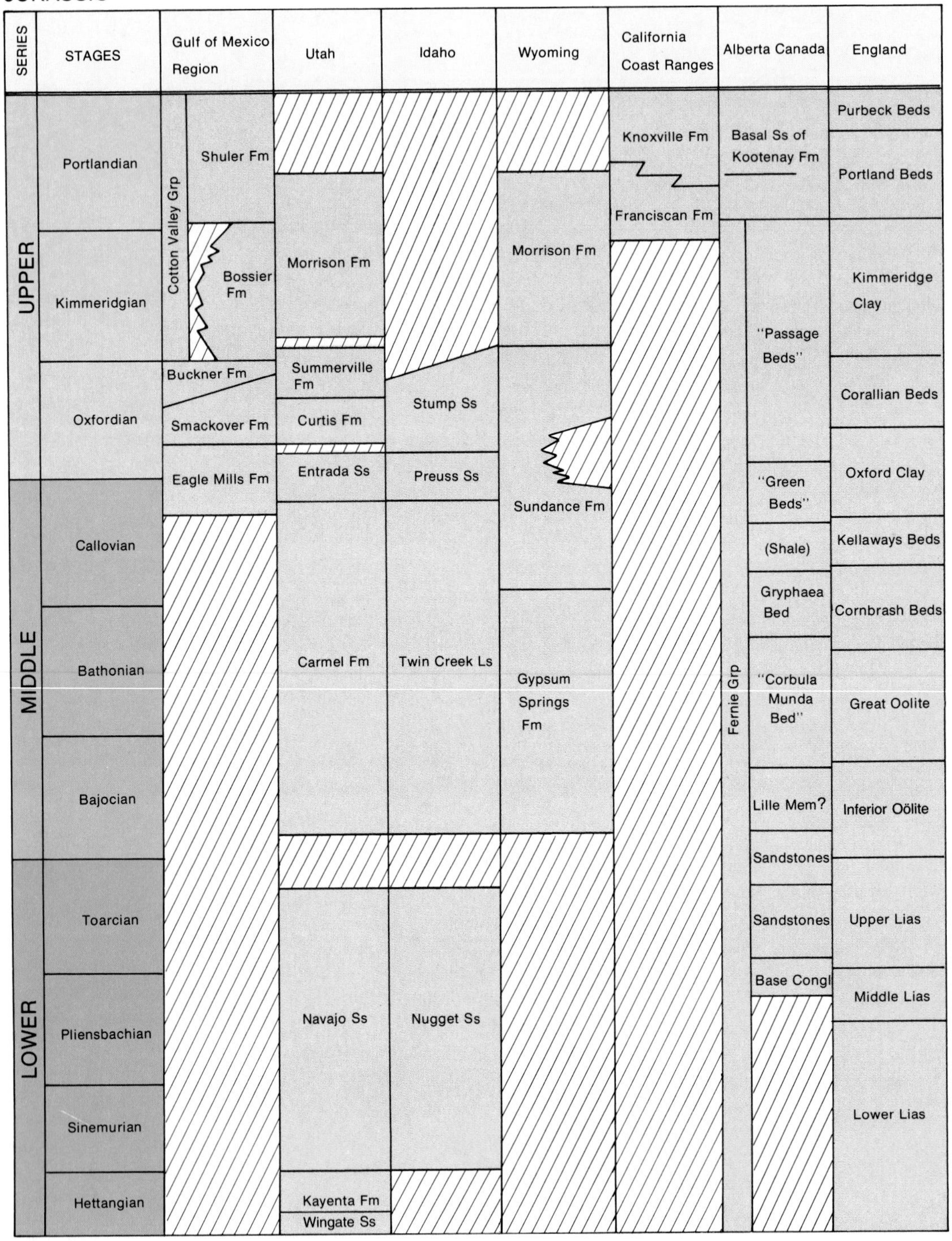
SERIES
STAGES
Gulf of Mexico Region
Utah
Idaho
Wyoming
California Coast Ranges
Alberta Canada
England
UPPER
MIDDLE
LOWER
Portlandian
Kimmeridgian
Oxfordian
Callovian
Bathonian
Bajocian
Toarcian
Pliensbachian
Sinemurian
Hettangian
Cotton Valley Grp
Shuler Fm
Bossier Fm
Buckner Fm
Smackover Fm
Eagle Mills Fm
Morrison Fm
Summerville Fm
Curtis Fm
Entrada Ss
Carmel Fm
Navajo Ss
Kayenta Fm
Wingate Ss
Stump Ss
Preuss Ss
Twin Creek Ls
Nugget Ss
Morrison Fm
Sundance Fm
Gypsum Springs Fm
Knoxville Fm
Franciscan Fm
Basal Ss of Kootenay Fm
Fernie Grp
"Passage Beds"
"Green Beds"
(Shale)
Gryphaea Bed
"Corbula Munda Bed"
Lille Mem?
Sandstones
Sandstones
Base Congl
Purbeck Beds
Portland Beds
Kimmeridge Clay
Corallian Beds
Oxford Clay
Kellaways Beds
Cornbrash Beds
Great Oolite
Inferior Oölite
Upper Lias
Middle Lias
Lower Lias

CRETACEOUS

SERIES	STAGES	New Jersey	Alabama (Tuscaloosa)	Texas	Kansas (Wallace Co.)	Colorado	Montana	England
UPPER CRETACEOUS	Maastrichtian	Monmouth Grp	Prairie Bluff / Ripley Fm	Navarro Grp — Escondido Fm	Pierre Sh	Animas Fm / McDermott Fm / Kirkland Fm / Fruitland Fm / Pictured Cliffs Ss / Lewis Sh / Mesaverde Grp	Hell Ck Fm / Fox Hills Ss / Bear Paw Sh	Upper Chalk
UPPER CRETACEOUS	Campanian	Matawan Grp: Wenonah Sand / Marshalltown Fm / Englishtown Sand / Woodbury Clay / Merchantville Clay	Selma Chalk	Navarro Grp — Corsicana Marl / Taylor Marl	Pierre Sh	Lewis Sh / Mesaverde Grp	Judith River Fm / Claggett Sh	Upper Chalk
UPPER CRETACEOUS	Santonian	Magothy Fm	Eutaw Fm	Austin Chalk		Mancos Sh	Eagle Ss / Telegraph Ck Fm	Upper Chalk
UPPER CRETACEOUS	Coniacian	Magothy Fm		Austin Chalk	Niobrara Fm	Mancos Sh		Upper Chalk
UPPER CRETACEOUS	Turonian	Raritan Fm	Tuscaloosa Fm	Eagle Ford Shale	Carlile Sh / Greenhorn Ls / Graneros Sh	Mancos Sh	Colorado Sh	Middle Chalk
UPPER CRETACEOUS	Cenomanian	Raritan Fm	Tuscaloosa Fm	Woodbine Ss	Dakota Ss	Dakota Ss	Colorado Sh	Lower Chalk
LOWER CRETACEOUS	Albian			Washita Grp / Fredericksburg Grp / Trinity Grp	Kiowa Sh / Cheyenne Ss			Upper Greensand
LOWER CRETACEOUS	Aptian			Trinity Grp		Burro Canyon Fm	Kootenai Fm	Lower Greensand
LOWER CRETACEOUS (Neocomian)	Barremian							Wealden Beds
LOWER CRETACEOUS (Neocomian)	Hauterivian							Wealden Beds
LOWER CRETACEOUS (Neocomian)	Valanginian							Wealden Beds
LOWER CRETACEOUS (Neocomian)	Berriasian							Wealden Beds

CENOZOIC

SYSTEMS	SERIES	STAGES	Gulf Coast	Atlantic Coast	Colorado Wyoming	S. Dakota Nebraska	California (S. Great Valley)	Western Washington
Quaternary	Quaternary / Pleistocene (Hol.)	Unnamed Holocene	Recent stream and lake deps	Pamlico Fm	Recent sediments	Recent alluvial deposits	Recent alluvial deposits	Recent alluvium
Quaternary	Pleistocene	Wisconsinan	Prairie Fm	Talbot Fm			Local terrace deposits	Marine & river terrace deposits
Quaternary	Pleistocene	Sangamon	Montgomery Fm	Wicomico Fm		Pleistocene sand & clay		Marine & river terrace deposits
Quaternary	Pleistocene	Illinoian / Yarmouth	Bentley Fm	Sunderland Fm		Pleistocene sand & clay		Marine & river terrace deposits
Quaternary	Pleistocene	Kansan / Aftonian / Nebraskan	Williana Fm	Brandywine Fm		Pleistocene sand & clay	Tulare Fm	Marine & river terrace deposits
Tertiary	Neogene / Pliocene	Astian	Citronelle Fm	Wacamaw Fm		Ogallala Grp	San Joaquin Fm	
Tertiary	Neogene / Pliocene	Piacenzan	Citronelle Fm	Wacamaw Fm		Ogallala Grp	Etchegoin Fm	
Tertiary	Neogene / Pliocene	Piacenzan	Citronelle Fm	Wacamaw Fm		Ogallala Grp	Jacalitos Fm	
Tertiary	Neogene / Miocene	Pontian	Choctawatchee Fm	Yorktown Fm		Sheep Ck Fm	Reef Ridge Sh	
Tertiary	Neogene / Miocene	Sarmatian	Choctawatchee Fm			Sheep Ck Fm	McClure Sh	
Tertiary	Neogene / Miocene	Tortonian	Choctawatchee Fm	St. Mary's Fm		Marsland Fm	McClure Sh	
Tertiary	Neogene / Miocene	Helvetian	Shoal River Fm	Choptank Fm		Marsland Fm	Temblor Fm	Astoria Fm
Tertiary	Neogene / Miocene	Burdigalian	Oak Grove Ss	Calvert Fm		Arikaree Grp	Temblor Fm	Astoria Fm
Tertiary	Neogene / Miocene	Burdigalian	Chipola Fm	Hawthorn Fm		Arikaree Grp	Temblor Fm	Astoria Fm
Tertiary	Neogene / Miocene	Aquitainian	Catahoula Ss	Trent Marl		Arikaree Grp		Twin Rivers Fm
Tertiary	Oligocene	Chattian		Flint River		White River Series		Blakely Fm
Tertiary	Oligocene	Rupelian	Vicksburg Ls		Florissant Fm	White River Series		Lincoln Fm
Tertiary	Oligocene	Tongrian			White River Grp	White River Series		Lincoln Fm
Tertiary	Paleogene / Eocene	Ludian	Jackson Fm		Duchesne River Fm		Tumey Ss	
Tertiary	Paleogene / Eocene	Bartonian	Jackson Fm	Castle Hayne Marl	Uinta Fm		Kreyenhagen Sh	Cowlitz Fm
Tertiary	Paleogene / Eocene	Auversian	Yegua Fm / Cook Mtn. Fm / Sparta Ss (Claiborne Grp)	Shark River Marl	Bridger Fm		Kreyenhagen Sh	Cowlitz Fm
Tertiary	Paleogene / Eocene	Lutetian	Mt. Selma Fm / Carrizo Ss (Claiborne Grp)	Manasquan Marl	Green River Fm		Domingine Fm	
Tertiary	Paleogene / Eocene	Cuisian	Wilcox Grp	Manasquan Marl	Green River Fm		Yokut Ss	
Tertiary	Paleogene / Eocene	Ypresian	Wilcox Grp	Aquia Fm	Cedar Breaks Fm		Lodo Fm	Metachosin Volcanics
Tertiary	Paleogene / Paleocene	Thanetian	Wills Pt. Fm (Midway Grp)	Clayton Ls	Fort Union Fm	Fort Union Fm	Lodo Fm	Metachosin Volcanics
Tertiary	Paleogene / Paleocene	Montian	Kincaid Fm (Midway Grp)	Clayton Ls	Fort Union Fm	Fort Union Fm		Metachosin Volcanics
Tertiary	Paleogene / Paleocene	Danian						

Physiographic Provinces of the United States

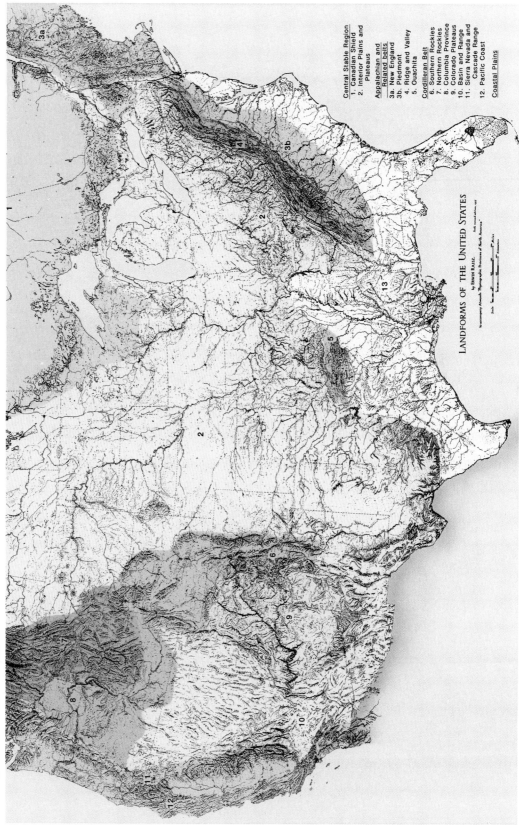

Central Stable Region
1. Canadian Shield
2. Interior Plains and Plateaus

Appalachian and Related belts
3a. New England
3b. Piedmont
4. Ridge and Valley
5. Ouachita

Cordilleran Belt
6. Southern Rockies
7. Northern Rockies
8. Columbia Province
9. Colorado Plateaus
10. Basin and Range
11. Sierra Nevada and Cascade Range
12. Pacific Coast

Coastal Plains

LANDFORMS OF THE UNITED STATES
by ERWIN RAISZ

ICELAND

GREENLAND
(Den.)

Reykjavik

RUSSIA

ALASKA
(U.S.
State)

Nuuk
(Godthåb)

CANADA

ST. PIERRE AND
MIQUELON (Fr.)

Ottawa

UNITED STATES

Washington D.C.

MIDWAY IS.
(U.S.)

BERMUDA (U.K.)

Tropic of Cancer

HAWAII
(U.S. State)

MEXICO

BAHAMAS
Nassau

TURKS AND
CAICOS IS. (U.K.)

REPUBLIC OF
THE MARSHALL
ISLANDS

Havana
CUBA
CAYMAN IS.
(U.K.)

Mexico City
GUATEMALA

Belmopan
BELIZE
HONDURAS

HAITI
JAMAICA

VIRGIN IS.
DR (U.S. & U.K.)
SJ
K
PP SD PR

ANGUILLA (U.K.)
ST. MARTIN (Fr. & Neth.)
ANTIGUA & BARBUDA
GUADELOUPE (Fr.)
DOMINICA
MARTINIQUE (Fr.)
BARBADOS
ST. VINCENT & THE GRENADINES
TRINIDAD & TOBAGO

CAPE
VERDE

ST. KITTS AND NEVIS

MONTSERRAT (U.K.)
ST. LUCIA
GRENADA

NETH. ANT.

Guatemala City
San Salvador
EL SALVADOR
Managua
San Jose
COSTA RICA
PANAMA
Panama City

Tegucigalpa
NICARAGUA

Caracas
VENEZUELA
Bogotá
COLOMBIA

Georgetown
Paramaribo
Cayenne

GUYANA
SURINAME
FRENCH GUIANA
(Fr.)

Equator

Quito
ECUADOR

GALAPAGOS IS.
(Ecuador)

NAURU

KIRIBATI

TUVALU

TOKELAU IS.
(N.Z.)

SOLOMON
ISLANDS

WALLIS &
FUTUNA
(Fr.)

WESTERN
SAMOA

AMERICAN
SAMOA
(U.S.)

FRENCH
POLYNESIA
(Fr.)

TONGA

COOK ISLANDS
(N.Z.)

Papeete

PERU

Lima

BRAZIL

BOLIVIA
La Paz
Sucre

Brasilia

Tropic of Capricorn

PITCAIRN I. (U.K.)

CHILE

PARAGUAY
Asunción

VANUATU

FIJI

LOYALTY
ISLANDS
NEW. (New Caledonia)
CALEDONIA
(Fr.)

EASTER I.
(Chile)

Santiago

ARGENTINA

URUGUAY
Montevideo
Buenos Aires

NEW ZEALAND

Wellington

FALKLAND
ISLANDS
(U.K.)

ABBREVIATIONS

CITIES

A	Amman	L	Ljubljana
Am	Amsterdam	P	Prague B Bratislava
B	Beirut	PP	Port-au-Prince
Bc	Bucharest	R	Rome
Bd	Budapest	S	Sarajevo
Bg	Belgrade	So	Sofia
Bl	Berlin	SD	Santo Domingo
Bn	Bern	SJ	San Juan
Br	Brussels	St	Stockholm
Co	Copenhagen	T	Tirana
D	Damascus	V	Vienna
TH	The Hague	W	Warsaw
J	Jerusalem	Y	Yamoussoukro
K	Kingston	Z	Zagreb

COUNTRIES & STATES

AU	AUSTRIA	POL.	POLAND
B	BULGARIA	PR	PUERTO RICO
B-H	BOSNIA-HERCEGOVINA	ROM.	ROMANIA
C	CROATIA	SL	SLOVENIA
CZ	CZECH R. S SLOVAKIA	SM	SAN MARINO
DR	DOMINICAN REPUBLIC	UAE	UNITED ARAB
GER.	GERMANY		EMIRATES
GR	GREECE	VC	VATICAN CITY
H	HUNGARY	Y	YUGOSLAVIA
LIECH.	LIECHTENSTEIN		
LUX.	LUXEMBOURG		
M	MACEDONIA		
NETH.	NETHERLANDS		
NETH. ANT.	NETHERLANDS ANTILLES		

ANTARCTICA

WORLD POLITICAL MAP

COUNTRIES GROUPED BY POPULATION SIZE
(MILLIONS, 1992)

LEGEND

Over 150
50–150
15–50
5–15
Under 5
★ Capitals
(Not shown for many small countries)
Territorial boundaries of island groups
are schematic

| 0 | 550 | 1100 | 1650 | 2200 | Miles |
| 0 | 875 | 1750 | 2625 | 3500 | Kilometers |

APPENDIX D

Periodic Table and Symbols for Chemical Elements

Scientists generally group or organize things that are similar to better understand them and determine how they relate to one another. An important step in the organization of elements was made in 1869 by Dimitri Mendeleev. Mendeleev showed that when elements are arranged in order of their increasing atomic weights, their physical and chemical properties tend to be repeated in cycles. The arrangement of elements was depicted on a chart called the periodic table of elements, a modern version of which is shown in Table D–1. In the periodic table, each box contains the symbol of the element and its atomic number. Except for two long sequences set apart at the bottom, the elements appear in the increasing order of their atomic numbers. The vertical columns are called *groups* and contain elements with similar properties. For example, in column VIIA, one finds fluorine, chlorine, bromine, and iodine. All of these elements are colored and highly reactive and share other similarities. Fluorine, however, is chemically the most active, chlorine somewhat less active, bromine still less active, and iodine the least active of the four. Except for hydrogen, all of the elements in group IA are soft, shiny metals and very reactive chemically. The horizontal rows on the table are called *periods* and contain sequences of elements having electron configurations that vary in characteristic patterns. It is thus apparent that the periodic table shows relationships among elements rather well and certainly better than an arbitrary listing.

Table D–2 provides the chemical names for the symbols of the elements used in the periodic table.

TABLE D–1 Periodic Table of the Elements

Key:

26
Fe
55.847

Atomic number (Z)
Element symbol
Atomic mass of naturally occurring isotopic mixture; for radioactive elements, numbers in parentheses are mass numbers of most stable isotopes

IA	IIA	IIIB	IVB	VB	VIB	VIIB	VIII	VIII	VIII	IB	IIB	IIIA	IVA	VA	VIA	VIIA	O
1 **H** 1.0079																	2 **He** 4.00260
3 **Li** 6.941	4 **Be** 9.01218											5 **B** 10.81	6 **C** 12.011	7 **N** 14.0067	8 **O** 15.9994	9 **F** 18.998403	10 **Ne** 20.179
11 **Na** 22.98977	12 **Mg** 24.305											13 **Al** 26.98154	14 **Si** 28.0855	15 **P** 30.97376	16 **S** 32.06	17 **Cl** 35.453	18 **Ar** 39.948
19 **K** 39.0983	20 **Ca** 40.08	21 **Sc** 44.9559	22 **Ti** 47.90	23 **V** 50.9415	24 **Cr** 51.996	25 **Mn** 54.9380	26 **Fe** 55.847	27 **Co** 58.9332	28 **Ni** 58.70	29 **Cu** 63.546	30 **Zn** 65.38	31 **Ga** 69.72	32 **Ge** 72.59	33 **As** 74.9216	34 **Se** 78.96	35 **Br** 79.904	36 **Kr** 83.80
37 **Rb** 85.4678	38 **Sr** 87.62	39 **Y** 88.9059	40 **Zr** 91.22	41 **Nb** 92.9064	42 **Mo** 95.94	43 **Tc** (98)	44 **Ru** 101.07	45 **Rh** 102.9055	46 **Pd** 106.4	47 **Ag** 107.868	48 **Cd** 112.41	49 **In** 114.82	50 **Sn** 118.69	51 **Sb** 121.75	52 **Te** 127.60	53 **I** 126.9045	54 **Xe** 131.30
55 **Cs** 132.9054	56 **Ba** 137.33	57 ***La** 138.9055	72 **Hf** 178.49	73 **Ta** 180.9479	74 **W** 183.85	75 **Re** 186.207	76 **Os** 190.2	77 **Ir** 192.22	78 **Pt** 195.09	79 **Au** 196.9665	80 **Hg** 200.59	81 **Tl** 204.37	82 **Pb** 207.2	83 **Bi** 208.9804	84 **Po** (209)	85 **At** (210)	86 **Rn** (222)
87 **Fr** (223)	88 **Ra** 226.0254	89 **†Ac** 227.0278	104 **Unq** (261)	105 **Unp** (262)	106 **Unh** (263)	107 **Uns**	109										

*Lanthanide Series

58 **Ce** 140.12	59 **Pr** 140.9077	60 **Nd** 144.24	61 **Pm** (145)	62 **Sm** 150.4	63 **Eu** 151.96	64 **Gd** 157.25	65 **Tb** 158.9254	66 **Dy** 162.50	67 **Ho** 164.9304	68 **Er** 167.26	69 **Tm** 168.9342	70 **Yb** 173.04	71 **Lu** 174.967

†Actinide Series

90 **Th** 232.0381	91 **Pa** 231.0359	92 **U** 238.029	93 **Np** 237.0482	94 **Pu** (244)	95 **Am** (243)	96 **Cm** (247)	97 **Bk** (247)	98 **Cf** (251)	99 **Es** (252)	100 **Fm** (257)	101 **Md** (258)	102 **No** (259)	103 **Lr** (260)

Note: Atomic masses shown here are 1977 IUPAC values.

TABLE D–2   Elements and Their Chemical Symbols

Actinium	Ac	Erbium	Er	Mercury	Hg	Samarium	Sm
Aluminum	Al	Europium	Eu	Molybdenum	Mo	Scandium	Sc
Americium	Am	Fermium	Fm	Neodymium	Nd	Selenium	Se
Antimony	Sb	Fluorine	F	Neon	Ne	Silicon	Si
Argon	Ar	Francium	Fe	Neptunium	Np	Silver	Ag
Arsenic	As	Gadolinium	Gd	Nickel	Ni	Sodium	Na
Astatine	At	Gallium	Ga	Niobium	Nb	Strontium	Sr
Barium	Ba	Germanium	Ge	Nitrogen	N	Sulfur	S
Berkelium	Bk	Gold	Au	Nobelium	No	Tantalum	Ta
Beryllium	Be	Hafnium	Hf	Osmium	Os	Technetium	Tc
Bismuth	Bi	Helium	He	Oxygen	O	Tellurium	Te
Boron	B	Holmium	Ho	Palladium	Pd	Terbium	Tb
Bromine	Br	Hydrogen	H	Phosphorus	P	Thallium	Tl
Cadmium	Cd	Indium	In	Platinum	Pt	Thorium	Th
Calcium	Ca	Iodine	I	Plutonium	Pu	Thulium	Tm
Californium	Cf	Iridium	Ir	Polonium	Po	Tin	Sn
Carbon	C	Iron	Fe	Potassium	K	Titanium	Ti
Cerium	Ce	Krypton	Kr	Praseodymium	Pr	Tungsten	W
Cesium	Cs	Lanthanum	La	Promethium	Pm	Uranium	U
Chlorine	Cl	Lawrencium	Lr	Protactinium	Pa	Vanadium	V
Chromium	Cr	Lead	Pb	Radium	Ra	Wolfram	
Cobalt	Co	Lithium	Li	Radon	Rn	Xenon	Xe
Copper	Cu	Lutetium	Lu	Rhenium	Re	Ytterbium	Yb
Curium	Cm	Magnesium	Mg	Rhodium	Rh	Yttrium	Y
Dysprosium	Dy	Manganese	Mn	Rubidium	Rb	Zinc	Zn
Einsteinium	Es	Mendelevium	Md	Ruthenium	Ru	Zirconium	Zr

APPENDIX E

Convenient Conversion Factors

To Convert from	To	Multiply by*
Centimeters	Feet	0.0328 ft/cm
	Inches	0.394 in/cm
	Meters	0.01 m/cm
	Microns (micrometers)	10,000 μm/cm
	Miles (statute)	6.214×10^{-6} mi/cm
	Millimeters	10 mm/cm
Feet	Centimeters	30.48 cm/ft
	Inches	12 in/ft
	Meters	0.3048 m/ft
	Microns (micrometers)	304800 μm/ft
	Miles (statute)	0.000189 mi/ft
	Yards	0.3333 yd/ft
Kilometers	Miles	0.6214 mi/km
Gallons (U.S., liq.)	Cu. centimeters	3785 cm³/gal
	Cu. feet	0.133 ft³/gal
	Liters	3.785 L/gal
	Quarts (U.S., liq.)	4 qt/gal
Grams	Kilograms	0.001 kg/g
	Micrograms	1×10^6 μg/g
	Ounces	0.03527 oz/g
	Pounds	0.002205 lb/g
Inches	Centimeters	2.54 cm/in
	Feet	0.0833 ft/in
	Meters	0.0254 m/in
	Yards	0.0278 yd/in
Kilograms	Ounces	35.27 oz/kg
	Pounds	2.205 lb/kg
Meters	Centimeters	100 cm/m
	Feet	3.2808 ft/m
	Inches	39.37 in/m
	Kilometers	0.001 km/m
	Miles (statute)	0.0006214 mi/m
	Millimeters	1000 mm/m
	Yards	1.0936 yd/m
Miles (statute)	Centimeters	160934 cm/mi
	Feet	5280 ft/mi
	Inches	63360 in/mi
	Kilometers	1.609 km/mi
	Meters	1609 m/mi
	Yards	1760 yd/mi
Ounces	Grams	28.3 g/oz
	Pounds	0.0625 lb/oz
Pounds	Grams	453.6 g/lb
	Kilograms	0.454 kg/lb
	Ounces	16 oz/lb

*Values are generally given to three or four significant figures.

Source: Turk, J., and Turk, A. *Physical Science*. Philadelpha: W. B. Saunders Company, 1977.

Exponential or Scientific Notation

Exponential or scientific notation is used by scientists all over the world. This system is based on exponents of 10, which are shorthand notations for repeated multiplications or divisions.

A positive exponent is a symbol for a number that is to be multiplied by itself a given number of times. Thus, the number 10^2 (read "ten squared" or "ten to the second power") is exponential notation for $10 \cdot 10 = 100$. Similarly, $3^4 = 3 \cdot 3 \cdot 3 \cdot 3 = 81$. The reciprocals of these numbers are expressed by negative exponents. Thus $10^{-2} = 1/10^2 = 1/(10 \cdot 10) = 1/100 = 0.01$.

To write 10^4 in longhand form you simply start with the number 1 and move the decimal four places to the right: 10000. Similarly, to write 10^{-4} you start with the number 1 and move the decimal four places to the left: 0.0001.

It is just as easy to go the other way—that is, to convert a number written in longhand form to an exponential expression. Thus, the decimal place of the number 1,000,000 is six places to the right of 1:

$$1\ 000\ 000 = 10^6$$
6 places

Similarly, the decimal place of the number 0.000001 is six places to the left of 1 and

$$0.000001 = 10^{-6}$$
6 places

Glossary

For terms not included in this glossary, students may wish to consult the Dictionary of Geologic Terms, *published in 1976 by Anchor/Doubleday under the direction of the American Geological Institute. A more comprehensive reference is the* Glossary of Geology, *1987 edited by R. L. Bates and J. A. Jackson, 3rd ed., Falls Church, Virginia, American Geological Institute.*

Absaroka sequence   A sequence of Permian–Pennsylvanian sediments bounded both above and below by a regional unconformity and recording an episode of marine transgression over an eroded surface, full flood level of inundation, and regression from the craton.

Absolute geologic age   The actual age, expressed in years, of a geologic material or event.

Acadian orogeny   An episode of mountain building in the northern Appalachians during the Devonian Period.

Acanthodians   The earliest known vertebrates (fishes) with a movable, well-developed lower jaw, or mandible; hence, the first jawed fishes.

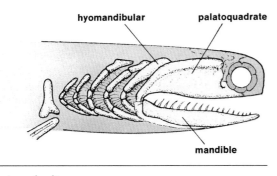

Acanthodian

Adaptation   A modification of an organism that better fits it for existence in its present environment or enables it to live in a somewhat different environment.

Adaptive radiation   The diversity that develops among species as each adapts to a different set of environmental conditions.

Adenosine diphosphate (ADP)   A product formed in the hydrolysis of adenosine triphosphate that is accompanied by release of energy and organic phosphate.

Adenosine triphosphate (ATP)   A compound that occurs in all cells and that serves as a source of energy for physiologic reactions such as muscle contraction.

Aerobic organism   An organism that uses oxygen in carrying out respiratory processes.

Age   The time represented by the time–stratigraphic unit called a stage. (Informally, may indicate any time span in geologic history, as "age of cycads.")

Agnatha   The jawless vertebrates, including extinct ostracoderms and living lampreys and hagfishes.

Algae   Any of a large group of simple plants (thallophyta) that contain chlorophyll and are capable of photosynthesis.

Allegheny orogeny   The late Palezoic episodes of mountain building along the present trend of the Appalachian Mountains.

Alluvium   Unconsolidated, poorly sorted detrital sediments ranging from clay to gravel sizes and characteristically fluvial in origin.

Alpha particle   A particle, equivalent to the nucleus of a helium atom, emitted from an atomic nucleus during radioactive decay.

Alpine orogeny   In general, the sequence of crustal disturbances beginning in the middle Mesozoic and continuing into the Miocene that resulted in the geologic structures of the Alps.

Amino acids   Nitrogenous hydrocarbons that serve as the building blocks of proteins and are thus essential to all living things.

Ammonites   Ammonoid cephalopods having more complex sutural patterns than either ceratites or goniatites.

Ammonoids   An extinct group of cephalopods, with coiled, chambered conch(s) and having septa with crenulated margins.

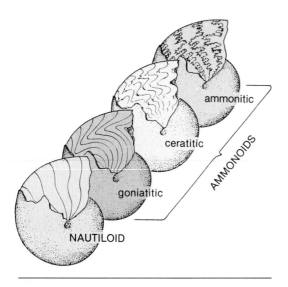

Ammonoids

Amniotic egg   That type of egg produced by reptiles, birds, and monotremes. In this type, the developing embryo is maintained and protected by an elaborate arrangement of shell membranes, yolk, sac, amnion, and allantois.

Amphibians   "Cold-blooded" vertebrates that utilize gills for respiration in the early life stages but that have air-breathing lungs as adults.

Amphibole   A ferromagnesium silicate mineral that occurs commonly in igneous and metamorphic rocks.

Anaerobic organism   An organism that does not require oxygen for respiration, but rather makes use of processes such as fermentation to obtain its energy.

Andesite   A volcanic rock that in chemical composition is intermediate between basalt and granite.

Angiosperms   An advanced group of plants having floral reproductive structures and seeds in a closed ovary. The "flowering plants."

Anthropoidea   The suborder of primates that includes monkeys, apes, and humans.

Anticline   A geologic structure in which strata are bent into an upfold or arch.

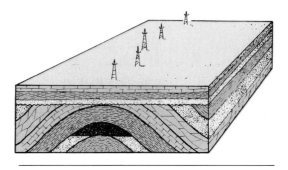

Anticline

Antler orogeny   A Late Devonian and Mississippian episode of mountain building involving folding and thrusting along a belt across Nevada to southwestern Alberta.

Archaeocyatha   A group of extinct marine organisms having double, perforated, calcareous, conical-to-cylindric walls. Archaeocyathids lived during the Cambrian.

Archean   Division of Precambrian time beginning 3.8 billion years ago and ending 2.5 billion years ago.

Archosaurs   Advanced reptiles of a group called diapsids, which includes thecodonts, "dinosaurs," pterosaurs, and crocodiles.

Arcoids   A group of bivalves (pelecypods) exemplified by species of *Arca*.

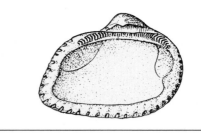

Arca

Artiodactyl   Hoofed mammals that typically have two or four toes on each foot.

Asteroid   One of numerous relatively small planetary bodies (less than 800 km in diameter) revolving around the sun in orbits lying between those of Mars and Jupiter.

Asthenosphere   A zone between 50 and 250 km below the surface of the Earth where shock waves of earthquakes travel at much reduced speeds, perhaps because of less rigidity. The asthenosphere may be a zone where convective flow of material occurs.

Atoll   A ringlike island or a series of islands formed by corals and calcareous algae around a central lagoon.

Atom   The smallest divisible unit retaining the characteristics of a specific element.

Atomic fission   A nuclear process that occurs when a heavy nucleus splits into two or more lighter nuclei, simultaneously liberating a considerable amount of energy.

Atomic fusion   A nuclear process that occurs when two light nuclei unite to form a heavier one. In the process, a large amount of energy is released.

Atomic mass   A quantity essentially equivalent to the number of neutrons plus the number of protons in an atomic nucleus.

Atomic number   The number of protons in the nuclei of atoms of a particular element. (An element is thus a substance in which all of the atoms have the same atomic number.)

Australopithecines   A general term applied loosely to Pliocene and early Pleistocene primates whose skeletal characteristics place them between typically apelike individuals and those more obviously human.

Autotroph   An organism that uses an external source of energy to produce organic nutrients from simple inorganic chemicals.

Basin   A depressed area that serves as a catchment area for sediments (basin of deposition). A structural basin is an area in which strata slope inward toward a central location. Structural basins tend to experience periodic downsinking and thus receive a thicker and more complete sequence of sediments than do adjacent areas.

Belemnites   Members of the molluscan class Cephalopoda, having straight internal shells.

Benioff seismic zone   An inclined zone along which frequent earthquake activity occurs and that marks the location of the plunging, forward edge of a lithospheric plate during subduction.

Benthic   A bottom-dwelling organism.

Bentonite   A layer of clay, presumably formed by the alteration of volcanic ash.

Beta particle   A charged particle, essentially equivalent to an electron, emitted from an atomic nucleus during radioactive disintegration.

Bivalvia   A class of the phylum Mollusca also known as the class Pelecypoda. (The term *Pelecypoda* is preferred in this text so that the pelecypods are not confused with other "bivalves," such as brachiopods and ostracods.)

Blastoids   Sessile (attached) Paleozoic echinoderms having a stem and an attached cup or calyx composed of relatively few plates.

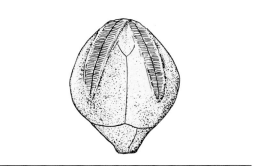

Blastoid

Bolide   A meteorite, asteroid, or comet that explodes on striking the Earth.

Brachiating   Swinging from branch to branch and tree to tree by using the limbs, as among monkeys.

Brachiopod   Bivalved (doubled-shelled) marine invertebrates. They were particularly common and widespread during the Paleozoic and persist in fewer numbers today.

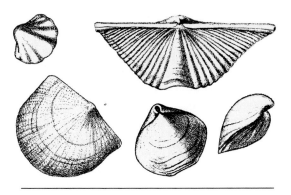

Brachiopods

Breccia A clastic sedimentary rock composed largely of angular fragments of granule size or larger.

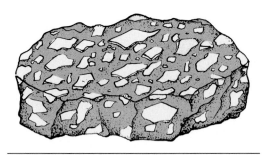

Breccia

Breeder reactor (nuclear) An atomic reactor that uses uranium-238 but that creates additional fuel by producing more fissionable material than it consumes.

Bryozoa A phylum of attached and incrusting colonial marine invertebrates.

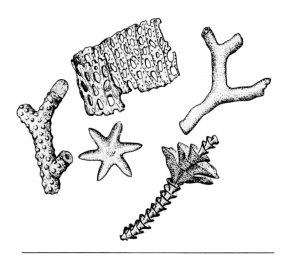

Fossil Bryozoa

Burgess Shale fauna A beautifully preserved fossil fauna of soft-bodied Cambrian animals discovered in 1910 by Charles Walcott in Kicking Horse Pass, Alberta.

Caledonian orogeny A major early Paleozoic episode of mountain building affecting Europe, which created an orogenic belt, the Caledonides, extending from Ireland and Scotland northwestward through Scandinavia.

Carbon-14 A radioactive isotope of carbon with an atomic mass of 14. Carbon-14 is frequently used in determining the age of materials less than about 50,000 years old.

Carbonate A general term for a chemical compound formed when carbon dioxide dissolved in water combines with oxides of calcium, magnesium, potassium, sodium, and iron. The most common carbonate minerals are calcite (which forms the carbonate rock limestone) and dolomite.

Carbonization The concentration of carbon during fossilization.

Cast (natural) A replica of an organic object, such as a fossil shell, formed when sediment fills a mold of that object.

Cataract Group A group of formations deposited during the Early Silurian and including the Whirlpool Sandstone, Manitoulin Dolomite, Cabot Head Shale, Dyer Bay Limestone, and Wingfield Shale of southern Ontario. The Cataract Group is correlative with the Albion Group of western New York.

Catskill delta A buildup of Middle and Upper Devonian clastic sediments as a broad, complex clastic wedge derived from the erosion of highland areas formed largely during the Acadian orogeny.

Ceboidea The New World monkeys, characterized by prehensile tails, and including the capuchin, marmoset, and howler monkeys.

Centrifugal force The apparent outward force experienced by an object moving in a circular path. Centrifugal force is a manifestation of inertia, the tendency of moving things to travel in straight lines.

Ceratites One of the three larger groups of ammonoid cephalopods having sutural complexity intermediate between goniatites and ammonites.

Ceratopsians The quadrupedal ornithischian dinosaurs characterized by the development of prominent horns on the head.

Cercopithecoidea The Old World monkeys (of Asia, southern Europe, and Africa), including macaques, guenons, langurs, baboons, and mandrills.

Cetaceans The group of marine mammals that includes whales and porpoises.

Chalicotheres Extinct perissodactyls having robust claws rather than hoofs.

Chalk A soft, white, fine-grained variety of limestone composed largely of the calcium carbonate skeletal remains of marine microplankton.

Chert A dense, hard sedimentary rock or mineral composed of submicrocrystalline quartz. Unless colored by impurities, chert is white, as opposed to flint, which is dark or black.

Chlorophyll The catalyst that makes possible the reaction of water and carbon dioxide in green plants to produce carbohydrates. Photosynthesis is the reaction.

Choanichthyes That group of fishes that includes both the dipnoans (lungfishes with weak pelvic and pectoral fins) and crossopterygians (lungfishes with stout lobe-fins).

Chondrichthyes The broad category of fishes with cartilaginous skeletons that is exemplified by sharks, skates, and rays.

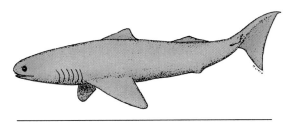

The Devonian chondrichthyan (shark)
Cladoselache

Chondrites Stony meteorites that contain rounded silicate grains or chondrules. Chondrules are believed to have formed by crystallization of liquid silicate droplets.

Chromosome Threadlike microscopic bodies composed of chromatin. Chromosomes appear in the nucleus of the cell at the time of cell division. They contain the genes. The number of chromosomes is normally constant for a particular species.

Chromosphere One of the concentric shells of the sun, lying above the photosphere and telescopically visible as a thin, brilliant red rim around the edge of the sun for a second or so at the beginning and end of a solar eclipse.

Clastic texture Texture that characterizes a rock made up of fragmental grains such as sand, silt, or parts of fossils. Conglomerates, sandstones, and siltstones are clastic rocks; the individual clastic grains are termed *clasts*.

Clastic wedge An extensive accumulation of largely clastic sediments deposited adjacent to uplifted areas. Sediments in the wedge become finer and the section becomes thinner in a direction away from the upland source areas. The Queenston and Catskill "deltas" are examples of clastic wedges.

Coccolithophorids Marine, planktonic, biflagellate, golden-brown algae that typically secrete coverings of discoidal calcareous platelets called *coccoliths.*

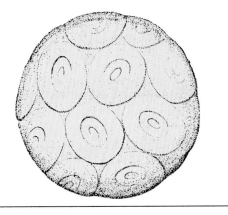

Coccolithophorid

Colorado mountains Highlands uplifted in Pennsylvanian time in Colorado. Sometimes inappropriately termed the "ancestral Rockies."

Conodonts Small, toothlike fossils composed of calcium phosphate and found in rocks ranging from Cambrian to Triassic in age. Although the precise nature of the conodont-producing organism has not been determined, this uncertainty does not detract from their usefulness as guide fossils.

Contact metamorphism The compositional and textural changes in a rock that result from heat and pressure emanating from an adjacent igneous intrusion.

Convergence (in evolution) The process by which similarity of form or structure arises among different organisms as a result of their becoming adapted to similar habitats.

Cordaites A primitive order of treelike plants with long, bladelike leaves and clusters of naked seeds. Cordaites were in some ways intermediate in evolutionary stage between seed ferns and conifers.

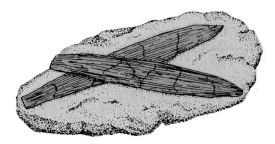

Fossil cordaite leaves

Core Central part of the Earth that lies beneath the mantle.

Coriolis effect   The deflection of winds and water currents to the right in the Northern Hemisphere and to the left in the Southern Hemisphere as a consequence of the Earth's rotation.

Cosmic rays   Extremely high-energy particles, mostly protons, which move through the galaxy and frequently strike the Earth's atmosphere.

Craton   The long-stable region of a continent, commonly with Precambrian rocks either at the surface or only thinly covered with younger sedimentary rocks.

Creodonta   Primitive, early, flesh-eating placental mammals.

Crinoids   Stalked echinoderms with a calyx composed of regularly arranged plates from which radiate arms for gathering food.

Cross-bedding (cross-stratification)   An arrangement of laminae or thin beds transverse to the planes of stratification. The inclined laminae are usually at inclinations of less than 30° and may be either straight or concave.

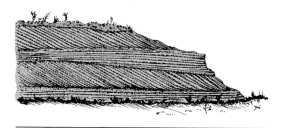

Cross-bedding

Crossopterygii   That group of choanichthyan fishes ancestral to earliest amphibians and characterized by stout pectoral and pelvic fins as well as lungs.

The Devonian crossopteryian *Eusthenopteron*

Crust (Earth)   The outer part of the lithosphere; it averages about 32 km in thickness.

Crustacea   A subphylum of the phylum Arthropoda that includes such well-known living animals as lobsters and crayfishes.

Curie temperature   The temperature at which a cooling mineral acquires permanent magnetic properties that record the surrounding magnetic field orientation and strength at the time of cooling. Magnetic properties of a mineral above the Curie

temperature will change as the surrounding field changes. Below the Curie temperature, magnetic characteristics of the mineral will not alter.

Cycadales   A group of seed plants that were especially common during the Mesozoic and were characterized by palmlike leaves and coarsely textured trunks marked by numerous leaf scars.

Cyclothem   A vertical succession of sedimentary units reflecting environmental events that occurred in a constant order. Cyclotherms are particularly characteristic of the Pennsylvania System.

Cystoids   Attached echinoderms with generally irregular arrangement and number of plates in the calyx and perforated by pores or slits.

Deccan traps   A thick sequence (3200 meters) of Upper Cretaceous basaltic lava flows that cover about 500,000 sq km of peninsular India.

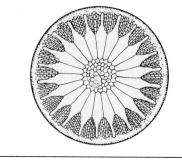

Deccan traps (outcrop map)

Décollement   Feature of stratified rocks in which upper formations may become "unstuck" from lower formations, deform, and slide thousands of meters over underlying beds.

Diatoms   Microscopic golden-brown algae (chrysophytes) that secrete a delicate siliceous frustule (shell).

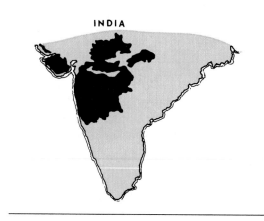

Diatom

Differentiation The process by which a planet becomes internally zoned, as when heavy materials sink toward its center and light materials accumulate near the surface.

Dinoflagellates Unicellular marine algae usually having two flagella and a cellulose wall.

Diploid cells Cells having two sets of chromosomes that form pairs, as in somatic cells.

Dipnoi An order of lungfishes with weak pectoral and pelvic fins; not considered ancestral to land vertebrates.

Disconformity A variety of unconformity in which bedding planes above and below the plane of erosion or nondeposition are parallel.

DNA The nucleic acid found chiefly in the nucleus of cells that functions in the transfer of genetic characteristics and in protein synthesis.

Docodonts A group of small, primitive Late Jurassic mammals possibly ancestral to the living monotremes.

Dolomite (or dolostone) A carbonate sedimentary rock that contains more than 50 percent of the mineral dolomite CaMg $(CO_3)_2$.

Dome An upfold in rocks having the general configuration of an inverted bowl. Strata in a dome dip outward and downward in all directions from a central area. An example is the Ozark dome.

Dryopithecine In general, a group of lightly built primates that lived during the Miocene and Pliocene in mostly open savannah country and that includes *Dryopithecus*, a form considered to be in the line leading to apes.

Dynamothermal (regional) metamorphism Metamorphism that has occurred over a wide region, caused by deep burial and high temperatures associated with pressures resulting from overburden and orogeny.

Echinoderms The large group (phylum Echinodermata) of marine invertebrates characterized by prominent pentamerous symmetry and skeleton frequently constructed of calcite elements and including spines. Cystoids, blastoids, crinoids, and echinoids are examples of echinoderms.

Edentata An order of placental mammals that includes extinct ground sloths and gyptodonts, as well as living armadillos, tree sloths, and South American anteaters.

Elastic limit The greatest amount of stress that can be applied to a body without causing permanent strain.

Endemic population The native fauna of any particular region.

Entelodonts A group of extinct artiodactyls bearing a superficial resemblance to giant wild boars.

Entelodont

Eon A major division of the geologic time scale. All of the geologic periods from the Cambrian to the Holocene comprise the Phanerozoic Eon. The term is also sometimes used to denote a span of 1 billion years.

Epifaunal organisms Organisms living *on*, as distinct from *in*, a particular body of sediment or another organism.

Epoch A chronologic subdivision of a geologic period. Rocks deposited or emplaced during an epoch constitute the *series* for that epoch.

Era A major division of geologic time, divisible into geologic periods.

Eukaryote A type of living cell containing a true nucleus, enclosed within a nuclear membrane, and having well-defined chromosomes and cell organelles.

Eupantotheres A group of Jurassic mammals with dentition similar to that of primitive representatives of later marsupial and placental mammals and thus thought to be ancestral to these groups.

Eurypterids Aquatic arthropods of the Paleozoic, superficially resembling scorpions, and probably carnivorous.

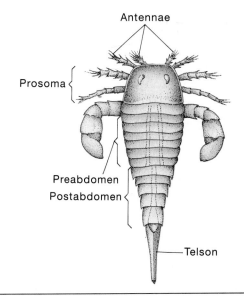

Eurypterid

Eustatic   Pertaining to worldwide simultaneous changes in sea level, such as might result from change in the volume of continental glaciers.

Evaporites   Sediments precipitated from a water solution as a result of the evaporation of that water. Evaporite minerals include anhydrite gypsum ($CaSO_4$) and halite ($NaCl$).

Evolution   The continuous genetic adaptation of organisms or species to the environment.

Facies   A particular aspect of sedimentary rocks that is a direct consequence of sedimentation in a particular depositional environment.

Fault   A fracture in the Earth's crust along which rocks on one side have been displaced relative to rocks on the other side.

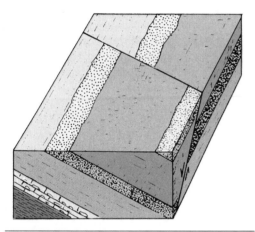

Fault in dipping strata

Fermentation   The partial breakdown of organic compounds by an organism in the absence of oxygen. The final product of fermentation is alcohol or lactic acid.

Filter-feeders (organisms)   An animal that obtains its food, which usually consists of small particles or organisms, by filtering it from the water.

Fissility   That property of rocks that causes them to split into thin slabs parallel to bedding. Fissility is particularly characteristic of shale.

Flood basalts   Regionally extensive layers of basalt that originated as low-viscosity lava pouring from fissure eruptions. The lavas of the Columbia Plateau and the Deccan Plateau are flood basalts.

Fluvial   Pertaining to sediments or other geologic features formed by streams.

Flysch   Term describing thick sequences of rapidly deposited, poorly sorted marine clastics.

Focus (earthquake)   The location at which rock rupture occurs so as to generate seismic waves.

Foliation   A textural feature especially characteristic of metamorphic rocks in which laminae develop by growth or realignment of minerals in parallel orientation.

Foraminifers   An order of mostly marine, unicellular protozoans that secrete tests (shells) that are usually composed of calcium carbonate.

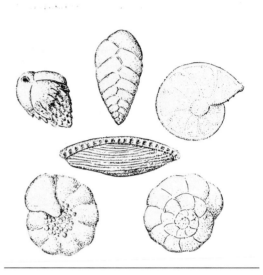

Foraminifers

Formation   A mappable, lithologically distinct body of rock having recognizable contacts with adjacent rock units.

Fossil   The remains or indications of an organism that lived in the geologic past.

Fractional crystallization   The separation of components of a cooling magma by sequential formation of particular mineral crystals at progressively lower temperatures.

Fusulinids   Primarily spindle-shaped foraminifers with calcareous, coiled tests divided into a complex of numerous chambers. Fusulinids were particularly abundant during the Pennsylvanian and Permian periods.

Galaxy   An aggregate of stars and planets, separated from other such aggregates by distances greater than those between member stars.

Gamete   Either of two cells (male or female) that must unite in sexual reproduction to initiate the development of a new individual.

Gamma rays   Very high-frequency electromagnetic waves.

Garnet   A family of aluminosilicates of iron and calcium that are particularly characteristic of metamorphic rocks.

Garnet crystal

Gene The unit of heredity transmitted in the chromosome.

Genus The major subdivision of a taxonomic family or subfamily of plants or animals, usually consisting of more than one species.

Geochronology The study of time as applied to Earth and planetary history.

Geologic range The geologic time span between the first and last appearance of an organism.

Glacier A large mass of ice, formed by the recrystallization of snow, that flows slowly under the influence of gravity.

Glauconite A green clay mineral frequently found in marine sandstones and believed to have formed at the site of deposition.

Glossopteris flora An assemblage of fossil plants found in rocks of Late Paleozoic and Early Triassic age in South Africa, India, Australia, and South America. The flora takes its name from the seed fern *Glossopteris.*

Gondwanaland The great Permo-Carboniferous Southern Hemisphere continent, comprising the assembled present continents of South Africa, India, Australia, Africa–Arabia, and Antarctica.

Goniatites One of the three large groups of ammonoid cephalopods with sutures forming a pattern of simple lobes and saddles and thus not as complex as either the ceratites or the ammonites.

Gowganda Conglomerate An apparent tillite of the Canadian Shield. The Gowganda rests on a surface of older rock that appears to have been polished by glacial action.

Graptolite facies A Paleozoic sedimentary facies composed of dark shales and fine-grained clastics that contain the abundant remains of graptolites and that are associated with volcanic rocks.

Graptolites Extinct colonial marine invertebrates considered to be protochordates. Graptolites range from the Late Cambrian to the Mississipian.

Gravity anomaly The differences between the observed value of gravity at any point on the Earth and the calculated theoretic value.

Greenhouse effect A process in which incoming solar radiation that is absorbed and reradiated cannot escape back into space because the Earth's atmosphere is not transparent to the reradiated energy, which is in the form of infrared radiation (heat).

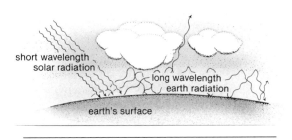

Greenhouse effect

Greenstones Low-grade metamorphic rocks containing abundant chlorite, epidote, and biotite and developed by metamorphism of basaltic extrusive igneous rocks. Great linear outcrops of greenstones are termed greenstone belts and are thought to mark the locations of former volcanic island arcs.

Guide fossil Fossil with a wide geographical distribution but narrow stratigraphic range and thus useful in correlating strata and for age determination.

Gutenberg discontinuity The boundary separating the mantle of the Earth from the core below. The Gutenberg discontinuity lies about 2900 km below the surface.

Gymnosperms An informal designation for flowerless seed plants in which seeds are not enclosed (hence "naked seeds"). Examples are conifers and cycads.

Hadrosaurs The ornithischian duck-billed dinosaurs of the Cretaceous.

Hadrosaur

Half-life The time in which one-half of an original amount of a radioactive species decays to daughter products.

Haploid cell One having a single set of chromosomes, as in gametes (see *diploid*).

Hercynian orogeny A major Late Paleozoic orogenic episode in Europe that formed the ancient Hercynian mountains. Today, only the eroded stumps of these mountains are exposed in areas where the cover of Mesozoic and Cenozoic strata has been removed by erosion.

Heterotroph An organism that depends on an external source of organic substances for its nutrition and energy.

Holocene A term sometimes used to designate the period of time since the last major episode of glaciation. The term is equivalent to Recent.

Homologous organs Organs having structural and developmental similarities due to genetic relationship.

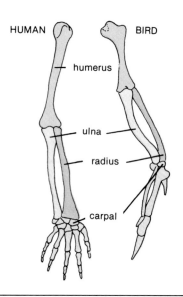

Homologous bones in forelimbs of human and bird

Hydrosphere The water and water vapor present at the surface of the Earth, including oceans, seas, lakes, and rivers.

Hylobatidae A group of persistently arboreal, small apes that are exemplified by the gibbons and siamangs.

Hyomandibular bone (in fishes) The modified upper bone of the hyoid arch, which functions as a connecting element between the jaws and the braincase in certain fishes.

Ichthyosaurs Highly specialized marine reptiles of the Mesozoic, recognized by their fishlike form.

Ictidosauria A group of extinct mammal-like reptiles or therapsids whose skeletal characteristics are considered to be very close to those of mammals.

Igneous rock A rock formed by the cooling and solidification of magma or lava.

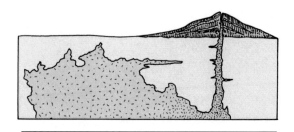

Igneous rocks

Infaunal organisms Organisms that live and feed within bottom sediments.

Ion An atom that, because of electron transfers, has excess positive or negative charges.

Island arcs Chains of islands, arranged in arcuate trends on the surface of the Earth. The arcs are sites of volcanic and earthquake activity and are usually bordered by deep oceanic trenches on the convex side.

Isopachous map A map depicting the thickness of a sedimentary unit.

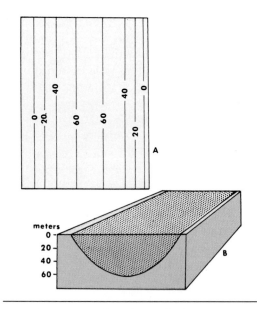

Isopach map (A) of sandstone body (B)

Isotope Atoms of an element that have the same

number of protons in the nucleus, the same atomic number, and the same chemical properties but different atomic masses because they have different numbers of neutrons in their nuclei.

Karoo system   A sequence of Permian to Lower Jurassic rocks, primarily continental formations, which outcrop in Africa and are approximately equivalent to the Gondwana system of peninsular India.

Kaskaskia sequence   A sequence of Devonian–Mississippian sediments, bounded above and below by regional unconformities and recording an episode of transgression followed by full flooding of a large part of the craton and by subsequent regression.

Lacustrine   Relating to lakes, as in lacustrine sediments (lake sediments).

Lagomorph   The order of small placental mammals that includes rabbits, hares, and pikas.

Lagomorph (pika)

Laramide orogeny   In general, those pulses of mountain building that were frequent in Late Cretaceous time and were in large part responsible for producing many of the structures of the Rocky Mountains.

Lateral fault   A fault in which the movement is largely horizontal and in the direction of the trend of the fault plane. Sometimes called a *strike–slip* fault.

Laurasia   A hypothetical supercontinent composed of what is now Europe, Asia, Greenland, and North America.

Limestone   A sedimentary rock consisting mainly of calcium carbonate.

Lithofacies map   A map that shows the areal variation in lithologic attributes of a stratigraphic unit.

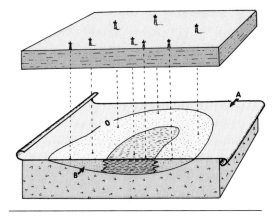

Lithofacies map (A) of time–rock unit (B)

Lithosphere   The outer shell of the Earth, lying above the asthenosphere and comprising the crust and upper mantle.

Litoptern   South American ungulates whose evolutionary history somewhat paralleled that of horses and camels.

Logan's line   A zone of thrust-faulting produced during the Taconic orogeny that extends from the west coast of Newfoundland along the trend of the St. Lawrence River to near Quebec and southward along Vermont's western border. (Named after the pioneer Canadian geologist Sir William Logan.)

Lophophore   An organ, located adjacent to the mouth of brachiopods and bryozoans, which bears ciliated tentacles and has as its primary function the capture of food particles.

Low-velocity zone   An interior zone of the Earth characterized by lower seismic wave velocities than the region immediately above it.

Lunar maria   Low-lying, dark lunar plains filled with volcanic rocks rich in iron and magnesium.

Lycopsids   Leafy plants with simple, closely spaced leaves bearing sporangia on their upper surfaces. They are represented by living club mosses and vast numbers of extinct late Paleozoic "scale trees."

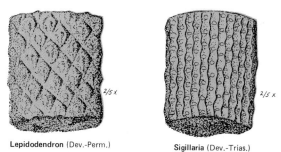

Lepidodendron (Dev.-Perm.)   Sigillaria (Dev.-Trias.)

Lycopsids

Magnetic declination   The horizontal angle between "true," or geographic, north and magnetic north, as indicated by the compass needle. Declination is the result of Earth's magnetic axis being inclined with respect to the Earth's rotational axis.

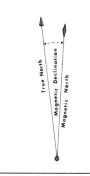

Magnetic declination

Magnetic inclination   The angle between the magnetic lines of force for the Earth and the Earth's surface; sometimes called "dip." Magnetic inclination can be demonstrated by observing a freely suspended magnetic needle. The needle will lie parallel to the Earth's surface at the equator but is increasingly inclined toward the vertical as the needle is moved toward the magnetic poles.

Mammoth   The name commonly applied to extinct elephants of the Pleistocene Epoch.

Marsupials   Mammals of the order Marsupialia. Female marsupials bear mammary glands and carry their immature young in a stomach pouch.

Mass spectrometer   An instrument that separates ions of different mass but equal charge and measures their relative quantities.

Mastodon   The group of extinct proboscideans (elephantoids), early forms of which were characterized by long jaws, tusks in both jaws, and low-crowned teeth.

Medusa   The free-swimming, umbrella-shaped jellyfish form of the phylum Cnidaria.

Meiosis   That kind of nuclear division, usually involving two successive cell divisions, that results in daughter cells having one-half the number of chromosomes that were in the original cell.

Mélange   A body of intricately folded, faulted, and severely metamorphosed rocks, examples of which can be seen in the Franciscan rocks of California.

Mesosphere   A zone of the Earth's mantle where pressures are sufficient to impart greater strength and rigidity to the rock.

Metamorphic rock   A rock formed from a previously existing rock by subjecting the parent rock to high temperature and pressure but without melting.

Metamorphism   The transformation of previously existing rocks into new types by the action of heat, pressure, and chemical solutions. Metamorphism usually takes place at depth in the roots of mountain chains or adjacent to large intrusive igneous bodies.

Metazoa   All multicellular animals whose cells become differentiated to form tissues (all animals except Protozoa).

Meteorites   Metallic or stony bodies from interplanetary space that have passed through the Earth's atmosphere and hit the Earth's surface.

Meteors   Sometimes called "shooting stars," meteors are generally small particles of solid material from interplanetary space that approach close enough to the Earth to be drawn into the Earth's atmosphere, where they are heated to incandescence. Most disintegrate, but a few land on the surface of the Earth as meteorites.

Micrite   A texture in carbonate rocks that, when viewed microscopically, appears as murky, fine-grained calcium carbonate. Micrite is believed to develop from fine carbonate mud or ooze.

Milankovitch effect   The hypothetical long-term effect on world climate caused by three known components of Earth motion. The combination of these components provides a possible explanation for repeated glacial to interglacial climatic swings.

Miliolids   A group of foraminifers with smooth, imperforate test walls and chambers arranged in various planes around a vertical axis. Miliolids are common in shallow marine areas.

Mineral   A naturally occurring element or compound formed by inorganic processes that has a definite chemical composition or range of compositions, as well as distinctive properties and form that reflect its characteristic atomic structure.

Mitosis   The method of cell division by means of which each of the two daughter nuclei receives exactly the same complement of chromosomes as had existed in the parent nucleus.

Mobile belt   An elongate region of the Earth's crust characterized by exceptional earthquake and volcanic activity, tectonic instability, and periodic mountain building.

Mohorovičić discontinuity   A plane that marks the boundary separating the crust of the Earth from the underlying mantle. The "moho," as it is sometimes called, is at a depth of about 70 km below the surface of continents and 6 to 14 km below the floor of the oceans.

Molasse Accumulations of primarily nonmarine, relatively light-colored, irregularly bedded conglomerates, shales, coal seams, and cross-bedded sandstones that are deposited subsequent to major orogenic events.

Mold An impression, or imprint, of an organism or part of an organism in the enclosing sediment.

Mollusk Any member of the invertebrate phylum Mollusca, including pelecypods, cephalopods, gastropods, scaphopods, and chitons.

Monoplacophorans Primitive marine molluscans with simple, cap-shaped shells.

Monotremes The egg-laying mammals.

Morganucodonts Early mammals found in Triassic beds of Europe and Asia and characterized by their small size and their retention of certain reptilian osteologic traits.

Mosasaurs Large marine lizards of the Late Cretaceous.

Mosasaur

Multituberculates An early group (Jurassic) of Mesozoic mammals with tooth cusps in longitudinal rows and other dental characteristics that suggest they may have been the earliest herbivorous mammals.

Mutation A stable and inheritable change in a gene.

Mytiloids Pelecypods having rather triangular shells that are most commonly identical but in some forms unequal. The edible mussel *Mytilus* is a representative form.

Nappe A large mass of rocks that have been moved a considerable distance over underlying formations by overthrusting, recumbent folding, or both.

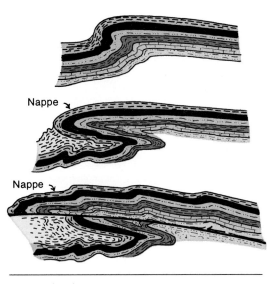

Nappe development

Nektic Pertaining to swimming organisms.

Neogene A subdivision of the Cenozoic that encompasses the Miocene, Pliocene, Pleistocene, and Holocene epochs.

Neutron Electrically neutral (uncharged) particles of matter existing along with protons in the atomic nucleus of all elements except the mass 1 isotope of hydrogen.

Nevadan orogeny In general, those pulses of mountain building, intrusion, and metamorphism that were most frequent during the Jurassic and Early Cretaceous along the western part of the Cordilleran orogenic belt.

Newark series A series of Upper Triassic, nonmarine red beds (shales, sandstones, and conglomerates), lava flows, and intrusions located within downfaulted basins from Nova Scotia to South Carolina.

New World monkeys Monkeys whose habitat is today confined to South America. They are thoroughly arboreal in habit and have prehensile tails, by which they hang and swing from tree limbs.

Nonconformity An unconformity developed between sedimentary rocks and older plutons or massive metamorphic rocks that had been exposed to erosion before the overlying sedimentary rocks were deposited.

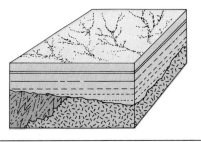

Nonconformity

Normal fault A fault in which the hanging wall appears to have moved downward relative to the footwall; normally occurring in areas of crustal tension.

Nothosaurs Relatively small early Mesozoic sauropterygians that were replaced during the Jurassic by the plesiosaurs.

Notochord A rod-shaped cord of cartilage cells forming the primary axial structure of the chordate body. In vertebrates, the notochord is present in the embryo and is later supplanted by the vertebral column.

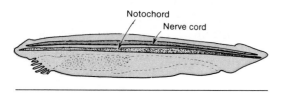

Notochord and dorsal nerve cord in the living lancelet

Notungulates A group of ungulates that diversified in South America and persisted until Plio-Pleistocene time.

Novaculites A term applied originally to rocks suitable for whetstones and, in America, to white chert found in Arkansas. Now applied to very tough, uniformly grained cherts composed of microcrystalline quartz.

Nuclear fission tracks Submicroscopic "tunnels" in minerals produced when high-energy particles from the nucleus of uranium are forcibly ejected during spontaneous fission.

Nucleic acid Any of a group of organic acids that control hereditary processes within cells and make possible the manufacture of proteins from the amino acids ingested by the cells as food.

Nuclides The different weight configurations of an element caused by atoms of that element having differing numbers of neutrons. Nuclides or isotopes of an element differ in number of neutrons but not in chemical properties.

Nummulites Large, coin-shaped foraminifers, especially common in Tertiary limestones.

Offlap A sequence of sediments resulting from a marine regression and characterized by an upward progression from offshore marine sediments (often limestones) to shales and finally sandstones (above which will follow an unconformity).

Oil shale A dark-colored shale rich in organic material that can be heated to liberate gaseous hydrocarbons.

Old World monkeys Monkeys of Asia, Africa, and southern Europe that include macaques, guenons, langurs, baboons, and mandrills.

Onlap A sequence of sediments resulting from a marine transgression. Normally, the sequence begins with a conglomerate or sandstone deposited over an erosional unconformity and followed upward in the vertical section by progressively more offshore sediments.

Oölites Limestones composed largely of small, round or ovate calcium carbonate bodies called oöids.

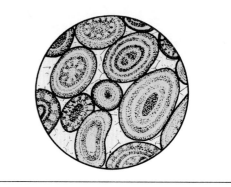

Oölites

Ophiolite suite An association of radiolarian cherts, pelagic muds, basaltic pillow and flow lavas, gabbros, and ultramafic rocks such as periodotite regarded as surviving masses of former oceanic crust largely destroyed in former subduction zones.

Oreodonts North American artiodactyls of the middle and late Tertiary.

The oreodont *Merycoidodon*

Ornithischia An order of dinosaurs characterized by birdlike pelvic structures and including such herbivores as the ornithopods, stegosaurs, ankylosaurs, and ceratopsians.

Orogenic belt Great linear tracts of deformed rocks, primarily developed near continental margins by compressional forces accompanying mountain building.

Orogeny The process by which great systems of mountains are formed. (*Orogenesis* means mountain building.)

Ostracoderms Extinct jawless fishes of the early Paleozoic.

Ostracods Small, bivalved, bean-shaped crustaceans.

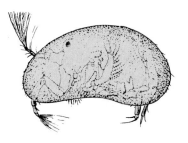

Ostracod

Ostreoids The "oyster family" of pelecypods.

The ostreoid *Exogyra* (Jur.-Cret.)

Outcrop An area where specific rock units are exposed at the Earth's surface or occur at the surface but are covered by surficial deposits.

Paleoecology The study of the relationship of ancient organisms to their environment.

Paleogene A subdivision of the Cenozoic that encompasses the Paleocene, Eocene, and Oligocene epochs.

Paleogeography The geography as it existed at some time in the geologic past.

Paleolatitudes The latitude that once existed across a particular region at a particular time in the geologic past.

Paleomagnetism The Earth's magnetic field and magnetic properties in the geologic past. Studies of paleomagnetism are helpful in determining positions of continents and magnetic poles.

Paleontology The study of all ancient forms of life, their interactions, and their evolution.

Pangea In Alfred Wegener's theory of continental drift, the supercontinent that included all present major continental masses.

Panthalassa The great universal ocean that surrounded the supercontinent Pangea prior to its breakup.

Paraconformity A rather obscure unconformity in which no erosional surface is discernible and in which beds above and below the break are parallel.

Partial melting Process by which a rock subjected to high temperature and pressure is partially melted, and the liquid fraction moved to another location. Partial melting results from the variations in melting points of different minerals in the original rock mass.

Pectenoids Pelecypods exemplified by the scallops. They have generally subcircular shells and straight hinge lines.

Pecten

Pelycosaurs Early mammal-like reptiles exemplified by the sail-back animals of the Permian Period.

Period A subdivision of an era.

Perissodactyl Progressive, hoofed mammals, characteristically having an odd number of toes on the hind feet and usually on the front feet as well.

Permineralization A manner of fossilization in which voids in an organic structure (such as bone) are filled with mineral matter.

Petrification The process of converting organic structures, such as bone, shell, or wood, into a stony substance, such as calcium carbonate or silica.

Phanerozoic The eon of geologic time during which the Earth has been populated by abundant and diverse life. The Phanerozoic Eon followed the Cryptozoic (or Precambrian) Eon and is divided into Paleozoic, Mesozoic, and Cenozoic eras.

Phosphorite A sediment composed largely of calcium phosphate.

Photosphere A relatively thin gaseous layer on the

sun that emits nearly all the light that the sun radiates into space.

Photosynthesis   The process of synthesizing carbohydrates from carbon dioxide and water, utilizing the radiant energy of light captured by the chlorophyll in plant cells.

Phytoplankton   Microscopic marine planktonic plants, most of which are various forms of algae.

Phytosaurs   Extinct aquatic, crocodile-like thecodonts of the Triassic.

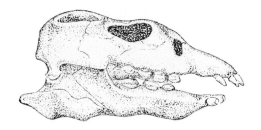

Skull of the placodont *Placodus*. Note blunt teeth for crushing shellfish.

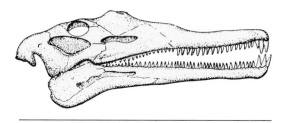

Phytosaur skull

Plankton   Minute, free-floating aquatic organisms.

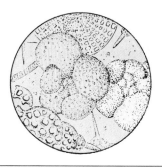

Plankton (foraminifers, diatoms, and radiolaria)

Pillow lava   Type of lava that is extruded under water and in which many pillow-shaped lobes break through the chilled surface of the flow and solidify. (Resembles a pile of pillows.)

Pinnipeds   Marine carnivores such as seals, sea lions, and walruses.

Placer deposit   An accumulation of sediment rich in a valuable mineral or metal that has been concentrated because of its greater density.

Placoderms   Extinct primitive jawed fishes of the Paleozoic Era.

Plate tectonics   The theory that explains the tectonic behavior of the crust of the Earth in terms of several moving plates that are formed by volcanic activity at oceanic ridges and destroyed along great ocean trenches.

Platform   That part of a craton covered thinly by layered sedimentary rocks and characterized by relatively stable tectonic conditions.

Plesiosaurs   The group of extinct Mesozoic marine reptiles (sauropterygians) characterized by large, paddle-shaped limbs and broad bodies, with either very long or relatively short necks.

Coccosteus, a Devonian placoderm

Plesiosaur

Placodonts   Extinct walrus-like marine reptiles that fed principally on shellfish.

Pluvial lake   A lake formed in an earlier climate when rainfall was greater than at present.

Polyp   The hydra-like form of some cnidaria in

which the mouth and tentacles are at the top of the body.

Porphyry A textural term used to describe an igneous rock in which some of the crystals, called phenocrysts, are distinctly larger than others.

Porphyry

Precambrian All of geologic time and its corresponding rocks before the beginning of the Paleozoic Era.

Primary earthquake waves Seismic waves that are propagated through solid rock as a train of compressions and dilations. Direction of vibration is parallel to direction of propagation.

Principle of cross-cutting relations The principle that states that geologic features such as faults, veins, and dikes must be younger than the rocks or features they cut across.

Principle of biologic succession The principle that states that the observed sequence of life forms has changed continuously through time, so that the total aspect of life (as recognized by fossil evidence) for a particular segment of time is distinct and different from that of life of earlier and later times.

Principle of temporal transgression The principle that stipulates that sediments of advancing (transgressing) or retreating (regressing) seas are not necessarily of the same geologic age throughout their lateral extent.

Proboscidea The elephants and their progenitors.

Prokaryotes Organisms that lack membrane-bounded nuclei and other membrane-bounded organelles.

Prosimii The less advanced primates, such as lemurs, tarsiers, and tree shrews.

Proteinoids Extra-large organic molecules containing most of the 20 amino acids of proteins and produced in laboratory conditions simulating those found in nature.

Proteins Giant molecules containing carbon, hydrogen, oxygen, nitrogen, and usually sulfur and phosphorus; composed of chains of amino acids and present in all living cells.

Proton An elemental particle found in the nuclei of all atoms that has a positive electric charge and a mass similar to that of a neutron.

Pyroclastics Fragments of volcanic debris that have usually been found fragmented during eruptions.

Pyroxene group A group of dark-colored iron and magnesium-rich silicate minerals.

Queenston delta A clastic wedge of red beds shed westward from highlands elevated in the course of the Taconic orogeny.

Radioactive decay The spontaneous emission of a particle from the atomic nucleus, thereby transforming the atom from one element to another.

Radiolaria Protozoa that secrete a delicate, often beautifully filigreed skeleton of opaline silica.

Recumbent fold A fold in which the axial plane is essentially horizontal; a fold that has been turned over by compressional forces so that it lies on its side.

Red beds Prevailing red, usually clastic sedimentary deposits.

Regolith Any solid materials, such as rock fragments or soil, lying on top of bedrock.

Regression A general term signifying that a shoreline has moved toward the center of a marine basin. Regression may be caused by tectonic emergence of the land, eustatic lowering of sea level, or prograding of sediments, as in deltaic build-outs.

Relative geologic age The placing of an event in a time sequence without regard to the absolute age in years.

Replacement A fossilization process in which the original skeletal substance is replaced after burial by inorganically precipitated mineral matter.

Reverse fault A fault formed by compression in which the hanging wall appears to move up relative to the foot wall.

Rhynchonellids A group of brachiopods having pronounced beaks, accordion-like plications, and triangular outline.

Rift valley A valley formed by faulting, usually involving a central fault block that moves downward in relation to adjacent blocks.

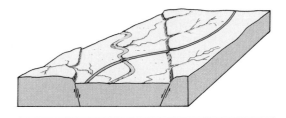

Rift valley

Rudists Peculiarly specialized Mesozoic bivalvia often having one valve in the shape of a horn coral, covered by the other valve in the form of a lid.

Rudistoid pelecypods

Rugosa The large group of solitary and colonial Paleozoic horn corals.

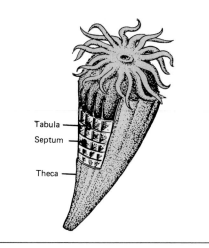

Tabula
Septum
Theca

Rugosa

Ruminant A herbivorous, cud-chewing ungulate.

Salt dome A structural dome in sedimentary strata resulting from the upward flow of a large body of salt.

Sauk sequence A sequence of upper Precambrian to Ordovician sediments bounded both above and below by a regional unconformity and recording an episode of marine transgression, followed by full flooding of a large part of the craton, and ending with a regression from the craton.

Saurischia An order of dinosaurs with triradiate pelvic structures, including both the gigantic herbivorous sauropods and the carnivorous theropods.

Scleractinid coral Coral belonging to the order Scleractinia, which includes most modern and post-Paleozoic corals.

Sea-floor spreading The process by which new sea-floor crust is produced along midoceanic ridges (divergence zones) and slowly conveyed away from the ridges.

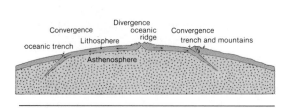

Convergence   Divergence   Convergence
oceanic trench   Lithosphere   oceanic   trench and mountains
                                    ridge
                    Asthenosphere

Sea-floor spreading

Secondary earthquake wave A seismic wave in which the direction of vibration of wave energy is at right angles to the direction the wave travels.

Sedimentary rock A rock that has formed as a result of the consolidation (lithification) of accumulations of sediment.

Sessile The bottom-dwelling habit of aquatic animals that live continuously in one place.

Series The time–rock term representing the rocks deposited or emplaced during a geologic epoch. Series are subdivisions of systems.

Shelly facies In general, sedimentary deposits consisting primarily of carbonate rocks containing the abundant fossil remains of marine invertebrates.

Silicoflagellates Unicellular, tiny, flagellate marine algae that secrete an internal skeleton composed of opaline silica.

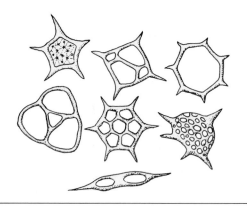

Silicoflagellates

Silicon tetrahedron An atomic structure in silicates consisting of a centrally located silicon atom linked to four oxygen atoms placed symmetrically around the silicon at the corners of a tetrahedron.

Sonoma orogeny Middle Permian orogenic movements, the structural effects of which are most evident in western Nevada.

Sorting A measure of the uniformity of the sizes of particles in a sediment or sedimentary rock.

Spar carbonate As viewed microscopically, the clear, crystalline carbonate that has been deposited in a carbonate rock as a cement between clasts or has developed by recrystallization of clasts.

Species A unit of taxonomic classification of organisms. In another sense, a species is a population of individuals that are similar in structural and functional characteristics and that in nature breed only with one another.

Sphenodon (tuatara) Large, lizard-like reptiles that have persisted from Triassic to the present and now inhabit islands off the coast of New Zealand.

Sphenopsids A group of spore-bearing plants that were particularly common during the Late Paleozoic and were characterized by articulated stems with leaves borne in whorls at nodes.

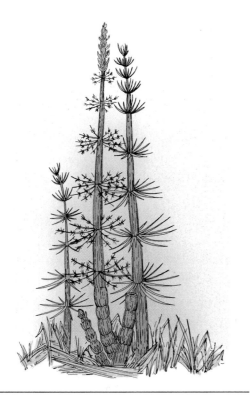

Sphenopsid calamites

Spiracle In cartilaginous fishes, a modified gill opening through which water enters the pharynx.

Spontaneous fission Spontaneous fragmentation of an atom into two or more lighter atoms and nuclear particles.

Spores A usually asexual reproductive body, such as occurs in bacteria, ferns, and mosses.

Stage The time–rock unit equivalent to an age. A stage is a subdivision of a series.

Stapes The innermost of the small bones in the middle ear cavity of mammals; also recognized in amphibians and reptiles.

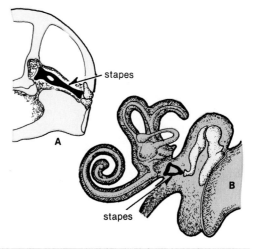

Stapes in reptile (A) and mammal (B)

Strain Deformation of a rock mass in response to stress.

Stratification The layering in sedimentary rocks that results from changes in texture, color, or rock type from one bed to the next.

Stratified drift Deposits of glacial clastics that have been sorted and stratified by the action of meltwater.

Stratigraphy The study of rock strata, with emphasis on their succession, age, correlation, form, distribution, lithology, fossil content, and all other characteristics useful in interpreting their environment of origin and geologic history.

Stromatolites Distinctly laminated accumulations of calcium carbonate having rounded, branching, or frondose shape and believed to form as a result of the metabolic activity of marine algae. They are usually found in the high intertidal to low supratidal zones.

Stromatoporoids An extinct group of reef-building organisms now believed to have affinities with the Porifera and noted for the large, often laminated masses constructed by the colonies.

Stromatopora

Subaerial   Formed or existing at or near a sediment surface significantly above sea level.

Subduction zone   An inclined planar zone, defined by high frequency of earthquakes, that is thought to locate the descending leading edge of a moving oceanic plate.

Sublittoral zone   The marine bottom environment that extends from low tide seaward to the edge of the continental shelf.

Surface earthquake waves   Seismic or earthquake waves that move only about the surface of the Earth.

Symmetrodonts   A group of primitive Mesozoic mammals characterized by a symmetric triangular arrangement of cusps on cheek teeth.

Syncline   A geologic structure in which strata are bent into a downfold.

System   The time–rock unit representing rock deposited or emplaced during a geologic period.

Taconic orogeny   A major episode of orogeny that affected the Appalachian region in Ordovician time. The northern and Newfoundland Appalachians were the most severely deformed during this orogeny.

Taxon (pl. taxa)   Any unit in the taxonomic classification, such as a phylum, class, order, or family.

Taxonomy   The science of naming, describing, and classifying organisms.

Tectonics   The structural behavior of a region of the Earth's crust.

Teleosts   The most advanced of the bony fishes, characterized by thin, rounded scales, completely bony internal skeleton, and symmetric tail. Teleosts range from Cretaceous to Recent.

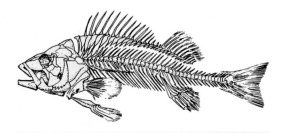

Teleost fish

Terebratulids   A group of Silurian to Recent mostly smooth-shelled brachiopods having a loop-shaped attachment for the lophomore. Terebratulids were most abundant during the Jurassic and Cenozoic.

Terrane   A three-dimensional block of crust having a distinctive assemblage of rocks (as opposed to *terrain,* implying topography, such as rolling hills or rugged mountains).

Tethys seaway   A great east–west trending seaway lying between Laurasia and Gondwanaland during Paleozoic and Mesozoic time and from which arose the Alpine–Himalayan Mountain ranges.

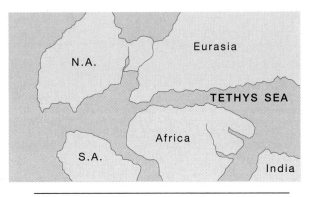

Tethys seaway

Thecodonts   An order of primarily Triassic reptiles considered to be the ancestral archosaurians.

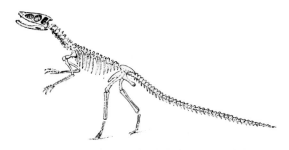

The Triassic thecodont *Hesperosuchus*

Therapsids   An order of advanced mammal-like reptiles.

Cynognathus, a Permian therapsid

Thermal plume A "hot spot" in the upper mantle believed to exist where a huge column of upwelling magma lies in a fixed position under the lithosphere. Thermal plumes are thought to cause volcanism in the overlying lithosphere.

Thermoremanent magnetism Permanent magnetization acquired by igneous rocks as they cool past the Curie point while in the Earth's magnetic field.

Theropods The carnivorous saurischian dinosaurs.

The small Jurassic theropod *Ornitholestes*

Thrust fault A low-angle reverse fault, with inclination of fault plane generally less than 45°.

Till Unconsolidated, unsorted, unstratified glacial debris.

Tillite Unsorted glacial drift (till) that has been converted into solid rock.

Time–stratigraphic unit (chronostratigraphic unit) The rocks formed during a particular unit of geologic time. Also called time–rock unit.

Tippecanoe sequence A sequence of Ordovician to Lower Devonian sediments bounded above and below by regional unconformities and recording an episode of marine transgression, followed by full flooding of a large region of the craton and subsequent regression.

Titanotheres Large, extinct perissodactyls (odd-toed ungulates) that attained the peak of their evolutionary development during the Oligocene. *Brontotherium* is a widely known titanothere.

Trace fossils Tracks, trails, burrows, and other markings made in now lithified sediments by ancient animals.

Transcontinental arch An elongate, uplifted region extending from Arizona northeastward toward Lake Superior. During the Cambrian, the arch was emergent, as indicated by the observation that Cambrian marine sediments are located on either side of the arch but are missing above it. (Along the crest of the arch, post-Cambrian rocks rest on Precambrian basement.)

Transform fault A strike–slip fault bounded at each end by an area of crustal spreading that tends to be more or less perpendicular to the trace of the fault.

Triconodonts A group of primitive Mesozoic mammals recognized primarily by the arrangement of three principal cheek tooth cusps in a longitudinal row.

Trilobites Paleozoic marine arthropods of the class Crustacea, characterized by longitudinal and transverse division of the carapace into three parts, or lobes. Trilobites were especially abundant during the early Paleozoic.

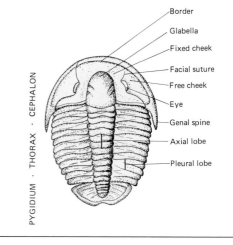

Trilobite

Tuff Volcanic ash that has become consolidated into rock.

Turbidites Sediment deposited from a turbidity current and characterized by graded bedding and moderate to poor sorting.

Turbidity current A mass of moving water that is denser than surrounding water because of its content of suspended sediment and that flows along slopes of the sea floor as a result of that higher density.

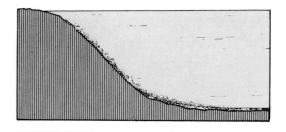

Turbidity current

Tylopod   The artiodactyl group to which camels and llamas belong.

Alticamelus, a Pliocene tylopod

Unconformity   A surface separating an overlying younger rock formation from an underlying formation and representing an episode of erosion or nondeposition. Because unconformities represent a lack of continuity in deposition, they are gaps in the geologic record.

Ungulate   Four-legged mammals whose toes bear hoofs.

Uniformitarianism   A general principle that suggests that the past history of the Earth can be interpreted and deciphered in terms of what is known about present natural laws.

Vagile   The bottom-dwelling habit of aquatic animals capable of locomotion.

Varve   A thin sedimentary layer or pairs of layers that represent the depositional record of a single year.

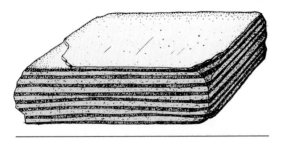

Varves

Vascular plants   Plants, including all higher land plants, that have a system of vessels and ducts for distributing moisture and nutrients.

Veneroids   A group of pelecypods exemplified by the common clam, *Mercenaria*.

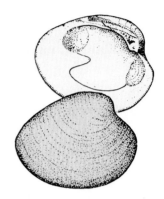

Veneroid bivalve (Venus)

Vestigial organ   An organ that is useless, small, or degenerate but representing a structure that was more fully developed or functional in an ancestral organism.

Vindelician arch   A highland area believed to have formed a barrier separating the Germanic and Alpine depositional areas during the Triassic and Jurassic.

Williston basin   A large structural basin extending from South Dakota and Montana northward into Canada; well known for the petroliferous Devonian formations deposited therein.

Zechstein sea   An arm of the Atlantic that extended across part of northern Europe during the late Permian and in which were deposited several hundred meters of evaporites, including the well-known potassium salts of Germany.

Zone   A bed or group of beds distinguished by a particular fossil content and frequently named after the fossil or fossils it contains. More formally known as a *biozone*.

Zygote   The cell formed by the union of two gametes. Thus, a zygote is a fertilized egg.

Index

In this index, an italic *f* following the page number indicates a *figure,* and an italic *t* after the page number indicates a *table.* Entries from the appendices are preceded by a capital A.

Common Rock-Forming Silicate Minerals

Silicate Mineral	Composition	Physical Properties
Quartz	Silicon dioxide (silica, SiO_2)	Hardness of 7 (on scale of 1 to 10)*; will not cleave (fractures unevenly); specific gravity: 2.65
Potassium feldspar group	Aluminosilicates of potassium	Hardness of 6.0–6.5; cleaves well in two directions; pink or white; specific gravity: 2.5–2.6
Plagioclase feldspar group	Aluminosilicates of sodium and calcium	Hardness of 6.0–6.5; cleaves well in two directions; white or gray; may show striations on cleavage planes; specific gravity: 2.6–2.7
Muscovite mica	Aluminosilicates of potassium with water	Hardness of 2–3; cleaves perfectly in one direction, yielding flexible thin plates; colorless; transparent in thin sheets; specific gravity: 2.8–3.0
Biotite mica	Aluminosilicates of magnesium, iron, potassium, with water	Hardness of 2.5–3.0; cleaves perfectly in one direction, yielding flexible, thin plates; black to dark brown; specific gravity: 2.7–3.2
Pyroxene group	Silicates of aluminum, calcium, magnesium, and iron	Hardness of 5–6; cleaves in two directions at 87° and 93°; black to dark green; specific gravity: 3.1–3.5
Amphibole group	Silicates of aluminum, calcium, magnesium, and iron	Hardness of 5–6; cleaves in two directions at 56° and 124°; black to dark green; specific gravity: 3.0–3.3
Olivine	Silicates of magnesium and iron	Hardness of 6.5–7.0; light green; transparent to translucent; specific gravity: 3.2–3.6
Garnet group	Aluminosilicates of iron, calcium, magnesium, and manganese	Hardness of 6.5–7.5; uneven fracture; red, brown, or yellow; specific gravity: 3.5–4.3

(Muscovite mica through Olivine are bracketed as Ferromagnesian minerals)

*The scale of hardness used by geologists was formulated in 1822 by Frederich Mohs. Beginning with diamond as the hardest mineral, he arranged the following table:

10	Diamond	5	Apatite
9	Corundum	4	Fluorite
8	Topaz	3	Calcite
7	Quartz	2	Gypsum
6	Feldspar	1	Talc